Lecture Notes in Mathematics

Volume 2360

Editors-in-Chief

Jean-Michel Morel, City University of Hong Kong, Kowloon Tong, China

Bernard Teissier, IMJ-PRG, Paris, France

Series Editors

Karin Baur, University of Leeds, Leeds, UK

Michel Brion, UGA, Grenoble, France

Rupert Frank, LMU, Munich, Germany

Annette Huber, Albert Ludwig University, Freiburg, Germany

Davar Khoshnevisan, The University of Utah, Salt Lake City, USA

Ioannis Kontoyiannis, University of Cambridge, Cambridge, UK

Angela Kunoth, University of Cologne, Cologne, Germany

Ariane Mézard, IMJ-PRG, Paris, France

Mark Podolskij, University of Luxembourg, Esch-sur-Alzette, Luxembourg

Mark Policott, Mathematics Institute, University of Warwick, Coventry, UK

László Székelyhidi, MPI for Mathematics in the Sciences, Leipzig, Germany

Gabriele Vezzosi, UniFI, Florence, Italy

Anna Wienhard, MPI for Mathematics in the Sciences, Leipzig, Germany

This series reports on new developments in all areas of mathematics and their applications - quickly, informally and at a high level. Mathematical texts analysing new developments in modelling and numerical simulation are welcome. The type of material considered for publication includes:

1. Research monographs
2. Lectures on a new field or presentations of a new angle in a classical field
3. Summer schools and intensive courses on topics of current research.

Texts which are out of print but still in demand may also be considered if they fall within these categories. The timeliness of a manuscript is sometimes more important than its form, which may be preliminary or tentative. Please visit the LNM Editorial Policy (https://drive.google.com/file/d/1MOg4TbwOSokRnFJ3ZR3ciEeKs9hOnNX_/view?usp=sharing)

Titles from this series are indexed by Scopus, Web of Science, Mathematical Reviews, and zbMATH.

Burglind Jöricke

Braids, Conformal Module, Entropy, and Gromov's Oka Principle

Springer

Burglind Jöricke
Berlin, Germany

ISSN 0075-8434 ISSN 1617-9692 (electronic)
Lecture Notes in Mathematics
ISBN 978-3-031-67287-3 ISBN 978-3-031-67288-0 (eBook)
https://doi.org/10.1007/978-3-031-67288-0

Mathematics Subject Classification: 20F36, 30F60, 32E10, 32G05, 32G08, 32G15, 32L05, 32Gxx, 32Q56, 37B40

This Springer imprint is published by the registered company Springer Nature Switzerland AG
The registered company address is: Gewerbestrasse 11, 6330 Cham, Switzerland

Preface

This book takes the reader on an excursion through a wide range of topics that are focused on the treatment of conformal invariants and dynamical invariants. The collection of conformal invariants of complex manifolds, considered in the book, was proposed by Gromov for the case of the twice punctured complex plane. They are called here the conformal modules of conjugacy classes of elements of the fundamental group of complex manifolds and play a role in understanding the obstructions to Gromov's Oka Principle. The fundamental group of the space of monic polynomials of degree n without multiple zeros is the braid group on n strands. The respective conformal invariants are also called the conformal modules of conjugacy classes of n-braids. They appeared much earlier in connection with the interest in the Thirteen's Hilbert Problem.

The interesting point is that the conformal invariant of a conjugacy class of braids, that is related to Gromov's Oka principle, is inversely proportional to a dynamical invariant, the entropy, which was studied in connection with Thurston's celebrated theory of surface homeomorphisms. This result can be considered as another instance of the numerous manifestations of the unity of mathematics. The relation has applications. Known results concerning the entropy can be applied to problems that use the concept of the conformal module, and vice versa; methods of complex analysis and quasiconformal mappings used for the study of the conformal invariants imply results on the entropy.

After prerequisites on Riemann surfaces, braids, mapping classes, and elements of Teichmüller theory, a detailed introduction to the entropy of braids and mapping classes is given with thorough, sometimes new proofs. The book contains the first published proof of the inverse proportionality of the conformal and the dynamical invariants of conjugacy classes of braids, and a number of applications of the concept of conformal module to Gromov's Oka Principle as well as to an older problem that appeared in connection with the Thirteen's Hilbert problem. Gromov's conformal invariants of the twice punctured complex plane are estimated from above and from below. The upper and the lower bounds differ by universal multiplicative constants. The estimates imply estimates of the entropy of any pure three-braid as well as quantitative statements on the limitation of Gromov's Oka Principle in the sense of finiteness theorems. For the finiteness theorems slightly more powerful conformal invariants related to elements of the fundamental group (not merely to conjugacy classes) are used.

Including detailed proofs of all the main results and proposing several research problems, the book is aimed at graduate students and researchers.

Acknowledgments

While working on the project of this book the author had the opportunity to gain from the wonderful working conditions at IHES, Max-Planck-Institute for Mathematics, Weizmann Institute, CRM Barcelona, and at Humboldt-University Berlin, in particular at the SFB "Raum-Zeit-Materie." Other parts of the work were done while visiting the École Normale Supérieure Paris, the Universities of Grenoble, Bern, Calais, Toulouse, Lille, Bloomington, and Canberra. Talks, or series of lectures, given on the topic of the book at various places encouraged research and helped to improve the explanation. A talk given at the Algebraic Geometry seminar at Courant Institute and a stimulating discussion with F. Bogomolov and M. Gromov had a special impact. The author is indebted to many other mathematicians for interesting and fruitful discussions and for information concerning references to the literature. Among them are B. Berndtsson, R. Bryant, D. Calegari, F. Dahmani, P. Eyssidieux, B. Farb, P. Kirk, M. Reimann, S. Orevkov, L. Stout, O. Viro and M. Zaidenberg. O. Viro asked about a notion of conformal module of braids rather than of conjugacy classes of braids. M. Reimann gave references to the literature related to the computation of the conformal module of special doubly connected domains and of quadrilaterals. The author is grateful to B. Farb who suggested to use the concept of conformal module and extremal length for a proof of finiteness theorems, and to B. Berndtsson for proposing the kernel for solving the $\bar{\partial}$-problem that arises in the proof of Proposition 11.1.

Special thanks go to Alexander Weiße for teaching how to draw figures and for producing the essential parts of some difficult figures. The author is much obliged to F. Dufour and Marie-Claude Vergne for drawing some other figures, and to Cécile Gourgues for typing a preliminary version of some chapters.

Berlin, Germany
Bures-sur-Yvette, France
June 2024

Burglind Jöricke

Introduction

I am the spirit that always denies!
And rightly so.
Mephistopheles, Faust I, verse 1338/1339

The book treats invariants from different topics of mathematics, namely conformal invariants and dynamical invariants. While the dynamical invariant, the entropy, is very popular and intensively studied, the conformal invariants were undeservedly almost forgotten.

The conformal invariants are important in connection with the heuristic principle, which is widely known as Oka's Principle. The Oka Principle goes back to J.P. Serre, who phrased it in 1951 as follows:

In the absence of topological obstructions analytic problems on Stein manifolds admit analytic solutions.

The Cartan-Serre Theorems A and B, and the equivalence between topological and holomorphic classification of principal fiber bundles over Stein manifolds are important examples of this principle.

In [35] M. Gromov offered an interpretation of the Oka Principle in terms of the h-principle for holomorphic mappings: "Oka Principle is an expression of an optimistic expectation with regard to the validity of the h-principle for holomorphic maps in the situation where the source manifold is Stein". The h-principle is said to hold for holomorphic mappings from a complex manifold X to a complex manifold Y, if each continuous mapping $X \to Y$ is homotopic to a holomorphic one. Gromov's Oka Principle became the subject of intensive and fruitful research on the "affirmative side", i.e. on situations, when the principle holds. For a comprehensive account see [29].

On the other side, what about the "negating side"? How interesting is a look at situations when Gromov's Oka Principle is violated? Are there interesting structures behind?

A basic problem concerning the "negating side" of Gromov's Oka Problem is the following.

Problem *Understand obstructions for Gromov's Oka Principle.*

Obstructions for Gromov's Oka Principle are based on the relation between conformal invariants of the source and the target, in case the relation between topological invariants does not completely describe the situation.

Gromov himself referred in his seminal paper [35] from 1989 to mappings from annuli into the twice punctured complex plane as the simplest example, where this principle fails and pointed out the special role of mappings from annuli for analyzing the failure of the principle. The occurring conformal invariants of the twice punctured complex plane (called here the conformal modules of conjugacy classes of elements of the fundamental group) are supposed to capture certain "conformal rigidity".

The conformal module of a conjugacy class of elements of the fundamental group $\pi_1(Y, y_0)$ of a complex manifold Y is defined as follows. The conjugacy class $\hat{e}$ of an element $e \in \pi_1(Y, y_0)$ can be identified with the free homotopy class of curves containing the curves of the class e. A continuous map g from an annulus $A = \{z \in \mathbb{C} : r < |z| < R\}$ into $\mathcal{X}$ is said to represent the conjugacy class $\hat{e}$ if for some (and hence for any) $\rho \in (r, R)$ the map $g : \{|z| = \rho\} \to \mathcal{X}$ represents $\hat{e}$.

The definition of the conformal module of $\hat{e}$ is based on Ahlfors' concept of conformal module and extremal length. According to Ahlfors the conformal module of an annulus $A = \{z \in \mathbb{C} : \ r < |z| < R\}$ is equal to $m(A) = \frac{1}{2\pi}\log(\frac{R}{r})$ and its extremal length equals $\lambda(A) = \frac{1}{m(A)}$. For any complex manifold $\mathcal{X}$ and any conjugacy class $\hat{e}$ of elements of the fundamental group of $\mathcal{X}$ the conformal module $\mathcal{M}(\hat{e})$ of $\hat{e}$ is defined as the supremum of the conformal modules of annuli that admit a holomorphic mapping into $\mathcal{X}$ representing $\hat{e}$. The extremal length $\Lambda(\hat{e})$ equals $\frac{1}{\mathcal{M}(\hat{e})}$.

An obstruction to Gromov's Oka principle is given by the following observation.

For a holomorphic mapping $f : X \to Y$ from a complex manifold X to a complex manifold Y and any element $e \in \pi_1(X, x_0)$ the conformal module of the free homotopy class of its image $\widehat{f_*(e)}$ is not smaller than the conformal module of the free homotopy class $\hat{e}$ of the element itself.

It was an encouraging observation, that for the twice punctured complex plane it is possible to give good estimates of the conformal invariants proposed by Gromov. More precisely, in Chap. 9 we give upper and lower bounds for the conformal module of conjugacy classes of elements of the fundamental group of the twice punctured complex plane. The upper and lower bound differ by multiplicative constants not depending on the conjugacy class.

Interestingly, the notion of the conformal module of conjugacy classes of elements of the fundamental group appeared already 1974 (without name) much before Gromov's work. It was used in the paper [33] which was motivated by the interest of the authors in Hilbert's Thirteen's Problem for algebraic functions. In Chap. 6 we present some related historical remarks. The complex manifold in this case was the space $\mathfrak{P}_n$, the space of monic polynomials of degree n without multiple zeros. (The adjective "monic" refers to the property that the coefficient by the highest power of the variable equals 1). The elements of the fundamental group of this space are braids on n strands. Thus, the invariant that was considered in [33] is the conformal module of conjugacy classes of braids. It was applied to the question of global reducibility of families of polynomials $\mathfrak{P}_n$ that depend holomorphically on a parameter varying over a connected open (i.e. non-compact) Riemann surface (in particular over an annulus).

Braids play a role in several mathematical fields. It is therefore not surprising that invariants of conjugacy classes of braids occurred in different areas of mathematics. There is a dynamical invariant, the entropy. The dynamical concept was considered in 1979 in [25] in the Asterisque volume dedicated to Thurston's theory of surface homeomorphisms. Thurston's theory of surface homeomorphisms was motivated by his celebrated geometrization conjecture. For the dynamical concept one assigns to each braid on n strands an isotopy class of self-homeomorphisms of the unit disc all elements of which fix the unit circle pointwise and permute a set E_n of n chosen points in the unit disc. The entropy of the braid is the infimum of the entropy of the mappings in the class. The entropy does not change under conjugation, in other words it is an invariant of conjugacy classes of braids. The entropy of braids, more generally, of classes of surface homeomorphisms has received a lot of attention.

It turns out that the two invariants related to the dynamical aspect and to the conformal aspect of braids are related. The following theorem holds.

Main Theorem *For each conjugacy class of braids $\hat{b} \in \hat{\mathcal{B}}_n$, $n \geq 2$, the following equality holds*

$$\mathcal{M}(\hat{b}) = \frac{\pi}{2} \frac{1}{h(\hat{b})} .$$

Here $h(\hat{b})$ denotes the entropy of the conjugacy class $\hat{b}$.

The relation between the invariants coming from different fields has applications. First, results related to the more intensively studied dynamical invariant entropy may be applied to problems that use the concept of conformal module. For instance, Chap. 6 contains a short conceptual proof (of a slightly stronger version) of the result of [33], based on a result [70] related to entropy.

Vice versa, methods of complex analysis including quasi-conformal mappings, that can be used for the study of the conformal module, may give results related to the entropy. For example, the relation between the conformal and the dynamical invariants together with the effective estimates of the conformal module gives effective upper and lower bounds for the entropy of pure 3-braids.

There is an invariant of braids (not merely an invariant of conjugacy classes of braids), the conformal module (or extremal length) of braids with totally real horizontal boundary values. It uses Ahlfors' extremal length of rectangles instead of the extremal length of annuli. This notion is more powerful for applications to problems related to Gromov's Oka Principle than the extremal length of conjugacy classes of braids. For instance, it is used to obtain quantitative statements concerning

Gromov's Oka Principle in the spirit of analogs of the Geometric Shafarevich Conjecture (see Chap. 11).

We tried to make the book as self-contained as possible and coherent for graduate students. Chapter 1 provides prerequisites from several topics, like Riemann surfaces, braids, and mapping classes, as well as Teichmüller theory. In some cases proofs are included, for instance, in the case when the material is less standard and not extensively treated in the literature, or when the idea behind a statement will appear to be important in the following. Readers who are familiar with the material may skip this chapter and consult it later for notation and for some less standard facts that are included in this chapter.

In Chap. 2 a self-contained thorough treatment of the entropy of braids and mapping classes is given. One of the theorems of [25] on irreducible mapping classes on closed surfaces is proved along the lines of [25]. The proof of another theorem is new. We include detailed prerequisites on foliations generated by quadratic differentials. A thorough treatment of the entropy of mapping classes on finite Riemann surfaces of second kind (i.e. on connected Riemann surfaces that are neither closed nor punctured and have finitely generated fundamental group) is presented. This is needed to treat the entropy of braids and the entropy of reducible mapping classes.

The proof of the Main Theorem for irreducible braids is given in Chap. 3. The chapter starts with a short introduction to Ahlfors' concept of the extremal length of families of curves in planar domains, and presents an extension of the concept to homotopy classes of curves in complex manifolds. The chapter contains explicit computations of the extremal length of some examples of conjugacy classes of 3-braids (which means by the Main Theorem, to give another way of computing some entropies), and it contains explicit computations of the extremal length with totally real horizontal boundary values of some examples of 3-braids.

In Chap. 4 we define the irreducible braid components of a conjugacy class of reducible pure braids and show that the conformal module of the conjugacy class of a pure braid equals the minimum of the conformal modules of the irreducible components. Similarly, we define the irreducible nodal components of the conjugacy class of mapping classes associated to the braid, and show that the entropy of the mapping class is equal to the maximum of the entropies of the irreducible nodal components. The Main Theorem for irreducible pure braids implies the Main Theorem for reducible pure braids. The conjugacy class of a reducible pure braid can be recovered from its irreducible components. The conjugacy class of the mapping class can be recovered up to products of commuting Dehn twists from the conjugacy classes of the irreducible nodal components. Similar statements for not necessarily pure braids are contained in Chap. 5.

As mentioned above, Chap. 6 discusses global reducibility of holomorphic families of polynomials. Since the conformal invariants are not much studied we present deeper applications of the concept of the conformal module of braids or their conjugacy classes (more generally, of elements of fundamental groups or their conjugacy classes) in Chaps. 7–11.

In Chap. 7 we study situations when Gromov's Oka Principle fails, using the concept of the conformal modules of conjugacy classes of elements of the fundamental group. For instance, we call a mapping $f : X \to Y$ from a connected finite open Riemann surface X to a complex manifold Y a Gromov-Oka mapping, if for each conformal structure on X with only thick ends the mapping is homotopic to a holomorphic mapping. (A conformal structure on a connected finite open Riemann surface has only thick ends if it makes the manifold a closed Riemann surface with a finite number of closed discs but no point removed.) The Gromov-Oka mappings from finite open Riemann surfaces to the twice punctured Riemann surface are completely described.

In Chap. 8 we prove theorems related to the failure of the Gromov-Oka Principle for bundles whose fibers are Riemann surfaces of type $(1, 1)$ (in other words, the fibers are once punctured tori). The problem can be reduced to the respective problem for $(0, 4)$-bundles (and, hence, for mappings into $\mathfrak{P}_3$) by considering double branched coverings.

Chapter 9 is devoted to effective upper and lower bounds for the extremal length with totally real horizontal boundary values of any element of the fundamental group of the twice punctured complex plane (not merely of the extremal length of conjugacy classes). The bounds differ by a multiplicative constant not depending on the element. This is done in terms of a natural syllable decomposition of the word representing the element. In the last section (Sect. 9.7) we give respective estimates for 3-braids. The extremal length with totally real horizontal boundary values is capable of giving more subtle information regarding Gromov's Oka Principle and limitations of its validity than the respective invariant of conjugacy classes. Moreover, the estimates for the extremal length with totally real horizontal boundary values of pure braids imply estimates of the extremal length of conjugacy classes of these braids, and hence, of their entropy.

In Chap. 10 we give effective estimates for the growth of the number of those elements of the braid group modulo center $\mathcal{B}_3 \diagup \mathcal{Z}_3$, whose extremal length with totally real boundary values is positive and does not exceed a positive number Y. As a corollary, we give an alternative proof of the result of Veech [80] on the exponential growth of the number of conjugacy classes of elements of $\mathcal{B}_3 \diagup \mathcal{Z}_3$ that have positive entropy not exceeding Y. The proof does not use deep techniques from Teichmüller theory.

In Chap. 11 we apply the concept of conformal module and extremal length to obtain quantitative statements on the limitation of Gromov's Oka Principle. More specifically, for any finite open Riemann surface X (maybe, of second kind) we give an effective upper bound for the number of irreducible holomorphic mappings up to homotopy from X to the twice punctured complex plane, and an effective upper bound for the number of irreducible holomorphic torus bundles up to isotopy on such a Riemann surface. These statements are analogs for certain objects on Riemann surfaces of second kind of the Geometric Shafarevich Conjecture and the Theorem of de Franchis, that state the finiteness of the number of certain holomorphic objects on closed or punctured Riemann surfaces.

Contents

Chapter 1
Riemann Surfaces, Braids, Mapping Classes, and Teichmüller Theory

In this chapter we provide prerequisites from several topics, like Riemann surfaces, braids and mapping classes, as well as Teichmüller theory. For completeness and convenience of the reader and for fixing notation and terminology we give here a recollection of basic facts, that will be used later frequently. References to the literature are given. In some cases proofs are included, for instance, in the case when the material is less standard and not extensively treated in the literature, or when the idea behind a statement will occur to be important in the following. Readers who are familiar with the material may skip this chapter and consult it later for notation and for some less standard facts that are include in this chapter.

1.1 Free Homotopy Classes of Mappings and Conjugacy Classes of Group Elements

The Change of the Base Point Let $\mathcal{X}$ be a connected smooth manifold with base point x_0 and non-trivial fundamental group $\pi_1(\mathcal{X}, x_0)$. Let α be an arc in $\mathcal{X}$ with initial point x_0 and terminal point x. Change the base point $x_0 \in \mathcal{X}$ along a curve α to the point $x \in \mathcal{X}$. This leads to an isomorphism $\mathrm{Is}_\alpha : \pi_1(\mathcal{X}, x_0) \to \pi_1(\mathcal{X}, x)$ of fundamental groups induced by the correspondence $\gamma \to \alpha^{-1}\gamma\alpha$ for any loop γ with base point x_0 and the arc α with initial point x_0 and terminal point x. We will denote the correspondence $\gamma \to \alpha^{-1}\gamma\alpha$ between curves also by Is_α.

We call two homomorphisms $h_j : G_1 \to G_2,\ j = 1, 2,$ from a group G_1 to a group G_2 conjugate if there is an element $g' \in G_2$ such that for each $g \in G_1$ the equality $h_2(g) = g'^{-1} h_1(g) g'$ holds. For two arcs α_1 and α_2 with initial point x_0 and terminal point x we have $\alpha_2^{-1}\gamma\alpha_2 = (\alpha_1^{-1}\alpha_2)^{-1}\alpha_1^{-1}\gamma\alpha_1(\alpha_1^{-1}\alpha_2)$. Hence, the two isomorphisms Is_{α_2} and Is_{α_1} differ by conjugation with the element of $\pi_1(\mathcal{X}, x)$ represented by $\alpha_1^{-1}\alpha_2$.

B. Jöricke, *Braids, Conformal Module, Entropy, and Gromov's Oka Principle*, Lecture Notes in Mathematics 2360, https://doi.org/10.1007/978-3-031-67288-0_1

Free homotopic curves are related by homotopy with fixed base point and an application of a homomorphism Is_α that is defined up to conjugation. Hence, free homotopy classes of curves can be identified with conjugacy classes of elements of the fundamental group $\pi_1(\mathcal{X}, x_0)$ of $\mathcal{X}$.

Let $\mathcal{X}$ and $\mathcal{Y}$ be oriented smooth connected non-compact surfaces with finitely generated fundamental group, with base points $x_0 \in \mathcal{X}$ and $y_0 \in \mathcal{Y}$. For a continuous mapping $F : \mathcal{X} \to \mathcal{Y}$ with $F(x_0) = y_0$ we denote by $F_* : \pi_1(\mathcal{X}, x_0) \to \pi_1(\mathcal{Y}, y_0)$ the induced map on fundamental groups. For each element $e_0 \in \pi_1(\mathcal{X}, x_0)$ the image $F_*(e_0)$ is called the monodromy along e_0, and the homomorphism F_* is called the monodromy homomorphism corresponding to F. The mapping $F \to F_*$ defines a one to one correspondence from the set of homotopy classes of continuous mappings $\mathcal{X} \to \mathcal{Y}$ with fixed base point x_0 in the source and fixed value y_0 at the base point to the set of homomorphisms $\pi_1(\mathcal{X}, x_0) \to \pi_1(\mathcal{Y}, y_0)$.

Consider a free homotopy F_t, $t \in (0, 1)$, of continuous mappings from $\mathcal{X}$ to $\mathcal{Y}$. Consider the curve $\alpha(t) \overset{def}{=} F_t(x_0)$, $t \in [0, 1]$. Suppose $F_0(x_0) = y_0$ and $F_1(x_0) = y_1$. Each curve β in $\mathcal{Y}$ with initial point y_0 and terminal point y_1 defines an isomorphism $\pi_1(\mathcal{Y}, y_0) \to \pi_1(\mathcal{Y}, y_1)$ by considering the compositions of curves $\beta^{-1}\gamma\beta$ for loops γ representing elements of $\pi_1(\mathcal{Y}, y_0)$. For different curves β the isomorphisms differ by conjugation. Identifying fundamental groups with different base point by a chosen isomorphism of the described kind we obtain a mapping that assigns to each free homotopy class of maps $\mathcal{X} \to \mathcal{Y}$ a conjugacy class of homomorphisms $\pi_1(\mathcal{X}, x_0) \to \pi_1(\mathcal{Y}, y_0)$. The mapping is surjective. It is also injective by the following reason. Let $F_j : \mathcal{X} \to \mathcal{Y}$, $j = 0, 1$, be continuous mappings, $F_j(x_0) = y_0$, such that for an element $e \in \pi_1(\mathcal{Y}, y_0)$ the equality $(F_1)_* = e^{-1}(F_0)_*\, e$ holds. Let α be a smooth curve in $\mathcal{Y}$ that represents e. There exists a free homotopy F^t, $t \in [0, 1]$, of mappings such that $F^0 = F_0$ and $F^t(x_0) = \alpha(t)$, $t \in [0, 1]$. Then $(F^1)_* = e^{-1}(F^0)_*\, e$. Hence, F_1 and F^1 define the same homomorphism $\pi_1(\mathcal{X}, x_0) \to \pi_1(\mathcal{Y}, y_0)$, and hence, they are homotopic. We obtain the following theorem (see also [36],[76], [75].)

Theorem 1.1 *The free homotopy classes of continuous mappings from $\mathcal{X}$ to $\mathcal{Y}$ are in one-to-one correspondence to the set of conjugacy classes of homomorphisms between the fundamental groups of $\mathcal{X}$ and $\mathcal{Y}$.*

1.2 Riemann Surfaces

We will use the common definition of a surface, of a smooth surface, of a surface with boundary, and of a Riemann surface. A homeomorphic mapping $\omega : S \to X$ from a surface S onto a Riemann surface X is called a conformal structure or a complex structure, having in mind, that the pull back of a complex analytic atlas on X provides a complex analytic atlas on S.

If the interior of a smooth surface with boundary is equipped with a complex structure, the surface with boundary is called (following Ahlfors) a bordered Riemann surface.

A surface (not a surface with boundary) is called open if no connected component is compact. A compact surface (without boundary) is called a closed surface. A closed surface with finitely many points removed is called a punctured surface. The removed points are called punctures. A connected closed oriented surface or a connected closed Riemann surface is called of type (g, m), if it has genus g and is equipped with m distinguished points.

The intersection number of an ordered pair of smooth loops with transversal intersection in a smooth oriented surface X is the sum of the intersection numbers over all intersection points. The intersection number at an intersection point equals $+1$ if the orientation determined by the tangent vector to the first curve followed by the tangent vector to the second curve is the orientation of X, and equals -1 otherwise. The intersection number does not change under homotopy, i.e. the intersection number of a homotopic pair of smooth curves with transversal intersection is equal to the intersection number of the original pair.

If the loops are not smooth or do not intersect transversally, their intersection number is defined as intersection number of a homotopic pair of smooth loops with transversal intersection.

We will consider only oriented surfaces, often without further mentioning. If not mentioned otherwise, the considered surfaces will be connected. A surface is called finite if its fundamental group is finitely generated. A finite surface is obtained from a closed surface by removing finitely many disjoint simply connected closed sets (called holes). The fundamental group of a surface of genus g with $m > 0$ holes is a free group in $2g + m - 1$ generators.

Each finite connected open Riemann surface X is conformally equivalent to a domain (denoted again by X) on a closed Riemann surface X^c such that each connected component of the complement $X^c \setminus X$ is either a point or a closed topological disc with smooth boundary [77].

A connected Riemann surface is said to be of first kind, if it is a closed or a punctured Riemann surface, otherwise it is said to be of second kind. If all holes of a finite open Riemann surface are closed topological discs, the Riemann surface is said to have only thick ends.

Standard Bouquets of Circles Let X be an oriented smooth open surface of genus $g \geq 0$ with $m > 0$ holes, equipped with a base point q_0. The union B of non-contractible circles in X with base point q_0 is called a bouquet of circles in X, if q_0 is the only common point of any pair of circles in B, and different circles in B with base point q_0 represent different elements of $\pi_1(X, q_0)$. Suppose B is the union of simple closed oriented curves α_j, β_j, $j = 1, \ldots, g'$, and γ_k, $k = 1, \ldots, m'$, with base point q_0 with the following property. Labeling the rays of the loops emerging from the base point q_0 by α_j^-, β_j^- γ_j^-, and the incoming rays by α_j^+, β_j^+ γ_j^+, and

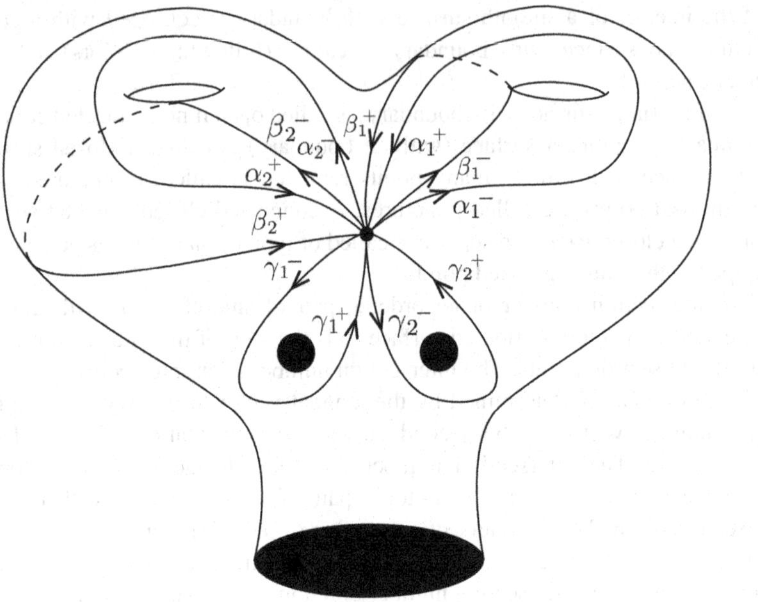

Fig. 1.1 A standard bouquet of circles for a connected finite open Riemann surface

moving in counterclockwise direction along a small circle around q_0 we meet the rays in the order

$$\ldots, \alpha_j^-, \beta_j^-, \alpha_j^+, \beta_j^+, \ldots, \gamma_k^-, \gamma_k^+, \ldots .$$

Then B is called a standard bouquet of circles in X.

Let B be a standard bouquet of circles in X with base point q_0. If the collection $\mathcal{E}$ of elements of the fundamental group $\pi_1(X, q_0)$ represented by the collection of curves in B is a system of generators of $\pi_1(X, q_0)$ (then in particular, $g' = g$, $m' = m$), we call B a standard bouquet of circles for X, and say that the system $\mathcal{E}$ of generators is associated to a standard bouquet of circles for X. See also Fig. 1.1.

Label the generators $\mathcal{E} \subset \pi_1(X, q_0)$ associated to a standard bouquet of circles for X as follows. The elements $e_{2j-1,0} \in \pi_1(X, q_0)$, $j = 1, \ldots, g$, are represented by α_j, the elements $e_{2j,0} \in \pi_1(X, q_0)$, $j = 1, \ldots, g$, are represented by β_j, and the elements $e_{2g+k,0} \in \pi_1(X, q_0)$, $k = 1, \ldots, m$, of $\pi_1(X, q_0)$ are represented by γ_k. The system $\mathcal{E} = \{e_j\}_{j=1}^{2g+m-1}$ labeled in this way is called a standard system of generators of $\pi_1(X, q_0)$.

A standard bouquet of circles for a connected finite open Riemann surface can be obtained as follows. Recall that the Riemann surface X is conformally equivalent to a domain in a closed Riemann surface X^c of genus g. Take a fundamental polygon of the compact Riemann surface X^c, so that the projection of its sides to X^c do not meet the holes of X. The projections of the sides provide a collection of simple closed curves α_j, β_j, $j = 1, \ldots, g$, with base point q_0, that are pairwise disjoint outside q_0, and represent the elements of the fundamental group $\pi_1(X^c, q_0)$. Each pair (α_j, β_j) corresponds to a handle of X^c. Cut X^c along the curves α_j, β_j, $j = 1, \ldots, g$. We obtain a simply connected domain $\overset{\circ}{X}$. Denote the connected components of $X^c \setminus X$ by $\mathcal{C}_1, \mathcal{C}_2, \ldots, \mathcal{C}_m$. (The $\mathcal{C}_\ell$ are also the connected components of $\overset{\circ}{X} \setminus \big(X \bigcup_j (\alpha_j \cup \beta_j)\big)$.) Each component is either a point or a closed disc. We associate to the components $\mathcal{C}_\ell$, $\ell = 1, \ldots, m-1$, a collection of $m-1$ simple closed curves γ_ℓ in X with base point q_0 that meet the union $\bigcup_j (\alpha_j \cup \beta_j)$ exactly at the point q_0 and are pairwise disjoint outside q_0. Moreover, each γ_ℓ is contractible in $X \cup \mathcal{C}_\ell$ and divides X into two connected components, one of them containing $\mathcal{C}_\ell$. We will say that γ_ℓ surrounds $\mathcal{C}_\ell$. We orient γ_ℓ so that $\mathcal{C}_\ell$ is on the left when walking along the curve γ_ℓ equipped with this orientation. We may choose the γ_ℓ so that the union $\bigcup_{j=1}^{g} (\alpha_j \cup \beta_j) \cup \bigcup_{\ell=1}^{m-1} \gamma_\ell$ is a standard bouquet of circles for X.

Let X be a connected finite open smooth surface with base point q_0, and B a standard bouquet of circles in X. Choose a neighbourhood $N(B)$ of B in X with the following properties. $N(B)$ contains a smoothly bounded open disc D in X around q_0, such that the intersection of each circle in the bouquet with $X \setminus D$ is a closed arc with endpoints on the boundary ∂D. Denote the arcs by s_j. Moreover, $N(B)$ is equal to the union $D \cup \bigcup_{j=1}^{2g+m-1} V_j$ of D with disjoint sets V_j, where each V_j has the following properties. There exists a diffeomorphism φ_j from a half-open rectangle $\{x + iy \in \mathbb{C} : |x| < \mathsf{a}, |y| \leq \mathsf{b}\}$ in the plane with sides parallel to the axes onto V_j, that maps the horizontal sides into ∂D, and the rest of the rectangle into $X \setminus D$. Moreover, its restriction to the vertical symmetry axis of the rectangle (with the direction of the imaginary axis) parameterizes the arc s_j (with the orientation of the respective circle of the bouquet). We call the V_j bands attached to D representing s_j, and we call a neighbourhood of the described form a standard neighbourhood of the standard bouquet B.

Lemma 1.1 *Suppose a standard bouquet B of circles for a connected open Riemann surface X consists of g pairs of curves α_j, β_j, and $m-1$ curves γ_k as above, that represent a system of generators of $\pi_1(X, q_0)$, and let $N(B)$ be a standard neighbourhood of S. Then there exists a continuous family of mappings $\varphi_t : X \to X$, $t \in [0, 1]$, that map X homeomorphically onto a domain X_t in X such that φ_0 is the identity and $X_1 = N(B)$. In particular, X has genus g and m holes.*

Proof For each j the pair of loops α_j, β_j in $N(B)$ has non-vanishing intersection number. Each of these pairs corresponds to a handle of $N(S)$. For any other pair of loops the intersection number equals zero. Each loop γ_k, $k = 1, \ldots, m-1$, divides $N(B)$. One of the connected components of $N(B) \setminus \gamma_k$ is an annulus A_k whose second boundary component ∂_k is a boundary component of $N(B)$.

The loop γ_k also divides X. Otherwise there would be a simple closed curve γ in X with base point q_0 whose only intersection point with γ_k is q_0, and that has points on A_k and in $N(B) \setminus A_k$. Then the intersection number of the curves γ and γ_k would be different from zero. But all elements of the fundamental group of $N(B)$, and hence all elements of the fundamental group of X, have vanishing intersection number with γ_k. Let Ω_k be the component of $X \setminus \gamma_k$ that contains ∂_k. Put $X_k = \Omega_k \setminus \overline{A_k}$.

The last boundary component ∂_m of $N(B)$ also divides X. Otherwise there would be an arc in X that joins two points that are close to a point in ∂_m and locally on different sides of ∂_m. Then there would exist a simple closed loop in X with base point q_0 that has non-zero intersection number with ∂_m. This is impossible, since all elements of the fundamental group $\pi_1(X, q_0)$ can be represented by loops that are contained in $N(B)$ and hence have zero intersection number with ∂_m. Let X_m be the connected component of $X \setminus \partial_m$, that contains the part of a neighbourhood of ∂_m which is contained in $X \setminus N(B)$.

Each of the X_k is an open Riemann surface of some genus with a number of holes. Since α_j, β_j, and γ_k represent generators of the fundamental group of both, X and $N(B)$, each X_k has genus zero. No X_k is a disc. If some X_k, $k \leq m-1$, was a disc, γ_k would be contractible in X, which is a contradiction. X_m is not a disc, because ∂_m is free homotopic to $\gamma_1 \dots \gamma_{m-1}\alpha_1\beta_1\alpha_1^{-1}\beta_1^{-1} \dots \alpha_g\beta_g\alpha_g^{-1}\beta_g^{-1}$ which is not the identity in the fundamental group of X. Since the fundamental groups of X and $N(B)$ are isomorphic, each X_k is an annulus bounded be a boundary component of X and ∂_k. We proved that X has genus g and m holes. Moreover, there exists a continuous family of mappings $\varphi_t : X \to X$, $t \in [0, 1]$, that map X diffeomorphically onto a domain X_t in X such that φ_0 is the identity and $X_1 = N(B)$. □

Coverings For the following facts see [23, 27]. By a covering $P : \mathcal{Y} \to \mathcal{X}$ we mean a continuous map P from a connected topological space $\mathcal{Y}$ to a connected topological space $\mathcal{X}$ such that for each point $x \in \mathcal{X}$ there is a neighbourhood $V(x)$ of x such that the mapping P maps each connected component of the preimage of $V(x)$ homeomorphically onto $V(x)$. (Note that in function theory sometimes these objects are called unlimited unramified coverings to reserve the notion "covering" for more general objects.) A covering $P : \mathcal{Y} \to \mathcal{X}$ is called a universal covering if for each covering $P_{\mathcal{Z}} : \mathcal{Z} \to \mathcal{X}$ of $\mathcal{X}$ by a connected topological space $\mathcal{Z}$ and for each pair of points $y \in \mathcal{Y}$ and $z \in \mathcal{Z}$ with $P(y) = P_{\mathcal{Z}}(z)$ there exists a unique fiber preserving continuous mapping $f : \mathcal{Y} \to \mathcal{Z}$ that takes y to z. A connected locally simply connected topological space has up to isomorphism at most one universal covering. If $\mathcal{X}$ and $\mathcal{Y}$ are connected manifolds, $\mathcal{Y}$ is simply connected, and $P : \mathcal{Y} \to \mathcal{X}$ is a covering, then P is the universal covering.

For each connected manifold $\mathcal{X}$ the universal covering exists and is denoted by $\mathsf{P} : \tilde{\mathcal{X}} \to \mathcal{X}$. It is constructed as follows. Fix a base point x_0 in $\mathcal{X}$. For each $x \in \mathcal{X}$ we denote by $\pi_1(\mathcal{X}; x_0, x)$ the set of homotopy classes of arcs in X with initial point x_0 and terminal point x. Consider the set that consists of the union of all elements $e_{x_0,x} \in \pi_1(\mathcal{X}; x_0, x)$ for all $x \in \mathcal{X}$. Consider the mapping that assigns to each element $e_{x_0,x} \in \pi_1(\mathcal{X}; x_0, x)$ the point $x \in \mathcal{X}$. We define a system of fundamental neighbourhoods of elements of the defined set as follows. For each element $e_{x_0,x} \in \pi_1(\mathcal{X}; x_0, x)$ we take any simply connected neighbourhood $U \subset \mathcal{X}$

and consider the set of all products $e_{x_0,x}e_{x,y} \in \pi_1(\mathcal{X}; x_0, y)$ for which $e_{x,y}$ can be represented by curves that are contained in U and join x with y. We obtain a simply connected manifold $\tilde{\mathcal{X}}$ and a covering $\mathsf{P} : \tilde{\mathcal{X}} \to \mathcal{X}$. (For a proof see [27].)

Consider the universal covering $\mathsf{P} : \tilde{X} \to X$ of a manifold X. A covering transformation is a homeomorphism $\varphi : \mathcal{Y} \circlearrowleft$ such that $\mathsf{P} \circ \varphi = \mathsf{P}$. The covering transformations form a group under composition, denoted by $\text{Deck}(\tilde{X}, X)$. By the definition of the universal covering for each pair of points z_1, z_2 of $\tilde{X}$ with $\mathsf{P}(z_1) = \mathsf{P}(z_2)$ there exists a unique element $\varphi \in \text{Deck}(\tilde{X}, X)$ with $z_2 = \varphi(z_1)$.

Equip X with a Riemannian metric d, and denote by the same letter d the lifted metric on $\tilde{X}$. In other words, the length of a curve in the metric d on $\tilde{X}$ is the length of its projection to X. Take any point $z_0 \in \tilde{X}$. The Dirichlet region containing z_0 is defined as

$$\mathrm{D}_{z_0} \overset{def}{=} \{z \in \tilde{X} : d(z, z_0) < d(z, g(z_0)) \text{ for all } g \in \text{Deck}(\tilde{X}, X)\}. \tag{1.1}$$

The Dirichlet region is an open set and $g_1(\mathrm{D}_{z_0}) \cap g_2(\mathrm{D}_{z_0}) = \emptyset$ if g_1 and g_2 are different covering transformations.

The union of the closures $\overline{g(\mathrm{D}_{z_0})}$ over all covering transformations g is equal to the whole $\tilde{X}$. Hence, the union of D_{z_0} with a suitable subset of its boundary is a fundamental region. A subset U of $\tilde{X}$ is called a fundamental region for the covering $\mathsf{P} : \tilde{X} \to X$ if each point in X has exactly one preimage under the mapping $\mathsf{P} : U \to X$. Further, for each point z of the boundary of D_{z_0} (taken in the topological space $\tilde{X}$) there is a covering transformation g such that $d(z, z_0) = d(z, g(z_0))$.

We saw that for each point $z_0 \in \tilde{X}$ there exists a neighbourhood U such that $U \cap g(U) = \emptyset$ for all covering transformations except the identity. In particular, no element of this group except the identity has a fixed point, and the group acts discontinuously on $\tilde{X}$, i.e. for two compact sets A and B in $\tilde{X}$ the set $A \cap g(B)$ is non-empty for at most finitely many elements g of the group.

Let X be a connected finite open Riemann surface with base point q_0 and let $\mathsf{P} : \tilde{X} \to X$ be the universal covering map. We fix a base point $q_0 \in X$ and a base point $\tilde{q}_0 \in \mathsf{P}^{-1}(q_0) \subset \tilde{X}$. The group of covering transformations of $\tilde{X}$ can be identified with the fundamental group $\pi_1(X, q_0)$ of X by the following correspondence (See e.g. [27]). Take a covering transformation $\sigma \in \text{Deck}(\tilde{X}, X)$. Let $\tilde{\gamma}_0$ be an arc in $\tilde{X}$ with initial point $\tilde{q}_0$ and terminal point $\sigma(\tilde{q}_0)$. Denote by $\text{Is}^{\tilde{q}_0}(\sigma)$ the element of $\pi_1(X, q_0)$ represented by the loop $\mathsf{P}(\tilde{\gamma}_0)$.

By the convention to write products of curves from left to right and compositions of mappings from right to left for the mapping $\text{Deck}(\tilde{X}, X) \ni \sigma \to \text{Is}^{\tilde{q}_0}(\sigma) \in \pi_1(X, q_0)$ the following relation holds

$$\text{Is}^{\tilde{q}_0}(\sigma_1\sigma_2) = \text{Is}^{\tilde{q}_0}(\sigma_2)\text{Is}^{\tilde{q}_0}(\sigma_1).$$

Hence, the mapping $\sigma \to \text{Is}^{\tilde{q}_0}(\sigma^{-1})$ is a group homomorphism. This homomorphism is injective and surjective, hence it is a group isomorphism. The inverse $(\text{Is}^{\tilde{q}_0})^{-1}$ of the mapping $\text{Is}^{\tilde{q}_0}$ is obtained as follows. Represent an element $e_0 \in$

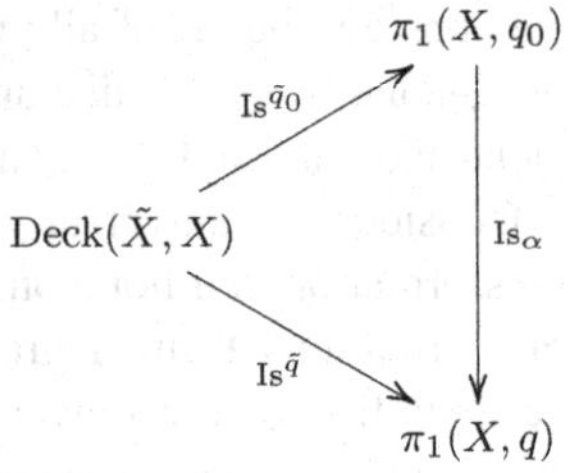

Fig. 1.2 A commutative diagram related to the change of the base point

$\pi_1(X, q_0)$ by a loop γ_0. Consider the lift $\tilde{\gamma}_0$ of γ_0 to $\tilde{X}$ that has initial point $\tilde{q}_0$. Then $(\mathrm{Is}^{\tilde{q}_0})^{-1}(e_0)$ is the covering transformation that maps $\tilde{q}_0$ to the terminal point of $\tilde{\gamma}_0$.

For another point $\tilde{q}$ of $\tilde{X}$ and the point $q \overset{def}{=} \mathsf{P}(\tilde{q}) \in X$ the isomorphism $\mathrm{Is}^{\tilde{q}}$: $\mathrm{Deck}(\tilde{X}, X) \to \pi_1(X, q)$ assigns to each $\sigma \in \mathrm{Deck}(\tilde{X}, X)$ the element of $\pi_1(X, q)$ that is represented by $\mathsf{P}(\tilde{\gamma})$ for a curve $\tilde{\gamma}$ in $\tilde{X}$ that joins $\tilde{q}$ with $\sigma(\tilde{q})$. $\mathrm{Is}^{\tilde{q}}$ is related to $\mathrm{Is}^{\tilde{q}_0}$ as follows. Let $\tilde{\alpha}$ be an arc in $\tilde{X}$ with initial point $\tilde{q}_0$ and terminal point $\tilde{q}$. Put $\alpha = \mathsf{P}(\tilde{\alpha})$. Then for the isomorphism $\mathrm{Is}_\alpha : \pi_1(X, q_0) \to \pi_1(X, q)$ the equation

$$\mathrm{Is}^{\tilde{q}}(\sigma) = \mathrm{Is}_\alpha \circ \mathrm{Is}^{\tilde{q}_0}(\sigma), \quad \sigma \in \mathrm{Deck}(\tilde{X}, X), \tag{1.2}$$

holds, i.e. the diagram Fig. 1.2 is commutative. Indeed, let $\tilde{\alpha}^{-1}$ denote the curve that is obtained from $\tilde{\alpha}$ by inverting the direction on $\tilde{\alpha}$, i.e. moving from $\tilde{q}$ to $\tilde{q}_0$. For a curve $\tilde{\gamma}_0$ in $\tilde{X}$ that joins $\tilde{q}_0$ with $\sigma(\tilde{q}_0)$, the curve $\tilde{\alpha}^{-1}\ \tilde{\gamma}_0\ \sigma(\tilde{\alpha})$ in $\tilde{X}$ has initial point $\tilde{q}$ and terminal point $\sigma(\tilde{q})$. Therefore $\mathsf{P}(\tilde{\alpha}^{-1}\ \tilde{\gamma}_0\ \sigma(\tilde{\alpha}))$ represents $\mathrm{Is}^{\tilde{q}}(\sigma)$. On the other hand

$$\mathsf{P}(\tilde{\alpha}^{-1}\ \tilde{\gamma}_0\ \sigma(\tilde{\alpha})) = \mathsf{P}(\tilde{\alpha}^{-1})\ \mathsf{P}(\tilde{\gamma}_0)\ \mathsf{P}(\sigma(\tilde{\alpha})) = \alpha^{-1}\gamma_0\,\alpha \tag{1.3}$$

represents $\mathrm{Is}_\alpha(e_0)$ with $e_0 = \mathrm{Is}^{\tilde{q}_0}(\sigma)$. In particular, if $\tilde{q}_0' \in \mathsf{P}^{-1}(q_0)$ is another preimage of the base point q_0 under the projection P, then the associated isomorphisms to the fundamental group $\pi_1(X, q_0)$ are conjugate, i.e. $\mathrm{Is}^{\tilde{q}_0'}(e_0) = (e_0')^{-1}\mathrm{Is}^{\tilde{q}_0}(e_0)e_0'$ for each $e_0 \in \pi_1(X, q_0)$. The element e_0' is represented by the projection of an arc in $\tilde{X}$ with initial point $\tilde{q}_0$ and terminal point $\tilde{q}_0'$.

Keeping fixed $\tilde{q}_0$ and q_0 we will say that a point $\tilde{q} \in \tilde{X}$ and a curve α in X are compatible if the diagram Fig. 1.2 is commutative, equivalently, if Eq. (1.2) holds. We may also start with choosing a curve α in X with initial point q_0 and terminal point q. Then there is a point $\tilde{q} = \tilde{q}(\alpha)$, such that $\tilde{q}$ and α are compatible. Indeed, let $\tilde{\alpha}$ be the lift of α, that has initial point $\tilde{q}_0$. Denote the terminal point of $\tilde{\alpha}$ by $\tilde{q}(\alpha)$, and repeat the previous arguments.

For two Riemann surfaces $\mathcal{X}$ and $\hat{\mathcal{X}}$ and a non-constant holomorphic mapping $p : \hat{\mathcal{X}} \to \mathcal{X}$ a point $\hat{x} \in \hat{\mathcal{X}}$ is called a branch point or ramification point if there is no neighbourhood of $\hat{x}$ such the restriction of p to it is injective. For each branch point $\hat{x}$ there exists a neighbourhood V of $\hat{x}$ and an integer number $k \geq 2$, such that $p \mid V$ is equal to the mapping $z \to z^k$ in suitable coordinates on $\hat{\mathcal{X}}$ near $\hat{x}$ and on $\mathcal{X}$ near $x = p(\hat{x})$.

A non-constant holomorphic mapping $p : \hat{\mathcal{X}} \to \mathcal{X}$ is called a branched covering, if the image E of the set $\hat{E}$ of branch points is discrete and the restriction $p \mid (\mathcal{X} \setminus \hat{E})$ is a covering. The set E is called the branch locus.

The branched covering is called simple if for each point $x \in E$ the fiber $p^{-1}(x)$ contains exactly one branch point and in a neighbourhood of this branch point the mapping is a double branched covering (i.e. in suitable coordinates it equals $z \to z^2$).

Hyperbolic Riemann Surfaces For the following collection of well-known results we refer e.g. to [58], and [59]. By the Uniformisation Theorem the universal covering space $\tilde{X}$ of each connected Riemann surface X (equipped with the complex structure that makes the projection $\mathsf{P} : \tilde{X} \to X$ holomorphic) is conformally equivalent to one of the following: the Riemann sphere $\mathbb{P}^1$, or the complex plane $\mathbb{C}$, or the upper half-plane $\mathbb{C}_+ \stackrel{def}{=} \{z \in \mathbb{C} : \mathrm{Im} z > 0\}$. The upper half-plane is equipped with the hyperbolic metric which is defined by the infinitesimal length element $\frac{|dz|}{2y}$. If the upper half-plane is replaced by the unit disc $\mathbb{D}$, the hyperbolic metric (that is induced by a conformal mapping $\mathbb{C}_+ \to \mathbb{D}$) is defined by $\frac{|dz|}{1-|z^2|}$. A Riemann surface whose universal covering is (conformally equivalent to) the upper half-plane will be called hyperbolic.

Let X be a finite Riemann surface with $\tilde{X}$ being the upper half-plane. Each covering transformation is a conformal mapping of the upper half-plane onto itself. Such a mapping extends to Möbius transformation on the Riemann sphere. Hence, the group of covering transformations is a subgroup of the projective special linear group $PSL_2(\mathbb{R}) = SL_2(\mathbb{R})/\langle -\mathrm{Id}\rangle$, where $SL_2(\mathbb{R})$ is the group of mappings $z \to \frac{az+b}{cz+d}$ with $ad - bc = 1$, for which a, b, c, d are real numbers, equivalently, that map the real axis to itself, and $\langle -\mathrm{Id}\rangle$ is the subgroup generated by $-\mathrm{Id}$ for the identity Id. The group of covering transformations acts properly discontinuously. Subgroups of $PSL_2(\mathbb{R})$ that act properly discontinuously are called Fuchsian groups. For each fixed point free Fuchsian group Γ the quotient $\mathbb{C}_+/\Gamma$ is a Riemann surface.

The limit set of a Fuchsian group Γ is the set of accumulation points of $\{g(z_0) : g \in \Gamma\}$, where z_0 is any point of the upper half-plane. The set is independent of the choice of the point z_0. The limit set of a fixed point free Fuchsian group is either the real axis $\mathbb{R}$ or a nowhere dense subset of the real axis without isolated points. A Fuchsian group is said to be of first kind if the limit set is the real axis, and is said to be of second kind if the limit set is a nowhere dense subset of the real axis without isolated points.

The following theorem motivates our definition of Riemann surfaces of first and second kind.

Theorem 1.2 *A connected Riemann surface is of first kind iff it is the quotient $\mathbb{C}_+/\Gamma$ of the upper half-plane by a Fuchsian group Γ of first kind.*

For a proof (in a more general situation) see e.g. [56], II Theorem 3.2 and [2].

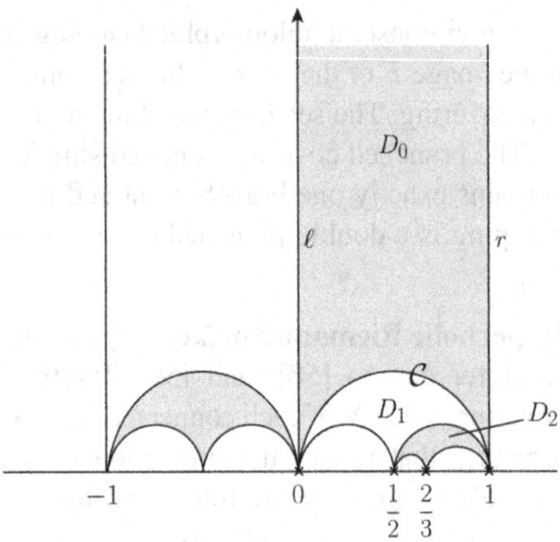

Fig. 1.3 The universal covering of the twice punctured complex plane

The Universal Covering of the Twice Punctured Complex Plane We recall the well-known explicit description of the universal covering of $\mathbb{C} \setminus \{0, 1\}$. We consider the domain $D_0 = \{z \in \mathbb{C}_+ : 0 < \text{Re} z < 1, |z - \frac{1}{2}| > \frac{1}{2}\}$ in the upper half-plane, which is bounded by a half-circle and two half-lines. These curves are geodesic curves in the hyperbolic metric of the upper half-plane. We call D_0 the geodesic triangle with vertices 0, 1, and ∞. Denote by P the conformal mapping from D_0 onto the upper half-plane, whose continuous extension to the boundary takes 0 to 0, 1 to 1 and ∞ to ∞. The extension of P to the boundary takes $\ell \overset{def}{=} (0, i\infty)$ to $(-\infty, 0)$, $\mathcal{C} \overset{def}{=} \{|z - \frac{1}{2}| = \frac{1}{2}\} \cap \mathbb{C}_+$ to $(0, 1)$ and $r \overset{def}{=} 1 + (0, +i\infty)$ to $(1, \infty)$.

We reflect D_0 in the circle $|z - \frac{1}{2}| = \frac{1}{2}$. By symmetry reasons the reflected domain is the geodesic triangle D_1 bounded by $\mathcal{C}$ (the reflection of itself), and the half-circles in $\mathbb{C}_+$ with diameter $(0, \frac{1}{2})$ (the reflection of ℓ), and diameter $(\frac{1}{2}, 1)$ (the reflection of r). In the same way we reflect D_0 in the half-lines in the boundary (see Fig. 1.3).

We reflect now D_1 in the circle $|z - \frac{3}{4}| = \frac{1}{4}$ with diameter $(\frac{1}{2}, 1)$. We obtain the geodesic triangle D_2 with vertices $\frac{1}{2}$, $\frac{2}{3}$, and 1. This can be seen by the following calculation. The mapping $z \to \zeta(z) = 4(z - \frac{3}{4})$ maps the circle $|z - \frac{3}{4}| = \frac{1}{4}$ to the unit circle $|\zeta| = 1$. The reflection map in $|\zeta| = 1$ is $\zeta \to \frac{1}{\bar{\zeta}}$, hence, the reflection in the circle $|z - \frac{3}{4}| = \frac{1}{4}$ is defined by the mapping

$$z = \frac{1}{4}(\zeta + 3) \to \frac{1}{4}(\frac{1}{\bar{\zeta}} + 3) = \frac{1}{4}(\frac{1}{4\bar{z} - 3} + 3) = \frac{3\bar{z} - 2}{4\bar{z} - 3}\,.$$

The mapping takes 1 to 1, 0 to $\frac{2}{3}$, and $\frac{1}{2}$ to $\frac{1}{2}$. We get further geodesic triangles by continuing reflecting in each geodesic arc of each already obtained geodesic triangle. We get a collection of open geodesic triangles whose closures in $\mathbb{C}_+$ fill the whole $\mathbb{C}_+$.

Extend the mapping P continuously to the closure of D_0, and by Schwarz' Reflection Principle through the geodesic arcs in the boundary of D_0. Extending P repeatedly through the geodesic boundary arcs of the images under repeated reflection, we obtain a holomorphic mapping form $\mathbb{C}_+$ to $\mathbb{C} \setminus \{0, 1\}$, denoted also by P. The extended P takes each geodesic triangle, obtained from D_0 by an even number of successive reflections, conformally onto the upper half-plane, and maps each geodesic triangle, obtained from D_0 by an odd number of successive reflections, conformally onto the lower half-plane. It is easy to see that $\mathsf{P} : \mathbb{C}_+ \to \mathbb{C} \setminus \{-1, 1\}$ is a covering, in fact it is the universal covering of the twice punctured complex plane.

Each reflection in a geodesic side of one of the geodesic triangles obtained from D_0 by reflection is an anti-holomorphic self-homeomorphism of $\mathbb{C}_+$. The composition φ of an even number of successive reflections is a conformal mapping of $\mathbb{C}_+$ onto itself, for which $\mathsf{P}\circ\varphi = \mathsf{P}$, in other words, it is a covering transformation. The covering transformation obtained by reflection in the imaginary axis followed by reflection in one of the half-circles $\{z \in \mathbb{C}_+ : |z \pm \frac{1}{2}| = \frac{1}{2}\}$ equals $z \to \frac{z}{1 \mp 2z}$. Indeed, the reflection in $\{z \in \mathbb{C}_+ : |z \pm \frac{1}{2}| = \frac{1}{2}\}$ is given by the mapping $z \to \frac{\mp \bar{z}}{2\bar{z} \pm 1}$, the reflection in the imaginary axis is given by $x + iy \to -x + iy$.

Reflection in the imaginary axis followed by reflection in $\{z \in \mathbb{C}_+ : \mathrm{Re} z = 1\}$ equals $z \to z + 2$. These covering transformation generate the group of covering transformations. In particular, each covering transformation is an element of $PSL_2(\mathbb{Z})$. Here $SL_2(\mathbb{Z})$ is the group of 2×2 matrices with integer entries and determinant 1, and $PSL_2(\mathbb{Z})$ is the quotient $SL_2(\mathbb{Z})/\langle -\mathrm{Id}\rangle$, for the subgroup $\langle -\mathrm{Id}\rangle$ where Id is the unit matrix.

Properties of Elements of $PSL_2(\mathbb{Z})$ Let $T(z) = \frac{az+b}{cz+d}$ represent an element of $PSL_2(\mathbb{Z})$. The fixed points of this mapping satisfy the equation $\frac{az+b}{cz+d} = z$, hence, if $c \neq 0$, $z^2 + \frac{d-a}{c} z - \frac{b}{c} = 0$. This implies that the fixed points are $z_\pm = -\frac{d-a}{2c} \pm \frac{\sqrt{(a+d)^2-4}}{2c}$. We used the fact that the determinant of the matrix $A = \begin{pmatrix} a & b \\ c & d \end{pmatrix}$ equals $ad - bc = 1$. The eigenvalues $t_\pm$ of the matrix A satisfy the equation $t^2 - (a+d)t + 1 = 0$, hence, $t_\pm = \frac{a+d}{2} \pm \frac{\sqrt{(a+d)^2-4}}{2}$. We used again that the determinant of the matrix equals 1.

If $|a + d| > 2$ there are two different real fixed points. The matrix A has two different real eigenvalues $t_\pm$, and can be conjugated by a real matrix to the matrix $\begin{pmatrix} t_+ & 0 \\ 0 & t_- \end{pmatrix}$. Hence, T can be conjugated by a holomorphic self-homeomorphism of $\mathbb{C}_+$ to the mapping $T'(z) = \frac{t_+ z}{t_-} = t_+^2 z$.

If $|a + d| = 2$, T has a double fixed point z_0. This point is real. Conjugating this fixed point to ∞ (for instance by the conformal self-mapping $z \to \frac{1}{z_0 - z}$ of $\mathbb{C}_+$) we arrive at a mapping of the form $z \to z + \mathsf{b}$ for a real number b. The eigenvalues are equal to $t_\pm = 1$. If $c = 0$ then $a = d = \pm 1$, hence $|a + d| = 2$.

If $|a+d| < 2$ the eigenvalues are equal to $t_\pm = \pm i$, if $|a+d| = 0$, and to $t_\pm = \frac{1}{2} \pm \frac{\sqrt{3}}{2} i$ or $t_\pm = -\frac{1}{2} \pm \frac{\sqrt{3}}{2} i$, if $|a+d| = 1$. In the first case the square of the mapping T^2 is the identity. In the second case T^3 is the identity.

1.3 Braids

In this section we outline basic facts on braids. For details the reader may consult [14] or [52].

Most intuitively, braids are described in terms of geometric braids. We will use here the complex plane $\mathbb{C}$ though the complex structure will play a role only later. Let E_n be a subset of $\mathbb{C}$ containing exactly n points. A geometric braid in $[0,1] \times \mathbb{C}$ with base point E_n is a collection of n mutually disjoint arcs in the cylinder $[0,1] \times \mathbb{C}$ which joins the set $\{0\} \times E_n$ in the bottom with the set $\{1\} \times E_n$ in the top of the cylinder and intersects each fiber $\{t\} \times \mathbb{C}$ along an unordered n-tuple $E_n(t)$ of points. The arcs are called the strands of the geometric braid.

There is a more conceptual point of view. We consider points $(z_1, \dots, z_n) \in \mathbb{C}^n$ as ordered tuples of points in $\mathbb{C}$. Usually the points $z_j \in \mathbb{C}$ will be required to be pairwise distinct, in other words, we require that $(z_1, \dots, z_n)$ belongs to the configuration space $C_n(\mathbb{C}) = \{(z_1, \dots, z_n) \in \mathbb{C}^n : z_i \neq z_j \text{ for } i \neq j\}$ of n particles moving in the plane without collision. Denote by $\mathcal{S}_n$ the symmetric group. Each permutation in $\mathcal{S}_n$ acts on $C_n(\mathbb{C})$ by permuting the coordinates. The quotient $C_n(\mathbb{C}) \diagup \mathcal{S}_n$ is called the symmetrized configuration space. We denote its elements by $\{z_1, \dots, z_n\}$ and consider them as unordered n-tuples of points in $\mathbb{C}$ or as subset E_n of $\mathbb{C}$ consisting of exactly n points. For a subset A of $\mathbb{C}$ we will put $C_n(A) = C_n(\mathbb{C}) \cap A^n$.

The configuration space inherits the topology and complex structure from $\mathbb{C}^n$. The symmetrized configuration space is given the quotient topology and quotient complex structure. Note that $\mathcal{S}_n$ acts freely and properly discontinuously on $C_n(\mathbb{C})$. The canonical projection $\mathcal{P}_{\text{sym}} : C_n(\mathbb{C}) \to C_n(\mathbb{C}) \diagup \mathcal{S}_n$ is holomorphic.

Identifying subsets of $\mathbb{C}$ consisting of n points with elements of $C_n(\mathbb{C}) \diagup \mathcal{S}_n$, we may regard a geometric braid with base point E_n as a set of the form

$$\{(t, f(t)),\ t \in [0,1]\} \subset [0,1] \times \mathbb{C} \tag{1.4}$$

for a continuous mapping $[0,1] \ni t \xrightarrow{f} C_n(\mathbb{C}) \diagup \mathcal{S}_n$ with $f(0) = f(1) = E_n$. We will often identify the geometric braid (1.4) with the mapping f defining it. Two geometric braids with base point E_n are called isotopic if there is a continuous family of geometric braids with base point E_n joining them. A braid on n strands with base point E_n is an isotopy class of geometric braids with base point E_n. Isotopy classes of geometric braids with base point E_n form a group. The operation is obtained by putting one geometric braid on the top of another.

Take a geometric braid with base point E_n. Assigning to each point in $\{0\} \times E_n$ the point in $\{1\} \times E_n$ which belongs to the same strand we obtain a permutation of the points of E_n. It depends only on the isotopy class which we denote by b. This permutation is denoted by $\tau_n(b)$. The braid b is called pure if $\tau_n(b)$ is the identity. The set of pure n-braids forms a group which we denote by $\mathcal{PB}_n$.

Algebraically, n-braids are represented as elements of the Artin braid group $\mathcal{B}_n$. This is the group with generators denoted by $\sigma_1, \ldots, \sigma_{n-1}$ and (if $n \geq 3$) with relations $\sigma_i \sigma_j = \sigma_j \sigma_i$ for $|i - j| \geq 2$, and $\sigma_i \sigma_{i+1} \sigma_i = \sigma_{i+1} \sigma_i \sigma_{i+1}$ for $i = 1, \ldots, n-2$. An isomorphism between $\mathcal{B}_n$ and isotopy classes of geometric braids with given base point E_n can be obtained as follows. Choose a projection Pr of $\mathbb{C}$ onto the real line $\mathbb{R}$ which is injective on E_n. Label the points in E_n so that $E_n = \{z_1, \ldots, z_n\}$ with $\Pr(z_j) < \Pr(z_{j+1})$. The strand of a (geometric) braid with initial point z_j is called the j-th strand. The projection $[0, 1] \times \mathbb{C} \ni (t, z) \to (t, \Pr z) \in [0, 1] \times \mathbb{R}$ assigns to each geometric braid a braid diagram (at intersection points of images of strands it is indicated which strands is "over" and which strand is "under"). Isotopy classes of geometric braids with base point E_n correspond to equivalence classes of diagrams. The equivalence classes of diagrams can be interpreted as elements of the Artin braid group. The braid diagram corresponding to the generator σ_j of the Artin braid group is the diagram with one intersection point, at this point the j-th strand is over the $j + 1$-st strand.

The isomorphism from the group of isotopy classes of geometric braids to the Artin group depends on the base point E_n and the projection Pr. A change of the base point and a change of the projection lead to conjugation with an element of the Artin group. Indeed, let α be a curve in $C_n(\mathbb{C})\diagup\mathcal{S}_n$ with initial point E_n and endpoint E_n'. A change of base point from E_n to E_n' along α assigns to each curve γ in $C_n(\mathbb{C})\diagup\mathcal{S}_n$ with initial point and endpoint equal to E_n (i.e. to each geometric braid with base point E_n) the curve $\alpha^{-1}\gamma\alpha$. The latter curve is obtained by travelling first along the arc α^{-1} in the symmetrized configuration space, that joins the new base point E_n' with E_n, then along the curve γ, and then along α. Looking at the associated braid diagram using a fixed projection (provided it is injective on both, E_n and E_n') and considering equivalence classes we obtain the statement concerning the change of base point. A change of the projection can be made by fixing the projection and considering an isotopy of the space $\mathbb{C}$. This leads to an isotopy of geometric braids, hence we obtain the statement for the change of projections.

The center $\mathcal{Z}_n$ of the Artin braid group $\mathcal{B}_n$ is generated by the element Δ_n^2, where $\Delta_n \overset{def}{=} (\sigma_1\sigma_2 \ldots \sigma_{n-1})(\sigma_1\sigma_2 \ldots \sigma_{n-2}) \ldots (\sigma_1\sigma_2)\sigma_1$ is called the Garside element. The braid Δ_n is represented by the following geometric braid. Choose a set $E_n \subset \mathbb{R}$ that is invariant under reflection in the imaginary axis. The representing geometric braid is obtained from the trivial geometric braid with base point E_n by a half-twist of the cylinder $\mathbb{C} \times [0, 1]$, i.e. by fixing the bottom of the cylinder and turning the top by the angle π. We will say that the braid Δ_n corresponds to a half-twist. The braid Δ_n^2 corresponds to a full twist. For an element e of a group we denote by $\langle e \rangle$ the subgroup that is generated by e. We put $\mathcal{Z}_n = \langle \Delta_n^2 \rangle$.

The pure braid group for $n = 3$ is related to a free group, more precisely the following lemma holds.

Lemma 1.2 *The group $\mathcal{PB}_3/\langle\Delta_3^2\rangle$ is isomorphic to the fundamental group of $\mathbb{C} \setminus \{-1, 1\}$ with base point 0. The generators $\sigma_j^2/\langle\Delta_3^2\rangle$ of $\mathcal{PB}_3/\langle\Delta_3^2\rangle$ correspond to the generators a_j of $\pi_1(\mathbb{C} \setminus \{-1, 1\}, 0)$.*

Here a_1 is represented by a loop that surrounds -1 counterclockwise, and a_2 is represented by a loop that surrounds 1 counterclockwise.

Proof For a point $z = (z_1, z_2, z_3) \in C_3(\mathbb{C})$ we denote by $M_z = M_{(z_1,z_2,z_3)}$ the Möbius transformation that maps z_1 to 0, z_3 to 1 and fixes ∞. Then $M_z(z_2)$ omits 0, 1 and ∞. Notice that $M_z(z_2)$ is equal to the cross ratio $(z_2, z_3; z_1, \infty) = \frac{z_2-z_1}{z_3-z_1} \cdot \frac{z_3-\infty}{z_2-\infty} = \frac{z_2-z_1}{z_3-z_1}$. We define $\mathfrak{C}(z) \overset{def}{=} 2M_z(z_2) - 1$ with $z = (z_1, z_2, z_3) \in C_3(\mathbb{C})$. The mapping $\mathfrak{C}$ takes $C_3(\mathbb{C})$ to $\mathbb{C} \setminus \{-1, 1\}$ and $C_3(\mathbb{R})^0 \overset{def}{=} \{(x_1, x_2, x_3) \in \mathbb{R}^3 : x_1 < x_2 < x_3\}$ to $(-1, 1)$. Associate to a curve $\tilde{\gamma}(t) = (\tilde{\gamma}_1(t), \tilde{\gamma}_2(t), \tilde{\gamma}_3(t))$, $t \in [0, 1]$, in $C_3(\mathbb{C})$ the curve $\mathfrak{C}(\tilde{\gamma})(t) \overset{def}{=} 2\,\frac{\tilde{\gamma}_2(t)-\tilde{\gamma}_1(t)}{\tilde{\gamma}_3(t)-\tilde{\gamma}_1(t)} - 1$, $t \in [0, 1]$, in $\mathbb{C}$ which omits the points -1 and 1. The mapping $\tilde{\gamma} \to \mathfrak{C}(\tilde{\gamma})$ defines a homomorphism $\mathfrak{C}_*$ from the fundamental group $\pi_1(C_3(\mathbb{C}), (-1, 0, 1))$ of $C_3(\mathbb{C})$ with base point $(-1, 0, 1)$ to the fundamental group $\pi_1 \overset{def}{=} \pi_1(\mathbb{C} \setminus \{-1, 1\}, 0)$ of the twice punctured complex plane with base point 0.

Identify the pure braid group $\mathcal{PB}_3$ with a subgroup of the fundamental group of $C_3(\mathbb{C})/\mathcal{S}_3$ with base point $\{-1, 0, 1\}$. For curves γ representing elements of this subgroup we consider the lift $\tilde{\gamma}$ under $\mathcal{P}_{\text{sym}}$ with base point $(-1, 0, 1) \in C_3(\mathbb{C})$. This gives an isomorphism from the pure braid group $\mathcal{PB}_3$ onto the fundamental group of $C_3(\mathbb{C})$ with base point $(-1, 0, 1)$.

The braids represented by the two loops $\tilde{\gamma}$ and $\mathring{\gamma}$, $\mathring{\gamma}(t) \overset{def}{=} (-1, \mathfrak{C}(\tilde{\gamma})(t), 1) = \big(2M_{\tilde{\gamma}(t)}(\tilde{\gamma}_1(t)) - 1,\, 2M_{\tilde{\gamma}(t)}(\tilde{\gamma}_2(t)) - 1,\, 2M_{\tilde{\gamma}_3(t)}(\tilde{\gamma}_1(t)) - 1\big)$, $t \in [0, 1]$, differ by a power Δ_3^{2N} of the full twist Δ_3^2. Indeed, to obtain $\mathring{\gamma}$ we act for each t by a complex linear mapping on the point $\tilde{\gamma}(t)$. The number N can be interpreted as the linking number of the first and the third strands of the geometric braid $\tilde{\gamma}(t)$, $t \in [0, 1]$. This linking number is obtained as follows. Discard the second strand. The resulting braid equals σ^{2N} where N is the mentioned linking number. For the geometric braid $\mathring{\gamma}$ the linking number of the first and third strand is zero. It follows that $\mathfrak{C}_*$ is surjective and its kernel equals $\langle\Delta_3^2\rangle$.

A word in the generators of a free group is called reduced, if neighbouring terms are powers of different generators. We saw that for each element $\boldsymbol{b}$ of the pure braid group modulo its center $\mathcal{PB}_3/\mathcal{Z}_3$ there is a unique element $b \in \mathcal{PB}_3$ that represents $\boldsymbol{b}$ and can be written as reduced word in σ_1^2 and σ_2^2. Indeed, this representative b of $\boldsymbol{b}$ is determined by the property that the linking number between the first and the third strand equals zero. Assigning to each element $\boldsymbol{b} \in \mathcal{PB}_3/\mathcal{Z}_3$ the mentioned element we obtain the isomorphism to the free group in two generators $\sigma_1^2/\mathcal{Z}_n$ and $\sigma_2^2/\mathcal{Z}_n$, or, equivalently to the fundamental group $\pi_1(\mathbb{C} \setminus \{-1, 1\}, 0)$ of the twice punctured complex plane with base point 0 with generators a_1 and a_2, respectively. □

Geometric braids with base point E_n were interpreted as paths in $C_n(\mathbb{C})/\mathcal{S}_n$ with initial and terminal point equal to E_n, in other words, as loops in this space with base point E_n. Isotopy classes of geometric braids with base point E_n correspond to homotopy classes of loops in $C_n(\mathbb{C})/\mathcal{S}_n$ with base point E_n, in other words, braids with base point E_n correspond to elements of the fundamental group $\pi_1(C_n(\mathbb{C})/\mathcal{S}_n, E_n)$ of $C_n(\mathbb{C})/\mathcal{S}_n$ with base point E_n. Thus the Artin braid group $\mathcal{B}_n$ is isomorphic to $\pi_1(C_n(\mathbb{C})/\mathcal{S}_n, E_n)$. We saw that a change of the base point leads to an automorphism of $\mathcal{B}_n$ defined by conjugation with an element of $\mathcal{B}_n$. Denote by $\widehat{\mathcal{B}_n}$ the set of conjugacy classes of $\mathcal{B}_n$. Its elements $\hat{b} \in \widehat{\mathcal{B}_n}$ can be interpreted as free homotopy classes of loops in $C_n(\mathbb{C})/\mathcal{S}_n$. In other words, two geometric braids $f_0 : [0,1] \to C_n(\mathbb{C})/\mathcal{S}_n$, $f_1 : [0,1] \to C_n(\mathbb{C})/\mathcal{S}_n$, represent the same class $\widehat{\mathcal{B}_n}$, iff there is a free homotopy joining them, i.e. if there exists a continuous mapping $h : [0,1] \times [0,1] \to C_n(\mathbb{C})/\mathcal{S}_n$ such that $h(t,0) = f_0(t)$, $h(t,1) = f_1(t)$, for $t \in [0,1]$, and $h(s,0) = h(s,1)$ for $s \in [0,1]$. We may consider the continuous family of braids $f_s : [0,1] \to C_n(\mathbb{C})/\mathcal{S}_n$ with variable base point as a free isotopy of braids.

We will also use the following terminology. A loop in $C_n(\mathbb{C})/\mathcal{S}_n$ (i.e. a continuous map of the circle into $C_n(\mathbb{C})/\mathcal{S}_n$) is called a closed geometric braid. A free homotopy class of loops in $C_n(\mathbb{C})/\mathcal{S}_n$ is called a closed braid. Closed braids correspond to elements of $\widehat{\mathcal{B}_n}$.

Arnold interpreted the symmetrized configuration space $C_n(\mathbb{C})/\mathcal{S}_n$ as the space of monic polynomials $\mathfrak{P}_n$ of degree n without multiple zeros. ("Monic" refers to the property that the coefficient by the highest power of the variable equals 1.) Denote by $\overline{\mathfrak{P}}_n$ the set of all monic polynomials of degree n. If we assign to each unordered n-tuple $E_n = \{z_1, \dots, z_n\}$ (this time the z_j are not necessarily pairwise distinct) the monic polynomial $\prod_{j=1}^{n}(z - z_j) = a_0 + a_1 z + \dots z^n$ whose set of roots equal E_n, we obtain a bijection onto the set $\overline{\mathfrak{P}}_n$. The set of monic polynomials of degree n can also be parameterized by the ordered n-tuple $(a_0, a_1, \dots, a_{n-1}) \in \mathbb{C}^n$ of the coefficients. Hence, we obtain a bijection between $C_n(\mathbb{C})/\mathcal{S}_n$ and the set of points $(a_0, a_1, \dots, a_{n-1}) \in \mathbb{C}^n$ which are coefficients of polynomials without multiple zeros. There is a polynomial D_n in the variables $a_0, \dots, a_{n-1}$, called the discriminant, which vanishes exactly if the polynomial with these coefficients has multiple zeros. Hence, the symmetrized configuration space $C_n(\mathbb{C})/\mathcal{S}_n$ can be identified with $\mathbb{C}^n \setminus V_{\mathsf{D}_n}$ where $V_{\mathsf{D}_n} \overset{\text{def}}{=} \{(a_0, \dots, a_{n-1}) \in \mathbb{C}^n : \mathsf{D}_n(a_0, \dots, a_{n-1}) = 0\}$. The identification is actually a biholomorphic map. Thus, $C_n(\mathbb{C})/\mathcal{S}_n$ is biholomorphic to a pseudoconvex domain in $\mathbb{C}^n$, even stronger, it is biholomorphic to the complement of an algebraic hypersurface in $\mathbb{C}^n$. The space $\mathfrak{P}_n \cong C_n(\mathbb{C})/\mathcal{S}_n$ received much attention in connection with problems of algebraic geometry. For instance, motivated by his interest in the Thirteen's Hilbert problem, Arnold [7] computed its topological invariants.

1.4 Mapping Class Groups

Mapping class groups are defined as follows. Let A be a topological space, not necessarily compact, but paracompact. Let A_1 and A_2 be disjoint closed subsets of A. Denote by $\mathrm{Hom}(A; A_1, A_2)$ the set of self-homeomorphisms of A that fix A_1 pointwise and A_2 setwise. We will also write $\mathrm{Hom}(A; A_1)$ for $\mathrm{Hom}(A; A_1, \emptyset)$, but in case we do not require that some points are fixed we write $\mathrm{Hom}(A; \emptyset, A_2)$. We will also write $\mathrm{Hom}(A)$ for $\mathrm{Hom}(A; \emptyset, \emptyset)$. Equip the set $\mathrm{Hom}(A; A_1, A_2)$ with compact open topology.

Let A be an oriented manifold. By $\mathrm{Hom}^+(A; A_1, A_2)$ we denote the set of orientation preserving self-homeomorphisms in $\mathrm{Hom}(A; A_1, A_2)$. It forms a group with respect to composition. The set of connected components of $\mathrm{Hom}^+(A; A_1, A_2)$ equipped with the inherited group structure is the mapping class group $\mathfrak{M}(A; A_1, A_2)$ corresponding to the triple (A, A_1, A_2).

Let $\bar{\mathbb{D}}$ be the closed unit disc in the complex plane $\mathbb{C}$ with boundary $\partial\mathbb{D}$ and interior $\mathbb{D}$. At this point we may forget the complex structure of $\mathbb{D}$, but we will need it later.

Let $E_n^0 = \left\{0, \frac{1}{n}, \dots, \frac{n-1}{n}\right\}$ be the "standard" subset of $\mathbb{D}$ containing n points. We also identify E_n^0 with the respective unordered n-tuple of points. The set $\mathfrak{M}(\bar{\mathbb{D}}; \partial\mathbb{D}, E_n^0)$ is commonly known as the mapping class group of the n-punctured disc. Note that $\mathfrak{M}(\bar{\mathbb{D}}; \partial\mathbb{D}, E_n^0)$ is isomorphic to $\mathfrak{M}(\bar{\mathbb{D}}\backslash E_n^0; \partial\mathbb{D})$ since each element of $\mathfrak{M}(\bar{\mathbb{D}}\backslash E_n^0; \partial\mathbb{D})$ extends continuously to each point of E_n^0 (the "punctures"). The points of E_n^0 are also called "distinguished points". The homeomorphisms in $\mathrm{Hom}^+(\bar{\mathbb{D}}; \partial\mathbb{D}, E_n^0)$ are called homeomorphisms of $\bar{\mathbb{D}}$ with distinguished points E_n^0. By an abuse of language we will also call them homeomorphisms of the n-punctured disc, having in mind the identification with their restrictions to $\mathbb{D} \setminus E_n^0$. The connected component $\mathrm{Hom}^0(\bar{\mathbb{D}}; \partial\mathbb{D}, E_n^0)$ of $\mathrm{Hom}^+(\bar{\mathbb{D}}; \partial\mathbb{D}, E_n^0)$ containing the identity consists of the self-homeomorphisms of $\bar{\mathbb{D}}$ which can be joined to the identity by a continuous family of homeomorphisms in $\mathrm{Hom}^+(\bar{\mathbb{D}}; \partial\mathbb{D}, E_n^0)$. In other words, it consists of homeomorphisms in $\mathrm{Hom}^+(\bar{\mathbb{D}}; \partial\mathbb{D}, E_n^0)$ which are isotopic to the identity through homeomorphisms fixing $\partial\mathbb{D}$ and E_n^0 pointwise. We have

$$\mathfrak{M}(\bar{\mathbb{D}}; \partial\mathbb{D}, E_n^0) = \mathrm{Hom}^+(\bar{\mathbb{D}}, \partial\mathbb{D}, E_n^0)/\mathrm{Hom}^0(\bar{\mathbb{D}}; \partial\mathbb{D}, E_n^0)\,. \tag{1.5}$$

The respective mapping class group $\mathfrak{M}(\bar{\mathbb{D}}; \partial\mathbb{D}, E_n)$ can be defined for any unordered n-tuple E_n of points in $\mathbb{D}$.

For a homeomorphism $\varphi \in \mathrm{Hom}^+(\bar{\mathbb{D}}; \partial\mathbb{D}, E_n)$ we consider its restriction $\varphi \mid \mathbb{D} \in \mathrm{Hom}^+(\mathbb{D}; \emptyset, E_n)$ and the corresponding mapping class in $\mathfrak{M}(\mathbb{D}; \emptyset, E_n)$. This mapping class is denoted by $\mathfrak{m}_\varphi^{\mathrm{free}}$ and is called the free isotopy class of φ. We also call the elements of $\mathfrak{M}(\bar{\mathbb{D}}; \partial\mathbb{D}, E_n)$ the relative mapping classes. The restriction map $\varphi \to \varphi \mid \mathbb{D}$ defines a surjective mapping from $\mathfrak{M}(\bar{D}; \partial\mathbb{D}, E_n)$ to $\mathfrak{M}(\mathbb{D}; \emptyset, E_n)$.

Indeed, each element of $\mathfrak{M}(\mathbb{D}; \emptyset, E_n)$ contains representatives which extend to the boundary $\partial\mathbb{D}$ as the identity mapping on $\partial\mathbb{D}$. There is a short exact sequence

$$0 \to \mathcal{K} \to \mathfrak{M}(\bar{\mathbb{D}}; \partial\mathbb{D}, E_n) \to \mathfrak{M}(\mathbb{D}; \emptyset, E_n) \to 0. \tag{1.6}$$

An element $\mathfrak{m} \in \mathfrak{M}(\bar{\mathbb{D}}; \partial\mathbb{D}, E_n)$ is in the kernel $\mathcal{K}$ iff each representative is isotopic to the identity through homeomorphisms of $\mathbb{D}$ which fix E_n setwise (and, hence, pointwise).

Let X be an oriented surface. A Dehn twist about a simple closed curve γ in an oriented surface S is a mapping that is isotopic to the following one. Take a neighbourhood of γ that can be parameterized as a round annulus $A = \{e^{-\varepsilon} < |z| < 1\}$ so that γ corresponds to $|z| = e^{-\frac{\varepsilon}{2}}$. The mapping is an orientation preserving self-homeomorphism of S which is the identity outside A and is equal to the mapping $e^{-\varepsilon s+2\pi i t} \to e^{-\varepsilon s+2\pi i(t+s)}$ for $e^{-\varepsilon s+2\pi i t} \in A$, i.e. $s \in (0, 1)$. Here ε is a small positive number.

For the following theorem see e.g. [42].

Theorem 1.3 *The kernel $\mathcal{K}$ is generated by a Dehn twist about a simple closed curve γ in $\mathbb{D}$ which is homologous to $\partial\mathbb{D}$ in $\mathbb{D}\setminus E_n$.*

Let again $E_n \subset \mathbb{D}$ be a set that contains exactly n points. The inclusion $\overline{\mathbb{D}} \hookrightarrow \mathbb{P}^1$ induces a surjective homomorphism $\mathcal{H}_\infty$

$$\mathcal{H}_\infty : \mathfrak{M}(\bar{\mathbb{D}}; \partial\,\mathbb{D}, E_n) \to \mathfrak{M}(\mathbb{P}^1; \infty, E_n), \tag{1.7}$$

which is described in detail as follows. Take a mapping class $\mathfrak{m} \in \mathfrak{M}(\overline{\mathbb{D}}; \partial\,\mathbb{D}, E_n)$. Represent it by a homeomorphism φ. Extend φ to a self-homeomorphism φ_∞ of $\mathbb{P}^1$ by putting $\varphi_\infty = \varphi$ on $\overline{\mathbb{D}}$ and $\varphi_\infty = \text{id}$ outside $\overline{\mathbb{D}}$. Let $\mathfrak{m}_\infty \in \mathfrak{M}(\mathbb{P}^1; \infty, E_n)$ be the mapping class of φ_∞. It depends only on the mapping class $\mathfrak{m}$ of φ in $\mathfrak{M}(\bar{\mathbb{D}}; \partial\,\mathbb{D}, E_n)$, not on the choice of the representative φ. Put

$$\mathcal{H}_\infty(\mathfrak{m}) \stackrel{def}{=} \mathfrak{m}_\infty \in \mathfrak{M}(\mathbb{P}^1; \infty, E_n). \tag{1.8}$$

The homomorphism is surjective since each element of $\mathfrak{M}(\mathbb{P}^1; \infty, E_n)$ can be represented by a homeomorphism that is equal to the identity on $\mathbb{P}^1 \setminus \mathbb{D}$. By Theorem 1.3 the kernel of the homomophism (1.8) consists of the class of powers of Dehn twists about a simple closed curve that is homologous to $\partial\mathbb{D}$.

Let A and A' be oriented manifolds, let A_1 and A_2 be subsets of A, and let A'_1 and A'_2 be subsets of A'. Take a homeomorphism $\psi : A' \to A$ that maps A'_j to A_j for $j = 1, 2$. The mapping $\varphi \to \psi^{-1} \circ \varphi \circ \psi$ is a one-to-one mapping from $\text{Hom}^+(A; A_1, A_2)$ onto $\text{Hom}^+(A'; A'_1, A'_2)$. It defines an isomorphism is_ψ from the mapping class $\mathfrak{M}(A; A_1, A_2)$ onto the mapping class $\mathfrak{M}(A'; A'_1, A'_2)$. The isomorphism is determined by ψ up to conjugation by an element of $\mathfrak{M}(A; A_1, A_2)$.

Indeed, let $\psi_1 : A' \to A$ be another homeomorphism that maps A'_1 to A_1 and A'_2 to A_2. Then

$$(\psi_1)^{-1} \circ \varphi \circ \psi_1 = (\psi^{-1} \circ \psi_1)^{-1} \circ \psi^{-1} \circ \varphi \circ \psi \circ (\psi^{-1} \circ \psi_1)\,.$$

The conjugacy class $\hat{\mathfrak{m}}$ of an element $\mathfrak{m}$ of the mapping class group $\mathfrak{M}(A; A_1, A_2)$ consists of all elements $\mathfrak{g}^{-1}\mathfrak{m}\mathfrak{g}$ with $\mathfrak{g} \in \mathfrak{M}(A; A_1, A_2)$.The set of conjugacy classes is denoted by $\widehat{\mathfrak{M}}(A; A_1, A_2)$. We will identify the conjugacy class of an element $\mathfrak{m} \in \mathfrak{M}(A; A_1, A_2)$ with the set

$$\big\{\psi^{-1}\varphi \circ \psi : \varphi \in \mathfrak{m},\ \psi : A' \to A \text{ a homeomorphism that maps } A'_1 \text{ to } A'_2 \text{ and } A'_2 \text{ to } A_2\big\}\,,$$

having in mind the isomorphisms is_{ψ}^{-1} that are determined up to conjugation by elements of $\mathfrak{M}(A; A_1, A_2)$.

Consider the set of conjugacy classes $\widehat{\mathfrak{M}}(\mathbb{P}^1; \infty, E_n)$. Any homeomorphism $h : \mathbb{D} \setminus E_n \to \mathbb{C} \setminus E_n$ that maps points close to $\partial\mathbb{D}$ to points close to ∞ defines an isomorphism

$$\mathfrak{M}(\mathbb{D}; \emptyset, E_n) \to \mathfrak{M}(\mathbb{P}^1; \infty, E_n)$$

by associating to each orientation preserving self-homeomorphism $\varphi : \mathbb{D} \setminus E_n \circlearrowleft$ the self-homeomorphism $h \circ \varphi \circ h^{-1}$ of $\mathbb{C} \setminus E_n$. The isomorphisms corresponding to different homomorphisms $\mathbb{D} \setminus E_n \to \mathbb{C} \setminus E_n$ differ by conjugation by a self-homeomorphism of $\mathbb{C} \setminus E_n$ whose extension to $\mathbb{P}^1$ fixes ∞. We obtain a one-to-one correspondence between the set of conjugacy classes $\widehat{\mathfrak{M}}(\mathbb{D} \setminus E_n)$ and the set of conjugacy classes $\widehat{\mathfrak{M}}(\mathbb{P}^1 \setminus E_n; \infty) \cong \widehat{\mathfrak{M}}(\mathbb{P}^1; \infty, E_n)$.

More generally, if X is a closed oriented surface and E_n a subset of X consisting of n points, the space $\mathfrak{M}(X; \emptyset, E_n) \cong \mathfrak{M}(X \setminus E_n)$ is called the mapping class group of the n-punctured surface. Let X be a compact connected oriented surface with boundary and let E_n be a subset of X that contains exactly n points. The set $\mathfrak{M}(X; \partial X, E_n) \cong \mathfrak{M}(X \setminus E_n; \partial X)$ is called the mapping class group of the n-punctured surface with boundary. Consider the mapping $\varphi \to \varphi|\mathrm{Int}X$ that assigns to each self-homeomorphism φ of X the restriction to the interior $\mathrm{Int}X$ of X. This mapping defines a homomorphism $\mathfrak{M}(X; \partial X,\ E_n) \to \mathfrak{M}(\mathrm{Int}X; \emptyset,\ E_n)$, whose kernel is generated by Dehn twists about closed curves in X that are homologous to a boundary component of X.

Let X be a bordered Riemann surface with smooth boundary. There is a compact Riemann surface X^c and a diffeomorphism from X onto the closure of a smoothly bounded domain in X^c with the following properties. The diffeomorphism is conformal on $\mathrm{Int}\,X$. The domain in X^c is obtained by removing from X^c a finite number of smoothly bounded topological discs. This follows from the fact, that the interior of X admits a conformal mapping onto a smoothly bounded domain on a closed Riemann surface ([77], see also Sect. 1.2), and the smooth analog of Caratheodory's Theorem (Theorem 4 Chapter 2, Section 3, in [32]) that states smooth extension to the boundary of conformal mappings between smoothly bounded domains (see e.g. [8]). Let $\partial_1, \ldots, \partial_N$ be the boundary components of X and let $\delta_1, \ldots, \delta_N$ be the open discs on X^c bounded by the ∂_j. Let $E_n \subset \operatorname{Int} X$ be a finite set. For each $j = 1, \ldots, N$ we pick a point $\zeta_j \in \delta_j$. Put $\zeta = \{\zeta_1, \ldots, \zeta_N\}$. There is a homomorphism

$$\mathcal{H}_\zeta : \mathfrak{M}(X; \partial X, E_n) \to \mathfrak{M}(X^c; \{\zeta_1, \ldots, \zeta_N\}, E_n) \tag{1.9}$$

which can be described as follows. Take $\mathfrak{m} \in \mathfrak{M}(X; \partial X, E_n)$. Let $\varphi \in \operatorname{Hom}^+(X; \partial X, E_n)$ be a representing homeomorphism. Let φ^c be the extension of φ to X^c which is the identity outside X. Then $\mathcal{H}_\zeta(\mathfrak{m}) = \mathfrak{m}_\zeta$ where $\mathfrak{m}_\zeta$ is the class of φ^c in $\mathfrak{M}(X^c; \{\zeta_1, \ldots, \zeta_N\}, E_n)$. The homomorphism (1.9) is surjective. Its kernel is generated by Dehn twists about closed curves that are homologous to a boundary component of X.

We remove now the requirement that the elements of a mapping class fix the boundary of a bordered Riemann surface X pointwise (or fixes each boundary component setwise), in other words we consider mapping classes in $\mathfrak{M}(X; \emptyset, E_n)$. We obtain a homomorphism

$$\mathfrak{M}(\operatorname{Int} X; \emptyset, E_n) \to \mathfrak{M}(X^c \setminus \{\zeta_1, \ldots, \zeta_N\}; \emptyset, E_n)\,.$$

We may identify the conjugacy classes

$$\widehat{\mathfrak{M}}(\operatorname{Int} X; \emptyset, E_n) \cong \widehat{\mathfrak{M}}(X^c \setminus \{\zeta_1, \ldots, \zeta_N\}; \emptyset, E_n)\,.$$

1.5 The Relation Between Braids and Mapping Classes

In this section we describe the isomorphism between the braid group $\mathcal{B}_n$ and the mapping class group of the n-punctured disc. We use complex notation.

For any subset E of the unit disc and any self-homeomorphism ψ in $\operatorname{Hom}^+(\overline{\mathbb{D}}, \emptyset, \emptyset)$ we put $\operatorname{ev}_E \psi = \psi(E)$. For $E_n^0 = \left\{0, \frac{1}{n}, \ldots, \frac{n-1}{n}\right\}$ (considered as unordered tuple of n points or as set) we obtain

$$\operatorname{ev}_{E_n^0} \psi = \left\{\psi(0), \psi\left(\frac{1}{n}\right), \ldots, \psi\left(\frac{n-1}{n}\right)\right\}. \tag{1.10}$$

We define the mapping e_n,

$$e_n(\psi) = \left(\psi(0), \ldots, \psi\left(\frac{n-1}{n}\right)\right) \tag{1.11}$$

that assigns to ψ the ordered n-tuple of values of ψ at the points $0, \frac{1}{n}, \ldots, \frac{n-1}{n}$. Note that $e_n(\psi) \in C_n(\mathbb{C})$, and $\mathcal{P}_{\text{sym}}\, e_n(\psi) = \text{ev}_{E_n^0}\, \psi$. The isomorphism between the mapping class group $\mathfrak{M}(\overline{\mathbb{D}}, \partial\,\mathbb{D}, E_n^0)$ and the group of isotopy classes of geometric braids with base point E_n^0 is obtained as follows.

Let $\varphi \in \text{Hom}^+(\overline{\mathbb{D}}; \partial\,\mathbb{D}, E_n^0)$. Consider a path $\varphi_t \in \text{Hom}^+(\overline{\mathbb{D}}, \partial\,\mathbb{D})$, $t \in [0, 1]$, which joins φ with the identity. In other words, φ_t is a continuous family of self-homeomorphisms of $\overline{\mathbb{D}}$ which fix the boundary $\partial\,\mathbb{D}$ pointwise, such that $\varphi_0 = \text{id}$ and $\varphi_1 = \varphi$. Notice that we do not require that φ_t maps E_n^0 to itself. By the Alexander-Tietze Theorem (see e.g. [52], Section 1.6.1) such a family exists for each $\varphi \in \text{Hom}^+(\overline{\mathbb{D}}; \partial\,\mathbb{D}, E_n^0)$. Consider the evaluation map

$$[0, 1] \ni t \to \text{ev}_{E_n^0}\, \varphi_t = \left\{\varphi_t(0), \ldots, \varphi_t\left(\frac{n-1}{n}\right)\right\} \in C_n(\mathbb{C})/\mathcal{S}_n\,. \tag{1.12}$$

This map defines a geometric braid in the cylinder $[0, 1] \times \mathbb{D}$ (i.e. $\text{ev}_{E_n^0}\, \varphi_t \in C_n(\mathbb{D})/\mathcal{S}_n$ for each t). Its base point is

$$E_n^0 = \varphi_0(E_n^0) = \varphi_1(E_n^0)\,.$$

Notice that the isotopy class of the obtained geometric braid depends only on the class of φ in $\mathfrak{M}(\overline{\mathbb{D}}; \partial\,\mathbb{D}, E_n^0)$. The obtained mapping from $\mathfrak{M}(\overline{\mathbb{D}}; \partial\,\mathbb{D}, E_n^0)$ to the group of braids with base point E_n^0, hence to $\mathcal{B}_n$, is a homomorphism. It is, in fact, an isomorphism. This is a consequence of Proposition 1.1 below.

Let E_n be an arbitrary unordered n-tuple of points in $\mathbb{D}$. A continuous family of homeomorphisms $\varphi_t \in \text{Hom}^+(\overline{\mathbb{D}}; \partial\,\mathbb{D})$, $t \in [0, 1]$, is called a parameterizing isotopy of a geometric braid $f : [0, 1] \to C_n(\mathbb{D})/\mathcal{S}_n$ with base point E_n if $\varphi_0 = \text{id}$ and $\text{ev}_{E_n}\, \varphi_t = f(t)$, $t \in [0, 1]$.

Proposition 1.1 *Let $f : [0, 1] \to C_n(\mathbb{D})/\mathcal{S}_n$ be a smooth geometric braid in $[0, 1] \times \mathbb{D}$ with any base point $E_n \in C_n(\mathbb{D})/\mathcal{S}_n$. Then there exists a smooth parameterizing isotopy φ_t for f. The mappings φ_t can be chosen so that the map $[0, 1] \times \overline{\mathbb{D}} \ni (t, z) \to (t, \varphi_t(z)) \in [0, 1] \times \overline{\mathbb{D}}$ is a diffeomorphism.*

For the continuous version see e.g. [52].

Proof of Proposition 1.1 Lift the mapping f to a mapping $\tilde{f} : [0, 1] \to C_n(\mathbb{D})$. Denote the coordinate functions of $\tilde{f}$ by $f_j : [0, 1] \to \mathbb{D}$. Choose $\delta > 0$ so that for each t the discs $\mathcal{U}_j(t)$ of radius δ around the points $f_j(t)$, $j = 1, \ldots, n$, are pairwise disjoint subsets of $\mathbb{D}$.

Consider for each j the tube $\mathfrak{T}_j = \bigcup_{t \in [0,1]} (t, \mathcal{U}_j(t))$. Define a smooth mapping v from the union $\bigcup_j \mathfrak{T}_j$ of the tubes to the complex plane $\mathbb{C}$ by putting

$$v(t, \zeta) \overset{def}{=} \frac{\partial}{\partial t} f_j(t), \quad (t, \zeta) \in \mathfrak{T}_j\,. \tag{1.13}$$

Notice that for $(t, \zeta) \in \mathfrak{T}_j$ the point ζ is in the δ-neighbourhood $\mathcal{U}_j(t)$ of $f_j(t)$ in $\mathbb{D}$. For each t the mapping v is constant on each $\mathcal{U}_j(t)$ as a function of ζ. The function f_j satisfies the differential equation

$$\frac{\partial}{\partial t} f_j(t) = v(t, f_j(t)),\ t \in [0, 1]. \tag{1.14}$$

Indeed, $(t, f_j(t))$ is contained in $\mathfrak{T}_j$, hence, $v(t, f_j(t)) = \frac{\partial}{\partial t} f_j(t)$.

Let $\mathfrak{T}_j^0 \Subset \mathfrak{T}_j$ be the $\frac{\delta}{2}$-tubes in $[0, 1] \times \mathbb{D}$ around the graphs of f_j. Let χ_0 be a C^∞-function on $\mathbb{C}$ with values in $[0, 1]$ that equals 1 on $|\zeta| \leq \frac{\delta}{2}$ and equals 0 outside $|\zeta| < \delta$. Define a mapping χ on $[0, 1] \times \mathbb{D}$ by the equations $\chi(t, \zeta) = \chi_0(\zeta - f_j(t))$ for $(t, \zeta) \in \mathfrak{T}_j$, $j = 1, \ldots, n$, and $\chi(t, \zeta) = 0$ else. Then χ is a C^∞-function on $[0, 1] \times \mathbb{D}$ with values in $[0, 1]$ that equals 1 in the union $\bigcup_j \mathfrak{T}_j^0$ of the smaller tubes and equals 0 outside the union of the bigger tubes $\bigcup_j \mathfrak{T}_j$. Put

$$V(t, \zeta) = \begin{cases} v(t, \zeta) \cdot \chi(t, \zeta) & \text{if } (t, \zeta) \in \bigcup_j \mathfrak{T}_j \\ 0 & \text{otherwise} \end{cases} \tag{1.15}$$

For $\zeta \in \mathbb{D}$ we denote by $\varphi_t(\zeta)$ the solution of the initial value problem

$$\frac{\partial}{\partial t} \varphi_t(\zeta) = V(t, \varphi_t(\zeta)),\ \varphi_0(\zeta) = \zeta, \tag{1.16}$$

on the maximal interval of existence. By the smooth version of the Picard-Lindelöf Theorem the initial value problem has a unique solution on a maximal interval of existence and the solution depends smoothly on all parameters (see e.g. [37], Theorem V, 3.1, Corollary V, 3.1 and the remark after it, and also Theorem V, 4.1). Near the ends of the maximal interval of existence the solution curve approaches the boundary of the domain. Since $V = 0$ in a neighbourhood of $[0, 1] \times \partial\mathbb{D}$, the solution curve approaches $\{1\} \times \mathbb{D}$, in other words the maximal interval of existence equals $[0, 1]$. Moreover, for each t the mapping $\zeta \to \varphi_t(\zeta)$ is a local diffeomorphism, and hence by the uniqueness theorem for solutions of initial value problems, this mapping is a global diffeomorphism. Since $V = v$ on the union $\bigcup_j \mathfrak{T}_j^0$ of the smaller tubes, Eq. (1.14) yields $\varphi_t(f_j(0)) = f_j(t)$ for all j and $t \in [0, 1]$. Since V vanishes in a neighbouhood of $[0, 1] \times \partial\mathbb{D}$, $\varphi_t(\zeta) = \zeta$ for ζ close to $\partial\mathbb{D}$. Hence, φ_t, $t \in [0, 1]$, is a parameterizing isotopy. □

Remark 1.1 *Let $f : [0, 1] \times [0, 1] \to C_n(\mathbb{D}) \diagup \mathcal{S}_n$ be a smooth isotopy of braids with fixed base point E_n, that joins the braids $t \to f(t, 0)$ and $t \to f(t, 1)$. Then there exists a diffeomorphism $[0, 1] \times [0, 1] \times \overline{\mathbb{D}} \ni (t, s, \zeta) \overset{\varphi_{t,s}(\zeta)}{\longrightarrow} [0, 1] \times [0, 1] \times \overline{\mathbb{D}}$ such that $\varphi_{t,s}(E_n) = f(t, s)$, $(t, s) \in [0, 1] \times [0, 1]$, and $\varphi_{t,s}$ is the identity on $[0, 1] \times [0, 1] \times \partial\mathbb{D}$. If φ_t^0 is a given smooth parameterizing isotopy for $t \to f(t, 0)$ and φ_t^1 is a given smooth parameterizing isotopy for $t \to f(t, 1)$, then the family $\varphi_{t,s}$ may be chosen so that $\varphi_{t,0} = \varphi_t^0$, $\varphi_{t,1} = \varphi_t^1$, and $\varphi_{0,s} = \mathrm{Id}$.*

The proof follows along the same lines as the proof of Proposition 1.1. We construct a vector field V on $[0, 1] \times [0, 1] \times \overline{\mathbb{D}}$, i.e. a vector field that depends on $(\zeta, t) \in \overline{\mathbb{D}} \times [0, 1]$ and the additional parameter $s \in [0, 1]$, such that

$$V(t, s, f_j(t, s)) = \frac{\partial}{\partial t} f_j(t, s) \text{ for } (t, s) \in [0, 1] \times [0, 1]$$

for the strands f_j of f, and solve the initial value problem

$$\frac{\partial}{\partial t}\varphi_{t,s}(\zeta) = V(t, s, \varphi_{t,s}(\zeta)),\ \varphi_{0,s}(\zeta) = \zeta\ ,\ (s, \zeta) \in [0, 1] \times \overline{\mathbb{D}}\,.$$

The families $\varphi_{t,0}$ and $\varphi_{t,1}$ define vector fields $\frac{\partial}{\partial t}\varphi_{t,0}$ and $\frac{\partial}{\partial t}\varphi_{t,1}$ for (t, s, ζ) in $[0, 1] \times \{0\} \times \overline{\mathbb{D}}$ and $[0, 1] \times \{1\} \times \overline{\mathbb{D}}$. The vector field V can be chosen to coincide with these vector fields on the respective sets. We omit the details.

By the proposition the inverse of the mapping $\mathfrak{M}(\overline{\mathbb{D}}; \partial\, \mathbb{D}, E_n^0) \to \mathcal{B}_n$ is obtained as follows. Take a braid $b \in \mathcal{B}_n$ and choose a representing smooth geometric braid in $[0, 1] \times \mathbb{D}$. Consider a parameterizing isotopy φ_t. Associate to b the mapping class of the homeomorphism φ_1.

Explicitly, the inverse mapping assigns to each generator $\sigma_j \in \mathcal{B}_n$ the class of the following homeomorphism which is called a half-twist around the interval $\left[\frac{j-1}{n}, \frac{j}{n}\right]$. Take two open discs D_1 and D_2 centered at the midpoint of the segment $\left[\frac{j-1}{n}, \frac{j}{n}\right]$, such that $\left[\frac{j-1}{n}, \frac{j}{n}\right] \subset D_1, \bar{D}_1 \subset D_2$, $\bar{D}_2$ does not contain points of E_n^0 other than $\frac{j-1}{n}$ and $\frac{j}{n}$. Define φ_{σ_j} to be the identity on $\overline{\mathbb{D}}\backslash D_2$ and to be counterclockwise rotation by the angle π on $\bar{D}_1$. Extend this mapping by a homeomorphism of $\bar{D}_2\backslash D_1$ which changes the argument of each point by a non-negative value at most equal to π (see Fig. 1.4). We will denote the mapping $\mathcal{B}_n \to \mathfrak{M}(\overline{\mathbb{D}}; \partial\, \mathbb{D}, E_n^0)$ by Θ_n.

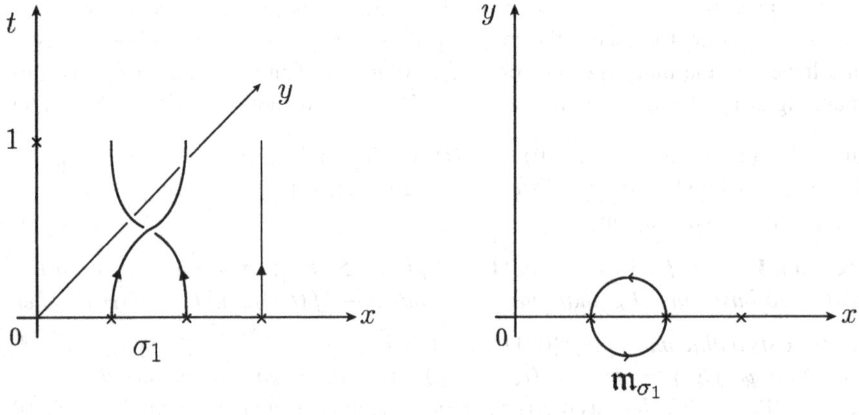

Fig. 1.4 The braid $\sigma_1 \in \mathcal{B}_3$ and the corresponding mapping class $\mathfrak{m}_{\sigma_1}$

The mapping class $\mathfrak{m}_{\Delta_n^2}$ corresponding to Δ_n^2 is the Dehn twist about a curve in $\mathbb{D}$ that is homologous to $\partial\mathbb{D}$ in $\mathbb{D} \setminus E_n^0$. Notice that the elements of the set of conjugacy classes $\widehat{\mathcal{B}}_n$ of braids are in one-to-one correspondence with the elements of the set of conjugacy classes $\widehat{\mathfrak{M}}(\overline{\mathbb{D}}; \partial\, \mathbb{D}, E_n^0)$ of mapping classes.

1.6 Beltrami Differentials and Quadratic Differentials

In this section we collect some prerequisites from Teichmüller theory, in particular we recall some facts related to Beltrami differentials and quadratic differentials. For more details we refer the reader for example to [4, 59, 67, 78].

Let X be a Riemann surface. A Beltrami differential μ on X assigns to each chart on X with holomorphic coordinates z an essentially bounded measurable function $\mu(z)$ so that $\mu(z)\,\frac{d\bar{z}}{dz}$ is invariant under holomorphic coordinate changes. In other words, the functions $\mu_1(z)$ and $\mu_2(\zeta)$ associated to local coordinates $z = z(\zeta)$ and ζ are related by the equation

$$\mu_1(z(\zeta)) \left(\frac{dz}{d\zeta}\right)^{-1} \overline{\left(\frac{dz}{d\zeta}\right)} = \mu_2(\zeta)\,. \tag{1.17}$$

We may interpret Beltrami differentials as sections in the bundle $\kappa^{-1} \otimes \bar{\kappa}$ for the cotangent bundle κ of X.

By an abuse of notation we will denote the Beltrami differential on X by μ and write $\mu = \mu(z)\,\frac{d\bar{z}}{dz}$, where the left hand side denotes the globally defined Beltrami differential and $\mu(z)$ on the right hand side is a representing function in local coordinates z. The value $|\mu(z)|$ is invariant under holomorphic coordinate change. Put $\|\mu\|_\infty \overset{\text{def}}{=} \sup\limits_X |\mu|$.

Let X and Y be Riemann surfaces and let $\varphi : X \to Y$ be a smooth orientation preserving homeomorphism. Let z be coordinates near a point of X and let ζ be coordinates on Y near its image under φ. We write $\zeta = \varphi(z)$ in these coordinates. Consider the function $\frac{\bar{\partial}\varphi}{\partial\varphi} = \frac{\frac{\partial}{\partial\bar{z}}\varphi}{\frac{\partial}{\partial z}\varphi}$ in these coordinates. Since for the Jacobian $\mathcal{J}(z)$ of $\varphi(z)$, $\mathcal{J}^2(z) = \left|\frac{\partial}{\partial z}\varphi(z)\right|^2 - \left|\frac{\partial}{\partial\bar{z}}\varphi(z)\right|^2 > 0$, the denominator $\partial\varphi$ does not vanish and $\left|\frac{\bar{\partial}\varphi}{\partial\varphi}\right| < 1$. The expression $\frac{\frac{\partial}{\partial\bar{z}}\varphi(z)\,d\bar{z}}{\frac{\partial}{\partial z}\varphi(z)\,dz}$ is invariant under holomorphic coordinate changes on X and on Y. It defines a Beltrami differential μ_φ on X. The mapping φ is called quasiconformal if $\|\mu_\varphi\|_\infty < 1$. If X and Y are compact this is automatically so. The condition that φ is differentiable can be weakened.

For the Beltrami differential of superpositions of quasiconformal mappings the following formulas hold.

$$\mu_g \circ f = \frac{\partial_z f}{\partial_{\bar{z}} \bar{f}} \frac{\mu_{g \circ f} - \mu_f}{1 - \overline{\mu_f} \, \mu_{g \circ f}} . \tag{1.18}$$

If g is conformal, then $\mu_g = 0$ and $\mu_{g \circ f} = \mu_f$. If f is conformal, then $\mu_f = 0$ and

$$\mu_g \circ f = \left(\frac{f'}{|f'|}\right)^2 \mu_{g \circ f} . \tag{1.19}$$

For a detailed account on quasiconformal mappings we refer to [4].

The quasiconformal dilatation of the mapping φ is defined as $K(\varphi) = \frac{1+\|\mu_\varphi\|_\infty}{1-\|\mu_\varphi\|_\infty}$ if φ is quasiconformal and is defined to be equal to ∞ otherwise.

A meromorphic (respectively, holomorphic) quadratic differential ϕ on X assigns to each chart on X with holomorphic coordinates z a meromorphic (respectively, holomorphic) function which by abusing notation we denote by $\phi(z)$, such that $\phi(z)(dz)^2$ is invariant under holomorphic changes of coordinates. In other words, the functions $\phi_1(z)$ and $\phi_2(\zeta)$ associated to local coordinates ζ and $z(\zeta)$ are related by the transformation rule

$$\phi_1(z(\zeta)) \left(\frac{dz}{d\zeta}\right)^2 = \phi_2(\zeta) . \tag{1.20}$$

Holomorphic quadratic differentials on X can be regarded as holomorphic sections of the bundle κ^2. By an abuse of notation we will write $\phi = \phi(z)\, dz^2$, where the left hand side denotes the quadratic differential and $\phi(z)$ on the right hand side denotes the meromorphic function that represents ϕ in local coordinates z. If the function $\phi_1(\zeta)$ corresponding to local coordinates ζ vanishes or equals ∞ at a point ζ_0 the function $\phi_2(z)$ corresponding to local coordinates $z(\zeta)$ vanishes at the point $z(\zeta_0)$ or equals ∞ there. The set of points on X where this happens is called the set of singular points of the quadratic differential ϕ. The other points are called regular points of the quadratic differential.

Suppose on a simply connected domain U with coordinates ζ the quadratic differential equals $\phi(\zeta)(d\zeta)^2$ where $\phi \neq 0$ is a nowhere vanishing analytic function (in particular ϕ has no pole on U). Consider the differential form $\sqrt{\phi(\zeta)}d\zeta$ on U for a branch of the square root of ϕ. There are local holomorphic coordinates z of the quadratic differential $\phi(\zeta)\, d\zeta^2$ on U in which the quadratic differential has the form dz^2. They are called flat coordinates. Such coordinates can be obtained as an integral of the mentioned differential form,

$$z(\zeta) = \int_{\zeta_0}^{\zeta} \sqrt{\phi(\zeta')}d\zeta' . \tag{1.21}$$

Together with the complex differential form $\sqrt{\phi(\zeta)}d\zeta$ on U we will consider the two real forms $dx(\zeta) = \mathrm{Re}(\sqrt{\phi(\zeta)}d\zeta)$ and $dy(\zeta) = \mathrm{Im}(\sqrt{\phi(\zeta)}d\zeta)$ on U.

The coordinates (1.21) vanish at ζ_0. Flat coordinates that vanish at the point ζ_0 are called distinguished coordinates at ζ_0. Different flat coordinates on U differ from each other by sign (since we may choose the other branch of the square root) and an additive constant (as the integral of the form is defined up to an additive constant). Denote by E' the set of singular points of the quadratic differential on X. The ϕ-regular set $X \setminus E'$ has a flat structure. This means that there is an atlas with transition functions being $z \to \pm z + b$ for complex numbers b.

The flat structure defines foliations on the ϕ-regular part of X as follows. Cover $X \setminus E'$ by simply connected open sets U_j each of which is equipped with flat coordinates z_j. Then the forms dx_j on U_j define a foliation on $X \setminus E'$ which is called horizontal foliation. The plaques (the local parts of leaves in U_j) are the connected components of the sets $x_j = \text{const}$ in U_j. The transition functions map plaques to plaques.

In the same way the forms dy_j on U_j define a foliation on $X \setminus E'$ which is called the vertical foliation. Actually, for each real θ the forms $\mathrm{Re}(e^{i\theta}dz_j)$ define a foliation on $X \setminus E'$.

The horizontal foliation on each U_j is generated by a vector field $v_h(\zeta)$, $\zeta \in U_j$. In other words, the leaves of this foliation are the integral curves of this vector field. In arbitrary coordinates ζ on U_j the vector field is defined by the conditions

$$\mathrm{Re}(\sqrt{\phi(\zeta)}\, d\zeta)(v_h(\zeta)) = 1\,, \quad \mathrm{Im}(\sqrt{\phi(\zeta)}\, d\zeta)(v_h(\zeta)) = 0\,. \tag{1.22}$$

Notice that if $\phi(\zeta) = |\phi(\zeta)|e^{i\theta}$ in coordinates $\zeta = \xi + i\eta$, then in these coordinates

$$\mathrm{Re}(\sqrt{\phi(\zeta)}\, d\zeta) = \sqrt{|\phi(\zeta)|}(cos\frac{\theta}{2}\, d\xi - \sin\frac{\theta}{2}\, d\eta)\,,$$

$$\mathrm{Im}(\sqrt{\phi(\zeta)}\, d\zeta) = \sqrt{|\phi(\zeta)|}(\sin\frac{\theta}{2}\, d\xi + \cos\frac{\theta}{2}\, d\eta)\,.$$

In distinguished coordinates z_j the conditions (1.22) become $dx_j(v_h) = 1$, $dy_j(v_h) = 0$. In other words, in these coordinates the vector field points in the direction of the positive real axis (or in the direction of the negative real axis if the other branch of the square root is chosen). Identifying a vector with a point in the complex plane we obtain the following. If $\phi(\zeta) = |\phi(\zeta)|e^{i\theta}$ in coordinates ζ then in these coordinates $v_h(\zeta) = |\phi(\zeta)|^{-\frac{1}{2}}\binom{\cos\frac{\theta}{2}}{-\sin\frac{\theta}{2}}$ which we identify with the point $|\phi(\zeta)|^{-\frac{1}{2}}e^{-\frac{1}{2}i\theta}$ in the complex plane. Moreover, $v_v(\zeta) = |\phi(\zeta)|^{-\frac{1}{2}}\binom{\sin\frac{\theta}{2}}{\cos\frac{\theta}{2}} = |\phi(\zeta)|^{-\frac{1}{2}}\binom{\cos\frac{\theta-\pi}{2}}{-\sin\frac{\theta-\pi}{2}}$ which can be identified with the point $|\phi(\zeta)|^{-\frac{1}{2}}e^{-\frac{1}{2}i(\theta-\pi)}$ in the complex plane.

In general the v_h (the v_v, respectively) in the coordinate patches do not define a global vector field on $X \setminus E'$ but merely a global line field. Indeed, the horizontal foliation is in general not orientable. The leaves of the horizontal foliation (of the vertical foliation, respectively) are also called horizontal trajectories (vertical trajectories, respectively). Arcs that are contained in horizontal (vertical, respectively) trajectories are called horizontal (vertical, respectively) arcs.

In a neighbourhood U of a singular point p_0 there are holomorphic coordinates z vanishing at the point in which the quadratic differential has the form

$$\phi(z)\,dz^2 = \left(\frac{a+2}{2}\right)^2 z^a\,dz^2 \tag{1.23}$$

for some integer a. The coordinates in which ϕ has the form (1.23) are uniquely defined up to multiplication by an $(a+2)$-nd root of unity and are called distinguished coordinates at the singular points. The number a is called the order of the point. For distinguished coordinates near regular points ϕ has the form (1.23) with $a = 0$. We will consider quadratic differentials with at worst poles of order one, i.e. $a \geq -1$.

Equip a simply connected subset U' of $U \setminus \{p_0\}$ with coordinates z satisfying (1.23). We assume that in these coordinates U has the form $U = \{|z| < r_0\}$ for a positive number r_0.

The vector field that equals $\frac{2}{a+2} r^{-\frac{a}{2}} e^{-\frac{ia\theta}{2}}$ at the point $z = e^{i\theta}$ in U' in the mentioned coordinates generates the horizontal foliation on U'. A ray $\{re^{i\theta} : 0 < r < r_0\}$ (in these coordinates) is a subset of a horizontal trajectory if at each point the unit tangent vector to the ray is equal to the unit vector in a horizontal direction, i.e $e^{i\theta} = \pm e^{-\frac{ia\theta}{2}}$, or equivalently, $\theta = \frac{2k\pi}{a+2}$, $k = 0, \dots, a+1$. The rays that are subsets of the vertical trajectories are $\{re^{\frac{(2k+1)\pi i}{a+2}} : r > 0 \text{ small}\}$, for $k = 0, \dots, a+1$ (we use the same coordinates satisfying (1.23)). We call these rays horizontal bisectrices (also horizontal separatrices), or vertical bisectrices (also vertical separatrices), respectively. Any other horizontal leaf in U is a subset of a sector between two neighbouring horizontal bisectrices and avoids a neighbourhood (depending on the leaf) of the singular point. The respective facts hold for the vertical leaves.

The same facts hold in a neighbourhood of each singular point. We obtained a singular foliation on X. On a sector between two consecutive horizontal bisectrices a branch of the power $z \to z^{\frac{a+2}{2}}$ provides flat coordinates. The image of the sector under the latter mapping is a subset of the upper or lower half-plane with horizontal leaves being the sets $x = \text{const}$. A sector between two consecutive vertical bisectrices is mapped to a subset of the right or left half-plane with vertical leaves being the sets $y = \text{const}$.

We want to associate a singular metric to ϕ. Consider again a simply connected domain U in $X \setminus E'$ equipped with coordinates ζ. The quadratic differential in these coordinates has the form $\phi(\zeta)d\zeta^2$. Let z be flat coordinates on U. Let $\gamma : [0, 1] \to U$ be a piecewise smooth curve. We define the length $\ell_\phi(\gamma)$ with respect to ϕ of the curve γ as its Euclidean length in some flat coordinates. This value is independent on the choice of the flat coordinates, and therefore it is well defined. If we denote by γ_1 the curve $t \to z(\gamma(t))$ in some flat coordinates z then

$$\ell_\phi(\gamma) = \int_{\gamma_1} |dz| = \int_0^1 \left|\frac{d}{dt}(\gamma_1(t)\right| dt\,. \tag{1.24}$$

In arbitrary coordinates ζ we obtain with $dz = \sqrt{\phi(\zeta)}d\zeta$

$$\ell_\phi(\gamma) = \int_\gamma |\sqrt{\phi(\zeta)}d\zeta| = \int_0^1 \sqrt{|\phi(\gamma(t))|}\left|\frac{d}{dt}(\gamma(t))\right| dt. \tag{1.25}$$

We call $|dz| = |\sqrt{\phi(\zeta)}d\zeta|$ the length element in the ϕ-metric and $\ell_\phi(\gamma)$ the ϕ-length of the curve γ. An arbitrary piecewise smooth curve in the regular part of ϕ can be subdivided into pieces contained in small discs around regular points. The ϕ-length of the curve is the sum of the ϕ-length of these pieces. The invariant form (1.25) makes sense also for curves passing through singular points. As the singular points of ϕ are at worst poles of order one, the ϕ-length of any piecewise smooth compact curve $\gamma : [0, 1] \to X$ is finite.

In the same way we define the horizontal ϕ-variation $\ell_{\phi,h}(\gamma)$ of a curve γ (the vertical ϕ-variation $\ell_{\phi,v}(\gamma)$, respectively). For a curve γ in a simply connected set U with flat coordinates z this is the total variation of the x coordinate (the y coordinate, respectively) along the curve γ,

$$\begin{aligned} \ell_{\phi,h}(\gamma) &= \int_\gamma |dx| = \int_0^1 \left|\mathrm{Re}\frac{d}{dt}(\gamma(t)\right| dt\,, \\ \ell_{\phi,v}(\gamma) &= \int_\gamma |dy| = \int_0^1 \left|\mathrm{Im}\frac{d}{dt}(\gamma(t)\right| dt\,. \end{aligned} \tag{1.26}$$

An arbitrary curve is divided into pieces contained in such sets and the variations of the pieces are added.

The area element in the ϕ-metric is defined as $dz \wedge d\bar{z}$ in flat coordinates z, or, equivalently, $\sqrt{\phi(\zeta)}d\zeta \wedge \overline{\sqrt{\phi(\zeta)}d\zeta} = |\phi(\zeta)| d\zeta \wedge \overline{d\zeta}$ in arbitrary local coordinates ζ. For a small open subset $U \subset X$ the integral $\iint\limits_U |\phi| = \frac{1}{2}\iint\limits_U |\phi(\zeta)|\, d\zeta \wedge d\bar{\zeta}$ is invariant under holomorphic coordinate changes. Hence, $\|\phi\|_1 \overset{\text{def}}{=} \iint\limits_X |\phi|$ is well defined. ϕ is called integrable on X if this integral is finite. The singularities of integrable meromorphic quadratic differentials are zeros or simple poles. The set of integrable meromorphic quadratic differentials on X with norm $\|\cdot\|_1$ forms a complex Banach space.

The following fact is an immediate corollary of the definitions. For each meromorphic quadratic differential ϕ and any number $k \in (0, 1)$ the object $k\,\frac{|\phi|}{\phi} =$

$k\,\frac{\bar{\phi}}{|\phi|}$ (given in local coordinates z by $k\,\frac{\bar{\phi}(z)}{|\phi(z)|}$ with a meromorphic function $\phi(z)$) is a Beltrami differential of norm $K = \frac{1+k}{1-k}$.

1.7 Elements of Teichmüller Theory

Teichmüller Theorem (Closed Surfaces) The following celebrated theorem was the key point for solving Riemann's problem on describing the classes of conformally equivalent Riemann surfaces. It was Teichmüller's vision that isotopy classes of conformally equivalent Riemann surfaces are easier to handle and their description provides a solution of Riemann's problem. For more details, see [2], [9].

Theorem 1.4 *Let X and Y be closed Riemann surfaces of genus $g \geq 2$, and let $\varphi : X \to Y$ be a homeomorphism. Then there is a unique homeomorphism isotopic to φ with smallest quasiconformal dilatation. This homeomorphism is either conformal or its Beltrami differential has the form $k \cdot \frac{|\phi|}{\phi}$ for a holomorphic quadratic differential ϕ on X and a constant $k \in (0, 1)$. ϕ is unique up to multiplication by a positive constant.*

A homeomorphism with Beltrami differential $k \cdot \frac{|\phi|}{\phi}$ for a holomorphic quadratic differential ϕ on X and a constant $k \in (0, 1)$ is called a Teichmüller mapping and ϕ is called its quadratic differential.

Teichmüller mappings are characterized by their local description in distinguished coordinates. Let X and Y be closed Riemann surfaces. Suppose $\varphi : X \to Y$ is a Teichmüller mapping with quadratic differential ϕ and constant k. Then the inverse mapping φ^{-1} is again a Teichmüller mapping with quadratic differential denoted by $-\psi$ and constant k. The order of ϕ at a point z is the same as the order of $-\psi$ at the image $\varphi(z)$. The quadratic differential ψ is called the terminal quadratic differential of φ. The quadratic differential ϕ of $\varphi(z)$ is also called the initial quadratic differential of $\varphi(z)$. For more insight see [11], or Theorem 8.1, V.8, of [59].

There are distinguished coordinates z for ϕ near a point $z_0 \in X$ that vanish at z_0 and distinguished coordinates ζ for $-\psi$ near the image point $\varphi(z_0) \in Y$ which vanish at $\varphi(z_0)$ so that the mapping φ has the form

$$\zeta = \left(\frac{z^{a+2} + 2\,k|z|^{a+2} + k^2\,\bar{z}^{a+2}}{1 - k^2} \right)^{\frac{1}{a+2}} \tag{1.27}$$

with $\zeta > 0$ for $z > 0$. If $a = 0$ this is equivalent to

$$\zeta = \xi + i\eta = K^{\frac{1}{2}} x + i\,K^{-\frac{1}{2}} y \tag{1.28}$$

with $z = x + iy$ and $K = \frac{1+k}{1-k}$ being the quasiconformal dilatation.

We will describe now an idea of Ahlfors ([2], section 4, pp. 19–20) which allows to reduce questions concerning self-homeomorphisms of punctured Riemann surfaces (self-homeomorphisms of closed Riemann surfaces with distinguished points, respectively) to the related questions concerning closed Riemann surfaces. Ahlfors used it for a proof of Teichmüller's theorem for punctured Riemann surfaces. It can also be used for reducing the study of the entropy of self-homeomorphisms of closed Riemann surfaces with distinguished points to the study of the entropy of self-homeomorphisms of closed Riemann surfaces.

Ahlfors' idea is the following. Let X and Y be closed Riemann surfaces, both with a set of m distinguished points. (Equivalently, we may think about two m-punctured Riemann surfaces by using the identification of $\mathrm{Hom}^+(X, \emptyset, E)$ with $\mathrm{Hom}^+(X\backslash E, \emptyset, \emptyset)$.) Assume $2g - 2 + m > 0$, so that the universal covering of the m-punctured surfaces equals $\mathbb{C}_+$. Except for $g = 0$ and $m \leq 4$ Ahlfors associates to X and Y closed Riemann surfaces $\hat{X}$ and $\hat{Y}$, which are holomorphic simple branched coverings $p_X : \hat{X} \to X$ of X and $p_Y : \hat{Y} \to Y$ of Y with branch locus being the set of distinguished points, and have genus at least two.

Such coverings can be obtained as follows. If the number of punctures is even (and not zero) and either $g \geq 1$, or $g = 0$ and $m \geq 6$, then one can take a double branched covering with branch locus being the set of distinguished points.

If $g > 0$ and the number of points in the set E of distinguished points is odd and bigger than 1, one first considers an intermediate double branched covering with branch locus being a non-empty set $E' \subset E$ that contains an even number of points,and takes then a double branched cover of the intermediate branched covering, this time with branch locus being the preimage of $E \setminus E'$ in the intermediate branched covering (which consists of an even number of points).

If the number of distinguished points equals 1 and $g > 0$, we take an intermediate unramified covering and proceed further as before.

If $g = 0$ and $m \geq 5$ is odd we consider an intermediate double branched covering with branch locus being $m - 1$ of the punctures, and then a double branched covering over the immediate covering branched at the preimages of the remaining puncture of X. This treats all cases except $g = 0$ and $m \leq 4$. If $m \leq 3$ there exist holomorphic self-mappings of $\mathbb{P}^1$ that map each distinguished point in X to the respective distinguished point in Y. If $m = 4$, the double branched covering with the required branch locus is a torus (i.e. a closed Riemann surface of genus 1), in which case the extremal mappings are well described, and also problems related to entropy are well-known in this case.

Consider a homeomorphism $\varphi : X \to Y$ which maps the set of distinguished points of X to the set of distinguished points of Y. There is a lift $\hat{\varphi}$ of φ, $\hat{\varphi} : \hat{X} \to \hat{Y}$, i.e. a homeomorphism between closed surfaces such that $p_Y \circ \hat{\varphi} = \varphi \circ p_X$. The homeomorphism $\hat{\varphi}$ has some "additional symmetries". Vice versa, a homeomorphism between $\hat{X}$ and $\hat{Y}$ with such symmetries is a lift of a homeomorphism between X and Y which maps the set of distinguished points in X to the set of distinguished points in Y.

Notice that a self-homeomorphism φ of a Riemann surface X which preserves a set E consisting of an even number of points, lifts to the double branched covering with branch locus E as follows. Cut X along a set Γ of disjoint arcs that join pairs of points of E. Consider two copies of $X \setminus \Gamma$. φ maps each copy to a copy of $X \setminus \varphi(\Gamma)$. Glue the copies in the source and the copies in the target crosswise along respective strands of the cuts and extend the mapping.

Teichmüller's theorem also applies in the situation of homeomorphisms with "additional symmetry" between closed Riemann surfaces. One obtains quadratic differentials on $\hat{X}$ with a "symmetry", and one obtains quadratic differentials on X which lift to the mentioned quadratic differentials on $\hat{X}$. This implies that the quadratic differentials on X have at most simple poles at the branch locus. This is a simple calculation using the behaviour of quadratic differentials under coordinate changes. Indeed, let $\hat{z}_0$ be a branch point in $\hat{X}$ and let z_0 be its image under the projection $\hat{X} \to X$. Let ζ be distinguished coordinates on $\hat{X}$ at $\hat{z}_0$, and let $z = z(\zeta) = \zeta^2$ be coordinates near z_0 on X. If ϕ is the quadratic differential on X and $\hat{\phi}$ its lift to $\hat{X}$ then by (1.20)

$$\hat{\phi}(\zeta) = \phi(z(\zeta))z'(\zeta)^2 = \phi(z(\zeta))(2\zeta)^2 = 4\phi(z(\zeta))z(\zeta) = 4\phi(z)z\,. \tag{1.29}$$

Since $\hat{\phi}$ does not have poles, ϕ has at most a simple pole at z_0.

The arguments imply that Teichmüller's theorem remains true for homeomorphisms between closed surfaces with distinguished points, if instead of holomorphic quadratic differentials one considers meromorphic quadratic differentials on closed surfaces with at most simple poles at distinguished points. These are exactly the integrable meromorphic quadratic differentials.

In a similar way Ahlfors treats finite Riemann surfaces of the second kind (i.e. Riemann surfaces with finitely many boundary components, some of which may be points, some are continua). He uses an extension of the homeomorphism to a homeomorphism between the doubles of the Riemann surfaces. We do not need this case here. For details see [2, 3].

The local characterization of Teichmüller maps implies the following useful fact. Let X, X_1 and X_2 be closed or punctured Riemann surfaces, and let $\varphi_1 : X \to X_1$ and $\varphi_2 : X_1 \to X_2$ be Teichmüller maps with quadratic initial differentials ϕ_1 and ϕ_2, respectively. Suppose ϕ_2 is equal to the terminal quadratic differential of φ_1. Then $\varphi_2 \circ \varphi_1$ is a Teichmüller map with quadratic differential ϕ_1.

Teichmüller Spaces Let X be a connected Riemann surface of genus g, with ℓ boundary continua and m punctures. We always assume that the universal covering of X equals $\mathbb{C}_+$. (This excludes the Riemann sphere, $\mathbb{C}$, $\mathbb{C}^* = \mathbb{C}\setminus\{0\}$, and tori (compact Riemann surfaces of genus 1).)

Let $w_j : X \to Y_j$, $j = 1, 2$, be two quasiconformal homeomorphisms onto Riemann surfaces (considered as conformal structures on X). They are called (Teichmüller) equivalent if there exists a conformal mapping $c : Y_1 \to Y_2$ such that $w_2^{-1} \circ c \circ w_1$ is isotopic to the identity by an isotopy which fixes the set of boundary continua pointwise (if any). We denote the equivalence class of a quasiconformal complex structure $w : X \to Y$ on X by $[w]$. The set of equivalence classes is the

Teichmüller space $\mathcal{T}(X)$. Equip the Teichmüller space with the Teichmüller metric $d_{\mathcal{T}}$,

$$d_{\mathcal{T}}([w_1],[w_2]) \stackrel{\text{def}}{=} \inf\left\{\frac{1}{2}\log K(v_2 \circ v_1^{-1}) : v_1 \in [w_1],\ v_2 \in [w_2]\right\}. \tag{1.30}$$

Note that the infimum is equal to the following

$$\inf\left\{\frac{1}{2}\log K(g) : g : Y_1 \to Y_2 \text{ such that } w_2^{-1} \circ g \circ w_1 : X \to X \text{ is isotopic}\right.$$
$$\left.\text{to the identity fixing the boundary continua pointwise}\right\}. \tag{1.31}$$

For a quasi-conformal homeomorphism w with $[w] \in \mathcal{T}(X)$ the space $\mathcal{T}(w(X))$ is canonically isometric to $\mathcal{T}(X)$. We choose a reference Riemann surface X with m punctures and ℓ boundary continua and write $\mathcal{T}(g, m, \ell)$ for the respective Teichmüller space.

We also need the Fuchsian model of the Teichmüller space. (For more details see [4], VI B.) Consider Riemann surfaces X and Y of first kind whose universal coverings equal $\mathbb{C}_+$. Represent X and Y as quotients of the upper half-plane $\mathbb{C}_+$ by the action of Fuchsian groups Γ and Γ_1 of first kind, i.e. $X \cong \mathbb{C}_+ \diagup \Gamma$, $Y \cong \mathbb{C}_+ \diagup \Gamma_1$. Then each homeomorphism w from X to Y lifts to a self-homeomorphism $\tilde{w}$ of the upper half-plane such that

$$\text{for each } \gamma \in \Gamma \text{ there is } \gamma_1^{\tilde{w}} \in \Gamma_1 \text{ such that } \tilde{w} \circ \gamma = \gamma_1^{\tilde{w}} \circ \tilde{w}\,, \tag{1.32}$$

and each self-homeomorphisms of $\mathbb{C}_+$ with this property projects to a homeomorphisms from X onto Y. Indeed, a mapping $\tilde{w}_1$ also lifts w if and only it differs from $\tilde{w}$ by precomposition with a covering transformation for the covering $\mathbb{C}_+ \to Y$.

Beltrami differentials and quadratic differentials on X lift to Beltrami differentials $\tilde{\mu}$ and quadratic differentials $\tilde{\phi}$ on $\mathbb{C}_+$ with the invariance property

$$\tilde{\mu} \circ \gamma = \tilde{\mu}\frac{\overline{\gamma'}}{\gamma'} \quad \text{for } \gamma \in \Gamma \qquad \text{(automorphic } (-1,1)\text{-forms)}, \tag{1.33}$$

$$\tilde{\phi} \circ \gamma \cdot \gamma'^2 = \tilde{\phi} \quad \text{for } \gamma \in \Gamma \qquad \text{(automorphic 2-forms)}. \tag{1.34}$$

A self-homeomorphism of the closed upper half-plane $\bar{\mathbb{C}}_+$ (a self-homeomorphism of the Riemann sphere $\mathbb{P}^1$, respectively) is called normalized if it maps 0 to 0, 1 to 1 and ∞ to ∞. Note that each quasiconformal self-homeomorphism of $\mathbb{C}_+$ extends to a self-homeomorphism of $\bar{\mathbb{C}}_+$ [2].

Let Γ be a Fuchsian group. Denote by $Q_{\text{norm}}(\Gamma)$ the set of normalized quasiconformal self-homeomorphisms of $\mathbb{C}_+$ that satisfy (1.32) for another Fuchsian group Γ_1. The Beltrami differentials on $\mathbb{C}_+$ that satisfy (1.33) are in one-to-one correspondence to elements of $Q_{\text{norm}}(\Gamma)$. Indeed, associate to each Beltrami differential μ on $\mathbb{C}_+$ the Beltrami differential $\hat{\mu}$ on $\mathbb{C}$, for which $\hat{\mu}(z) = \mu(z)$, $z \in \mathbb{C}_+$, $\hat{\mu}(z) = \bar{\mu}(\bar{z})$, $z \in \mathbb{C}_-$ ($\mathbb{C}_-$ denotes the lower half-plane). There is a unique normalized solution w of the equation $w_{\bar{z}} = \hat{\mu}(z)\, w_z$ on the complex plane. It maps $\mathbb{C}_+$ onto itself. Its restriction to $\mathbb{C}_+$ is denoted by w^μ. Let $\Gamma_1 = \Gamma^\mu$ be the group $w^\mu \circ \gamma \circ (w^\mu)^{-1}$, $\gamma \in \Gamma$. This is a Fuchsian group. If μ satisfies (1.33) then w^μ satisfies (1.32) for $\Gamma_1 = \Gamma^\mu$. Hence, w^μ induces a quasiconformal mapping $W^\mu : \mathbb{C}_+/\Gamma \to \mathbb{C}_+/\Gamma_1$.

Two elements of $Q_{\text{norm}}(\Gamma)$ are called equivalent iff their restrictions to the real axis coincide. The Teichmüller space $\mathcal{T}(\Gamma)$ is defined as set of equivalence classes of elements of $Q_{\text{norm}}(\Gamma)$. Two mappings $w^\mu, w^\nu \in Q_{\text{norm}}(\Gamma)$ are equivalent iff the mappings w^μ, w^ν of $\mathbb{C}_+$ induce Teichmüller equivalent mappings W^μ, W^ν on $X = \mathbb{C}_+/\Gamma$ ([4], VI B, Lemma 2).

Let μ be a Beltrami differential on X, let $\tilde{\mu}$ be its lift to $\mathbb{C}_+$ and let $w^{\tilde{\mu}}$ be the normalized solution of the Beltrami equation on $\mathbb{C}_+$ for $\tilde{\mu}$. The projection of $w^{\tilde{\mu}}$ to X is denoted by W^μ. For later use we give the following definition.

Definition 1.1 *For each Beltrami differential μ on X the homeomorphism W^μ is called the normalized solution of the Beltrami equation on X for the Beltrami differential μ.*

Also, we assign to μ the element $[W^\mu]$ of the Teichmüller space $\mathcal{T}(X)$. We use the notation $\{\mu\}$ for $[W^\mu]$. The obtained mapping from $\mathcal{T}(\Gamma)$ to $\mathcal{T}(X)$ is a bijection.

Let now X be a Riemann surface of genus g with m punctures and no boundary continuum, i.e. $X = X^c \setminus E$ for a closed Riemann surface X^c with set of distinguished points E. We assume that the universal covering of X equals $\mathbb{C}_+$. The Teichmüller space is denoted by $\mathcal{T}(X) \cong \mathcal{T}(g, m, 0)$. Instead of $\mathcal{T}(g, m, 0)$ we will write $\mathcal{T}(g, m)$. Teichmüller's theorem implies the following.

Denote by $QC(X)$ the set of quasiconformal homeomorphisms of X onto another Riemann surface. The mapping $QC(X) \xrightarrow{[\,]} \mathcal{T}(X)$ assigns to each element $w \in QC(X)$ its class $[w]$ in the Teichmüller space $\mathcal{T}(X)$. The Teichmüller space is equipped with the Teichmüller metric $d_{\mathcal{T}}$. Associate to each non-trivial class $[w] \in \mathcal{T}(X)$ a (unique up to composition with conformal mappings) extremal quasiconformal homeomorphism in this class. We obtain a bijection from non-trivial elements of the Teichmüller space to Beltrami differentials of the form $k\frac{|\phi|}{\phi}$ on X, where k is a constant in $(0, 1)$ and ϕ is a holomorphic quadratic differential on X, that extends to a meromorphic quadratic differential on X^c with at most simple poles at the points of E. Notice that a meromorphic quadratic differential on X^c is integrable iff it has at most simple poles. Assign to the trivial class in the Teichmüller space (corresponding to a conformal extremal mapping) the zero quadratic differential. We obtain a bijection of the Teichmüller space to the space of integrable holomorphic quadratic differentials of norm less than 1. The real dimension of the space of integrable holomorphic quadratic differentials on

X is equal to $6g - 6 + 2m$. It can be proved that the Teichmüller space $\mathcal{T}(X)$ is homeomorphic to the unit ball in the Banach space of such quadratic differentials.

There is a unique conformal structure on $\mathcal{T}(g, m)$ with the following property. Take any family of complex structures in $QC(X)$ whose Beltrami differentials depend holomorphically on certain complex parameters. More precisely, for almost all $x \in X$ the Beltrami differentials at x of the members of the family depend holomorphically on the complex parameters. Then the equivalence classes in $\mathcal{T}(X)$ depend holomorphically on the complex parameters.

The explicit construction uses the Fuchsian model (see [10]). Write $X = \mathbb{C}_+/\Gamma$. Let μ be a Beltrami differential on $\mathbb{C}_+$ satisfying (1.33). Let w_μ be the unique normalized solution of the Beltrami equation on the Riemann sphere with Beltrami coefficient equal to μ on $\mathbb{C}_+$ and equal to 0 on $\mathbb{C}_-$. The mapping w_μ does not map $\mathbb{C}_+$ onto $\mathbb{C}_+$, but if the Beltrami differentials depend holomorphically on complex parameters then the mappings w_μ depend holomorphically on them. The mappings w_μ are conformal on the lower half-plane $\mathbb{C}_-$. Moreover, for two Beltrami differentials ν and μ the equality $w^\mu = w^\nu$ holds on $\mathbb{R}$ iff the equality $w_\mu = w_\nu$ holds on $\mathbb{R}$ (and hence the latter equality holds on $\mathbb{C}_-$) ([4], VI B, Lemma 1). Consider the Schwarzian derivative $\mathfrak{S}(w_\mu \mid \mathbb{C}_-)$. (For a locally conformal mapping f on an open set in $\mathbb{P}^1$ the Schwarzian derivative is defined as $\mathfrak{S}(f) = \left(\frac{f''}{f'}\right)' - \frac{1}{2}\left(\frac{f''}{f'}\right)^2$.)

The Schwarzian derivatives of Möbius transformations equal zero and $\mathfrak{S}(f \circ g) = \mathfrak{S}(f) \circ g \cdot (g')^2 + \mathfrak{S}(g)$ for locally conformal mappings f and g. This implies that $\phi_\mu = \mathfrak{S}(w_\mu \mid \mathbb{C}_-)$ is a quadratic differential on $\mathbb{C}_-$ which satisfies (1.34) (with respect to Γ acting on $\mathbb{C}_-$). If μ depends holomorphically on parameters then so does ϕ_μ. We obtained a map $\mu \to \phi_\mu = \mathfrak{S}(w_\mu \mid \mathbb{C}_-)$ from the set of Beltrami differentials on $\mathbb{C}_+$ satisfying (1.33) to the space of holomorphic quadratic differentials on $\mathbb{C}_-$. The mapping satisfies the condition

$$\phi_\mu = \phi_\nu \quad \text{if} \quad w^\mu, w^\nu \in Q_{\text{norm}}(\Gamma) \text{ are equivalent.} \tag{1.35}$$

Note that for each holomorphic function f on a simply connected domain in the complex plane there is a meromorphic function w in the domain, unique up to a Möbius transformation for which $\mathfrak{S}(w) = f$. There is an explicit way to find such a function.

Consider the Banach space $B(\mathbb{C}_-, \Gamma)$ of holomorphic quadratic differentials in $\mathbb{C}_-$ with norm $\|\varphi\| = \sup |y^2 \varphi(z)|$. By the condition (1.35) the mapping

$$\mathcal{T}(X) \cong \mathcal{T}(\Gamma) \ni \{\mu\} \to \phi_\mu \tag{1.36}$$

is well defined. It can be proved that it defines a homeomorphism of $\mathcal{T}(X)$ onto an open subset of the unit ball of $B(\mathbb{C}_-, \Gamma)$ [4]. This homeomorphism is called Bers embedding. The complex structure on $\mathcal{T}(X)$ induced by this homeomorphism from $B(\mathbb{C}_-, \Gamma)$ is the desired one. For Beltrami differentials μ on X depending holomorphically on parameters, the Teichmüller classes $[W^\mu] = \{\mu\}$ also depend holomorphically on parameters, but the homeomorphisms W^μ do possibly not have this property.

Teichmüller Discs Let X be a Riemann surface of genus g with m punctures with universal covering $\mathbb{C}_+$. Write $X = \mathbb{C}_+/\Gamma$ for a Fuchsian group Γ. Let ϕ be a meromorphic quadratic differential on X which is holomorphic on X^c except, maybe, at some punctures, where it may have simple poles. For each $z = r\,e^{i\theta} \in \mathbb{D}$ we consider the Beltrami differential $\mu_z = z\,\frac{|\phi|}{\phi} = r\,\frac{|e^{-i\theta}\phi|}{e^{-i\theta}\phi}$. Notice that the absolute value $|\mu_z|$ of μ_z is the constant function $|z|$ on $\mathbb{C}_+$. Each Beltrami differential μ_z defines a unique element $w^{\mu_z} \in \mathcal{Q}_{\text{norm}}(\Gamma)$, equivalently a unique normalized solution $W^{\mu_z} \in QC(X)$ associated to μ_z, and a unique Teichmüller class $[W^{\mu_z}] = \{\mu_z\} \in \mathcal{T}(X) \cong \mathcal{T}(\Gamma)$. The mapping

$$\mathbb{D} \ni z \to \{\mu_z\} \in \mathcal{T}(X) \tag{1.37}$$

is a holomorphic embedding. For each $z \neq 0$ the homeomorphism W^{μ_z} is a Teichmüller map. Let $z_1, z_2 \in \mathbb{D}$, $z_1 \neq z_2$. Up to positive multiplicative constants the quadratic differential of $W^{\mu_{z_j}}$ equals $e^{-i\theta_j}\phi$, $j = 1, 2$, i.e. the quadratic differentials differ by a constant factor. Hence, by Lemma 9.1 of [59] the composition $W^{\mu_{z_2}} \circ (W^{\mu_{z_1}})^{-1}$ is a Teichmüller mapping of the form $W^{\mu_{z'}}$ for a point $z' \in \mathbb{D}$. By formula (9.5) of [59] the point z' is real, if z_1 and z_2 are real. One can show that the absolute value of the Beltrami differential of $W^{\mu_{z_2}} \circ (W^{\mu_{z_1}})^{-1}$ is a constant function on $\mathbb{C}_+$ that equals

$$k = |\mu_{W^{\mu_{z_2}} \circ (W^{\mu_{z_1}})^{-1}}| = \left| \frac{z_2 - z_1}{1 - z_2\,\bar{z}_1} \right| . \tag{1.38}$$

Hence, with $K = \frac{1+k}{1-k}$ the value

$$\frac{1}{2} \log K = \frac{1}{2} \log \frac{|1 - z_2\overline{z_1}| + |z_2 - z_1|}{|1 - z_2\overline{z_1}| - |z_2 - z_1|}$$

is equal to the distance of z_1 and z_2 in the Poincaré metric on the unit disc. (For details see, e.g. [59].) Thus the mapping $\mathbb{D} \ni z \to \{\mu_z\} \in \mathcal{T}(X)$ is an isometric proper holomorphic embedding of the disc with Poincaré metric into Teichmüller space with Teichmüller metric. It is called the Teichmüller disc associated to ϕ. We will denote its image in $\mathcal{T}(X)$ by $\mathcal{D}_{X,\phi}$.

The Modular Group Consider a closed Riemann surface of genus g with m punctures, denoted by $X = X^c \setminus E$. Here X^c is a closed Riemann surface with a set E of m distinguished points. A quasiconformal self-homeomorphism φ of X induces a mapping φ^* of the Teichmüller space $\mathcal{T}(X) \cong \mathcal{T}(g, m)$ to itself. It is defined as follows. For each homeomorphism $w \in QC(X)$, $w : X \to Y$ the composition $w \circ \varphi : X \to Y$ is another quasiconformal homeomorphism. Its class $[w \circ \varphi] \in \mathcal{T}(g, m)$ depends only on the class $[w]$ of w. Put $\varphi^*([w]) = [w \circ \varphi]$. For each quasiconformal self-homeomorphism φ of X the mapping φ^* is an isometry on the Teichmüller space $\mathcal{T}(g, m)$. Moreover, it maps $\mathcal{T}(g, m)$ biholomorphically onto itself. The mapping φ^* is called the modular transformation of φ.

Notice that φ^* depends only on the mapping class of φ in the mapping class group $\mathfrak{M}(X; \emptyset) = \mathfrak{M}(X^c, \emptyset, E)$, in other words, isotopic self-homeomorphisms of X have the same modular transformation. The modular group is isomorphic to the mapping class group $\mathfrak{M}(X) = \mathfrak{M}(X^c, \emptyset, E)$. For two quasiconformal self-homeomorphism φ_1 and φ_2 of the punctured Riemann surface X the equality $(\varphi_1 \circ \varphi_2)^* = \varphi_1^* \varphi_2^*$ holds, because $[w \circ (\varphi_1 \circ \varphi_2)] = \varphi_2^*([w \circ \varphi_1]) = \varphi_1^* \varphi_2^*([w])$. Hence, the set of modular transformations forms a group, called the modular group. It is often denoted by $\text{Mod}(g, m)$. The quotient $\mathcal{T}(g, m) / \text{Mod}(g, m)$ can be identified with the Riemann space of conformally equivalent complex structures, i.e. with the moduli space of Riemann surfaces of genus g with m punctures.

Royden's Theorem The following deep theorem of Royden [71] has many applications.

Theorem 1.5 (Royden) *The Teichmüller metric on the Teichmüller space $\mathcal{T}(g, m)$ is equal to the Kobayashi metric. In other words, a holomorphic mapping from the unit disc $\mathbb{D}$ with Poincaré metric $\frac{|dz|}{1-|z|^2}$ into $\mathcal{T}(g, m)$ with Teichmüller metric is a contraction. (Equivalently, a holomorphic map from $\mathbb{C}_+$ with hyperbolic metric $\frac{|dz|}{2y}$ into $\mathcal{T}(g, m)$ is a contraction.)*

Configuration Space and Teichmüller Space We will relate geometric n-braids to paths in the Teichmüller space $\mathcal{T}(0, n+1)$ of the Riemann sphere with $n+1$ punctures. The reference Riemann surface will be denoted by $X_0 = \mathbb{C} \backslash E_n^0 = \mathbb{P}^1 \backslash (\{\infty\} \cup E_n^0)$, where $E_n^0 = \left\{0, \frac{1}{n}, \ldots \frac{n-1}{n}\right\}$. Let $E_n^1 \subset \mathbb{C}$ be another set containing exactly n points. For a homeomorphism from $\mathbb{C} \backslash E_n^0$ onto $\mathbb{C} \backslash E_n^1$ we will use the same notation for this homeomorphism and for the extension of this homeomorphism to a self-homeomorphism of $\mathbb{P}^1$ which maps E_n^0 to E_n^1 and ∞ to ∞.

Consider the set $QC_\infty(0, n+1)$ of orientation preserving quasiconformal homeomorphisms $w : \mathbb{C} \setminus E_n^0 \to \mathbb{C} \setminus E_n^1$ whose extensions to $\mathbb{P}^1$ fix ∞. We equip this set with the topology for which a neighbourhood basis of an element w_0 is given by the sets

$$
\begin{aligned}
N_\varepsilon(w_0) = \big\{ & w \in QC_\infty(0, n+1) : \\
& \|\mu_w - \mu_{w_0}\|_\infty < \varepsilon,\ |w(0) - w_0(0)| < \varepsilon,\ |w(\frac{1}{n}) - w_0(\frac{1}{n})| < \varepsilon \big\}.
\end{aligned}
$$

Convergence in this topology implies uniform convergence on compact subsets of $\mathbb{C}$ (of the extensions of the mappings across the punctures). Indeed, let $\mathfrak{a}_{w,w_0} : \mathbb{C} \circlearrowleft$ be the complex affine mapping that takes $w(0)$ to $w_0(0)$ and $w(\frac{1}{n})$ to $w_0(\frac{1}{n})$. Then $\mathfrak{a}_{w,w_0} \circ w$ takes 0 to $w_0(0)$ and $\frac{1}{n}$ to $w_0(\frac{1}{n})$, and for any sequence of elements $w_n \in N_{\varepsilon_n}(w_0)$ with $\varepsilon_n \to 0$ the sequence $\mathfrak{a}_{w_n,w_0}$ converges to the identity uniformly on compacts. By formulas (1.18) and (1.19) the L^∞-norm of the Beltrami coefficient of the mapping $\mathfrak{a}_{w_n,w_0} \circ w_n \circ w_0^{-1}$ converges to 0 for any sequence $w_n \in N_{\varepsilon_n}(w_0)$, $\varepsilon_n \to 0$. The mappings $\mathfrak{a}_{w_n,w_0} \circ w_n \circ w_0^{-1}$ fix the points $w_0(0)$, $w_0(\frac{1}{n})$, and ∞, i.e. after conjugating by a complex linear mapping they are normalized solutions of Beltrami equations on $\mathbb{C}$ with Beltrami coefficients converging to 0 in the L_∞-norm. Hence, these mappings converge to the identity uniformly on compacts (see [4], Ch. V, B, Lemma 1 and the proof of Theorem 3).

Each self-homeomorphism ψ of $\mathbb{C}$ acts diagonally on the configuration space $C_n(\mathbb{C})$. Denote this action again by ψ:

$$\psi : (z_1, \dots, z_n) \to (\psi(z_1), \dots, \psi(z_n)), \quad (z_1, \dots, z_n) \in C_n(\mathbb{C}). \tag{1.39}$$

The action descends to an action on the symmetrized configuration space $C_n(\mathbb{C})\diagup\mathcal{S}_n$.

Let $\mathcal{A}$ be the set of complex affine mappings on the complex plane (equivalently, the set of Möbius transformations on the Riemann sphere that fix ∞). Each element $\mathfrak{a} \in \mathcal{A}$ has the form $\mathfrak{a}(z) = az + b$, $z \in \mathbb{C}$. Here $b \in \mathbb{C}$ and $a \in \mathbb{C}^* = \mathbb{C}\setminus\{0\}$ are constants. $\mathcal{A}$ has the complex structure of $\mathbb{C}^* \times \mathbb{C}$. It forms a group under composition.

For $z = (z_1, z_2, \dots, z_n)$ we put $\mathfrak{a}_z(\zeta) = \frac{1}{n}\frac{\zeta - z_1}{z_2 - z_1}$. Then

$$\begin{aligned} \mathfrak{a}_z(\zeta)((z_1, z_2, \dots, z_n)) =& \Big(0, \frac{1}{n}, \frac{1}{n}\frac{z_3 - z_1}{z_2 - z_1}, \dots, \frac{1}{n}\frac{z_n - z_1}{z_2 - z_1}\Big) \\ \in& \{0\} \times \{\frac{1}{n}\} \times C_{n-2}(\mathbb{C}\setminus\{0, \frac{1}{n}\}). \end{aligned}$$

Recall that

$$C_{n-2}\big(\mathbb{C}\setminus\{0, \frac{1}{n}\}\big) = \Big\{(z_3, \dots, z_n) \in \big(\mathbb{C}\setminus\{0, \frac{1}{n}\}\big)^{n-2} : z_i \neq z_j \text{ for } i \neq j\Big\}.$$

The mapping

$$(z_1, z_2, \dots, z_n) \to \Big(z_1,\ z_2 - z_1,\ \frac{1}{n}\frac{z_3 - z_1}{z_2 - z_1}, \dots, \frac{1}{n}\frac{z_n - z_1}{z_2 - z_1}\Big)$$

is a holomorphic isomorphism from $C_n(\mathbb{C})$ onto $\mathbb{C} \times \mathbb{C}^* \times C_{n-2}(\mathbb{C}\setminus\{0, \frac{1}{n}\})$. We denote by $\mathcal{P}_{\mathcal{A}}$ the projection from $C_n(\mathbb{C})$ onto the quotient $C_n(\mathbb{C})\diagup\mathcal{A}$ by the action of $\mathcal{A}$. The quotient $C_n(\mathbb{C})\diagup\mathcal{A}$ with the inherited complex structure is isomorphic to $\{0\} \times \{\frac{1}{n}\} \times C_{n-2}(\mathbb{C}\setminus\{0, \frac{1}{n}\})$. Denote by $\mathrm{Is_n}$ the isomorphism $\mathrm{Is_n} : C_n(\mathbb{C})\diagup\mathcal{A} \to \{0\} \times \{\frac{1}{n}\} \times C_{n-2}(\mathbb{C}\setminus\{0, \frac{1}{n}\})$ that assigns to each element of the quotient $C_n(\mathbb{C})\diagup\mathcal{A}$ the normalized element of $C_n(\mathbb{C})$ whose first two coordinates are 0 and $\frac{1}{n}$. The composition $\mathcal{P}_{\mathcal{A},\mathrm{n}} \stackrel{def}{=} \mathrm{Is_n} \circ \mathcal{P}_{\mathcal{A}}$ is equal to the mapping $(z_1, z_2, \ldots, z_n) \xrightarrow{\mathrm{Is_n}\circ\mathcal{P}_{\mathcal{A}}} (0, \frac{1}{n}, \frac{1}{n}\frac{z_3-z_1}{z_2-z_1}, \ldots, \frac{1}{n}\frac{z_n-z_1}{z_2-z_1})$.

Recall that for each self-homeomorphism w of $\mathbb{P}^1$ the map $e_n(w)$ is defined by $e_n(w) = \left(w(0), w\left(\frac{1}{n}\right), \ldots, w\left(\frac{n-1}{n}\right)\right)$. Recall also that two complex structures w_1, w_2 on $\mathbb{C}\backslash E_n^0$ are Teichmüller equivalent if there is a conformal mapping $c : w_1(\mathbb{C}\backslash E_n^0) \to w_2(\mathbb{C}\backslash E_n^0)$ such that $w_2^{-1} \circ c \circ w_1 : \mathbb{C}\backslash E_n^0 \to \mathbb{C}\backslash E_n^0$ is isotopic to the identity on $\mathbb{C}\backslash E_n^0$. Denote by the same letters c, w_1, w_2 the extensions of the previous mappings to $\mathbb{P}^1$. The mapping c is a Möbius transformation that fixes ∞, hence $c \in \mathcal{A}$. This implies that for two Teichmüller equivalent conformal structures w_1 and w_2 on $\mathbb{C}\backslash E_n^0$ there exists a mapping $c \in \mathcal{A}$ such that $e_n(w_2) = c(e_n(w_1))$. Indeed, if $w_2^{-1} \circ c \circ w_1$ is isotopic to the identity on $\mathbb{C}\backslash E_n^0$ its extension to $\mathbb{C}$ fixes each point in E_n^0, hence w_2 and $c \circ w_2$ take the same values at each point of E_n^0.

The arguments imply that to each Teichmüller class $[w] \in \mathcal{T}(0, n+1)$ corresponds a unique class in $C_n(\mathbb{C})\diagup\mathcal{A}$. It is obtained as follows. Take an element $\tau \in \mathcal{T}(0, n+1)$. Represent τ by a self-homeomorphism $w \in \tau$. The element $e_n(w) \in C_n(\mathbb{C})$ is defined by τ modulo a diagonal action of the group $\mathcal{A}$ of complex affine self-mappings of $\mathbb{C}$. We denote by $\mathcal{P}_{\mathcal{T}}(\tau)$ the unique element of the class in $C_n(\mathbb{C})\diagup\mathcal{A}$ corresponding to τ, that has the form $(0, \frac{1}{n}, z_3, \ldots, z_n)$.

In the following simple but useful lemma we again identify homomorphisms between punctured surfaces and their extensions to $\mathbb{P}^1$.

Lemma 1.3 *Let E_n^1 and E_n^2 be subsets of $\mathbb{C}$, each containing exactly n points. Let $w_1 : \mathbb{C}\backslash E_n^0 \to \mathbb{C}\backslash E_n^1$ and $w_2 : \mathbb{C}\backslash E_n^0 \to \mathbb{C}\backslash E_n^2$ be Teichmüller equivalent homeomorphisms. If w_1 and w_2 take equal values on two different points $\frac{j_1}{n}$ and $\frac{j_2}{n}$ of E_n^0, i.e. $w_1\left(\frac{j_1}{n}\right) = w_2\left(\frac{j_1}{n}\right)$ and $w_1\left(\frac{j_2}{n}\right) = w_2\left(\frac{j_2}{n}\right)$, then $E_n^1 = E_n^2$ and w_1, w_2 are isotopic as mappings from $\mathbb{C}\backslash E_n^0$ onto $\mathbb{C}\backslash E_n^1$ (i.e. $w_1 \circ w_2^{-1} : \mathbb{C} \setminus E_n^2 \circlearrowleft$ is isotopic to the identity.) In particular, $e_n \circ w_1 = e_n \circ w_2$.*

Proof Under the conditions of the lemma there is an affine map $c \in \mathcal{A}$ such that the mapping $w_2^{-1} \circ c \circ w_1$ (see Fig. 1.5) is isotopic to the identity through self-homeomorphisms of $\mathbb{C}\diagup E_n^0$. Then $w_2^{-1} \circ c \circ w_1$ fixes E_n^0 pointwise, i.e. $c \circ w_1 \mid E_n^0 = w_2 \mid E_n^0$. Since $w_1\left(\frac{j_1}{n}\right) = w_2\left(\frac{j_1}{n}\right)$ and $w_1\left(\frac{j_2}{n}\right) = w_2\left(\frac{j_2}{n}\right)$, c fixes two points in $\mathbb{C}$, hence c is the identity. Thus $E_n^1 = w_1(E_n^0) = w_2(E_n^0) = E_n^2$ and $w_2^{-1} \circ w_1$ is isotopic to the identity through self-homeomorphisms of $\mathbb{C}\backslash E_n^0$. In other words, there is a continuous family φ^t, $t \in [0, 1]$, of self-homeomorphisms of

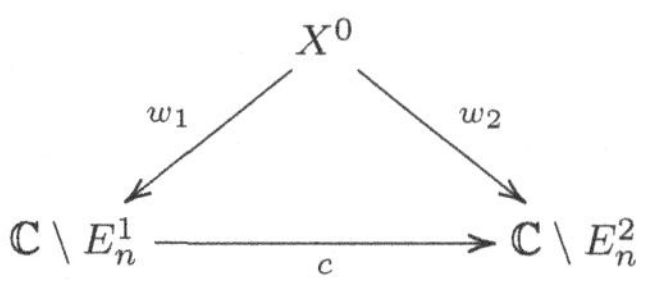

Fig. 1.5 Teichmüller equivalent mappings

$\mathbb{C} \mid E_n^0$ with $\varphi^0 = w_2^{-1} \circ w_1$ and $\varphi^1 = \text{id}$. The required isotopy is $w_2 \circ \varphi^t$, $t \in [0, 1]$. The equality $e_n \circ w_1 = e_n \circ w_2$ follows. □

The following theorem was proved by Kaliman [49, 50] and later independently by Bers and Royden [12]. Our proof follows the approach of Kaliman.

Theorem 1.6 *The mapping $\mathcal{P}_{\mathcal{T}} : \mathcal{T}(0, n+1) \to \{0\} \times \{\frac{1}{n}\} \times C_{n-2}(\mathbb{C}\backslash\{0, \frac{1}{n}\})$ is a holomorphic covering. Hence, the Teichmüller space $\mathcal{T}(0, n+1)$ is the holomorphic universal covering space of the space $\{0\} \times \{\frac{1}{n}\} \times C_{n-2}(\mathbb{C}\backslash\{0, \frac{1}{n}\})$.*

Remark *Theorem 1.6 implies in particular that the Teichmüller space $\mathcal{T}(0, 4)$ is isomorphic to $\mathbb{C}_+$. For the case $n = 3$ it will be convenient to consider instead of the covering $\mathcal{P}_{\mathcal{T}} : \mathcal{T}(0, 4) \to \{0\} \times \{\frac{1}{3}\} \times \mathbb{C}\backslash\{0, \frac{1}{3}\}$ the covering $\mathcal{P}'_{\mathcal{T}} : \mathcal{T}(0, 4) \to \mathbb{C}\backslash\{0, 1\}$ that is obtained as follows. Consider the mapping $C_3(\mathbb{C}) \to \{0\} \times (\mathbb{C} \setminus \{0, 1\}) \times \{1\}$ that assigns to each class $(z_1, z_2, z_3)/\mathcal{A}$ the unique element of $C_3(\mathbb{C})$ whose first entry equals 0, and whose last entry equals 1. This element equals $M_z\big((z_1, z_2, z_3)\big)$, where for $z = (z_1, z_2, z_3)$ the mapping M_z is the Möbius transformation $M_z(\zeta) = \frac{\zeta - z_1}{z_3 - z_1}$, $\zeta \in \mathbb{C}$, and $M_z\big((z_1, z_2, z_3)\big) = \big(M_z(z_1), M_z(z_2), M_z(z_3)\big)$. Instead of $\mathcal{P}_{\mathcal{A},\mathsf{n}}$ we consider the mapping $\mathcal{P}'_{\mathcal{A},\mathsf{n}} : C_3(\mathbb{C}) \to \mathbb{C} \setminus \{0, 1\}$, $\mathcal{P}'_{\mathcal{A},\mathsf{n}}\big((z_1, z_2, z_3)\big) = M_z(z_2)$. The mapping $\mathcal{P}'_{\mathcal{T}} : \mathcal{T}(0, 4) \to \mathbb{C} \setminus \{0, 1\}$ is defined by choosing for each $\tau \in \mathcal{T}(0, 4)$ a self-homeomorphism w of $\mathbb{P}^1$ representing τ and putting $\mathcal{P}'_{\mathcal{T}}(\tau) = \mathcal{P}'_{\mathcal{A},\mathsf{n}}(e_3(w)) = \frac{z_2 - z_1}{z_3 - z_1}$.*

The covering $\mathcal{P}'_{\mathcal{T}} : \mathcal{T}(0, 4) \to \mathbb{C}\backslash\{0, 1\}$ and the covering $\mathsf{P} : \mathbb{C}_+ \to \mathbb{C}\backslash\{0, 1\}$ described in Sect. 1.2 are canonically identified as follows. Consider the points $[\text{Id}] \in \mathcal{T}(0, 4)$ and $\frac{1}{2} \in \mathbb{C}\backslash\{0, 1\}$. Then $\mathcal{P}'_{\mathcal{T}}([\text{Id}]) = \mathsf{P}(\frac{1+i}{2}) = \frac{1}{2}$. Hence, there is a holomorphic isomorphism $\omega : \mathcal{T}(0, 4) \to \mathbb{C}_+$ that takes $[\text{Id}]$ to $\frac{1+i}{2}$ such that $\mathsf{P} = \mathcal{P}'_{\mathcal{T}} \circ \omega^{-1}$.

Proof of Theorem 1.6. The theorem is a consequence of the following lemma. □

In the Lemma 1.4 we consider $\{0\} \times \{\frac{1}{n}\} \times C_{n-2}(\mathbb{C}\backslash\{0, \frac{1}{n}\})$ as subspace of $C_n(\mathbb{C})$ and denote points of this space by $\tilde{E}_n$.

Lemma 1.4 *For any point $\tilde{E}_n^* \in \{0\} \times \{\frac{1}{n}\} \times C_{n-2}(\mathbb{C}\backslash\{0, \frac{1}{n}\})$ there exists a neighbourhood U in $\{0\} \times \{\frac{1}{n}\} \times C_{n-2}(\mathbb{C}\backslash\{0, \frac{1}{n}\})$, such that for any preimage $\tau_0 \in \mathcal{P}_{\mathcal{T}}^{-1}(\tilde{E}_n^*)$ there exists a holomorphic mapping $g_U : U \to \mathcal{T}(0, n+1)$ that takes the value τ_0 at the point $\tilde{E}_n^*$ and is a local inverse of $\mathcal{P}_{\mathcal{T}}$.*

Moreover, for any given homeomorphism $w^0_{\tilde{E}_n^*} \in \mathcal{QC}_\infty(0, n+1)$, such that $e_n(w^0_{\tilde{E}_n^*}) = \tilde{E}_n^*$ and $[w^0_{\tilde{E}_n^*}] = \tau_0$, there exists a smooth mapping

$$U \ni \tilde{E}_n \xrightarrow{g_U^1} w_{\tilde{E}_n} \in \mathcal{QC}_\infty(0, n+1) \text{ with } w_{\tilde{E}_n^*} = w^0_{\tilde{E}_n^*} \text{ and } e_n(w_{\tilde{E}_n}) = \tilde{E}_n, \text{ such}$$

that $g_U(\tilde{E}_n) = [w_{\tilde{E}_n}] = [g_U^1(\tilde{E}_n)]$, and hence, $\mathcal{P}_{\mathcal{T}}([w_{\tilde{E}_n}]) = \mathcal{P}_{\mathcal{T}}(g_U(\tilde{E}_n)) = \tilde{E}_n$.

Proof We consider the family $v_z^{\mathbb{D}} : \overline{\mathbb{D}} \circlearrowleft$ of self-homeomorphisms of the unit disc,

$$v_z^{\mathbb{D}}(\zeta) = \frac{\rho(\zeta)z + \zeta}{1 + \rho(\zeta)z\overline{\zeta}}, \zeta \in \mathbb{D}, \tag{1.40}$$

where the parameter z runs over a disc $\varepsilon\mathbb{D}$ with center 0 and small radius ε. Here $\zeta \to \rho(\zeta)$ is a (real) non-negative smooth function on $\overline{\mathbb{D}}$, that vanishes near the unit circle, equals 1 in a neighbourhood of the closed disc $\varepsilon\overline{\mathbb{D}}$, and satisfies the inequality $|\rho(\zeta)| \leq 1$ on $\mathbb{D}$. For each $z \in \varepsilon\mathbb{D}$ the restriction $v_z^{\mathbb{D}}|\varepsilon\overline{\mathbb{D}}$ equals $\zeta \to \frac{z+\zeta}{1+z\overline{\zeta}}$. This mapping takes 0 to z. The mapping $v_z^{\mathbb{D}}$ is equal to the identity on the unit circle. For each $\zeta \in \overline{\mathbb{D}}$ the mapping $z \to v_z^{\mathbb{D}}(\zeta)$, $z \in \mathbb{D}$, is holomorphic. The following equalities hold.

$$\begin{aligned}
v_z^{\mathbb{D}}(\zeta) - \zeta &= \frac{\rho(\zeta)z(1 - |\zeta|^2)}{1 + z\rho(\zeta)\overline{\zeta}}, \\
\partial_{\overline{\zeta}}(v_z^{\mathbb{D}}(\zeta) - \zeta) &= \frac{z\partial_{\overline{\zeta}}\rho(\zeta)(1 - |\zeta|^2) - \rho(\zeta)z\zeta}{1 + z\rho(\zeta)\overline{\zeta}} \\
&\quad - \frac{\rho(\zeta)z(1 - |\zeta|^2)(z\partial_{\overline{\zeta}}\rho(\zeta)\overline{\zeta} + z\rho(\zeta))}{(1 + z\rho(\zeta)\overline{\zeta})^2}, \\
\partial_\zeta(v_z^{\mathbb{D}}(\zeta) - \zeta) &= \frac{z\partial_\zeta\rho(\zeta)(1 - |\zeta|^2) - \rho(\zeta)z\overline{\zeta}}{1 + z\rho(\zeta)\overline{\zeta}} - \frac{\rho(\zeta)z(1 - |\zeta|)^2\, z(\partial_\zeta\rho)(\zeta)\overline{\zeta}}{(1 + z\rho(\zeta)\overline{\zeta})^2}.
\end{aligned} \tag{1.41}$$

Put $C \overset{def}{=} \max_{|\zeta| \leq 1} |(\partial_\zeta\rho)(\zeta)| = \max_{|\zeta| \leq 1} |(\partial_{\overline{\zeta}}\rho)(\zeta)|$. For each $z \in \varepsilon\mathbb{D}$ we extend $v_z^{\mathbb{D}}$ to $\mathbb{P}^1$ by the identity outside the unit disc. Denote the extended function again by $v_z^{\mathbb{D}}$. By Eq. (1.41) for each $z \in \varepsilon\mathbb{D}$ the mapping $v_z^{\mathbb{D}}(\zeta) - \zeta$, $\zeta \in \mathbb{P}^1$, is smooth and vanishes outside the unit disc. Its supremum norm and the supremum norm of its differential do not exceed $C'\varepsilon$ for a constant C' depending on C. These facts imply, that if ε is small, for each $z \in \varepsilon\mathbb{D}$ the mapping $v_z^{\mathbb{D}} : \mathbb{P}^1 \circlearrowleft$ is a diffeomorphism. Indeed, $v_z^{\mathbb{D}}$ is a local diffeomorphism and it is $C'\varepsilon$-close to the identity in the supremum norm, hence it is injective, if ε is small. Hence, $v_z^{\mathbb{D}}$ is a diffeomorphism from $\mathbb{P}^1$ onto its image. Since it is the identity outside the unit disc, it is a diffeomorphism from $\mathbb{P}^1$ onto itself, and it maps the unit disc onto itself. Moreover, the Beltrami coefficient $\mu_{v_z^{\mathbb{D}}}(\zeta) = \frac{\partial_{\overline{\zeta}} v_z^{\mathbb{D}}(\zeta)}{\partial_\zeta v_z^{\mathbb{D}}(\zeta)}$ of $v_z^{\mathbb{D}}$ for $|\zeta| < 1$ satisfies the inequality $\sup_{\zeta \in \mathbb{D}} |\mu_z^{\mathbb{D}}(\zeta)| \leq C''\varepsilon$, and, hence, is small for each $z \in \varepsilon\mathbb{D}$ if ε small. Moreover, the mapping $z \to \mu_{v_z^{\mathbb{D}}}(\zeta)$, $z \in \varepsilon\mathbb{D}$, is holomorphic for each $\zeta \in \mathbb{D}$.

Write $\tilde{E}_n^* = (0, \frac{1}{n}, z_3^0, \ldots, z_n^0)$, and let D_j be disjoint discs in $\mathbb{C} \setminus \{0, \frac{1}{n}\}$ with center z_j^0, $j = 3, \ldots, n$. The set $U' \overset{def}{=} \{0\} \times \{1\} \times D_3 \times \ldots \times D_n \subset$

$\{0\} \times \{\frac{1}{n}\} \times C_{n-2}(\mathbb{C}\backslash\{0, \frac{1}{n}\})$ is a neighbourhood of $(0, \frac{1}{n}, z_3^0, \ldots, z_n^0)$ in $\{0\} \times \{\frac{1}{n}\} \times C_{n-2}(\mathbb{C}\backslash\{0, \frac{1}{n}\})$. Let D_j^ε, $j = 3, \ldots, n$, be the disc with the same center as D_j and radius equal to the radius of D_j multiplied by the small number ε chosen above. We consider the neighbourhood $U \overset{def}{=} \{0\} \times \{1\} \times D_3^\varepsilon \times \ldots \times D_n^\varepsilon$, $U \subset U'$, of $(0, \frac{1}{n}, z_3^0, \ldots, z_n^0)$. Take any $\tau_0 \in \mathcal{T}(0, n+1)$ for which $\mathcal{P}_\mathcal{T}(\tau_0) = (0, \frac{1}{n}, z_3^0, \ldots, z_n^0) = \tilde{E}_n^*$ and any $w^0_{\tilde{E}_n^*}$ with $[w^0_{\tilde{E}_n^*}] = \tau_0$. We will define a continuous family $w_{\tilde{E}_n} \in QC_\infty(0, n+1)$, $\tilde{E}_n \in U$, of quasiconformal homeomorphisms of $\mathbb{P}^1$ with the following properties. $e_n(w_{\tilde{E}_n}) = \tilde{E}_n$, $g_U(\tilde{E}_n) = [w_{\tilde{E}_n}]$, and $\mathcal{P}_\mathcal{T}([w_{\tilde{E}_n}]) = \tilde{E}_n$. Moreover, $w_{\tilde{E}_n^*} = w^0_{\tilde{E}_n^*}$, and for all $\zeta \in \mathbb{P}^1$ the Beltrami coefficient $\mu_{w_{\tilde{E}_n}}(\zeta)$ depends holomorphically on $\tilde{E}_n \in U$.

For $j = 3, \ldots, n$ we denote by $\mathfrak{a}_j \in \mathcal{A}$ a complex affine mapping that maps the unit disc onto D_j and maps 0 to z_j^0. Let $\tilde{E}_n = (0, \frac{1}{n}, z_3, \ldots, z_n) \in U$. Put $v_{\tilde{E}_n}(\zeta)$ equal to $\mathfrak{a}_j \circ v^{\mathbb{D}}_{\mathfrak{a}_j^{-1}(z_j)} \circ \mathfrak{a}_j^{-1}(\zeta)$ if $\zeta \in D_j$, $j = 3, \ldots, n$, and equal to ζ on the rest of $\mathbb{P}^1$. Since $v_z^{\mathbb{D}}$ is the identity near $\partial\mathbb{D}$, for each $\tilde{E}_n = (1, \frac{1}{n}, z_3, \ldots, z_n) \in U$ the mapping $(\zeta, \tilde{E}_n) \to v_{\tilde{E}_n}(\zeta)$ is smooth on $\mathbb{P}^1$, takes z_j^0 to z_j, $j = 3, \ldots, n$, and fixes $0, \frac{1}{n}$, and ∞. $v_{\tilde{E}_n^*}$ is equal to the identity. Moreover, for each $\zeta \in \mathbb{P}^1$ each $v_{\tilde{E}_n}$ in the family is differentiable at ζ. By Eqs. (1.18) and (1.19) (see also [4], 1D, equations (7) and (8)) for the Beltrami differential of compositions, the absolute value of its Beltrami coefficient is bounded by $C''\varepsilon$. Put $g_U^1(\tilde{E}_n) = w_{\tilde{E}_n} = v_{\tilde{E}_n} \circ w^0_{\tilde{E}_n^*}$. Then $w_{\tilde{E}_n^*} = w^0_{\tilde{E}_n^*}$. The mapping $\tilde{E}_n \to \mu_{w_{\tilde{E}_n}}$ is uniformly continuous in $\tilde{E}_n$ and $\zeta \in \mathbb{P}^1$ (recall that each $\mu_{w_{\tilde{E}_n}}$ vanishes outside $\mathbb{D}$). Hence, the mapping $\tilde{E}_n \to v_{\tilde{E}_n} \circ w^0_{\tilde{E}_n^*} \in QC_\infty(0, n+1)$ is continuous. Moreover, the mapping $\tilde{E}_n \to \mu_{w_{\tilde{E}_n}}(\zeta)$ is holomorphic for every $\zeta \in \mathbb{P}^1$. By the choice of the complex structure on the Teichmüller space $\mathcal{T}(0, n+1)$ the mapping $\tilde{E}_n \to [w_{\tilde{E}_n}] \in \mathcal{T}(0, n+1)$ is holomorphic. It is clear that $\mathcal{P}_\mathcal{T}([g_U^1(\tilde{E}_n)]) = \mathcal{P}_\mathcal{T}([w_{\tilde{E}_n}]) = \tilde{E}_n = (0, \frac{1}{n}, z_3, \ldots, z_n)$. The lemma is proved. □

For later use we state the following tautological lemma.

Lemma 1.5 *The following diagram is commutative.*

$$\begin{array}{ccccc}
QC_\infty(0,n+1) & & \xrightarrow{\quad [\,] \quad} & & \mathcal{T}(0,n+1) \\
\downarrow{\scriptstyle e_n} & & & & \downarrow{\scriptstyle \mathcal{P}_\mathcal{T}} \\
C_n(\mathbb{C}) & \xrightarrow{\mathcal{P}_\mathcal{A}} & C_n(\mathbb{C})/\mathcal{A} & \xrightarrow{\mathfrak{Is}_n} & \{0\} \times \{\frac{1}{n}\} \times C_{n-2}(\mathbb{C} \setminus \{0, \frac{1}{n}\})
\end{array}$$

All mappings in the diagram are continuous.

The Modular Transformation Corresponding to Braids For each n-braid $b \in \mathcal{B}_n$ we consider the mapping class

$$\mathfrak{m}_{b,\infty} = \mathcal{H}_\infty(\mathfrak{m}_b) \in \mathfrak{M}(\mathbb{P}^1; \{\infty\}, E_n^0) \subset \mathfrak{M}(\mathbb{P}^1 \setminus (E_n^0 \cup \{\infty\})) .$$

We associate to $\mathfrak{m}_{b,\infty}$ the modular transformation $\varphi_{b,\infty}^*$ on the Teichmüller space $\mathcal{T}(n+1,0)$, that is induced by a homeomorphism $\varphi_{b,\infty}$ that represents the mapping class $\mathfrak{m}_{b,\infty}$. The modular transformation does not depend on the choice of the representing homeomorphism $\varphi_{b,\infty}$. It is convenient to choose for $\varphi_{b,\infty}$ the mapping φ_1 of a parameterizing isotopy for a geometric braid representing b. We denote the mapping $\varphi_{b,\infty}^*$ on $\mathcal{T}(n+1,0)$ by T_b and call it the modular transformation T_b of b (though the mapping T_b depends only on $\mathfrak{m}_{b,\infty}$, i.e. on $b/\mathcal{Z}_n$). Since braids are composed as elements of a fundamental group and modular transformations are composed as mappings, the relation

$$T_{b_1 b_2} = T_{b_2} T_{b_1} \tag{1.42}$$

holds for two braids b_1 and b_2.

Vice versa, take a modular transformation T on the Teichmüller space $\mathcal{T}(n+1,0)$, i.e. a mapping $T : \mathcal{T}(n+1,0) \circlearrowleft$ such that for a representative φ of a class in $\mathfrak{M}(\mathbb{P}^1; \infty, E_n^0)$ for each element $[w] \in \mathcal{T}(n+1,0)$ the equality $T([w]) = [w \circ \varphi]$ holds. We associate to T an element $b/\mathcal{Z}_n$ as follows.

Let $\tilde{q}_0 = [\mathrm{Id}] \in \mathcal{T}(n+1,0)$. Then $T([\mathrm{Id}]) = [\varphi]$. Let φ be a representative of $[\varphi]$ and let φ_t be a continuous family of self-homeomorphisms of $\mathbb{P}^1$ that fix ∞ such that $\varphi_0 = \mathrm{Id}$ and $\varphi_1 = \varphi$. Then $e_n(\varphi_0) = e_n(\mathrm{Id})$ and $e_n(\varphi_1) = e_n(\varphi) = S_\varphi(e_n(\mathrm{Id}))$ for a permutation S_φ of the set E_n^0, since φ maps E_n^0 onto itself permuting the points of E_n^0. Then the curve $\mathcal{P}_{\mathrm{sym}} e_n(\varphi_t)$, $t \in [0,1]$, in $C_n(\mathbb{C})/\mathcal{S}_n$ defines a geometric braid. Let b be the braid represented by it. Then $T = T_b$. Indeed, φ_t, $t \in [0,1]$, is a parameterizing isotopy for the geometric braid $\mathcal{P}_{\mathrm{sym}} e_n(\varphi_t)$, $t \in [0,1]$, and $\varphi_1 = \varphi = \varphi_{b,\infty}$.

We proved that the mapping $\mathcal{B}_n/\mathcal{Z}_n \ni b/\mathcal{Z}_n \to T_b \in \mathcal{T}(n+1,0)$ is a bijection that satisfies (1.42).

1.8 Thurston's Classification of Mapping Classes

Thurston's interest in surface homeomorphisms was motivated by his celebrated geometrization conjecture. He considered a closed surface S and a self-homeomorphism φ of S. The mapping torus

$$([0,1] \times S)/(0,x) \sim (1, \varphi(x))$$

is obtained by gluing the fiber over the point 0 of the cylinder $[0, 1] \times S$ to the fiber over 1 using the homeomorphism φ. Thurston observed that for a class of homeomorphisms, which he called pseudo-Anosov, the mapping torus admits a complete hyperbolic metric of finite volume. This was one of the eight geometric structures. Moreover, Thurston gave a classification of mapping classes of surface homeomorphisms. Here is Thurston's theorem on classification of mapping classes.

Theorem 1.7 (Thurston [79]) *A self-homeomorphism of a closed surface which is not isotopic to a periodic one is either isotopic to a pseudo-Anosov homeomorphism or is reducible, but not both.*

We explain now Thurston's notion of pseudo-Anosov homeomorphisms. Let S be a connected finite smooth oriented surface. It is either closed or homeomorphic to a surface with a finite number of punctures. We will assume from the beginning that S is either closed or punctured.

A finite non-empty set of mutually disjoint Jordan curves $\{C_1, \dots, C_\alpha\}$ on a connected closed or punctured oriented surface S is called admissible if no C_i is homotopic to a point in X, or to a puncture, or to a C_j with $i \neq j$. Thurston calls an isotopy class $\mathfrak{m}$ of self-homeomorphisms of S (in other words, a mapping class on S) reducible if there is an admissible system of curves $\{C_1, \dots, C_\alpha\}$ on S such that some (and, hence, each) element in $\mathfrak{m}$ maps the system to an isotopic system. In this case we say that the system $\{C_1, \dots, C_\alpha\}$ reduces $\mathfrak{m}$. A mapping class which is not reducible is called irreducible. A conjugacy class is called reducible if the representing mapping classes are reducible.

Thurston calls an individual self-homeomorphism φ of a closed or punctured surface S reduced by the set $\{C_1, \dots, C_\alpha\}$, if this set is admissible and

$$\varphi(C_1 \cup C_2 \cup \ldots \cup C_\alpha) = C_1 \cup C_2 \cup \ldots C_\alpha .$$

A self-homeomorphism of S is reduced by an admissible system of curves if and only if the mapping class of φ is reducible.

Let S be a closed or punctured surface with set E of distinguished points. We say that φ is a self-homeomorphism of S with distinguished points E, if φ is a self-homeomorphism of S that maps the set of distinguished points E to itself. Notice that each self-homeomorphism of the punctured surface $S \setminus E$ extends to a self-homeomorphism of the surface S with set of distinguished points E. We will sometimes identify self-homeomorphisms of $S \setminus E$ and self-homeomorphism of S with set E of distinguished points.

For a (connected oriented closed or punctured) surface S and a finite subset E of S a finite non-empty set of mutually disjoint Jordan curves $\{C_1, \dots, C_\alpha\}$ in $S \setminus E$ is called admissible for S with set of distinguished points E if it is admissible for $S \setminus E$. An admissible system of curves for S with set of distinguished points E is said to reduce a mapping class $\mathfrak{m}$ on S with set of distinguished points E, if the induced mapping class on $S \setminus E$ is reduced by this system of curves.

Similarly, a braid $b \in \mathcal{B}_n$ is called reducible, if its associated mapping class $\mathfrak{m}_b = \Theta_n(b) \in \mathfrak{M}(\overline{\mathbb{D}}; \partial\,\mathbb{D}, E_n^0)$ is reducible and is called irreducible otherwise.

Bers [11] gave a proof of Thurston's Theorem from the point of view of Teichmüller theory and obtained a description of reducible mappings. The proof of our Main Theorem makes explicit use of the technique developed by Bers. We will outline now the results of Bers' approach to Thurston's theory which we need.

Consider a smooth closed or punctured surface S. We do not require that it is connected. We allow S to be the union of more than one, but at most finitely many, connected components. A conformal structure on S is a homeomorphism w of S onto a Riemann surface. Let φ be a self-homeomorphism of S. Consider the extremal problem to find the following infimum

$$I(\varphi) \stackrel{\text{def}}{=} \inf\{K(w \circ \tilde{\varphi} \circ w^{-1}) : w : S \to w(S) \text{ is a conformal structure on } S,$$

$$\tilde{\varphi} \text{ is free isotopic to} \varphi\}. \tag{1.43}$$

This extremal problem differs from Teichmüller's extremal problem by varying also the conformal structure. Notice that in (1.43) we do not required that the conformal structures are quasiconformal. If the infimum is realized on a pair (w_0, φ_0), then $w_0 \circ \varphi_0 \circ w_0^{-1}$ is called absolutely extremal and w_0 is called a φ_0-minimal conformal structure.

Denote by $\mathfrak{m}_\varphi$ the mapping class of φ and by $\widehat{\mathfrak{m}_\varphi}$ its conjugacy class,

$$\widehat{\mathfrak{m}_\varphi} = \{\psi = w \circ \tilde{\varphi} \circ w^{-1} : w \text{ is a conformal structure on } S,\ \tilde{\varphi} \text{ is isotopic to } \varphi\}.$$

Then $I(\varphi)$ can be written as follows

$$I(\varphi) = \inf\{K(\psi) : \psi \in \widehat{\mathfrak{m}_\varphi}\}. \tag{1.44}$$

In the following we will again consider connected surfaces S unless said otherwise. A self-homeomorphism φ of S is periodic if φ^n is the identity on S for a natural number n. For mappings that are isotopic to periodic self-homeomorphisms the infimum is attained. More precisely, the following theorem holds.

Theorem 1.8 (Bers, see [11]) *A self-homeomorphism φ of a surface S is (free) isotopic to a periodic self-homeomorphism iff there is a conformal structure w on S and a self-homeomorphism $\tilde{\varphi}$ of S such that $w \circ \tilde{\varphi} \circ w^{-1}$ is conformal (thus $K(w \circ \tilde{\varphi} \circ w^{-1}) = 0$). Moreover, if φ is free isotopic to a periodic self-homeomorphism then there is a φ-minimal conformal structure of first kind.*

For self-homeomorphisms of S that are not isotopic to periodic ones the following theorem holds.

Theorem 1.9 (Bers, see [11]) *A conformal structure w of second kind on a surface S cannot be φ-minimal for a self-homeomorphism φ of S that is not free isotopic to a periodic one. Moreover, there exists a conformal structure w_1 of first kind and a self-homeomorphism $\tilde{\varphi}$ of S which is isotopic to φ and such that $K(w_1 \circ \tilde{\varphi} \circ w_1^{-1}) < K(w \circ \varphi \circ w^{-1})$.*

In the light of the two theorems it is sufficient to consider the extremal problem only for conformal structures of first kind. Moreover, we may fix a reference conformal structure of first kind and replace S by the obtained Riemann surface X. Composing the mappings w in (1.43) with the inverse of the reference conformal structure we may consider conformal structures on X rather than on S. Assume that X is of genus g with m punctures and $3g-3+m > 0$. (We require that the universal covering of X is $\mathbb{C}_+$ and, in case of genus 0, that the number of punctures is at least 4, to avoid trivial cases.)

The Teichmüller space of a Riemann surface $\mathcal{T}(X)$ of first kind can equivalently be described as follows. Consider the set of *all* homeomorphisms of X onto another Riemann surface Y of first kind (instead of *quasiconformal* homeomorphisms). Call two (not necessarily quasiconformal) homeomorphisms $w_j : X \to Y_j$, $j = 1, 2$, Teichmüller equivalent if there is a conformal mapping $c : Y_1 \to Y_2$ such that $w_2^{-1} \circ c \circ w_1$ is isotopic to the identity. The thus obtained equivalence classes are the same as the Teichmüller classes. Indeed, if the Riemann surfaces X and Y are of first kind and w is an arbitrary homeomorphism from X onto Y then w can be extended to a homeomorphism between closed Riemann surfaces. The extended homeomorphism can be uniformly approximated by smooth, and thus, quasiconformal homeomorphisms. For Riemann surfaces of first kind this implies also that the Teichmüller distance (1.30) can be computed by letting all homeomorphisms in (1.30) be arbitrary homeomorphisms instead of quasiconformal homeomorphisms. In particular, for a Riemann surface X of first kind and for homeomorphisms $w_1 : X \to X_1$, $w_2 : X \to X_2$, the Teichmüller distance $d_{\mathcal{T}}([w_1], [w_2])$ can also be written as follows:

$$d_{\mathcal{T}}([w_1],[w_2]) = \inf\Big\{\frac{1}{2}\log K(g) : g : X_1 \to X_2 \text{ a surjective homeomorphism which is isotopic to } w_2 \circ w_1^{-1} \text{ through such homeomorphisms}\Big\}. \tag{1.45}$$

Let X be a Riemann surface of first kind and let φ be a self-homeomorphism of X. Denote by φ^* the modular transformation induced by (the mapping class of) φ on $\mathcal{T}(g, m)$. Put

$$L(\varphi^*) \stackrel{\text{def}}{=} \inf_{\tau \in \mathcal{T}(g,m)} d_{\mathcal{T}}(\tau, \varphi^*(\tau)). \tag{1.46}$$

The quantity $L(\varphi^*)$ is called the translation length of φ^*. Write $\tau = [w]$. Then $\varphi^*([w]) = [w \circ \varphi]$, and by (1.30), (1.43) and (1.45)

$$\frac{1}{2}\log I(\varphi) = \inf_{\tau \in \mathcal{T}(g,m)} d_{\mathcal{T}}(\tau, \varphi^*(\tau)). \tag{1.47}$$

Bers uses the following terminology in analogy to the classification of elements of PSL$(2, \mathbb{Z})$. Recall that PSL$(2, \mathbb{Z})$ is the quotient of the group SL$(2, \mathbb{Z})$ of quadratic matrices of determinant 1 with integer entries by the subgroup consisting of the identity Id and $-$Id. Consider a modular transformation φ^*, i.e. an element of the modular group Mod(g, m). A point $\tau \in \mathcal{T}(g, m)$ is called φ^*-minimal, if $d_{\mathcal{T}}(\tau, \varphi^*(\tau)) = L(\varphi^*)$. A modular transformation $\varphi^* \in$ Mod(g, m) is elliptic, if it has a fixed point in $\mathcal{T}(g, m)$, parabolic, if it has no fixed point but $L(\varphi^*) = 0$, hyperbolic, if $L(\varphi^*) > 0$ and $L(\varphi^*)$ is attained, and pseudohyperbolic, if $L(\varphi^*) > 0$ but $d_{\mathcal{T}}(\tau, \varphi^*(\tau)) > L(\varphi^*)$ for all $\tau \in \mathcal{T}(g, m)$.

A conformal structure w is φ-minimal for a self-homeomorphism φ of X iff $[w]$ is φ^*-minimal.

Theorem 1.10 (Bers, see [11]) *An element $\varphi^* \in$ Mod(g, m) is elliptic iff it is periodic. This happens iff the absolutely extremal map in the isotopy class of self-homeomorphisms containing φ is conformal.*

The following theorem is a reformulation of Thurston's result.

Theorem 1.11 (See [11]) *For an irreducible self-homeomorphism φ of a Riemann surface X of first kind the modular transformation φ^* of φ is either elliptic or hyperbolic.*

Corollary 1.1 *An irreducible self-homeomorphism φ of a Riemann surface X of first kind leads to an absolutely extremal self-homeomorphism $\tilde{\varphi}$ of a Riemann surface Y by isotopy and conjugation with a homeomorphism. For the quasiconformal dilatation $K(\tilde{\varphi})$ of the absolutely extremal mapping $\tilde{\varphi}$ we have*

$$\frac{1}{2} \log K(\tilde{\varphi}) = \frac{1}{2} \log I(\varphi) = L(\varphi^*). \tag{1.48}$$

The following theorem characterizes the absolutely extremal maps with hyperbolic modular transformation in terms of Teichmüller mappings and quadratic differentials.

Theorem 1.12 (Bers, see [11]) *Let X be a Riemann surface of genus g with m punctures, $3g - 3 + m > 0$. A mapping $\varphi : X \to X$ is absolutely extremal, iff it is either conformal or a Teichmüller mapping satisfying the following two equivalent conditions*

(i) the mapping $\varphi \circ \varphi$ is also a Teichmüller mapping with $K(\varphi \circ \varphi) = (K(\varphi))^2$,
(ii) the initial and terminal quadratic differentials of φ coincide.

Remark 1.2 *Suppose φ is an absolutely extremal self-homeomorphism of X with quadratic differential ϕ. Then its modular transformation maps the Teichmüller disc $\mathcal{D}_{X,\phi}$ homeomorphically onto itself.*

Indeed, the mapping φ has Beltrami differential $k\frac{|\phi|}{\phi}$, and hence it has initial quadratic differential ϕ. The inverse φ^{-1} has Beltrami differential $-k\frac{|\psi|}{\psi}$, where ψ is the terminal quadratic differential of φ. Since φ is absolutely extremal, $\psi = \phi$.

For each $z \in \mathbb{D}$ and $\mu_z = z\frac{|\phi|}{\phi}$ the element $\varphi^*([W^{\mu_z}]) \in \mathcal{T}(X)$ is represented by $W^{\mu_z} \circ \varphi$. Since $\varphi^{\pm 1}$ has Beltrami differential $\pm k\frac{\phi}{|\phi|}$ for a number $k \in (0, 1)$, the homeomorphism $\varphi^{\pm 1}$ has the form $W^{\mu_{\pm k}}$. Hence the composition $W^{\mu_z} \circ \varphi = W^{\mu_z} \circ (W^{\mu_{-k}})^{-1}$ equals $W^{\mu_{z'}}$ for a number $z' \in \mathbb{D}$. Since also for the inverse φ^{-1} the composition $W^{\mu_z} \circ \varphi^{-1} = W^{\mu_z} \circ (W^{\mu_k})^{-1}$ equals $W^{\mu_{z''}}$ for a number $z'' \in \mathbb{D}$, φ maps the Teichmüller disc $\mathcal{D}_{X,\phi}$ homeomorphically onto itself. In case $z = x \in \mathbb{R}$ is a positive real number, the composition $W^{\mu_z} \circ (\varphi)^{\pm 1}$ is equal to $W^{\mu_{z_\pm}}$ for real numbers $z_\pm$. In other words, φ^* maps real points $\{\mu_z\}$, $z \in \mathbb{R}$, in the Teichmüller disc to real points in the Teichmüller disc.

The following definition of pseudo-Anosov mappings is equivalent to Thurston's definition (see [79], [25]).

Definition 1.2 *Let S be an oriented smooth connected surface. A self-homeomorphism φ of S is called a pseudo-Anosov mapping, if there exists a homeomorphism $w : S \to X$ onto a Riemann surface X of first kind such that $w \circ \varphi \circ w^{-1}$ is an absolutely extremal self-homeomorphism of X with hyperbolic modular transformation (equivalently, a non-periodic absolutely extremal self-homeomorphism of X).*

The absolutely extremal self-homeomorphisms φ of a Riemann surface of first kind with hyperbolic modular transformation (equivalently, the non-periodic absolutely extremal self-homeomorphisms φ of a Riemann surface of first kind) are characterized by the following properties. There is a quadratic differential ϕ such that φ maps singular points to singular points, it maps the leaves of the horizontal foliation to leaves of the horizontal foliation and leaves of the vertical foliation to leaves of the vertical foliation. These two foliations are orthogonal outside the common singularity set and are measured foliations in the sense of Thurston by using the metric $ds = |\phi|^{\frac{1}{2}}$ to measure the distance between leaves. The mapping φ decreases the distance between horizontal trajectories by the factor $K^{-\frac{1}{2}}$ and increases the distance between vertical trajectories by the factor $K^{\frac{1}{2}}$. Here K is the quasiconformal dilatation of φ.

The image of $(-1, 1)$ under an isometry from $(-1, 1)$ with metric $\frac{dx}{1-x^2}$ into the Teichmüller space $\mathcal{T}(g, m)$ with Teichmüller metric is called a geodesic line. Each pair of distinct points $\tau_1, \tau_2 \in \mathcal{T}(g, m)$ lies on a unique geodesic line. This geodesic line contains each point τ for which $d_{\mathcal{T}}(\tau_1, \tau) + d_{\mathcal{T}}(\tau, \tau_2) = d_{\mathcal{T}}(\tau_1, \tau_2)$ (see [11], [57]).

Self-homeomorphisms of Riemann surfaces with parabolic or pseudohyperbolic modular transformation are reducible. Notice that self-homeomorphisms with elliptic modular transformation also may be reducible but those with hyperbolic modular transformation are irreducible (see Theorem 1.13, statement (3) below).

Following Bers [11] we consider now the extremal problem for the quasiconformal dilatation in the case of reducible self-homeomorphisms of Riemann surfaces.

Suppose again X is a Riemann surface of first kind with $3g - 3 + m > 0$ and a self-homeomorphism $\varphi : X \to X$ is reduced by a non-empty admissible system of curves $\{C_1, \ldots, C_\alpha\}$. φ is called maximally reduced by this system if there is no admissible system of more than α curves which reduces a mapping which is isotopic to φ. φ is called completely reduced by this system of curves if for each connected component X_j of the complement $X \setminus \bigcup_{\ell=1}^{\alpha} C_\ell$ and the smallest positive integer N_j, for which $\varphi^{N_j}(X_j) = X_j$, the map $\varphi^{N_j} \mid X_j$ is irreducible.

Lemma 1.6 ([11]) *A reducible mapping is isotopic to a maximally reduced mapping.*

Lemma 1.7 ([11]) *If φ is maximally reduced by a system of curves $\{C_1, \ldots, C_\alpha\}$ it is completely reduced by this system.*

Let φ be a self-homeomorphism of a Riemann surface X of first kind, $3g - 3 + m > 0$, which is completely reduced by a non-empty admissible system $\{C_1, \ldots, C_\alpha\}$ of curves. If φ is not periodic then an absolutely extremal self-mapping of a Riemann surface related to φ by isotopy and conjugation does not exist (see Theorem 1.13 below). But there exists an absolutely extremal self-mapping ψ of a *nodal* Riemann surface Y. The nodal Riemann surface Y is the image of X by a continuous mapping w which collapses each curve C_j to a point and maps $X \setminus \bigcup_1^{\alpha} C_j$ homeomorphically onto its image. The absolutely extremal mapping ψ is related to φ by isotopy on X and semi-conjugation with w. The nodal Riemann surface Y and the absolutely extremal mapping ψ on it can be regarded as a "limit" of a sequence of non-singular Riemann surfaces X_j and self-homeomorphisms φ_j of X_j. For the X_j we have $X_j = w_j(X)$ for a quasiconformal complex structure w_j on X. The φ_j are related to φ by isotopy and conjugation: $w_j^{-1} \circ \varphi_j \circ w_j$ is isotopic to φ on X. The quasiconformal dilatations $K(\varphi_j)$ converge to the infimum

$$\inf\{K(w \circ \tilde{\varphi} \circ w^{-1}) : w \in QC(X),\ \tilde{\varphi} \text{ isotopic to } \varphi\} = e^{2L(\varphi^*)}$$

and are strictly larger than the infimum.

We will need the details later and give them here. A nodal Riemann surface X (or Riemann surface with nodes) is a one-dimensional complex space, each point of which has a neighbourhood which is either biholomorphic to the unit disc $\mathbb{D}$ in the complex plane or to the set $\{z = (z_1, z_2) \in \mathbb{D}^2 : z_1 z_2 = 0\}$. (A mapping on the latter set is holomorphic if its restriction to either of the sets, $\{(z_1, 0) : z_1 \in \mathbb{D}\}$, and $\{(0, z_2) : z_2 \in \mathbb{D}\}$, is holomorphic.) In the second case the point $(0, 0)$ is called a node. We assume that X is connected and has finitely many nodes. Let $\mathcal{N}$ be the set of nodes. The connected components of $X \setminus \mathcal{N}$ are called the parts of the nodal Riemann surface. We do not require that the set $\mathcal{N}$ of nodes is non-empty. If it is empty we also call the Riemann surface non-singular. Thus, non-singular Riemann surfaces are particular cases of nodal Riemann surfaces.

We will say that a non-singular Riemann surface is of finite type and stable if it has no boundary continuum, has genus g and m punctures with $2g - 2 + m > 0$ (hence the universal covering is $\mathbb{C}_+$). A connected nodal Riemann surface with finitely many parts, each of which is a stable Riemann surface of finite type, is called of finite type and stable.

Let X and Y be stable nodal Riemann surfaces of finite type. A surjective homeomorphism $\varphi : X \to Y$ is orientation preserving if its restriction to each part of X is so. Notice that φ defines a bijection between the parts of X and the parts of Y. The quasiconformal dilatation of φ is defined as

$$K(\varphi) = \max_{X_j} K(\varphi \mid X_j), \tag{1.49}$$

where X_j runs over all parts of X.

Let φ be an orientation preserving self-homeomorphism of a stable nodal Riemann surface X of first kind. The mapping φ permutes the parts of X along cycles. Let X_0 be a part of X and let n be the smallest number for which $\varphi^n(X_0) = X_0$. Put $X_j = \varphi^j(X_0)$ for $j = 1, \dots, n-1$,, and call $(X_0, X_1, ..., X_{n-1})$ a φ-cycle of length n. The mapping φ is called absolutely extremal if for any φ-cycle $(X_0, X_1, ..., X_{n-1})$ the restriction $\varphi | X_0 \cup ... \cup X_{n-1}$ to the Riemann surface $X_0 \cup ... \cup X_{n-1}$ (which is not connected if $n > 1$) is absolutely extremal. In other words the following holds. Let $w : X \to Y$ be a homeomorphism onto another nodal Riemann surface Y (considered as conformal structure on X). Let $\hat{\varphi}$ be a self-homeomorphism of X which is isotopic to φ. Let $(X_0, ..., X_{n-1})$ be a cycle of parts for φ. Then

$$\max_{0 \le j \le n-1} K(\varphi \mid X_j) \le \max_{0 \le j \le n-1} K(w \circ \hat{\varphi} \circ w^{-1} \mid w(X_j)). \tag{1.50}$$

If the mapping $\varphi | X_0 \cup ... \cup X_{n-1}$ is absolutely extremal, then the restriction $\varphi^n | X_0$ is an absolutely extremal self-homeomorphism of the connected Riemann surface X_0. Notice that if φ fixes all parts of X then (1.50) is equivalent to the condition that $\varphi \mid X_j$ is an absolutely extremal self-homeomorphism of X_j for each part X_j of X. In this case Theorem 1.11 applied to the parts gives a description of the absolutely extremal self-homeomorphisms.

Let $X_0, X_1, \dots X_n$ be Riemann surfaces, and let $\hat{\varphi}$ be a self-homeomorphism of the disjoint union $X \stackrel{def}{=} X_0 \cup ... \cup X_{n-1}$ that commutes the X_j along the cycle

$$X_0 \xrightarrow{\hat{\varphi}} X_1 \xrightarrow{\hat{\varphi}} \dots \xrightarrow{\hat{\varphi}} X_n \stackrel{\text{def}}{=} X_0. \tag{1.51}$$

Let w be a conformal structure on X, in other words, consider Riemann surfaces Y_j, their disjoint union $Y \stackrel{def}{=} \bigcup_{j=0}^{n-1} Y_j$, and a conformal mapping $w : X \to Y$. We may assume that w maps X_j conformally onto Y_j for each j. We define $\psi = w \circ \hat{\varphi} \circ w^{-1}$ on $Y = w(X)$. Let $\hat{\varphi}_j = \hat{\varphi} | X_j$, $\psi_j = \psi | Y_j$ for $j = 0, \dots, n-1$, and $w_j = w \mid X_j$, $j = 0, \dots, n-1$, $X_n = X_0$, and $w_n = w | X_n = w_0$. Then we get the commutative diagram Fig. 1.6.

$$\begin{array}{ccccccccc}
X_0 & \xrightarrow{\hat{\varphi}} & X_1 & \xrightarrow{\hat{\varphi}} & \dots & \xrightarrow{\hat{\varphi}} & X_{n-1} & \xrightarrow{\hat{\varphi}} & X_n = X_0\ . \\
\downarrow{\scriptstyle w_0} & & \downarrow{\scriptstyle w_1} & & & & \downarrow{\scriptstyle w_{n-1}} & & \downarrow{\scriptstyle w_n = w_0} \\
Y_0 & \xrightarrow{\psi} & Y_1 & \xrightarrow{\psi} & \dots & \xrightarrow{\psi} & Y_{n-1} & \xrightarrow{\psi} & Y_n = Y_0
\end{array}$$

Fig. 1.6 A commutative diagram for self-homeomorphisms of non-connected surfaces

The commutativity of the diagram implies the following equations

$$\begin{aligned}
\psi_j \quad &= w_{j+1} \circ \hat{\varphi}_j \circ w_j^{-1}, \ j = 0, \dots n-1\,, \\
\psi^j | Y_0 &= w_j \circ \hat{\varphi}^j \circ w_0^{-1}, \ j = 1, \dots n\,, \\
\psi^n | Y_0 &= w_0 \circ \hat{\varphi}^n \circ w_0^{-1}\,.
\end{aligned} \tag{1.52}$$

It will be convenient to have in mind the following lemma which describes the absolutely extremal self-homeomorphisms on the cycles of parts of the nodal surface X.

Lemma 1.8 *Let $X_0, X_1, ..., X_{n-1}, X_n \overset{\text{def}}{=} X_0$ be non-singular stable Riemann surfaces of finite type. Suppose φ is a self-homeomorphism of $X_0 \cup ... \cup X_{n-1}$ which permutes the X_j along the n-cycle*

$$X_0 \xrightarrow{\varphi} X_1 \xrightarrow{\varphi} \dots \xrightarrow{\varphi} X_n \overset{\text{def}}{=} X_0. \tag{1.53}$$

Then φ is absolutely extremal iff the following two conditions hold.

(1) *The mapping $F \overset{\text{def}}{=} \varphi^n | X_0$ is absolutely extremal.*
(2) *The quasiconformal dilations $K(\varphi_j)$ of the mappings $\varphi_j \overset{def}{=} \varphi | X_j$ satisfy the equality*

$$\frac{1}{2} \log K(\varphi) = \max_j \frac{1}{2} \log K(\varphi_j) = \frac{1}{2n} \log K(\varphi^n \mid X_0)\,. \tag{1.54}$$

Moreover, if φ is absolutely extremal then one of the following two situations occurs.

(2a) *The mapping F is conformal and φ is conformal.*
(2b) *The mapping $F : X_0 \circlearrowleft$ is a Teichmüller mapping whose initial and terminal quadratic differentials are equal, and the Teichmüller classes $[\varphi^j] \in \mathcal{T}(X_0)$, $j = 1, ..., n-1$, divide the bounded segment with endpoints $[id]$ and $[F]$*

on the unique geodesic line through these two points into n segments, each of which has $d_{\mathcal{T}}$-length equal to

$$\frac{1}{2n} d_{\mathcal{T}(X_0)}([id], [F]) = \frac{1}{2n} \log K(F) .$$

Moreover, for each j, $j = 1, ..., n-1$, the mapping $\varphi_j \stackrel{def}{=} \varphi|X_j$ is a Teichmüller map from $\varphi^j(X_0)$ onto $\varphi^{j+1}(X_0)$.

Proof Notice first that for all homeomorphisms $\hat{\varphi}$ that permute the X_j along the cycle (1.53) the inequalities

$$n \max_j \frac{1}{2} \log K(\hat{\varphi}_j) \geq \sum_0^{n-1} \frac{1}{2} \log K(\hat{\varphi}_j) \geq \frac{1}{2} \log K(\hat{\varphi}^n | X_0) \tag{1.55}$$

hold.

We prove first that φ is absolutely extremal if $F = \varphi^n|X_0$ is absolutely extremal and equality (1.54) holds. Let $\hat{\varphi}$ be a self-homeomorphism of $X_0 \cup \ldots \cup X_{n-1}$ that is isotopic to φ, and let ψ be related to φ by the diagram Fig. 1.6. If $F = \varphi^n|X_0$ is absolutely extremal, then

$$\frac{1}{2} \log K(\varphi^n | X_0) \leq \frac{1}{2} \log K(\psi^n | X_0) .$$

If also equality (1.54) holds, then

$$\frac{n}{2} \log K(\varphi) = \frac{1}{2} K(\varphi^n | X_0) \leq \frac{1}{2} K(\psi^n | X_0) \leq n \max_j \frac{1}{2} \log K(\psi_j) = \frac{n}{2} \log K(\psi) .$$

We proved that φ is absolutely extremal.

Vice versa, suppose φ is absolutely extremal. Then $F = \varphi^n|X_0$ is absolutely extremal. It remains to show that condition (2) holds. If F is conformal, we associate to F the self-homeomorphism

$$X_0 \xrightarrow{\text{Id}} X_0 \xrightarrow{\text{Id}} \ldots \xrightarrow{F} X_0$$

of the formal disjoint union of n copies of X_0. This homeomorphism is conformal, hence its quasiconformal dilation equals zero. By the Lemma on Conjugation (see Appendix A.2) this homeomorphism is conjugate to φ. Since φ is absolutely extremal, φ must have vanishing quasiconformal dilation, hence φ is conformal (see Eq. (1.50)).

Consider the remaining case when F is a Teichmüller mapping whose initial and terminal quadratic differentials are equal. Let ϕ be the quadratic differential of F and $\mathcal{D}_{X_0,\phi}$ the Teichmüller disc in X_0 corresponding to ϕ. Suppose $\mu_\phi = k\frac{|\phi|}{\phi}$. The set $\left\{\{x\frac{|\phi|}{\phi}\} : x \in \mathbb{R}\right\}$, equipped with the metric $d_{\mathcal{T}}$, is the unique geodesic line through $[F] = \{k\frac{|\phi|}{\phi}\}$ and $[\mathrm{Id}] = \{0\}$. Consider the points $\tau_j = \{x_j\frac{|\phi|}{\phi}\}$ on this line that divide the segment $\left\{\{x\frac{|\phi|}{\phi}\} : x \in [0,k]\right\}$ on this line into n segments of equal $d_{\mathcal{T}}$-length $\frac{1}{2n}\log K(F)$. The class τ_j, $j = 1, \dots, n-1$, is represented by the absolutely extremal self-homeomorphism $W^{\mu_{x_j}}$ of X_0, τ_0 is represented by the identity Id and τ_n is represented by $W^{\mu_{x_n}} = W^{\mu_k} = F$. Put $\mathring{X}_0 = \mathring{X}_n = X_0$ and $\mathring{X}_j = W^{\mu_{x_j}}(X_0)$, $j = 1, \dots, n-1$. By Lemma 9.1 of [59] for each $j = 0, \dots, n-1$, the homeomorphism $\mathring{\varphi}_j \overset{def}{=} W^{\mu_{x_{j+1}}} \circ (W^{\mu_{x_j}})^{-1} : \mathring{X}_j \to \mathring{X}_{j+1}$ is a Teichmüller mapping with quadratic differential ϕ. For its quasi-conformal dilatation the equality $\frac{1}{2}\log K(\mathring{\varphi}_j) = d_{\mathcal{T}}(\{x_j\frac{|\phi|}{\phi}\}, \{x_{j+1}\frac{|\phi|}{\phi}\}) = \frac{1}{2n}\log K(F)$ holds.

Consider the mapping $\mathring{\varphi}$ on $\mathring{X} \overset{def}{=} \bigcup_0^{n-1}\mathring{X}_j$ that equals $\mathring{\varphi}_j$ on $\mathring{X}_j$. Then $\mathring{\varphi}_0^n \mid \mathring{X}_0 = F$, $\frac{1}{2}\log K(\mathring{\varphi}) = \frac{1}{2n}\log K(\mathring{\varphi}^n \mid \mathring{X}_0)$. Since $\mathring{X}_0 = X_0$ and $\mathring{\varphi}^n \mid X_0 = \varphi^n \mid X_0$ the mapping $\mathring{\varphi}$ on $\mathring{X}$ is conjugate to the mapping φ on X. Since φ is absolutely extremal, the inequality

$$\frac{1}{2}\log K(\varphi) \le \frac{1}{2}\log K(\mathring{\varphi}) = \frac{1}{2n}\log K(\mathring{\varphi}^n \mid \mathring{X}_0) = \frac{1}{2n}\log K(\varphi^n \mid X_0)$$

holds. Condition (2) follows from (1.55).

Conditions (1) and (2) imply that φ is conformal if F is conformal, and if F is a Teichmüller mapping, then equality (1.54) implies that the elements τ_j of the Teichmüller space represented by the φ_j divide the segment on the geodesic line between $[Id]$ and $[F]$ into n segments of equal length. The lemma is proved. □

Now we describe in more detail the solution of the extremal problem (1.43) in the reducible case. Let X be a connected Riemann surface, which is closed or of first kind, with universal covering equal to $\mathbb{C}_+$. Let $\mathfrak{m} \in \mathfrak{M}(X)$ be an isotopy class of orientation preserving self-homeomorphisms. Let $\mathcal{C}$ be an admissible system of curves which completely reduces an element φ of $\mathfrak{m}$. By an isotopy we may assume that $\mathcal{C}$ is real analytic (or even geodesic). Associate to X and the system of curves $\mathcal{C}$ a nodal surface Y and a continuous surjection $w : X \to Y$. This can be done as follows.

Surround each connected component C_j of $\mathcal{C}$ by an annulus $A(C_j)$, which admits a conformal mapping $\mathfrak{c}_j$ onto a round annulus $A_{r_j} = \{\frac{1}{r_j} < |z| < r_j\}$ for some $r_j > 1$, such that $\mathfrak{c}_j$ maps the curve C_j to the unit circle. For each j we define the set $V_j \overset{def}{=} \{(z_1, z_2) \in \mathbb{C}^2 : z_1 z_2 = 0,\ |z_j| < r_j \text{ for } j = 1, 2\}$ and the holomorphic mapping $\psi_j : A_{r_j} \setminus \{|z| = 1\} \to V_j$

$$\psi_j(z) = \begin{cases} (0, z) & \text{if } 1 < |z| < r_j\,, \\ (\frac{1}{z}, 0) & \text{if } \frac{1}{r_j} < |z| < 1\,. \end{cases}$$

Notice that $\psi_j(A_{r_j} \setminus \{|z|=1\}) = V_j \setminus \overline{\mathbb{D}^2}$. The nodal surface Y is obtained by gluing each V_j to $X \setminus \bigcup \mathcal{C}_j$ along the sets $V_j \setminus \overline{\mathbb{D}^2}$ and $A(\mathcal{C}_j) \setminus \mathcal{C}_j$, respectively, using the holomorphic homeomorphism $\mathfrak{c}_j^{-1} \circ \psi_j^{-1} : V_j \setminus \overline{\mathbb{D}^2} \to A(\mathcal{C}_j) \setminus \mathcal{C}_j$. The set of nodes $\mathcal{N}$ of Y is the set of nodes of the V_j, hence, $\mathcal{N}$ is in bijective correspondence to the set of connected components of $\mathcal{C}$.

By construction the nodal surface Y contains the "copy" $Y \setminus \cup_j V_j$ of $X \setminus \cup_j A(\mathcal{C}_j)$. The restriction $w \mid X \setminus \cup_j A(\mathcal{C}_j)$ is defined to be the canonical mapping from $X \setminus \cup_j A(\mathcal{C}_j)$ onto its copy in Y.

For all j we let $\chi_j : [1, r_j] \to [0, 1]$ be a diffeomorphism such that $\chi_j(1) = 0$ and $\chi_j(r_j) = 1$. We define for each j a continuous surjective mapping $w_j : A_{r_j} \setminus \mathcal{C}_j \to V_j \setminus \{(0,0)\}$,

$$w_j(z) = \begin{cases} \chi_j(|z|)\psi_j(z) & \text{if } 1 < |z| < r_j\,, \\ \chi_j(\frac{1}{|z|})\psi_j(z) & \text{if} \frac{1}{r_j} < |z| < 1\,. \end{cases}$$

The mapping $w_j \circ \mathfrak{c}_j$ maps $A(\mathcal{C}_j) \setminus \mathcal{C}_j$ homeomorphically onto $V_j \setminus \{(0,0)\}$. It extends to a continuous surjection from $A(\mathcal{C}_j)$ onto V_j that collapses $\mathcal{C}_j$ to a point. The mapping w that is equal to $w \mid X \setminus \cup_j A(\mathcal{C}_j)$ on $X \setminus \cup_j A(\mathcal{C}_j)$ and is for each j equal to the extension of $w_j \circ \mathfrak{c}_j$ to $A(\mathcal{C}_j)$ is the required surjection $w : X \to Y$.

Since φ maps the complement $X \setminus \bigcup_{C \in \mathcal{C}} C$ of the set of curves homeomorphically onto itself, we may consider the mapping $w \circ \varphi \circ w^{-1}$ on $Y \setminus \mathcal{N}$. It extends continuously to the nodes. Denote the obtained mapping on Y by $\varphi_{\odot}$ and its isotopy class on Y by $\mathfrak{m}_{\odot}$.

Note that the nodal surface Y is defined up to homeomorphism by the isotopy class of $\mathcal{C}$. (We consider isotopies of the system of curves $\mathcal{C}$ within the class of real analytic systems of curves). The class $\mathfrak{m}_{\odot}$ is determined by $\mathfrak{m}$, and by the system of curves $\mathcal{C}$. The conjugacy class $\widehat{\mathfrak{m}_{\odot}}$ of $\mathfrak{m}_{\odot}$ is defined by the isotopy class of $\mathcal{C}$ and the class $\widehat{\mathfrak{m}}$.

The mapping $\varphi_{\odot}$, and, hence its isotopy class $\mathfrak{m}_{\odot}$, permutes the parts Y_k of $Y \setminus \mathcal{N}$ along cycles denoted by $\mathring{\text{cyc}}_j$. Let $\mathfrak{m}_{\odot,j}$ be the restrictions of the class $\mathfrak{m}_{\odot}$ to the cycles $\mathring{\text{cyc}}_j$. We call the conjugacy classes $\widehat{\mathfrak{m}_{\odot,j}}$ of the restrictions to the cycles $\mathring{\text{cyc}}_j$ the irreducible components of $\widehat{\mathfrak{m}}$ related to the class of $\mathcal{C}$. Notice that the class $\widehat{\mathfrak{m}_{\odot}}$ determines the class $\widehat{\mathfrak{m}}$ only modulo products of powers of Dehn twists around curves which are homotopic to curves of the system $\mathcal{C}$. It is not hard to see that the $\widehat{\mathfrak{m}_{\odot}}$ are irreducible (see also Remark 5.1).

Notice also that the isotopy class of a system of curves $\mathcal{C}$ which completely reduces an element of $\mathfrak{m}$ is not uniquely determined by X and $\mathfrak{m}$, even if we require that the system maximally reduces the homeomorphism. In particular, the type of the nodal surface Y is not uniquely determined. This may occur, for instance for reducible homeomorphisms with elliptic modular transformation.

It is a remarkable fact that the extremal problem for the quasiconformal dilatation has a solution in terms of nodal Riemann surfaces and irreducible parts of mapping classes.

The following theorem is due to Bers.

Theorem 1.13 ([11]) *Let $\mathfrak{m}$ be a mapping class of orientation preserving self-homeomorphisms of a Riemann surface X of first kind with universal covering $\mathbb{C}_+$. Let $\mathcal{C}$ be an admissible system of real analytic curves which completely reduces an element $\varphi \in \mathfrak{m}$. Choose a stable nodal Riemann surface Y of first kind with set of nodes $\mathcal{N}$ and a continuous surjection $w : X \to Y$ which contracts each curve of $\mathcal{C}$ to a point and whose restriction to the complement of $\mathcal{C}$ is a homeomorphism onto $Y \setminus \mathcal{N}$. Denote by $\mathfrak{m}_\odot$ the mapping class on Y induced by $\mathfrak{m}$ and w, and by $\widehat{\mathfrak{m}_\odot}$ its conjugacy class. Then the following holds.*

1. *There exists a conformal structure $\tilde{w}$ on Y, $\tilde{w} : Y \to \tilde{w}(Y) = \tilde{Y}$, and an absolutely extremal self-homeomorphism $\tilde{\varphi}$ of $\tilde{Y}$, representing the class $\widehat{\mathfrak{m}_\odot}$.*

The self-homeomorphism $\tilde{\varphi}$ has the following stronger extremal properties.

2. *The equality*

$$I(\varphi) = e^{2L(\varphi^*)} = K(\tilde{\varphi}) \tag{1.56}$$

holds for the modular transformation φ^* of φ. Moreover, for any continuous surjection $w' : X \to Y'$ onto a nodal Riemann surface Y', such that the preimages of the nodes are disjoint Jordan curves of the system $\mathcal{C}$ and the restriction of w' to the complement of the curves is a homeomorphism, and any self-homeomorphism φ' in the class $\widehat{\mathfrak{m}'_\odot}$ induced by $\mathfrak{m}$ and w', we have

$$K(\tilde{\varphi}) \le K(\varphi').$$

3. *If $\mathfrak{m}$ is reducible and not periodic then for each non-singular Riemann surface Y', each surjective homeomorphism $w' : X \to Y'$, and each self-homeomorphism φ' of Y' such that $(w')^{-1} \circ \varphi' \circ w' \in \widehat{\mathfrak{m}}$, we have strict inequality*

$$K(\tilde{\varphi}) < K(\varphi').$$

4. *If $\mathfrak{m}$ is reducible then there exists a sequence $Y^{(j)}$ of non-singular Riemann surfaces $Y^{(j)}$, surjective homeomorphisms $w^{(j)} : X \to Y^{(j)}$, and self-homeomorphisms $\varphi^{(j)}$ of $Y^{(j)}$ with the following property:*

$$(w^{(j)})^{-1} \circ \varphi^{(j)} \circ w^{(j)} \in \mathfrak{m} \text{ and } K(\varphi^{(j)}) \to K(\tilde{\varphi}).$$

Recall that a self-homeomorphism $\tilde{\varphi}$ of a nodal Riemann surface $\tilde{Y}$ is absolutely extremal if the quantity $K(\tilde{\varphi}) = \max_{\tilde{Y}_j} K(\tilde{\varphi}|Y_j)$ is smallest among the respective quantities for all nodal Riemann surfaces that are homeomorphic to Y and all self-homeomorphisms of Y that are obtained from $\tilde{\varphi}$ by isotopy and conjugation. Here $\tilde{Y}_j$ are the parts of $\tilde{Y}$. Recall also that the quantity $I(\varphi)$ for a self-homeomorphism φ of a non-singular Riemann surface equals the infimum of

$K(\varphi')$ over all self-homeomorphisms φ' of non-singular Riemann surfaces in the class of φ. Equality (1.56) relates the value $I(\varphi)$ to the quasiconformal dilatation of the absolutely extremal self-homeomorphism of the associated nodal surface.

The second part of Statement 2 is strictly stronger than statement 1. In Statement 1 the nodal surface up to homeomorphism and the class of $\tilde{\varphi}$ are fixed. In statement 2 the type of the nodal surface Y' is not prescribed. It may be different from the type of Y. We require only that Y' is the image of X under a continuous surjection that maps some subcollection of the set of admissible curves to nodes.

The deepest part of the theorem is its particular case concerning irreducible homeomorphisms and Statement 3.

Chapter 2
The Entropy of Surface Homeomorphisms

Here we give a self-contained proof of the Theorem of Fathi and Shub on the entropy minimizing property of pseudo-Anosov self-homeomorphisms of closed Riemann surfaces of genus at least two following mainly along the lines of the original proof. A proof of the theorem for the case of punctured surfaces is also included. The exposition contains a thorough account on the necessary prerequisites on the trajectories of quadratic differentials. The present proof of the Fathi and Shub Theorem on the relation between the entropy and the quasi-conformal dilatation is new. The chapter concludes with a thorough treatment of the entropy of self-homeomorphisms of Riemann surfaces of second kind which is a prerequisite for the treatment of reducible braids.

2.1 The Topological Entropy of Mapping Classes

The topological entropy of continuous surjective mappings of a compact topological space to itself is defined as follows.

Let X be a compact topological space and φ a continuous mapping from X onto itself. Let $\mathcal{A}$ be a collection of open subsets of X which cover X (for short, $\mathcal{A}$ is an open cover of X). For two open covers $\mathcal{A}$ and $\mathcal{B}$ we define

$$\mathcal{A} \vee \mathcal{B} = \{A \cap B : A \in \mathcal{A}, B \in \mathcal{B}\}.$$

Let $\mathcal{N}(\mathcal{A})$ be the minimal cardinality of a subset $\mathcal{A}_1$ of $\mathcal{A}$ which is a cover of X. The entropy $h(\varphi, \mathcal{A})$ of φ with respect to $\mathcal{A}$ is defined as

$$h(\varphi, \mathcal{A}) \stackrel{\text{def}}{=} \overline{\lim}_{N\to\infty} \frac{1}{N} \log \mathcal{N}(\mathcal{A} \vee \varphi^{-1}(\mathcal{A}) \vee \ldots \vee \varphi^{-N}(\mathcal{A}))\,. \tag{2.1}$$

B. Jöricke, *Braids, Conformal Module, Entropy, and Gromov's Oka Principle*, Lecture Notes in Mathematics 2360, https://doi.org/10.1007/978-3-031-67288-0_2

(Here $\varphi^{-1}(\mathcal{A}) = \{\varphi^{-1}(A) : A \in \mathcal{A}\}$, $\varphi^{-k-1}(\mathcal{A}) = \varphi^{-1}(\varphi^{-k}(\mathcal{A}))$.) The entropy $h(\varphi)$ of φ is defined as

$$h(\varphi) \stackrel{def}{=} \sup_{\mathcal{A}} h(\varphi, \mathcal{A}). \tag{2.2}$$

An open cover $\mathcal{A}$ is a refinement of an open cover $\mathcal{B}$ if each set $A \in \mathcal{A}$ is contained in a set $B \in \mathcal{B}$. We write $\mathcal{A} \prec \mathcal{B}$. A sequence $\{\mathcal{A}_n\}$, $n = 1, 2, \ldots$, of open covers is refining if for each n the cover $\mathcal{A}_{n+1}$ is a refinement of $\mathcal{A}_n$ and each open cover $\mathcal{B}$ has a refinement among the $\mathcal{A}_n$. For a refining sequence of open covers $\mathcal{A}_n$ the equality

$$h(\varphi) = \lim_{n\to\infty} h(\varphi, \mathcal{A}_n)$$

holds. (The sequence of numbers on the right is non-decreasing.) See [1], Proposition 12.

An example of a refining sequence of covers is the following. Suppose X is equipped with a metric. Let $\mathcal{A}_n$ be the cover consisting of all open balls of radius ε_n. If ε_n decreases to zero for $n \to \infty$ then the sequence is refining.

The entropy is a conjugacy invariant (see Theorem 1 of [1]):

If $\psi : X \to Y$ is a homeomorphism of topological spaces, then $h(\psi \circ \varphi \circ \psi^{-1}) = h(\varphi)$.

Let φ be a self-homeomorphism of X. The following relation is not difficult to prove (see Theorem 2 of [1]).

For any non-zero integral number n, $h(\varphi^n) = |n|\, h(\varphi)$.

Moreover, the following result of [1] (see Theorem 4 there) holds.

Let X_1 and X_2 be two closed subsets of a topological space X such that $X = X_1 \cup X_2$. If a self-homeomorphism φ of X fixes each X_j, j=1, 2, setwise, then $h(\varphi) = \max_{j=1,2} h(\varphi \mid X_j)$.

For more details see [1].

We also need the following property of entropies (see Theorem 5 of [1]).

Suppose X is a compact topological space and $\sim$ is an equivalence relation on X. Let $p : X \to X/\sim$ be the projection of X to the quotient. If $\tilde{\varphi}$ is a continuous mapping from X into itself such that $\tilde{\varphi}(x) \sim \tilde{\varphi}(y)$ if $x \sim y$, then for the mapping φ on the quotient that is defined by $\varphi \circ p = p \circ \tilde{\varphi}$ the inequality $h(\varphi) \le h(\tilde{\varphi})$ holds.

We will be concerned with the topological entropy of self-homeomorphisms of compact surfaces (with or without boundary). Let X be a compact surface (with or without boundary) with a finite set E_n of distinguished points.

Let $\mathfrak{m} \in \mathfrak{M}(X; \partial X, E_n)$ (or $\mathfrak{m} \in \mathfrak{M}(X; \emptyset, E_n)$ if $\partial X = \emptyset$) be a mapping class. The entropy of the mapping class $\mathfrak{m}$ is defined as follows

$$h(\mathfrak{m}) = \inf\{h(\varphi) : \varphi \in \mathfrak{m}\}. \tag{2.3}$$

Since the topological entropy of a homeomorphism is invariant under conjugation the following holds:

$$h(\mathfrak{m}) = h(\widehat{\mathfrak{m}}) = \inf\{h(\psi \circ \varphi \circ \psi^{-1}) : \varphi \in \mathfrak{m}\,,\ \psi : X \to Y \text{ a homeomorphism}\}. \tag{2.4}$$

For a braid $b \in \mathcal{B}_n$ with base point E_n we define the entropy $h(b)$ as the entropy of the mapping class $\mathfrak{m}_b = \Theta_n(b) \in \mathfrak{M}(\bar{D}; \partial D, E_n)$ corresponding to b:

$$h(b) \overset{\text{def}}{=} h(\mathfrak{m}_b). \tag{2.5}$$

By (2.4) the entropy $h(b)$ does not depend on the choice of the base point E_n and is a conjugacy invariant. Hence,

$$h(b) = h(\hat{b}) = h(\widehat{\mathfrak{m}_b}) \overset{\text{def}}{=} \inf\{h(\varphi) : \varphi \in \widehat{\mathfrak{m}_b}\}. \tag{2.6}$$

The entropy of self-homeomorphisms of closed surfaces was studied first in Exposé 10 of the Asterisque volume dedicated to Thurston's work [25].

In Sects. 2.2 and 2.4 of this chapter we give the proof of the following two theorems.

Theorem 2.1 *Let X be a (closed connected) Riemann surface of genus g with a set E_m of $m \geq 0$ distinguished points, $3g - 3 + m > 0$. Let $\varphi_0 \in \mathrm{Hom}(X; \emptyset, E_m)$ be a non-periodic absolutely extremal self-homeomorphism of X with set of distinguished points E_m (by an abuse of language we speak about a non-periodic absolutely extremal self-homeomorphism of $X \setminus E_m$). Then*

$$h(\varphi_0) = \frac{1}{2}\log K(\varphi_0) = L(\varphi_0^*). \tag{2.7}$$

Here $K(\varphi_0)$ is the quasiconformal dilatation of φ_0, and φ_0^* is the modular transformation on the Teichmüller space $\mathcal{T}(g, m)$ induced by φ_0 and $L(\varphi_0^*)$ is its translation length.

The theorem was first proved for closed surfaces by Fathi and Shub in the exposé 10 of the volume [25]. The present proof of Theorem 2.1 (see Sect. 2.3) differs from that of Fathi and Shub and works also for punctured surfaces.

Theorem 2.2 ([25]) *Let X be a closed connected Riemann surface of genus $g \geq 2$. Then any pseudo-Anosov self-homeomorphism of X is entropy minimizing in its isotopy class.*

Theorem 2.2 is proved in [25] but the ingredients of the proof of Theorem 2.2 in [25] are spread over several chapters. In Sects. 2.2 and 2.4 we present a self-contained proof of Theorem 2.2 following mainly the ideas of Fathi and Shub. Ahlfors' trick (see [2], section 4, pp. 19–20), and the Lemma 2.1 below on the entropy of lifts to simple branched coverings are needed to prove the analog of Theorem 2.2 for punctured Riemann surfaces. Theorem 2.2 for punctured Riemann surfaces is proved in Sect. 2.3. The result of Sect. 2.5 on the entropy of self-homeomorphisms of Riemann surfaces of second kind is needed to treat the entropy of reducible braids.

Recall that a branched covering is called simple, if over each point there is at most one branch point and the mapping is a double branched covering in a neighbourhood of this branch point.

Lemma 2.1 *Let φ be a self-homeomorphism of a compact Riemann surface X with set of distinguished points E. Let $\hat{X} \xrightarrow{p} X$ be a simple branched covering of X with branch locus in E and let $\hat{\varphi} : \hat{X} \to \hat{X}$ be a self-homeomorphism of $\hat{X}$ such that $p \circ \hat{\varphi} = \varphi \circ p$ (for short, $\hat{\varphi}$ is a lift of φ). Then $h(\hat{\varphi}) = h(\varphi)$.*

Proof of the Lemma 2.1 We assume that the covering multiplicity m is bigger than 1.

I. The Inequality $h(\varphi) \le h(\hat{\varphi})$ follows from Theorem 5 [1]. Here is a proof for convenience of the reader. Let $\mathcal{A}$ be an open cover of X. Put $p^{-1}(\mathcal{A}) \stackrel{\text{def}}{=} \{p^{-1}(A) : A \in \mathcal{A}\}$. $p^{-1}(\mathcal{A})$ is an open cover of $\hat{X}$. Since $p \circ \hat{\varphi} = \varphi \circ p$ it follows that $p^{-1}(\varphi^{-1}(\mathcal{A})) = \hat{\varphi}^{-1}(p^{-1}(\mathcal{A}))$. Further, $p^{-1}(\mathcal{A} \vee \mathcal{B}) = p^{-1}(\mathcal{A}) \vee p^{-1}(\mathcal{B})$ for two open covers $\mathcal{A}$ and $\mathcal{B}$ of X. Hence

$$\begin{aligned} \mathcal{N}\Big(p^{-1}(\mathcal{A}) \vee \ldots \vee \hat{\varphi}^{-N}(p^{-1}(\mathcal{A}))\Big) &= \mathcal{N}\Big(p^{-1}(\mathcal{A}) \vee \ldots \vee p^{-1}(\varphi^{-N}(\mathcal{A}))\Big) \\ = \mathcal{N}\Big(p^{-1}(\mathcal{A} \vee \ldots \vee \varphi^{-N}(\mathcal{A}))\Big) &= \mathcal{N}\Big(\mathcal{A} \vee \ldots \vee \varphi^{-N}(\mathcal{A})\Big). \end{aligned}$$

Hence

$$h(\varphi) = \sup_{\mathcal{A}} h(\varphi, \mathcal{A}) = \sup_{\substack{\hat{\mathcal{A}} = p^{-1}(\mathcal{A}) \text{ for} \\ \text{a cover } \mathcal{A} \text{ of } X}} h(\hat{\varphi}, \hat{\mathcal{A}}) \le \sup_{\substack{\tilde{\mathcal{A}} \text{ an arbitrary} \\ \text{cover of } \hat{X}}} h(\hat{\varphi}, \tilde{\mathcal{A}}) = h(\hat{\varphi}).$$

Notice that whatever sequence of covers $\mathcal{A}_n$ of X we take, the sequence of covers $p^{-1}(\mathcal{A}_n)$ is not refining if the covering multiplicity is bigger than 1.

II. The Opposite Inequality $h(\varphi) \ge h(\hat{\varphi})$ We will choose a suitable refining sequence of coverings of X and associate to it a refining sequence of coverings of $\hat{X}$. Equip X with a metric d and denote by d also the induced metric on $\hat{X}$ (which is defined by putting the length of any smooth curve in $\hat{X}$ equal to the length of its projection). Let $\sigma > 0$ be the minimal distance between two points in the branch locus of the covering.

A connected open subset V of X is called nicely covered if $p^{-1}(V)$ contains no more than one branch point. For a nicely covered connected open set V the preimage $p^{-1}(V)$ consists either of m or $m-1$ connected components. In the first case p is a homeomorphism from each connected component of $p^{-1}(V)$ onto V. In the second case this statement is true for $m-2$ of the connected components of $p^{-1}(V)$. The restriction of p to the remaining component of $p^{-1}(V)$ is a double branched covering of V with a single branch point. Each open set of diameter not exceeding σ is nicely covered.

Remarks Preceding the Proof of the Opposite Inequality Suppose we have three simply connected sets A_1, A_2, and A_3 belonging to a certain open cover $\mathcal{A}$ of X. Suppose for integer numbers k_1, k_2, and k_3 the set $A_1 \cap \varphi^{-k_2}(A_2) \cap \varphi^{-k_3}(A_3)$ is not empty and the union $A_1 \cup \varphi^{-k_2}(A_2) \cup \varphi^{-k_3}(A_3)$ is contained in a simply connected set B, that does not intersect the branch locus. Since φ is a homeomorphism with set of distingiushed points E, none of the sets A_j, $j = 1, 2, 3$, intersects the branch locus. For $k' = 1, 2, 3$, we let $A^l_{k'}$ be the m connected components of $p^{-1}(A_{k'})$, and $\hat{\varphi}$ a self-homeomorphism of $\hat{X}$ for which $p \circ \hat{\varphi} = \varphi \circ p$. Then the intersection of two or three sets among the $A_1^{j_1}$, $\hat{\varphi}^{-k_2}(A_2^{j_2})$, and $\hat{\varphi}^{-k_3}(A_2^{j_3})$ is not empty if and only if the sets lie in the same of the m connected components of $p^{-1}(B)$. Hence, there are exactly m non-empty sets of the form $A_1^{j_1} \cap \hat{\varphi}^{-k_2}(A_2^{j_2}) \cap \hat{\varphi}^{-k_3}(A_2^{j_3})$.

Consider the case when A_1 and A_3 do not meet the branch locus but A_2 contains a single point of the branch locus. Suppose the intersection $A_1 \cap \varphi^{-k_2}(A_2) \cap \varphi^{-k_3}(A_3)$ is not empty and the union $A_1 \cup \varphi^{-k_2}(A_2) \cup \varphi^{-k_3}(A_3)$ is contained in a simply connected set B with a single point in the branch locus. The previous arguments remain true for the $m-2$ connected components of $p^{-1}(B)$ that do not contain a branch point. The connected component B^{m-1}, for which $p : B^{m-1} \to B$ is a double branched covering contains two preimages of A_1, and two preimages of $\varphi^{-k_3}(A_3)$, and we cannot exclude that each of the two components of $p^{-1}(A_1)$ intersects each of the two components of $p^{-1}(\varphi^{-k_3}(A_3))$. We may get more than m, but no more than $m-2+4$ non-empty sets of the form $A_1^{j_1} \cap \hat{\varphi}^{-k_2}(A_2^{j_2}) \cap \hat{\varphi}^{-k_3}(A_3^{j_3})$. More detailed, there are at most two of the connected components of $p^{-1}(A_1)$, such that each of them may contain 2 different sets of the form $A_1^{j_1} \cap \hat{\varphi}^{-k_2}(A_2^{j_2}) \cap \hat{\varphi}^{-k_3}(A_3^{j_3})$, and all other components contain exactly one set of this form. Equivalently, there are at most two of the connected components of $p^{-1}(\varphi^{-k_3}(A_3))$, such that each of them may contain 2 sets of the form $A_1^{j_1} \cap \hat{\varphi}^{-k_2}(A_2^{j_2}) \cap \hat{\varphi}^{-k_3}(A_3^{j_3})$, and all other components contain exactly one set of this form.

Beginning of the Formal Proof We make the following choices of positive numbers by induction on the natural number ℓ. For each ℓ there exists a positive number $\varepsilon_\ell < \varepsilon_{\ell-1}$, $\varepsilon_1 < \frac{\sigma}{2}$, such that for any pair of points $z_1, z_2 \in X$ the following implication holds

$$d(z_1, z_2) < \varepsilon_\ell \Rightarrow d(\varphi^{\pm l}(z_1), \varphi^{\pm l}(z_2)) < \frac{\sigma}{2} \text{ for } l = 1, 2, \ldots, \ell. \tag{2.8}$$

Further, for each ℓ there exists a positive number $\varepsilon'_\ell < \frac{\varepsilon_\ell}{2}$, $\varepsilon'_\ell < \varepsilon'_{\ell-1}$, such that for any pair of points $z_1, z_2 \in X$ the implication

$$d(z_1, z_2) \geq \frac{\varepsilon_\ell}{2} \Rightarrow d(\varphi^{\pm l}(z_1), \varphi^{\pm l}(z_2)) \geq \varepsilon'_\ell \text{ for } l = 1, 2, \ldots, \ell \tag{2.9}$$

holds. Finally, there exists a positive number $\delta_\ell < \frac{\varepsilon'_\ell}{4}$, $\delta_\ell < \delta_{\ell-1}$, such that such that for any pair of points $z_1, z_2 \in X$

$$d(z_1, z_2) \leq 2\delta_\ell \Rightarrow d(\varphi^{\pm l}(z_1), \varphi^{\pm l}(z_2)) < \frac{\varepsilon'_\ell}{2} \text{ for } l = 1, 2, \ldots, \ell\,. \tag{2.10}$$

The statements follow from the uniform continuity of $\varphi^{\pm 1}$ on the compact space X. The implication (2.9) is equivalent to the implication

$$d(z_1, z_2) < \frac{\varepsilon_\ell}{2} \Leftarrow d(\varphi^{\pm l}(z_1), \varphi^{\pm l}(z_2)) < \varepsilon'_\ell\,.$$

Consider the cover $\mathcal{A}_\ell$ of X whose elements are the open balls of radius ε_ℓ in the metric d centered at each point of the branch locus and the open balls of radius δ_ℓ centered at each point of the complement of all the described ε_ℓ-balls.

Associate to $\mathcal{A}_\ell$ the cover $\hat{\mathcal{A}}'_\ell$ of $\hat{X}$ that consists of all connected components of $p^{-1}(A)$ for all $A \in \mathcal{A}_\ell$. For $\mathcal{A}$ being one of the $\mathcal{A}_\ell$ the cover $\hat{\varphi}^{-1}(\hat{\mathcal{A}}')$ consists of all sets of the following form. For a set $A \in \mathcal{A}$ we take a connected component $\hat{A}^j$ of $p^{-1}(A)$ and consider $\hat{\varphi}^{-1}(\hat{A}^j)$.

The cover $\hat{\mathcal{A}}' \vee \hat{\varphi}^{-1}(\hat{\mathcal{A}}') \vee \ldots \vee \hat{\varphi}^{-n+1}(\hat{\mathcal{A}}')$ of $\hat{X}$ consists of all sets of the form

$$\hat{A}_1^{j_1} \cap \hat{\varphi}^{-1}(\hat{A}_2^{j_2}) \cap \ldots \cap \hat{\varphi}^{-n+1}(\hat{A}_n^{j_n}) \tag{2.11}$$

for $A_1, \ldots, A_n \in \mathcal{A}$ and $\hat{A}_{k'}^{j_{k'}}$ being a connected component of $p^{-1}(A_{k'})$, $k' = 1, 2, \ldots, n$. We want to compare $\mathcal{N}(\mathcal{A} \vee \varphi^{-1}(\mathcal{A}) \vee \ldots \vee \varphi^{-n+1}(\mathcal{A}))$ and $\mathcal{N}(\hat{\mathcal{A}}' \vee \hat{\varphi}^{-1}(\hat{\mathcal{A}}') \vee \ldots \vee \hat{\varphi}^{-n+1}(\hat{\mathcal{A}}'))$.

1. The Case When No A_j Meets the Branch Locus Consider the case $n = 2$. Notice that $A_1 \cap \varphi^{-1}(A_2) \neq \emptyset$ if the intersection $\hat{A}_1^{j_1} \cap \hat{\varphi}^{-1}(\hat{A}_2^{j_2})$ is not empty. Suppose A_1 and A_2 do not meet the branch locus. Then there are local inverses of $p : \hat{X} \to X$, $i_{l,A_1} : A_1 \to \hat{X}$, and $i_{l,\varphi^{-1}(A_2)} : \varphi^{-1}(A_2) \to \hat{X}$, $l = 1, \ldots, m$. Recall that m is the covering multiplicity of the covering $p : \hat{X} \to X$. We claim that the sets of the form

$$\hat{A}_1^{j_1} \cap \hat{\varphi}^{-1}(\hat{A}_2^{j_2}) \tag{2.12}$$

are the images

$$i_{l,A_1}(A_1 \cap \varphi^{-1}(A_2)) = i_{l,\varphi^{-1}(A_2)}(A_1 \cap \varphi^{-1}(A_2)),\ l = 1, \ldots, m\,.$$

The claim implies that in this case the number of sets of the form (2.12) equals m.

The claim can be obtained as follows. We write for short ε instead of ε_ℓ, ε' instead of ε'_ℓ, and δ instead of δ_ℓ. In the considered case the center of A_j, $j = 1, 2$, has distance in the metric d at least ε from the branch locus, the diameter of A_1 equals $\mathrm{diam}(A_1) = 2\delta$, and by the implication (2.10) the diameter of $\varphi^{-1}(A_2)$ equals $\mathrm{diam}(\varphi^{-1}(A_2)) < \frac{\varepsilon'}{2}$. Since the sets A_1 and $\varphi^{-1}(A_2)$ intersect, their union has diameter $\mathrm{diam}(A_1 \cup \varphi^{-1}(A_2)) < 2\delta + \frac{\varepsilon'}{2} < \varepsilon'$. Since the center of A_1 has distance bigger than ε from the branch locus, the union $A_1 \cup \varphi^{-1}(A_2)$ is contained in an ε'-ball B that has distance at least ε' from the branch locus. The preimage $p^{-1}(B)$ consists of m connected components (called sheets) on which p is a homeomorphism. Let $i_{l,B}$ be the inverse of the restriction of p to the l-th sheet. Then $p^{-1}(A_1)$ is the union of the $i_{l,B}(A_1)$, and $p^{-1}(\varphi^{-1}(A_2))$ is the union of the $i_{l,B}(\varphi^{-1}(A_2))$. Hence, the non-empty sets of the form (2.11) are exactly the sets $i_{l,B}(A_1 \cap \varphi^{-1}(A_2))$, $l = 1, \ldots, m$, i.e. the sheets over $A_1 \cap \varphi^{-1}(A_2)$. We obtained the claim.

For all $k \geq 2$ we have the following

Claim I_k Suppose the intersection

$$A_1 \cap \varphi^{-1}(A_2) \cap \ldots \cap \varphi^{-k+1}(A_k) \tag{2.13}$$

is not empty and none of the $A_{k'}$ intersects the branch locus. Then the non-empty sets of the form

$$\hat{A}_1^{j_1} \cap \hat{\varphi}^{-1}(\hat{A}_2^{j_2}) \cap \ldots \cap \hat{\varphi}^{-k+1}(\hat{A}_k^{j_k}) \tag{2.14}$$

for $A_{k'} \in \mathcal{A}$ and $\hat{A}_{k'}^{j_{k'}}$ a connected component of $p^{-1}(A_{k'})$, $k' = 1, 2, \ldots, k$, are the images

$$i_{l,\varphi^{-k+1}(A_k)}(A_1 \cap \varphi^{-1}(A_2) \cap \ldots \cap \varphi^{-k+1}(A_k)) \,, l = 1, \ldots, m \,, \tag{2.15}$$

under the m local inverses $i_{l,\varphi^{-k+1}(A_k)}$ of $p \mid \varphi^{-k+1}(A_k)$.

For all $k' = 1, \ldots, k$ we have the following

Claim $I_{(k,k')}$ In (2.15) we may replace the local inverses $i_{l,\varphi^{-k+1}(A_k)}$ of $p \mid \varphi^{-k+1}(A_k)$ by the local inverses $i_{l,\varphi^{-k'+1}(A_{k'})}$ of $p \mid \varphi^{-k'+1}(A_{k'})$ for any $k' = 1, \ldots, k$.

The Claim I_k was proved for $k = 2$.

The Claim $I_{(k,k')}$ is easy to see. Indeed, if Claim I_k is true for some $k' > 1$ in (2.15), then it is true also for $k' - 1$, since $A_{k'-1} \cup \varphi^{-1}(A_{k'})$ is simply connected and, hence, $\varphi^{-k'}(A_{k'-1}) \cup \varphi^{-k'+1}(A_{k'})$ is simply connected and on $\varphi^{-k'}(A_{k'-1}) \cap \varphi^{-k'+1}(A_{k'})$ the equality $i_{l,\varphi^{-k'+1}(A_{k'})} = i_{l,\varphi^{-k'}(A_{k'-1})} = i_{l,\varphi^{-k'+1}(A_{k'}) \cup \varphi^{-k'}(A_{k'-1})}$

holds. In the same way one can see that Claim I_k is true for $k' + 1$, if it is true for k' and the inequality $k' + 1 \leq k$ holds.

We now prove the Claim I_k by induction for $k > 2$. Suppose it is proved for some number $k \geq 2$. The collection of sets (2.14) computed for the number $k + 1$ is the collection of intersections of the sets (2.14) computed for the number k with the sets $\hat{\varphi}^{-k}(\hat{A}_{k+1}^{j_{k+1}})$. By the statement for $k = 2$ the collection of sets of the form $\hat{A}_k^{j_k} \cap \hat{\varphi}^{-1}(\hat{A}_{k+1}^{j_{k+1}})$ equals the collection $i_{l,A_k^{j_k}}(A_k \cap \varphi^{-1}(A_{k+1}))$, $l = 1, \ldots, m$. Since φ^{-k+1} is a homeomorphism, the collection $\hat{\varphi}^{-k+1}(\hat{A}_k^{j_k}) \cap \hat{\varphi}^{-k}(\hat{A}_{k+1}^{j_{k+1}})$ equals the collection $i_{l,\varphi^{-k+1}(A_k)}(\varphi^{-k+1}(A_k) \cap \varphi^{-k}(A_{k+1}))$. Together with the statement for k this gives the statement for $k + 1$ and $k' = k$. Claim I_k follows from Claim $I_{(k,k')}$.

We proved in particular, that there are exactly m non-empty sets of form (2.14) if the $A_{k'}$ do not meet the branch locus and the set (2.13) is not empty.

2. All A_j Intersect the Branch Locus Let again $n = 2$. Suppose now, that both, A_1 and $\varphi^{-1}(A_2)$, intersect the branch locus. Then the intersection of each set with the branch locus consists of a single point which is the center of A_1 and also the preimage under φ of the center of A_2. By implication (2.8) the set $A_1 \cup \varphi^{-1}(A_2)$ is contained in a disc B_1 of radius not exceeding $\frac{\sigma}{2}$ centered at a point in the branch locus. Consider the connected components ${B_1}^j$, $j = 1, \ldots, m - 1$, of $p^{-1}(B_1)$, where $p : {B_1}^{m-1} \to B_1$ is a double branched covering, and $p : B_1^j \to B_1$ is a homeomorphism for $j < m - 1$. Since $\hat{A}_1^{j_1}$ and $\hat{\varphi}^{-1}(\hat{A}_2^{j_2})$ may intersect only if they are in the same component B_1^j, there are $m - 1$ non-empty sets of the form (2.12) each contained in a ${B_1}^j$ and each ${B_1}^j$ contains a set of form (2.12). In this case there are $m - 1$ non-empty sets of the form (2.12).

Consider now k sets $A_{k'}$ each of which intersects the branch locus. Suppose that $A_1 \cap \varphi^{-1}(A_2) \cap \ldots \cap \varphi^{-k+1}(A_k)$ is not empty. The same induction argument as in the case when none of the $A_{k'}$ meets the branch locus gives the following. Take a set $A_{k'}$ of the collection. The set $\varphi^{-k'+1}(A_{k'})$ contains a single point in the branch locus, namely the center of the ε-disc A_1. The preimage $p^{-1}(\varphi^{-k'+1}(A_{k'}))$ consists of $m - 1$ connected components B_l, such that the restriction of p to the first $m - 2$ of them is a homeomorphism, and the restriction of p to the last component is a double branched covering. For each $l = 1, \ldots, m - 1$ we consider the preimage of $A_1 \cap \varphi^{-1}(A_2) \cap \ldots \cap \varphi^{-k+1}(A_k)$ under the projection p, that is contained in the connected component B^l of $p^{-1}(B)$. The collection of non-empty sets of the form (2.11) is equal to this collection.

3. The Mixed Case, Two Sets Suppose now that A_1 intersects the branch locus, A_2 does not, and $A_1 \cap \varphi^{-1}(A_2) \neq \emptyset$. Then A_1 is a disc of radius ε with center at a point in the branch locus. The ball B_1 of radius 2ε with the same center also has a single point in the branch locus. Since A_2 does not intersect the branch locus, it is a disc of radius δ around a point z that has distance bigger than ε to the branch locus. By implication (2.9) the preimage $\varphi^{-1}(z)$ has distance at least ε' from the branch locus. By (2.10) the diameter of $\varphi^{-1}(A_2)$ is less than $\frac{\varepsilon'}{2}$. Hence, the distance of $\varphi^{-1}(A_2)$ from the branch locus is not smaller than $\varepsilon' - \frac{1}{2}\varepsilon' = \frac{1}{2}\varepsilon'$. Hence, $\varphi^{-1}(A_2)$ is contained in a sector S_1 of an angle not bigger than π of the disc B_1. The preimages of S_1 under p are m sheets that are mapped by p homeomorphically onto S_1. It is now clear that the collection of sets (2.12) is equal to the collection of the m images of the local inverses of p on S_1.

4. The Building Block for the General Case Consider now elements $A_{k_0}, A_{k_0+1}, \ldots, A_{k_0+k_1}, A_{k_0+k_1+1}$ of the cover $\mathcal{A}_\ell$ such that A_{k_0} and $A_{k_0+k_1+1}$ do not intersect the branch locus, but all other sets are discs around a point in the branch locus. Suppose $\varphi^{-k_0+1}(A_{k_0}) \cap \ldots \cap \varphi^{-k_0-k_1}(A_{k_0+k_1+1}) \neq \emptyset$ and $k_1 \leq \ell - 1$. (The chosen number ℓ is the label of the cover $\mathcal{A}_\ell$.)

Claim The collection of sets of the form

$$\hat{\varphi}^{-k_0+1}(\hat{A}_{k_0}^{j_{k_0}}) \cap \ldots \cap \hat{\varphi}^{-k_0-k_1}(\hat{A}_{k_0+k_1+1}^{j_{k_0+k_1+1}}) \tag{2.16}$$

equals the collection of the m images of the set

$$\varphi^{-k_0+1}(A_{k_0}) \cap \ldots \cap \varphi^{-k_0-k_1}(A_{k_0+k_1+1})$$

under the local inverses of p on the set $\varphi^{-k_0-k_1}(A_{k_0+k_1+1})$.

By the preceding arguments the respective statement is true for the sets of this collection that do intersect the branch locus. Namely, the collection of sets of the form

$$\hat{\varphi}^{-k_0}(\hat{A}_{k_0+1}^{j_{k_0+1}}) \cap \ldots \cap \hat{\varphi}^{-k_0-k_1+1}(\hat{A}_{k_0+k_1}^{j_{k_0+k_1}}) \tag{2.17}$$

equals the collection of preimages of $\varphi^{-k_0}(A_{k_0+1}) \cap \ldots \cap \varphi^{-k_0-k_1+1}(A_{k_0+k_1})$ contained in the $m-1$ connected components of the preimage $p^{-1}(\varphi^{-k'+1}(A_{k'}))$ for any k' between k_0+1 and k_0+k_1.

Recall that the two sets A_{k_0} and $A_{k_0+k_1+1}$, that do not intersect the branch locus, are discs of radius δ_ℓ with center at distance more than ε_ℓ from the branch locus. Also, $A_{k_0} \cap \varphi^{-k_1-1}(A_{k_0+k_1+1}) \neq \emptyset$. By the implication (2.10) the diameter of $\varphi^{-k_1-1}(A_{k_0+k_1+1})$ is smaller then $\frac{1}{2}\varepsilon'_\ell$. Hence, $A_{k_0} \cup \varphi^{-k_1-1}(A_{k_0+k_1+1})$ is contained in a disc B'_{k_0} of radius not exceeding $\delta_\ell + \frac{1}{2}\varepsilon'_\ell \leq \varepsilon'_\ell$ around the center of A_{k_0}. Since the distance of the center of A_{k_0} to the branch locus is at least ε_ℓ, the distance of $B'_{k_0} \supset A_{k_0} \cup \varphi^{-k_1-1}(A_{k_0+k_1+1})$ to the branch locus is at least $\varepsilon_\ell - \varepsilon'_\ell \geq \frac{\varepsilon_\ell}{2} \geq \varepsilon'_\ell$.

By (2.8) $\varphi^{-1}(A_{k_0+1})$ is contained in the disc of radius $\frac{\sigma}{2}$ around the point in the branch locus that is contained in $\varphi^{-1}(A_{k_0+1})$. Hence, the union $A_{k_0} \cup \varphi^{-k_1-1}(A_{k_0+k_1+1}) \cup \varphi^{-1}(A_{k_0+1})$ is contained in the σ-disc around a point of the branch locus. It is now clear, that the collection of sets $\hat{A}_{k_0}^{j_{k_0}} \cap \hat{\varphi}^{-k_1-1}(\hat{A}_{k_0+k_1+1}^{j_{k_0+k_1+1}}) \cap \hat{\varphi}^{-1}(\hat{A}_{k_0+1}^{j_{k_0+1}})$ is equal to the collection of the m images of the set $A_{k_0} \cap \varphi^{-k_1-1}(A_{k_0+k_1+1}) \cap \varphi^{-1}(A_{k_0+1})$ under the local inverses of p

on the disc B'_{k_0} of radius ε'_ℓ that contains $A_{k_0} \cup \varphi^{-k_1-1}(A_{k_0+k_1+1})$. Apply the diffeomorphism φ^{-k_0+1}. We see that the collection of the non-empty sets of the form $\hat{\varphi}^{-k_0+1}(\hat{A}_{k_0}^{j_{k_0}}) \cap \hat{\varphi}^{-k_0}(\hat{A}_{k_0+1}^{j_{k_0+1}}) \cap \hat{\varphi}^{-k_0-k_1}(\hat{A}_{k_0+k_1+1}^{j_{k_0+k_1+1}})$ is the collection of the m images of the set $\varphi^{-k_0+1}(A_{k_0}) \cap \varphi^{-k_0}(A_{k_0+1}) \cap \varphi^{-k_0-k_1}(A_{k_0+k_1+1})$ under the local inverses of p on the set $\varphi^{-k_0+1}(A_{k_0})$ (equivalently, under the local inverses of p on the set $\varphi^{-k_0-k_1}(A_{k_0+k_1+1})$). Hence the same is true for the collection of sets (2.16). The claim is proved.

5. The General Case Take now a maximal collection of consecutive sets $A_{k_0+1}, \ldots, A_{k_0+k_1}$, that intersect the branch locus. We assume that $k_0 + 1 > 1$ and $k_0 + k_1 < n$ and $k_1 \geq \ell$. Notice that A_{k_0} and $A_{k_0+k_1+1}$ do not meet the branch locus. By the remark in the beginning of the proof there are no more than $m - 2 + 4 < 2m$ different non-empty sets of the form $\hat{\varphi}^{-k_0+1}(\hat{A}_{k_0}^{j_{k_0}}) \cap \ldots \cap \hat{\varphi}^{-k_0-k_1}(\hat{A}_{k_0+k_1+1}^{j_{k_0+k_1+1}})$, and each set is contained in a connected component of $p^{-1}(\varphi^{-k_0-k_1}(A_{k_0+k_1+1}^{j_{k_0+k_1+1}}))$ for the set $A_{k_0+k_1+1}$ that does not intersect the branch locus. Moreover, each connected component of $p^{-1}(\hat{\varphi}^{-k_0-k_1}(A_{k_0+k_1+1}))$ contains at most two sets of the form $\hat{\varphi}^{-k_0+1}(\hat{A}_{k_0}^{j_{k_0}}) \cap \ldots \cap \hat{\varphi}^{-k_0-k_1}(\hat{A}_{k_0+k_1+1}^{j_{k_0+k_1+1}})$.

Consider now any collection $A_1, A_2, \ldots, A_n \in \mathcal{A}_\ell$. Divide it into maximal collections of consecutive sets, so that either all sets of a collection intersect the branch locus or all sets of the collection do not intersect the branch locus. For each integer number k we consider all the maximal collections of consecutive sets $A_{k_0+1}, \ldots, A_{k_0+k_1}$, that intersect the branch locus and for which $k_1 \geq \ell$ and $k_0 + k_1 \leq k$. Its number is denoted by $\mathfrak{n}(k)$.

If $A_{k_0+1}, \ldots, A_{k_0+k_1}$ is a maximal collection of consecutive sets among the $A_1, A_2, \ldots, A_n \in \mathcal{A}_\ell$, that do not intersect the branch locus, then each connected component of the preimage $p^{-1}(\varphi^{-k_0-k_1-1}(A_{k_0+k_1}^{j_{k_0+k_1}}))$ contains at most $2^{\mathfrak{n}(k_0)} = 2^{\mathfrak{n}(k_0+k_1)}$ different sets of the form $\hat{A}_1^{j_1} \cap \ldots \cap \hat{\varphi}^{-k_0-k_1-1}(\hat{A}_{k_0+k_1}^{j_{k_0+k_1}})$.

If $A_{k_0+1}, \ldots, A_{k_0+k_1}$ is a maximal collection of consecutive sets that intersect the branch locus and $k_0 + k_1 \leq n$, then, if $k_1 < \ell$, each connected component of $p^{-1}(\varphi^{-k_0-k_1+1}(A_{k_0+k_1}^{j_{k_0+k_1}}))$ contains at most $2^{\mathfrak{n}(k_0)}$ different sets of the form $\hat{A}_1^{j_1} \cap \ldots \cap \hat{\varphi}^{-k_0-k_1-1}(\hat{A}_{k_0+k_1}^{j_{k_0+k_1}})$, and contains at most $2 \cdot 2^{\mathfrak{n}(k_0)} = 2^{\mathfrak{n}(k_0+k_1)}$ different sets of the form $\hat{A}_1^{j_1} \cap \ldots \cap \hat{\varphi}^{-k_0-k_1+1}(\hat{A}_{k_0+k_1}^{j_{k_0+k_1}})$, if $k_1 \geq \ell$.

Since $\mathfrak{n}(n) \leq [\frac{n}{\ell}]$, we obtain

$$\mathcal{N}\Big(\hat{\mathcal{A}}_\ell \vee \ldots \vee \hat{\varphi}^{-n}(\hat{\mathcal{A}}_\ell)\Big) \leq 2^{[\frac{n}{\ell}]} \cdot m \cdot \mathcal{N}\Big(\mathcal{A}_\ell \vee \ldots \vee \varphi^{-n}(\mathcal{A}_\ell)\Big).$$

Hence,

$$h(\hat{\mathcal{A}}_\ell, \hat{\varphi}) \leq \frac{\log 2}{\ell} + h(\mathcal{A}_\ell, \varphi) \tag{2.18}$$

for each of the described covers $\mathcal{A}_\ell$ of X. Since the sequences $\mathcal{A}_\ell$ and $\hat{\mathcal{A}}_\ell$ are refining we obtain

$$h(\hat{\varphi}) \le h(\varphi) . \tag{2.19}$$

The lemma is proved. □

Theorem 2.2 and Lemma 2.1 together with Ahlfors' trick (see Sect. 1.7) will imply the following.

Theorem 2.3 *Let X be a closed connected Riemann surface of genus g with a set E_m of $m \ge 0$ distinguished points, $3g-3+m > 0$. Then any non-periodic absolutely extremal self-homeomorphism of X with set of distinguished points E_m is entropy minimizing in its isotopy class.*

Together with Corollary 1.1 we obtain the following statement, which includes homeomorphisms with elliptic or hyperbolic modular transformation.

Corollary 2.1 *Let φ be an irreducible self-homeomorphism of a closed connected Riemann surface X with a set E_m of $m \ge 0$ distinguished points, $3g-3+m > 0$. Let φ_0 be an absolutely extremal self-homeomorphism of a Riemann surface Y with m distinguished points, which is obtained from φ by isotopy and conjugation. Then*

$$\begin{aligned} h(\varphi_0) &= \frac{1}{2}\log K(\varphi_0) = L(\varphi^*) = h(\widehat{\mathfrak{m}_{\varphi_0}}) \\ &= \inf\{h(\varphi_1) : \varphi_1 \textit{ is obtained from } \varphi \textit{ by isotopy and conjugation}\} . \end{aligned}$$

Proof of the Implication Theorem 2.2 ⇒ Theorem 2.3 Let X be a connected closed Riemann surface with set of distinguished points $\{z_1, \dots, z_m\}$, $3g-3+m > 0$. According to [2] (see also Sect. 1.7) there is a closed Riemann surface $\hat{X}$ of genus at least two which is a simple branched covering of X so that the set of branch points projects onto the set $\{z_1, \dots, z_m\}$. The non-periodic absolutely extremal self-homeomorphism φ_0 lifts to a non-periodic absolutely extremal self-homeomorphism $\hat{\varphi}_0$ on $\hat{X}$ with $K(\varphi_0) = K(\hat{\varphi}_0)$.

By Lemma 2.1

$$\begin{aligned} &\inf\{h(\varphi_1) : \varphi_1 \text{ is a self-homeomorphism of } X \text{which is isotopic to } \varphi\} \\ &\quad = \inf\{h(\hat{\varphi}_1) : \hat{\varphi}_1 \text{ is the lift of a self-homeomorphism of } X \\ &\qquad \text{which is isotopic to } \varphi\} \\ &\quad \ge \inf\{h(\mathcal{F}_1) : \mathcal{F}_1 \text{ is a self-homeomorphism of } \hat{X} \text{ which is isotopic to } \hat{\varphi}\}. \end{aligned}$$

Since $\hat{\varphi}_0$ is absolutely extremal on $\hat{X}$ and isotopic to $\hat{\varphi}$, the Fathi-Shub Theorem 2.2 implies that the last infimum equals $h(\hat{\varphi}_0)$. Thus, the last infimum is attained on a homeomorphism of $\hat{X}$ which is a lift. Hence the inequality between the second and the third infimum is an equality too. Since by Lemma 2.1 we have

$h(\varphi_0) = h(\hat{\varphi}_0)$, the first infimum equals $h(\varphi_0)$. Also, $K(\varphi_0) = K(\hat{\varphi}_0)$. It remains to use Corollary 1.1, and the invariance of the entropy under conjugation. □

2.2 Absolutely Extremal Homeomorphisms and the Trajectories of the Associated Quadratic Differentials

We collect here some facts concerning trajectories of quadratic differentials that are related to non-periodic absolutely extremal self-homeomorphisms. For a more comprehensive exposition we refer the reader to the book of Strebel [78].

Let ϕ be a quadratic differential associated to a non-periodic absolutely extremal self-homeomorphism φ on a closed Riemann surface X possibly with set of distinguished points E. Recall that ϕ is a meromorphic quadratic differential. It is analytic except, maybe, at some distinguished points where it has at worst first order poles. We need some information about the horizontal and vertical trajectories of ϕ.

For a horizontal trajectory γ and a point p on it we call the union of $\{p\}$ with any of the connected components of $\gamma \setminus \{p\}$ a horizontal trajectory ray. If the set of limit points of a trajectory ray (that is the closure of the ray in the ϕ-metric minus the ray itself) contains a regular point it contains a horizontal arc through this point. If the limit set of a trajectory ray consists of a single point this point must be critical and the trajectory ray is called critical.

Proposition 2.1 *For a quadratic differential ϕ associated to a pseudo-Anosov self-homeomorphism φ of a closed Riemann surface (possibly with distinguished points) there are no leaf of the horizontal foliation and point on it with both trajectory rays critical (and no leaf of the vertical foliation and point on it with this property).*

Proof Suppose there is a horizontal leaf joining two critical points. There are only finitely many critical points and in a neighbourhood of each critical point there are only finitely many horizontal bisectrices. Since φ maps the set of singular points of ϕ onto itself (perhaps, permuting the points) and φ maps horizontal trajectories to itself, a power φ^n of φ maps the leaf onto itself. Adding the two endpoints to the leaf we obtain a compact arc which has finite ϕ-length. But by (1.28) the homeomorphism φ^n expands the length of the horizontal leaves by the factor $K^{\frac{n}{2}}$ which is impossible. □

Proposition 2.2 *There is no closed leaf of the horizontal foliation and no closed leaf of the vertical foliation of a quadratic differential ϕ associated to a pseudo-Anosov homeomorphism φ.*

Proof Suppose there is a closed horizontal leaf γ. Choose an orientation for γ. We claim that there is an annulus A around this leaf which is contained in the regular part of X and is the union of closed horizontal leaves. Indeed, each point x of the leaf γ is regular. Hence, there is a neighbourhood V of each point x, that is equipped with distinguished coordinates centered at the point, such that $V \cap \gamma$ is the set of points in V that are real in distinguished coordinates, and the direction of the positive orientation of γ corresponds to the positive direction of the real axis. We may assume that, perhaps after shrinking, the neighbourhood of each x is a square $R(x)$ with center x written in the distinguished coordinates centered at x as $\{|\mathrm{Re}\, z)| < \varepsilon(x),\ |\mathrm{Im}\, z| < \varepsilon(x)\}$. Cover γ by a finite number of such squares labeled by R_j, $j = 1, \ldots, N$, of side length ε_j. We may do this so, that considering an infinite sequence of squares with $R_{j+N} = R_j$ for each j, any R_j intersects exactly the squares $R_{j\pm 1}$. Let ε be the minimum of the ε_j.

The squares R_j cover an annulus A around γ. By the choice of the distinguished coordinates on the R_j each transition function is a translation by a real number. The arcs given in distinguished coordinates on R_j by $(-\varepsilon_j, \varepsilon_j)$ cover the closed horizontal curve γ. Hence, for any real parameter t, $|t| < \epsilon \stackrel{def}{=} \min \varepsilon_j$, the arcs $(-\varepsilon_j + it, \varepsilon_j + it)$ cover a closed horizontal curve. The union of these curves is an annulus A around γ whose existence was claimed.

Take any boundary point of the annulus A in the sense, that the point is a limit point of A in the ϕ-metric and is not contained in A. If this point is a regular point then there is a horizontal arc contained in ∂A that contains the point. If all points of a boundary component of A are regular, then by the preceding arguments there is a bigger annulus that contains the previous annulus as well as the boundary component, and is the union of closed horizontal curves. Take the maximal annulus containing γ that is the union of closed horizontal curves. Then each boundary component must contain a critical point and the complement in each boundary component of the critical points consists of horizontal leaves with both rays being critical. This is not possible by Proposition 2.1. □

We call a trajectory ray divergent if its limit set contains more than one point. A trajectory ray with initial point p is called recurrent if p is contained in the limit set of the ray. We will consider trajectory rays parameterized so that the initial point has the smallest parameter and write $p_2 > p_1$ for two points p_1 and p_2 on the ray if the parameter of p_1 is smaller than that of p_2. The parametrization defines an orientation of the ray.

Let as before X be a Riemann surface with quadratic differential ϕ, and let R be an open rectangle (and let $\bar{R}$ be a closed rectangle, respectively) in the complex plane with sides parallel to the axes. A mapping $\mathcal{F} : R \to X$ is called an open ϕ-rectangle (and a mapping $\mathcal{F} : \bar{R} \to X$ is called a closed ϕ-rectangle, respectively,) if $\mathcal{F}$ takes horizontal segments in R (in $\bar{R}$, respectively) to ϕ-horizontal arcs and takes vertical segments in R (in $\bar{R}$, respectively) to ϕ-vertical arcs, so that the Euclidean length of the horizontal (vertical arc, respectively) is equal to the ϕ-length of its image.

Recall that trajectories do not contain ϕ-singular points, hence the images of ϕ-rectangles are contained in the ϕ-regular part of X. We call a ϕ-rectangle embedded if the mapping $\mathcal{F}$ is injective.

Theorem 2.4 *Every divergent trajectory ray is recurrent. More precisely, let α^+ be a divergent horizontal trajectory ray with initial point p_0. Let β be any oriented open arc through p_0 contained in a vertical leaf, that is cut positively by α^+ at p_0. Then for each point p_1 on the ray α^+ there exists a point $p_2 > p_1$ on α^+, such that α^+ cuts β positively at p_2. The ϕ-length of the arc on α^+ between p_1 and p_2 is bounded from below by a constant c that depends only on the quadratic differential ϕ.*

Proof Let $\beta' \subset \beta$ be a vertical arc which has center p_0 and ϕ-length $\ell_\phi(\beta')$, such that the vertical arc with center p_0 and ϕ-length $3\ell_\phi(\beta')$ is contained in β. We call a point $p \in \beta'$ exceptional, if the horizontal trajectory ray, which cuts β' positively at p, is critical and does not cut β' positively at a point greater than p. Notice that a point p is not exceptional, if the trajectory ray which cuts β' positively at p is critical but cuts β' positively at a point greater than p before running into the critical point. We claim that only finitely many points of β' are exceptional. Indeed, each such point p and the respective critical ray are obtained as follows. Take a critical point, take any oriented trajectory emerging from this critical point, consider its part between the critical point and its first negative intersection point with β' (if there is such), and invert orientation. There are only finitely many such rays, hence the claim.

Recall that the ray α^+ is not critical. We take a closed arc β_0, that is contained in β', has midpoint p_0, and does not contain exceptional points. We will prove now the following

Claim There is a horizontal trajectory ray that emerges at a point $p \in \beta_0$ and intersects β' positively.

For positive numbers t we consider the closed rectangle $\overline{R_t} \subset \mathbb{C}$ with sides parallel to the axes, with horizontal Euclidean side length equal to t, vertical Euclidean side length $\ell_\phi(\beta_0)$, and with midpoint of the left side equal to the origin.

For small t there is a closed ϕ-rectangle $\mathcal{F}_{\phi,t} : \overline{R_t} \to X$ which maps the left side of $\overline{R_t}$ onto β_0. Such a ϕ-rectangle can be defined for parameters t as long as each horizontal trajectory ray that emerges at a point $p \in \beta_0$ in positive direction does not run into a critical point before running along a closed arc of ϕ-length t with initial point p.

Assume by contradiction that the claim is not true. Since β_0 does not contain exceptional points, each trajectory ray that emerges at a point in β_0 in positive direction is non-critical (otherwise it would intersect β' positively at a point after its initial point). Hence, a closed ϕ-rectangle $\mathcal{F}_{\phi,t} : \overline{R_t} \to X$ which maps the left side of $\overline{R_t}$ onto β_0 exists for all $t > 0$.

If $\mathcal{F}_{\phi,t}$ is injective on $\overline{R_t}$, the image $\mathcal{F}_{\phi,t}(\overline{R_t})$ has ϕ-area $t \cdot \ell_\phi(\beta_0)$. Since the ϕ-area $\|\phi\|_1$ of X is finite, there exists a smallest number $t_0 > 0$ for which the ϕ-rectangle $\mathcal{F}_{\phi,t_0} : \overline{R_{t_0}} \to X$ is not injective (see Fig. 2.1). Denote by s_{t_0} the right side of $\overline{R_{t_0}}$. The mapping $\mathcal{F}_{\phi,t_0}$ is injective on $\overline{R_{t_0}} \setminus \mathsf{s}_{t_0}$. Hence, $\mathcal{F}_{\phi,t_0}(\mathsf{s}_{t_0})$ intersects the rest of the boundary of $\mathcal{F}_{\phi,t_0}(\overline{R_{t_0}})$. We claim, that the vertical segment

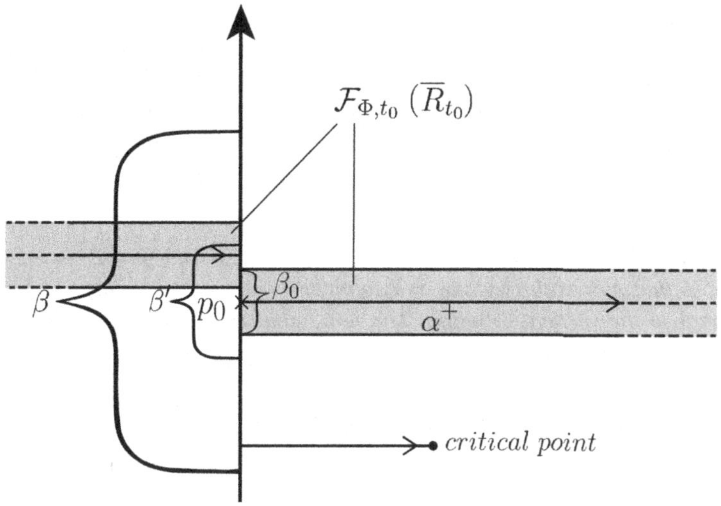

Fig. 2.1 The smallest number t_0 with non-injective ϕ-rectangle $\overline{R_{t_0}} \to \mathcal{F}_{\phi,t_0}(\overline{R_{t_0}})$

$\mathcal{F}_{\phi,t_0}(\mathsf{s}_{t_0})$ cannot intersect the image of the open horizontal sides of $\overline{R_{t_0}}$. Suppose, in contrary, $\mathcal{F}_{\phi,t_0}(\mathsf{s}_{t_0})$ intersects the image of an open horizontal side s_h of $\overline{R_{t_0}}$. Let p' be the intersection point. Then $p' = \mathcal{F}_{\phi,t_0}(z') = \mathcal{F}_{\phi,t_0}(z'')$, where z' is on an open horizontal side s_h of $\overline{R_{t_0}}$, and $z'' = t_0 + iy$ is contained in s_{t_0}. But then for all $t < t_0$ and close to t_0 the point $\mathcal{F}_{\phi,t_0}(t + iy)$ is also contained in the image $\mathcal{F}_{\phi,t_0}(\mathsf{s}_h)$ of the open horizontal side s_h. But this means, that $\mathcal{F}_{\phi,t_0}$ is not injective on $\overline{R_t}$ for $t < t_0$ and close to t, contrary to the definition of t_0.

Hence, $\mathcal{F}_{\phi,t_0}(\mathsf{s}_{t_0})$ intersects the image of the left side of $\overline{R_{t_0}}$, i.e $\mathcal{F}_{\phi,t_0}(\mathsf{s}_{t_0})$ intersects $\beta_0 \subset \beta'$. Since $\mathcal{F}_{\phi,t_0}$ is injective on $\overline{R_{t_0}} \setminus \mathsf{s}_{t_0}$, the points on $\overline{R_{t_0}}$ that are close to different vertical sides of the rectangle, but not on the vertical sides, are mapped to different (local) sides of the vertical arc containing β_0. This means, that there is a horizontal trajectory that starts at a point β_0 and cuts $\beta_0 \subset \beta'$ positively after running along a segment of length t_0. This contradicts our assumption that the claim is not true. The claim is proved.

Let t_0 be the infimum of positive numbers t for which a horizontal trajectory ray that emerges at a point of β_0 in positive direction intersects β' positively after running along a segment of length t. Then, since β_0 does not contain exceptional points, the closed ϕ-rectangle $\mathcal{F}_{\phi,t_0} : \overline{R_{t_0}} \to X$ exists. Moreover, the image of the right side $\mathcal{F}_{\phi,t_0}(\mathsf{s}_{t_0})$ of $\overline{R_{t_0}}$ intersects β'. Then $\mathcal{F}_{\phi,t_0}(\mathsf{s}_{t_0})$ is a vertical arc of ϕ-length $\ell_\phi(\beta_0)$ that intersects β', hence, it is contained in β. This means that each trajectory ray that emerges at a point of β_0 in positive direction, in particular α_+, intersects β positively after running along a segment of length t_0.

We proved that α_+ cuts β positively after p_0. Denote the intersection point by p_1. The same argument applied to p_1 instead of p_0 shows that α_+ cuts β positively at a point p_2 after p_1.

There is a positive constant c such that the ϕ-length of an arc on α^+ between two distinct positive intersection points with β exceeds c. Indeed, such an arc must leave a simply connected neighbourhood of the closure of β, that depends only on the quadratic differential ϕ. Repeating the argument for the already found intersection points with β we obtain a sequence of positive intersection points $p_0 < p_1 < \ldots < p_n < \ldots$ of α_+ with β such that the ϕ-length of the arc on α^+ between p_0 and p_n tends to ∞. The theorem is proved. □

Theorem 2.5 *For each ϕ-regular point $x_0 \in X$ there exists a simple closed smooth curve in X that passes through x_0, is contained in the ϕ-regular part of X, and is transversal to the vertical foliation. The horizontal ϕ-length $\ell_{\phi,h}$ of this curve is positive.*

Proof By Proposition 2.1 one of the horizontal trajectory rays that start at x_0 is not critical and therefore recurrent. We consider this recurrent trajectory ray, denoted by α^+, oriented in the direction away from x_0. Let β be a small vertical arc through x_0 that is oriented so that α^+ intersects β positively at x_0. By Theorem 2.4 α^+ intersects β positively at a point x_1 after x_0. Consider an open Φ-rectangle V with small horizontal side length whose right open vertical side contains the closed arc $\beta^* \subset \beta$ that joins the points x_0 and x_1 by. Let p_1 be the last (before x_1) intersection point of α_+ with ∂V. We take the closed curve which is the union of the part of α_+ between x_0 and p_1 and a curve in V that joins p_1 with x_0 and is transversal to the vertical foliation (see Fig. 2.2). The curve can be chosen to be smooth and simple closed. Its horizontal length equals the length of the part of α_+ between x_0 and x_1.

The theorem is proved. □

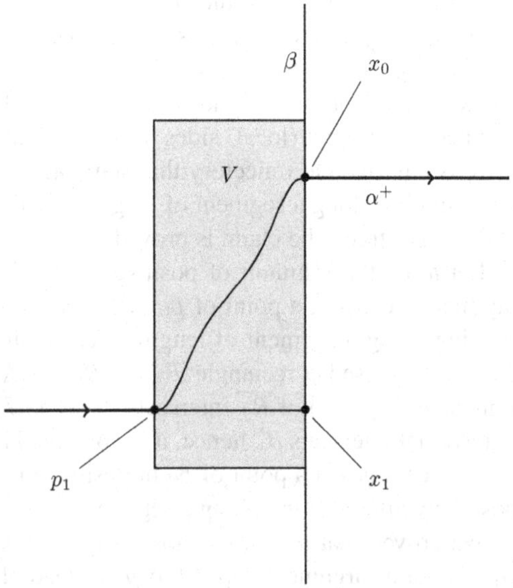

Fig. 2.2 A recurrent horizontal trajectory gives to a simple closed curve that is transversal to the horizontal foliation

2.3 The Entropy of Pseudo-Anosov Self-homeomorphisms of Surfaces with Distinguished Points

In this section we will prove Theorem 2.1. Our proof will be different from that in [25]. We begin with an especially simple example which is the entropy of a pseudo-Anosov self-homeomorphism of a torus with a distinguished point. This example is considered in [1]. The arguments given here will be adapted later to give a proof of the general case of the theorem.

Consider the standard torus $X = \mathbb{C}/(\mathbb{Z} + i\mathbb{Z})$ with distinguished point $0/(\mathbb{Z} + i\mathbb{Z})$. Let φ be a self-homeomorphism of X that lifts to a real linear self-homeomorphism of the complex plane which maps the integer lattice onto itself. Then φ fixes the distinguished point $0/(\mathbb{Z} + i\mathbb{Z})$ of X. The lift $\tilde{\varphi}$ of φ to the universal covering $\mathbb{C}$ corresponds to a matrix $\begin{pmatrix} a & b \\ c & d \end{pmatrix}$ with integer entries and determinant equal to one, i.e.

$$\tilde{\varphi}\begin{pmatrix} \xi \\ \eta \end{pmatrix} = \begin{pmatrix} a & b \\ c & d \end{pmatrix}\begin{pmatrix} \xi \\ \eta \end{pmatrix}$$

where $\zeta = \xi + i\eta \in \mathbb{C}$ are complex coordinates on $\tilde{X} = \mathbb{C}$, and $ad - bc = 1$. Suppose the eigenvalues of the matrix are positive and not equal to each other. Denote the bigger eigenvalue by λ and the corresponding eigenvector by v_+. The other eigenvalue equals $\frac{1}{\lambda}$. Denote the corresponding eigenvector by v_-. Then $\tilde{\varphi}(xv_+ + yv_-) = \lambda x v_+ + \lambda^{-1} y v_-$. Consider the conformal structure $w : X \to w(X)$ on X that lifts to a mapping $\tilde{w} : \tilde{X} \to \tilde{w}(\tilde{X})$ that takes the point $xv_+ + yv_-$ to the point $z = x + iy$. Then

$$\tilde{w} \circ \tilde{\varphi} \circ \tilde{w}^{-1}(x + iy) = \lambda x + i\lambda^{-1} y\,. \tag{2.20}$$

Consider the self-homeomorphism $\varphi_0 \stackrel{def}{=} w \circ \varphi \circ w^{-1}$ of $w(X)$ with distinguished point $0/\tilde{w}(\mathbb{Z} + i\mathbb{Z})$. Take local flat coordinates z vanishing at a point z_0 and local flat coordinates ζ vanishing at the point $\varphi_0(z_0)$. The mapping φ_0 in these local coordinates becomes

$$\zeta = w \circ \varphi \circ w^{-1}(x + iy) = \lambda x + i\frac{1}{\lambda} y\,. \tag{2.21}$$

The Beltrami differential of the mapping is

$$\frac{\bar{\partial}\varphi_0(z)}{\partial\varphi_0(z)} = \frac{(\lambda - \lambda^{-1})\, d\bar{z}}{(\lambda + \lambda^{-1})\, dz} = \frac{(\lambda - \lambda^{-1})\, d\bar{z}\, dz}{(\lambda + \lambda^{-1})\, (dz)^2}\,. \tag{2.22}$$

The equation shows the following. Let the quadratic differential ϕ be given in local coordinates by $(dz)^2$. The Beltrami differential of the mapping φ_0 equals $k\frac{|\phi|}{\phi} = k\frac{d\bar{z}}{dz}$

with $k = \frac{(\lambda-\lambda^{-1})}{(\lambda+\lambda^{-1})}$. Hence, the quadratic differential of φ_0 is the globally defined quadratic differential $(dz)^2$ on $w(X)$, and φ_0 has quasiconformal dilatation $K = \frac{1+k}{1-k} = \lambda^2$.

The quadratic differential of the inverse mapping $\xi + i\eta \to \frac{\xi}{\lambda} + i\lambda\eta$ has Beltrami differential

$$\frac{(\lambda^{-1} - \lambda)d\bar{z}\,dz}{(\lambda^{-1} + \lambda)(dz)^2}\,.$$

Hence, the quadratic differential of the inverse mapping is $-(dz)^2$, and therefore the terminal quadratic differential of φ_0 is equal to its initial quadratic differential ϕ. We proved that φ_0 is absolutely extremal on $w(X)$ with distinguished point and its quadratic differential equals $\phi = (dz)^2$. Notice that the quadratic differential ϕ does not have singular points, the distinguished point is also a regular point. Such a situation is exceptional and especially easy to treat.

The following proposition on the entropy of the mapping holds.

Proposition 2.3 $h(\varphi) = h(\varphi_0) = \log \lambda\,.$

Proof Choose a small positive number ε_0 so that the projection $\mathbb{C} \cong \tilde{w}(\tilde{X}) \to w(X)$ is injective on open squares of side length $2\lambda\epsilon_0$ (in coordinates $x + iy$ on $\mathbb{C}$). Take $\varepsilon < \varepsilon_0$. Choose the open cover $\mathcal{A}_\varepsilon$ of $w(X)$ that consists of projections to $w(X)$ of all open squares in $\mathbb{C}$ of side length ϵ with sides parallel to the axes. The minimal cardinality $\mathcal{N}(\mathcal{A}_\varepsilon)$ of a subcover of $\mathcal{A}_\varepsilon$ does not exceed the minimal cardinality of a cover of the closure of a fundamental domain on $\tilde{X}$ by squares of side length ε. This number does not exceed $c_+\,\varepsilon^{-2}$ for a positive number c_+. On the other hand $\mathcal{N}(\mathcal{A}_\varepsilon)$ is bounded from below by the maximal number of disjoint squares of side length ε that can be put in the interior of a fundamental domain on $\tilde{X}$. This number is bounded from below by $c_-\,\varepsilon^{-2}$ for a positive constant c_-.

The cover $\varphi_0^{-1}(\mathcal{A}_\varepsilon)$ consists of the projection to $w(X)$ of all rectangles in $\tilde{w}(\tilde{X})$ with sides parallel to the axes of horizontal side length $\lambda^{-1}\epsilon$ and vertical side length $\lambda\varepsilon$. If such a rectangle in $\tilde{w}(\tilde{X})$ intersects an ε-square in $\tilde{w}(\tilde{X})$, then by the choice of ε the projection is injective on the union of the square and the rectangle. Hence, the cover $\mathcal{A}_\varepsilon \vee \varphi_0^{-1}(\mathcal{A}_\varepsilon)$ consists of the projection to $w(X)$ of all rectangles in $\tilde{w}(\tilde{X})$ with sides parallel to the axes of horizontal side length at most $\lambda^{-1}\varepsilon$ and vertical side length at most ε.

By induction $\mathcal{A}_\varepsilon \vee \varphi_0^{-1}(\mathcal{A}_\varepsilon) \vee \ldots \vee \varphi_0^{-n}(\mathcal{A}_\varepsilon)$ consists of projections to $w(X)$ of all rectangles in $\tilde{w}(\tilde{X})$ of horizontal side length at most $\lambda^{-n}\varepsilon$ and vertical side length at most ε.

Indeed, if this is proved for a natural number n it is obtained for $n+1$ as follows. The cover $\varphi_0^{-1}(\mathcal{A}_\varepsilon) \vee \ldots \vee \varphi_0^{-n-1}(\mathcal{A}_\varepsilon)$ is the image under $(\varphi_0)^{-1}$ of the cover $\mathcal{A}_\varepsilon \vee \ldots \vee \varphi_0^{-n}(\mathcal{A}_\varepsilon)$. Hence, it consists of rectangles of horizontal side length at most $\lambda^{-n-1}\varepsilon$ and vertical side length at most $\lambda\varepsilon$. The cover $\mathcal{A}_\varepsilon \vee \varphi_0^{-1}(\mathcal{A}_\varepsilon) \vee \ldots \vee \varphi_0^{-n-1}(\mathcal{A}_\varepsilon)$ consists of intersections of such rectangles with ε-squares.

For each n a lift of each set of a minimal subcover of $\mathcal{A}_\varepsilon$ can be covered by $\lambda^n + 1$ rectangles of horizontal side length $\lambda^{-n}\varepsilon$ and vertical side length ε. On the other hand at least λ^n rectangles of horizontal side length at most $\lambda^{-n}\varepsilon$ and vertical side length at most ε are needed to cover a lift a single ε-square of a minimal subcover of $\mathcal{A}_\varepsilon$.

We obtain

$$\lambda^n \leq \mathcal{N}(\mathcal{A}_\varepsilon \vee \varphi_0^{-1}(\mathcal{A}_\varepsilon) \vee \ldots \vee \varphi_0^{-n}(\mathcal{A}_\varepsilon)) \leq c_+ \varepsilon^{-2}(\lambda^n + 1)\,. \tag{2.23}$$

Hence,

$$h(\varphi_0, \mathcal{A}_\varepsilon) = \limsup_{n\to\infty} \frac{1}{n} \log \mathcal{N}(\mathcal{A}_\varepsilon \vee \varphi_0^{-1}(\mathcal{A}_\varepsilon) \vee \ldots \vee \varphi_0^{-n}(\mathcal{A}_\varepsilon)) = \log \lambda\,.$$

For any decreasing sequence of small numbers ε_n the sequence of coverings $\mathcal{A}_{\varepsilon_n}$ is refining. Indeed, it is clear that $\mathcal{A}_{\varepsilon_1} \prec \mathcal{A}_{\varepsilon_2}$ for $\varepsilon_1 < \varepsilon_2$. Any open cover $\mathcal{B}$ has a finite subcover. For small enough $\varepsilon > 0$ each set in $\mathcal{A}_\varepsilon$ is contained in a set of the finite subcover, hence, $\mathcal{A}_\varepsilon \prec \mathcal{B}$.

Hence,

$$h(\varphi_0) = \log \lambda\,.$$

The proposition is proved. □

We will now prove Theorem 2.1 in the general case. In the rest of the section X will be a closed connected Riemann surface of genus g with a set E_m of $m \geq 0$ distinguished points, $3g - 3 + m > 0$. Let φ_0 be an absolutely extremal self-homeomorphism of X with set of distinguished points E_m. Suppose φ_0 is not conformal and has quasiconformal dilatation $K(\varphi_0) = \lambda^2 > 1$. Then there exists a quadratic differential ϕ on X with the following properties. φ_0 maps the set of critical points of ϕ to itself, it maps horizontal leaves of ϕ to horizontal leaves and vertical leaves to vertical leaves.

Let $\varepsilon > 0$ be a small number. We prepare the definition of an open cover $\mathcal{A}_\varepsilon$ of X. We will call an open ϕ-rectangle with both side lengths equal to ε an open ε-square. An open ϕ-rectangle with length of the horizontal sides equal to ε_1 and length of the vertical sides equal to ε_2 will be called an open $(\varepsilon_1, \varepsilon_2)$-rectangle.

An ε-star at a singular point of ϕ is defined as follows. Consider a simply connected neighbourhood U of a singular point of ϕ and distinguished coordinates z on it in which ϕ has the form $\phi(z) = (\frac{a+2}{2})^2 z^a (dz)^2$ where a is the order of the singular point. Since there are only finitely many singular points, we may assume that in these coordinates U has the form $\{|z| < c\}$ for a constant c that does not depend on the singular point. The horizontal bisectrices $\{r \exp(\frac{2\pi j i}{a+2}),\ 0 < r < c\}$, $j = 0, \ldots, a+1$, are segments of horizontal trajectories, the vertical bisectrices $\{r \exp(\frac{2\pi (j+\frac{1}{2}) i}{a+2}),\ 0 < r < c\}$, $j = 0, \ldots, a+1$, are segments of vertical trajectories. Consider the "half-sectors" between a horizontal bisectrix and a nearest

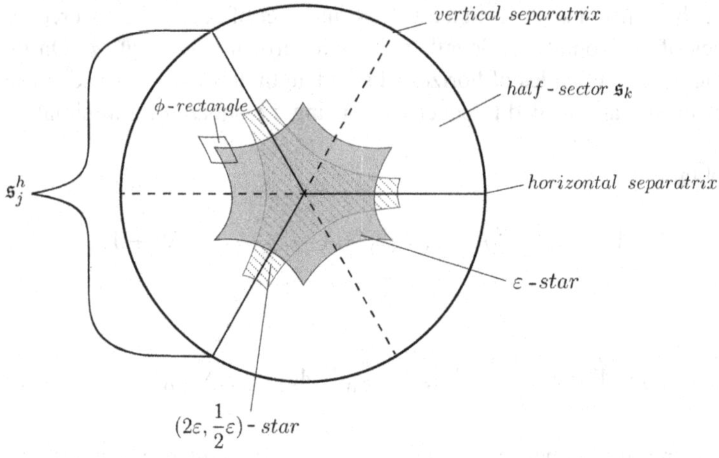

Fig. 2.3 An ε-star and a $(2\varepsilon, \frac{\varepsilon}{2})$-star at a ϕ-singular point

vertical bisectrix. There are $2(a + 2)$ such "half-sectors". They have the form $\mathfrak{s}_j = \{r \exp(xi) : x \in (\frac{2\pi ji}{2(a+2)}, \frac{2\pi(j+1)i}{2(a+2)}),\ 0 < r < c\}, j = 0, \ldots, 2(a + 2) - 1$. Denote by $\mathfrak{s}_j^h \overset{def}{=} \mathrm{Int}(\overline{\mathfrak{s}_{2j} \cup \mathfrak{s}_{2j+1}})$ the sectors between two horizontal bisectrices.

Each half-sector $\mathfrak{s}_j$ is taken by a branch g_j of the mapping $z \to z^{\frac{a+2}{2}}$ conformally onto a quarter-disc with center 0 in one of the quarters $\{\pm\mathrm{Im}z > 0, \pm\mathrm{Re}z > 0\}$ of the complex plane. For $\sqrt{2}\varepsilon < c^{\frac{a+2}{2}}$ we intersect the quarter disc with an open square in the complex plane with center 0 and side length 2ε. We obtain an open square $Q_j(\varepsilon)$ of side length ε. The union of the preimages $g_j^{-1}(Q_j(\varepsilon))$, $j = 1, \ldots, 2(a+2) - 1$, with the $2(a+2)$ open segments of ϕ-length ε emerging from the singular point and contained in a bisectrix, is a punctured neighbourhood of the singular point. Its union with the singular point is the required ε-star at the singular point (see Fig. 2.3).

The ε-star is bounded by $a + 2$ horizontal segments of ϕ-length 2ε and $a + 2$ vertical segments of ϕ-length 2ε. The distance in the ϕ-metric from the center of the ε-star to the horizontal and vertical sides in the boundary of the star is equal to ε. Hence, the ϕ-distance from the center of an ε-star to its boundary equals ε. The ε-star is contained in a $\sqrt{2}\varepsilon$-neighbourhood (in the ϕ-metric) of its center.

The definition of a star at a distinguished point implies immediately that for any positive $\varepsilon < \frac{1}{\sqrt{2}} c^{\frac{a+2}{2}}$ any ε-star punctured at its singular point of order a can be covered by $4(a + 2)$ ε-squares contained in the ε-star. Indeed, the part of the star contained in the half-sectors $\mathfrak{s}_j$ is covered by $2(a+2)$ ε-squares, and each bisectrix is covered by an open ε-square. The latter follows from the fact that e.g. each horizontal bisectrix is contained in the sector between two neighbouring vertical bisectrices, and a branch of the mapping $z \to z^{\frac{a+2}{2}}$ takes this sector conformally onto a half-disc with center 0 in the complex plane.

An $(\varepsilon_1, \varepsilon_2)$-star at a singular point is defined similarly. Let $\sqrt{\varepsilon_1^2 + \varepsilon_2^2} < c^{\frac{a+2}{2}}$. Consider again the half-sectors $\mathfrak{s}_j$ at the given singular point and the respective branches g_j of the mapping $z \to z^{\frac{a+2}{2}}$ on the half-sectors. Intersect each image $g_j(\mathfrak{s}_j)$ (which is a quarter disc at 0) with an open rectangle with center 0, horizontal side length $2\varepsilon_1$, and vertical side length $2\varepsilon_2$. The intersection is a rectangle $Q_j(\varepsilon_1, \varepsilon_2)$ of horizontal side length ε_1, and vertical side length ε_2. We take the preimage $g_j^{-1}(Q_j(\varepsilon_1, \varepsilon_2))$. Consider the closure of the union of the obtained preimages. Its interior is the ϕ-rectangle of horizontal side length ε_1 and vertical side length ε_2 at the given singular point. (See Fig. 2.3.)

Lemma 2.2 *Let ε_1 and ε_2 be positive numbers such that $\sqrt{2}(\varepsilon_1+\varepsilon_2) < c^{\frac{a+2}{2}}$ for the number c introduced above. If a ϕ-rectangle with both side length's not exceeding ε_2 intersects an ε_1-star $\mathfrak{S}_{\varepsilon_1}$, then the $(\varepsilon_1 + \varepsilon_2)$-star $\mathfrak{S}_{\varepsilon_1+\varepsilon_2}$ at the singular point of $\mathfrak{S}_{\varepsilon_1}$ exists, and the rectangle is contained in the intersection of $\mathfrak{S}_{\varepsilon_1+\varepsilon_2}$ with a sector between two consecutive horizontal bisectrices (or two consecutive vertical bisectrices).*

Proof Since the distance (in the ϕ-metric) of each point of the ε_1-star from the center does not exceed $\sqrt{2}\varepsilon_1$ and the diameter of the rectangle does not exceed $\sqrt{2}\varepsilon_2$, the rectangle is contained in the $\sqrt{2}(\varepsilon_1+\varepsilon_2)$-neighbourhood (in distinguished coordinates) of the distinguished point. Suppose the rectangle intersects a vertical bisectrix at a point x. Let $\mathfrak{s}_j^h$ be the sector between two neighbouring horizontal bisectrices that contains the vertical bisectrix. The rectangle can be obtained as follows. Take the maximal open vertical segment s through x that is contained in it. It has ϕ-length at most ε_2. For each point in s we take the maximal open horizontal segment through this point that is contained in the rectangle. Each has ϕ-length at most ε_2. Notice that the open rectangle does not contain singular points. Take the union of all these segments. The points on these segments are reachable starting from the singular point along a piece of a vertical trajectory of length smaller than $\varepsilon_1 + \varepsilon_2$ followed by a piece of a horizontal trajectory of length at most ε_2. Hence, the rectangle is contained in the $(\varepsilon_1+\varepsilon_2)$-star. If the rectangle intersects a horizontal bisectrix or does not intersect any bisectrix, the proof is the same. □

Lemma 2.3 *There exists a positive number δ_0 such that there exist disjoint δ_0-stars at the singular points. Moreover, there exists a positive number $\varepsilon_0 < \delta_0$ such that for each $\varepsilon < \varepsilon_0$ and for each point $x \in X$ which is not in an ε-star at a singular point there exists an ε-square with center x.*

Recall that a square in X is required to be contained in the ϕ-regular part of X.

Proof It is clear, that for some $\delta_0 > 0$ there exist disjoint δ_0-stars at the singular points. For each point x in the complement X_{δ_0} of all $\frac{1}{2}\delta_0$-stars there is an $\varepsilon(x)$-square with center at this point for some positive number $\varepsilon(x)$ depending on x. Since X_{δ_0} is compact it can be covered by a finite number of $\frac{1}{2}\varepsilon(x_j)$-squares, $j = 1, \ldots, N$. Let ε_0 be the minimum of the numbers $\frac{1}{2}\varepsilon(x_j)$. Then for each $x \in X_{\delta_0}$ there is an ε_0-square with center at this point.

Let x be in a δ_0-star at a singular point but not in an ε-star. Suppose the distance of the point x to a vertical bisectrix is equal to its distance to the union of all bisectrices. Let $\mathfrak{s}_j^h$ be the sector of the δ_0-star, that is bounded by two consecutive horizontal bisectrices and contains the latter vertical bisectrix. Let g_j^h be a branch of the mapping $z \to z^{\frac{a+2}{2}}$ that maps $\mathfrak{s}_j^h$ to the upper or lower half-plane. It defines flat coordinates on $\mathfrak{s}_j^h$. The point $g_j^h(x)$ is contained in the intersection of the upper or lower half-plane with a square of side length $2\delta_0$ and center 0, but not in the square of side length 2ε and center 0. Moreover, the distance of $g_j^h(x)$ to the imaginary axis does not exceed its distance to the real axis. Hence, the intersection of the half-plane with the square of side length $2\delta_0$ and center 0 contains a square of side length ε with center $g_j^h(x)$. The existence of an ε-square with center x is proved. □

We prove now the following proposition which is one of the estimates of Theorem 2.1.

Proposition 2.4 *Let X be a (closed connected) Riemann surface of genus g with a set E_m of $m \geq 0$ distinguished points, $3g - 3 + m > 0$. Let $\varphi_0 \in \mathrm{Hom}(X; \emptyset, E_m)$ be a non-periodic absolutely extremal self-homeomorphism of X with set of distinguished points E_m. Then*

$$h(\varphi_0) \leq \frac{1}{2} \log K(\varphi_0)$$

Proof Let ε_0 be as in Lemma 2.3. Take a positive number $\varepsilon < \min\{\frac{\varepsilon_0}{\lambda}, \frac{c^{\frac{a+2}{2}}}{4\sqrt{2}\lambda}\}$. Let $\mathcal{A}_\varepsilon$ be the set which consists of all ε-squares and all ε-stars at distinguished points. Then $\mathcal{A}_\varepsilon$ is an open cover of X.

The cover $\varphi_0^{-1}(\mathcal{A}_\varepsilon)$ consists of all $(\lambda^{-1}\varepsilon, \lambda\varepsilon)$-stars at distinguished points and all $(\lambda^{-1}\varepsilon, \lambda\varepsilon)$-rectangles. Indeed, the image under φ_0^{-1} of an ε-star at a singular point of ϕ is a $(\lambda^{-1}\varepsilon, \lambda\varepsilon)$-star at some singular point of ϕ. Vice versa, take a $(\lambda^{-1}\varepsilon, \lambda\varepsilon)$-star $\mathfrak{S}$. Its image $\varphi_0(\mathfrak{S})$ under φ_0 is an ε-star and $\mathfrak{S}$ is the image of the ε-star $\varphi_0(\mathfrak{S})$ under φ_0^{-1}. The argument for rectangles is the same.

Each $(\lambda^{-1}\varepsilon, \varepsilon)$-rectangle is the intersection of an ε-square with a $(\lambda^{-1}\varepsilon, \lambda\varepsilon)$-rectangle. Indeed, if the midpoint of the $(\lambda^{-1}\varepsilon, \varepsilon)$-rectangle is outside the $\lambda\varepsilon$-stars, then by Lemma 2.3 there exists a $\lambda\varepsilon$-square with center at this point. This square contains an ε-square and a $(\lambda^{-1}\varepsilon, \lambda\varepsilon)$-rectangle centered at this point. The intersection of the latter two sets is the required $(\lambda^{-1}\varepsilon, \varepsilon)$-rectangle.

Suppose the center x of a $(\lambda^{-1}\varepsilon, \varepsilon)$-rectangle R is in a $\lambda\varepsilon$-star $\mathfrak{S}_{\lambda\varepsilon}$. Lemma 2.2 with $\varepsilon_1 = \lambda\varepsilon$, $\varepsilon_2 = \varepsilon$, and $\sqrt{2}(\varepsilon_1 + \varepsilon_2) = \sqrt{2}(\lambda + 1)\varepsilon < \frac{1}{2}c^{\frac{a+2}{2}}$ implies that the rectangle R is contained in a sector of the $(\lambda+1)\varepsilon$-star $\mathfrak{S}_{(\lambda+1)\varepsilon}$ between consecutive horizontal bisectrices, or between consecutive vertical bisectrices. We assume that R is contained in a sector $\mathfrak{s}_j^h \cap \mathfrak{S}_{(\lambda+1)\varepsilon}$ of the $(\lambda+1)\varepsilon$-star between two consecutive horizontal bisectrices. The remaining case is treated similarly and is even slightly simpler.

By our choice of ε the $2(\lambda+1)\varepsilon$-star $\mathfrak{S}_{2(\lambda+1)\varepsilon}$ at the given singular point exists. The image $g_j^h(\mathfrak{s}_j^h \cap \mathfrak{S}_{2(\lambda+1)\varepsilon})$ is the intersection of the upper or lower half-plane with a Euclidean square of side length $4(\lambda+1)\varepsilon$. The image $g_j^h(R)$ is a Euclidean rectangle of horizontal side length $\lambda^{-1}\varepsilon$ and vertical side length ε contained in the Euclidean rectangle $g_j^h(\mathfrak{s}_j^h \cap \mathfrak{S}_{2(\lambda+1)\varepsilon})$ of horizontal side length $4(\lambda+1)\varepsilon$ and vertical side length $2(\lambda+1)\varepsilon$. Hence, $g_j^h(R)$ can be written as intersection of a Euclidean rectangle of side lengths $\lambda^{-1}\varepsilon$ and $\lambda\varepsilon$, and a Euclidean ε-square, both contained in the Euclidean rectangle $g_j^h(\mathfrak{s}_j^h \cap \mathfrak{S}_{2(\lambda+1)\varepsilon})$. Hence, R can be written as the desired intersection.

We showed that each $(\lambda^{-1}\varepsilon, \varepsilon)$-rectangle can be written as $A_0 \cap \varphi_0^{-1}(A_1) \in \mathcal{A}_\varepsilon \vee \varphi_0^{-1}(\mathcal{A}_\varepsilon)$ for ε-squares A_0 and A_1.

Any $(\lambda^{-n}\varepsilon, \varepsilon)$-rectangle can be written as $A_0 \cap \varphi_0^{-1}(A_1) \cap \ldots \cap \varphi_0^{-n}(A_n) \in \mathcal{A}_\varepsilon \vee \varphi_0^{-1}(\mathcal{A}_\varepsilon) \vee \ldots \vee \varphi_0^{-n}(\mathcal{A}_\varepsilon)$ for ε-squares A_j, $j = 0, 1, \ldots, n$. The proof goes by induction. Assume the claim is true for a natural number n. We prove that it is true for $n+1$. Suppose first the midpoint of the $(\lambda^{-n-1}\varepsilon, \varepsilon)$-rectangle A is outside the $\lambda\varepsilon$-stars. Then by Lemma 2.3 the $\lambda\varepsilon$-square around this point exists. Hence, there is an ε-square A' and a $(\lambda^{-n-1}\varepsilon, \lambda\varepsilon)$-rectangle A'' with the same center so that the intersection $A' \cap A''$ is equal to A. The image $\varphi_0(A'')$ is a $(\lambda^{-n}\varepsilon, \varepsilon)$-rectangle which is an element of $\mathcal{A}_\varepsilon \vee \varphi_0^{-1}(\mathcal{A}_\varepsilon) \vee \ldots \vee \varphi_0^{-n}(\mathcal{A}_\varepsilon)$ by the claim for the number n. Hence, its preimage A'' under φ_0 is in $\varphi_0^{-1}(\mathcal{A}_\varepsilon) \vee \varphi_0^{-2}(\mathcal{A}_\varepsilon) \vee \ldots \vee \varphi_0^{-1-n}(\mathcal{A}_\varepsilon)$. Since $A = A' \cap A''$ the claim is proved in this case.

In the remaining case the midpoint of the $(\lambda^{-n-1}\varepsilon, \varepsilon)$-rectangle A is inside a $\lambda\varepsilon$-star. The statement was proved for $n = 1$. For $n > 1$ we may proceed by induction similarly as above.

Each $(\lambda^{-n}\varepsilon, \varepsilon)$-star can be written as $A_0 \cap \varphi_0^{-1}(A_1) \cap \ldots \cap \varphi_0^{-n}(A_n) \in \mathcal{A}_\varepsilon \vee \varphi_0^{-1}(\mathcal{A}_\varepsilon) \vee \ldots \vee \varphi_0^{-n}(\mathcal{A}_\varepsilon)$ for ε-stars A_j, $j = 0, \ldots, n$. This is true for $n = 1$ since each $(\lambda^{-1}\varepsilon, \varepsilon)$-star is the intersection of a $(\lambda^{-1}\varepsilon, \lambda\varepsilon)$-star with an ε-star. We suppose that the claim is true for n, and prove it for $n+1$. Each $(\lambda^{-n-1}\varepsilon, \varepsilon)$-star is the intersection of an ε-star with a $(\lambda^{-n-1}\varepsilon, \lambda\varepsilon)$-star. The latter is a preimage under φ_0 of a $(\lambda^{-n}\varepsilon, \varepsilon)$-star, hence, by the induction hypothesis it can be written as $\varphi_0^{-1}(A_1) \cap \ldots \cap \varphi_0^{-n-1}(A_{n+1}) \in \varphi_0^{-1}(\mathcal{A}_\varepsilon) \vee \ldots \vee \varphi_0^{-n-1}(\mathcal{A}_\varepsilon)$. This proves the claim.

Cover X by a finite subcover of $\mathcal{A}_\varepsilon$. We may assume that this subcover consists of all ε-stars and a finite number of ε-squares that cover the complement of the singular points in X. Let C_1 be the number of squares in the subcover, and C_2 the number of stars. Each square can be covered by no more than $\lambda^n + 1$ sets that are $(\lambda^{-n}\varepsilon, \varepsilon)$-rectangles. These rectangles are in $\mathcal{A}_\varepsilon \vee \varphi_0^{-1}(\mathcal{A}_\varepsilon) \vee \ldots \vee \varphi_0^{-n}(\mathcal{A}_\varepsilon)$. Each singular point can be covered by a $(\lambda^{-n}\varepsilon, \varepsilon)$-star which is also contained in $\mathcal{A}_\varepsilon \vee \varphi_0^{-1}(\mathcal{A}_\varepsilon) \vee \ldots \vee \varphi_0^{-n}(\mathcal{A}_\varepsilon)$. We obtain a subcover of cardinality at most $C_1(\lambda^n + 1) + C_2 \le C(\lambda^n + 1)$. Hence, for any $\varepsilon < \varepsilon_0$

$$h(\varphi_0, \mathcal{A}_\varepsilon) \le \limsup \frac{1}{n} \log(C(\lambda^n + 1)) = \log \lambda \,. \tag{2.24}$$

Take a sequence of ε's decreasing to 0. We get (2.24) for a refining sequence of coverings. Hence,

$$h(\varphi_0) \leq \log \lambda = \frac{1}{2} \log K(\varphi_0) .$$

The proposition is proved. □

The following proposition states the opposite inequality.

Proposition 2.5 *Let X be a (closed connected) Riemann surface of genus g with a set E_m of $m \geq 0$ distinguished points, $3g - 3 + m > 0$. Let $\varphi_0 \in \mathrm{Hom}(X; \emptyset, E_m)$ be a non-periodic absolutely extremal self-homeomorphism of X with set of distinguished points E_m. Then*

$$h(\varphi_0) \geq \frac{1}{2} \log K(\varphi_0)$$

Proof Let $\varepsilon < \min\{\frac{\varepsilon_0}{3\lambda+1}, \frac{c^{\frac{a+2}{2}}}{4\sqrt{2}\lambda}\}$. We consider the cover $\mathcal{B}_\varepsilon$ which consist of all ε-squares and all $\frac{\varepsilon}{2}$-stars at singular points. We prove first that each non-empty set of the cover $\mathcal{B}_\varepsilon \vee \varphi_0^{-1}(\mathcal{B}_\varepsilon)$ is either a $(\lambda^{-1}\frac{\varepsilon}{2}, \frac{\varepsilon}{2})$-star at a singular point of ϕ or an $(\varepsilon_1, \varepsilon_2)$-rectangle with $\varepsilon_1 \leq \lambda^{-1}\varepsilon$ and $\varepsilon_2 \leq \varepsilon$.

We have to consider non-empty intersections $A_1 \cap \varphi_0^{-1}(A_2)$ for two sets in $\mathcal{B}_\varepsilon$. Suppose both, A_1 and A_2 are ε-squares, hence, $\varphi_0^{-1}(A_2)$ is an $(\lambda^{-1}\varepsilon, \lambda\varepsilon)$-rectangle. If the center x_0 of A_1 is not in a $2\lambda\varepsilon$-star, then by Lemma 2.3 there exists a $2\lambda\varepsilon$-square with center at the same point x_0. $\varphi_0^{-1}(A_2)$ is a $(\lambda^{-1}\varepsilon, \lambda\varepsilon)$-rectangle and intersects A_1, hence $A_1 \cap \varphi_0^{-1}(A_2)$ is contained in the $2\lambda\varepsilon$-square centered x_0. It is now clear that $A_1 \cap \varphi_0^{-1}(A_2)$ is an $(\varepsilon_1, \varepsilon_2)$-rectangle with $\varepsilon_1 \leq \lambda^{-1}\varepsilon$ and $\varepsilon_2 \leq \varepsilon$.

Suppose A_1 and A_2 are ε-rectangles and the centre of A_1 is in a $2\lambda\varepsilon$-star. Then by Lemma 2.2 A_1 is contained in the $(2\lambda + 1)\varepsilon$-star and $\varphi_0^{-1}(A_2)$ intersects the $(2\lambda + 1)\varepsilon$-star. Hence, by the same Lemma 2.2 $\varphi_0^{-1}(A_2)$ is contained in the $(3\lambda + 1)\varepsilon$-star. Each of the A_1 and $\varphi_0^{-1}(A_2)$ is contained in a sector between two consecutive horizontal or between two consecutive vertical bisectrices. Moreover, the two sectors intersect at least along a half-sector. The union of the two sectors, equipped with flat coordinates, is an open subset of a half-plane or a three-quarter plane with Euclidean coordinates. It is now clear that the intersection of A_1 and $\varphi_0^{-1}(A_2)$ is an $(\varepsilon_1, \varepsilon_2)$-rectangle with $\varepsilon_1 \leq \lambda^{-1}\varepsilon$ and $\varepsilon_2 \leq \varepsilon$.

If A_1 is an $\frac{\varepsilon}{2}$-star, further A_2 is an ε-square, and $A_1 \cap \varphi_0^{-1}(A_2)$ is non empty, then by Lemma 2.2 $A_1 \cup \varphi_0^{-1}(A_2)$ is contained in the $(\frac{\varepsilon}{2} + \lambda\varepsilon)$-star. Moreover, $\varphi_0^{-1}(A_2)$ is contained in a sector of this star between two consecutive horizontal or two consecutive vertical bisectrices. It is now clear that the intersection of A_1 and $\varphi_0^{-1}(A_2)$ is an $(\varepsilon_1, \varepsilon_2)$-rectangle with $\varepsilon_1 \leq \lambda^{-1}\varepsilon$ and $\varepsilon_2 \leq \varepsilon$.

Let A_1 be an ε-square and let A_2 be an $\frac{\varepsilon}{2}$-star. Then $\varphi_0^{-1}(A_2)$ is a $(\lambda^{-1}\frac{\varepsilon}{2}, \lambda\frac{\varepsilon}{2})$-star. If the intersection $A_1 \cap \varphi_0^{-1}(A_2)$ is non-empty, then by Lemma 2.2 A_1 is contained in a sector of a $(\frac{\lambda}{2}+1)\varepsilon$-star between two consecutive separatrices of the same kind (both horizontal or both vertical) with the same center. In flat coordinates on this set we obtain the intersection of an ε-square with an $(\varepsilon_1, \varepsilon_2)$-rectangle with $\varepsilon_1 \le \lambda^{-1}\frac{\varepsilon}{2}$ and $\varepsilon_2 \le \lambda\frac{\varepsilon}{2}$. The claim is obtained also in this case.

If both, A_1 and A_2 are stars they have the same center if the intersection is not empty, and the claim is clear.

We prove now by induction that each non-empty set of the form

$$A_0 \cap \varphi_0^{-1}(A_1) \cap \ldots \cap \varphi_0^{-n}(A_n) \tag{2.25}$$

(with A_j, $j = 0, \ldots n$, being sets in $\mathcal{B}_\varepsilon$), is either an $(\varepsilon_1, \varepsilon_2)$-rectangle with $\varepsilon_1 \le \lambda^{-n}\varepsilon$ and $\varepsilon_2 \le \varepsilon$ or a $(\lambda^{-n}\frac{\varepsilon}{2}, \frac{\varepsilon}{2})$-star at a singular point of ϕ. Suppose this is true for a natural number n. Prove it for $n+1$. We have to consider intersections $A_0 \cap \varphi_0^{-1}(B)$ with $A_0 \in \mathcal{B}_\varepsilon$ and $B \in \mathcal{B}_\varepsilon \vee \varphi_0^{-1}(\mathcal{B}_\varepsilon) \vee \ldots \vee \varphi_0^{-n}(\mathcal{B}_\varepsilon)$. Let A_0 be an ε-square. We have to intersect it either with an $(\varepsilon_1, \varepsilon_2)$-rectangle with $\varepsilon_1 \le \lambda^{-n-1}\varepsilon$ and $\varepsilon_2 \le \lambda\varepsilon$, or with a $(\lambda^{-n-1}\frac{\varepsilon}{2}, \lambda\frac{\varepsilon}{2})$-star at singular a point of ϕ. In the same way as before we obtain that the intersection is a $(\varepsilon_1, \varepsilon_2)$-rectangle with $\varepsilon_1 \le \lambda^{-n-1}\varepsilon$ and $\varepsilon_2 \le \varepsilon$.

Let A_0 be an $\frac{\varepsilon}{2}$-star at a singular point of ϕ. We have to intersect it either with a $(\lambda^{-n-1}\frac{\varepsilon}{2}, \lambda\frac{\varepsilon}{2})$-star at the same singular point, or with an $(\varepsilon_1, \varepsilon_2)$-rectangle with $\varepsilon_1 \le \lambda^{-n-1}\varepsilon$ and $\varepsilon_2 \le \lambda\varepsilon$. In the first case we obtain a $(\lambda^{-n-1}\frac{\varepsilon}{2}, \frac{\varepsilon}{2})$-star. In the second case the rectangle is contained in a sector between two consecutive bisectrices of the same kind of a larger star and we obtain in flat coordinates the intersection of the $(\varepsilon_1, \varepsilon_2)$-rectangle with an $(\varepsilon_1', \varepsilon_2')$-rectangle, where ε_1' and ε_2' do not exceed ε. The smaller horizontal side length is $\lambda^{-n-1}\varepsilon$, the smaller vertical side length is ε. We proved the statement.

The proof of the proposition is completed as follows. Consider an ε-square Q with ϕ-distance bigger than $\sqrt{2}\varepsilon$ from any singular point. Then for any non-negative integer n no star in the cover $\mathcal{B}_\varepsilon \vee \varphi_0^{-1}(\mathcal{B}_\varepsilon) \vee \ldots \vee \varphi_0^{-n}(\mathcal{B}_\varepsilon)$ intersects Q. Fix a natural number n. We want to estimate $\mathcal{N}(\mathcal{B}_\varepsilon \vee \varphi_0^{-1}(\mathcal{B}_\varepsilon) \vee \ldots \vee \varphi_0^{-n}(\mathcal{B}_\varepsilon))$ from below. The sets in the cover of X that intersect Q are $(\varepsilon_1, \varepsilon_2)$-rectangles with $\varepsilon_1 \le \lambda^{-n}\varepsilon$ and $\varepsilon_2 \le \varepsilon$. To cover Q at least λ^n sets are needed. Hence,

$$\begin{aligned} h(\varphi_0, \mathcal{B}_\epsilon) &\ge \limsup_{n\to\infty} \frac{1}{n} \log \mathcal{N}(\mathcal{B}_\varepsilon \vee \varphi_0^{-1}(\mathcal{B}_\varepsilon) \vee \ldots \vee \varphi_0^{-n}(\mathcal{B}_\varepsilon)) \\ &\ge \limsup_{n\to\infty} \frac{1}{n} \log \lambda^n = \lambda . \end{aligned} \tag{2.26}$$

Take a sequence of ε's decreasing to 0 we obtain a refining sequence of coverings. We proved that $h(\varphi_0) \ge \lambda$. □

The first equality in Theorem 2.1 is proved. The second one is Corollary 1.1. Theorem 2.1 is proved. □

2.4 Pseudo-Anosov Self-homeomorphisms of Closed Surfaces Are Entropy Minimizing

In this section we will prove Theorem 2.2 along the lines of [25]. Let X be a closed surface of genus at least two and let x_0 be a chosen base point of X. Let f be an arbitrary self-homeomorphism of X. If $f(x_0) = x_0$ the mapping f induces a homomorphism $f_* : \pi_1(X, x_0)\circlearrowleft$. Indeed, for each loop γ with base point x_0 the image $f \circ \gamma$ is a loop with base point x_0, whose class in $\pi_1(X, x_0)$ depends only on the class of γ in $\pi_1(X, x_0)$.

In the general situation the plan is the following. We will show that the mapping f induces a conjugacy class of homomorphisms $\widehat{f_\#}$ from the group $\mathrm{Deck}(\tilde{X}, X)$ of covering transformations (identified with the fundamental group of X) into itself. We will define a quantity $\Gamma_{\widehat{f_\#}}$ that depends only on the conjugacy class $\widehat{f_\#}$ and provides a lower bound for the entropy of f. Moreover, the quantity is more geometric than the entropy. In the case when f is a non-periodic absolutely extremal homeomorphism the quantity can be estimated from below by $\frac{1}{2}\log K(f)$.

Before coming to the fundamental group we consider any group G. Let $\mathcal{G} = \{g_1, \ldots, g_r\}$ be any subset of G that generates G. For an element $g \in G$ we denote by $\mathcal{L}_{\mathcal{G}}(g)$ the minimal length of a word in the g_j's and g_j^{-1}'s that represents g. (The length of a word in the g_j is the sum of the absolute values of the powers of exponents of the g_j that appear in the word.) For any other subset $\mathcal{G}' = \{g'_1, \ldots, g'_{r'}\}$ of G that generates G the inequality $\mathcal{L}_{\mathcal{G}'}(g) \le (\max \mathcal{L}_{\mathcal{G}'}(g_j))\mathcal{L}_{\mathcal{G}}(g)$ holds.

For a group homomorphism $A : G \to G$ we put

$$\Gamma_A = \sup_{g \in G} \limsup_{n\to\infty} \frac{1}{n} \log \mathcal{L}_{\mathcal{G}}(A^n g) \tag{2.27}$$

The following lemma is a straightforward consequence of the definition.

Lemma 2.4

$$\Gamma_A = \max_{g_j \in \mathcal{G}} \limsup_{n\to\infty} \frac{1}{n} \log \mathcal{L}_{\mathcal{G}}(A^n g_j)\,.$$

Hence, Γ_A is finite.

For an element $g \in G$ we consider the group homomorphism $gAg^{-1} : G \to G$ defined by $gAg^{-1}(x) = gA(x)g^{-1}$.

For convenience of the reader we formulate the following technical lemma from [25] (see Lemma 10.6 in [25]). The proof is a straightforward calculation and is left to the reader.

Lemma 2.5 *Let $(a_n)_{n=1}^{\infty}$ and $(b_n)_{n=1}^{\infty}$ be two sequences of positive numbers. Then*

(1) $\limsup \frac{1}{n}\log(a_n + b_n) = \max\big(\limsup \frac{1}{n}\log(a_n)), \limsup \frac{1}{n}\log(b_n)$;
(2) $\limsup \frac{1}{n}\log(a_n) \le \limsup \frac{1}{n}\log(a_1 + \ldots + a_n) \le \max\big(0, \limsup \frac{1}{n}\log(a_n)\big)$.

The following proposition is proved in [25] (see Proposition 10.5 of [25] and its proof).

Proposition 2.6 *The equality*

$$\Gamma_A = \Gamma_{gAg^{-1}}$$

holds.

Proof By the definition we have $(gAg^{-1})(x) = gA(x)g^{-1},\ g \in G$. Hence,

$$(gAg^{-1})^2(x) = gA(gA(x)g^{-1})g^{-1} = gA(g)A^2(x)A(g^{-1})g^{-1}.$$

By induction

$$(gAg^{-1})^n(x) = gA(g)\dots A^{n-1}(g)A^n(x)A^{-n+1}(g)\dots A(g^{-1})g^{-1}.$$

Hence, since $\mathcal{L}_{\mathcal{G}}(y) = \mathcal{L}_{\mathcal{G}}(y^{-1})$ for each $y \in G$ we obtain

$$\begin{aligned}\Gamma_{gAg^{-1}} &= \sup_{x\in G} \limsup \frac{1}{n} \log(\mathcal{L}_{\mathcal{G}}(g\dots A^{n-1}(g)A^n(x)A(g^{-n+1})\dots g^{-1})) \\ &\leq \sup_{x\in G} \limsup \frac{1}{n} \log\left(2(\mathcal{L}_{\mathcal{G}}(g) + \dots + \mathcal{L}_{\mathcal{G}}(A^{n-1}(g))) + \mathcal{L}_{\mathcal{G}}(A^n(x))\right).\end{aligned} \tag{2.28}$$

If $A^{n_0}(g)$ is the identity for some n_0, then it follows from Statement (1) of Lemma 2.5, that the right hand side of (2.28) does not exceed

$$\limsup \frac{1}{n} \log \mathcal{L}_{\mathcal{G}}(A^n(x)).$$

In the remaining case $\mathcal{L}_{\mathcal{G}}(A^n(g)) \geq 1$ for each n, and by Statements (1) and (2) of Lemma 2.5

$$\begin{aligned}&\limsup \frac{1}{n} \log\left(\mathcal{L}_{\mathcal{G}}(gAg^{-1})(x)\right) \\ &\quad\leq \max\left(\limsup \frac{1}{n} \log \mathcal{L}_{\mathcal{G}}(A^n(g)), \limsup \frac{1}{n} \log \mathcal{L}_{\mathcal{G}}(A^n(x))\right).\end{aligned}$$

Taking the maximum over g running along a finite set g_j of elements that generates G, we obtain $\Gamma_{gAg^{-1}} \leq \Gamma_A$. The opposite inequality follows by symmetry. □

Let X be a closed surface and $x_0 \in X$ a base point. Denote by $\mathsf{P} : \tilde{X} \to X$ the universal covering of X. Recall that each point of $\tilde{X}$ that projects to a point $x \in X$ can be identified with an element $e_{x_0,x} \in \pi_1(X; x_0, x)$ (see e.g. [27] and Sect. 1.2). Here $\pi_1(X; x_0, x)$ denotes the set of homotopy classes of arcs in X with initial point

x_0 and terminal point x. We write $\{e_{x_0,x}\}$ if we mean the point of $\tilde{X}$ associated to the element $e_{x_0,x} \in \pi_1(X; x_0, x)$. The projection assigns to each point $\{e_{x_0,x}\}$ the point x.

Let $f : X \to X$ be a continuous mapping. For each pair of points x_1, x_2 in X the mapping f induces a mapping $f_* : \pi_1(X; x_1, x) \to \pi_1\big(X; f(x_1), f(x)\big)$. Namely, $f_*(e_{x_1,x})$ is the class represented by $f \circ \gamma_{x_1,x}$ for a curve $\gamma_{x_1,x}$ that represents $\{e_{x_1,x}\}$.

Let $x_0 \in X$ be the base point of X. A lift of f to $\tilde{X}$ is a continuous mapping $\tilde{f}$ such that $f \circ \mathsf{P} = \mathsf{P} \circ \tilde{f}$. A lift of f to $\tilde{X}$ can be described as follows (see also [27]). Choose and fix a homotopy class of curves $e^0_{x_0, f(x_0)} \in \pi_1(X; x_0, f(x_0))$ with fixed endpoints x_0 and $f(x_0)$. Put

$$\tilde{f}(\{e_{x_0,x}\}) = \{e^0_{x_0, f(x_0)}\, f_*(e_{x_0,x})\},\ \{e_{x_0,x}\} \in \tilde{X}\,. \tag{2.29}$$

Here $e^0_{x_0, f(x_0)}\, f(e_{x_0,x})$ is the homotopy class of arcs represented by first traveling along an arc representing $e^0_{x_0, f(x_0)}$ and then along an arc representing the image $f_*(e_{x_0,x})$ of $e_{x_0,x}$ under the induced mapping $f_* : \pi_1(X; x_0, x) \to \pi_1(X; f(x_0), f(x))$. All other lifts are obtained by replacing $e^0_{x_0, f(x_0)}$ by another element of $\pi_1(X; x_0, f(x_0))$. Equivalently, for two lifts $\tilde{f}_1$ and $\tilde{f}_2$ of f there exists a covering transformation $\theta \in \pi_1(X, x_0)$ such that $\tilde{f}_1 = \theta \circ \tilde{f}_2$. Indeed, if $\mathsf{P} \circ \tilde{f}_1 = \mathsf{P} \circ \tilde{f}_2 = f \circ \mathsf{P}$, then for each $\tilde{x} \in \tilde{X}$ the points $\tilde{f}_1(\tilde{x})$ and $\tilde{f}_2(\tilde{x})$ differ by a covering transformation. The covering transformations depend continuously on $\tilde{x}$, hence, there is a single covering transformation θ such that $\tilde{f}_1 = \theta \circ \tilde{f}_2$.

The following lemma gives a lower bound for the entropy of a mapping f in terms of metric properties of the lift $\tilde{f}$ of f.

Lemma 2.6 *Let X be a closed surface with metric d, $\tilde{X}$ its universal covering with the lifted metric also denoted by d, $f : X \to X$ a self-homeomorphism of X and $\tilde{f}$ a lift of f to $\tilde{X}$. Then for any points $\tilde{x}, \tilde{y} \in \tilde{X}$*

$$\limsup_{n\to\infty} \frac{1}{n} d(\tilde{f}^n(\tilde{x}), \tilde{f}^n(\tilde{y})) \le h(f). \tag{2.30}$$

Proof Take any curve $\tilde{\gamma}$ in $\tilde{X}$ joining $\tilde{x}$ and $\tilde{y}$. Let γ be the projection of $\tilde{\gamma}$ to X. Since $\tilde{\gamma}$ is a compact subset of $\tilde{X}$, there is a finite constant l such that each point of γ is covered at most l times by points in $\tilde{\gamma}$. For small $\varepsilon > 0$ we denote by $\mathcal{A}_\varepsilon$ the cover of X by all discs of radius ε in the metric d. For any decreasing sequence of ε's we obtain a refining sequence of covers. Fix any small $\varepsilon > 0$ and any positive number δ. If n is large there is a cover $\mathcal{A}_{\varepsilon,n}$ of X by at most $\exp(n(h(f, \mathcal{A}_\varepsilon) + \delta))$ elements of the cover $\mathcal{A}_\varepsilon \vee f^{-1}(\mathcal{A}_\varepsilon) \vee \ldots \vee f^{-n}(\mathcal{A}_\varepsilon)$. Note that $\mathcal{A}_{\varepsilon,n}$ is a subcover of $\mathcal{A}_\varepsilon \vee f^{-1}(\mathcal{A}_\varepsilon) \vee \ldots \vee f^{-n}(\mathcal{A}_\varepsilon)$. For each element of the subcover $\mathcal{A}_{\varepsilon,n}$ there is an ε-disc $A_{\varepsilon,n}$ such that the element is contained in $f^{-n}(A_{\varepsilon,n})$. Since f^{-n} is a homeomorphism we see that $f^n(\gamma)$ can be covered by $\exp(n(h(f, \mathcal{A}_\varepsilon)+\delta))$ ε-discs.

The mapping $(\tilde{f})^n$ is a lift of f^n to $\tilde{X}$. Indeed, $\tilde{f}$ lifts f, i.e. $\mathsf{P} \circ \tilde{f}(\tilde{x}) = f \circ \mathsf{P}(\tilde{x})$ for $\tilde{x} \in \tilde{X}$. By induction $\mathsf{P} \circ \tilde{f}^{n+1}(\tilde{x}) = \mathsf{P} \circ \tilde{f}^n(\tilde{f}(\tilde{x})) = f^n \circ \mathsf{P}(\tilde{f}(\tilde{x})) = f^{n+1} \circ \mathsf{P}(\tilde{x})$

for $\tilde{x} \in \tilde{X}$. The curve $\tilde{f}^n(\tilde{\gamma})$ is a lift of $f^n(\gamma)$ to $\tilde{X}$. Hence, each point of the curve $\tilde{f}^n(\tilde{\gamma})$ is covered by a lift of one of the ε-discs. Since each point of $f^n(\gamma)$ is the projection of at most l points of $\tilde{f}^n(\tilde{\gamma})$ the latter curve can be covered by $l \exp(n(h(f, \mathcal{A}_\varepsilon) + \delta))$ ε-discs. This implies that the distance between its two endpoints $d(\tilde{f}^n(x), \tilde{f}^n(y))$, does not exceed $2\varepsilon l \exp(n(h(f, \mathcal{A}_\varepsilon) + \delta))$. Hence,

$$\limsup_{n\to\infty} \frac{1}{n} \log d(\tilde{f}^n(x), \tilde{f}^n(y)) \le h(f, \mathcal{A}_\varepsilon) + \delta. \tag{2.31}$$

Since $\delta > 0$ is arbitrary, Eq. (2.31) holds with $h(f, \mathcal{A}_\varepsilon) + \delta$ replaced by $h(f, \mathcal{A}_\varepsilon)$. For a sequence of ε's we obtain a refining sequence of coverings. The statement of the lemma follows. □

We will consider now a lift $\tilde{f}$ of f, and associate to it an action $\tilde{f}_\#$ on the fundamental group. Recall how to associate to each element α of the fundamental group $\pi_1(X, x_0)$ a covering transformation. We take a point $\tilde{x}_0 \in \tilde{X}$ that projects to x_0. Consider the lift $\tilde{\alpha}$ of a representative of α with initial point $\tilde{x}_0$. Associate to α the covering transformation $(\mathrm{Is}^{\tilde{x}_0})^{-1}(\alpha)$ that maps $\tilde{x}_0$ to the terminal point of $\tilde{\alpha}$. The point $\tilde{x}_0$ will be fixed in this section. We denote the covering transformation $(\mathrm{Is}^{\tilde{x}_0})^{-1}(\alpha)$ associated to $\alpha \in \pi_1(X, x_0)$ by α^{cov}. Recall that $(\alpha_1\alpha_2)^{cov} = \alpha_2^{cov}\alpha_1^{cov}$. The covering transformation α^{cov}, acts as follows. Write the elements of $\tilde{X}$ as $\tilde{x} \cong \{e_{x_0,x}\}$. Then

$$\alpha^{cov}(\{e_{x_0,x}\}) = \{\alpha\, e_{x_0,x}\}, \ \{e_{x_0,x}\} \in \tilde{X}\,. \tag{2.32}$$

Fix a lift $\tilde{f}$ of f. Since for each $\alpha \in \pi_1(X, x_0)$ the mapping $\tilde{f} \circ \alpha^{cov}$ is a lift of f, there is an element $(\tilde{f})_\#(\alpha) \in \pi_1(X, x_0)$ depending on α such that

$$\tilde{f} \circ \alpha^{cov} = \big((\tilde{f})_\#(\alpha)\big)^{cov} \circ \tilde{f}\,. \tag{2.33}$$

Since

$$\begin{aligned}\tilde{f} \circ (\alpha_1\alpha_2)^{cov}(\tilde{x}) &= \tilde{f} \circ (\alpha_2^{cov} \circ \alpha_1^{cov})(\tilde{x}) \\ &= \big((\tilde{f})_\#(\alpha_2)\big)^{cov} \circ \tilde{f}(\alpha_1^{cov}(\tilde{x})) \\ &= \Big(\big((\tilde{f})_\#(\alpha_2)\big)^{cov} \circ \big((\tilde{f})_\#(\alpha_1)\big)^{cov}\Big) \circ \tilde{f}(\tilde{x})\end{aligned}$$

the mapping $\alpha \to (\tilde{f})_\#(\alpha)$ is a group isomorphism of $\pi_1(X, x_0)$.

This isomorphism can be given explicitly as follows. Let $\tilde{f}$ be given by (2.29). For any element $\alpha \in \pi_1(X, x_0)$ we get the equality

$$\begin{aligned}\tilde{f} \circ \alpha^{cov}(\tilde{x}) = \tilde{f}(\{\alpha e_{x_0,x}\}) &= \{e^0_{x_0,f(x_0)} f_*(\alpha e_{x_0,x})\} \\ &= \{e^0_{x_0,f(x_0)} f_*(\alpha) f_*(e_{x_0,x})\}.\end{aligned} \tag{2.34}$$

Recall that $\tilde{f}(\tilde{x}) = \{e^0_{x_0,f(x_0)} f_*(e_{x_0,x})\}$. Then

$$
\begin{aligned}
(\tilde{f})_\#(\alpha) &= e^0_{x_0,f(x_0)}\, f_*(\alpha e_{x_0,x})\big(e^0_{x_0,f(x_0)}\, f_*(e_{x_0,x})\big)^{-1} \\
&= e^0_{x_0,f(x_0)}\, f_*(\alpha)(e^0_{x_0,f(x_0)})^{-1} .
\end{aligned}
\tag{2.35}
$$

Equality (2.33) implies by induction

$$
\tilde{f}^n \circ \alpha^{cov}(\tilde{x}) = \big((\tilde{f})^n_\#(\alpha)\big)^{cov}(\tilde{f}^n(\tilde{x})),\ \tilde{x} \in \tilde{X} . \tag{2.36}
$$

Indeed,

$$
\tilde{f}\big(\tilde{f}^{n-1} \circ \alpha^{cov}(\tilde{x})\big) = \tilde{f}\Big(\big((\tilde{f})^{n-1}_\#(\alpha)\big)^{cov}(\tilde{f}^{n-1}(\tilde{x}))\Big) = \Big((\tilde{f})^n_\#(\alpha)\Big)^{cov}(\tilde{f}^n(\tilde{x})) .
$$

Lemma 2.7 *If f_0 and f_1 are two self-homeomorphisms of a closed surface X that are isotopic to each other then $(\tilde{f}_0)_\# = (\tilde{f}_1)_\#$.*

Proof Let f_t be a homotopy of self-homeomorphisms of X joining f_0 and f_1. Denote by $f : [0,1]\times X \to [0,1]\times X$ the mapping for which $f(t,x) = f_t(x)$, $t \in [0,1]$, $x \in X$. Put $f^{-1}(t,x) = (f_t)^{-1}(x)$, $t \in [0,1]$, $x \in X$. Denote by $\tilde{f}$ a lift of f to $[0,1]\times \tilde{X}$. Fix a covering transformation $\alpha \in \pi_1(X, x_0)$. Let $\tilde{X}_0$ be a connected compact subset of $\tilde{X}$ whose interior covers X.

The mappings $\tilde{f}$ and $\tilde{f}^{-1}$, $\tilde{f}^{-1}(t,x) = (\tilde{f}_t)^{-1}(\tilde{x})$, $t \in [0,1], \tilde{x} \in X$, are uniformly continuous on $[0,1] \times (\tilde{X}_0 \cup \alpha^{cov}(\tilde{X}_0))$. Hence, for each $\varepsilon > 0$ there exists $\delta > 0$ such that the implication

$$
\begin{aligned}
&(t_1,\tilde{x}_1), (t_2,\tilde{x}_2) \in [0,1]\times(\tilde{X}_0 \cup \alpha^{cov}(\tilde{X}_0)), \\
&|t_1 - t_2| < \delta,\ d(\tilde{x}_1,\tilde{x}_2) < \delta \Rightarrow d(\tilde{f}_{t_1}(\tilde{x}_1), \tilde{f}_{t_2}(\tilde{x}_2)) < \varepsilon
\end{aligned}
\tag{2.37}
$$

holds, and the respective implication holds for $\tilde{f}^{-1}$.

Put $\xi(\tilde{x}) = \tilde{f}_{t_1}(\tilde{x})$. We want to show that

$$
d\Big(\big((\tilde{f}_{t_2})_\#(\alpha)\big)^{cov}(\xi(\tilde{x})), \big((\tilde{f}_{t_1})_\#(\alpha)\big)^{cov}(\xi(\tilde{x})\Big) < 2\varepsilon \tag{2.38}
$$

for $\tilde{x} \in \tilde{X}_0$ and $|t_1 - t_2| < \delta$. Equation (2.38) shows that the values of the covering transformations $\big((\tilde{f}_{t_1})_\#(\alpha)\big)^{cov}$ and $\big((\tilde{f}_{t_2})_\#(\alpha)\big)^{cov}$ are 2ε-close on the set $\tilde{f}_{t_1}(\tilde{X}_0)$. Since $\tilde{f}_{t_1}$ is a lift of a self-homeomorphism of X this set covers X. If ε is small, inequality (2.38) can only hold, if the covering transformations coincide. This implies that $\big((\tilde{f}_t)_\#(\alpha)\big)^{cov}$ is locally constant for $t \in [0,1]$, hence, it is constant and $\big((\tilde{f}_0)_\#(\alpha)\big)^{cov} = \big((\tilde{f}_1)_\#(\alpha)\big)^{cov}$. Since α was an arbitrary covering transformation the lemma will be proved if inequality (2.38) is proved.

We prove now inequality (2.38). By the implication (2.37) the inequality $d(\tilde{f}_{t_1} \circ \alpha^{cov}(\tilde{x}), \tilde{f}_{t_2} \circ \alpha^{cov}(\tilde{x})) < \varepsilon$ holds for $\tilde{x} \in \tilde{X}_0$ and $t_1, t_2 \in [0,1]$, $|t_1 - t_2| < \delta$.

Hence, by (2.33)

$$d\Big(\big((\tilde{f}_{t_1})_\#(\alpha)\big)^{cov}(\tilde{f}_{t_1}(\tilde{x})), \big((\tilde{f}_{t_2})_\#(\alpha)\big)^{cov}(\tilde{f}_{t_2}(\tilde{x}))\Big) < \varepsilon \tag{2.39}$$

for those $\tilde{x}$, t_1 and t_2. Since the covering transformation $\big((\tilde{f}_{t_2})_\#(\alpha)\big)^{cov}$ is an isometry, the implication (2.37) implies the inequality

$$d\Big(\big((\tilde{f}_{t_2})_\#(\alpha)\big)^{cov}(\tilde{f}_{t_2}(\tilde{x})), \big((\tilde{f}_{t_2})_\#(\alpha)\big)^{cov}(\tilde{f}_{t_1}(\tilde{x}))\Big) < \varepsilon\,. \tag{2.40}$$

for $\tilde{x} \in \tilde{X}_0$, $|t_1 - t_2| < \delta$ Inequalities (2.39) and (2.40) imply inequality (2.38). The lemma is proved. □

For two lifts $\tilde{f}_1$ and $\tilde{f}_2$ of f the associated isomorphisms $(\tilde{f}_1)_\#$ and $(\tilde{f}_2)_\#$ of $\pi_1(X, x_0)$ are conjugate. Indeed, for a covering transformation θ

$$\begin{aligned}\tilde{f}_1 \circ \alpha^{cov}(\tilde{x}) &= \theta \circ \tilde{f}_2 \circ \alpha^{cov}(\tilde{x}) = \theta \circ \big((\tilde{f}_2)_\#(\alpha)\big)^{cov}(\tilde{f}_2(\tilde{x}))\\ &= \theta \circ \big((\tilde{f}_2)_\#(\alpha)\big)^{cov}(\theta^{-1}\tilde{f}_1(\tilde{x}))\,.\end{aligned}$$

We obtain the equality $\Big((\tilde{f}_1)_\#(\alpha)\Big)^{cov} = \theta\big((\tilde{f}_2)_\#(\alpha)\big)^{cov}\theta^{-1}$. In other words, the mapping f defines a conjugacy class of isomorphisms of $\pi_1(X, x_0)$, denoted by $\widehat{f_\#}$. By Proposition 2.6 the quantities $\Gamma_{(\tilde{f}_j)_\#}$ defined by Eq. (2.27) for the isomorphisms $(\tilde{f}_j)_\#$ of $\pi_1(X, x_0)$ satisfy the equality $\Gamma_{(\tilde{f}_1)_\#} = \Gamma_{(\tilde{f}_2)_\#}$. Hence, there is a quantity $\Gamma_{\widehat{f_\#}}$ related to f, that is correctly defined by putting it equal to $\Gamma_{\tilde{f}_\#}$ for any lift $\tilde{f}$ of f. Moreover, by Lemma 2.7 $\Gamma_{\widehat{f_\#}}$ depends only on the isotopy class of f.

The following lemma relates the word length $\mathcal{L}_{\pi_1(X,x_0)}(g)$ of an element $g \in \pi_1(X, x_0)$ to the distance from a point in the universal covering $\tilde{X}$ to its image under the related covering transformation g^{cov}. The proof of this lemma is due to J.Milnor [65]. (See also [25], Lemma 10.7 there.) Again, d is a metric on X and its lift to the universal covering $\tilde{X}$ is denoted by the same letter d. Notice that the metric d on $\tilde{X}$ is invariant under covering transformations.

Lemma 2.8 *Fix a point $\tilde{x} \in \tilde{X}$ and a set $\mathcal{G}$ of elements of the fundamental group $\pi_1(X, x_0)$ that generates $\pi_1(X, x_0)$. There exist two positive constants C_1 and C_2 such that for each $g \in \pi_1(X, x_0)$ the inequality*

$$C_1\mathcal{L}_{\mathcal{G}}(g) \le d(\tilde{x}, g^{cov}(\tilde{x})) \le C_2\mathcal{L}_{\mathcal{G}}(g) \tag{2.41}$$

holds.

Proof Let δ be the diameter of X. Put $\tilde{N} = \{\tilde{y} \in \tilde{X} : d(\tilde{y}, \tilde{x}) \le \delta\}$. Then for the projection $\mathsf{P} : \tilde{X} \to X$ we have $\mathsf{P}(N) = X$. The sets $\{g^{cov}(\tilde{N}) : g \in \pi_1(X, x_0)\}$ cover $\tilde{X}$. Indeed, for each point $\tilde{y}$ in $\tilde{X}$ there is a point $\tilde{y}' \in \tilde{N}$ with $p(\tilde{y}') = p(\tilde{y}) = y$. There exists a covering transformation g such that $g(\tilde{y}') = (\tilde{y})$.

The family $g^{cov}(\tilde{N})$, $g \in \pi_1(X, x_0)$, is a locally finite cover of $\tilde{X}$ (by compact subsets). Indeed, let $\tilde{U}(\tilde{x}) \subset \tilde{N}$ be a neighbourhood of $\tilde{x}$ in $\tilde{X}$ such that $g^{cov}(\tilde{U}(\tilde{x})) \cap \tilde{U}(\tilde{x}) = \emptyset$ for each $g \in \pi_1(X, x)$. If $g^{cov}(\tilde{N})$ intersects $\tilde{N}$ then, since g is an isometry, $g^{cov}(\tilde{N})$ is contained in $\{\tilde{y} \in \tilde{X} : d(\tilde{y}, \tilde{x}) \le 3\delta\}$. If there were infinitely many different $g_j \in \pi_1(X, x_0)$ for which $g_j^{cov}(\tilde{N}) \cap \tilde{N} \neq \emptyset$, then since all $g_j^{cov}(\tilde{U}(\tilde{x}))$ are disjoint, have the same non-zero area in the metric d and are contained in $\{\tilde{x} \in \tilde{X} : d(\tilde{y}, \tilde{x}) \le 3\delta\}$, the area of the set $\{\tilde{y} \in \tilde{X} : d(\tilde{y}, \tilde{x}) \le 3\delta\}$ would be infinite. This is impossible.

We saw that the set

$$\mathcal{G}' \stackrel{def}{=} \{g \in \pi_1(X, x_0) : g^{cov}(\tilde{N}) \cap \tilde{N} \neq \emptyset\}$$

is finite. Since the sets $g^{cov}(\tilde{N})$, $g \in \pi_1(X, x_0)$, cover $\tilde{X}$, $\mathcal{G}'$ generates $\pi_1(X, x_0)$. Notice that $\mathcal{G}'$ contains the inverse of each of its elements.

Moreover, there is a positive number ν such that $d(\tilde{N}, g^{cov}(\tilde{N})) \ge \nu$ if $g^{cov}(\tilde{N}) \cap \tilde{N} = \emptyset$. Indeed, the union of all $(g')^{cov}(\tilde{N})$, $g' \in \pi_1(X, x_0)$, that intersect $\tilde{N}$, contains an open neighbourhood of the closure $\overline{\tilde{N}}$. If for some $g \in \pi_1(X, x_0)$ the set $g^{cov}(\tilde{N})$ does not intersect this open neighbourhood of $\overline{\tilde{N}}$, the distance of $g^{cov}(\tilde{N})$ to $\tilde{N}$ is bigger than a positive constant not depending on g. There are only finitely many g for which $g^{cov}(\tilde{N})$ intersects the open neighbourhood of $\tilde{N}$, and each $g^{cov}(\tilde{N})$ has positive distance to $\tilde{N}$, if it does not intersect $\tilde{N}$.

The second inequality in (2.41) is now easy to prove. If for an element $g \in \pi_1(X, x_0)$ we have $\mathcal{L}_{\mathcal{G}}(g) = n$, then $\mathcal{L}_{\mathcal{G}'}(g) = n'$ for $n' \le (\max \mathcal{L}_{\mathcal{G}'}(g_j))\, \mathcal{L}_{\mathcal{G}}(g) = C_2' n$ with $C_2' = \max \mathcal{L}_{\mathcal{G}'}(g_j)$. We can write $g = g_1 g_2 \dots g_{n'}$ with $g_j^{cov}(\tilde{N}) \cap \tilde{N} \neq \emptyset$. Then $d(\tilde{x}, g^{cov}(\tilde{x})) \le 2\delta n' \le 2\delta C_2' n = 2\delta C_2' \mathcal{L}_{\mathcal{G}}(g)$.

The first inequality in (2.41) is obtained as follows. Let $g \in \pi_1(X, x_0)$. Take the smallest positive integer number k for which $d(\tilde{x}, g^{cov}(\tilde{x})) < k\nu$. Consider a sequence of points $\tilde{y}_0 = \tilde{x}, \dots, \tilde{y}_{k-1}, \tilde{y}_k = g^{cov}(\tilde{x})$, such that $d(\tilde{y}_j, \tilde{y}_{j+1}) < \nu$ for $j = 0, 1, \dots, k-1$. Choose for $j = 1, \dots, k-1$ a point $\tilde{y}_j' \in \tilde{N}$ and an element $g_j \in \mathcal{G}'$ such that $\tilde{y}_j = g_j^{cov}(\tilde{y}_j')$ and put g_0 equal to the identity and $g_k = g$. We obtain $d(g_j^{cov}(\tilde{y}_j'), g_{j+1}^{cov}(\tilde{y}_{j+1}')) < \nu$. Hence, $g_j^{-1} g_{j+1} \in \mathcal{G}'$. Since $g = (g_0^{-1} g_1) \dots (g_{k-1}^{-1} g_k)$, we obtain $L_{\mathcal{G}'}(g) < k$.

Since k is minimal, we have with $\mu = \min\{d(\tilde{x}, g^{cov}(\tilde{x})) : g \neq \mathrm{Id}, g \in \pi_1(X, x_0)\}$

$$L_{\mathcal{G}'}(g) \le \frac{1}{\nu} d(\tilde{x}, g^{cov}(\tilde{x})) + 1 \le (\frac{1}{\nu} + \frac{1}{\mu}) d(\tilde{x}, g^{cov}(\tilde{x})).$$

The lemma follows by the inequality $L_{\mathcal{G}}(g) \le (\max L_{\mathcal{G}}(g_j')) L_{\mathcal{G}'}(g)$. □

We give now the proof of the following theorem from Exposè 10 of [25] which is interesting in itself. It gives a lower bound of the entropy by the quantity $\Gamma_{\widehat{f_\#}}$.

Theorem 2.6 *For a self-homeomorphism f of a closed surface X the inequality*

$$h(f) \geq \Gamma_{\tilde{f}_\#} \tag{2.42}$$

holds.

Proof Let $\tilde{f}$ be a lift of f to the universal covering $\tilde{X}$. By Milnor's Lemma with g replaced by $\tilde{f}^n_\#(g) \in \pi_1(X, x_0)$ for each $\tilde{x} \in \tilde{X}$

$$\begin{aligned}\Gamma_{\tilde{f}_\#} &= \sup_{g\in\pi_1(X,x_0)} \limsup \frac{1}{n} \log \mathcal{L}_{\mathcal{G}}(\tilde{f}^n_\#(g)) \\ &= \sup_{g\in\pi_1(X,x_0)} \limsup \frac{1}{n} \log d\Big(\tilde{x}, \big((\tilde{f}_\#)^n(g)\big)^{cov}(\tilde{x})\Big)\end{aligned} \tag{2.43}$$

for a metric d on X and its lift to $\tilde{X}$ denoted also by d. Take in Lemma 2.6 $\tilde{y} = g^{cov}(\tilde{x})$. Since by Eq. (2.36) $\tilde{f}^n(g^{cov}(\tilde{x})) = \big((\tilde{f}_\#)^n(g)\big)^{cov}(\tilde{f}^n(\tilde{x}))$, Lemma 2.6 gives

$$\limsup \frac{1}{n} d\Big(\tilde{f}^n(\tilde{x}), \big((\tilde{f}_\#)^n(g)\big)^{cov}(\tilde{f}^n(\tilde{x}))\Big) \leq h(f). \tag{2.44}$$

Note that

$$\begin{aligned} d\Big(\tilde{x}, \big((\tilde{f}_\#)^n(g)\big)^{cov}(\tilde{x})\Big) \leq\; & d\Big(\tilde{x}, \tilde{f}^n(\tilde{x})\Big) + d\Big(\tilde{f}^n(\tilde{x}), \big((\tilde{f}_\#)^n(g)\big)^{cov}(\tilde{f}^n(\tilde{x}))\Big) \\ & + d\Big(\big((\tilde{f}_\#)^n(g)\big)^{cov}(\tilde{f}^n(\tilde{x})), \big((\tilde{f}_\#)^n(g)\big)^{cov}(\tilde{x})\Big). \end{aligned} \tag{2.45}$$

The first term on the right is estimated by

$$d(\tilde{x}, \tilde{f}^n(\tilde{x})) \leq \sum_{\ell=0}^{n-1} d(\tilde{f}^\ell(\tilde{x}), \tilde{f}^\ell(\tilde{f}(\tilde{x}))). \tag{2.46}$$

Apply Lemma 2.6 with $\tilde{y} = \tilde{f}(\tilde{x})$, and Lemma 2.5, Statement (2). We obtain from (2.46)

$$\limsup \frac{1}{n} \log d(\tilde{x}, \tilde{f}^n(\tilde{x})) \leq h(f). \tag{2.47}$$

The last quantity $d\Big(\big((\tilde{f}_\#)^n(g)\big)^{cov}(\tilde{f}^n(\tilde{x})), \big((\tilde{f}_\#)^n(g)\big)^{cov}(\tilde{x})\Big)$ on the right of (2.45) is equal to the left hand side of Eq. (2.46) since the covering transformation $(\tilde{f}_\#)^n(g)$ is an isometry in the metric d. Hence, $d(\tilde{x}, (\tilde{f}_\#)^n(g)(\tilde{x}))$ can be estimated from above by the sum of the three positive terms $I_j(n)$, $j = 1, 2, 3$, on the right of (2.45) with $\limsup \frac{1}{n} \log I_j(n) \leq h(f)$ for each j. By Lemma 2.5, Statement (1), inequality (2.42) follows. □

We need the Poincaré-Hopf (Theorem 2.7). We consider a compact differentiable manifold X of dimension n with or without boundary and a smooth tangent vector field v on X. If X has a boundary then v is required to point out of X at points of ∂X. The singularities of v are its zeros on X. We assume that the zeros are isolated and contained in the interior of X. The index $i_v(p)$ of v at an isolated singularity p is defined as follows. Consider local coordinates in a small neighbourhood of p. Then the mapping $z \to \frac{v(z)}{|v(z)|}$ takes a small sphere around p in these coordinates to the unit sphere in $\mathbb{R}^{n-1}$. The degree of this mapping, considered as mapping from the unit sphere to itself, is called the index if v at p. The index of any regular (i.e. non-singular) point equals zero.

Theorem 2.7 (Poincaré-Hopf) *Let X be a compact differentiable manifold (or manifold with boundary) of dimension n, and v a smooth tangent vector field on X with isolated zeros. If X has a boundary then we require that v has no singularities on ∂X, and v points out of X at points of ∂X. Then*

$$\sum i_v(p) = \chi(X), \tag{2.48}$$

where $\chi(X)$ is the Euler characteristic of X.

For a proof see [64] or [16].

The Poincaré-Hopf (Theorem 2.7) implies the following theorem (see Proposition 5.6 in [25]). Actually, we need only the version of the Poincaré-Hopf (Theorem 2.7) for the dimension 2 which was proved already by Poincaré.

Theorem 2.8 *Let X be a Riemann surface, and let ϕ be a holomorphic quadratic differential on X (equivalently, a meromorphic quadratic differential without poles). Then there is no relatively compact topological disc in X with piecewise smooth boundary consisting of the union of a (perhaps empty) closed smooth vertical arc and a (perhaps empty) closed smooth arc that is contained in the regular part of X and is transversal to the vertical foliation.*

Proof Suppose, in contrary, that such a disc Δ exists. Then by Proposition 2.2 the whole boundary $\partial\Delta$ cannot be a vertical curve. If the whole boundary $\partial\Delta$ is transversal to the vertical foliation, we may deform the boundary of Δ slightly within the regular part of X, so that the boundary of the new disc is piecewise smooth and equals the union of a non-empty smooth vertical arc and a non-empty smooth arc that is transversal to the vertical foliation. We will assume from the beginning that $\partial\Delta$ is piecewise smooth and equals the union of a non-empty smooth closed vertical arc γ_v and a non-empty smooth closed arc γ_t that is contained in the ϕ-regular part and is transversal to the vertical foliation.

Consider a conformal mapping of Δ onto the unit half-disc $\mathbb{D}_r = \{z \in \mathbb{C}_r : |z| < 1\}$ in the right half-plane $\mathbb{C}_r$, whose continuous extension to the boundary maps the vertical arc γ_v onto the interval $[-i, i]$. Since the vertical segment γ_v is real analytic, the conformal mapping extends holomorphically by the reflection principle across the interval $(-i, i)$. The extension maps a neighbourhood N of the arc γ_v to a neighbourhood V of the segment $(-i, i)$.

Consider the push-foreword ϕ' to $\mathbb{D}_r \cup V$ of the quadratic differential $\phi \mid (\Delta \cup N)$ under the conformal mapping. For each singular point of ϕ' that is contained in $\mathbb{D}_r$ we consider a closed arc that joins the singular point with a point in the boundary half-circle, such that all points of the arc except one endpoint are contained in $\mathbb{D}_r$. We may choose the arcs pairwise disjoint. Let Ω' be the complement of the arcs in $\mathbb{D}_r$, let Ω'_{refl} be its reflection in the imaginary axis, and let Ω be the union of Ω', Ω'_{refl} and the interval $(-i, i)$.

After possibly shrinking V, the domain $V \cup \Omega'$ is simply connected, and the quadratic differential $\phi' \mid V \cup \Omega'$ has the form $\phi'(\zeta)d\zeta^2$ for a non-vanishing holomorphic function ϕ' on $V \cup \Omega'$. Consider the form $\sqrt{\phi'}d\zeta$ on $\Omega' \cup V$ for a branch of the square root of ϕ'. Since the segment $(-i, i)$ is a vertical trajectory of the quadratic differential ϕ', the real part of the vertical vector field v_v associated to ϕ' vanishes at points of $(-i, i)$. With $\sqrt{\phi'(\zeta)} = \sqrt{|\phi'(\zeta)|}e^{\frac{i\theta}{2}}$ and $v_v(\zeta) = |\phi'(\zeta)|^{-\frac{1}{2}}\binom{\sin\frac{\theta}{2}}{\cos\frac{\theta}{2}}$ we obtain $\text{Im}(\sqrt{\phi'}) \mid (-i, i) = 0$ (see equality (1.22) and the following calculations).

By Schwarz Reflection Principle the function $\sqrt{\phi'} \mid \Omega'$ extends to a holomorphic function on Ω. Hence $\phi' \mid \Omega'$ extends to a holomorphic function on Ω. Since $\phi' d\zeta^2 \mid \Omega'$ extends across the cuts to a holomorphic quadratic differential on $\mathbb{D}_r$ whose singular points are the initial points of the cuts, the reflection provides a holomorphic quadratic differential on $\mathbb{D}$, denoted by $\phi_{\mathbb{D}}$, whose singular points are the singular points of the push-forward of $\phi \mid \Delta$ and their reflections in the imaginary axis.

Denote by E_{even} the set of singular points of $\phi_{\mathbb{D}}$ of even order and by E_{odd} the set of singular points of odd order. We want to apply the Poincaré-Hopf (Theorem 2.7). If the set E_{odd} is not empty, $\phi_{\mathbb{D}}$ does not define a vector field on $\mathbb{D}$. We consider the double branched covering S of $\mathbb{D}$ with branch locus E_{odd}. Recall that the points of E_{odd} come in pairs. Let $\tilde{\phi}_{\mathbb{D}}$ be a lift of $\phi_{\mathbb{D}}$ to S. In distinguished coordinates z at any point in E_{even} the quadratic differential $\phi_{\mathbb{D}}$ is written as $(\frac{a+2}{2})^2 z^a (dz)^2$ with a positive even number a, and the lift of $\phi_{\mathbb{D}}$ has the same behavior near the lifts of points in E_{even}.

At any point in E_{odd} the quadratic differential $\phi_{\mathbb{D}}$ is written in distinguished coordinates z as $(\frac{a+2}{2})^2 z^a (dz)^2$ with a positive odd number a. The lift $\tilde{\phi}_{\mathbb{D}}$ can be written in coordinates ζ on S with $z(\zeta) = \zeta^2$ as $\tilde{\phi}_{\mathbb{D}}(\zeta) = \phi_{\mathbb{D}}(z(\zeta))(z'(\zeta))^2 = (a+2)^2 \zeta^{2a+2}$.

The quadratic differential $\tilde{\phi}_{\mathbb{D}}$ defines a vector field on S. Indeed, cut $\mathbb{D}$ along disjoint arcs that join points E_{odd} with the circle $\partial\mathbb{D}$. We obtain a simply connected domain $\hat{\Omega}$ which is contained in the unit disc and contains only singular points of $\phi_{\mathbb{D}}$ of even order. Hence, each point in the obtained domain $\hat{\Omega}$ has a neighbourhood on which two branches of the differential $\sqrt{\phi_{\mathbb{D}}(z)}dz$ are defined. The two branches differ by sign. Since the domain is simply connected there are two branches of

$\sqrt{\phi_{\mathbb{D}}(z)}dz$ that are globally defined on $\hat{\Omega}$. The surface S is obtained as follows. Take two copies of $\hat{\Omega}$ which we call sheets. Consider for each cut in the disc the respective copies of the two sheets and do cross-gluing of the edges. Take a global branch $\sqrt{\phi_{\mathbb{D}}(z)}dz$ on one sheet and $-\sqrt{\phi_{\mathbb{D}}(z)}dz$ on the other sheet. The behaviour along small closed curves surrounding any preimage of a point in E_{odd} shows that we obtain a globally defined differential which we denote by $\sqrt{\phi_{\mathbb{D}}(z)}dz$. Consider the vector field $v(z)$ on S that is defined by $\sqrt{\phi_{\mathbb{D}}(z)}dz(v(z)) = i$ for the non-singular points z (a vector field in the vertical direction). The vector field v is transversal to the boundary. Changing v to $-v$ if necessary we may assume that v points out of S at all boundary points.

Apply the Poincaré-Hopf (Theorem 2.7). Take a point in S whose projection to $\mathbb{D}$ is in E_{even}. In distinguished coordinates the quadratic differential $\tilde{\phi}_1$ near this point has the form $(\frac{a+2}{2})^2 z^a (dz)^2$ with a even. The vertical vector field near this point is given by $v(z) = \frac{2}{a+2} i z^{-\frac{a}{2}}$. The index of the vector field v at this point equals $-\frac{a}{2}$. For a point in S whose projection to $\mathbb{D}$ is in E_{odd}. the quadratic differential equals $(a+2)^2 \zeta^{2a+2} d\zeta^2$ and the vertical vector field can be written as $v(\zeta) = \frac{1}{a+2} i z^{-(a+1)}$. The index equals $-(a+1)$. For a point $p \in \mathbb{D}$ we denote by $a(p)$ the order of ϕ_1 at this point. Then

$$\sum_{p\in S} i(p) = 2 \sum_{p\in E_{even}} -\frac{a(p)}{2} + 2 \sum_{p\in E_{odd}} -(1 + a(p)) . \tag{2.49}$$

For the Euler characteristic of the double branched covering S we have

$$\chi(S) = 2\chi(\mathbb{D}) - B = 2 - B .$$

Here B is the number of branch points which is equal to $2\,|E_{odd}|$, where $|E_{odd}|$ is the number of points in E_{odd}. Hence, by the Poincaré-Hopf (Theorem 2.7)

$$-\sum_{p\in E_{even}} a(p) - 2 \sum_{p\in E_{odd}} a(p) - 2\,|E_{odd}| = 2 - 2\,|E_{odd}| . \tag{2.50}$$

This is impossible. The contradiction proves the theorem. □

The following lemma is proved by Douady (see Lemma 9.22 in [25]).

Lemma 2.9 *Let X be a compact Riemann surface with a quadratic differential ϕ and let ϱ be any Riemannian metric on X. There exist positive constants c_1 and c_2 such that for any free homotopy class $\hat{\alpha}$ of closed curves on X the inequality*

$$c_1\, \ell_\phi(\hat{\alpha}) \le \ell_\varrho(\hat{\alpha}) \le c_2\, \ell_\phi(\hat{\alpha}) \tag{2.51}$$

holds. Here $\ell_\phi(\hat{\alpha})$ is the infimum of the ϕ-length over all piecewise smooth loops in $\hat{\alpha}$, and $\ell_\varrho(\hat{\alpha})$ is the infimum of the length in the metric ϱ over all piecewise smooth loops in $\hat{\alpha}$.

Proof Take a small enough number $\delta > 0$, such that the δ-neighbourhood $V_\delta(z_j)$ in the ϱ-metric of any singular point z_j of ϕ is simply connected. Any loop that is contained in such a neighbourhood $V_\delta(z_j)$ is contractible to the singular point z_j. Hence, if $\hat{\alpha}$ contains a loop contained in $V_\delta(z_j)$, than both, $\ell_\phi(\hat{\alpha})$ and $\ell_\varrho(\hat{\alpha})$ are equal to zero. In this case the inequalities (2.51) hold.

On the other hand, there exist positive numbers c_1' and c_2' depending on δ such that for any curve γ' avoiding the $\frac{\delta}{2}$-neighbourhoods in the ϱ-metric of the singular points the inequalities

$$c_1' \,\ell_\phi(\gamma') \leq \ell_\varrho(\gamma') \leq c_2' \,\ell_\phi(\gamma') \tag{2.52}$$

hold. Moreover, there exists a constant $c'' \geq 1$ such that for each j

$$\begin{aligned} \ell_\varrho(\partial V_\delta(z_j)) &\leq 2c'' d_\varrho\big(\partial V_\delta(z_j), \partial V_{\frac{\delta}{2}}(z_j)\big), \\ \ell_\phi(\partial V_\delta(z_j)) &\leq 2c'' d_\phi\big(\partial V_\delta(z_j), \partial V_{\frac{\delta}{2}}(z_j)\big). \end{aligned} \tag{2.53}$$

Suppose the free homotopy class of curves $\hat{\alpha}$ cannot be represented by a loop contained in one of the $V_\delta(z_j)$. Take a curve $\gamma \in \hat{\alpha}$ which almost minimizes the ϱ-length, $\ell_\varrho(\gamma) \leq \ell_\varrho(\hat{\alpha}) + \varepsilon$ for a small positive number ε. For each part of γ that is contained in $X \setminus \cup V_{\frac{\delta}{2}}(z_j)$ (i.e. lies outside the $\frac{\delta}{2}$-neighbourhoods of the critical points) the estimates (2.52) hold.

Suppose there is a j and a connected component $\tilde{\gamma}$ of $\gamma \cap V_\delta(z_j)$ which intersects $V_{\frac{\delta}{2}}(z_j)$. The connected component $\tilde{\gamma}$ is an arc with the two endpoints on $\partial V_\delta(z_j)$. Replace $\tilde{\gamma}$ by an arc $\tilde{\gamma}'$ on $\partial V_\delta(z_j)$ which is homotopic to $\tilde{\gamma}$ with fixed endpoints. By (2.53)

$$\ell_\varrho(\tilde{\gamma}') \leq \ell_\varrho(\partial V_\delta(z_j)) \leq 2c'' d_\varrho(\partial V_\delta(z_j)), \partial V_{\frac{\delta}{2}}(z_j)) \leq c'' \ell_\varrho(\tilde{\gamma}). \tag{2.54}$$

Consider for each singular point z_j all connected components of the intersection $\gamma \cap V_\delta(z_j)$ which have points in $V_{\frac{\delta}{2}}(z_j)$. Replace in the same way as above each such component by an arc which avoids the sets $V_{\frac{\delta}{2}}(z_j)$. We obtain a curve γ' that is homotopic to γ, avoids the $\frac{\delta}{2}$-neighbourhoods in the ϱ-metric of the singular points, and satisfies the estimates

$$\ell_\varrho(\gamma') \leq c'' \ell_\varrho(\gamma) \leq c''(\ell_\varrho(\hat{\alpha}) + \epsilon). \tag{2.55}$$

Hence, by the inequalities (2.52) the inequality $\ell_\phi(\hat{\alpha}) \leq \frac{c''}{c_1'}(\ell_\varrho(\hat{\alpha}) + \varepsilon)$ holds. Since ε can be taken to be an arbitrary positive number, the first inequality in (2.51) holds. The second inequality follows by interchanging the role of ϱ and ϕ. □

The following proposition is proved in [25] (see Proposition 5.7. there).

Proposition 2.7 *Let X be a closed Riemann surface, and let ϕ be a holomorphic quadratic differential on X. Suppose γ is a simple closed (connected and smooth) curve in X which is contained in the ϕ-regular part of X and is transversal to the vertical foliation. Let γ' be any closed (connected and piecewise smooth) curve in X which is free homotopic to γ. Then the following estimate for the horizontal ϕ-length holds*

$$\ell_{\phi,h}(\gamma) \leq \ell_{\phi,h}(\gamma'). \tag{2.56}$$

Proof Let x_0 be the base point of γ. Denote by α the class of γ in the fundamental group $\pi_1(X, x_0)$, and let $\mathsf{P} : \tilde{X} \to X$ be the universal covering of X. Consider the quotient $\hat{X} \stackrel{def}{=} \tilde{X}/\alpha^{cov}$ for the covering transformation α^{cov} associated to α and the fixed base point $\tilde{x}_0 \in \mathsf{P}^{-1}(x_0)$. The quotient is conformally equivalent to an annulus. The universal covering $\mathsf{P} : \tilde{X} \to X$ induces a covering $\hat{p} : \hat{X} \to X$. (For more details see also Sect. 11.2). The holomorphic quadratic differential ϕ on X lifts to a holomorphic quadratic differential $\hat{\phi}$ on $\hat{X}$ and the free homotopic loops γ and γ' on X lift to free homotopic loops $\hat{\gamma}$ and $\hat{\gamma}'$ on $\hat{X}$. Notice that $\ell_{\hat{\phi},h}(\hat{\gamma}) = \ell_{\phi,h}(\gamma)$ and $\ell_{\hat{\phi},h}(\hat{\gamma}') = \ell_{\phi,h}(\gamma')$. Hence, it is sufficient to prove the theorem in the case when X is an annulus with a holomorphic quadratic differential, and both loops are free homotopic to the positive generator of the fundamental group of the annulus. We may assume that γ' is also simple closed, removing otherwise some parts of γ' which does not increase its horizontal ϕ-length. We will now prove the Theorem under these conditions.

Assume first that γ and γ' are disjoint. Then the set $\gamma \cup \gamma'$ consists of all boundary points of an annulus $A \subset X$. Indeed, after a conformal mapping we may assume that X is a subset of the plane. Then the claim follows from the Jordan Curve Theorem applied to each of the curves.

For each point $p \in \gamma$ we consider the vertical trajectory ray r_p with initial point p that enters A at p. We call a trajectory ray r_p exceptional if it stays in A and runs into a critical point contained in A. We claim that each non-exceptional vertical trajectory ray r_p, $p \in \gamma$, meets the boundary ∂A after p. Indeed, if the non-exceptional ray r_p is critical, it meets ∂A at a point after p by the definition of exceptional rays.

Suppose r_p is non-critical. Consider an open ϕ-rectangle $\mathcal{R}$ that contains $p \in \gamma$. Shrinking $\mathcal{R}$ in the horizontal direction if necessary, we may assume that the connected component γ_p, that contains p, of the intersection $\gamma \cap \mathcal{R}$ is an arc with endpoints on opposite vertical sides of $\mathcal{R}$.

Denote by γ_p^h the connected component containing p of the intersection of $\mathcal{R}$ with the horizontal trajectory through p. By (the proof of) Theorem 2.4 the divergent vertical trajectory ray r_p intersects γ_p^h after p infinitely often in the same direction. Hence, the ray r_p contains points that are not in A and therefore r_p intersects ∂A. The claim is proved.

By Proposition 2.8 for any non-exceptional trajectory ray r_p, $p \in \gamma$, the first point after p on the boundary ∂A is contained in γ'. Indeed, suppose it is a point

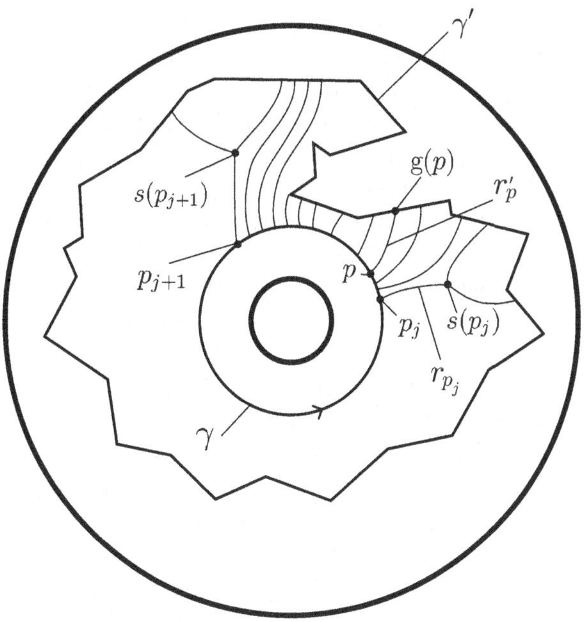

Fig. 2.4 Curves that are transversal to the vertical foliation are $\ell_{\phi,h}$-minimizing in their free homotopy class

p' on γ. The union of one of the arcs of γ with endpoints p and p' and the arc of r_p between p and p' bounds a disc. But the first arc is transversal to the vertical foliation and the second is an arc of a vertical leaf. This is impossible.

We assign to each non-exceptional ray r_p, $p \in \gamma$, the first point $g(p) > p$ of the trajectory ray r_p which is on γ'. Let r'_p be the segment on r_p between p and $g(p)$. There are only finitely many exceptional trajectory rays. Let p_j, $j = 1, \ldots, N$, be the initial points of the exceptional trajectory rays, ordered cyclically with respect to some orientation of γ. Let $s(p_j)$ be the singular point in A which is the limit point of r_{p_j}. Then for $p_j \neq p_{j'}$ the singular points $s(p_j)$ and $s(p_{j'})$ are different. Indeed, suppose not. Then there are two neighbouring points p_j and p_{j+1} with $s(p_j) = s(p_{j+1})$. (If the two neighbouring points are p_N and p_1 we relabel the p_j.) Then r_{p_j} and $r_{p_{j+1}}$, are segments on neighbouring bisectrices at the singular point $s(p_j) = s(p_{j'})$, and the two segments together with an arc of γ between p_j and p_{j+1} bound a sector which is laminated by vertical arcs with the two endpoints on the arc of γ between p_j and p_{j+1}. This is impossible by Theorem 2.8 (see Fig. 2.4).

It is enough to prove the proposition for the case when γ' does not contain singular points of the quadratic differential and is piecewise linear in flat coordinates. Indeed, one can achieve this by a deformation of γ' that changes the ϕ-length no more than by an a priory given amount.

Consider the open arc γ_j contained in γ whose endpoints are two neighbouring points p_j and p_{j+1} (p_N and p_1 for $j = N$). We consider the set $\gamma'_j = g(\gamma_j)$.

The obtained subset γ_j' of γ' is the union of a finite number of open linear segments contained in γ' and a finite number of points on γ' (that are endpoints of some of the just mentioned linear segments). Hence, the set γ_j' is measurable for the horizontal ϕ-length. Moreover, each maximal open linear piece of γ_j' has the same horizontal ϕ-length as the corresponding part on γ_j which is the preimage of this linear piece under g. Indeed, each point in the linear piece of γ' is joined with its preimage under g by a vertical segment. Hence, $\ell_{\phi,h}(\gamma_j) = \ell_{\phi,h}(\gamma_j')$. The union of the γ_j equals $\gamma \setminus \cup\{p_j\}$. The union of the γ_j', $j = 1, \dots N$, may be much smaller than γ'. The proposition is proved for the case when γ and γ' are disjoint.

In the general case we may assume again that the curve γ' does not contain singular points of the quadratic differential and is piecewise linear in flat coordinates. Consider the set $\gamma' \setminus \gamma$ of points that are in γ' but not in γ. It consists of finitely many connected components, each being a piecewise linear open arc. Take any component γ_j'. There is an arc γ_j on γ so that the union of the closures of γ_j and γ_j' (each with suitable orientation) bounds a disc in A. The same argument as before shows that $\ell_{\phi,h}(\gamma_j) \le \ell_{\phi,h}(\gamma_j')$. Instead of γ' we consider the closed piecewise smooth curve γ'' obtained by replacing each γ_j' by γ_j oriented so that γ_j' and γ_j are homotopic with fixed endpoints. The horizontal ϕ-length $\ell_{\phi,h}(\gamma'')$ of the new curve γ'' does not exceed $\ell_{\phi,h}(\gamma')$. All points of γ'' lie on γ. Each point of γ is covered by at least one point of γ'', but some parts of γ may be covered multiply. Hence $\ell_{\phi,h}(\gamma) \le \ell_{\phi,h}(\gamma'') \le \ell_{\phi,h}(\gamma')$. □

Proof of Theorem 2.2. Let φ_0 be a non-periodic absolutely extremal self-homeomorphism of a closed Riemann surface X of genus $g \ge 2$, and let φ be a self-homeomorphism of X that is isotopic to φ_0. We will prove that the entropy of φ is not smaller than $\frac{1}{2}\log K(\varphi_0)$. Theorem 2.2 will then be a consequence of Theorem 2.1.

Denote by ϕ the quadratic differential of φ_0. It is holomorphic on X. Let γ be a simple closed curve in X with base point x_0 that is contained in the ϕ-regular part of X, is transversal to the vertical foliation of ϕ, and has positive $\ell_{\phi,h}$-length. By Theorem 2.5 such a curve exists. By Theorem 2.8 γ is not contractible. Denote by α the class of γ in $\pi_1(X, x_0)$. Let d_{hyp} be the hyperbolic metric on the universal covering $\tilde{X} \cong \mathbb{C}_+$. We use the same notation for the induced metric on X. By Lemma 2.9 and Proposition 2.7 the inequalities

$$\ell_{hyp}(\widehat{\alpha}) \ge c_1\ell_{\phi}(\widehat{\alpha}) \ge c_1\ell_{\phi,h}(\widehat{\alpha}) = c_1\ell_{\phi,h}(\gamma) > 0 \tag{2.57}$$

hold for the free homotopy class $\widehat{\alpha}$ represented by α.

Denote by $\tilde{x}_0 \in \tilde{X}$ the point corresponding to the identity Id_{x_0,x_0} in $\pi_1(X, x_0, x_0)$. Let $\tilde{\alpha}$ be the lift of α to the universal covering with initial point $\tilde{x}_0$. Let as before α^{cov} be the covering transformation that maps $\tilde{x}_0$ to $\tilde{\alpha}(\tilde{x}_0)$. Since $d_{hyp}(\tilde{x}_0, \alpha^{cov}(\tilde{x}_0))$ is the infimum of the hyperbolic lengths $\ell_{hyp}(\tilde{\gamma}')$ over all curves $\tilde{\gamma}'$ in $\tilde{X}$ with initial point $\tilde{x}_0$ and terminal point $\alpha^{cov}(\tilde{x}_0)$ and $\ell_{hyp}(\tilde{\gamma}') = \ell_{hyp}(\gamma')$, we obtain by (2.57)

$$d_{hyp}(\tilde{x}_0, \alpha^{cov}(\tilde{x}_0)) \ge c_1\ell_{\phi,h}(\gamma)\,. \tag{2.58}$$

Since φ_0 is a non-periodic absolutely extremal self-homeomorphism of X with quadratic differential ϕ, for any natural number n the curve $(\varphi_0)^n(\gamma)$ is a simple closed curve in X that avoids the ϕ-singular points and is transversal to the vertical foliation. The mapping $(\varphi_0)^n$ expands the horizontal ϕ-length of arcs by the factor $K^{\frac{n}{2}}$. This implies the following estimate of the horizontal ϕ-length of $(\varphi_0)^n(\gamma)$

$$\ell_{\phi,h}((\varphi_0)^n(\gamma)) \geq K^{\frac{n}{2}} \ell_{\phi,h}(\gamma). \tag{2.59}$$

The loop $(\varphi_0)^n(\gamma)$ represents the free homotopy class of $((\varphi_0)_\#)^n(\alpha)$ (considered as element of the fundamental group $\pi_1(X, x_0)$). Indeed, $(\varphi_0)^n(\gamma)$ lifts to the curve $(\tilde{\varphi}_0)^n(\tilde{\gamma})$ in $\tilde{X}$ with initial point $(\tilde{\varphi}_0)^n(\tilde{x}_0)$ and terminal point $(\tilde{\varphi}_0)^n(\alpha^{cov}(\tilde{x}_0)) = \left(\left((\varphi_0)_\#\right)^n(\alpha)\right)^{cov}\left((\tilde{\varphi}_0)^n(\tilde{x}_0)\right)$. The last equality follows from equality (2.36).

Lemma 2.9 and Proposition 2.7 apply to the loop $(\varphi_0)^n(\gamma)$ in the same way as to the loop γ. Hence, inequality (2.57) with $(\varphi_0)^n(\gamma)$ instead of γ and $((\varphi_0)_\#)^n(\alpha)$ instead of α holds. Moreover, inequality (2.58) is true with α^{cov} replaced by $(((\varphi_0)_\#)^n(\alpha))^{cov}$ and γ replaced by $(\varphi_0)^n(\gamma)$. Using inequality (2.59), we obtain

$$d_{hyp}(\tilde{x}_0, (((\varphi_0)_\#)^n(\alpha))^{cov}(\tilde{x}_0)) \geq c_1 \ell_{\phi,h}((\varphi_0)^n(\gamma)) \geq c_1 K^{\frac{n}{2}} \ell_{\phi,h}(\gamma). \tag{2.60}$$

Consider the self-homeomorphism φ of X that is isotopic to φ_0. Since by Theorem 2.6, Lemma 2.7, and equality (2.43)

$$h(\varphi) \geq \Gamma_{\varphi_\#} = \Gamma_{(\varphi_0)_\#} \geq \limsup \frac{1}{n} \log d_{hyp}(\tilde{x}_0, (((\tilde{\varphi}_0)_\#)^n(\alpha))^{cov}(\tilde{x}_0)),$$

we obtain $h(\varphi) \geq \frac{1}{2} \log K$. The Theorem is proved. □

2.5 The Entropy of Mapping Classes on Second Kind Riemann Surfaces

The following Theorem 2.9 allows us to treat the entropy of mapping classes of Riemann surfaces of second kind.

Theorem 2.9 *Let X be a connected closed Riemann surface with a set E of distinguished points. Assume that $X \setminus E$ is hyperbolic (i.e. covered by $\mathbb{C}_+$). Let φ be an absolutely extremal non-periodic self-homeomorphism of X that fixes E setwise. Suppose $z_0 \in E$ is a fixed point of φ, $\varphi(z_0) = z_0$.*

Then φ is isotopic through self-homeomorphisms of X that fix E to a homeomorphism φ_0 that fixes a topological disc Δ' around z_0 pointwise and has the same entropy $h(\varphi_0) = h(\varphi)$ as φ. Moreover, the homeomorphism φ_0 can be chosen to represent any a priory given preimage under the mapping $\mathcal{H}_{z_0} : \mathfrak{M}(X \setminus \Delta'; \partial\Delta', E \setminus \{z_0\}) \to \mathfrak{M}(X; \{z_0\}, E \setminus \{z_0\})$ of the class of φ in $\mathfrak{M}(X; \{z_0\}, E \setminus \{z_0\})$.

(For the notation see Sect. 1.4.)

Notice that the complement X_0 in X of an open disc around z_0 is a bordered Riemann surface (the closure of a Riemann surface of second kind), and the restriction $\varphi_0 \mid X_0$ is a self-homeomorphism of X_0 with entropy $h(\varphi_0 \mid X_0) = h(\varphi)$.

We will now prepare the proof of Theorem 2.9.

In a neighbourhood of the distinguished point z_0 of X we consider distinguished coordinates ζ with $\zeta(z_0) = 0$. In these coordinates the quadratic differential of φ has the form $\phi(\zeta)(d\zeta)^2 = (\frac{a+2}{2})^2\zeta^a(d\zeta)^2$. Take a disc $\Delta = \{|\zeta| < c\}$ in these coordinates. The following lemma says that the restriction $\varphi \mid X \setminus \{z_0\}$ has an extension to a self-homeomorphism of the "real blowup" of X at z_0.

Lemma 2.10 *In distinguished coordinates $\zeta = re^{i\theta}$ the quotient $\frac{\varphi(re^{i\theta})}{|\varphi(re^{i\theta})|}$ does not depend on $r \in (0, c)$ and defines a homeomorphism*

$$\mathfrak{h}(e^{i\theta}) \overset{def}{=} \frac{\varphi(re^{i\theta})}{|\varphi(re^{i\theta})|} \tag{2.61}$$

of the unit circle onto itself.

Proof The distinguished coordinates are unique up to multiplication by an $(a+2)^{\text{nd}}$ root of unity. In other words

$$\left(\zeta\, e^{\frac{2\pi i}{a+2}\cdot l}\right)^a \left(d\zeta\, e^{\frac{2\pi i}{a+2}\cdot l}\right)^2 = \zeta^a(d\zeta)^2 \tag{2.62}$$

for any integer l, and these are the only holomorphic changes of coordinates under which the quadratic differential has again the canonical form.

The initial quadratic differential of φ coincides with its terminal quadratic differential. Hence in the distinguished coordinates ζ the mapping φ has the form

$$\varphi(\zeta) = \mathcal{Z}_a(\zeta)\cdot e^{\frac{2\pi i}{a+2}\cdot \ell} \quad \text{for some integer } \ell, \tag{2.63}$$

where

$$\mathcal{Z}_a(\zeta) = \left(\frac{\zeta^{a+2} + 2k\,|\zeta|^{a+2} + k^2\,\bar{\zeta}^{a+2}}{1-k^2}\right)^{\frac{1}{a+2}}. \tag{2.64}$$

Here $k = \frac{K-1}{K+1}$, and $K = K(\varphi)$ is the quasiconformal dilatation of φ. We take the root which is positive on the positive real axis. (See e.g. [11], formula (2.3) and Theorem 6 of [11] or Chap. 1, Eq. (1.27)).

The mapping $\mathcal{Z}_a$ fixes each horizontal bisectrix $\{r\exp(\frac{2\pi ji}{a+2}),\ 0 < r < c\}$, $j = 0, \dots, a+1$, and each vertical bisectrix $\{r\exp(\frac{2\pi(j+\frac{1}{2})i}{a+2}),\ 0 < r < c\}$, $j = 0, \dots, a+1$. Consider a sector $\mathfrak{s}_j^h = \{r\exp(\theta i) : \theta \in (\frac{2\pi ji}{a+2}, \frac{2\pi(j+1)i}{a+2}),\ 0 < r < c\}$, $j = 0, \dots, a+1$, of Δ between two consecutive horizontal bisectrices. The flat coordinates g_j^h on the sector $\mathfrak{s}_j^h$ map the sector to a half-disc in the upper or lower half-plane. Recall that g_j^h is a suitable branch of the mapping $\zeta \to \zeta' \stackrel{def}{=} \zeta^{\frac{a+2}{2}}$, $g_j^h(re^{i\theta}) = r^{\frac{a+2}{2}} e^{i\frac{a+2}{2}\theta}$ for $re^{i\theta} \in \mathfrak{s}_j^h$. The image is in the upper half-plane if j is even and in the lower half-plane if j is odd.

Put $\zeta' = \xi' + i\eta' = r'e^{i\theta'}$. In coordinates ζ' the mapping $\mathcal{Z}_a$ has the form

$$\zeta' = \xi' + i\eta' \to K^{\frac{1}{2}}\xi' + i\,K^{-\frac{1}{2}}\eta' = \mathcal{Z}'_a(\zeta') \tag{2.65}$$

with $K = \frac{1+k}{1-k}$. Hence, for $\zeta' = r'e^{i\theta'}$, $\theta' \in (0, \pi)$ or $\theta' \in (\pi, 2\pi)$, respectively, $0 < r' < c^{\frac{2}{a+2}}$, the equality

$$\begin{aligned}\frac{\mathcal{Z}'_a(r'e^{i\theta'})}{|\mathcal{Z}'_a(r'e^{i\theta'})|} &= \frac{K^{\frac{1}{2}}r'\cos\theta' + iK^{-\frac{1}{2}}r'\sin\theta'}{\sqrt{Kr'^2\cos^2(\theta') + K^{-1}r'^2\sin^2(\theta')}}\\ &= \frac{K^{\frac{1}{2}}\cos\theta' + iK^{-\frac{1}{2}}\sin\theta'}{\sqrt{K\cos^2(\theta') + K^{-1}\sin^2(\theta')}}\end{aligned} \tag{2.66}$$

holds. The quotient $\frac{\mathcal{Z}'_a(r'e^{i\theta'})}{|\mathcal{Z}'_a(r'e^{i\theta'})|}$ does not depend on r' and is denoted by $\mathfrak{h}'(e^{i\theta'})$. The mapping $\mathfrak{h}'$ defines a homeomorphism of $(0, \pi)$ ($(\pi, 2\pi)$, respectively) onto itself and extends to a homeomorphism of $[0, \pi]$ ($[\pi, 2\pi]$, respectively) onto itself. In coordinates ζ we obtain a homeomorphism $\mathfrak{h}$ from $[\frac{2\pi ji}{a+2}, \frac{2\pi(j+1)i}{a+2}]$ onto itself, that fixes each endpoint. Moreover, $\mathcal{Z}_a(re^{i\theta}) = |\mathcal{Z}_a(re^{i\theta})|\mathfrak{h}(e^{i\theta})$. The lemma follows from the Eq. (2.63). □

We will now define a homeomorphism from the "real blowup" of X at z_0 onto a subset of X. We put $\Delta_0 = \{\zeta \in \Delta : |\zeta| < \frac{c}{2}\}$, and $X^{z_0} = X\backslash\overline{\Delta_0}$. Then $X^{z_0} \subset X$ is a Riemann surface of second kind which is homeomorphic to $X\backslash\{z_0\}$. The closure $\overline{X^{z_0}}$ (the closure is taken in X) is a bordered Riemann surface and can be identified with the "real blowup" of X at z_0. We choose a diffeomorphism $\psi^{z_0} : X^{z_0} \to X\backslash\{z_0\}$ which is the identity on $X\backslash\Delta$ and is defined on $X^{z_0} \cap \Delta$ in coordinates ζ by the mapping

$$\left\{\frac{1}{2}c < |\zeta| < c\right\} \ni \zeta \to \psi^{z_0}(\zeta) = \alpha(|\zeta|) \cdot \frac{\zeta}{|\zeta|} \in \Delta\backslash\{0\} \tag{2.67}$$

for a smooth strictly increasing function $\alpha : \left(\frac{1}{2}c, c\right) \to (0, c)$ with $\lim_{t\to\frac{c}{2}} \alpha(t) = 0$ and $\alpha(t) = t$ for t close to c. Then the self-homeomorphism

$$(\psi^{z_0})^{-1} \circ \varphi \circ \psi^{z_0} : X^{z_0} \circlearrowleft$$

of X^{z_0} extends to a self-homeomorphism of $\overline{X^{z_0}}$ denoted by φ^{z_0}.

Lemma 2.11 *The following equality for entropies hold:* $h(\varphi^{z_0}) = h(\varphi)$.

Proof of Lemma 2.11

The Upper Bound The inequality $h(\varphi) \le h(\varphi^{z_0})$ follows from Theorem 5 of [1]. Indeed, the mapping $\psi^{z_0} : X^{z_0} \to X \setminus \{z_0\}$ extends to a continuous mapping $\widetilde{\psi^{z_0}} : \overline{X^{z_0}} \to X$ that maps the circle $\partial\Delta$ to z_0. We introduce the equivalence relation $\sim$ on $\overline{X^{z_0}}$, for which $z \sim z'$ for different points $z, z' \in \overline{X^{z_0}}$ if and only if z and z' are points of $\partial\Delta_0$. We denote by π the canonical projection $\pi : \overline{X^{z_0}} \to \overline{X^{z_0}}/\sim$. The quotient $\overline{X^{z_0}}/\sim$ is homeomorphic to X. A homeomorphism is given by the mapping $\mathring{\psi}^{z_0}$, $\mathring{\psi}^{z_0}(x/\sim) = \psi^{z_0}(x)$ for $x \in X^{z_0}$ and $\mathring{\psi}^{z_0}(\partial X^{z_0}/\sim) = z_0$. Then $\widetilde{\psi^{z_0}} = \mathring{\psi}^{z_0} \circ \pi$. The equality $\varphi \circ \widetilde{\psi^{z_0}} = \widetilde{\psi^{z_0}} \circ \varphi^{z_0}$ holds on X^{z_0} by the definition of ψ^{z_0}. It is clear that it holds on the whole $\overline{X^{z_0}}$. We obtain the equality

$$\left((\mathring{\psi}^{z_0})^{-1} \circ \varphi \circ \mathring{\psi}^{z_0}\right) \circ \pi = \pi \circ \varphi^{z_0} .$$

By Theorem 5 of [1] the inequality $h(\varphi) = h\left((\mathring{\psi}^{z_0})^{-1} \circ \varphi \circ \mathring{\psi}^{z_0}\right) \le h(\varphi^{z_0})$ holds. (See also the beginning of the proof of Lemma 2.1 of Sect. 2.1.)

The Lower Bound We need to prove the opposite inequality $h(\varphi^{z_0}) \le h(\varphi)$. The difficulty is that there is no refining sequence of covers of $\overline{X^{z_0}}$ of the form $(\widetilde{\psi^{z_0}})^{-1}(\mathcal{A}_j))$ for covers $\mathcal{A}_j$ of X. Each cover $(\widetilde{\psi^{z_0}})^{-1}(\mathcal{A})$ with $\mathcal{A}$ a cover of X contains an element that covers the whole circle ∂X^{z_0}.

Refining Sequences of Covers of $\overline{X^{z_0}}$ We will find refining sequences of covers for $\overline{X^{z_0}}$ as follows. Let N be a large natural number. For a positive number $\varepsilon < \varepsilon_0$ we take the open cover $\mathcal{A}_\varepsilon$ of X consisting of ε-squares and ε-stars. We associate to $\mathcal{A}_\varepsilon$ a cover $\mathcal{A}'_{\varepsilon,N}$ of $X \setminus \{z_0\}$ which is described as follows. Consider N radii in the disc $\Delta \subset X$ around z_0 which divide the disc into N open sectors S_j of equal angle. We may choose the number N and the radii so that each sector S_j is contained in a sector between a horizontal bisectrix and one of the closest vertical bisectrices. (Thus they divide each such sector into sectors of equal angle.) For each j we let S'_j be an open sector which has an angle not exceeding twice the angle of S_j and contains $\overline{S_j} \setminus \{0\}$. Then the S'_j cover the punctured disc $\Delta \setminus \{z_0\}$. We may assume that the maximum over $j = 0, \dots N$, of the (Euclidean) length of the arcs $\{e^{i\theta} : \frac{c}{2}e^{i\theta} \in S'_j\}$ of the unit circle goes to 0 for $N \to \infty$.

Let ε be small and N big. The cover $\mathcal{A}'_{\varepsilon,N}$ of $X \setminus \{z_0\}$ consists of all sets in $\mathcal{A}_\varepsilon$ that do not intersect the disc $\Delta_0 \overset{def}{=} \{|\zeta| < \frac{c}{2}\}$ around z_0, and all sets of the form $A \cap S'_j$ for any sector S'_j and any set $A \in \mathcal{A}_\varepsilon$ that intersects Δ_0 (see Fig. 2.5).

We define an open cover $\mathcal{A}_{\varepsilon,N}(z_0)$ of $\overline{X^{z_0}}$ as follows. For each element $A \in \mathcal{A}'_{\varepsilon,N}$ that is not equal to the intersection of the ε-star at z_0 with a sector S'_j we let the set $(\psi^{z_0})^{-1}(A)$ be an element of the cover $\mathcal{A}_{\varepsilon,N}(z_0)$. For each element $A \in \mathcal{A}'_{\varepsilon,N}$ that is an intersection of the ε-star at z_0 with a sector S'_j the boundary of the set $(\psi^{z_0})^{-1}(A)$ intersects the circle $\{|z - z_0| = \frac{c}{2}\}$ along an arc γ_A. We let the union $(\psi^{z_0})^{-1}(A) \cup \gamma_A$ be an element of the cover $\mathcal{A}_{\varepsilon,N}(z_0)$. These are all elements of the cover.

Fig. 2.5 The cover $\mathcal{A}'_{\varepsilon,N}$

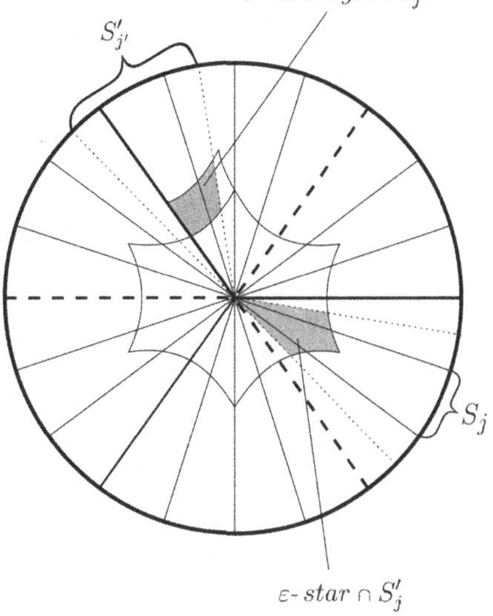

Take a sequence of ε_k's that decrease to 0, a sequence of N_k's with N_k dividing N_{k+1} and $N_k \to \infty$. Choose radii and sectors $S_{j_k}^{N_k}$ successively so that the $S_{j_1}^{N_1}$ subdivide the $\mathfrak{s}_j$ and for all $k > 1$ the $S_{j_{k+1}}^{N_{k+1}}$ subdivide the $S_{j_k}^{N_k}$. Define sectors $(S_{j_k}^{N_k})' \supset \overline{S_{j_k}^{N_k}} \setminus \{z_0\}$ as above and so that each $(S_{j_{N+1}}^{N_{k+1}})'$ is contained in some $(S_{j_k}^{N_k})'$. Consider for each k the cover $\mathcal{A}_{\varepsilon_k,N_k}(z_0)$ of $\overline{X^{z_0}}$ obtained for the choice of ε_k, N_k, and $(S_j^{N_k})'$. Then the $\mathcal{A}_{\varepsilon_k,N_k}(z_0)$ provide a refining sequence of open covers of $\overline{X^{z_0}}$. This is not difficult to see, taking into account the choice of the $(S_j^{N_k})'$, the fact that the sequence of coverings $\mathcal{A}_{\varepsilon_k}$ of X is refining, and the following observation. Fix k and consider the elements of the cover $\mathcal{A}_{\varepsilon_k,N_k}(z_0)$ contained in $\overline{X^{z_0}} \cap \Delta$. Let σ_k be the supremum of the diameters (in coordinates ζ on Δ) of such sets. Then the σ_k tend to zero for $k \to \infty$, since in polar coordinates $\zeta = \varrho e^{i\theta}$ on $\Delta \cap \overline{X^{x_0}}$ each set $(\psi^{z_0})^{-1}(A)$ with $A \in \mathcal{A}_{\varepsilon_k,N_k}(z_0)'$ has width comparable to ε_k in the ϱ-direction and width equal to the maximal angle of the $(S_j^{N_k})'$ in the θ direction.

An Upper Bound for the Entropy $h(\varphi^{z_0}, \mathcal{A}_{\varepsilon_k,N_k}(z_0))$ We will take a large number k, put $N = N_k$, $\varepsilon = \varepsilon_k$, and give an upper bound of the entropy $h(\varphi^{z_0}, \mathcal{A}_{\varepsilon,N}(z_0))$ of φ^{z_0} with respect to the cover $\mathcal{A}_{\varepsilon,N}(z_0)$ of $\overline{X^{z_0}}$.

For each set A in the above defined cover $\mathcal{A}_\varepsilon$ of X and one of the N sectors S_j^N of $\Delta \setminus \{z_0\}$ we define $A \cap^* S_j^N$ to be equal to A if A does not intersect Δ_0 and equal to $A \cap S_j^N$ if A intersects Δ_0. We prove first the following claim.
For each $(\lambda^{-n}\varepsilon, \varepsilon)$-rectangle and each $(\lambda^{-n}\varepsilon, \varepsilon)$-star written in the form

$$A_0 \cap \ldots \cap \varphi^{-n}(A_n) \in \mathcal{A}_\varepsilon \vee \ldots \vee \varphi^{-n}(\mathcal{A}_\varepsilon) \tag{2.68}$$

with either all A_j being ε-squares or all A_j being ε-stars, there are at most $1 + \frac{(n+1)N((n+1)N+1)}{2}$ *different non-empty sets of the form*

$$A \stackrel{def}{=} (A_0 \cap^* S^N_{j0}) \bigcap \dots \bigcap \varphi^{-n}(A_n \cap^* S^N_{jn}) \tag{2.69}$$

where each S^N_{jl} is one of the N sectors of Δ.

The claim is proved as follows. Suppose first that the set (2.68) is a $(\lambda^{-n}\varepsilon, \varepsilon)$-rectangle, i.e. it is the image of a ϕ-rectangle $\mathcal{F} : R \to A_0 \cap \dots \cap \varphi^{-n}(A_n)$ for a rectangle R in the complex plane with sides parallel to the axes. If a set A_l intersects Δ_0 then by a similar argument as used in the proof of Lemma 2.2 it is contained in a sector $\mathfrak{s}^h_{jl}$ (or $\mathfrak{s}^v_{jl}$) of Δ between two consecutive horizontal bisectrices (or between two consecutive vertical bisectrices). A branch of the mapping $\zeta \to \zeta^{\frac{2}{a+2}}$ defines flat coordinates on the sector and maps each of the N radii that are contained in the sector to a relatively closed straight line segment in the image. The mapping φ is real linear in flat coordinates. Hence the mapping φ^{-l} takes each of the N radii that are contained in the sector to a relatively closed curve in the image which is a straight line segment in flat coordinates. This implies that each set of the form (2.69) is obtained as follows.

Take the rectangle R in the plane which is the preimage $\mathcal{F}^{-1}(A)$ of a set A of the form (2.68). For each l the collection of preimages $\mathcal{F}^{-1}(A \bigcap^* \varphi^{-l}(A_l \cap S^N_{jl}))$ is the collection of intersections of R with connected components of the complement of the union of no more than N lines in the complex plane. We have to count non-empty sets each of which is the intersection over $l = 0, \dots, n$, of an element of the collection of preimages $\mathcal{F}^{-1}(A \bigcap \varphi^{-l}(A_l \cap^* S^N_{jl}))$.

Each such intersection Q is a connected component of the intersection of R with a connected component Q' of the complement of the union of no more than $(n+1)N$ lines in the complex plane. Indeed, Q is contained in such a component Q', and since the boundary of Q is the union of segments of some of the lines, it has no boundary points in Q', hence it coincides with Q'.

The estimate of the number of connected components of the complement of the union of N different real lines in the complex plane can be given by induction on the number N of lines. If N equals 1 the number is obviously equal to 2. Suppose an upper bound for the number is found for N lines. The observation for $N + 1$ lines is the following. The $(N + 1)$-st line intersects a number of lines among the N previous lines, and each of them is intersected once. The intersection points with the N previous lines divide the $(N + 1)$-st line into no more than $N + 1$ connected components. Each connected component of the last line is contained in exactly one connected component of the complement of the first N lines in $\mathbb{C}$. This component is divided into two parts by the last line. Hence, adding an $(N + 1)$-st line increases the number of connected components of the lines at most by $N + 1$. By induction, the number of connected components of the complement of N lines in $\mathbb{C}$ does not exceed $1 + \frac{N(N+1)}{2}$. We proved the claim for the case when A is a $(\lambda^{-n}\varepsilon, \varepsilon)$-rectangle.

If the set (2.68) is a $(\lambda^{-n}\varepsilon, \varepsilon)$-star and intersects Δ_0, then all A_j are stars at z_0, and for all A_j that are stars at z_0 we obtain at most N radii in each A_j. Hence, for the stars in $\mathcal{A}_\varepsilon \vee \varphi^{-1}(\mathcal{A}_\varepsilon) \vee \ldots \vee \varphi^{-n}(\mathcal{A}_\varepsilon)$ there are at most $(n+1)N$ different non-empty sets of the form (2.69). The claim is proved.

For each small $\varepsilon > 0$ there is a finite number C_1 depending on ε, such that the set $X \setminus \{z_0\}$ can be covered by C_1 sets, each either an ε-square or an ε-star. We may cover each ε-square by $[\lambda^n + 1]$ sets that are $(\lambda^{-n}\varepsilon, \varepsilon)$-rectangles. Consider an ε-star centered at a singular point of order a'. Each of the $2a' + 2$ sectors of the ε-star between a horizontal and the nearest vertical bisectrices is an open ε-square. Hence, the punctured star can be covered by no more than $(2a' + 2) \cdot \big(\lambda^n + 1\big)$ sets that are $(\lambda^{-n}\varepsilon, \varepsilon)$-rectangles, and the puncture is covered by a $(\lambda^{-n}\varepsilon, \varepsilon)$-star.

By the proof of Proposition 2.4 each $(\lambda^{-n}\varepsilon, \varepsilon)$-rectangle can be written in the form (2.68) for ε-squares A_l, and each $(\lambda^{-n}\varepsilon, \varepsilon)$-star can be written in the form (2.68) for ε-stars A_l. Hence, there is a finite constant C_2, such that for each natural number n the set X can be covered by no more than $C_2 \cdot (\lambda^n + 1)$ sets of the form (2.68). By the claim, $X \setminus \{z_0\}$ can be covered by no more than $C_2\big(1 + \frac{(n+1)(N(n+1)+1)}{2}\big) \cdot (\lambda^n + 1)$ different non-empty sets of the form (2.69) with each S_l^N replaced by the slightly bigger $(S_l^N)'$. These sets form a subcover of $\mathcal{A}'_{\varepsilon,N} \vee \varphi^{-1}(\mathcal{A}'_{\varepsilon,N}) \vee \ldots \vee \varphi^{-n}(\mathcal{A}'_{\varepsilon,N})$.

We obtain a cover of $\overline{X^{z_0}}$ by considering for each element of $\mathcal{A}'_{\varepsilon,N}$ the respective element of $\mathcal{A}_{\varepsilon,N}(z_0)$. This gives the inequality

$$\begin{aligned} &\mathcal{N}\big(\mathcal{A}_{\varepsilon,N}(z_0) \vee \varphi^{-1}(\mathcal{A}_{\varepsilon,N}(z_0)) \vee \ldots \vee \varphi^{-n}(\mathcal{A}_{\varepsilon,N}(z_0))\big) \\ &\quad \le C_2\Big(1 + \frac{(n+1)N((n+1)N+1)}{2}\Big) \cdot (\lambda^n + 1)\,. \end{aligned}$$

Hence,

$$\begin{aligned} &h(\varphi^{z_0}) \\ &= \limsup \frac{1}{n} \log\Big(\mathcal{N}\big(\mathcal{A}_{\varepsilon,N}(z_0) \vee \varphi^{-1}(\mathcal{A}_{\varepsilon,N}(z_0)) \vee \ldots \vee \varphi^{-n}(\mathcal{A}_{\varepsilon,N}(z_0))\big)\Big) \\ &\le \limsup \frac{1}{n} \log\Big(C_2\big(1 + \frac{N(n+1)(N(n+1)+1)}{2}\big) \cdot (\lambda^n + 1)\Big) = \log\lambda\,. \end{aligned} \tag{2.70}$$

Lemma 2.11 is proved. □

The restriction of the homeomorphism φ^{z_0} to ∂X^{z_0} is not equal to the identity mapping on ∂X^{z_0}. The Lemmas 2.12 and 2.14 below produce a self-homeomorphism of a slightly larger bordered Riemann surface whose restriction to the boundary is the identity mapping.

Lemma 2.12 *For each integer ℓ there exists a self-homeomorphism $\tilde{\varphi}$ of $\{\frac{r}{4} \le |\zeta| \le \frac{r}{2}\} \subset \Delta \subset X$ of entropy zero which equals φ^{z_0} on $\{|\zeta| = \frac{r}{2}\}$ and is equal to multiplication by $e^{\frac{2\pi i \ell}{2(a+2)}}$ on $\{|\zeta| = \frac{r}{4}\}$ for the integer ℓ.*

Recall that $r \exp(\frac{2\pi ji}{2(a+2)})$, $j = 1, \ldots, 2(a+2)$, $0 < r < c$, are the points on the vertical and horizontal bisectrices. Consider the arc $\gamma_j = \left\{\frac{r}{2} e^{i\theta} : \frac{2\pi j}{2(a+2)} \le \theta \le \frac{2\pi(j+1)}{2(a+2)}\right\}$, $j = 0, 1, \ldots, 2(a+2) - 1$. We will obtain Lemma 2.12 from the following lemma.

Lemma 2.13 *Let ψ_j be a self-homeomorphism of γ_j which fixes the endpoints of γ_j pointwise and does not fix any other point. Then there is a self-homeomorphism $\tilde{\psi}_j$ of the truncated sector $\bar{\Omega}_j = \{\rho\, e^{i\theta} : \frac{r}{4} \le \rho \le \frac{r}{2}, \frac{r}{2} e^{i\theta} \in \gamma_j\}$ which equals the identity on the boundary part $\partial\, \Omega_j \setminus \mathrm{Int}\, \gamma_j$ of $\partial\, \Omega_j$, equals ψ_j on γ_j, and has entropy zero.*

Proof We may assume that $j = 0$. Put $\varrho(t) = \frac{1}{2}t + \frac{1}{4}(1-t)$, $t \in [0, 1]$. Then $\varrho(0) = \frac{1}{4}$, $\varrho(1) = \frac{1}{2}$, and ϱ is a linear homeomorphism from $[0, 1]$ onto $[\frac{1}{4}, \frac{1}{2}]$. Put $I \overset{def}{=} [0, \frac{2\pi}{2a+2}]$. The mapping

$$[0, 1] \times I \ni (t, \theta) \overset{g}{\to} \varrho(t) e^{i\theta} \in \overline{\Omega_0}$$

is a homeomorphism. The segment $\{1\} \times I$ is mapped homeomorphically onto γ_0. Put $\Psi = g^{-1} \circ \psi_0 \circ g \mid \{1\} \times I$. The mapping Ψ is a self-homeomorphism of $\{1\} \times I$ that is strictly increasing in the variable $\theta \in I$. We need to find a self-homeomorphism $\tilde{\Psi}$ of $[0, 1] \times I$ such that $\tilde{\Psi} \mid \{1\} \times I = \Psi$, the restriction of $\tilde{\Psi}$ to the rest of the boundary of $[0, 1] \times I$ is the identity, and the entropy of $\tilde{\Psi}$ equals zero. Then the mapping $\tilde{\psi}_0 \overset{def}{=} g \circ \tilde{\Psi} \circ g^{-1}$ satisfies the requirement of the lemma for $j = 0$.

We put

$$\tilde{\Psi}(t, \theta) = (t,\ t\, \Psi(\theta) + (1-t)\,\theta), \quad (t, \theta) \in [0, 1] \times I\,, \tag{2.71}$$

and $\Psi_t(\theta) = t\, \Psi(\theta) + (1-t)\,\theta$. The mapping $\tilde{\Psi}$ fixes the coordinate t, hence, it fixes each maximal vertical segment in $[0, 1] \times I$, and maps the horizontal line segment with endpoints $(0, \theta)$ and $(1, \theta)$ onto the line segment with endpoints $(0, \theta)$ and $(1, \Psi(\theta))$.

We will prove now that the homeomorphism $\tilde{\Psi}$ has entropy zero. We choose open covers of $[0, 1] \times I$ as follows. For a natural number N we consider N vertical line segments in $[0, 1] \times I$ which divide $[0, 1] \times I$ into $N+1$ vertical strips of equal width, and N horizontal line segments in $[0, 1] \times I$ which divide $[0, 1] \times I$ into $N+1$ horizontal strips of equal width. The complement of the lines in $[0, 1] \times I$ consists of $(N+1)^2$ rectangles R_l. Let $\mathcal{A}_N$ be the cover of $[0, 1] \times I$ by open $(N+1)^2$ rectangles each containing the closure of one of the R_l and having diameter close to that of R_l. For a suitable sequence of N's we get a refining sequence of covers of $[0, 1] \times I$. We prove now that for each N the entropy $h(\tilde{\Psi}, \mathcal{A}_N)$ is equal to zero.

The mapping Ψ has no fixed point on the interior $\mathrm{Int}I$ of I. Hence, either $\Psi(\theta) > \theta$ for each $\theta \in \mathrm{Int}I$, or $\Psi(\theta) < \theta$ for each θ in $\mathrm{Int}I$. We assume that the first case holds. (In the second case we conjugate the mapping by $(t, \theta) \to (t, -\theta)$.) Then for each $\theta \in \mathrm{Int}I$ the mapping $t \to \Psi_t(\theta)$, $t \in I$, is strictly increasing. Since Ψ is a strictly increasing function on $\{1\} \times I$, also each the mapping $\theta \to \Psi_t(\theta)$, $\theta \in I$, is strictly increasing for each $t \in [0, 1]$.

Let Ψ_t^k be the k-th iterate of Ψ_t (considered as function of θ for fixed parameter t). Note that each iterate Ψ_t^k is strictly increasing in θ. With $\tilde{\Psi}(t, \theta) = (t, \Psi_t(\theta))$, for the k-th iterate $(\tilde{\Psi})^k$ we have by induction

$$\tilde{\Psi}^k\big(t, \theta\big) = \tilde{\Psi}\big(t, \Psi_t^{k-1}(\theta)\big) = \big(t, \Psi_t^k(\theta)\big) .$$

We claim that for each $k \geq 1$ the mapping $t \to \Psi_t^k(\theta))$ is strictly increasing for each θ in the interior of I. This is clear for the linear mapping $t \to \Psi_t(\theta)$ for $\theta \in \mathrm{Int}I$. By induction the claim for $k + 1$ is obtained as follows.

Suppose we know that for all $0 < t_1 < t_2 < 1$ and $\theta \in \mathrm{Int}I$ the inequality $\Psi_{t_2}^k(\theta) > \Psi_{t_1}^k(\theta)$ holds. We have for all $t \in [0, 1]$

$$\Psi_t^{k+1}(\theta) = t\,\Psi\big(\Psi_t^k(\theta)\big) + (1-t)\,\Psi_t^k(\theta) . \tag{2.72}$$

Since Ψ strictly increases and the mapping $t \to \Psi_t^k(\theta)$ strictly increases, we obtain

$$t\,\Psi\big(\Psi_{t_1}^k(\theta))\big) + (1-t)\,(\Psi_{t_1})^k(\theta) < t\,\Psi\big(\Psi_{t_2}^k(\theta)\big) + (1-t)\,\Psi_{t_2}^k(\theta) \tag{2.73}$$

for each $t \in [0, 1]$, $\theta \in \mathrm{Int}I$, and $t_1, t_2 \in [0, 1]$, $t_1 < t_2$. Since $\Psi(\theta') > \theta'$ for each θ' in the interior of I, we obtain the inequality

$$t_1\,\Psi\big(\Psi_{t_2}^k(\theta)\big) + (1-t_1)\,\Psi_{t_2}^k(\theta) < t_2\,\Psi\big(\Psi_{t_2}^k(\theta)\big) + (1-t_2)\,\Psi_{t_2}^k(\theta) . \tag{2.74}$$

Combining the last inequality (2.74) with inequality (2.73) for $t = t_1$ and equality (2.72) we obtain the claim.

The claim implies that for each natural number k and for each θ in the interior of I the curve

$$\mathcal{C}_k^\theta \overset{def}{=} \{\big(t, \Psi_t^k(\theta)\big), t \in [0, 1]\} = \tilde{\Psi}^k\big([0, 1] \times \{\theta\}\big) \tag{2.75}$$

intersects each vertical line and each horizontal line in $[0, 1] \times I$ at most once. Since $\mathcal{C}_{k+l}^\theta = \tilde{\Psi}^k(\mathcal{C}_l^\theta)$ the intersection of two curves of form (2.75) consists of at most one point. The intersection behaviour of the union of the set of vertical and horizontal lines and the set curves $\mathcal{C}_l^\theta$ is the same as the intersection behaviour of straight lines.

A similar argument as in the proof of Lemma 2.11 proves that the entropy $h(\tilde{\Psi}, \mathcal{A}_N)$ equals zero for each N. Hence, $h(\tilde{\Psi}) = 0$. Lemma 2.13 is proved. □

The following lemma states in particular that for a bordered Riemann surface $\overline{X}$ one can change a mapping class $\mathfrak{m} \in \mathfrak{M}(\overline{X}, \partial X)$ without increasing entropy by a product of Dehn twists about circles that are homologous to boundary curves.

Lemma 2.14 *Let $\hat{\varphi}$ be a twist on an annulus $\{\frac{r}{8} \le |\zeta| \le \frac{r}{4}\}$, more precisely, for some real constant ν we have $\hat{\varphi}(\zeta) = \zeta \cdot e^{i\nu\cdot(\log|\zeta| - \log\frac{r}{8})}$. Then $h(\hat{\varphi}) = 0$.*

Proof Denote by A the annulus $A = \left\{\frac{r}{8} \le |\zeta| \le \frac{r}{4}\right\}$. Let N be a natural number. Divide A by $N-1$ circles $\mathrm{Circ}_j = \left\{|\zeta| = \frac{r}{8}\cdot(1+\frac{j}{N})\right\}$, $j = 1, \ldots, N-1$, and N radii $\mathrm{Rad}_j = \left\{\rho \cdot e^{\frac{2\pi i}{N}j} : \frac{r}{8} \le \rho \le \frac{r}{4}\right\}$, $j = 1, \ldots, N$ into a collection $\mathcal{Q}^N$ of connected open sets Q_l^N. Locally the circles and the radii are straight line segments in logarithmic coordinates, and the Q_l^N are open rectangles in these coordinates. Similarly as in the proof of Lemma 2.11 we take for each l an open rectangle $(Q_l^N)'$ in local logarithmic coordinates that contains the closure $\overline{Q_l^N}$ and has diameter not exceeding twice the diameter of Q_l^N. We obtain an open cover $\mathcal{A}_N$ of the annulus A. For an increasing sequence N_k of natural numbers with N_{k+1} being an integer multiple of N_k and a suitable choice of the $(Q_l^{N_k})'$ we obtain a refining sequence of open covers.

For each N the annulus can be covered by the collection of sets $(Q_0^N)' \cap \ldots \cap \hat{\varphi}^{-n}((Q_n^N)')$ for which the sets $Q_0^N \cap \ldots \cap \hat{\varphi}^{-n}(Q_n^N)$ ($Q_l^N \in \mathcal{Q}^N$) are not empty. These sets $(Q_0)'^N \cap \ldots \cap \hat{\varphi}^{-n}((Q_n^N)')$ are elements of a subcover of the cover $\mathcal{A}_N \vee \ldots \vee \hat{\varphi}^{-n}(\mathcal{A}_N)$. We will estimate the number of these sets for each N and n.

Each iterate $\hat{\varphi}^{-\ell}$, $\ell = 1, \ldots, n$, is real linear in local logarithmic coordinates. It maps each circle Circ_j, $j = 1, \ldots, N-1$, onto itself and each radius Rad_j, $j = 1, \ldots N$, onto the curve

$$\hat{\varphi}^{-\ell}(\mathrm{Rad}_j) = \left\{e^x \cdot e^{i\frac{2\pi}{N}j + i\cdot\ell\cdot\nu(x-\log\frac{r}{8})} : \log\frac{r}{8} \le x \le \log\frac{r}{4}\right\}. \tag{2.76}$$

Each curve (2.76) intersects each given set $Q_l^N \in \mathcal{Q}^N$ along at most $c(\nu, N)\cdot\ell$ connected components, each a straight line segment in logarithmic coordinates on Q_l^N. Hence, as in the proof of Lemma 2.11 the union of the $\hat{\varphi}^{-\ell}(\mathrm{Rad}_j)$, $j = 1, \ldots, N$, $\ell = 1, \ldots, n$, intersects the given set Q_l^N along at most $c(\nu, N)(N + \ldots + Nn) \le c'(\nu, N)n^2$ straight line segments in logarithmic coordinates. They divide Q_l^N into at most $C'(\nu, N)n^4$ connected components. Hence, the number of non-empty sets of the form $Q_0^N \cap \ldots \hat{\varphi}^{-n}(Q_n^N)$ for sets $Q_l^N \in \mathcal{Q}^N$ does not exceed $C(\nu, N)n^4$. Here $c(\nu, N)$, $c'(\nu, N)$, $C'(\nu, N)$, and $C(\nu, N)$ are positive constants depending only on ν and N.

Hence, $\mathcal{N}(\mathcal{A}_N \vee \ldots \vee \hat{\varphi}^{-n}(\mathcal{A}_N)) \le C(\nu, N)n^4$. This implies that the entropy of $\hat{\varphi}$ with respect to the cover $\mathcal{A}_N$ does not exceed $\overline{\lim}_{n\to\infty}\frac{1}{n}\log(C(\nu, N)\cdot n^4) = 0$.

The estimate is obtained for each element of a refining sequence of open covers of the annulus. Hence $h(\hat{\varphi}) = 0$. $\square$

Proof of Lemma 2.12. Let ℓ be the integer from equality (2.63) and let k be the smallest positive integer for which $k \cdot \ell$ is a multiple of $2(a+2)$. Then φ^{z_0} permutes the γ_j in cycles of length k and $(\varphi^{z_0})^k$ fixes each γ_j setwise. In particular, for each j the mapping $(\varphi^{z_0})^k \mid \gamma_j$ satisfies the conditions stated for the function ψ_j of Lemma 2.13. Let $\gamma_{\ell_1} \xrightarrow{\varphi^{z_0}} \gamma_{\ell_2} \longrightarrow \dots \longrightarrow \gamma_{\ell_k} \xrightarrow{\varphi^{z_0}} \gamma_{\ell_1}$ be one of the cycles for φ^{z_0}. For each ℓ_j, $j = 1, \dots, k-1$, we take any homeomorphism $\tilde{\varphi}_{\ell_j}$: $\bar{\Omega}_{\ell_j} \to \bar{\Omega}_{\ell_{j+1}}$ which equals $\varphi^{z_0} \mid \gamma_{\ell_j}$ on γ_{ℓ_j} and equals multiplication by $e^{\frac{2\pi i \ell}{2(a+2)}}$ on the rest of the boundary of $\bar{\Omega}_{\ell_j}$. Let ψ_{ℓ_1} be the self-homeomorphism of $\bar{\Omega}_{\ell_1}$ obtained by applying Lemma 2.13 to $(\varphi^{z_0})^k \mid \gamma_{\ell_1}$. We put $\tilde{\varphi}_{\ell_k} = \psi_{\ell_1} \circ (\tilde{\varphi}_{\ell_{k-1}} \circ \dots \circ \tilde{\varphi}_{\ell_1})^{-1}$. Then on γ_{ℓ_k} the equality $\tilde{\varphi}_{\ell_k} = (\varphi^{z_0})^k \circ (\varphi^{z_0})^{-k+1} = \varphi^{z_0}$ holds. Denote by $\tilde{\varphi}_{cyc}$ the mapping that is defined by $\tilde{\varphi}_{\ell_j}$ on $\bar{\Omega}_{\ell_j}$ for $j = 1, \dots, k$.

The mapping that equals $\psi_{\ell_1} = \tilde{\varphi}_{\ell_k} \circ \dots \circ \tilde{\varphi}_{\ell_2} \circ \tilde{\varphi}_{\ell_1}$ on $\bar{\Omega}_{\ell_1}$ and $\tilde{\varphi}_{\ell_{j-1}} \circ \dots \circ \tilde{\varphi}_{\ell_1} \circ \tilde{\varphi}_{\ell_k} \circ \dots \circ \tilde{\varphi}_{\ell_{j+1}} \circ \tilde{\varphi}_{\ell_j}$ on $\bar{\Omega}_{\ell_j}$ has entropy zero, since by Lemma 2.14 $h(\psi_{\ell_1}) = 0$ and the mapping on $\bar{\Omega}_{\ell_j}$ is conjugate to ψ_{ℓ_1}. Moreover, the mapping is equal to the k-th iterate $\tilde{\varphi}_{cyc}$. Hence, $h(\tilde{\varphi}_{cyc}) = 0$.

Proceed in the same way with all cycles of γ_j under iteration by φ^{z_0}.

The obtained mappings $\tilde{\varphi}_j$ map $\bar{\Omega}_j$ homeomorphically onto $e^{\frac{2\pi i \ell}{2(a+2)}} \bar{\Omega}_j$. They match together to give a well-defined self-homeomorphism $\tilde{\varphi}$ of the annulus $\left\{\frac{r}{4} \le |\zeta| \le \frac{r}{2}\right\}$ which equals φ^{z_0} on $\left\{|\zeta| = \frac{r}{2}\right\}$ and equals rotation by $e^{\frac{2\pi i \ell}{2(a+2)}}$ on $\left\{|\zeta| = \frac{r}{4}\right\}$. The ψ_j match together to give a self-homeomorphism ψ of the annulus such that $\tilde{\varphi}^k = \psi$. Since $h(\psi) = 0$ we have $h(\tilde{\varphi}) = 0$. $\square$

Proof of Theorem 2.9. Put $\varphi_0 = \varphi^{z_0}$ on $\overline{X^{z_0}}$, take $r = c$, and put φ_0 equal to the mapping $\tilde{\varphi}$ of Lemma 2.12 on the annulus $\left\{\frac{c}{4} \le |\zeta| \le \frac{c}{2}\right\}$ with distinguished coordinates ζ around z_0. Let φ_0 be equal to the homeomorphism $\hat{\varphi}$ of Lemma 2.14 with $\nu = -\frac{2\pi \ell}{2(a+2) \cdot \log 2}$ on the annulus $\left\{\frac{c}{8} \le |\zeta| \le \frac{c}{4}\right\}$ and equal to the identity on $\left\{|\zeta| \le \frac{c}{8}\right\}$. Then φ_0 is isotopic to φ through self-homeomorphisms of X that fix E, φ_0 fixes the disc $\Delta' \overset{def}{=} \left\{|\zeta| \le \frac{c}{8}\right\}$ around z_0 and has the same entropy as φ. It follows that the isotopy classes of the two mappings φ and φ_0 differ by a Dehn twist about a curve that is homologous to $\partial \Delta'$. By Lemma 2.14 the homeomorphism φ_0 can be chosen to represent any a priory given preimage of the class of φ in $\mathfrak{M}(X; \{z_0\}, E \setminus \{z_0\})$ under the mapping $\mathcal{H}_{z_0} : \mathfrak{M}(X \setminus \Delta'; \partial \Delta', E \setminus \{z_0\}) \to \mathfrak{M}(X; \{z_0\}, E \setminus \{z_0\})$. Theorem 2.9 is proved. $\square$

The following theorem concerns the slightly more general case when the self-homeomorphism φ of a Riemann surface of first kind is changed without increasing entropy on the union of discs around points of a subset of the set of distinguished points (rather than on a single disc around a fixed distinguished point).

Theorem 2.10 *Let X be a connected closed Riemann surface with a set E of distinguished points. Assume that $X \backslash E$ is hyperbolic. Let φ be a non-periodic* *absolutely extremal self-homeomorphism of X which fixes E setwise. Suppose there is a φ-invariant subset $E' \subset E$.*

Then φ is isotopic through self-homeomorphisms which fix E setwise to a self-homeomorphism φ_0 of the same entropy $h(\varphi_0) = h(\varphi)$ with the following property.

For each $z \in E'$ there is a closed round disc $\overline{\delta}_z$ in distinguished coordinates for the quadratic differential of φ, such that z is the center of δ_z, and φ_0 maps each δ_z, $z \in E'$, conformally onto another disc of the collection and maps the center of the source disc to the center of the target disc. Moreover, if a φ-cycle of points in E' has length k, then the iterate φ_0^k fixes pointwise the disc δ_z around each point z of the cycle.

Denote by X' the bordered Riemann surface which is the complement of the union of the open discs $\cup_{z \in E'} \delta_z$. The isotopy class of the restriction $\varphi_0 | X'$ of the mapping φ_0 of Theorem 2.10 is determined by φ up to a product of powers of Dehn twists around simple closed curves that are homologous to the boundary circles of X'. The mapping φ_0 of Theorem 2.10 can be chosen so that the restriction $\varphi_0|X'$ represents the class corresponding to an a priory chosen product of powers of Dehn twists about the boundary circles. The proof of Theorem 2.10 follows along the same lines as the proof of Theorem 2.9 and is left to the reader.

We have the following corollaries. We formulate Corollary 2.2 concerning mapping classes of braids separately because it is simple and useful, although it is a particular case of Corollary 2.3. We call a braid b irreducible if the associated mapping class $\mathfrak{m}_{b,\infty}$ is irreducible.

Corollary 2.2 *Let $b \in \mathcal{B}_n$ be an irreducible braid and let $\mathfrak{m}_b = \Theta_n(b) \in \mathfrak{M}(\mathbb{D}; \partial\,\mathbb{D}, E_n)$ be its mapping class. Then*

$$h(\widehat{\mathfrak{m}_b}) = h(\widehat{\mathfrak{m}_b \cdot \mathfrak{m}_{\Delta_n^{2k}}}) = h(\widehat{\mathfrak{m}_{b,\infty}})$$

for each integer k.

Recall that $\mathfrak{m}_{b,\infty} = \mathcal{H}_\infty(\mathfrak{m}_b)$ (see Sect. 1.5), Δ_n is the Garside element in $\mathcal{B}_n$ and by $\widehat{\mathfrak{m}}$ we denote the conjugacy class of a mapping class $\mathfrak{m}$.

Proof Identify the set of elements of $\mathfrak{m}_b$ with the set of elements $\mathfrak{m}_b' \subset \mathfrak{m}_{b,\infty}$ which are equal to the identity outside the unit disc $\overline{\mathbb{D}}$. The entropy of a class is the infimum of entropies of mappings in the class. We obtain the inequality $h(\widehat{\mathfrak{m}_b}) = h(\widehat{\mathfrak{m}_b'}) \geq h(\widehat{\mathfrak{m}_{b,\infty}})$.

On the other hand Theorem 2.9 assigns to each absolutely extremal representative of $\widehat{\mathfrak{m}_{b,\infty}}$ a representative of $\widehat{\mathfrak{m}_{b,\infty}}$ which has the same entropy and equals the identity in a neighbourhood of infinity and, thus represents $\widehat{\mathfrak{m}_b'}$. Hence $h(\widehat{\mathfrak{m}_{b,\infty}}) \geq h(\widehat{\mathfrak{m}_b'}) = h(\widehat{\mathfrak{m}_b})$. The first equality of the statement of the corollary is Lemma 2.14 and the fact that the mapping class corresponding to Δ_n^2 is the Dehn twist about a curve that is homologous in $\mathbb{D} \setminus E_n$ to $\partial\mathbb{D}$. □

Corollary 2.3 *Let X be a bordered Riemann surface, and let E be a finite set of distinguished points in* Int *X. Let $\mathfrak{m}$ be an irreducible relative isotopy class of mappings, $\mathfrak{m} \in \mathfrak{M}(X; \partial X, E)$. Then for the mapping $\mathcal{H}_\zeta$ defined by Eq.* (1.9) *in*

Sect. 1.6 the following equalities hold

$$h(\widehat{\mathfrak{m}}) = h(\widehat{\mathfrak{m} \cdot \mathfrak{m}_D}) = h(\widehat{\mathcal{H}_\zeta \mathfrak{m}})$$

where $\mathfrak{m}_D$ *is the mapping class of an arbitrary product of powers of Dehn twists about simple closed curves which are free homotopic to boundary curves of* X.

The proof follows along the same lines as the proof of Corollary 2.2. It relies on Theorem 2.10.

Chapter 3
Conformal Invariants of Homotopy Classes of Curves. The Main Theorem

In this chapter we define the entropy and the conformal module of conjugacy classes of braids. We formulate our Main Theorem, stating that the entropy of each conjugacy class of braids is inversely proportional with factor $\frac{\pi}{2}$ to its conformal module. Here we prove this theorem in the case of irreducible braids. The general case will be proved in the next two chapters.

The Main Theorem allows to apply methods and known results concerning the more intensively studied entropy to problems whose solutions are based on the concept of the conformal module. Vice versa, for instance methods of quasi-conformal mappings related to the concept of extremal length and conformal module can be applied to give upper and lower bounds for the entropy of 3-braids, the bounds differing by universal multiplicative constants (see Chap. 9).

We also define the conformal module of conjugacy classes of elements of the fundamental group of complex manifolds, in particular, of the twice punctured complex plane. A version of the definition for elements of fundamental groups, not merely of their conjugacy classes, appears more effective for applications to quantitative versions of the Gromov-Oka Principle that will be given in later Chapters. In this chapter we compute these quantities explicitly in a number of examples.

3.1 Ahlfors' Extremal Length and Conformal Module

Ahlfors defined the extremal length of a family of curves in the complex plane as follows. Let Γ be a family each member of which consists of the union of no more than countably many connected locally rectifiable (open, half-open, or closed) arcs or connected closed curves (loops) in the complex plane. (We do not require that this union reparametrizes to a single connected curve.) In this context we will call also the elements of Γ "curves". Ahlfors defined the extremal length of the family

B. Jöricke, *Braids, Conformal Module, Entropy, and Gromov's Oka Principle*, Lecture Notes in Mathematics 2360, https://doi.org/10.1007/978-3-031-67288-0_3

Γ as follows. For a non-negative measurable function ϱ in the complex plane he defines $A(\varrho) = \iint_{\mathbb{C}} \varrho^2$. For an element $\gamma \in \Gamma$ and such a function ϱ he puts $L_\gamma(\varrho) = \int_\gamma \varrho |dz|$, if ϱ is measurable on γ with respect to arc length and $L_\gamma(\varrho) = \infty$ otherwise. Put $L(\varrho) = \inf_{\gamma \in \Gamma} L_\gamma(\varrho)$. The extremal length of the family Γ is the following value

$$\lambda(\Gamma) = \sup_{\varrho} \frac{L(\varrho)^2}{A(\varrho)},$$

where the supremum is taken over all non-negative measurable functions ϱ for which $A(\varrho)$ is finite and does not vanish.

It is not hard to see from this definition that the extremal length is invariant under conformal mappings (Theorem 3 in [4], Chapter I.D).

The conformal module $\mathcal{M}(\Gamma)$ of the family Γ is defined to be

$$\mathcal{M}(\Gamma) = \frac{1}{\lambda(\Gamma)}.$$

Example 3.1 *Let R be an open rectangle in the complex plane $\mathbb{C}$. Unless said otherwise the considered rectangles will always have sides parallel to the coordinate axes. Denote the length of the horizontal sides of R by b and the length of the vertical sides by a. (For instance, we may consider $R = \{z = x + iy : 0 < x < \mathsf{b},\ 0 < y < \mathsf{a}\}$.) The extremal length of the rectangle is defined to be the extremal length of the family of connected open arcs in the rectangle that join the two horizontal sides. An open arc is said to join the two horizontal sides if its limit sets at the two endpoints of the arc are contained in the closure of different horizontal sides. A small computation shows that the extremal length of $R = \{z = x + iy : 0 < x < \mathsf{b},\ 0 < y < \mathsf{a}\}$ equals $\lambda(R) = \frac{\mathsf{a}}{\mathsf{b}}$ and the conformal module equals $m(R) = \frac{\mathsf{b}}{\mathsf{a}}$ (see Example 1 in [4], Chapter I.D).*

For a conformal mapping $\omega : R \to U$ of the rectangle R onto a domain $U \subset \mathbb{C}$ the image U is called a curvilinear rectangle, if ω extends to a continuous mapping on the closure $\bar{R}$, and the restriction to each (open) side of R is a homeomorphism onto its image. The images of the vertical (horizontal, respectively) sides of R are called the vertical (horizontal, respectively) curvilinear sides of the curvilinear rectangle $\omega(R)$. The extremal length of the curvilinear rectangle U is the extremal length of the family of open arcs in U that join the two horizontal curvilinear sides. Thus the extremal length of U equals the extremal length of R. (See [4], Chapter I.D).

Example 3.2 *Ahlfors [4], Chapter I.D, defined the extremal length of an annulus $A = \{z \in \mathbb{C} : r_1 < |z| < r_2\}$ in the complex plane as the extremal length of the family of closed curves which are contained in the annulus and represent the conjugacy class of the positively oriented generator of the fundamental group of the annulus. A simple computation (Example 3 in [4], Chapter I.D) shows that the* *extremal length of an annulus* $A = \{z \in \mathbb{C} : r_1 < |z| < r_2\}$ *is equal to* $\lambda(A) = \frac{2\pi}{\log(\frac{r_2}{r_1})}$, *and the conformal module of this annulus equals* $m(A) = \frac{1}{2\pi}\log(\frac{r_2}{r_1})$. *Two annuli of finite conformal module are conformally equivalent iff they have equal conformal module. If a manifold Ω is conformally equivalent to an annulus A in the complex plane, its conformal module is defined to be $m(A)$. Recall that any domain in the complex plane with fundamental group isomorphic to the group of integer numbers $\mathbb{Z}$ is conformally equivalent to an annulus.*

Example 2 in [4], Chapter I.D, shows that the extremal length of the family of open arcs γ in $A = \{z \in \mathbb{C} : r_1 < |z| < r_2\}$ that join the two boundary circles is equal to $\frac{\log \frac{r_2}{r_1}}{2\pi}$.

Example 3.3 *The following example is a generalization of Example 3.1. Let Φ be a real C^1-function Φ on the closure of an open interval J, and let* b *be a positive number. We consider the curvilinear rectangle $R_{J,\Phi,\mathrm{b}} \overset{def}{=} \{x + iy \in \mathbb{C} : y \in J,\ x \in (\Phi(y), \Phi(y) + \mathrm{b})\}$ whose horizontal sides are the horizontal pieces of its boundary. The following lemma holds.*

Lemma 3.1 *Let Φ be a real C^1-function on the closure of an open interval J and let* b *be a positive number. Denote by Γ_Φ the set of curves in the curvilinear rectangle $R_{J,\Phi,\mathrm{b}} = \{x+iy \in \mathbb{C} : y \in J,\ x \in (\Phi(y), \Phi(y)+\mathrm{b})\}$ which join the two horizontal curvilinear sides. Suppose the absolute value $|\Phi'|$ of the derivative of Φ is bounded by the constant C. Then*

$$\lambda(\Gamma_\Phi) \le (1 + C^2)\lambda(\Gamma_0),$$

where Γ_0 is the family corresponding to the function Φ_0 which is identically equal to zero.

Proof The proof is similar to the arguments used in Example 1 in Chapter 1 of [4], Chapter I.D. For any measurable function ϱ on $\mathbb{C}$ and any $x \in (0, \mathrm{b})$ we have

$$\int_J \varrho(x + \Phi(y) + iy)\sqrt{1 + \Phi'(y)^2}dy \ge L_{\Gamma_\Phi}(\varrho).$$

Integrate over the interval $(0, \mathrm{b})$ and apply Fubini's Theorem and Hölder's inequality. Using the bound for $|\Phi'|$ we obtain

$$\Big(\int\int_{R_{\Phi,\mathrm{b}}} dm_2 \int\int_{R_{\Phi,\mathrm{b}}} \varrho^2 \cdot (1 + C^2)dm_2\Big)^{\frac{1}{2}} \ge \mathrm{b}L_{\Gamma_\Phi}(\varrho).$$

Denote by $|J|$ the length of the interval J. We obtain

$$\mathrm{b}|J|(1 + C^2)A(\varrho) \ge \mathrm{b}^2 L_{\Gamma_\Phi}(\varrho)^2.$$

Hence,

$$\frac{L_{\Gamma_\Phi}(\varrho)^2}{A(\varrho)} \le \frac{|J|}{\mathrm{b}} \cdot (1 + C^2) = \lambda(\Gamma_0)(1 + C^2).$$

Taking the supremum over all measurable functions ϱ with finite non-vanishing integral we obtain

$$\lambda(\Gamma_\Phi) \le (1 + C^2)\lambda(\Gamma_0).$$

The lemma is proved. □

For later use we formulate three theorems of Ahlfors.

For two families Γ_1 and Γ_2 as above the following relation is introduced by Ahlfors: $\Gamma_1 < \Gamma_2$ if each "curve" $\gamma_2 \in \Gamma_2$ contains a "curve" $\gamma_1 \in \Gamma_1$.

Suppose Γ_1 and Γ_2 are contained in disjoint measurable sets. Ahlfors defines the sum $\Gamma_1 + \Gamma_2$ of two such families as follows. Consider a "curve" $\gamma_1 \in \Gamma_1$ and a "curve" $\gamma_2 \in \Gamma_2$. The sum $\gamma_1 + \gamma_2$ is the " curve" consisting of the union of γ_1 and γ_2. The set $\Gamma_1 + \Gamma_2$ is the set consisting of all sums $\gamma_1 + \gamma_2$ for $\gamma_1 \in \Gamma_1$ and $\gamma_2 \in \Gamma_2$.

The following theorems were proved by Ahlfors.

Theorem A ([4], Ch.1.D Theorem 2) *If* $\Gamma' < \Gamma$ *then* $\lambda(\Gamma') < \lambda(\Gamma)$.

Theorem B ([4], Ch.1.D Theorem 4) *If the families* Γ_j *are contained in disjoint measurable sets then* $\sum \lambda(\Gamma_j) \le \lambda(\sum \Gamma_j)$.

Corollary 3.1 *Let* R *be a rectangle and let* α *be an open arc in* R*, that extends to a closed arc with endpoints on different open vertical sides of* R*. The arc* α *cuts* R *into two curvilinear rectangles* R_1 *and* R_2*. The following estimate holds:*

$$\lambda(R) \ge \lambda(R_1) + \lambda(R_2)\,. \tag{3.1}$$

Further, let $A = \{z \in \mathbb{C} : r_1 < |z| < r_2\}$ *be an annulus in the complex plane and* α *an open arc in* A *that extends to a closed arc in* $\bar{A}$ *with endpoints on different boundary circles. The arc cuts* A *into a curvilinear rectangle* R*. The following inequality holds*

$$\lambda(A) \ge \lambda(R)\,. \tag{3.2}$$

Proof To prove inequality (3.1) we let R_1 be the curvilinear rectangle below of α and R_2 the curvilinear rectangle above of α. Let Γ_j, $j = 1, 2$, be the family of open arcs in R_j that join the two open horizontal curvilinear sides of R_j, and let Γ be the family of open arcs in R that join the two open horizontal sides of R. Then by Theorem A (see [4], Ch.1 Theorem 2) $\lambda(\Gamma) \ge \lambda(\Gamma_1 + \Gamma_2)$. By Theorem B (see [4], Ch.1 Theorem 4) $\lambda(\Gamma_1 + \Gamma_2) \ge \lambda(\Gamma_1) + \lambda(\Gamma_2)$. By Example 3.1 of Sect. 3.1 the equalities $\lambda(\Gamma) = \lambda(R)$, $\lambda(\Gamma_j) = \lambda(R_j)$, $j = 1, 2$ hold.

For the proof of inequality (3.2) we let $\Gamma(A)$ be the family of curves in A that represent the positively oriented generator of the fundamental group of A. By Example 3.2 of Sect. 3.1 the extremal length $\lambda(A)$ is equal to the extremal length $\lambda(\Gamma(A))$. Further, the set $A \setminus \alpha$ is a curvilinear rectangle with horizontal sides being the strands of α that are reachable from $A \setminus \alpha$ by moving clockwise or counterclockwise, respectively. Its extremal length $\lambda(A \setminus \alpha)$ is the extremal length $\lambda(\Gamma(A \setminus \alpha))$ in the sense of Ahlfors [4] of the family $\Gamma(A \setminus \alpha)$ of curves in the curvilinear rectangle $A \setminus \alpha$ that join the two horizontal sides of the curvilinear rectangle. By Theorem A (see [4], Ch.1 Theorem 2) the inequality

$$\lambda(\Gamma(A \setminus \alpha)) \leq \lambda(\Gamma(A)) \tag{3.3}$$

holds. □

3.2 Another Notion of Extremal Length

We will use another notion of the extremal length of families of curves. This notion is in the spirit of the definition of the Kobayashi metric and applies also to arbitrary families of curves, that are homotopy classes of curves contained in complex manifolds, maybe, of dimension bigger than one.

Definition 3.1 *Let $\mathcal{X}$ be a complex manifold, X a domain in $\mathcal{X}$, and let $\mathcal{E}_1$ and $\mathcal{E}_2$ be relatively closed subsets of X. Let $\mathsf{h} = _{\mathcal{E}_1}\mathsf{h}_{\mathcal{E}_2}$ be a homotopy class of curves in X with initial point in $\mathcal{E}_1$ and terminal point in $\mathcal{E}_2$. If $\mathcal{E}_1 = \mathcal{E}_2$ we write $\mathsf{h}_{\mathcal{E}_1}$ instead of $\mathsf{h} = _{\mathcal{E}_1}\mathsf{h}_{\mathcal{E}_2}$. A continuous mapping f from an open rectangle into X which admits a continuous extension to the closure of the rectangle (denoted again by f) is said to represent h if the lower open horizontal side is mapped to $\mathcal{E}_1$, the upper horizontal side is mapped to $\mathcal{E}_2$ and the restriction of f to the closure of each maximal vertical segment in the rectangle represents h.*

The extremal length of homotopy classes of such curves is defined as follows.

Definition 3.2 *For a complex manifold $\mathcal{X}$, a domain X in $\mathcal{X}$, two relatively closed connected subsets $\mathcal{E}_1$ and $\mathcal{E}_2$ of X and a homotopy class $\mathsf{h} = _{\mathcal{E}_1}\mathsf{h}_{\mathcal{E}_2}$ of curves in X with initial point in $\mathcal{E}_1$ and terminal point in $\mathcal{E}_2$ the extremal length $\Lambda(\mathsf{h})$ is defined as*

$$\Lambda(\mathsf{h}) = \inf\{\lambda(R) : R \text{ a rectangle which admits a holomorphic mapping to } X \text{ that represents } \mathsf{h}\}. \tag{3.4}$$

The conformal module $\mathcal{M}(\mathsf{h})$ of the class h is defined as

$$\mathcal{M}(\mathsf{h}) = \frac{1}{\Lambda(\mathsf{h})} = \sup\{m(R) : R \text{ a rectangle which admits a holomorphic mapping to } X \text{ that represents } \mathsf{h}\}. \tag{3.5}$$

A similar definition can be given for free homotopy classes of curves in a complex manifold, in other words for curves representing a conjugacy class of elements of the fundamental group of the manifold.

Definition 3.3 *Let X be a complex manifold, and $\hat{e}$ a conjugacy class of elements of the fundamental group of X. A continuous mapping from an annulus $A = \{z \in \mathbb{C} : \frac{1}{r} < |z| < r\}$ with $1 < r \leq \infty$ into X is said to represent $\hat{e}$, if the restriction to the circle $\{|z| = 1\}$ with positive orientation (and, hence, the restriction to each curve that is free homotopic to this circle) represents $\hat{e}$.*

The extremal length and the conformal module of free homotopy classes of curves is defined as follows.

Definition 3.4 *Let X be a complex manifold, and $\hat{e}$ a conjugacy class of elements of the fundamental group of X. The extremal length $\Lambda(\hat{e})$ of $\hat{e}$ is defined as*

$$\Lambda(\hat{e}) \stackrel{def}{=} \inf\{\lambda(A) : A \textit{ an annulus that admits a holomorphic mapping to } X \textit{ that represents } \hat{e}\}\,. \tag{3.6}$$

The conformal module $\mathcal{M}(\hat{e})$ of $\hat{e}$ is defined as

$$\mathcal{M}(\hat{e}) \stackrel{def}{=} \frac{1}{\Lambda(\hat{e})} = \sup\{m(A) : A \textit{ an annulus that admits a holomorphic mapping to } X \textit{ that represents } \hat{e}\}\,. \tag{3.7}$$

For a curve γ we consider the curve γ^{-1} obtained from γ by inverting orientation. For a family of curves Γ we let Γ^{-1} be the family of curves γ^{-1} with $\gamma \in \Gamma$. It follows immediately from the definitions that $\Lambda(\mathsf{h}) = \Lambda(\mathsf{h}^{-1})$ and $\Lambda(\hat{e}) = \Lambda(\widehat{e^{-1}})$. Indeed, suppose a mapping $f : R \to X$ represents h. Put $-R = \{-z : z \in R\}$ and $f_-(z) = f(-z)$. Then $f_- : -R \to X$ represents h^{-1}. The argument for $\hat{e}$ is similar.

We need the following two lemmas.

Lemma 3.2 *Let R and R' be rectangles with sides parallel to the axes. Suppose S' is the vertical strip bounded by the two vertical lines which are prolongations of the vertical sides of the rectangle R'. Let $f : R \to S'$ be a holomorphic mapping with continuous extension to the closure that takes the two horizontal sides of R into different horizontal sides of R'. Then*

$$\lambda(R) \geq \lambda(R')\,.$$

Equality holds if and only if the mapping is a surjective conformal mapping from R onto R'.

The following lemma concerns holomorphic mappings between annuli.

Lemma 3.3 *Let A and A' be two annuli and let f be a holomorphic mapping from A into A' which induces an isomorphism on fundamental groups. Then*

$$\lambda(A) \geq \lambda(A').$$

Equality holds if and only if the mapping is a conformal mapping from A onto A'.

Proof of Lemma 3.2 Normalize the rectangles and the mapping so that $R = \{x + iy : x \in (0,1), y \in (0, \mathsf{a})\}$ and $R' = \{x + iy : x \in (0,1), y \in (0, \mathsf{a}')\}$. Denote the continuous extension of f to the closure of R again by f.

We may assume that f maps the upper side of R to the upper side of R' and the lower side of R to the lower side of R'. Put $u = \operatorname{Re} f$ and $v = \operatorname{Im} f$. Then

$$\begin{aligned} \mathsf{a}' &= \int_0^1 \mathsf{a}'\,dx = \int_0^1 (v(x,\mathsf{a}) - v(x,0))dx = \int_0^1 dx \int_0^{\mathsf{a}} dy \frac{\partial}{\partial y} v(x,y) \\ &= \int_0^{\mathsf{a}} dy \int_0^1 dx\,\frac{\partial}{\partial x} u(x,y) = \int_0^{\mathsf{a}} dy\,(u(1,y) - u(0,y)) \\ &\leq \int_0^{\mathsf{a}} 1\,dy = \mathsf{a}. \end{aligned} \tag{3.8}$$

We used the Cauchy-Riemann equations. To justify, for instance, the third equality we take for each $x \in (0,1)$ the limit of the equality $v(x, \mathsf{a} - \varepsilon) - v(x, \varepsilon) = \int_\varepsilon^{\mathsf{a}-\varepsilon} dy \frac{\partial}{\partial y} v(x,y)$ for $\varepsilon \to +0$ and use that v is continuous on the closure of R.

The relation (3.8) implies the inequality $\mathsf{a}' \leq \mathsf{a}$.

If $\mathsf{a}' = \mathsf{a}$ then $u(1,y) - u(0,y) = 1$ for each $y \in (0,a)$. Hence, the left side of R is mapped to the left side of R' and the right side of R is mapped to the right side of R'. Since also the lower side of R is mapped to the lower side of R' and the upper side of R is mapped to the upper side of R' the image of the positively oriented boundary curve of R has index 1 with respect to any point of R' and index 0 with respect to each point in $\mathbb{C} \setminus \bar{R}$. By the argument principle $f(R) = R'$ and f takes each value in R' exactly once. Hence, f is a conformal map of R onto R'. □

Proof of Lemma 3.3 Assume the annuli A and A' have center 0, smaller radius 1 and larger radius r and r', respectively. The set $A \setminus (0, \infty)$ is conformally equivalent to the rectangle $R = \{\xi + i\eta : \xi \in (0, \log r), \eta \in (0, 2\pi)\}$. The exponential function covers the annulus A' by the strip $S' = \{\xi + i\eta : \xi \in (0, \log r'), \eta \in \mathbb{R}\}$. We obtain a holomorphic mapping $g = U + iV$ from R to S' for which either $V(\xi, 2\pi) = V(\xi, 0) + 2\pi$ or $V(\xi, 2\pi) = V(\xi, 0) - 2\pi$ for $\xi \in (0, \log r)$. Assume without loss of generality that the first option holds. Then

$$2\pi \log r = \int_0^{\log r} (V(\xi, 2\pi) - V(\xi, 0))d\xi = \int_0^{\log r} d\xi \int_0^{2\pi} d\eta \frac{\partial}{\partial \eta} V(\xi, \eta) \tag{3.9}$$

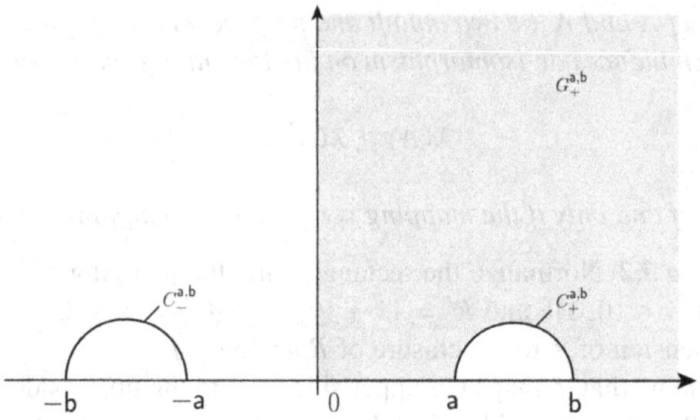

Fig. 3.1 The domains $G^{a,b}$ and $G_{\pm}^{a,b}$

$$= \int_0^{\log r} d\xi \int_0^{2\pi} d\eta \frac{\partial}{\partial \xi} U(\xi, \eta) = \int_0^{2\pi} (U(\log r, \eta) - U(0, \eta)) d\eta$$
$$\leq \int_0^{2\pi} \log r' d\eta = 2\pi \log r'.$$

Equality $r = r'$ holds iff f maps the bigger circle of A to the bigger circle of A' and maps the smaller circle of A to the smaller circle of A'. Since the map f induces an isomorphism of fundamental groups an application of the argument principle shows that f is a conformal mapping of A onto A'. □

Example 3.4 *Let* $0 < a < b$ *be two positive numbers. Denote by* $\Gamma^{a,b}$ *the family of curves in the upper half-plane* $\mathbb{C}_+$ *that join the two half-circles* $C_{\pm}^{a,b} = \{z \in \mathbb{C}_+ : |z \mp \frac{a+b}{2}| = \frac{b-a}{2}\}$ *with diameter* (a, b) *and* $(-b, -a)$*, respectively. Similarly, let* $\Gamma_+^{a,b}$ *be the family of curves in the upper half-plane* $\mathbb{C}_+$ *that join* $C_+^{a,b}$ *with the imaginary half-axis.*

Let $G_+^{a,b}$ be the domain $G_+^{a,b} = \{z \in \mathbb{C} : \text{Re} z > 0, \text{Im} z > 0, |z - \frac{a+b}{2}| > b - a\}$, and $G^{a,b} = \{z \in \mathbb{C}_+ : |z \pm \frac{a+b}{2}| > \frac{b-a}{2}\}$. We consider $G^{a,b}$ ($G_+^{a,b}$, respectively) as curvilinear rectangles with curvilinear horizontal sides equal to $C_{\pm}^{a,b}$ (equal to $C_+^{a,b}$ and $\{z \in \mathbb{C}_+ : \text{Re} z > 0\}$, respectively). See Fig. 3.1. Notice, that the elements of $\Gamma_+^{a,b}$ are not supposed to be contained in $G_+^{a,b}$, and the respective remark concerns $\Gamma^{a,b}$ and $\Gamma_-^{a,b}$.

The following lemma holds.

Lemma 3.4

$$\lambda(G_{\pm}^{\mathsf{a,b}}) = \frac{1}{\pi}\log\frac{(\sqrt{\mathsf{a}}+\sqrt{\mathsf{b}})^2}{\mathsf{b}-\mathsf{a}}, \qquad \lambda(G^{\mathsf{a,b}}) = \frac{2}{\pi}\log\frac{(\sqrt{\mathsf{a}}+\sqrt{\mathsf{b}})^2}{\mathsf{b}-\mathsf{a}}. \tag{3.10}$$

Moreover,

$$\Lambda(\Gamma_{\pm}^{\mathsf{a,b}}) = \frac{1}{\pi}\log\frac{(\sqrt{\mathsf{a}}+\sqrt{\mathsf{b}})^2}{\mathsf{b}-\mathsf{a}}, \qquad \Lambda(\Gamma^{\mathsf{a,b}}) = \frac{2}{\pi}\log\frac{(\sqrt{\mathsf{a}}+\sqrt{\mathsf{b}})^2}{\mathsf{b}-\mathsf{a}}. \tag{3.11}$$

Recall that $\Lambda(\Gamma_{\pm}^{\mathsf{a,b}})$ ($\Lambda(\Gamma^{\mathsf{a,b}})$, respectively) denotes the extremal length of $\Gamma_{\pm}^{\mathsf{a,b}}$ ($\Gamma^{\mathsf{a,b}}$, respectively) in the sense of Definition 3.2.

Proof To prove equality (3.10) we consider the Möbius transformation $T^{\mathsf{a,b}}$ that maps 0 to 0, ∞ to 1, a to a positive real number $t < \frac{1}{2}$, and b to $1-t$. $T^{\mathsf{a,b}}$ is a conformal mapping of the Riemann sphere $\mathbb{P}^1$ onto itself, that takes $G_{+}^{\mathsf{a,b}}$ to a half-annulus and maps the horizontal curvilinear sides of $G_{+}^{\mathsf{a,b}}$ to the half-circles. Each Möbius transformation has the form $z \to \frac{az+b}{cz+d}$. The first two conditions for $T^{\mathsf{a,b}}$ imply that $b = 0$ and $\frac{a}{c} = 1$. To determine the Möbius transformation we ignore the normalization by the condition that the matrix with entries a, b, c, d has determinant 1 and put $a = c = 1$. The remaining two conditions give the equations $\frac{\mathsf{a}}{\mathsf{a}+d} = t$, $\frac{\mathsf{b}}{\mathsf{b}+d} = 1-t$, hence by eliminating d,

$$\mathsf{a}(1-t)^2 = \mathsf{b}t^2 .$$

We obtain the quadratic equation

$$(\mathsf{b}-\mathsf{a})t^2 + 2\mathsf{a}t - \mathsf{a} = 0$$

in t. Its solutions are

$$t_{\pm}(= t_{\pm}^{\mathsf{a,b}}) = \frac{-\mathsf{a} \pm \sqrt{\mathsf{ab}}}{\mathsf{b}-\mathsf{a}}. \tag{3.12}$$

The value t_- is negative, so the required solution is $t_+ < 1 - t_+$. The half-annulus, whose boundary circles have center $\frac{1}{2}$ and radii $\frac{1}{2} - t_+ = \frac{1}{2} - \frac{-\mathsf{a}+\sqrt{\mathsf{ab}}}{\mathsf{b}-\mathsf{a}}$ and $\frac{1}{2}$, is a curvilinear rectangle which is the conformal image of a true rectangle with horizontal side length π and vertical side length $\log\frac{1}{1-2t_+}$. Hence, $\lambda(G_{+}^{\mathsf{a,b}}) = \frac{1}{\pi}\log\frac{1}{1-2t_+}$. Since

$$\frac{1}{1-2t_+} = \frac{1}{1-2\frac{-\mathsf{a}+\sqrt{\mathsf{ab}}}{\mathsf{b}-\mathsf{a}}} = \frac{\mathsf{b}-\mathsf{a}}{\mathsf{b}+\mathsf{a}-2\sqrt{\mathsf{ab}}} = \frac{(\sqrt{\mathsf{a}}+\sqrt{\mathsf{b}})^2}{\mathsf{b}-\mathsf{a}}, \tag{3.13}$$

the extremal length $\lambda(G_+^{\mathsf{a},\mathsf{b}})$ of the curvilinear rectangle $G_+^{\mathsf{a},\mathsf{b}}$ with horizontal sides being $C_+^{\mathsf{a},\mathsf{b}} = \{z \in \mathbb{C}_+ : |z - \frac{\mathsf{a}+\mathsf{b}}{2}| = \frac{\mathsf{b}-\mathsf{a}}{2}\}$ and the imaginary half-axis is equal to $\frac{1}{\pi} \log \frac{(\sqrt{\mathsf{a}}+\sqrt{\mathsf{b}})^2}{\mathsf{b}-\mathsf{a}}$. Hence, we obtain equality (3.10) for $\lambda(G_+^{\mathsf{a},\mathsf{b}})$. Equalities (3.10) for $\lambda(G_-^{\mathsf{a},\mathsf{b}})$ and for $\lambda(G^{\mathsf{a},\mathsf{b}})$ are obtained by the same reasoning.

For the proof of equality (3.11) we consider the conformal mapping $R \to G^{\mathsf{a},\mathsf{b}}$ of a true rectangle R onto $G^{\mathsf{a},\mathsf{b}}$ that maps the lower horizontal side of R onto $C_-^{\mathsf{a},\mathsf{b}}$ and upper side to $C_+^{\mathsf{a},\mathsf{b}}$. Apply Schwarz Reflection Principle to each horizontal side of R. We obtain a conformal mapping from a true rectangle, denoted by $3R$, that, has the same horizontal side length as R and three times the vertical side length of R, onto the upper half-plane with two smaller half-discs removed. The removed half-discs are symmetric with respect to the imaginary axis.

After repeating n times the application of the reflection principle we obtain a conformal mapping from a rectangle $3^n R$ onto a domain Ω_n which is equal to the left half-plane with two half-discs removed. The half-discs are symmetric with respect to the imaginary axis. The domains Ω_n are increasing. The diameter of the removed half-discs at step n tends to zero for $n \to \infty$ since the extremal length of $3^n R$ tends to ∞. We obtain a conformal mapping, denoted by $\mathfrak{c}$, of an infinite strip S' onto the left half-plane.

Let f be any holomorphic mapping from a rectangle to the left half-plane whose extension to the closure takes the upper side to $C_+^{\mathsf{a},\mathsf{b}}$ and the lower side to $C_-^{\mathsf{a},\mathsf{b}}$. Apply Lemma 3.2 to the mapping $\mathfrak{c}^{-1} \circ f$. Equality (3.11) for $\Lambda(\Gamma^{\mathsf{a},\mathsf{b}})$ follows. The equality for $\Lambda(\Gamma_\pm^{\mathsf{a},\mathsf{b}})$ is proved in the same way. □

The Fundamental Group and the Relative Fundamental Group of the Twice Punctured Complex Plane The fundamental group $\pi_1(\mathbb{C} \setminus \{0, 1\}, \frac{1}{2})$ of the twice punctured complex plane with base point is a free group in two generators a_1 and a_2, were a_1 is represented by a curve that surrounds 0 counterclockwise, and a_2 is represented by a curve that surrounds 1 counterclockwise. The fundamental group $\pi_1(\mathbb{C} \setminus \{0, 1\}, \frac{1}{2})$ is canonically isomorphic to the relative fundamental group $\pi_1(\mathbb{C} \setminus \{0, 1\}, (0, 1))$. The elements of the latter group are homotopy classes of arcs in $\mathbb{C} \setminus \{0, 1\}$ with initial point and terminal point in the interval $(0, 1)$. (The initial point may differ from the terminal point.) The isomorphism assigns to the element $e \in \pi_1(\mathbb{C} \setminus \{0, 1\}, \frac{1}{2})$ represented by a curve γ with base point $\frac{1}{2}$ the element in the relative fundamental group represented by γ. For an element $e \in \pi_1(\mathbb{C} \setminus \{0, 1\}, \frac{1}{2})$ we denote be $e_{(0,1)}$ the image of e under the mentioned canonical isomorphism.

Example 3.5 *The extremal length of $(a_1^k)_{(0,1)}$ and $(a_2^k)_{(0,1)}$ for $k \in \mathbb{Z}$ equals zero, and the extremal length of $(a_1^{-1}a_2)_{(0,1)}$ equals $\frac{2}{\pi}\log(2 + \sqrt{5})$.*

This can be seen as follows. The class $(a_1)_{(0,1)}$ can be represented by the holomorphic mapping $z \to e^z$ from the rectangle $R = \{z = x + iy : x \in (-\infty, 0), y \in (0, 2\pi)\}$, of extremal length $\lambda(R) = 0$ into $\mathbb{C} \setminus \{0, 1\}$. There is a similar representation for all powers of $(a_1)_{(0,1)}$ and for all powers of $(a_2)_{(0,1)}$, showing that the extremal length of these elements of the relative fundamental group equals zero.

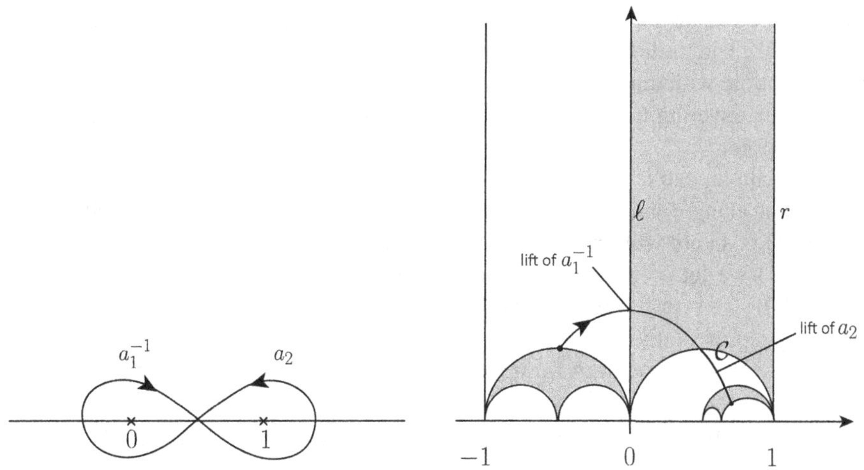

Fig. 3.2 A lift of $a_1^{-1}a_2$ to the universal covering

A representing arc for $(a_2)_{(0,1)}$ lifts under the universal covering map $\mathsf{P} : \mathbb{C}_+ \to \mathbb{C} \setminus \{0, 1\}$ to an arc in $\mathbb{C}_+$ that joins the half-circle with diameter $(0, 1)$ with the half-circle with diameter$(\frac{2}{3}, 1)$ (see Sect. 1.2). A representing arc for $(a_1)_{(0,1)}$ lifts under the universal covering map $\mathsf{P} : \mathbb{C}_+ \to \mathbb{C} \setminus \{0, 1\}$ to an arc in $\mathbb{C}_+$ that joins the half-circle with diameter $(0, 1)$ with the half-circle with diameter $(-1, 0)$. Hence, a representing arc for $(a_1^{-1}a_2)_{(0,1)}$ lifts to an arc in $\mathbb{C}_+$ that joins the half-circle with diameter $(-1, 0)$ with the half-circle with diameter $(\frac{2}{3}, 1)$. (See Fig. 3.2.) Any holomorphic mapping f from a rectangle R into $\mathbb{C} \setminus \{0, 1\}$, that represents $(a_1^{-1}a_2)_{(0,1)}$, lifts to a holomorphic mapping from R into $\mathbb{C}_+$ that takes the horizontal sides to the mentioned circles. The Möbius transformation $z \to \frac{z}{z+1}$ maps $\mathbb{C}_+$ onto itself and takes -1 to ∞ and 0 to 0. Moreover it maps $\frac{2}{3}$ to $\frac{2}{5}$ and 1 to $\frac{1}{2}$. Compose this mapping with the mapping $z \to 10z$. The composition $\mathfrak{c}$ is a conformal self-mapping of $\mathbb{C}_+$ that takes $-1, 0, \frac{2}{3}, 1$ to $\infty, 0, 4, 5$. Composing the lift of f to $\mathbb{C}_+$ with this conformal self-mapping $\mathfrak{c}$ of $\mathbb{C}_+$, we obtain a holomorphic mapping $\mathfrak{c} \circ f$ from R into $\mathbb{C}_+$ that takes the lower horizontal side to the imaginary half-axis and the upper horizontal side to the circle with diameter $(4, 5)$. Vice versa, for each holomorphic mapping g from a rectangle into $\mathbb{C}_+$, that takes the lower horizontal side into the imaginary axis and the upper horizontal side into the circle with diameter $(4, 5)$, the composition $\mathfrak{c}^{-1} \circ g$ is a holomorphic mapping of a rectangle into the twice punctured complex plane that represents $(a_1^{-1}a_2)_{(0,1)}$. Lemma 3.4 implies the equality $\Lambda((a_1^{-1}a_2)_{(0,1)}) = \lambda(G_+^{4,5}) = \frac{1}{\pi}\log(2+\sqrt{5})^2 = \frac{2}{\pi}\log(2+\sqrt{5})$.

Example 3.6 *The extremal length of the free homotopy class of closed curves* $\widehat{a_1^{-1}a_2}$ *equals* $\frac{2}{\pi}\log(3+2\sqrt{2})$.

This can be seen as follows. We lift a representative γ of $a_2 \in \pi_1(\mathbb{C} \setminus \{0, 1\}, \frac{1}{2})$ to a curve in $\mathbb{C}_+$ with initial point q on the half-circle with diameter $(0, 1)$. The lift joins the half-circle with diameter $(0, 1)$ with the half-circle with diameter $(\frac{2}{3}, 1)$. There is a unique covering transformation T_{a_2} that maps the initial point of this lift to its terminal point.

To obtain an explicit expression for the mapping T_{a_2}, we make a change of the base point along a curve α in $\mathbb{C} \setminus \{0, 1\}$. More precisely, suppose $\alpha : [0, 1] \to \mathbb{C} \setminus \{0, 1\}$ is an arc with initial point $\frac{1}{2}$, such that $\alpha((0, 1])$ is contained in $\mathbb{C}_+$. For $t \in [0, 1]$ we let α_t be the curves obtained by moving along α from the initial point $\alpha(0) = \frac{1}{2}$ to the point $\alpha(t)$. Consider the continuous family $\mathrm{Is}_{\alpha_t}(\gamma)$, $t \in [0, 1]$, of representatives of the free homotopy class $\widehat{a_2}$, and their lift to the universal covering with initial point in D_0 if $t \in (0, 1]$. The terminal point of each such lift is contained in the geodesic triangle D_2 with vertices $\frac{2}{3}, 1, \frac{1}{2}$ (see Sect. 1.2). The set of covering transformations is discrete, and the family of covering transformations, that map the initial point of the considered lift of $\mathrm{Is}_{\alpha_t}(\gamma)$ to its terminal point, depends continuously on the parameter. Hence, each transformation of the family equals T_{a_2}. Hence, T_{a_2} maps D_0 into D_2 and maps the half-circle with diameter $(0, 1)$ to the half-circle with diameter $(\frac{2}{3}, 1)$. Consider in the same way the mapping $T_{a_2^{-1}}$, that is associated to the inverse a_2^{-1}, and its lift with initial point in D_2. We see that the covering transformation $T_{a_2^{-1}}$ maps the geodesic triangle D_2 with vertices $\frac{2}{3}, 1, \frac{1}{2}$ into D_0, and maps the half-circle with diameter $(\frac{2}{3}, 1)$ into the half-circle with diameter $(0, 1)$. The composition $T_{a_2^{-1}} T_{a_2}$ is the identity. Hence, T_{a_2} maps D_0 conformally onto the geodesic triangle D_2 with vertices $\frac{2}{3}, 1, \frac{1}{2}$, and maps the half-circle with diameter $(0, 1)$ onto the half-circle with diameter $(\frac{2}{3}, 1)$. Hence, $T_{a_2} | D_0$ coincides with the double reflection. This can be used for explicit computation of T_{a_2}. One can also use the fact that T_{a_2} is a conformal self-mapping of $\mathbb{C}_+$, hence extends to a Möbius transformation, that takes the vertex 0 (adjacent to the sides ℓ and C) to $\frac{2}{3}$, the vertex ∞ (adjacent to the sides ℓ and r) to $\frac{1}{2}$, and the vertex 1 (adjacent to r and C) to 1. This Möbius transformation equals $T_{a_2} = \frac{z-2}{2z-3}$. Its inverse equals $T_{a_2}^{-1} = T_{a_2^{-1}} = \frac{-3z+2}{-2z+1}$.

In the same way we will now associate to $a_1 \in \pi_1(\mathbb{C} \setminus \{0, 1\}, \frac{1}{2})$ a covering transformation T_{a_1}. We lift a representative of a_1 to a curve with initial point in the half-circle with diameter $(0, 1)$. This lift joins the half-circle with diameter $(0, 1)$ with the half-circle with diameter $(-1, 0)$. The covering transformation T_{a_1} that takes the initial point of the lift to its terminal point, maps D_0 conformally onto the geodesic triangle with vertices $-1, -\frac{1}{2}, 0$, and maps the half-circle with diameter $(0, 1)$ to the half-circle with diameter $(-1, 0)$. Hence, T_{a_1} maps 0 to 0, 1 to -1, and ∞ to $-\frac{1}{2}$. A small computation gives $T_{a_1} = \frac{z}{1-2z}$. The inverse mapping equals $T_{a_1}^{-1} = T_{a_1^{-1}} = \frac{z}{2z+1}$.

The covering transformation corresponding to $a_1^{-1}a_2$ equals $T_{a_1^{-1}a_2}(z) = T_{a_2} \circ T_{a_1}^{-1}(z) = \frac{3z+2}{4z+3}$. The matrix $A = \begin{pmatrix} 3 & 2 \\ 4 & 3 \end{pmatrix}$ has two real eigenvalues $t_\pm = 3 \pm 2\sqrt{2}$ whose product equals the determinant 1. The matrix A can be conjugated to a diagonal matrix $B^{-1}AB = \begin{pmatrix} t_+ & 0 \\ 0 & t_- \end{pmatrix}$ by a matrix B with real entries whose columns are eigenvectors. The Möbius transformation corresponding to B maps the real axis to itself. Changing if needed the direction of one of the eigenvectors we may assume that the Möbius transformation maps $\mathbb{C}_+$ to itself. We see that the mapping $T_{a_1^{-1}a_2} : \mathbb{C}_+ \circlearrowleft$ can be conjugated by a conformal self-mapping of $\mathbb{C}_+$ to the mapping T_λ, $T_\lambda(z) \overset{def}{=} \frac{t_+ z}{t_-} = t_+^2 z = (3+2\sqrt{2})^2 z$ with $\lambda = (3+2\sqrt{2})^2$. The quotient $\mathbb{C}_+ / \langle T_{a_1^{-1}a_2} \rangle$ is conformally equivalent to the quotient $\mathbb{C}_+ / T_\lambda$. This quotient is an annulus of extremal length $\frac{1}{\pi}\log(3+2\sqrt{2})^2 = \frac{2}{\pi}\log(3+2\sqrt{2})$.

Consider a holomorphic mapping f of an annulus $A = \{z \in \mathbb{C} : r < |z| < R\}$ into $\mathbb{C} \setminus \{0, 1\}$ that represents the free homotopy class $\widehat{a_1^{-1}a_2}$. Cut the annulus along the arc $\ell = (r, R)$, and lift the restriction $f|A \setminus \ell$ to a mapping $\tilde{f}$ into $\mathbb{C}_+$. Let ℓ_- and ℓ_+ be the strands of ℓ that are accessible from a point in $A \setminus \ell$ by moving clockwise, or counterclockwise, respectively, and for each point $p \in \ell$ we let $p_\pm \in \ell_\pm$ be the points corresponding to p. Then the equality $\tilde{f}(p_+) = T_{a_1^{-1}a_2}(\tilde{f}(p_-))$ holds for the continuous extension of $\tilde{f}$ to $\ell_\pm$. This means, that the mapping $\tilde{f}$ from $(A \setminus \ell) \cup \ell_- \cup \ell_+$ descends to a holomorphic mapping from A into $\mathbb{C}_+ / T_\lambda$ that represents $\widehat{a_1^{-1}a_2}$. By Lemma 3.3 the extremal length of A is not smaller than $\lambda(\mathbb{C}_+ / T_\lambda) = \frac{2}{\pi}\log(3+2\sqrt{2})$, and there is a conformal mapping of an annulus A of extremal length $\frac{2}{\pi}\log(3+2\sqrt{2})$ onto $\mathbb{C}_+ / T_\lambda$ that represents $\widehat{a_1^{-1}a_2}$.

The Extremal Length of Conjugacy Classes of Braids and the Extremal Length of Braids with Totally Real Horizontal Boundary Values

These are conformal invariants of braids and conjugacy classes of braids which will play a key role later.

Recall that for a subset A of the complex plane $\mathbb{C}$ we defined the configuration space $C_n(A) = \{(z_1, \ldots, z_n) \in A^n : z_i \neq z_j \text{ for } i \neq j\}$ of n particles moving along A without collision. Each permutation in the symmetric group $\mathcal{S}_n$ acts on $C_n(A)$ by permuting the coordinates. The quotient $C_n(A)/\mathcal{S}_n$ is called the symmetrized configuration space related to A. Recall that the natural projection $C_n(\mathbb{C}) \to C_n(\mathbb{C})/\mathcal{S}_n$ is denoted by $\mathcal{P}_{\text{sym}}$.

Choose a base point $E_n \in C_n(\mathbb{R})/\mathcal{S}_n$. Recall that braids on n strands (n-braids for short) with base point E_n are homotopy classes of loops with base point E_n in the symmetrized configuration space, equivalently, they are elements of the fundamental group $\pi_1(C_n(\mathbb{C})/\mathcal{S}_n, E_n)$ of the symmetrized configuration space with base point E_n.

Recall that conjugacy classes of n-braids are free homotopy classes of loops in $C_n(\mathbb{C})/\mathcal{S}_n \cong \mathfrak{P}_n$. According to Definition 3.4 the extremal length $\Lambda(\hat{b})$ of a conjugacy class of n-braids $\hat{b}$ is defined as $\Lambda(\hat{b}) = \inf_{A \in \mathcal{A}} \lambda(A)$, where $\mathcal{A}$ denotes the set of all annuli which admit a holomorphic mapping into $C_n(\mathbb{C})/\mathcal{S}_n$ that represents $\hat{b}$.

To define the extremal length of braids with totally real horizontal boundary values we consider the totally real subspace $\mathcal{E}^n_{tr} \overset{def}{=} C_n(\mathbb{R})/\mathcal{S}_n$ of $C_n(\mathbb{C})/\mathcal{S}_n$. The totally real subspace $C_n(\mathbb{R})/\mathcal{S}_n$ of $C_n(\mathbb{C})/\mathcal{S}_n$ is connected and simply connected. Indeed, the totally real subspace $C_n(\mathbb{R})$ of $C_n(\mathbb{C})$ is the union of the connected components $\{(x_1, \ldots, x_n) \in \mathbb{R}^n : x_{\sigma(1)} < x_{\sigma(2)} < \ldots < x_{\sigma(n)}\}$ over all permutations $\sigma \in \mathcal{S}_n$. Thus $C_n(\mathbb{R})$ is invariant under the action of $\mathcal{S}_n$ and the quotient is homeomorphic to $\{(x_1, \ldots, x_n) \in \mathbb{R}^n : x_1 < x_2 < \ldots < x_n\}$. Hence the claim.

The fundamental group $\pi_1(C_n(\mathbb{C})/\mathcal{S}_n, E_n)$ is isomorphic to the relative fundamental group $\pi_1(C_n(\mathbb{C})/\mathcal{S}_n, C_n(\mathbb{R})/\mathcal{S}_n)$. The elements of the latter group are homotopy classes of arcs in $C_n(\mathbb{C})/\mathcal{S}_n$ with endpoints in $C_n(\mathbb{R})/\mathcal{S}_n$.

The isomorphism between the two groups is obtained as follows. Since the fundamental groups with different base point are isomorphic, we may assume that E_n is contained in the totally real subspace $C_n(\mathbb{R})/\mathcal{S}_n$. Each element of $\pi_1(C_n(\mathbb{C})/\mathcal{S}_n, E_n)$ is a subset of an element of $\pi_1(C_n(\mathbb{C})/\mathcal{S}_n, C_n(\mathbb{R})/\mathcal{S}_n)$. Vice versa, since $C_n(\mathbb{R})/\mathcal{S}_n$ is connected and E_n is contained in $C_n(\mathbb{R})/\mathcal{S}_n$, each class in $\pi_1(C_n(\mathbb{C})/\mathcal{S}_n, C_n(\mathbb{R})/\mathcal{S}_n)$ contains a class in $\pi_1(C_n(\mathbb{C})/\mathcal{S}_n, E_n)$. Since $C_n(\mathbb{R})/\mathcal{S}_n$ is simply connected, each class in $\pi_1(C_n(\mathbb{C})/\mathcal{S}_n, C_n(\mathbb{R})/\mathcal{S}_n)$ contains no more than one class of $\pi_1(C_n(\mathbb{C})/\mathcal{S}_n, E_n)$. Indeed, if two loops in $C_n(\mathbb{C})/\mathcal{S}_n$ with base point E_n are homotopic as loops in $C_n(\mathbb{C})/\mathcal{S}_n$ with varying base point in $C_n(\mathbb{R})/\mathcal{S}_n$ then they are homotopic as loops in $C_n(\mathbb{C})/\mathcal{S}_n$ with fixed base point E_n.

Let $b \in \mathcal{B}_n$ be a braid. Denote its image in the relative fundamental group $\pi_1(C_n(\mathbb{C})/\mathcal{S}_n, C_n(\mathbb{R})/\mathcal{S}_n)$ by b_{tr}.

We are now ready to define for any braid its extremal length with totally real boundary values (and the conformal module with totally real boundary values, respectively).

Definition 3.5 *Let $b \in \mathcal{B}_n$ be an n-braid. The extremal length $\Lambda_{tr}(b)$ with totally real horizontal boundary values is defined as*

$$\Lambda_{tr}(b) = \inf\{\lambda(R) : R \text{ a rectangle which admits a holomorphic mapping to } C_n(\mathbb{C})/\mathcal{S}_n \text{ that represents } b_{tr}\}.$$

The conformal module $\mathcal{M}_{tr}(b)$ of b with totally real horizontal boundary values, respectively, is defined as

$$\mathcal{M}_{tr}(b) = \sup\{m(R) : R \text{ a rectangle which admits a holomorphic mapping to } C_n(\mathbb{C})/\mathcal{S}_n \text{ that represents } b_{tr}\}.$$

We will also use the notation $\Lambda(b_{tr})$ for $\Lambda_{tr}(b)$ and the notation $\mathcal{M}(b_{tr})$ for $\mathcal{M}_{tr}(b)$. Note that the two invariants are inverse to each other. Sometimes it is more convenient to work with the extremal length, some other times it is more appropriate to speak about the conformal module.

Recall that Δ_n denotes the Garside element in the braid group $\mathcal{B}_n$.

Lemma 3.5 *For each braid $b \in \mathcal{B}_n$ the equalities $\Lambda(\hat{b}) = \Lambda(\widehat{b\Delta_n^2})$ and $\Lambda_{tr}(b) = \Lambda_{tr}(b\Delta_n) = \Lambda_{tr}(\Delta_n b)$ hold.*

Proof Let $R = \{x + iy : x \in (0, 1),\ y \in (0, \mathsf{a})\}$, and suppose a holomorphic mapping $f = \{f_1, f_2, \ldots, f_n\} : R \to C_n(\mathbb{C}^n)/\mathcal{S}_n$ represents b_{tr}. The mapping $\zeta \to e^{\frac{\pi}{\mathsf{a}}\zeta} f(\zeta) = e^{\frac{\pi}{\mathsf{a}}\zeta}\{f_1(\zeta), f_2(\zeta), \ldots, f_n(\zeta)\}$ is holomorphic on R and represents $(b\,\Delta_n)_{tr}$. Since $\Lambda_{tr}(b^{-1}) = \Lambda_{tr}(b)$ for each braid b, we obtain $\Lambda_{tr}(\Delta_n b) = \Lambda_{tr}(b)$. The stated relation for $\Lambda(\hat{b})$ is proved in the same way. □

Example 3.7 *The extremal length with totally real horizontal boundary values of $\sigma_1^{-1}\sigma_2$ equals $\Lambda((\sigma_1^{-1}\sigma_2)_{tr}) = \frac{\log(3+2\sqrt{2})}{\pi}$.*

To prove this fact, we let γ be an arc in $C_3(\mathbb{C})/\mathcal{S}_3$ that represents the homotopy class $(\sigma_1^{-1}\sigma_2)_{tr} \in \pi_1(C_3(\mathbb{C})/\mathcal{S}_3, C_3(\mathbb{R})/\mathcal{S}_3)$. The initial point (and also the terminal point) of γ is an unordered triple of distinct points, each contained in the real axis. Lift γ to an arc $\tilde{\gamma} = (\tilde{\gamma}_1, \tilde{\gamma}_2, \tilde{\gamma}_3)$ in $C_3(\mathbb{C})$, $\mathcal{P}_{\text{sym}}(\tilde{\gamma}) = \gamma$, so that the initial point of the lift $\tilde{\gamma}$ is an ordered triple (x_1, x_2, x_3) with $x_2 < x_1 < x_3$.

Recall that we put $\mathcal{P}'_{\mathcal{A},\mathrm{n}}\big((z_1, z_2, z_3)\big) = M_z(z_2) = \frac{z_2 - z_1}{z_3 - z_1}$. We associate to $\tilde{\gamma}$ the curve $M_{\tilde{\gamma}}$, $M_{\tilde{\gamma}}(t) = M_{(\tilde{\gamma}_1(t), \tilde{\gamma}_2(t), \tilde{\gamma}_3(t))} = \mathcal{P}'_{\mathcal{A},\mathrm{n}}\big(\tilde{\gamma}_1(t), \tilde{\gamma}_2(t), \tilde{\gamma}_3(t)\big)$, $t \in [0, 1]$, in $\mathbb{C} \setminus \{0, 1\}$. The initial point of this curve is in $(-\infty, 0)$. Figure 3.3 shows the curve $M_{\tilde{\gamma}}$ for a curve γ in $C_3(\mathbb{C})/\mathcal{S}_3$ representing $(\sigma_1^{-1}\sigma_2)_{tr}$.

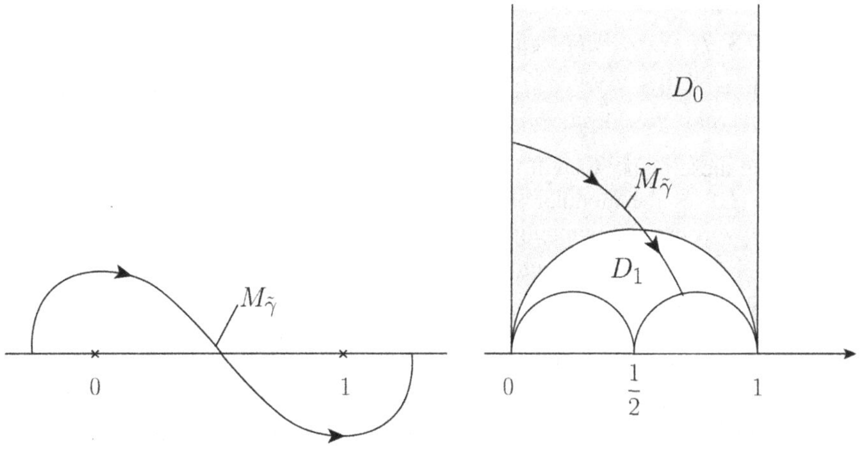

Fig. 3.3 The curve $\tilde{M}_{\tilde{\gamma}}$ on the Teichmüller space $\mathcal{T}(0, 4)$, that corresponds to a representative of $(\sigma_1^{-1}\sigma_2)_{tr}$

Lift the curve $M_{\tilde{\gamma}}$ under the covering $\mathsf{P} : \mathbb{C}_+ \to \mathbb{C} \setminus \{0, 1\}$ described in Sect. 1.2. to a curve $\tilde{M}_{\tilde{\gamma}}$ in the universal covering $\mathbb{C}_+$. (We identify the covering $\mathsf{P} : \mathbb{C}_+ \to \mathbb{C} \setminus \{0, 1\}$ described in Sect. 1.2 and the covering $\mathcal{P}'_{\mathcal{T}} : \mathcal{T}(0, 4) \to \mathbb{C} \setminus \{0, 1\}$ according to the remark after the statement of Theorem 1.6.) We choose the lift $\tilde{M}_{\tilde{\gamma}}$ that has initial point on the imaginary half-axis, see Fig. 3.3. Its terminal point is contained in the half-circle $\mathbb{C}_+ \cap \{|z - \frac{3}{4}| = \frac{1}{4}\}$.

Consider a holomorphic mapping g of a rectangle to $C_3(\mathbb{C})/\mathcal{S}_3$ that represents $(\sigma_1^{-1}\sigma_2)_{tr}$. Lift this mapping to a holomorphic mapping $\tilde{g}$ into $C_3(\mathbb{C})$, consider the associated mapping $M_{\tilde{g}} : R \to \mathbb{C} \setminus \{0, 1\}$, and lift it to a mapping $\tilde{M}_{\tilde{g}} : R \to \mathbb{C}_+$, so that $\tilde{M}_{\tilde{g}}$ represents the curve $\tilde{M}_{\tilde{\gamma}}$ in Fig. 3.3. The mapping $\tilde{M}_{\tilde{g}}$ is holomorphic and maps the lower side of R to the imaginary half-axis, and the upper side to the half-circle $\mathbb{C}_+ \cap \{|z - \frac{3}{4}| = \frac{1}{4}\}$. Since $\frac{(\sqrt{\frac{1}{2}}+\sqrt{1})^2}{\frac{1}{2}} = 2(\frac{1}{2} + 2\sqrt{\frac{1}{2}} + 1) = 3 + 2\sqrt{2}$, by Lemmas 3.2 and 3.4 the extremal length $\lambda(R)$ is not smaller than $\frac{\log(3+2\sqrt{2})}{\pi}$.

Vice versa, a rectangle of extremal length $\frac{\log(3+2\sqrt{2})}{\pi}$ admits a conformal mapping $\mathfrak{c}$ onto $\{\mathrm{Re} z > 0, \mathrm{Im} z > 0, |z - \frac{3}{4}| > \frac{1}{4}\}$, such that the horizontal sides of R correspond to the imaginary half-axis, and to the half-circle $\mathbb{C}_+ \cap \{|z - \frac{3}{4}| = \frac{1}{4}\}$. The composition $\mathsf{P} \circ \mathfrak{c}$ of $\mathfrak{c}$ with the universal covering map P is a holomorphic mapping from the rectangle into $\mathbb{C} \setminus \{0, 1\}$ that represents the curve $M_{\tilde{\gamma}}$ of Fig. 3.3 and takes the horizontal sides of the rectangle into $(-\infty, 0)$ and $(1, \infty)$, respectively. The mapping $R \ni z \to (0, \mathsf{P} \circ \mathfrak{c}(z), 1) \in C_3(\mathbb{C})$ projects under $\mathcal{P}_{\mathrm{sym}}$ to a mapping into $C_3(\mathbb{C})/\mathcal{S}_3$ that represents $(\sigma_1^{-1}\sigma_2)_{tr}$.

In the same way we see that the extremal length $\Lambda((\sigma_1^{-1}\sigma_2^2\sigma_1^{-1})_{tr})$ equals $\frac{2\log(3+2\sqrt{2})}{\pi}$. For a curve γ representing $(\sigma_1^{-1}\sigma_2^2\sigma_1^{-1})_{tr}$ the curve $\tilde{M}_{\tilde{\gamma}}$ joins the two half-circles $\{|z \pm \frac{3}{4}| = \frac{1}{4}\} \cap \mathbb{C}_+$.

Example 3.8 *The extremal length of the conjugacy class of 3-braids $\widehat{\sigma_1^{-1}\sigma_2}$ equals $\Lambda(\widehat{\sigma_1^{-1}\sigma_2}) = \frac{2}{\pi}\log\frac{3+\sqrt{5}}{2}$. This is the smallest non-vanishing extremal length among conjugacy classes of 3-braids.*

To prove these facts, we will first describe the action on the Teichmüller space $\mathcal{T}(0, 4) \cong \mathbb{C}_+$ of the modular transformations T_{σ_1} and T_{σ_2} corresponding to σ_1 and σ_2. We will use the mapping $\mathcal{P}'_{\mathcal{A},\mathsf{n}} : C_3(\mathbb{C}) \to (\mathbb{C} \setminus \{0, 1\}$ and the covering $\mathcal{P}'_{\mathcal{T}} : \mathcal{T}(0, 4) \to \mathbb{C} \setminus \{0, 1\}$ identified with $\mathsf{P} : \mathbb{C}_+ \to \mathbb{C} \setminus \{0, 1\}$ as in Example 3.7 of this section.

Recall that the modular transformation T_b associated to an n-braid b is the modular transformation φ_b^* of a self-homeomorphism φ_b of $\mathbb{P}^1$ that represents the mapping class $\mathfrak{m}_{b,\infty}$ corresponding to b. We consider the modular transformation T_b for the 3-braids $b = \sigma_1$ and $b = \sigma_2$. With the same normalization as in Example 3.7 of this section the mapping φ_b fixes ∞ and maps the set $E_3 = \{0, \frac{1}{2}, 1\}$ onto itself. Such a homeomorphism φ_b can be obtained using a parameterizing isotopy φ_t for a geometric braid γ representing b, i.e. a continuous family of self-homeomorphisms of $\mathbb{P}^1$ such that $\varphi_t(E_3) = \gamma(t)$, $t \in [0, 1]$. We may take $\varphi_b = \varphi^1$ and $T_b = (\varphi^1)^*$.

Lift the base point $E_3 = \{0, \frac{1}{2}, 1\} \in C_3(\mathbb{C})\diagup \mathcal{S}_3$ under $\mathcal{P}_{\text{sym}}$ to the point $\tilde{E}_3 = (0, \frac{1}{2}, 1)$ of $C_3(\mathbb{C})$. Then $\mathcal{P}'_{\mathcal{A},\mathsf{n}}(\tilde{E}_3) = \frac{1}{2}$. Take the preimage $\frac{1+i}{2}$ of $\frac{1}{2}$ under P. The image of $\frac{1+i}{2}$ under T_{σ_1} can be obtained as follows.

Consider a curve γ_1 in $C_3(\mathbb{C})\diagup \mathcal{S}_3$ that represents σ_1 with base point E_3 and its lift $\tilde{\gamma}_1$ under $\mathcal{P}_{\text{sym}}$ with initial point $\tilde{E}_3$. The projection $\mathcal{P}'_{\mathcal{A},\mathsf{n}}(\tilde{\gamma}_1)$ (identified with the curve $t \to M_{\tilde{\gamma}_1(t)}$) is a curve in $\mathbb{C} \setminus \{0, 1\}$ with initial point $\frac{1}{2}$. The curve γ_1 may be chosen so that $\mathcal{P}'_{\mathcal{A},\mathsf{n}}(\tilde{\gamma}_1)$ joins $\frac{1}{2}$ with a point in the negative real axis, and the interior points of the curve $\mathcal{P}'_{\mathcal{A},\mathsf{n}}(\tilde{\gamma}_1)$ are contained in upper half-plane $\mathbb{C}_+$.

Consider the lift $\tilde{M}_{\tilde{\gamma}_1}$ of $M_{\tilde{\gamma}_1}$ under $\mathsf{P} : \mathbb{C}_+ \to \mathbb{C} \setminus \{0, 1\}$ with initial point $\frac{1+i}{2}$. The lift $\tilde{M}_{\tilde{\gamma}_1}$ of $M_{\tilde{\gamma}_1}$ joins $\frac{1+i}{2}$ with a point on the positive imaginary axis. The interior of $\tilde{M}_{\tilde{\gamma}_1}$ is contained in D_0.

The terminal point of $\tilde{M}_{\tilde{\gamma}_1}$ is the image of $\frac{1+i}{2}$ under T_{σ_1}. Indeed, the following commutative diagram (see also Lemma 1.5)

$$\begin{array}{ccc} \varphi^t & \xrightarrow{\quad [\,] \quad} & \tilde{\mathcal{M}}_{\tilde{\gamma}_t} \\ \Big\downarrow{\scriptstyle e_3} & & \Big\downarrow{\scriptstyle \mathsf{P}} \\ \tilde{\gamma}_1(t) & \xrightarrow{\quad \mathcal{P}'_{\mathcal{A},\mathsf{n}} \quad} & \mathcal{M}_{\tilde{\gamma}_t} \end{array}$$

implies the equality

$$T_{\sigma_1}(\frac{1+i}{2}) = \varphi^*_{\sigma_1}([\mathrm{Id}]) = [\varphi_{\sigma_1}] = [\varphi^1] = \tilde{M}_{\tilde{\gamma}_1(1)} \,.$$

We proved that $T_{\sigma_1}(\frac{1+i}{2})$ is contained in the imaginary half-axis.

We will now prove that T_{σ_1} maps the half-circle $C_+^{0,1} = \{|z - \frac{1}{2}| = \frac{1}{2}\} \cap \mathbb{C}_+$ into the positive imaginary half-axis. Take any point in $C_+^{0,1}$ and join it with $\frac{1+i}{2}$ by an arc $\alpha(s)$, $s \in [0, 1]$, in $C_+^{0,1}$. Let w_s, $s \in [0, 1]$, be a continuous family of self-homeomorphisms of $\mathbb{P}^1$, that fix ∞, such that $w_0 = \mathrm{Id}$ and $\tau_s \overset{def}{=} [w_s] = \alpha(s)$. (See also Lemma 1.4.) Notice that for $\tilde{E}_3^s \overset{def}{=} w_s(\tilde{E}_3)$ the point $x(s) \overset{def}{=} \mathcal{M}_{\tilde{E}_3^s}$ is contained in $(0, 1)$. The equality

$$T_{\sigma_1}(\tau_s) = (\varphi^1_{\sigma_1})^*([w_s]) = [w_s \circ \varphi^1_{\sigma_1}]$$

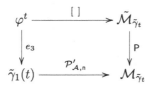

holds. Since $\varphi^1_{\sigma_1}$ acts on $\tilde{E}_3$ by permuting the first two coordinates, the homeomorphism $w_s \circ \varphi^1_{\sigma_1}$ acts on $w_s(\tilde{E}_n) = \tilde{E}_n^s$ by permuting the first two coordinates. This means that $x'(s) \overset{def}{=} \mathcal{M}_{w_s \circ \varphi^1_{\sigma_1}(\tilde{E}_3)} \in (-\infty, 0) \subset \mathbb{C}\setminus\{0, 1\}$. Hence, the lift $[w_s \circ \varphi^1_{\sigma_1}]$ of $\mathcal{M}_{w_s \circ \varphi^1_{\sigma_1}(\tilde{E}_3)}$ under P is contained in the imaginary half-axis.

Each member of the continuous family of curves $\{\mathcal{M}_{w_s \circ \varphi^t_{\sigma_1}(\tilde{E}_3)},\ t \in [0, 1]\}_{s\in[0,1]}$ in $\mathbb{C}\setminus\{0, 1\}$ has initial point on $(0, 1)$ and terminal point in $(-\infty, 0)$. The family lifts to a continuous family of curves in $\mathbb{C}_+$ with initial point in the half-circle $C_+^{0,1}$ and terminal point in the imaginary half-axis, and the interior $\{\mathcal{M}_{w_0 \circ \varphi^t_{\sigma_1}(\tilde{E}_3)},\ t \in (0, 1)\}$ of the curve corresponding to the parameter $s = 0$, is contained in the upper half-plane. Hence, T_{σ_1} maps the half-circle $C_+^{0,1}$ into the positive imaginary half-axis.

The same argument for the braid σ_1^{-1} shows that the modular transformation T_{σ_1} maps the half-circle $\{z \in \mathbb{C}_+ : |z - \frac{1}{2}| = 1\}$ onto the positive imaginary half-axis.

We show now that T_{σ_1} maps the geodesic triangle D_1 with vertices $0, \frac{1}{2}, 1$ onto D_0. Let $\alpha(s),\ s \in [0, 1]$, be an arc in $\mathbb{C}_+$ with $\alpha(0) = \frac{1+i}{2}$ and $\alpha((0, 1]) \subset D_1$. The projection $\mathsf{P}(\alpha((0, 1]))$ to $\mathbb{C}\setminus\{0, 1\}$ is contained in $\mathbb{C}_-$ and $\mathsf{P}(\alpha(0)) = \mathsf{P}(\frac{1+i}{2}) = \frac{1}{2}$.

Let again w_s be a continuous family of self-homeomorphisms of $\mathbb{P}^1$ for which $w_0 = \mathrm{Id}$ and $\tau_s \overset{def}{=} [w_s] = \alpha_s$. Consider again the continuous family of curves $\{\mathcal{M}_{w_s \circ \varphi^t_{\sigma_1}(\tilde{E}_3^0)},\ s \in [0, 1]\}$ in $\mathbb{C} \setminus \{0, 1\}$ with initial points $\{\mathcal{M}_{w_s(\tilde{E}_3^0)},\ s \in [0, 1]\}$ in $(0, 1) \cup \mathbb{C}_-$. The terminal points $\{\mathcal{M}_{w_s \circ \varphi^1_{\sigma_1}(\tilde{E}_3^0)},\ s \in [0, 1]\}$ are contained in $(-\infty, 0) \cup \mathbb{C}_+$, since $\varphi^1_{\sigma_1}$ acts on $\tilde{E}_3^0$ by permuting the first two coordinates. Indeed, for a point $(z_1, z_2, z_3) \in C_3(\mathbb{C})$ the inclusion $M_{(z_1,z_2,z_3)} \in \mathbb{C}_-$ means, that the point z_2 lies on the right of the line through z_1 and z_3, oriented in the direction from z_1 to z_3. Then z_1 lies on the left of the line through z_2 and z_3, oriented in the direction from z_2 to z_3. This means, $M_{(z_2,z_1,z_3)} \in \mathbb{C}_+$. The lift under P of the family of curves $\{\mathcal{M}_{w_s \circ \varphi^t_{\sigma_1}(\tilde{E}_3^0)},\ t \in [0, 1]\}_{s\in[0,1]}$, in $\mathbb{C}\setminus\{0, 1\}$ provides a continuous family of curves in $\mathbb{C}_+$ with initial points in $D_1 \cup C_+^{0,1}$ and terminal points in $D_0 \cup \{\mathrm{Re} z = 0\} \cap \mathbb{C}_+$. The terminating point of each curve is obtained from its initial point by applying the modular transformation T_{σ_1}. We showed that T_{σ_1} maps the geodesic triangle with vertices $0, \frac{1}{2}, 1$ into D_0. Similar arguments for the braid σ_1^{-1} imply that T_{σ_1} maps the geodesic triangle with vertices $0, \frac{1}{2}, 1$ onto D_0. Since T_{σ_1} maps the half-circle $\{z \in \mathbb{C}_+ : |z - \frac{1}{2}| = 1\}$ onto the positive imaginary half-axis, T_{σ_1} takes 0 to 0, 1 to ∞, and $\frac{1}{2}$ to 1. Hence, $T_{\sigma_1}(z) = \frac{z}{1-z}$ and $T_{\sigma_1^{-1}}(z) = \frac{z}{1+z}$.

The modular transformation T_{σ_2} is computed similarly. Represent σ_2 by a closed curve $\gamma_2 : [0, 1] \to C_3(\mathbb{C})\diagup\mathcal{S}_3$ with initial point $E_3^x = \{0, x, 1\}$, where $x \in (0, 1)$. We associate to it a curve $\tilde{M}_{(\tilde{\gamma}_2)_t}$ in $\mathbb{C}_+$ with initial point contained in the half-circle $\{|z - \frac{1}{2}| = \frac{1}{2}\}$. The terminal point is contained in the half-circle $\{|z - \frac{3}{4}| = \frac{1}{4}\}$. By the same arguments as used for σ_1 we see that T_{σ_2} is a Möbius transformation that maps the half-circle $\{|z - \frac{1}{2}| = \frac{1}{2}\}$ onto the half-circle $\{|z - \frac{3}{4}| = \frac{1}{4}\}$, and maps D_0 conformally onto the geodesic triangle with vertices $0, \frac{1}{2}, 1$. Hence, T_{σ_2} maps 0 to $\frac{1}{2}$, 1 to 1, and ∞ to 0. This implies that the mapping equals $T_{\sigma_2}(z) = \frac{1}{-z+2}$. (See also Fig. 3.3.)

The modular transformation corresponding to $\sigma_1^{-1}\sigma_2$ equals $T_{\sigma_1^{-1}\sigma_2} = T_{\sigma_2} \circ T_{\sigma_1^{-1}}(z) = \frac{z+1}{z+2}$. The eigenvalues of the matrix $\begin{pmatrix} 1 & 1 \\ 1 & 2 \end{pmatrix}$ are $t_\pm = \frac{3\pm\sqrt{5}}{2}$. The same arguments as in Example 3.6 of this section show that $\Lambda(\widehat{\sigma_1^{-1}\sigma_2}) = \frac{2}{\pi}\log\frac{3+\sqrt{5}}{2}$.

For each 3-braid b the respective modular transformation T_b is a finite product of powers of the T_{σ_j}. Hence, each T_b is an element of $PSL_2(\mathbb{Z})$. If for the trace $a+d$ of the matrix corresponding to T_b the equality $|a+d| = 2$ holds, then T_b is conjugate to a translation by a real number, and the quotient $\mathbb{C}_+ \diagup \langle T_b \rangle$ is conformally equivalent to the once punctured plane, hence has extremal length 0. If $|a+d| < 2$, then a power of the mapping is the identity. Hence, in a neighbourhood of the fixed point the mapping is conjugate to the mapping $z \to e^{i\frac{2\pi}{k}} z$. The quotient $\{0 < |z| < \delta\} \diagup (z \sim e^{i\frac{2\pi}{k}} z)$ is a punctured disc and has extremal length 0. Hence, $\mathbb{C}_+ \diagup T_b$ has extremal length 0. If $|a+d| > 2$ the extremal length of $\mathbb{C}_+ \diagup T_b$ equals $\frac{2}{\pi}\log|t_+|$ for the larger absolute value $|t_+| = \frac{|a+d|}{2} + \frac{\sqrt{(a+d)^2-4}}{2}$ among the eigenvalues of the matrix. The smallest among these values over all matrices in $SL_2(\mathbb{Z})$ is obtained for $|a+d| = 3$. Hence, $\frac{2}{\pi}\log\frac{3+\sqrt{5}}{2}$ is the smallest non-vanishing extremal length among conjugacy classes of 3-braids. □

Notice that the trace of the matrix corresponding to T_{σ_j}, $j = 1, 2$, equals 2, and both eigenvalues are equal to 1. Hence, the matrix is conjugate by a matrix in $SL_2(\mathbb{R})$ to the sum of the unit matrix and an upper diagonal matrix. Hence, both eigenvalues of all non-trivial powers of T_{σ_1} and of all non-trivial powers of T_{σ_2} are equal to 1. This implies that $\Lambda(\widehat{\sigma_1^k}) = \Lambda(\widehat{\sigma_2^k}) = 0$ for all integers $k \neq 0$. Moreover, for $T_{\sigma_1\sigma_2}(z) = T_{\sigma_2} \circ T_{\sigma_1}(z) = \frac{z-1}{3z-2}$ the associated matrix has trace 1. The mapping has a fixed point in $\mathbb{C}_+$. The eigenvalues of the associated matrix are $\frac{1}{2} \pm \frac{\sqrt{3}}{2}i$ (see Sect. 1.2). The power $T^3_{\sigma_1\sigma_2}$ is equal to the identity and $\Lambda(\widehat{\sigma_1\sigma_2}) = 0$.

3.3 Invariants of Conjugacy Classes of Braids: Statement of the Main Theorem

The Entropy of Braids Recall that the entropy $h(b)$ of a braid $b \in \mathcal{B}_n$ is defined as the infimum of the entropies of self-homeomorphims of the closed disc $\overline{\mathbb{D}}$ that are contained in the mapping class $\mathfrak{m}_b$ associated to b,

$$h(b) = \inf\{h(\varphi) : \varphi \in \mathfrak{m}_b\}. \tag{3.14}$$

The value $h(\varphi)$ is invariant under conjugation with self-homeomorphisms of the closed disc $\overline{\mathbb{D}}$, hence it does not depend on the position of the set of distinguished points and on the choice of the representative of the conjugacy class $\hat{b}$. We write

$h(\hat{b}) = h(b)$. In Chap. 2 we studied the entropy of irreducible mapping classes and braids.

The entropy of a mapping class is an important dynamical invariant which measures the complexity of its behaviour in terms of iterations. It has received a lot of attention and has been studied intensively. As for braids (and the associated mapping classes), it has been known that the entropy of any irreducible braid is the logarithm of an algebraic number. Further, the lowest non-vanishing entropy h_n among irreducible braids on n strands, $n \geq 3$, has been estimated from below by $\frac{\log 2}{4} n^{-1}$ ([70]) and has been computed for small n. The entropy of a few more braids has been computed explicitly. It has been known that the smallest non-vanishing entropy among 3-braids equals $\log \frac{3+\sqrt{5}}{2}$ and is attained on the braid $\sigma_1^{-1}\sigma_2$. Further, there is an algorithm which detects in principle whether a braid (respectively, a mapping class) is pseudo-Anosov and in this case it gives a computer assisted computation of the entropy ([13]). This approach uses so-called train tracks. Further, fluid mechanics related to stirring devises uses the entropy of the arising closed braids as a measure of complexity.

The Conformal Module of Conjugacy Classes of Braids On the other hand, the conformal module (or its inverse, the extremal length) is a conformal invariant of conjugacy classes of braids, and is even older than the entropy. The conformal module of conjugacy classes of braids appeared first (without name) in the paper [33] in connection with the interest of the authors in Hilbert's Thirteen's Problem. This invariant of conjugacy classes of braids is undeservedly almost forgotten, although it appeared again in Gromov's seminal paper [35]. Gromov observed that the conformal module of conjugacy classes of braids defines restrictions for the validity of his Oka Principle concerning homotopies of continuous or smooth objects involving braids to the respective holomorphic objects.

We recall the definition of the conformal module of conjugacy classes of braids. (See Definition 3.4 for the definition of the conformal module of the conjugacy classes of elements of the fundamental group of a complex manifold.)

We say that a continuous mapping f of a round annulus $A = \{z \in \mathbb{C} : r < |z| < R\}$, $0 \leq r < R \leq \infty$, into $C_n(\mathbb{C})\diagup \mathcal{S}_n$ represents an element $\hat{b} \in \widehat{\mathcal{B}_n}$ if for some (and hence for any) circle $\{|z| = \rho\} \subset A$ the loop $f : \{|z| = \rho\} \to C_n(\mathbb{C})\diagup \mathcal{S}_n$ represents $\hat{b}$.

Let $\hat{b}$ be a conjugacy class of n-braids, $n \geq 2$. The conformal module $M(\hat{b})$ of $\hat{b}$ is defined as $M(\hat{b}) = \sup_{\mathcal{A}(\hat{b})} m(A)$, where $\mathcal{A}(\hat{b})$ denotes the set of all annuli which admit a holomorphic mapping into $C_n(\mathbb{C})\diagup \mathcal{S}_n$ which represents $\hat{b}$. The extremal length $\Lambda(\hat{b})$ of $\hat{b}$ is defined as $\Lambda(\hat{b}) = \inf_{\mathcal{A}(\hat{b})} \lambda(A)$.

The Relation Between Entropy and Conformal Module The interesting point is that the two invariants of conjugacy classes of braids, the entropy, which is a dynamical invariant, and the conformal module, which is a conformal invariant related to complex and algebraic geometry, carry the same information on the braid class. More precisely, the following theorem holds.

Main Theorem *For each conjugacy class of braids $\hat{b} \in \hat{\mathcal{B}}_n$, $n \geq 2$, the following equality holds*

$$\mathcal{M}(\hat{b}) = \frac{\pi}{2} \frac{1}{h(\hat{b})}.$$

Equivalently,

$$\Lambda(\hat{b}) = \frac{2}{\pi} h(\hat{b}).$$

The Main Theorem allows to apply methods and results known for the more intensively studied entropy to problems whose solutions are based on the concept of the conformal module. For instance, the Main Theorem together with the lower bounds for the entropy of irreducible n-braids allow to give a new conceptual proof of a slightly improved version of the Theorem of [33], which was the first theorem that used the concept of conformal module (see Chap. 6). The Main Theorem together with Example 3.8 of Sect. 3.2 give another proof of the fact that the smallest non-vanishing entropy among 3-braids equals $\log \frac{3+\sqrt{5}}{2}$. Methods of quasi-conformal mappings related to the concept of extremal length and conformal module are applied to give upper and lower bounds, differing by a universal multiplicative constant, for the entropy of 3-braids (see Chap. 9).

In the present chapter we will give a proof of the Main Theorem for the case of irreducible braids. In Chap. 4 we will give its proof for arbitrary pure braids, and in Chap. 5 the proof of the general case will be completed.

In Chaps. 6, 7, 8, 9, 10, and 11 we will give applications of the concept of the conformal module. Chapters 7, 8, and 11 are devoted to Gromov's Oka Principle. In these Chapters we apply the concept of the conformal module to describe restrictions for the existence of homotopies of continuous objects involving braids to the respective holomorphic objects.

3.4 Main Theorem. The Upper Bound for the Conformal Module. The Irreducible Case

Let $\hat{b} \in \hat{\mathcal{B}}_n$ be a conjugacy class of (maybe, reducible) braids and let $\mathrm{f} : A \to C_n(\mathbb{C})/\mathcal{S}_n$ be a holomorphic mapping of an annulus A into the symmetrized configuration space which represents $\hat{b}$. Our goal is to give an upper bound for the conformal module $m(A)$.

It will be convenient to identify the annulus with the quotient of the upper half-plane by an automorphism of the upper half-plane as follows. For a number $\rho > 1$ we denote by Λ_ρ the linear map $z \to \rho z$ on the upper half-plane $\mathbb{C}_+ = \{z \in \mathbb{C} : \operatorname{Im} z > 0\}$. The quotient $\mathbb{C}_+ / \Lambda_\rho$ is conformally equivalent to an annulus of conformal module $\frac{\pi}{\log \rho}$. Indeed, the half-open curvilinear rectangle $\{re^{i\theta} : 1 \le r < \rho\,,\ 0 < \theta < \pi\}$ is a fundamental polygon for the covering $\Lambda_\rho : \mathbb{C}_+ \to \mathbb{C}_+ \diagup \Lambda_\rho$. The logarithm maps it to $\{x + iy : 0 \le x < \log \rho\,,\ 0 < y < \pi\}$. Identifying points on the vertical sides with equal y-coordinate we obtain an annulus of conformal module $\frac{\pi}{\log \rho}$.

Royden's Theorem and Translation Length The key ingredient for obtaining the upper bound for the conformal module is Royden's Theorem 1.5 on equality of the Kobayashi and the Teichmüller metric on the Teichmüller space $\mathcal{T}(0, n+1)$. Let d_{hyp} be the hyperbolic metric $\frac{|dz|}{2y}$ on $\mathbb{C}_+$. By Royden's Theorem 1.5 any holomorphic mapping $\mathcal{F} : \mathbb{C}_+ \to \mathcal{T}(0, n+1)$ is a contraction from $(\mathbb{C}_+, d_{\text{hyp}})$ to $(\mathcal{T}(0, n+1), d_{\mathcal{T}})$. In particular, for any positive number ρ

$$d_{\mathcal{T}}(\mathcal{F}(i), \mathcal{F}(\rho i)) \le d_{\text{hyp}}(i, \rho i) = \frac{1}{2} \log \rho\,. \tag{3.15}$$

Let φ be a self-homeomorphism of $\mathbb{P}^1$ with set of distinguished points $E_n^0 \cup \{\infty\}$, and φ^* its modular transformation on $\mathcal{T}(0, n+1)$. Suppose $\mathcal{F}$ has the following invariance property

$$\mathcal{F}(\rho z) = \varphi^*(\mathcal{F}(z)),\ z \in \mathbb{C}_+\,. \tag{3.16}$$

The invariance property allows to associate to $\mathcal{F}$ a holomorphic map of the annulus $\mathbb{C}_+ \diagup z \sim \rho z$ of conformal module $\frac{\pi}{\log \rho}$ to the quotient $\mathcal{T}(0, n+1) \diagup (\tau \sim \varphi^*(\tau))$. By Royden's Theorem

$$\frac{1}{2} \log \rho \ge L(\varphi^*)\,. \tag{3.17}$$

Indeed, for the term on the left hand side of (3.15) the inequality

$$\begin{aligned} d_{\mathcal{T}}(\mathcal{F}(i), \mathcal{F}(\rho\, i)) &= d_{\mathcal{T}}(\mathcal{F}(i), \varphi^*(\mathcal{F}(i))) \\ &\ge \inf_{\tau \in \mathcal{T}(0,n+1)} d_{\mathcal{T}}(\tau, \varphi^*(\tau)) = L(\varphi^*) \end{aligned} \tag{3.18}$$

holds.

The Mapping to $\mathcal{T}(0, n{+}1)$ Associated to a Mapping to $C_n(\mathbb{C})\diagup\mathcal{S}_n$ The relation between the symmetrized configuration space $C_n(\mathbb{C})\diagup\mathcal{S}_n$ and the Teichmüller space of the $(n+1)$-punctured Riemann sphere (see Chap. 1) will allow us to apply Teichmüller theory.

Take any holomorphic mapping $\mathsf{f} : \mathbb{C}\diagup\Lambda_\rho \to C_n(\mathbb{C})\diagup\mathcal{S}_n$. The set $\mathbb{C}\diagup\Lambda_\rho$ is identified with an annulus of conformal module $\frac{\pi}{\log \rho}$. Lift f to a Λ_ρ-equivariant mapping from $\mathbb{C}_+$ to $C_n(\mathbb{C})\diagup\mathcal{S}_n$ which we denote by f.

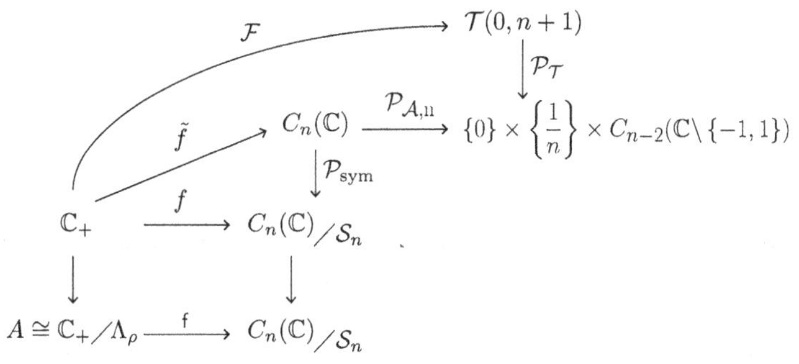

Fig. 3.4 The holomorphic mapping to Teichmüller space associated to a braid

Take a lift $\tilde{f} : \mathbb{C}_+ \to C_n(\mathbb{C})$ of f, i.e. a mapping $\tilde{f}$ for which $\mathcal{P}_{\mathrm{sym}}\tilde{f} = f$. The mapping $\mathcal{P}_{\mathcal{A}}\tilde{f}$ is a holomorphic mapping from $\mathbb{C}_+$ to $C_n(\mathbb{C})\diagup\mathcal{A}$. For the holomorphic isomorphism $\mathrm{Is}_{\mathrm{n}} : C_n(\mathbb{C})\diagup\mathcal{A} \to \{0\} \times \{\frac{1}{n}\} \times C_{n-2}(\mathbb{C} \setminus \{-1, 1\})$ we lift the mapping $\mathcal{P}_{\mathcal{A},\mathrm{n}}\tilde{f} = \mathrm{Is}_{\mathrm{n}} \circ \mathcal{P}_{\mathcal{A}}\tilde{f}$ with respect to the holomorphic projection $\mathcal{P}_{\mathcal{T}} : \mathcal{T}(0, n + 1) \to \{0\} \times \{\frac{1}{n}\} \times C_{n-2}(\mathbb{C} \setminus \{-1, 1\})$ and obtain a holomorphic map $\mathcal{F} : \mathbb{C}_+ \to \mathcal{T}(0, n + 1)$, such that $\mathcal{P}_{\mathcal{T}}\mathcal{F} = \mathcal{P}_{\mathcal{A},\mathrm{n}}\tilde{f}$. We have the commutative diagram Fig. 3.4. (For the definition of $\mathcal{P}_{\mathrm{sym}}$, $\mathcal{P}_{\mathcal{A},\mathrm{n}}$ and Is_{n} see Chap. 1.)

Denote by E_n the point $E_n = f(i) = f(i\rho)$ and by $(z_1, \ldots, z_n) \in C_n(\mathbb{C})$ the point $\tilde{f}(i)$. Note that $\mathcal{P}_{\mathrm{sym}}((z_1, \ldots, z_n)) = E_n$. For notational convenience we put $E_n' = E_n \cup \{\infty\}$. Choose a smooth self-homeomorphism ψ_i of $\mathbb{P}^1$ that is the identity outside a large disc containing $E_n^0 = \{0, \frac{1}{n}, \ldots, \frac{n-1}{n}\}$ and E_n, and maps ∞ to ∞, and $\frac{j-1}{n}$ to z_j for $j = 1, \ldots, n$, and has the property $\mathcal{F}(i) = [\psi_i]$.

Recall that each braid b with base point E_n^0 corresponds to a mapping class $\mathfrak{m}_b$ of self-homeomorphisms of the unit disc which fix the set E_n^0 setwise and the boundary circle pointwise. We denoted by $\varphi_{b,\infty}^*$ the modular transformation on $\mathcal{T}(0, n+1)$ of an element of the mapping class $\mathfrak{m}_{b,\infty} \in \mathcal{T}(0, n + 1)$. Recall that a representative of the mapping class $\mathfrak{m}_{b,\infty}$ may be obtained by extending an element of $\mathfrak{m}_b$ to the Riemann sphere putting it equal to the identity outside the disc. For a self-homeomorphism ψ of $\mathbb{P}^1$ that fixes ∞ and an element $\{\zeta_1, \ldots, \zeta_n\} \in C_n(\mathbb{C})\diagup\mathcal{S}_n$ we put as before $\psi(\{\zeta_1, \ldots, \zeta_n\}) = \{\psi(\zeta_1), \ldots, \psi(\zeta_n)\}$.

The following lemma holds.

Lemma 3.6 *Let $\tilde{f} : \mathbb{C}_+ \to C_n(\mathbb{C})$ be a smooth mapping such that $f = \mathcal{P}_{\mathrm{sym}}(\tilde{f})$ is Λ_ρ-invariant. Let $\mathcal{F} : \mathbb{C}_+ \to \mathcal{T}(0, n + 1)$ be a smooth mapping such that the diagram Fig. 3.4 commutes. Denote by b the braid with base point E_n^0 represented by the geometric braid*

$$\psi_i^{-1} \circ f \mid [i, i\rho],\ t \in [1, \rho] \tag{3.19}$$

where $\psi_i \in \mathcal{F}(i)$ is the self-diffeomorphism of $\mathbb{P}^1$ that was chosen above. Then for all $z \in \mathbb{C}_+$

$$\mathcal{F}(\rho\, z) = \varphi^*_{b,\infty}(\mathcal{F}(z)). \tag{3.20}$$

Proof Let z be an arbitrary point in $\mathbb{C}_+$, and let α_z be a smooth curve in $\mathbb{C}_+$ that joins the points i and z. By the proof of Proposition 1.1 (applied to $\psi_i^{-1} \circ f$) there exists a smooth family of self-diffeomorphisms ψ_ζ of $\mathbb{P}^1$, $\zeta \in \alpha_z$, that are equal to the identity outside a large disc, such that $\psi_\zeta(E_n^0) = f(\zeta)$, and for $\zeta = i$ the diffeomorphism of the family coincides with the diffeomorphism ψ_i in the statement of the lemma.

As a consequence $[\psi_\zeta] = \mathcal{F}(\zeta)$, $\zeta \in \alpha_z$, since both mappings, $\zeta \to [\psi_\zeta]$ and $\zeta \to \mathcal{F}(\zeta)$ lift $\mathcal{P}_{\mathcal{A},\mathrm{n}}(\tilde{f}(\zeta))$ by the commutative diagram of Lemma 1.5.

The family

$$\zeta \to \Big\{\big(t,\ \psi_\zeta^{-1}(f(\zeta t))\big),\ t \in [1,\rho]\Big\}, \zeta \in \alpha_z\,, \tag{3.21}$$

is an isotopy of geometric braids with base point E_n^0 that joins the geometric braid (3.19) with the geometric braid

$$\Big\{\big(t,\ \psi_z^{-1}(f(zt))\big),\ t \in [1,\rho]\Big\}. \tag{3.22}$$

Fix $z \in \mathbb{C}_+$ and take a parameterizing isotopy for the geometric braid (3.22). More precisely, for a large positive number R we take a smooth family of homeomorphisms $\varphi_t^z \in \mathrm{Hom}^+(\mathbb{P}^1;\ \mathbb{P}^1 \setminus R\mathbb{D})$, $t \in [1,\rho]$, such that

$$\mathrm{ev}_{E_n^0}\, \varphi_t^z = \psi_z^{-1}(f(z\,t)), \quad t \in [1,\rho], \tag{3.23}$$

and φ_1^z is the identity. Then the mapping φ_ρ^i represents the mapping class $\mathfrak{m}_{b,\infty} \in \mathfrak{M}(\mathbb{P}^1;\ \{\infty\},\ E_n^0)$. Write $\varphi_\rho^z = \varphi^z_{b,\infty}$. Since the $\varphi^z_{b,\infty}$ are isotopic with base point E_n^0, their modular transformations do not depend on z. Let $\varphi^*_{b,\infty}$ be the modular transformation induced on $\mathcal{T}(0, n+1)$ by the $\varphi^z_{b,\infty}$.

Equation (3.23) can also be written as $\mathrm{ev}_{E_n^0}(\psi_z \circ \varphi_t^z) = f(z\,t), t \in [1,\rho]$. Hence, the mapping $t \to e_n(\psi_z \circ \varphi_t^z)$ is a lift of the mapping $t \to f(zt), t \in [0,1]$. Clearly also $t \to \tilde{f}(zt), t \in [0,1]$, lifts the mapping $t \to f(zt), t \in [0,1]$. Moreover, both mappings, $e_n(\psi_z \circ \varphi_t^z)$ and $\tilde{f}(z\,t)$, take the value $\tilde{f}(i) = (z_1, \ldots, z_n)$ for $t = 1$. For $e_n(\psi_z \circ \varphi_t^z)$ this follows from the definition of ψ_i and the fact that $\varphi_1^z = \mathrm{id}$. Hence, $e_n(\psi_z \circ \varphi_t^z) = \tilde{f}(z\,t), t \in [1,\rho]$. Using also Lemma 1.5 we obtain the commutative diagram Fig. 3.5.

The diagrams Figs. 3.4 and 3.5 show that both mappings, $t \to \mathcal{F}(z\,t)$ and $t \to [\psi_z \circ \varphi_t^z]$, lift the mapping $t \to \mathcal{P}_{\mathcal{A},\mathrm{n}}(\tilde{f}(z\,t))$ to a mapping from $[1,\rho]$ to the Teichmüller space $\mathcal{T}(0, n+1)$. Moreover, since $\mathcal{F}(z) = [\psi_z]$ and $\varphi_1^z = \mathrm{id}$, we may

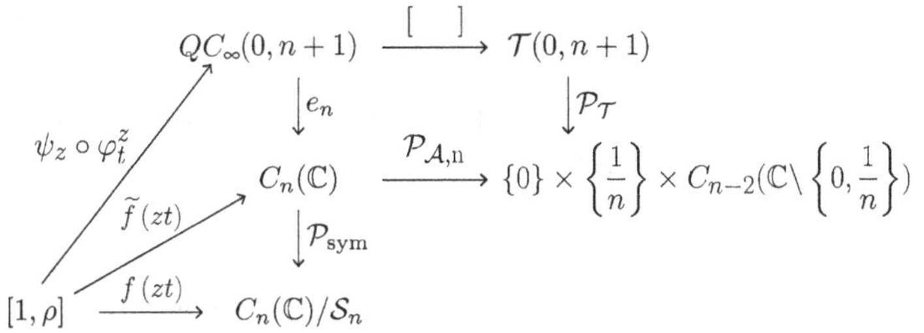

Fig. 3.5 Another lift to the Teichmüller space

write $\mathcal{F}(z) = [\psi_z \circ \varphi_1^z]$, which is the value of $[\psi_z \circ \varphi_t^z]$ for $t = 1$. Therefore, the two lifts coincide:

$$\mathcal{F}(z\,t) = [\psi_z \circ \varphi_t^z], \quad t \in [1, \rho]. \tag{3.24}$$

In particular,

$$\mathcal{F}(\rho\, z) = [\psi_z \circ \varphi_\rho^z] = [\psi_z \circ \varphi_{b,\infty}^z] = (\varphi_{b,\infty})^*([\psi_z]) = \varphi_{b,\infty}^*(\mathcal{F}(z)). \tag{3.25}$$

We obtained (3.20) for the arbitrarily chosen point z. The lemma is proved. □

Translation Length and the Upper Bound for the Conformal Module: The General Case The following proposition gives an upper bound for the conformal module of a conjugacy class of braids in terms of the translation length of the associated modular transformation that holds for all (maybe, reducible) n-braids.

Proposition 3.1 *Let b be an n-braid with base point E_n^0 and $\mathfrak{m}_b$ its mapping class (considered as mapping class on a large n-punctured disc). For the associated mapping class $\mathfrak{m}_{b,\infty} = \mathcal{H}_\infty(\mathfrak{m}_b) \in \mathfrak{M}(\mathbb{P}^1;\ \infty,\ E_n^0)$ on the Riemann sphere we consider a representative $\varphi_{b,\infty}$ of $\mathfrak{m}_{b,\infty}$, its modular transformation $\varphi_{b,\infty}^*$ on $\mathcal{T}(0, n+1)$, and the translation length $L(\varphi_{b,\infty}^*)$. Then*

$$L(\varphi_{b,\infty}^*) \le \frac{\pi}{2}\,\frac{1}{\mathcal{M}(\hat{b})}. \tag{3.26}$$

Proof As before we take an arbitrary holomorphic mapping f that takes the annulus $A = \mathbb{C}_+/\Lambda_\rho$ of conformal module $m(A) = \frac{\pi}{\log \rho}$ to the symmetrized configuration space, and represents the conjugacy class $\hat{b}$ of the braid b. Then by Lemma 3.6 and Royden's Theorem the inequality (3.17) holds for the modular transformation $\varphi_{b,\infty}^*$ and the mapping $\mathcal{F}$ associated to f by the diagram Fig. 3.4. Hence, in terms of the conformal module of A

$$\frac{1}{2}\frac{\pi}{m(A)} \ge L(\varphi_{b,\infty}^*).$$

Taking the supremum over the conformal modules of all annuli admitting a holomorphic mapping representing $\hat{b}$, we obtain (3.26). □

Entropy and the Upper Bound for the Conformal Module: The Irreducible Case Proposition 3.1 implies the upper bound for the conformal module of *irreducible* braids in terms of the entropy. More precisely, the following proposition holds.

Proposition 3.2 *For each irreducible conjugacy class of braids $\hat{b} \in \hat{\mathcal{B}}_n$ the inequality*

$$h(\hat{b}) \leq \frac{\pi}{2} \frac{1}{\mathcal{M}(\hat{b})}$$

holds.

Proof If a braid b with base point E_n^0 is irreducible, then, $\varphi_{b,\infty}$ is irreducible and therefore $\varphi_{b,\infty}^*$ is either elliptic or hyperbolic. By Corollary 1.1 there is an absolutely extremal self-homeomorphism $\tilde{\varphi}_{b,\infty}$ of a Riemann surface $\mathbb{C} \backslash \tilde{E}_n$ which is obtained from $\varphi_{b,\infty}$ by isotopy and conjugation. More precisely, there is a homeomorphism $w : \mathbb{C} \backslash E_n^0 \to \mathbb{C} \backslash \tilde{E}_n$ and a self-homeomorphism $\hat{\varphi}_{b,\infty}$ of $\mathbb{C} \backslash E_n^0$ which is isotopic to $\varphi_{b,\infty}$ on $\mathbb{C} \backslash E_n^0$ so that $\tilde{\varphi}_{b,\infty} = w \circ \hat{\varphi}_{b,\infty} \circ w^{-1} : \mathbb{C} \backslash \tilde{E}_n \to \mathbb{C} \backslash \tilde{E}_n$ is absolutely extremal. By Theorems 1.8 and 1.10 the mapping $\tilde{\varphi}_{b,\infty}$ is pseudo-Anosov if $\varphi_{b,\infty}^*$ is hyperbolic, and is a periodic conformal mapping if $\varphi_{b,\infty}^*$ is elliptic. The entropy of a periodic mapping and its quasi-conformal dilatation vanish. Hence, by Theorem 2.1 and Corollary 1.1 for the absolutely extremal mapping $\tilde{\varphi}_{b,\infty}$

$$\frac{1}{2} \log K(\tilde{\varphi}_{b,\infty}) = L(\varphi_{b,\infty}^*) = h(\tilde{\varphi}_{b,\infty}) = h(\widehat{\mathfrak{m}_{b,\infty}}) . \tag{3.27}$$

In the pseudo-Anosov case Theorem 2.9 provides an isotopy of $\tilde{\varphi}_{b,\infty}$ through self-homeomorphisms of $\mathbb{C} \backslash \tilde{E}_n$ to a homeomorphism $\tilde{\varphi}_b$ which is the identity outside the disc $R\,\overline{\mathbb{D}}$ and has the same entropy as $\tilde{\varphi}_{b,\infty}$:

$$h(\tilde{\varphi}_b) = h(\tilde{\varphi}_{b,\infty}) . \tag{3.28}$$

Suppose $\varphi_{b,\infty}^*$ is elliptic. Then there is a periodic conformal map $\tilde{\varphi}_{b,\infty} \in \text{Hom}^+(\mathbb{C}; \emptyset, E_n)$ which is obtained from $\varphi_{b,\infty}$ by isotopy and conjugation. Here $E_n \subset \mathbb{C}$ is a set consisting of n points. The mapping $\tilde{\varphi}_{b,\infty}$ is a Möbius transformation that fixes ∞ and, being periodic, has a fixed point on $\mathbb{P}^1$ different from ∞. Conjugate this fixed point to zero. We obtain that $\tilde{\varphi}_{b,\infty}$ is conjugate to multiplication by a root of unity $\omega = e^{2\pi i \frac{\mathsf{p}}{\mathsf{q}}}$ for integer numbers p and q, $\mathsf{q} \neq 0$. Since $\tilde{\varphi}_{b,\infty}(E_n) = E_n$, the set E_n is either equal to $\left\{r, r\,e^{\frac{2\pi i}{n}}, \ldots, r\,e^{\frac{2\pi i(n-1)}{n}}\right\}$ for some $r > 0$, and in this case $\omega = e^{\frac{2\pi i \mathsf{p}}{n}}$; or it is equal to $E_n = \left\{0, r, r e^{\frac{2\pi i}{n-1}}, \ldots, r\,e^{\frac{2\pi i(n-2)}{n-1}}\right\}$ for some $r > 0$, and in this case

$\omega = e^{\frac{2\pi i p}{n-1}}$. Lemma 2.14 provides a self-homeomorphism $\tilde{\varphi}_b$ of $\mathbb{P}^1$ that is isotopic to $\tilde{\varphi}_{b,\infty}$ by an isotopy that fixes E_n, equals $\tilde{\varphi}_{b,\infty}$ on the disc $R_1\mathbb{D}$ for a large number R_1, equals the identity outside $R\mathbb{D}$ for some $R > R_1$, is a twist on the annulus $R\mathbb{D} \setminus R_1\mathbb{D}$, and, finally, it has the same entropy as $\tilde{\varphi}_{b,\infty}$.

In both cases for the obtained homeomorphism $\tilde{\varphi}_b$ the mapping $w^{-1} \circ \tilde{\varphi}_b \circ w$ is isotopic to $\varphi_{b,\infty}$ through self-homeomorphisms of $\mathbb{C}\setminus E_n^0$ and is the identity outside the disc $R\,\overline{\mathbb{D}}$. By Theorem 1.3 the mapping class $\mathfrak{m}_b$ associated to b, and the mapping class of $w^{-1} \circ \tilde{\varphi}_b \circ w \mid R\,\overline{\mathbb{D}}$ in $\mathfrak{M}(R\,\overline{\mathbb{D}}; \partial(R\,\overline{\mathbb{D}}), E_n^0)$ differ by a power of a Dehn twist about a circle in $R\,\mathbb{D}$ of large radius. By Corollary 2.2 the entropies of the two mapping classes are equal. We obtain

$$h(\hat{b}) = h(b) = \inf\{h(\varphi) : \varphi \in \mathfrak{m}_b\} = \inf\{h(\varphi) : \varphi \in \mathfrak{m}_{w^{-1} \circ \tilde{\varphi}_b \circ w | R\mathbb{D}}\}$$
$$\leq h(w^{-1} \circ \tilde{\varphi}_b \circ w \mid R\,\mathbb{D})\,. \tag{3.29}$$

Hence we obtain from (3.27), (3.28), and (3.29)

$$h(\hat{b}) \leq h(w^{-1} \circ \tilde{\varphi}_b \circ w) = h(\tilde{\varphi}_b) = h(\tilde{\varphi}_{b,\infty}) = L(\varphi^*_{b,\infty})\,.$$

By (3.26) we have $h(\hat{b}) \leq \frac{\pi}{2}\frac{1}{\mathcal{M}(\hat{b})}$.

The proposition is proved. □

3.5 The Lower Bound for the Conformal Module. The Irreducible Case

We will prove here the following proposition for the irreducible case.

Proposition 3.3 *Let $\hat{b} \in \hat{\mathcal{B}}_n$ be an irreducible conjugacy class of braids. Then for any annulus A of conformal module $m(A) = \frac{\pi}{2}\frac{1}{h(\hat{b})}$ the class $\hat{b}$ can be represented by a holomorphic map $f : A \to C_n(\mathbb{C})\diagup \mathcal{S}_n$.*

Proposition 3.3 is stronger than the statement of the Main Theorem in the irreducible case. Indeed, the Proposition 3.3 asserts furthermore that in the irreducible case the supremum in Definition 3.4 is attained. Moreover, Proposition 3.3 says, that in case the supremum is infinite, it is attained on the punctured complex plane, not merely on the punctured disc.

Take an irreducible conjugacy class of n-braids $\hat{b} \in \hat{\mathcal{B}}_n$. Let $b \in \mathcal{B}_n$ be a braid with base point E_n^0 representing $\hat{b}$. Let φ_b be a self-homeomorphism of $\overline{\mathbb{D}}\setminus E_n^0$ which represents the mapping class $\mathfrak{m}_b = \Theta_n(b) \in \mathfrak{M}(\overline{\mathbb{D}}; \partial\,\mathbb{D}, E_n^0)$ of b. Let $\varphi_{b,\infty} \in \mathrm{Hom}^+(\mathbb{C}; \emptyset, E_n^0)$ be the extension of φ_b to the whole plane which is the identity outside the unit disc.

Since the class $\hat{b}$ is irreducible, the mapping $\varphi_{b,\infty}$ is irreducible. By Theorem 1.11 the induced modular transformation $\varphi^*_{b,\infty}$ on the Teichmüller space $\mathcal{T}(0, n+1)$ is either elliptic or hyperbolic.

Proof of Proposition 3.3

The Elliptic Case In this case the mapping $\tilde{\varphi}_{b,\infty}$ is conjugate to $\tilde{\varphi}_{b,\infty}(z) = e^{2\pi i \frac{\mathsf{p}}{\mathsf{q}}} z$ on $\mathbb{P}^1$ with set of distinguished points E_n either equal to $\left\{r, r\,e^{\frac{2\pi i}{n}}, \dots, r\,e^{\frac{2\pi i(n-1)}{n}}\right\}$ for some $r > 0$, and in this case $\omega = e^{\frac{2\pi i \mathsf{p}}{n}}$; or it is equal to $E_n = \left\{0, r, r e^{\frac{2\pi i}{n-1}}, \dots, r\,e^{\frac{2\pi i(n-2)}{n-1}}\right\}$, and in this case $\omega = e^{\frac{2\pi i \mathsf{p}}{n-1}}$. As above p and q are integer numbers. We may assume that the mapping itself has this form.

Consider the universal covering

$$\mathbb{C} \ni \zeta = \xi + i\eta \to e^{\xi + i\eta} \in \mathbb{C}^*$$

of the annulus $\mathbb{C}^* = \mathbb{C}\backslash\{0\} = \{z \in \mathbb{C} : 0 < |z| < \infty\}$.

Denote by $\tilde{\mathcal{F}}$ the following holomorphic mapping from $\mathbb{C}$ into the space $\mathcal{A}$ of complex affine mappings,

$$\tilde{\mathcal{F}}(\zeta) = \mathfrak{a}(\zeta) \in \mathcal{A}\,, \text{ where } \quad \mathfrak{a}(\zeta)(z) = e^{\frac{\mathsf{p}}{\mathsf{q}} \cdot \zeta} \cdot z\,. \tag{3.30}$$

Notice that

$$\mathfrak{a}(0)(z) = z \quad \text{and} \quad \mathfrak{a}(2\pi i)(z) = e^{2\pi i \frac{\mathsf{p}}{\mathsf{q}}} z = \tilde{\varphi}_{b,\infty}(z)\,. \tag{3.31}$$

The evaluation map

$$\mathbb{C} \ni \zeta \to \mathrm{ev}_{E_n}\, \mathfrak{a}(\zeta) \in C_n(\mathbb{C})\diagup\mathcal{S}_n \tag{3.32}$$

is holomorphic and is periodic with period $2\pi i$. Hence, the evaluation map induces a holomorphic map f from $\mathbb{C}^*$ to $C_n(\mathbb{C})\diagup\mathcal{S}_n$. f represents a conjugacy class $\hat{b}' \in \hat{\mathcal{B}}_n$. By Theorem 1.3 the mapping classes corresponding to the braids $b' \in \hat{b}'$, and $b \in \hat{b}$ differ by a power of a Dehn twist. Replacing p by $\mathsf{p} + 2\pi k \mathsf{q}$ for a suitable integer number k we may achieve that $f : \mathbb{C}^* \to C_n(\mathbb{C})\diagup\mathcal{S}_n$ represents the conjugacy class $\hat{b} \in \hat{\mathcal{B}}_n$. The proposition is proved for the elliptic case. □

Note that in the elliptic case $r = \infty$, and we may take $A = \mathbb{C}^*$.

The Hyperbolic Case. Teichmüller Discs The plan of the proof in the hyperbolic case is the following. We will associate to the irreducible braid b a Teichmüller disc, i.e. a holomorphic map $\mathcal{F} : \mathbb{C}_+ \to \mathcal{T}(0, n+1)$ (see Fig. 3.4). Given $\mathcal{F}$, we need to find a mapping $\tilde{f} : \mathbb{C}_+ \to C_n(\mathbb{C})$ for which $\mathcal{P}_{\mathcal{A},n}\tilde{f} = \mathcal{P}_{\mathcal{T}}\mathcal{F}$ (which means to find a holomorphic section), such that $\mathcal{P}_{\mathrm{sym}}\tilde{f}$ has the equivariance property for which the Diagram Fig. 3.4 is commutative.

Associate to the irreducible braid $b \in \hat{b}$ with base point $E_n^0 = \left\{0, \frac{1}{n}, \dots, \frac{n-1}{n}\right\}$ a representative φ_b of the mapping class $\mathfrak{m}_b \in \mathfrak{M}(\overline{\mathbb{D}}; \partial\,\mathbb{D}, E_n^0)$. Let $\varphi_{b,\infty} \in \mathrm{Hom}^+(\mathbb{C}; \emptyset, E_n^0)$ be the extension of φ_b to $\mathbb{C}$ which is the identity outside the unit disc $\mathbb{D}$. Consider the modular transformation $\varphi^*_{b,\infty}$ on $\mathcal{T}(0, n+1)$ that is associated to b, and let $\tilde{\varphi}_{b,\infty}$ be the absolutely extremal map that is obtained from $\varphi_{b,\infty}$ by isotopy and conjugation. Put $X_0 = \mathbb{C} \setminus E_n^0 = \mathbb{P}^1 \setminus (E_n^0)'$. Denote by $w_0 : X_0 \to w_0(X_0) = X$ the $\tilde{\varphi}_{b,\infty}$-minimal conformal structure on X_0. Let $X = w_0(X_0)$ and $E_n = w_0(E_n^0)$. The mapping $\tilde{\varphi}_{b,\infty}$ is a self-homeomorphism of X, equivalently, $\tilde{\varphi}_{b,\infty}$ defines a self-homeomorphism of $\mathbb{P}^1$ with set of distinguished points $E_n \cup \infty$, and $\tilde{\varphi}_{b,\infty}$ is isotopic to $w_0 \circ \varphi_{b,\infty} \circ w_0^{-1}$.

By Corollary 2.2 $h(\tilde{\varphi}_{b,\infty}) = h(\widehat{\mathfrak{m}_{\varphi_{b,\infty}}}) = h(\hat{b})$. Again we identify self-homeomorphisms of punctured surfaces with self-homeomorphisms of closed surfaces with distinguished points obtained by extension to the punctures.

The Beltrami differential of the self-homeomorphism $\tilde{\varphi}_{b,\infty}$ of X has the form $k\frac{\bar{\phi}}{|\phi|}$ for a number $k \in (0, 1)$ and a meromorphic quadratic differential ϕ on $\mathbb{P}^1$ which is holomorphic on $\mathbb{C} \setminus E_n$ and has at worst simple poles at the points of $E_n \cup \{\infty\}$. The quasiconformal dilatation of $\tilde{\varphi}_{b,\infty}$ equals $K = \frac{1+k}{1-k}$. Recall that by Theorem 2.1 we have $\frac{1}{2}\log K = h(\tilde{\varphi}_{b,\infty})$.

For $\mu_z = z\frac{\bar{\phi}}{|\phi|}$, $z \in \mathbb{D}$, we let W^{μ_z} be the normalized solution on X of the Beltrami equation for μ_z (see Definition 1.1). Consider the Teichmüller disc $\mathbb{D} \ni z \to [W^{\mu_z}] \in \mathcal{D}_{X,\phi}$ in $\mathcal{T}(X)$ which is associated to the quadratic differential ϕ on X (see Sect. 1.7). The modular transformation $\tilde{\varphi}^*_{b,\infty}$ maps the disc $\mathcal{D}_{X,\phi}$ onto itself. Indeed, $\tilde{\varphi}_{b,\infty}$ has Beltrami differential $k\frac{\bar{\phi}}{|\phi|}$. For each z in the open unit disc the mapping $W^{\mu_z} \circ \tilde{\varphi}_{b,\infty}$ is a Teichmüller mapping with quadratic differential $\mathrm{const} \cdot \phi$ (see Sect. 1.7). Hence $[W^{\mu_z} \circ \tilde{\varphi}_{b,\infty}] = \tilde{\varphi}^*_{b,\infty}([W^{\mu_z}])$ has the form $\{\mu_{z'}\}$ for some $z' \in \mathbb{D}$. If z is real then also z' is real. Each element $\{\mu_{z'}\}$, $z' \in \mathbb{D}$, is in the image under $\tilde{\varphi}^*_{b,\infty}$ of the Teichmüller disc . Since W^0 is the identity mapping, $\tilde{\varphi}^*_{b,\infty}$ maps the point $\{\mu_0\}$ to $\{\mu_k\}$.

The conformal structure w_0 realizes a canonical isomorphism between the Teichmüller space $\mathcal{T}(X)$ and the canonical Teichmüller space $\mathcal{T}(X_0) = \mathcal{T}(0, n+1)$. Indeed, associate to each conformal structure w on X the conformal structure $w \circ w_0$ on X_0. Its class $[w \circ w_0]$ in $\mathcal{T}(0, n+1)$ depends only on the class $[w]$ in $\mathcal{T}(X)$. Put $w_0^*([w]) = [w \circ w_0]$, $[w] \in \mathcal{T}(X)$. The mapping $w_0^* : \mathcal{T}(X) \to \mathcal{T}(0, n+1)$ gives the canonical isomorphism. w_0^* is a holomorphic mapping between Teichmüller spaces. We denote the image $w_0^*(\mathcal{D}_{X,\phi})$ in $\mathcal{T}(0, n+1)$ by $\mathcal{D}_\phi^0$. The mapping w_0^* takes the Teichmüller disc in $\mathcal{T}(X)$ to a Teichmüller disc

$$\mathbb{D} \ni z \xrightarrow{\mathfrak{e}} w_0^*(\{\mu^z\}) \in \mathcal{D}_\phi^0 \subset \mathcal{T}(0, n+1) \tag{3.33}$$

in $\mathcal{T}(0, n+1)$.

It will be convenient to reparameterize the Teichmüller disc by a mapping from the upper half-plane to Teichmüller space. For this purpose we consider the composition

$$\mathcal{F} : \mathbb{C}_+ \to \mathcal{D}_\phi^0, \quad \mathcal{F} = \mathfrak{e} \circ \mathfrak{c}^{-1} \tag{3.34}$$

with the conformal mapping $\mathfrak{c}(z) = i\,\frac{1+z}{1-z}$ from $\mathbb{D}$ onto $\mathbb{C}_+$.

The mapping $\varphi^*_{b,\infty} = w^*_0 \circ \tilde{\varphi}^*_{b,\infty} \circ (w^*_0)^{-1}$ takes $\mathcal{D}^0_\phi$ onto itself. Conjugate the restriction of the modular transformation $\varphi^*_{b,\infty} \mid \mathcal{D}^0_\phi$ by $\mathcal{F}$. We obtain a holomorphic homeomorphism Λ of the upper half-plane onto itself,

$$\Lambda = \mathcal{F}^{-1} \circ \varphi^*_{b,\infty} \circ \mathcal{F} : \mathbb{C}_+ \to \mathbb{C}_+ . \tag{3.35}$$

Since $\tilde{\varphi}^*_{b,\infty}$ maps each point $\{\mu_x\}$ with x real to a point $\{\mu_{x'}\}$ with real x', and maps $\{\mu_0\}$ to $\{\mu_k\}$, the mapping Λ fixes the imaginary axis and maps i to $i\,\frac{1+k}{1-k} = iK$. Hence

$$\Lambda(\zeta) = \Lambda_K(\zeta) \overset{def}{=} K\zeta\,, \quad \zeta \in \mathbb{C}_+. \tag{3.36}$$

By (3.35) and (3.36)

$$\mathcal{F}(K\zeta) = \varphi^*_{b,\infty}(\mathcal{F}(\zeta)). \tag{3.37}$$

The annulus $A = \mathbb{C}_+ \diagup \Lambda_K$ has conformal module $\frac{\pi}{2\log K^{\frac{1}{2}}}$ and is conformally equivalent to the quotient $\mathcal{D}^0_\phi \diagup \langle \varphi^*_{b,\infty} \rangle$.

The Mapping Representing the Braid Class Using a mapping $\mathcal{F} : \mathbb{C}_+ \to \mathcal{D}^0_\phi$ that satisfies (3.37), we will now represent the braid class $\hat{b}$ by a holomorphic mapping from an annulus to $C_n(\mathbb{C}) \diagup \mathcal{S}_n$. Associate to $\mathcal{F}$ the mapping $\mathcal{P}_\mathcal{T} \circ \mathcal{F}$ from $\mathbb{C}_+$ into $\{0\} \times \{\frac{1}{n}\} \times C_{n-2}(\mathbb{C} \setminus \{1, \frac{1}{n}\}) = \mathrm{Is}_n(C_n(\mathbb{C}) \diagup \mathcal{A})$ (see Sect. 1.7). The diagram Fig. 3.4 and the diagram of Lemma 1.5 suggest to find a suitable holomorphic section for the projection

$$\mathbb{C}_+ \times C_n(\mathbb{C}) \ni (\zeta, z) \to (\zeta, z \diagup \mathcal{A}) \in \mathbb{C}_+ \times C_n(\mathbb{C}) \diagup \mathcal{A}\,.$$

We need a suitable section only for $\mathbb{C}_+$ replaced by a neighbourhood $\mathcal{U}(\varepsilon) \overset{def}{=} \{\zeta \in \mathbb{C}_+ : 1 - \varepsilon < |\zeta| < (1+\varepsilon)K\}$ of the closure $\{\zeta \in \mathbb{C}_+ : 1 \le |\zeta| \le K\}$ of a fundamental domain in the universal covering $\mathbb{C}_+$ of the annulus $\mathbb{C}_+ \diagup \Lambda$. Each such holomorphic section has the following form

$$\mathcal{U}(\varepsilon) \times C_n(\mathbb{C}) \diagup \mathcal{A} \ni (\zeta, Z) \to (\zeta, \mathfrak{A}(\zeta)(\mathrm{Is}_n(Z))) \in \mathcal{U}(\varepsilon) \times C_n(\mathbb{C})$$

for a holomorphic map $\mathfrak{A} : \mathcal{U}(\varepsilon) \to \mathcal{A}$ that assigns to each point $\zeta \in \mathcal{U}(\varepsilon)$ a complex linear self-mapping $\mathfrak{A}(\zeta) \in \mathcal{A}$ of $\mathbb{C}$. Recall that for an element $\mathfrak{A}$ of $\mathcal{A}$ the point

$$\mathfrak{A}\Big((0, \frac{1}{n}, z_3, \ldots, z_n)\Big) \overset{def}{=} \Big(\mathfrak{A}(0), \mathfrak{A}(\frac{1}{n}), \mathfrak{A}(z_3), \ldots, \mathfrak{A}(z_n)\Big) \tag{3.38}$$

is the result of the diagonal action of the complex affine mapping $\mathfrak{A}$ on the element $(0, \frac{1}{n}, z_3, \ldots, z_n) \in \{0\} \times \{\frac{1}{n}\} \times C_{n-2}(\mathbb{C} \setminus \{0, \frac{1}{n}\})$.

For $\zeta \in \mathbb{C}_+$ we write

$$\mathcal{P}_{\mathcal{T}}(\mathcal{F}(\zeta)) = \big(0, \frac{1}{n}, z_3(\zeta), \ldots, z_n(\zeta)\big). \tag{3.39}$$

For a holomorphic mapping $\mathfrak{A} : \mathcal{U}(\varepsilon) \to \mathcal{A}$ we define the holomorphic mapping $f_{\mathfrak{A}} : \mathcal{U}(\varepsilon) \to C_n(\mathbb{C})$ by

$$f_{\mathfrak{A}}(\zeta) \overset{def}{=} \mathfrak{A}(\zeta)(\mathcal{P}_{\mathcal{T}}\Big(\mathcal{F}(\zeta))\Big).$$

The following lemma guarantees the existence of a mapping $\mathfrak{A}$ that is suitable for our purposes.

Lemma 3.7 *There exists a holomorphic map $\mathfrak{A} : \mathcal{U}(\varepsilon) \to \mathcal{A}$ such that the mapping $f(\zeta) \overset{\text{def}}{=} \mathcal{P}_{\text{sym}}\, f_{\mathfrak{A}}(\zeta)$, $\zeta \in \mathcal{U}(\varepsilon)$, has the following properties:*

$$f(K\zeta) = f(\zeta) \quad for \quad \zeta \in \mathcal{U}'(\varepsilon) \overset{def}{=} \{\zeta \in \mathbb{C} : 1 - \varepsilon < |\zeta| < 1 + \varepsilon\}, \tag{3.40}$$

and the family $f(it)$, $t \in [1, K]$, defines a closed path in $C_n(\mathbb{C})\diagup \mathcal{S}_n$ in the free isotopy class $\widehat{\Delta_n^{2\ell}\, b}$ for some $\ell \in \mathbb{Z}$.

Proof of Lemma 3.7 By Lemma 1.4 in a neighbourhood of each point ζ in $\mathbb{C}_+$ we may choose a smooth family $w_\zeta \in QC_\infty(0, n+1)$, such that w_ζ fixes 0 and $\frac{1}{n}$, $[w_\zeta] = \mathcal{F}(\zeta)$, and the equality $\mathcal{P}_{\mathcal{T}}(\mathcal{F}(\zeta)) = e_n(w_\zeta)$ holds.

The mapping $\varphi_{b,\infty}$ is a self-homeomorphism of $\mathbb{C} \setminus E_n^0$. Its extension to $\mathbb{C}$ acts on E_n^0 by permutation:

$$\varphi_{b,\infty}(\frac{j}{n}) = S_b(\frac{j}{n}),\ j = 0, 1, \ldots, n-1, \tag{3.41}$$

for the permutation $S_b = \tau(b)$ that is related to the braid b and maps the set $\{0, \frac{1}{n}, \ldots, \frac{n-1}{n}\}$ onto itself. Equation (3.41) is equivalent to the equation $e_n(\varphi_{b,\infty}) = S_b(e_n(\text{Id}))$.

Since w_ζ represents $\mathcal{F}(\zeta)$, the mapping $w_\zeta \circ \varphi_{b,\infty}$ represents $\mathcal{F}(K\zeta) = \varphi_{b,\infty}^*(\mathcal{F}(\zeta))$. Hence, the point $e_n(w_\zeta \circ \varphi_{b,\infty})$ represents the class of $\mathcal{P}_{\mathcal{T}}(\mathcal{F}(K\zeta))$ in the quotient $C_n(\mathbb{C})\diagup\mathcal{A}$ of $C_n(\mathbb{C})$ by the action of the group $\mathcal{A}$.

The locally defined family $\zeta \to e_n(w_\zeta)$ does not depend on the local choice of the ω_ζ and is holomorphic since it coincides with the holomorphic mapping $\zeta \to \mathcal{P}_{\mathcal{T}}(\mathcal{F}(\zeta))$. For $e_n(w_\zeta) = \big(0, \frac{1}{n}, \ldots, z_n(\zeta)\big) = \big(z_1(\zeta), \ldots, z_n(\zeta)\big)$ we have $e_n(w_\zeta \circ \varphi_{b,\infty}) = \big(z_{S(1)}(\zeta), \ldots, z_{S(n)}(\zeta)\big)$ for the permutation S, $S(j) = nS_b(\frac{j-1}{n}) + 1$ on $\{1, \ldots, n\}$, that is induced by S_b. Put $S(e_n(w_\zeta)) = S\big((z_1(\zeta), \ldots, z_n(\zeta))\big) = (z_{S(1)}(\zeta), \ldots, z_{S(n)}(\zeta))$. We obtained

$$e_n(w_\zeta \circ \varphi_{b,\infty}) = S(e_n(w_\zeta)). \tag{3.42}$$

Since w_ζ represents $\mathcal{F}(\zeta)$, and by Eq. (3.37) $w_\zeta \circ \varphi_{b,\infty}$ represents $\mathcal{F}(K\zeta) = \varphi^*_{b,\infty}(\mathcal{F}(\zeta))$, the equality (3.42) means that $\mathcal{P}_{\mathcal{T}}(\mathcal{F}(K\zeta))$ differs from $S(\mathcal{P}_{\mathcal{T}}\big(\mathcal{F}(\zeta))\big)$ by the diagonal action of a Möbius transformation depending on $\zeta \in \mathbb{C}_+$. This Möbius transformation is determined by its action on the first two coordinates of $\mathcal{P}_{\mathcal{T}}(\mathcal{F}(K\zeta))$ and $S(\mathcal{P}_{\mathcal{T}}\big(\mathcal{F}(\zeta))\big)$. □

The following lemma holds.

Lemma 3.8 *There exists a holomorphic mapping* $\mathfrak{A} : \mathcal{U}(\varepsilon) \to \mathcal{A}$ *such that for each* $\zeta \in \mathcal{U}'(\varepsilon)$ *the first two coordinates of the two n-tuples* $\mathfrak{A}(\zeta)\Big(S(\mathcal{P}_{\mathcal{T}}(\mathcal{F}(\zeta)))\Big)$ *and* $\mathfrak{A}(K\zeta)\Big(\mathcal{P}_{\mathcal{T}}\big(\mathcal{F}(K\zeta)\big)\Big)$ *coincide on* $\mathcal{U}'(\varepsilon)$.

We will prove this lemma after the proof of the present Lemma 3.7 is finished.

Continuation of Proof of Lemma 3.7 Let $\mathfrak{A}$ be the mapping of Lemma 3.8. We put $f_{\mathfrak{A}}(\zeta) \overset{def}{=} \mathfrak{A}(\zeta)(\mathcal{P}_{\mathcal{T}}\big(\mathcal{F}(\zeta))\big)$, $f_{\mathfrak{A}} : \mathcal{U}(\varepsilon) \to C_n(\mathbb{C})$. Since the first two coordinates of $\mathfrak{A}\big(K\zeta\big)\Big(\mathcal{P}_{\mathcal{T}}\big(\mathcal{F}(K\zeta)\big)\Big)$ and $\mathfrak{A}(\zeta)\Big(S(\mathcal{P}_{\mathcal{T}}\big(\mathcal{F}(\zeta))\big)\Big)$ coincide on $\mathcal{U}(\varepsilon)$, Eq. (3.42) implies

$$\mathfrak{A}(K\zeta)\Big(\mathcal{P}_{\mathcal{T}}\big(\mathcal{F}(K\zeta)\big)\Big) = \mathfrak{A}(\zeta)\Big(S\big(\mathcal{P}_{\mathcal{T}}(\mathcal{F}(\zeta))\big)\Big), \quad \zeta \in \mathcal{U}'(\varepsilon). \tag{3.43}$$

Since the action of $\mathfrak{A}(\zeta)$ commutes with S, we have

$$\mathfrak{A}(K\zeta)\Big(\mathcal{P}_{\mathcal{T}}\big(\mathcal{F}(K\zeta)\big)\Big) = S\Big(\mathfrak{A}(\zeta)\big(\mathcal{P}_{\mathcal{T}}(\mathcal{F}(\zeta))\big)\Big), \quad \zeta \in \mathcal{U}'(\varepsilon). \tag{3.44}$$

Equation (3.44) means that $f_{\mathfrak{A}}(K\zeta) = S(f_{\mathfrak{A}}(\zeta))$ for $\zeta \in \mathcal{U}'(\varepsilon)$, hence,

$$\mathcal{P}_{\text{sym}}\, f_{\mathfrak{A}}(K\zeta) = \mathcal{P}_{\text{sym}}\, f_{\mathfrak{A}}(\zeta), \; \zeta \in \mathcal{U}'(\varepsilon). \tag{3.45}$$

Equality (3.40) is proved.

The second statement of Lemma 3.7 is proved as follows. Take the smooth family $t \to w_{it} \in QC_\infty(0, n+1)$, $t \in [1, K]$, of the beginning of the proof of Lemma 3.7, for which $\mathcal{F}(it) = [w_{it}]$, and $\mathcal{P}_{\mathcal{T}}(\mathcal{F}(it)) = \mathcal{P}_{\mathcal{T}}([w_{it}]) = e_n(w_{it})$.

We need to prove that $f(t) = \mathcal{P}_{\text{sym}}(\mathfrak{A}(it)e_n(w_{it}))$, $t \in [1, K]$, represents $\hat{b}$ up to a power of a Dehn twist. By Eq. (3.44) the equality

$$\mathfrak{A}(Kit)\big(e_n(w_{Kit})\big) = S\big(\mathfrak{A}(it)(e_n(w_{it})\big)$$

holds. The mapping $\tilde{f}(it) \overset{def}{=} \mathfrak{A}(it)\big(e_n(w_{it})\big) = e_n\big(\mathfrak{A}(it)(w_{it})\big)$, $t \in [1, K]$, is a lift to $C_n(\mathbb{C})$ of the mapping $f(it) = \mathcal{P}_{\text{sym}}\big(\mathfrak{A}(it)e_n(w_{it})\big) = \text{ev}_{E_n^0}\big(\mathfrak{A}(it)(w_{it})\big)$. Hence,

$t \to \mathfrak{A}(it)(w_{it})$ is a parameterizing isotopy of the geometric braid $t \to f(it)$. Since $[w_{iK}] = [w_i \circ \varphi_{b,\infty}]$, also the equality

$$[\mathfrak{A}(iK)(w_{iK})] = [\mathfrak{A}(i)(w_i \circ \varphi_{b,\infty})]$$

holds. Since

$$\begin{aligned} e_n\big(\mathfrak{A}(iK) \circ w_{iK}\big) &= \mathfrak{A}(iK)\big(e_n(w_{iK})\big) = \mathfrak{A}(iK)\big(\mathcal{P}_{\mathcal{T}}(\mathcal{F}(iK))\big)\,, \\ e_n(\mathfrak{A}(i) \circ w_i \circ \varphi_{b,\infty}) &= \mathfrak{A}(i)(S(e_n(w_i))) = S(\mathfrak{A}(i)(\mathcal{P}_{\mathcal{T}}(\mathcal{F}(i)))\,, \end{aligned}$$

Eq. (3.43) and Lemma 1.3 imply that the Teichmüller equivalent mappings $\mathfrak{A}(iK)(w_{iK})$ and $\mathfrak{A}(i)(w_i \circ \varphi_{b,\infty})$ are isotopic. Hence, the class of $\varphi_{b,\infty}$ is the mapping class in $\mathfrak{M}(\mathbb{P}^1; \infty, E_n^0)$ of the geometric braid $t \to f(it)$, $t \in [1, K]$. Hence, the conjugacy classes of the braid b and of the braid represented by $t \to f(it)$, $t \in [1, K]$, differ by a power of a Dehn twist. Lemma 3.7 is proved. □

Proof of Lemma 3.8 Write

$$\mathcal{P}_{\mathcal{T}}(\mathcal{F}(\zeta)) = \big(z_1(\zeta), z_2(\zeta), z_3(\zeta), \ldots, z_n(\zeta)\big)$$

with $z_1(\zeta) \equiv 0$, $z_2(\zeta) \equiv \frac{1}{n}$. For the proof of the lemma we have to find a holomorphic mapping $\mathfrak{A} : \mathcal{U}(\varepsilon) \to \mathcal{A}$ to the space $\mathcal{A}$ of affine mappings so that the following condition is satisfied

$$\begin{aligned} \mathfrak{A}(K\zeta)\left(0\right) &\equiv \mathfrak{A}(K\zeta)(z_1(K\zeta)) = \mathfrak{A}(\zeta)(z_{S(1)}(\zeta))\,, \\ \mathfrak{A}(K\zeta)\left(\frac{1}{n}\right) &\equiv \mathfrak{A}(K\zeta)(z_2(K\zeta)) = \mathfrak{A}(\zeta)(z_{S(2)}(\zeta)), \quad \zeta \in \mathcal{U}'(\varepsilon)\,. \end{aligned} \tag{3.46}$$

We write $\mathfrak{A}(\zeta)$ in the form $\mathfrak{A}(\zeta)\,(z) = \frac{z-b(\zeta)}{a(\zeta)}$, $z \in \mathbb{C}$, for holomorphic functions a and b in $\mathbb{C}_+$ with $a \neq 0$ in $\mathbb{C}_+$. For a number $K > 1$ and any function q on $\mathcal{U}(\varepsilon)$ we use the notation $q^K(\zeta)$, $\zeta \in \mathcal{U}'(\varepsilon)$, for the function $q^K(\zeta) = q(K\zeta)$. The Eq. (3.46) can be written on $\mathcal{U}'(\varepsilon)$ as

$$\begin{aligned} \frac{z_{S(1)} - b}{a} &= \frac{-b^K}{a^K}\,, \\ \frac{z_{S(2)} - b}{a} &= \frac{\frac{1}{n} - b^K}{a^K}\,. \end{aligned} \tag{3.47}$$

Put

$$\chi \overset{\text{def}}{=} \frac{1}{n(z_{S(2)} - z_{S(1)})}\,.$$

χ is an analytic function on $\mathcal{U}'(\varepsilon)$, $\chi \neq 0$ on $\mathcal{U}'(\varepsilon)$. The Eq. (3.47) can be rewritten as

$$\frac{a^K}{a} = \chi\,,$$
$$\frac{b^K}{a^K} - \frac{b}{a} = -\frac{z_{S(1)}}{a}\,, \quad \zeta \in \mathcal{U}'(\varepsilon)\,. \tag{3.48}$$

The first equation leads to a Second Cousin Problem for the coefficient a on the annulus $\mathbb{C}_+ \diagup \Lambda_K$ (see [31], [40], or Appendix A.1). Indeed, let $\varepsilon > 0$ be small compared to K. Cover the annulus $\mathbb{C}_+ \diagup \Lambda_K$ by the open sets

$$\mathcal{U}_1 = \{\zeta \in \mathbb{C}_+ : 1 - \varepsilon < |\zeta| < 1 + 3\varepsilon\} \diagup \Lambda_K\,,$$
$$\mathcal{U}_2 = \{\zeta \in \mathbb{C}_+ : 1 + 2\varepsilon < |\zeta| < 1 + 5\varepsilon\} \diagup \Lambda_K\,, \text{ and}$$
$$\mathcal{U}_3 = \{\zeta \in \mathbb{C}_+ : 1 + 4\varepsilon < |\zeta| < (1 + \varepsilon)K\} \diagup \Lambda_K\,. \tag{3.49}$$

Then $\mathcal{U}_1 \cap \mathcal{U}_2 \cap \mathcal{U}_3 = \emptyset$, and $\mathcal{U}_3 \cap \mathcal{U}_1 = \mathcal{U}'(\varepsilon) \diagup \Lambda_K$. On $\mathcal{U}_{3,1} = \mathcal{U}_3 \cap \mathcal{U}_1$ we take the holomorphic transition function $g_{3,1}(\zeta \diagup \Lambda_K) = \chi(\zeta)$, $\zeta \in \mathcal{U}'(\varepsilon)$, on all other intersections of covering sets we take the holomorphic functions that are identically equal to one. We obtained a Cousin II distribution. Since the intersections of two covering sets are simply connected and an annulus is a Stein manifold, the Second Cousin Problem has a solution. This means, there exist nowhere vanishing holomorphic functions g_j on $\mathcal{U}_j$ such that $g_{j,k} = \frac{g_j}{g_k}$. Cover $\mathcal{U}_\varepsilon$ by lifts $\tilde{\mathcal{U}}_j$ of $\mathcal{U}_j$, $j = 1, 2, 3$. Let $\tilde{g}_j$ be the lift of g_j to $\mathcal{U}_j$ The function a on $\mathcal{U}(\varepsilon)$ which is equal to $\tilde{g}_j$ on $\tilde{\mathcal{U}}_j$, $j = 1, 2, 3$, satisfies the first equation of (3.48).

After we found a function a satisfying the first equation, we are looking for a function $\frac{b}{a}$ which satisfies the second equation. This leads to a First Cousin Problem on A which is solvable. The lemma is proved. □

End of Proof of Proposition 3.3 To prove the Proposition 3.3 we consider the mapping $\mathfrak{a}_\ell(\zeta)(f(\zeta))$, $\zeta \in \mathcal{U}(\varepsilon)$. Here f is the mapping of Lemma 3.7 and

$$\mathfrak{a}_\ell(\zeta)(z) = e^{\frac{2\pi i \ell}{\log K} \cdot \log \frac{\zeta}{i}} \cdot z\,, \quad \zeta \in \mathcal{U}(\varepsilon)\,, \quad z \in \mathbb{C}\,. \tag{3.50}$$

Equation (3.50) defines a holomorphic mapping from $A = \mathbb{C}_+ \diagup \Lambda_K$ to $C_n(\mathbb{C}) \diagup \mathcal{S}_n$. When ζ ranges over $[i, iK]$ the function $\log \frac{\zeta}{i}$ ranges over $[1, \log K]$, and $\mathfrak{a}_\ell(\zeta)$, $\zeta \in [i, iK]$, defines the ℓ-th power of the full twists about the circle $[i, iK] \diagup \Lambda_K$. Hence, for a suitable integer number ℓ the path $\mathfrak{a}_\ell(\zeta)(f(\zeta))$, $\zeta \in [i, iK]$, represents the free isotopy class $\hat{b}$. The conformal module $m(A) = m(\mathbb{C}_+ \diagup \Lambda_K)$ equals

$$\frac{\pi}{2} \frac{1}{\log \frac{K}{2}} = \frac{\pi}{2} \frac{1}{h(\hat{b})}.$$

Proposition 3.3 is proved in the hyperbolic case. □

Chapter 4
Reducible Pure Braids. Irreducible Nodal Components, Irreducible Braid Components, and the Proof of the Main Theorem

The purpose of this chapter is the proof of the Main Theorem for conjugacy classes of reducible pure braids. The idea is to reduce the proof to the case of irreducible braids and mapping classes. We will describe a decomposition of reducible elements of $\widehat{\mathfrak{M}}(\mathbb{P}^1; \{\infty\} \cup E_n)$ into irreducible nodal components (See Sect. 4.1). The elements of $\widehat{\mathfrak{M}}(\mathbb{P}^1; \{\infty\} \cup E_n)$ can be reconstructed from the irreducible nodal components up to a family of commuting powers of Dehn twists. On the other hand we provide a decomposition of conjugacy classes of reducible pure braids into irreducible braid components. The conjugacy class of a reducible pure braid can be recovered from its irreducible braid components.

In Sect. 4.4 we will establish the relation between the irreducible braid components of the conjugacy class of pure braids and the irreducible nodal components of the conjugacy class of mapping classes associated to the pure braid.

In Sect. 4.5 we prove that the conformal module of a conjugacy class $\hat{b}$ of pure braids is equal to the minimum of the conformal modules of the irreducible braid components, and the entropy of the class $\hat{b}$ is equal to the maximum of the entropies of the irreducible nodal components of the mapping class associated to $\hat{b}$. The Main Theorem for reducible pure braids is proved by applying the version of the Main Theorem for irreducible braids to the irreducible braid components and the related irreducible nodal components.

4.1 Irreducible Nodal Components. Pure Braids

Throughout this chapter $b \in \mathcal{B}_n$ will be a reducible pure braid and $\mathfrak{m}_b \in \mathfrak{M}(\overline{\mathbb{D}}; \partial\mathbb{D} \cup E_n)$ its mapping class. Here $E_n \subset \mathbb{D}$ is a set consisting of n points. We will describe the decomposition of the conjugacy class $\widehat{\mathfrak{m}_b}$ into irreducible components. For this purpose we again represent the mapping class $\mathfrak{m}_{b,\infty} = \mathcal{H}_\infty(\mathfrak{m}_b) \in$

B. Jöricke, *Braids, Conformal Module, Entropy, and Gromov's Oka Principle*, Lecture Notes in Mathematics 2360, https://doi.org/10.1007/978-3-031-67288-0_4

$\mathfrak{M}(\mathbb{P}^1; \{\infty\} \cup E_n)$ by a self-homeomorphism $\varphi_{b,\infty}$ of $\mathbb{P}^1$ which is the identity outside the unit disc $\mathbb{D}$. We assume that $\varphi_{b,\infty}$ is completely reduced by an admissible system of curves $\mathcal{C} = \{C_1, \dots, C_k\}$ in $\mathbb{D} \setminus E_n \subset \mathbb{C} \setminus E_n$. In particular, $\varphi_{b,\infty}$ leaves the union $\bigcup_{C \in \mathcal{C}} C$ invariant, and also leaves the complement $\mathbb{P}^1 \setminus \bigcup_{C \in \mathcal{C}} C$ invariant. Notice that the decomposition into irreducible components in general depends on the isotopy class of the admissible system $\mathcal{C}$. For simplicity of notation we put $E_n' = E_n \cup \infty$.

The Jordan Curve Theorem induces a partial order on the set of connected components of $\mathbb{P}^1 \setminus \bigcup_{C \in \mathcal{C}} C$. The partial order is described as follows. Let S be a connected component of $\mathbb{P}^1 \setminus \bigcup_{C \in \mathcal{C}} C$. Its boundary ∂S is the union of some curves that are elements of $\mathcal{C}$. By the Jordan Curve Theorem each connected component ∂_j of ∂S divides $\mathbb{P}^1$ into two connected components. If S does not contain ∞ then there exists exactly one boundary component of S denoted by $\partial_{\mathfrak{E}} S$ for which S is contained in the bounded component of $\mathbb{C} \setminus \partial_{\mathfrak{E}} S$. We call $\partial_{\mathfrak{E}} S$ the exterior boundary of S. If S is contained in the component of $\mathbb{P}^1 \setminus \partial_j$ that contains ∞ the component is called an interior boundary component of S . The union $\partial S \setminus \partial_{\mathfrak{E}} S$ of the interior boundary components of S is called the interior boundary of S and denoted by $\partial_{\mathfrak{I}} S$.

Denote by $S^{1,1}$ the connected component of $\mathbb{P}^1 \setminus \bigcup_{C \in \mathcal{C}} C$ that contains the point ∞. The exterior boundary of $S^{1,1}$ is empty. We call $S^{1,1}$ the outermost connected component of $\mathbb{P}^1 \setminus \bigcup_{C \in \mathcal{C}} C$.

The components of the interior boundary $\partial_{\mathfrak{I}} S^{1,1}$ are called the curves of generation 2. By the Jordan Curve Theorem each connected component of the interior boundary $\partial_{\mathfrak{I}} S^{1,1}$ bounds a disc contained in $\mathbb{D}$. (Recall that all curves of the system $\mathcal{C}$ are contained in $\mathbb{D}$.) Denote the discs by $\delta^{2,j}$, $j = 1, \dots, k_2$. For each j there is exactly one connected component $S^{2,j}$ of $\mathbb{C} \setminus \bigcup_{C \in \mathcal{C}} C$ which is contained in $\delta^{2,j}$ and has a boundary component in common with $S^{1,1}$. (For short, $S^{2,j}$ is adjacent to a boundary component of $S^{1,1}$.) This boundary component is the exterior boundary component $\partial_{\mathfrak{E}} S^{2,j}$, $j = 1, \dots, k_2$ of $S^{2,j}$. The components $S^{2,j}$ are called the components of second generation.

The connected components of $\mathbb{P}^1 \setminus \bigcup_{j=1}^{k} C_j$ of generation ℓ are defined by induction as follows. Consider the union of all connected components of $\mathbb{P}^1 \setminus \bigcup_{j=1}^{k} C_j$ of generation not exceeding $\ell - 1$. Take its closure

$$\overline{Q_{\ell-1}} \stackrel{def}{=} \overline{\bigcup_{1 \le \ell' \le \ell-1} \bigcup_{1 \le j \le k_{\ell'}} S^{\ell',j}}\,, \tag{4.1}$$

which is the closure of a domain $Q_{\ell-1} \subset \mathbb{P}^1$. $Q_{\ell-1}$ is the union of all components of generation not exceeding $\ell - 1$ and the exterior boundaries of all components of generation between 2 and $\ell - 1$. Its boundary $\partial Q_{\ell-1}$ is the union of the interior boundary components of all components $S^{\ell-1,j}$ $j = 1, \dots, k_{\ell-1}$, of generation $\ell - 1$. The connected components of $\mathbb{P}^1 \setminus \bigcup_{j=1}^{k} C_j$ which share a boundary component with $Q_{\ell-1}$ are called the components of generation ℓ and are labeled by $S^{\ell,j}$, $j = 1, \dots, k_\ell$. Each $S^{\ell,j}$ has exactly one boundary component in common with $Q_{\ell-1}$, the exterior boundary $\partial_{\mathfrak{E}} S^{\ell,j}$. The exterior boundaries of the components $S^{\ell,j}$ are also called the curves of generation ℓ. We put

$$\overline{Q_\ell} \stackrel{def}{=} \overline{\bigcup_{1 \le \ell' \le \ell} \bigcup_{1 \le j \le k_{\ell'}} S^{\ell',j}}\,. \tag{4.2}$$

The process terminates after we obtained components $S^{N,j}$ of some generation N that all have empty interior boundary. The boundary of Q_N is empty, i.e. Q_N coincides with $\mathbb{P}^1$.

The mapping $\varphi_{b,\infty}$ fixes each connected component $S^{\ell,j}$ setwise. Indeed, it fixes $S^{1,1}$ setwise since it fixes $\infty \in S^{1,1}$. Hence it fixes the interior boundary of $S^{1,1}$ setwise and therefore, it fixes the union of the discs $\delta^{2,j}$ setwise. Further, since the system of curves $\mathcal{C}$ is admissible, each disc $\delta^{2,j}$ contains a point of E_n. Hence, since b is a pure braid, the mapping $\varphi_{b,\infty}$ fixes the point. Hence, it fixes each disc $\delta^{2,j}$ setwise, and therefore it fixes setwise the component $S^{2,j}$ whose exterior boundary component is the boundary of the disc $\delta^{2,j}$. By an induction on the number of generation we see that $\varphi_{b,\infty}$ fixes each connected component $S^{\ell,j}$ setwise.

Let

$$w : \mathbb{C} \setminus E_n' \to Y$$

be a continuous surjection onto a nodal surface Y associated to the isotopy class of the curve system $\mathcal{C}$ in $\mathbb{P}^1$ (see Sect. 1.8). Notice that the nodal surface Y has punctures. The mapping w maps each component $S^{\ell,j} \setminus E_n'$ of $\mathbb{P}^1 \setminus (E_n' \cup \bigcup C_j)$ homeomorphically onto a part $Y^{\ell,j}$ of Y. (Recall that a part of a nodal surface Y with set of nodes $\mathcal{N}$ is a connected component of $Y \setminus \mathcal{N}$.) The correspondence between the components $S^{\ell,j} \setminus E_n'$ and the parts $Y^{\ell,j}$ is a bijection.

We conjugate the restriction $\varphi_{b,\infty} | \mathbb{P}^1 \setminus (E_n' \cup \mathcal{C})$ of the self-homeomorphism $\varphi_{b,\infty}$ with the inverse of w. We obtain a self-homeomorphism of $Y \setminus \mathcal{N}$ that extends to a self-homeomorphism of Y. The isotopy class of this extension is denoted by $\mathfrak{m}_{b,\odot}$ and is called the isotopy class of self-mappings of the nodal surface Y determined by $\mathfrak{m}_{b,\infty}$ and the isotopy class of the admissible system $\mathcal{C}$. Let $\widehat{\mathfrak{m}_{b,\odot}}$ be the conjugacy class of $\mathfrak{m}_{b,\odot}$.

Let Y^c be the compact nodal surface obtained by filling the punctures of Y. The surjection w extends to a continuous surjection $w : \mathbb{P}^1 \to Y^c$ which by an abuse of notation we denote again by w. The elements of the class $\mathfrak{m}_{b,\odot}$ extend across the punctures of Y to self-homeomorphisms of Y^c with set of distinguished points

$\mathcal{N} \cup w(E'_n)$. By an abuse of notation we will also denote the class of these extensions by $\mathfrak{m}_{b,\odot}$. The conjugacy class $\widehat{\mathfrak{m}_{b,\odot}}$ is called the nodal conjugacy class associated to $\widehat{\mathfrak{m}_{b,\infty}}$ and the admissible system of curves $\mathcal{C}$.

More detailed, the mapping $\varphi_{b,\infty}$ fixes each $S^{\ell,j} \setminus E'_n$ setwise, hence each mapping in $\mathfrak{m}_{b,\odot}$ fixes each nodal component $Y^{\ell,j}$ setwise. For each (ℓ, j) the restriction $\mathfrak{m}_{b,\odot}^{\ell,j} \overset{def}{=} \mathfrak{m}_{b,\odot}|Y^{\ell,j}$ is obtained in the following way. Put $E'^{\ell,j} \overset{def}{=} E'_n \cap S^{\ell,j}$. We conjugate the restriction of $\varphi_{b,\infty}$ to $S^{\ell,j} \setminus E'^{\ell,j}$ by the inverse of the restriction $w^{\ell,j} \overset{def}{=} w|S^{\ell,j} \setminus E'^{\ell,j}$. The class $\mathfrak{m}_{b,\odot}^{\ell,j}$ is the mapping class of the conjugated mapping which is a self-homeomorphism of $Y^{\ell,j}$.

The elements of $\mathfrak{m}_{b,\odot}|Y^{\ell,j}$ extend continuously across the punctures of $Y^{\ell,j}$. The extensions are self-homeomorphisms of the closed Riemann surface $(Y^{\ell,j})^c \cong \mathbb{P}^1$ with distinguished points $w(E'^{\ell,j}) \cup (\mathcal{N} \cap (Y^{\ell,j})^c)$. The set of extensions is denoted by the same letter $\mathfrak{m}_{b,\odot}^{\ell,j}$. The conjugate of $\varphi_{b,\infty} \mid S^{\ell,j} \setminus E'^{\ell,j}$ by $(w^{\ell,j})^{-1}$ represents the class $\mathfrak{m}_{b,\odot}^{\ell,j}$ on $Y^{\ell,j}$, and, hence, $\varphi_{b,\infty} \mid S^{\ell,j} \setminus E'^{\ell,j}$ represents the conjugacy class $\widehat{\mathfrak{m}_{b,\odot}^{\ell,j}}$.

The conjugacy classes $\widehat{\mathfrak{m}_{b,\odot}^{\ell,j}}$ will be called the irreducible nodal components of the class $\widehat{\mathfrak{m}_{b,\infty}}$ with respect to the system $\mathcal{C}$. The conjugacy classes $\widehat{\mathfrak{m}_{b,\odot}^{\ell,j}}$ of the restrictions $\mathfrak{m}_{b,\odot}^{\ell,j}$ depend only on the conjugacy class $\widehat{\mathfrak{m}_{b,\infty}}$ and the isotopy class of the curve system $\mathcal{C}$. Each such class is indeed irreducible (see Remark 5.1 for the case of not necessarily pure braids and its proof). Notice, that given the isotopy class of curve system $\mathcal{C}$, the irreducible nodal components determine the class $\widehat{\mathfrak{m}_{b,\infty}}$ up to products of powers of some Dehn twists. (See Remark 4.1 below.)

4.2 The Irreducible Braid Components. Pure Braids

In this section we will describe the decomposition of reducible braids into irreducible braid components. The decomposition depends again on the isotopy class of an admissible system of curves that completely reduces the mapping class associated to the braid.

We start with defining a few objects. This will be done here for the case of arbitrary (not necessarily pure) braids. A tube in the "infinite" cylinder $[0, 1] \times \mathbb{C}$ is the image of a "finite" cylinder $[0, 1] \times \overline{\mathbb{D}}$ under a diffeomorphism of $[0, 1] \times \mathbb{C}$ onto itself which preserves each fiber $\{t\} \times \mathbb{C}$, $t \in [0, 1]$. Let $\Omega^{n_2} = \{\Omega_1, \dots, \Omega_{n_2}\}$ be an unordered n_2-tuple of pairwise disjoint closed topological discs in the complex plane, and $E_{n_1} = \{z_1, \dots, z_{n_1}\}$ an unordered n_1-tuple of points in $\mathbb{C} \setminus \cup\Omega_j$. A tubular geometric (n_1, n_2)-braid with base point (E_{n_1}, Ω^{n_2}) consists of the following two objects. The first object is a collection of n_2 mutually disjoint tubes in the cylinder $[0, 1] \times \mathbb{C}$, each of which intersects the fibers $\{0\} \times \mathbb{C}$ and $\{1\} \times \mathbb{C}$

along a copy of Ω^{n_2}. The second object is a geometric n_1-braid with base point E_{n_1} whose strands do not meet the n_2 tubes.

Take a point $E_{n_1} \in C_{n_1}(\mathbb{C})/\mathcal{S}_{n_1}$ and a point $\boldsymbol{E}_{n_2} \in C_{n_2}(\mathbb{C})/\mathcal{S}_{n_2}$ such that $E_{n_1} \cup \boldsymbol{E}_{n_2} \in C_{n_1+n_2}(\mathbb{C})/\mathcal{S}_{n_1+n_2}$. A geometric fat braid (of type (n_1, n_2)) with base point $E_{n_1} \cup \boldsymbol{E}_{n_2}$ is a geometric $(n_1 + n_2)$-braid such that all strands with initial point in $\boldsymbol{E}_{n_2}$ have also their endpoints in $\boldsymbol{E}_{n_2}$ and are declared to be fat. The strands with initial point in E_{n_1} will be called ordinary strands. A tubular geometric braid is associated to a fat braid if there is a one-to-one correspondence between the tubes and the fat strands so that each fat strand is contained in the corresponding tube and is a deformation retraction of this tube.

A fat braid is an isotopy class of geometric fat braids with fixed base point. Two geometric fat braids are called isotopic if they are isotopic as geometric braids and the isotopy moves the fat strands to fat strands and the ordinary strands to ordinary strands. A geometric tubular fat braid is a tubular geometric braid some of whose strands are declared to be fat.

We consider now again pure braids. Take a reducible pure braid $b \in \mathcal{B}_n$ with base point E_n^0, and describe the decomposition into irreducible braid components. As in Sect. 4.1 we consider a mapping $\varphi_{b,\infty} \in \mathfrak{m}_{b,\infty} \in \mathfrak{M}(\mathbb{P}^1; \{\infty\} \cup E_n^0)$ that is the identity outside the unit disc and is completely reduced by an admissible system $\mathcal{C}$ of curves contained in the unit disc. (Notice that in general the decomposition into irreducible braid components depends on the isotopy class of the admissible system $\mathcal{C}$.) Replacing $\varphi_{b,\infty}$ by an isotopic mapping which differs from the original mapping only in small neighbourhoods of the admissible curves we may assume that $\varphi_{b,\infty}$ fixes each curve C pointwise.

Consider a path $\varphi_t \in \mathrm{Hom}^+(\mathbb{P}^1; \mathbb{P}^1 \setminus \mathbb{D})$, $t \in [0, 1]$, which joins $\varphi_{b,\infty}$ with the identity. More detailed, φ_t is a continuous family of self-homeomorphisms of $\mathbb{P}^1$ which fix the complement $\mathbb{P}^1 \setminus \mathbb{D}$ of the unit disc pointwise, such that $\varphi_0 = \mathrm{id}$ and $\varphi_1 = \varphi_{b,\infty}$. The evaluation map

$$[0, 1] \ni t \to \mathrm{ev}_{E_n} \varphi_t = \{\varphi_t(z_1), \ldots, \varphi_t(z_n)\} \in C_n(\mathbb{C})/\mathcal{S}_n \tag{4.3}$$

with $E_n = \{z_1, \ldots, z_n\}$, gives a geometric braid that is contained in the cylinder $[0, 1] \times \mathbb{D}$ (i.e. $\mathrm{ev}_{E_n} \varphi_t \in \mathbb{D}$ for each t), has base point $E_n = \varphi_0(E_n) = \varphi_1(E_n)$, and represents b. The family φ_t, $t \in [0, 1]$ is a parameterizing isotopy of the geometric braid (4.3). We will write the geometric braid (4.3) as continuous mapping $g : [0, 1] \to C_n(\mathbb{C})/\mathcal{S}_n$,

$$g(t) = \varphi_t(E_n),\ t \in [0, 1], \tag{4.4}$$

and label the coordinate functions of g so that for $z' \in E_n$ the coordinate function $g_{z'}$ of g has initial point $g_{z'}(0) = g_{z'}(1) = z'$.

We will associate to the parameterizing isotopy tubular geometric braids and geometric fat braids. The classes obtained from the fat braids by isotopy and conjugation are the irreducible braid components. This is done as follows.

The Outermost Braid Component $\widehat{b(1,1)}$ of a Pure Braid b Recall that we regard a point in $C_n(\mathbb{C})$ sometimes as ordered subset of $\mathbb{C}$, and sometimes as a point in $\mathbb{C}^n$.

We will associate now a fat braid $\boldsymbol{b}(1,1)$ to the outermost component $S^{1,1}$. Put $E^{1,1} \overset{def}{=} E_n \cap S^{1,1}$. Let $C^{2,j}$, $j = 1, \ldots, k_2$, be the interior boundary components of $S^{1,1}$, labeled so that $C^{2,j}$ is the exterior boundary of the component $S^{2,j}$ of $\mathbb{P}^1 \setminus \cup_{C \in \mathcal{C}} C$, $j = 1, \ldots, k_2$. Denote by $\delta^{2,j} \subset \mathbb{D}$ the topological disc bounded by $C^{2,j}$.

The tubular geometric braid associated to the parameterizing isotopy φ_t and the component $S^{1,1}$ is

$$\{t\} \times \varphi_t \left(E^{1,1} \cup \bigcup_{j=1}^{k_2} \overline{\delta^{2,j}} \right), \quad t \in [0,1]. \tag{4.5}$$

For each tube $\{t\} \times \varphi_t(\overline{\delta^{2,j}})$, $t \in [0,1]$, we consider a strand contained in it and call it a fat strand. More detailed, for each j we pick a point $z^{2,j} \in \delta^{2,j} \cap E_n$ and denote it by the fat letter $\boldsymbol{z}^{2,j}$. Such a point exists by the definition of an admissible system of curves and an induction argument. Since the braid is pure we have $\varphi_{b,\infty}(\boldsymbol{z}^{2,j}) = \boldsymbol{z}^{2,j}$. The respective fat strand is $\varphi_t(\boldsymbol{z}^{2,j})$, $t \in [0,1]$. Put $\boldsymbol{E}^{1,1} = \{\boldsymbol{z}^{2,1}, \ldots, \boldsymbol{z}^{2,k_2}\}$. Note that the choice of the points in $\boldsymbol{E}^{1,1}$ may not be unique. We associate to $S^{1,1}$ the set of points $E^{1,1} \cup \boldsymbol{E}^{1,1}$.

The geometric fat braid associated to $S^{1,1}$ and the parameterizing isotopy equals

$$\varphi_t(E^{1,1} \cup \boldsymbol{E}^{1,1}), \quad t \in [0,1]. \tag{4.6}$$

The strands with initial point in $E^{1,1}$ are the ordinary strands, and the strands with initial point in $\boldsymbol{E}^{1,1}$ are the fat strands. The geometric fat braid (4.6) is obtained from the geometric tubular braid (4.5) by taking a deformation retract of each tube to a fat strand.

Let $n(1,1)$ be the number of points of $E^{1,1} \cup \boldsymbol{E}^{1,1}$. We will write the geometric fat braid (4.6) as mapping

$$\boldsymbol{g}^{1,1} : [0,1] \to C_{n(1,1)}(\mathbb{C}) \diagup \mathcal{S}_{n(1,1)}, \quad \boldsymbol{g}^{1,1}(0) = \boldsymbol{g}^{1,1}(1) = E^{1,1} \cup \boldsymbol{E}^{1,1}. \tag{4.7}$$

If needed, the "strands" of $\boldsymbol{g}^{1,1}$ will be labeled by points of $E^{1,1} \cup \boldsymbol{E}^{1,1}$, so that for $z' \in E^{1,1} \cup \boldsymbol{E}^{1,1}$ the strand corresponding to $\boldsymbol{g}^{1,1}_{z'}$ has initial point $\boldsymbol{g}^{1,1}_{z'}(0) = z'$. The fat strands are those with initial points in $\boldsymbol{E}^{1,1}$.

We define the fat braid $\boldsymbol{b}(1,1)$ with base point $E^{1,1} \cup \boldsymbol{E}^{1,1}$ as the isotopy class of the geometric fat braid (4.6) (equivalently, (4.7)) with given base point. The fat strands are those with initial point in $\boldsymbol{E}^{1,1}$. Notice that $\boldsymbol{b}(1,1)$ is obtained from the braid b by discarding all strands with initial point not in $E^{1,1} \cup \boldsymbol{E}^{1,1}$ and indicating the set of fat strands.

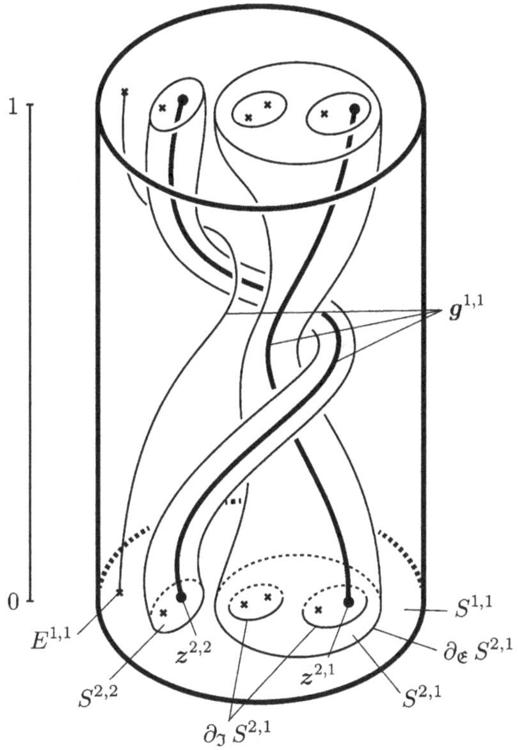

Fig. 4.1 The irreducible braid component $\widehat{\boldsymbol{b}(1,1)}$ of a pure braid b

Finally we take the conjugacy class $\widehat{\boldsymbol{b}(1,1)}$ of the fat braid $\boldsymbol{b}(1,1)$. This is the irreducible braid component of $\hat{b}$ associated to b and the component $S^{1,1}$. The class $\widehat{\boldsymbol{b}(1,1)}$ depends only on the conjugacy class $\hat{b}$ and on the isotopy class of the curve system $\mathcal{C}$.

See Fig. 4.1 for an example of the geometric fat braid $\boldsymbol{g}^{1,1}$ and the respective tubular geometric braid. In the figure the set $E^{1,1}$ consists of a single point. The strand with this initial point is the only ordinary strand. There are two components of $\mathbb{C}\setminus \bigcup_{C\in\mathcal{C}} C$ of generation 2, $S^{2,1}$ and $S^{2,2}$. The respective distinguished points are $z^{2,1}$ and $z^{2,2}$. These are the points in $\boldsymbol{E}^{1,1}$. The strands with these initial points are the fat strands of $\boldsymbol{g}^{1,1}$.

The Irreducible Braid Component $\widehat{\boldsymbol{b}(\ell,j)}$ associated to $S^{\ell,j}$ with $\ell > 1$ Take a component $S^{\ell,j}$ of generation ℓ. Recall that $E^{\ell,j} = E \cap S^{\ell,j}$. Consider the interior boundary components of $S^{\ell,j}$. Each such boundary component is the exterior boundary of a component $S^{\ell+1,j'}$ of $\mathbb{P}^1\setminus \bigcup_{C\in\mathcal{C}} C_i$ of generation $\ell+1$. Let $\delta^{\ell+1,j'} \subset \mathbb{C}$ be the topological disc bounded by $\partial_{\mathfrak{E}}\, S^{\ell+1,j'}$, and let $z^{\ell+1,j'}$ be a point in $E \cap \delta^{\ell+1,j'}$. Let $\boldsymbol{E}^{\ell,j}$ be the collection of $z^{\ell+1,j'}$ obtained in this way (one such point is assigned to each interior boundary component of $S^{\ell,j}$).

We obtain the tubular geometric braid

$$\varphi_t\left(E^{\ell,j}\cup\bigcup_{j=1}^{k_{\ell+1}}\overline{\delta^{\ell+1,j'}}\right),\quad t\in[0,1],\tag{4.8}$$

and the geometric fat braid

$$\varphi_t\left(E^{\ell,j}\cup \boldsymbol{E}^{\ell,j}\right),\quad t\in[0,1].\tag{4.9}$$

The geometric fat braid is obtained by taking a deformation retraction of each of the cylinders $\varphi_t\left(\overline{\delta^{\ell+1,j'}}\right)$, $t\in[0,1]$, to the fat strand of b contained in this cylinder.

Let $n(\ell,j)$ be the number of points of $E^{\ell,j}\cup\boldsymbol{E}^{\ell,j}$. We will write the geometric fat braid (4.9) as a mapping $\boldsymbol{g}^{\ell,j}:[0,1]\to C_{n(\ell,j)}(\mathbb{C})\diagup\mathcal{S}_{n(\ell,j)}$ from the unit interval into $C_{n(\ell,j)}(\mathbb{C})\diagup\mathcal{S}_{n(\ell,j)}$.

Denote the isotopy class of the geometric fat braid by $\boldsymbol{b}(\ell,j)$. Intuitively, the fat braid $\boldsymbol{b}(\ell,j)$ is obtained from the braid b by discarding all strands with initial point not in $E^{\ell,j}\cup\boldsymbol{E}^{\ell,j}$ and indicating the set of fat strands.

The conjugacy class $\widehat{\boldsymbol{b}(\ell,j)}$ of the fat braid $\boldsymbol{b}(\ell,j)$ is the irreducible component of the braid b associated to $S^{\ell,j}$. It depends only on the conjugacy class $\hat{b}$ of the braid and the isotopy class of the curve system $\mathcal{C}$.

The decomposition of the conjugacy class of a braid into irreducible braid components is described.

4.3 The Relation Between Irreducible Nodal Components and Irreducible Braid Components

Recall that for each (ℓ,j) we denoted by $\widehat{\mathfrak{m}_{b,\odot}^{\ell,j}}$ the irreducible nodal component that is associated to $\mathfrak{m}_{b,\infty}$ and the isotopy class of the admissible system of curves $\mathcal{C}$. For the irreducible braid component $\widehat{\boldsymbol{b}(\ell,j)}$, that is a conjugacy class of fat braids which was associated to b and the isotopy class of $\mathcal{C}$, we now consider the respective conjugacy class of ordinary braids $\widehat{b(\ell,j)}$ by declaring all strands to be ordinary. Let $\widehat{\mathfrak{m}_{b(\ell,j),\infty}}$ be the conjugacy class of mapping classes corresponding to the braid class $\widehat{b(\ell,j)}$. The following lemma holds.

Lemma 4.1 *For each (ℓ,j) the equality*

$$\widehat{\mathfrak{m}_{b,\odot}^{\ell,j}}=\widehat{\mathfrak{m}_{b(\ell,j),\infty}}\tag{4.10}$$

holds.

In the statement of the lemma either both conjugacy classes are considered as classes of self-homeomorphisms of a punctured Riemann surface or both are considered as classes of self-homeomorphisms of a closed Riemann surface with distinguished points.

Proof of Lemma 4.1 Recall that $E'_n = E_n \cup \{\infty\}$, and the mapping $\varphi_{b,\infty}|S^{\ell,j} \setminus E'_n$ represents the class $\widehat{\mathfrak{m}_{b,\odot}^{\ell,j}}$ considered as element of $\widehat{\mathfrak{M}}(Y^{\ell,j})$. We will write again $E'^{\ell,j} = E'_n \cap S^{\ell,j}$. As before we will identify the conjugacy class $\widehat{\mathfrak{M}}(Y^{\ell,j})$ on the punctured surface with the conjugacy class $\widehat{\mathfrak{M}}\Big((Y^{\ell,j})^c;\ (\mathcal{N} \cap (Y^{\ell,j})^c) \cup w^{\ell,j}(E'^{\ell,j})\Big)$ on the closed surface.

The irreducible braid component $\widehat{b(\ell,j)}$ is represented by the geometric braid (4.9) after forgetting that some strands are declared to be fat. The parameterizing family φ_t that was chosen for the geometric braid (4.4) representing the braid b serves also as parameterizing family for the latter geometric braid.

The mapping class $\mathfrak{m}_{b(\ell,j),\infty} \in \mathfrak{M}(\mathbb{P}^1;\ E'^{\ell,j} \cup \boldsymbol{E}^{\ell,j})$ that is associated to the braid $b(\ell,j)$ (see (4.9)) is the class of $\varphi_1 = \varphi_{b,\infty}$ in $\mathfrak{M}(\mathbb{P}^1;\ E'^{\ell,j} \cup \boldsymbol{E}^{\ell,j})$ identified with $\mathfrak{M}(\mathbb{P}^1 \setminus (E'^{\ell,j} \cup \boldsymbol{E}^{\ell,j}))$. This follows from the fact that φ_t is a parameterizing isotopy also for the geometric braid (4.9) that represents $b(\ell,j)$.

The mapping $\varphi_{b,\infty} \mid (\mathbb{P}^1 \setminus (E'^{\ell,j} \cup \boldsymbol{E}^{\ell,j}))$ is obtained from the mapping $\varphi_{b,\infty} \mid (S^{\ell,j} \setminus E'_n)$ by isotopy and conjugation. Indeed, consider all holes of $S^{\ell,j}$. These are all discs $\delta^{\ell+1,i'}$ that are adjacent to an interior boundary component of $S^{\ell,i}$. Each disc $\delta^{\ell+1,i'}$ contains a point $z^{\ell+1,i'} \in \boldsymbol{E}^{\ell,j}$ and each point in $\boldsymbol{E}^{\ell,j}$ is contained in one of the discs. If $(\ell,j) \neq (1,1)$ then there is one more hole of $S^{\ell,j}$, namely the complement of $\overline{\delta^{\ell,j}}$. This disc contains ∞ and no other point of $E'^{\ell,j}$. We denote it by $\delta_\infty^{\ell,j}$.

For each hole $\delta^{\ell+1,i'}$ (and $\delta_\infty^{\ell,j}$, if $\ell \neq 1$) we take an annulus $A^{\ell+1,i'}$ (and $A_\infty^{\ell,j}$ respectively, if $\ell \neq 1$), that is contained in $S^{\ell,j}$ and shares a boundary component with the boundary of the hole. The annuli are chosen to have disjoint closure. For each disc $\delta^{\ell+1,i'}$ ($\delta_\infty^{\ell,j}$, respectively) we consider the disc $\delta'^{\ell+1,i'} = \overline{\delta^{\ell+1,i'}} \cup A^{\ell+1,i'}$ (and $\delta'^{\ell,j}_\infty = \overline{\delta_\infty^{\ell,j}} \cup A_\infty^{\ell,j}$, respectively).

There is a homeomorphism $\tilde{w}^{\ell,j}$ from $S^{\ell,j} \setminus E'^{\ell,j}$ onto $\mathbb{P}^1 \setminus (E'^{\ell,j} \cup \boldsymbol{E}^{\ell,j})$ with the following properties. $\tilde{w}^{\ell,j}$ is equal to the identical injection on the complement of all annuli $A^{\ell+1,i'}$ (and $A_\infty^{\ell,j}$ respectively, if $\ell \neq 1$) in $S^{\ell,j}$. Then $\tilde{w}^{\ell,j}$ maps each annulus $A^{\ell+1,i'}$ ($A_\infty^{\ell,j}$, respectively) homeomorphically onto the respective punctured disc $\delta'^{\ell+1,i'} \setminus \{z^{\ell+1,i'}\}$ ($\delta'^{\ell,j}_\infty \setminus \{\infty\}$, respectively) and its continuous extension to the boundary component $\partial\delta'^{\ell+1,i'}$ of $A^{\ell+1,i'}$ (to the boundary component $\partial\delta'^{\ell,j}_\infty$ of $A_\infty^{\ell,j}$, respectively) equals the identity on the boundary component.

The mapping $(\tilde{w}^{\ell,j})^{-1}$ conjugates $\varphi_{b,\infty} \mid S^{\ell,j} \setminus E'^{\ell,j}$ to a self-homeomorphism $\varphi'_{b,\infty}$ of $\mathbb{P}^1 \setminus (E'^{\ell,j} \cup \boldsymbol{E}^{\ell,j})$ that differs from $\varphi_{b,\infty} \mid \mathbb{P}^1 \setminus (E'^{\ell,j} \cup \boldsymbol{E}^{\ell,j})$ only on the union of the punctured discs $\delta'^{\ell+1,i'} \setminus \{z^{\ell+1,i'}\}$ ($\delta'^{\ell,j}_\infty \setminus \{\infty\}$, respectively) and maps each punctured disc onto itself. Moreover, $\varphi'_{b,\infty} \circ \varphi_{b,\infty}^{-1}$ is equal to the identity on the boundary of each disc. Since two self-homeomorphisms of a punctured disc that fix the boundary circle pointwise are isotopic to the identity through self-homeomorphisms of the punctured disc that fix the boundary circle pointwise, the conjugated homeomorphism is isotopic to $\varphi_{b,\infty} \mid \mathbb{P}^1 \setminus E'^{\ell,j}$. The equality (4.10) is proved. □

Lemma 4.1 (together with Remark 5.1) shows in particular that the irreducible braid components are indeed irreducible conjugacy classes of braids.

4.4 Recovery of the Conjugacy Class of a Pure Braid from the Irreducible Braid Components

We will consider now again the isotopy class of an admissible system of curves that completely reduces the class $\mathfrak{m}_{b,\infty}$ of the braid and consider the connected components $S^{\ell,j}$ of the complement of the curves. We will show how to recover the conjugacy class of the pure braid b from knowing the irreducible braid components $\widehat{b(\ell,j)}$ associated to the $S^{\ell,j}$. For this purpose we will take arbitrary geometric fat braids, that are contained in $[0,1] \times \mathbb{D}$ and represent the irreducible braid components $\widehat{b(\ell,j)}$. By induction on the generation ℓ we will "insert the geometric fat braids representing the irreducible braid components of generation ℓ into tubes around fat strands of the geometric fat braids representing the irreducible braid components of generation $\ell - 1$". We have to show that indeed for arbitrary choices of geometric fat braids in $[0,1] \times \mathbb{D}$ representing the $\widehat{b(\ell,j)}$ the geometric braid obtained by the inductive construction represents the class $\hat{b}$. For this purpose we will compare an arbitrary inductively constructed geometric braid with the geometric braid (4.3) that represents $\hat{b}$.

We will use the following operation on points of symmetrized configuration spaces. Consider a point $E \cup \boldsymbol{E} \in C_n(\mathbb{C})\diagup \mathcal{S}_n$ (which is also considered as a subset of $\mathbb{C}$). The points in the set $\boldsymbol{E}$ are declared to be fat points. Let $z' \in \boldsymbol{E}$ be one of the fat points. Take a subset E' of $\mathbb{C}$ that contains n' points (also considered as point in $C_{n'}(\mathbb{C})\diagup \mathcal{S}_{n'}$.) E' may contain fat points or not. Suppose that E' does not intersect $E \cup (\boldsymbol{E} \setminus \{z'\})$. Remove the point z' from the set $E \cup \boldsymbol{E}$ and add the points of the set E'. The resulting set is denoted by

$$(E \cup \boldsymbol{E}) \sqcup_{z'} E' \overset{def}{=} (E \cup \boldsymbol{E}' \setminus \{z'\}) \cup E'. \tag{4.11}$$

The set (4.11) may be considered as subset of $\mathbb{C}$ consisting of $n + n' - 1$ points, or as point in $C_{n+n'-1}(\mathbb{C})\diagup \mathcal{S}_{n+n'-1}$.

The respective operation on mappings into symmetrized configuration spaces is defined pointwise. For instance, let $\boldsymbol{g}(t), t \in [0,1]$, be a geometric fat braid, let $\boldsymbol{g}_1(t), t \in [0,1]$, be a fat strand of $\boldsymbol{g}$, and let $\mathfrak{T} = \cup_{t\in[0,1]}\mathfrak{T}(t)$ be a tube that contains the fat strand $\boldsymbol{g}_1$ (i.e. $\boldsymbol{g}_1(t) \in \mathfrak{T}(t), t \in [0,1]$,) and does not intersect any other strand (i.e. $\mathfrak{T}(t)$ does not intersect the set $g(t) \setminus g_1(t)$, $t \in [0, 1]$). Suppose $g'(t)$, $t \in [0, 1]$, is another geometric fat braid, such that g' is contained in the tube $\mathfrak{T}$, i.e. $g'(t) \in \mathfrak{T}(t)$, $t \in [0, 1]$. By removing the fat strand g_1 and inserting the geometric fat braid g' which is contained in the tube $\mathfrak{T}$, we obtain a new geometric fat braid $g(t) \sqcup_{g_1(t)} g'(t)$, $t \in [0, 1]$, that is denoted by $g \sqcup_{g_1} g'$.

We will need also the following operation on points in $C_n(\mathbb{C})\diagup\mathcal{S}_n$ and on mappings from an interval to $C_n(\mathbb{C})\diagup\mathcal{S}_n$. For an element $E \in C_n(\mathbb{C})\diagup\mathcal{S}_n$ and a number ε we denote by εE the element $E \in C_n(\mathbb{C})\diagup\mathcal{S}_n$ which is obtained by multiplying each point in E (considered as unordered set of points in $\mathbb{C}$) by ε. For a point $z \in \mathbb{C}$ we let

$$z \boxplus \varepsilon E \tag{4.12}$$

be the set that is obtained from εE by adding z to each point in εE. Similarly, for a set Ω and a mapping $f_1 : \Omega \to \mathbb{C}$, a number ε, and a mapping $f^2 : \Omega \to C_n(\mathbb{C})\diagup\mathcal{S}_n$ we denote by

$$f_1 \boxplus \varepsilon f^2 \tag{4.13}$$

the mapping from Ω to symmetrized configuration space defined for each $t \in \Omega$ by $f_1(t) \boxplus \varepsilon f^2(t)$.

The following two lemmas on isotopies of geometric fat braids will be needed for the recovery.

Lemma 4.2 *Let g^0 and g^1 be two geometric fat braids contained in a closed tube $\mathfrak{T}$ (i.e. for each $t \in [0, 1]$ the points $g^0(t)$ and $g^1(t)$ are contained in $\mathfrak{T} \cap (\{t\} \times \mathbb{C})$). If g^0 and g^1 are isotopic then there exists an isotopy of geometric fat braids which is contained in $\mathfrak{T}$ and joins the two braids.*

Proof Let g^s, $s \in [0, 1]$, be the smooth family of geometric fat braids defined by the isotopy. Let $R > 0$ be a number such that all geometric braids g^s, $s \in [0, 1]$, together with the tube $\mathfrak{T}$, are contained in $[0, 1] \times R\mathbb{D}$. Let $\mathfrak{T}^0 \subset \mathfrak{T}$ be a tube which contains both geometric braids g^0 and g^1, such that the fiber $\mathfrak{T}^0 \cap (\{t\} \times \mathbb{C})$ is relatively compact in the fiber $\mathfrak{T} \cap (\{t\} \times \mathbb{C})$ for each $t \in [0, 1]$.

Let χ be a self-homeomorphism of $[0, 1] \times \mathbb{C}$ which preserves each fiber $\{t\} \times \mathbb{C}$, is equal to the identity on $\mathfrak{T}^0$ and outside $[0, 1] \times R_1\overline{\mathbb{D}}$ for a large number $R_1 > R$, and maps the set $[0, 1] \times R\overline{\mathbb{D}}$ onto $\mathfrak{T}$.

Then the equalities $\chi \circ g^0 = g^0$ and $\chi \circ g^1 = g^1$ hold, and $\chi \circ g^s$, $s \in [0, 1]$, is a smooth family of geometric fat braids contained in $\mathfrak{T}$ that joins g^0 with g^1. The lemma is proved. □

Lemma 4.3 *Let f^1, f^2, g^1, and g^2 be pure geometric fat braids, such that f^1 is free isotopic to f^2, and g^1 is free isotopic to g^2. For $j = 0, 1$ we let f_1^j be fat strands of f^j that correspond to each other under the isotopy. Further, for $j = 1, 2$*

we let $\mathfrak{T}(f_1^j)$ be a tube around f_1^j which does not meet the other strands of f^j and contains g^j. Then $f^1 \sqcup_{f_1^1} g^1$ and $f^2 \sqcup_{f_1^2} g^2$ are free isotopic geometric fat braids.

Proof Note first that for any sufficiently small number ε the braids defined by the mappings g^1 and $f_1^1 \boxplus \varepsilon g^1$ are free isotopic. Further, for small ε the fat braid $f_1^1 \boxplus \varepsilon g^1$ is contained in the tube $\mathfrak{T}(f_1^1)$ around the fat strand f_1^1 of f^1. Since, g^1 is also contained in $\mathfrak{T}(f_1^1)$, Lemma 4.2 (which applies to fat braids as well) provides an isotopy that is contained in $\mathfrak{T}(f_1^1)$ and joins g^1 and $f_1^1 \boxplus \varepsilon g^1$. As a consequence, there is a free isotopy joining $f^1 \sqcup_{f_1^1} g^1$ and $f^1 \sqcup_{f_1^1} (f_1^1 \boxplus \varepsilon g^1)$. In the same way for small ε the braids $f^2 \sqcup_{f_1^2} g^2$ and $f^2 \sqcup_{f_1^2} (f_1^2 \boxplus \varepsilon g^2)$ are free isotopic. Hence, for any a priory given small enough positive number ε we may assume from the beginning that $g^1 = f_1^1 \boxplus \varepsilon \tilde{g}^1$, and $g^2 = f_1^2 \boxplus \varepsilon \tilde{g}^2$, where $\tilde{g}^1$, and $\tilde{g}^2$ are geometric fat braids that are contained in the cylinder $[0, 1] \times \mathbb{D}$.

Let f^s, $s \in [1, 2]$, be an isotopy of braids joining f^1 with f^2, and let f_1^s be the family of strands joining f_1^1 with f_1^2. Also, let $\tilde{g}^s$, $s \in [1, 2]$, be an isotopy of braids contained in the cylinder $[0, 1] \times \mathbb{D}$, that join $\tilde{g}^1$ and $\tilde{g}^2$. Choose $\varepsilon > 0$ so that for each $s \in [1, 2]$ the ε-tubes around the strand of f^s are disjoint. Then $f^s \sqcup_{f_1^s} (f_1^s \boxplus \varepsilon \tilde{g}^s)$, $s \in [1, 2]$, is an isotopy of braids that joins $f^1 \sqcup_{f_1^1} (f_1^1 \boxplus \varepsilon \tilde{g}^1)$ with $f^2 \sqcup_{f_1^2} (f_1^2 \boxplus \varepsilon \tilde{g}^2)$. The Lemma is proved. □

We describe now in detail the recovery procedure for the conjugacy class $\hat{b}$ knowing the isotopy class of the admissible set of curves $\mathcal{C} \subset \mathbb{D}$, and, hence, up to isotopy, the sets $S^{\ell,j}$, and knowing for all ℓ and j the sets of the fat braids $\widehat{b(\ell, j)}$ associated to the $S^{\ell,j}$.

For this purpose we will take arbitrary representatives of the $\widehat{b(\ell, j)}$ contained in the cylinder $[0, 1] \times \mathbb{D}$ and describe an inductive procedure to put them together after isotopy and normalization. To show that any geometric braid obtained by this procedure represents the conjugacy class $\hat{b}$, we will use Lemmas 4.2 and 4.3 to compare the geometric fat braid obtained at each inductive step with a geometric fat braid related to $t \to \varphi_t(E)$, $t \in [0, 1]$, where $E \subset \mathbb{D}$ and φ_t is a continuous family of self-homeomorphisms of $\mathbb{P}^1$ such that $\varphi_0 = \text{id}$, and φ_1 represents $\mathfrak{m}_{b,\infty}$ and fixes the admissible set of curves $\mathcal{C} \subset \mathbb{D}$.

Take a connected component $S^{\ell,j}$ of $\mathbb{P}^1 \backslash \mathcal{C}$. Recall that the conjugacy class of fat braids $\widehat{b(\ell, j)}$ was defined in terms of the set E, and the family φ_t, as the class of the geometric fat braid $g^{\ell,j} : [0, 1] \to C_{n(\ell,j)}(\mathbb{C}) \diagup \mathcal{S}_{n(\ell,j)}$ with

$$g^{\ell,j}(t) = \varphi_t(E^{\ell,j} \cup \boldsymbol{E}^{\ell,j}) \tag{4.14}$$

for the set $E^{\ell,j} = E \cap S^{\ell,j}$ and the set $\boldsymbol{E}^{\ell,j}$. The set $\boldsymbol{E}^{\ell,j} = \{z^{\ell,1}, \ldots\}$ was constructed so that for each interior boundary component ∂' of $S^{\ell,j}$ it contains exactly one point of E that is contained in the bounded connected component of the complement of ∂' in $\mathbb{C}$. Here $n(\ell, j)$ is the total number of points in $E^{\ell,j} \cup \boldsymbol{E}^{\ell,j}$.

We represent each $\widehat{\boldsymbol{b}(\ell, j)}$ by an arbitrary geometric fat braid

$$\boldsymbol{f}^{\ell,j} : [0, 1] \to C_{n(\ell,j)}(\mathbb{C})\diagup \mathcal{S}_{n(\ell,j)}$$

which is contained in the tube $[0, 1] \times \mathbb{D}$ and has base point $\tilde{E}^{\ell,j} \cup \tilde{\boldsymbol{E}}^{\ell,j}$ for some sets $\tilde{E}^{\ell,j}$ and $\tilde{\boldsymbol{E}}^{\ell,j}$. The set of initial points of the fat strands equals $\tilde{\boldsymbol{E}}^{\ell,j}$ and $n(\ell, j)$ is the total number of strands (fat strands and ordinary strands together) of $\boldsymbol{f}^{\ell,j}$.

Let $\tilde{\boldsymbol{E}}^{1,1} = \{\tilde{z}^{2,1}, \dots, \tilde{z}^{2,k_2}\}$ and $\boldsymbol{E}^{1,1} = \{z^{2,1}, \dots, z^{2,k_2}\}$. Denote by $\boldsymbol{f}_{2,j}$ the strand of $\boldsymbol{f}^{1,1}$ with initial point $\boldsymbol{f}_{2,j}(0) = \tilde{z}^{2,j}$. We may choose the label so that for each j the strand $\varphi_t(z^{2,j})$, $t \in [0, 1]$, of $\boldsymbol{g}^{1,1}$ corresponds to $\boldsymbol{f}_{2,j}$ under the isotopy of $\boldsymbol{g}^{1,1}$ and $\boldsymbol{f}^{1,1}$. Take a small positive number ε_2 such that the ε_2-neighbourhoods of the strands of $\boldsymbol{f}^{1,1}$ are pairwise disjoint.

We consider the pure geometric fat braid

$$\boldsymbol{f}^{1,1} \sqcup_{\boldsymbol{f}_{2,1}} (\boldsymbol{f}_{2,1} \boxplus \varepsilon_2\, \boldsymbol{f}^{2,1}) : [0, 1] \to C_{n(1,1)+n(2,1)-1}(\mathbb{C})\diagup \mathcal{S}_{n(1,1)+n(2,1)-1}\,, \tag{4.15}$$

that is obtained by removing the strand $\boldsymbol{f}_{2,1}$ from $\boldsymbol{f}^{1,1}$ and inserting into a tube around $\boldsymbol{f}_{2,1}$ a geometric fat braid that is obtained from $\boldsymbol{f}^{2,1}$ by the operations $\sqcup$ and $\boxplus$. The base point of the obtained geometric braid (4.15) is the point

$$(\tilde{E}^{1,1} \cup \tilde{\boldsymbol{E}}^{1,1}) \sqcup_{\tilde{z}^{2,1}} (\tilde{z}^{2,1} \boxplus \varepsilon_2\, (\tilde{E}^{2,1} \cup \tilde{\boldsymbol{E}}^{2,1}))\,. \tag{4.16}$$

Since the geometric fat braids $\boldsymbol{f}^{1,1}$ and $\boldsymbol{g}^{1,1}$ (see (4.14)) both represent $\widehat{\boldsymbol{b}(1, 1)}$, and the geometric fat braids $\boldsymbol{f}^{2,1}$ and $\boldsymbol{g}^{2,1}$ (see (4.14)) both represent $\widehat{\boldsymbol{b}(2, 1)}$, Lemma 4.3 shows that for small ε_2 the geometric fat braids (4.15) and $\boldsymbol{g}^{1,1} \sqcup_{\boldsymbol{g}_{2,1}} \boldsymbol{g}^{2,1}$ are free isotopic. The geometric braid $\boldsymbol{g}^{1,1} \sqcup_{\boldsymbol{g}_{2,1}} \boldsymbol{g}^{2,1}$ equals

$$\begin{aligned} &\varphi_t((E^{1,1} \cup \boldsymbol{E}^{1,1}) \quad \sqcup_{z^{2,1}} \quad (E^{2,1} \cup \boldsymbol{E}^{2,1})) \\ &= \varphi_t((E^{1,1} \cup \boldsymbol{E}^{1,1}) \setminus z^{2,1}) \cup (E^{2,1} \cup \boldsymbol{E}^{2,1})),\ t \in [0, 1]\,. \end{aligned} \tag{4.17}$$

We proceed in the same way by induction on $j = 2, \dots, k_2,$. We arrive at a geometric fat braid

$$\boldsymbol{f}^{1,1} \sqcup_{\boldsymbol{f}_{2,1}} (\boldsymbol{f}_{2,1} \boxplus \varepsilon_2 \boldsymbol{f}^{2,1}) \sqcup \dots \sqcup_{\boldsymbol{f}_{2,k_2}} (\boldsymbol{f}_{2,k_2} \boxplus \varepsilon_2 \boldsymbol{f}^{2,k_2})\,. \tag{4.18}$$

Since $\boldsymbol{E}^{1,1} = \{z^{2,1}, \dots, z^{2,k_2}\}$, the equality

$$(E^{1,1} \cup \boldsymbol{E}^{1,1}) \sqcup_{z^{2,1}} (E^{2,1} \cup \boldsymbol{E}^{2,1}) \sqcup \dots \sqcup_{z^{2,k_2}} (E^{2,k_2} \cup \boldsymbol{E}^{2,k_2}) =$$

$$E^{1,1} \cup (E^{2,1} \cup \boldsymbol{E}^{2,1}) \cup \dots \cup (E^{2,k_2} \cup \boldsymbol{E}^{2,k_2}) = (E \cap Q_2) \cup (\boldsymbol{E}^{2,1} \cup \dots \cup \boldsymbol{E}^{2,k_2})$$

holds. The respective equality holds for $\tilde{E}^{1,1}$, $\tilde{\boldsymbol{E}}^{1,1}$, $\tilde{E}^{2,j}$, and $\tilde{\boldsymbol{E}}^{2,j}$. Hence, repeated application of Lemma 4.3 shows that (4.18) is free isotopic to the geometric fat braid

$$\varphi_t\big((E \cap Q_2) \cup (\boldsymbol{E}^{2,1} \cup \ldots \cup \boldsymbol{E}^{2,k_2})\big),\ t \in [0,1]\,, \tag{4.19}$$

all whose fat strands have initial points in discs $\delta^{3,i}$ bounded by curves of third generation.

We make an induction on the number of generation $\ell = 2, \ldots, N$. Suppose we found by induction small positive numbers ε_l and obtained a geometric fat braid

$$\begin{aligned} \boldsymbol{f}_{Q_\ell} \stackrel{def}{=} \boldsymbol{f}_1^{1,1} \sqcup_{f_{2,1}} (\boldsymbol{f}_{2,1} \boxplus \varepsilon_2 \boldsymbol{f}^{2,1}) \sqcup \ldots \sqcup_{f_{2,k_2}} (\boldsymbol{f}_{2,k_2} \boxplus \varepsilon_2 \boldsymbol{f}^{2,k_2}) \\ \sqcup \ldots \sqcup_{f_{\ell,k_\ell}} (\boldsymbol{f}_{\ell,k_\ell} \boxplus \varepsilon_\ell \boldsymbol{f}^{\ell,k_\ell}) \end{aligned} \tag{4.20}$$

that is free homotopic to the geometric fat braid

$$\varphi_t\big((E \cap Q_\ell) \cup (\boldsymbol{E}^{\ell,1} \cup \ldots \cup \boldsymbol{E}^{\ell,k_\ell})\big),\ t \in [0,1]\,, \tag{4.21}$$

Denote the fat strands of (4.20) by $\boldsymbol{f}_{\ell+1,j}$. Choose a positive number $\varepsilon_{\ell+1}$ such that the $\varepsilon_{\ell+1}$-neighbourhoods of the strands of the geometric fat braid (4.20) are pairwise disjoint. Add strands to (4.20) by the operation

$$\boldsymbol{f}_{Q_\ell} \sqcup_{f_{\ell+1,1}} (\boldsymbol{f}_{\ell+1,1} \boxplus \varepsilon_{\ell+1} \boldsymbol{f}^{\ell+1,1}) \sqcup \ldots \sqcup_{f_{\ell+1,k_{\ell+1}}} (\boldsymbol{f}_{\ell+1,k_{\ell+1}} \boxplus \varepsilon_{\ell+1} \boldsymbol{f}^{\ell+1,k_{\ell+1}})\,. \tag{4.22}$$

We obtain a geometric fat braid of the same form as (4.20) with ℓ replaced by $\ell+1$. Repeated application of Lemma 4.3 shows that (4.22) is isotopic to the geometric fat braid (4.21) with ℓ replaced by $\ell + 1$. Since the boundary of Q_N is empty, the geometric fat braid $\boldsymbol{f}_{Q_N}$ is an ordinary braid that is isotopic to

$$\varphi_t(E),\ t \in [0,1]\,. \tag{4.23}$$

We described a procedure that gives for arbitrary geometric fat braids $\boldsymbol{f}_{i,j}$ in $[0,1] \times \mathbb{D}$, that represent $\widehat{\boldsymbol{b}(i,j)}$, the ordinary geometric braid f_{Q_N} (see (4.20) with ℓ replaced by N), that is free isotopic to (4.22), hence, represents $\hat{b}$. The recovery procedure is described.

4.5 Pure Braids, the Reducible Case. Proof of the Main Theorem

In this section we will prove the Main Theorem for reducible pure braids. The theorem will follow from the Propositions 4.1 and 4.2 below which are of independent interest.

We will represent a conjugacy class of pure braids $\hat{b}$ by a holomorphic map from an annulus A to the symmetrized configuration space $C_n(\mathbb{C})\diagup\mathcal{S}_n$ (or by a holomorphic map on A that extends continuously to the closure $\bar{A}$). The set $\{(z, f(z));\ z \in A\}$ (or $\{(z, f(z));\ z \in \bar{A}\}$, respectively) is the union of n connected components which we also call strands. A map with some strands declared to be fat will be called a fat map and denoted by $\boldsymbol{f}$.

Proposition 4.1 *The conformal module of a conjugacy class $\hat{b}$ of pure n-braids is equal to the smallest conformal module among the irreducible braid components of $\hat{b}$. In other words, the equality*

$$\mathcal{M}(\hat{b}) = \min_{\ell,j} \mathcal{M}\left(\widehat{b(\ell, j)}\right) \tag{4.24}$$

holds.

Proof Let A be any annulus of conformal module $m(A) < \mathcal{M}(\hat{b})$. Then there exists a holomorphic mapping

$$f : A \to C_n(\mathbb{C})\diagup\mathcal{S}_n \tag{4.25}$$

representing $\hat{b}$. Recall that for all (ℓ, j) the class $\widehat{b(\ell, j)}$ is obtained from $\hat{b}$ by forgetting some strands. Forgetting the respective strands of the mapping (4.25) provides us a holomorphic map $f^{\ell,j} : A \to C_{n(\ell,j)}(\mathbb{C})\diagup\mathcal{S}_{n(\ell,j)}$ that represents $\widehat{b(\ell, j)}$. Hence, for each annulus A with $m(A) < \mathcal{M}(\hat{b})$ the inequality $\mathcal{M}(\widehat{b(\ell, j)}) > m(A)$ holds for all (ℓ, j). Hence,

$$\mathcal{M}(\hat{b}) \le \min_{\ell,j} \mathcal{M}\left(\widehat{b(\ell, j)}\right).$$

To prove the opposite inequality, let A be an annulus of conformal module $m(A) < \min_{\ell,j} \mathcal{M}\left(\widehat{b(\ell, j)}\right)$. Then for each (ℓ, j) there is a continuous fat map $\boldsymbol{f}^{\ell,j} : \bar{A} \to C_{n(\ell,j)}(\mathbb{D})\diagup\mathcal{S}_{n(\ell,j)}$ that is holomorphic on A and represents the conjugacy class of fat braids $\widehat{\boldsymbol{b}(\ell, j)}$. We will denote by $\boldsymbol{f}_{\ell+1,j'}$ the strands of the mappings $\boldsymbol{f}^{\ell,j}$ that correspond to fat strands of $\widehat{\boldsymbol{b}(\ell, j)}$. As in the recovery procedure of the conjugacy class of braids from the irreducible braid components we successively choose small numbers ε_ℓ, $\ell \le N$, and consider the mapping

$$\begin{aligned} &\boldsymbol{f}^{1,1} \sqcup_{\boldsymbol{f}_{2,1}} (\boldsymbol{f}_{2,1} \boxplus \varepsilon_2 \boldsymbol{f}^{2,1}) \sqcup \ldots \sqcup_{\boldsymbol{f}_{2,k_2}} (\boldsymbol{f}_{2,k_2} \boxplus \varepsilon_2 \boldsymbol{f}^{2,k_2}) \\ &\sqcup \ldots \sqcup_{\boldsymbol{f}_{\ell,k_\ell}} (\boldsymbol{f}_{\ell,k_\ell} \boxplus \varepsilon_\ell \boldsymbol{f}^{\ell,k_\ell}) \end{aligned} \tag{4.26}$$

that is defined pointwise for $z \in \bar{A}$ in the same way as the respective objects were defined in Sect. 4.4 for geometric fat braids. The ε_ℓ are chosen successively so that the ε_ℓ-neighbourhoods of the strands of the fat map defined by (4.26) for ℓ replaced by $\ell - 1$ are pairwise disjoint. For $\ell = N$ the mapping (4.26) is an ordinary mapping and represents the conjugacy class $\hat{b}$. Since all mappings $\boldsymbol{f}^{\ell,j}$ and $\boldsymbol{f}_{\ell',j'}$ are holomorphic on A, the mapping (4.26) is holomorphic by construction. Hence, $\mathcal{M}(\hat{b}) \geq m(A)$ for any A with $m(A) < \min_{\ell,j} \mathcal{M}\left(\widehat{b(\ell,j)}\right)$. We obtained the inequality

$$\mathcal{M}(\hat{b}) \geq \min_{\ell,j} \mathcal{M}\left(\widehat{b(\ell,j)}\right).$$

The proposition is proved. □

Proposition 4.2 *For the mapping class $\mathfrak{m}_{b,\infty}$ associated to a pure n-braid b and its irreducible nodal components $\widehat{\mathfrak{m}_{b,\odot}^{\ell,j}}$ the equality*

$$h(\widehat{\mathfrak{m}_{b,\infty}}) = \max_{\ell,j} h(\widehat{\mathfrak{m}_{b,\odot}^{\ell,j}}) \tag{4.27}$$

holds.

In the lemma all conjugacy classes are considered as classes of self-homeomorphisms of compact Riemann surfaces with distinguished points.

Proof We prove first the inequality

$$h(\widehat{\mathfrak{m}_{b,\infty}}) \geq \max_{\ell,j} h(\widehat{\mathfrak{m}_{b,\odot}^{\ell,j}}). \tag{4.28}$$

We take again a self-homeomorphism $\varphi_{b,\infty}$ of $\mathbb{P}^1$ with distinguished points $E_n' = E_n \cup \{\infty\}$ that fixes pointwise the complement of the unit disc and represents the mapping class $\mathfrak{m}_{b,\infty} \in \mathfrak{M}(\mathbb{P}^1; E_n')$ of the braid b. We require that $\varphi_{b,\infty}$ is completely reduced by the admissible system of curves $\mathcal{C}$ that is contained in the unit disc $\mathbb{D}$. As in Sect. 4.2 we choose the set $\boldsymbol{E}^{\ell,j} \subset E_n$ as follows. For each hole of $S^{\ell,j}$ bounded by an interior boundary component of $S^{\ell,j}$ we chose a point of E contained in it. These points constitute the set $\boldsymbol{E}^{\ell,j}$. Then $\mathfrak{m}_{b(\ell,j),\infty}$ is the mapping class of $\varphi_{b,\infty}$ in $\mathfrak{M}(\mathbb{P}^1; E'^{\ell,j} \cup \boldsymbol{E}^{\ell,j})$. We obtain

$$h(\mathfrak{m}_{b(\ell,j),\infty}) = \inf\{h(\varphi) : \varphi \text{ is } \mathrm{Hom}^+(\mathbb{P}^1; E'^{\ell,j} \cup \boldsymbol{E}^{\ell,j}) - \text{isotopic to } \varphi_{b,\infty}\}. \tag{4.29}$$

On the other hand,

$$h(\mathfrak{m}_{b,\infty}) = \inf\{h(\varphi) : \varphi \text{ is } \mathrm{Hom}^+(\mathbb{P}^1; E_n') - \text{isotopic to } \varphi_{b,\infty}\}. \tag{4.30}$$

Since the space which appears in (4.30) is contained in the space which appears in (4.29), the infimum in (4.30) is not smaller than the infimum in (4.29):

$$h(\mathfrak{m}_{b,\infty}) \geq h(\mathfrak{m}_{b(\ell,j),\infty}) \quad \text{for each } \ell,\, j. \tag{4.31}$$

Since by Lemma 4.1 the equality $\widehat{\mathfrak{m}_{b,\odot}^{\ell,j}} = \widehat{\mathfrak{m}_{b(\ell,j),\infty}}$ holds, the inequality (4.28) is proved.

The proof of the opposite inequality is based on Theorem 2.10. Recall that the class $\widehat{\mathfrak{m}_{b,\odot}^{\ell,j}}$ (considered as element of $\widehat{\mathfrak{M}}(Y^{\ell,j})$ for a punctured Riemann sphere $Y^{\ell,j}$) is represented by the mapping $\varphi_{b,\infty}|(S^{\ell,j} \setminus E_n')$ on a Riemann surface of second kind. Take an absolutely extremal representative of this class. This is a self-homeomorphism of a punctured Riemann surface $\tilde{Y}^{\ell,j}$. It extends across the punctures to a self-homeomorphism $\tilde{\varphi}^{\ell,j}$ of $\mathbb{P}^1$ with distinguished points $\tilde{E}'^{\ell,j} \cup \tilde{\boldsymbol{E}}^{\ell,j}$. (The set $\tilde{E}'^{\ell,j}$ corresponds to $S^{\ell,j} \cap E_n'$, the set $\tilde{\boldsymbol{E}}^{\ell,j}$ corresponds to the holes of $S^{\ell,j}$.) $\tilde{\varphi}^{\ell,j}$ is entropy minimizing in the class of self-homeomorphisms of $\mathbb{P}^1$ with distinguished points that is obtained from $\varphi_{b,\infty}|(S^{\ell,j} \setminus E_n')$ by isotopy, conjugation and extension across punctures, i.e.

$$h(\tilde{\varphi}^{\ell,j}) = \inf\{h(\varphi) : \varphi \in \widehat{\mathfrak{m}_{b,\odot}^{\ell,j}}\}.$$

By Theorem 2.10 for each (ℓ, j) there is a closed topological disc around each distinguished point in $\tilde{\boldsymbol{E}}^{\ell,j}$ and a self-homeomorphism $\tilde{\varphi}_0^{\ell,j}$ of $\mathbb{P}^1$ with distinguished points $\tilde{E}'^{\ell,j} \cup \tilde{\boldsymbol{E}}^{\ell,j}$ that is isotopic (with distinguished points $\tilde{E}'^{\ell,j} \cup \tilde{\boldsymbol{E}}^{\ell,j}$) to $\tilde{\varphi}^{\ell,j}$, has the same entropy $h(\tilde{\varphi}_0^{\ell,j}) = h(\tilde{\varphi}^{\ell,j})$, and equals the identity in a neighbourhood of each of the discs.

Since $\tilde{\varphi}^{\ell,j} \in \widehat{\mathfrak{m}_{b,\odot}^{\ell,j}}$ there exists a homeomorphism $\tilde{w}^{\ell,j} : S^{\ell,j} \setminus E'^{\ell,j} \to \tilde{Y}^{\ell,j}$ that conjugates $\tilde{\varphi}^{\ell,j} \mid \mathbb{P}^1 \setminus (E'^{\ell,j} \cup \tilde{\boldsymbol{E}}^{\ell,j})$ to a self-homeomorphism of $S^{\ell,j} \setminus E'^{\ell,j}$ that is isotopic on this set to $\varphi_{b,\infty} \mid S^{\ell,j} \setminus E'^{\ell,j}$. The conjugate $(\tilde{w}^{\ell,j})^{-1} \circ \tilde{\varphi}_0^{\ell,j} \circ \tilde{w}^{\ell,j}$ of the isotopic homeomorphism $\tilde{\varphi}_0^{\ell,j}$ equals the identity in a neighbourhood of all boundary components of $S^{\ell,j}$. Hence, it extends to the boundary of $S^{\ell,j}$ as a self-homeomorphism $\varphi_0^{\ell,j}$ of $\overline{S^{\ell,j}} \setminus E_n'$ that fixes the boundary pointwise. The isotopy class in $\mathfrak{M}(\overline{S^{\ell,j}} \setminus E_n'; \partial S^{\ell,j})$ of the extension $\varphi_0^{\ell,j}$ differs from that of the restriction $\varphi_{b,\infty} \mid \overline{S^{\ell,j}} \setminus E_n'$ by products of powers of Dehn twists about curves in $S^{\ell,j}$ that are homologous to one of the boundary components of $S^{\ell,j}$. By Lemma 2.14 for each (ℓ, j) the entropy minimizing homeomorphism $\tilde{\varphi}_0^{\ell,j}$ may be chosen from the beginning so that $\varphi_0^{\ell,j}$ and $\varphi_{b,\infty} \mid \overline{S^{\ell,j}} \setminus E_n'$ are $\mathfrak{M}(\overline{S^{\ell,j}} \setminus E_n'; \partial S^{\ell,j})$-isotopic.

Consider the self-homeomorphism φ_0 of $\mathbb{P}^1 \setminus E_n'$ whose restriction to $\overline{S^{\ell,j}} \setminus E_n'$ equals $\varphi_0^{\ell,j}$ for all ℓ and j. (The self-homeomorphism is well defined since each $\varphi_0^{\ell,j}$ is equal to the identity near the boundary $\partial S^{\ell,j}$.) The mapping φ_0 is isotopic to $\varphi_{b,\infty}$ in $\mathfrak{M}(\mathbb{P}^1 \setminus E_n')$, hence it represents $\mathfrak{m}_{b,\infty}$ (considered as class of self-homeomorphisms of a punctured surface). For the extensions $\hat{\varphi}_0$ and $\hat{\varphi}_0^{\ell,j}$ of the mappings φ_0 and $\varphi_0^{\ell,j}$ across the punctures the inequality $h(\hat{\varphi}_0) \leq \max_{\ell,j} h(\hat{\varphi}_0^{\ell,j})$ holds. Hence, the opposite inequality

$$h(\widehat{\mathfrak{m}_{b,\infty}}) \leq \max_{\ell,j} h(\widehat{\mathfrak{m}_{b,\odot}^{\ell,j}}) \tag{4.32}$$

holds and the proposition is proved. □

The Proof of the Main Theorem for Reducible Pure Braids is an immediate consequence of Propositions 4.1 and 4.2, Lemma 4.1, and the Main Theorem for irreducible braids. □

Remark 4.1 *The mapping class $\widehat{\mathfrak{m}_{b,\infty}}$ can be recovered from the irreducible nodal components of the braid up to a product of Dehn twists about simple closed curves that are homologous to the admissible curves.*

Indeed, this statement was obtained by the proof of inequality (4.32).

Chapter 5
The General Case. Irreducible Nodal Components, Irreducible Braid Components, and the Proof of the Main Theorem

In this chapter we prove the Main Theorem for reducible non-pure braids. While the strategy of the proof in this case is similar to that in the case of pure braids, there are some difficulties that need additional considerations. For instance, for each conjugacy class of pure braids the irreducible braid components can be obtained by forgetting strands. This approach does not work for non-pure braids. We define irreducible braid components and irreducible nodal components in the non-pure case, and describe the respective decompositions into irreducible components and the recovery procedures.

5.1 Irreducible Nodal Components. The General Case

In this Chapter we consider a reducible, not necessarily pure, braid $b \in \mathcal{B}_n$ and its mapping class $\mathfrak{m}_b \in \mathfrak{M}(\overline{\mathbb{D}}; \partial\mathbb{D}, E_n)$, $E_n \subset \mathbb{D}$. We represent the mapping class $\mathfrak{m}_{b,\infty} = \mathcal{H}_\infty(\mathfrak{m}_b) \in \mathfrak{M}(\mathbb{P}^1; \infty, E_n)$ (see equality (1.8)) by a self-homeomorphism $\varphi_{b,\infty}$ of $\mathbb{P}^1$ which is the identity outside the unit disc $\mathbb{D}$. We may choose $\varphi_{b,\infty}$ so that it is completely reduced by an admissible system of curves $\mathcal{C} = \{C_1, \ldots, C_k\}$ in $\mathbb{D}\setminus E_n \subset \mathbb{P}^1\setminus E_n'$. As before $E_n' = E_n \cup \{\infty\}$. In particular, $\varphi_{b,\infty}$ leaves the union $\bigcup\limits_{C\in\mathcal{C}} C$ invariant, and also leaves the complement $\mathbb{P}^1\setminus \bigcup\limits_{C\in\mathcal{C}} C$ invariant. The nodal components and the irreducible braid components will be associated to the isotopy class of the chosen admissible system of curves.

As in Chap. 4 the connected component of $\mathbb{P}^1 \setminus \bigcup_{C\in\mathcal{C}} C$ that contains ∞ is denoted by $S^{1,1}$ and is said to be of generation 1. By induction for $\ell \geq 2$ the interior of the union of the closure of all components of generation not exceeding $\ell - 1$ is denoted by $Q_{\ell-1}$. The components of $\mathbb{P}^1 \setminus \bigcup_{C\in\mathcal{C}} C$ that share a boundary component with $Q_{\ell-1}$ are called the components of generation ℓ. However, since b is not required to be a pure braid, the connected components of $\mathbb{C} \setminus \mathcal{C}$ of generation

B. Jöricke, *Braids, Conformal Module, Entropy, and Gromov's Oka Principle*, Lecture Notes in Mathematics 2360, https://doi.org/10.1007/978-3-031-67288-0_5

ℓ are not necessarily fixed by $\varphi_{b,\infty}$. But the components of each generation ℓ are permuted along cycles. Indeed, since φ fixes $\infty \in Q_{\ell-1}$ it fixes $Q_{\ell-1}$ setwise and, hence, permutes its boundary components, and therefore permutes the components of $\mathbb{P}^1 \setminus \bigcup_{C \in \mathcal{C}} C$ that share a boundary component with $Q_{\ell-1}$.

We label the connected components of $\mathbb{P}^1 \setminus \bigcup_{C \in \mathcal{C}} C$ of generation $\ell \geq 2$ as follows. Choose for each cycle of components of $\mathbb{P}^1 \setminus \bigcup_{C \in \mathcal{C}} C$ of generation ℓ a set of the cycle and denote it by $S_1^{\ell,i}$. Let $k(\ell, i)$ be the length of the cycle, i.e. the smallest positive integer for which $\varphi_{b,\infty}^{k(\ell,i)}$ maps $S_1^{\ell,i}$ onto itself. Put

$$S_{j+1}^{\ell,i} \overset{def}{=} \varphi_{b,\infty}(S_j^{\ell,i}), \quad j = 1, \dots, k(\ell, i) - 1. \tag{5.1}$$

For all ℓ, and i

$$\varphi_b(S_{k(\ell,i)}^{\ell,i}) = S_1^{\ell,i}, \tag{5.2}$$

The $\varphi_{b,\infty}$-cycles of connected components of generation ℓ of $\mathbb{P}^1 \setminus \bigcup_{C \in \mathcal{C}} C$ are denoted by $\mathrm{cyc}_0^{\ell,i} = \left(S_1^{\ell,i}, \dots, S_{k(\ell,i)}^{\ell,i}\right),\ i = 1, \dots, k_\ell$,

$$S_1^{\ell,i} \overset{\varphi_{b,\infty}}{\longrightarrow} S_2^{\ell,i} \overset{\varphi_{b,\infty}}{\longrightarrow} \dots \overset{\varphi_{b,\infty}}{\longrightarrow} S_{k(\ell,i)}^{\ell,i} \overset{\varphi_{b,\infty}}{\longrightarrow} S_1^{\ell,i}.$$

Note that the number of cycles k_ℓ does not exceed the number of components k'_ℓ of generation ℓ.

The set of distinguished points $E'^{\ell,i}_j = S_j^{\ell,i} \cap E'_n$ is permuted in the same way:

$$E'^{\ell,i}_1 \overset{\varphi_{b,\infty}}{\longrightarrow} E'^{\ell,i}_2 \overset{\varphi_{b,\infty}}{\longrightarrow} \dots \overset{\varphi_{b,\infty}}{\longrightarrow} E'^{\ell,i}_{k(\ell,i)} \overset{\varphi_{b,\infty}}{\longrightarrow} E'^{\ell,i}_1.$$

The cycles of connected components of $\mathbb{P}^1 \setminus \left((\bigcup_{C \in \mathcal{C}} C) \bigcup E'_n \right)$ of generation ℓ are denoted by $\mathrm{cyc}^{\ell,i} = \left(S_1^{\ell,i} \setminus E'_n, \dots, S_{k(\ell,i)}^{\ell,i} \setminus E'_n\right),\ i = 1, \dots, k_\ell$.

Consider the $\varphi_{b,\infty}$-cycle of curves $\partial_{\mathbb{C}} S_j^{\ell,i}$, $j = 1, \dots, k(\ell, i)$, of generation ℓ and the cycle of discs $\delta_j^{\ell,i}$ in $\mathbb{C}$ that are bounded by $\partial \delta_j^{\ell,i} = \partial_{\mathbb{C}} S_j^{\ell,i}$. The homeomorphism $\varphi_{b,\infty}$ permutes the discs along the cycle $(\delta_1^{\ell,i}, \delta_2^{\ell,i}, \dots, \delta_{k(\ell,i)}^{\ell,i})$,

$$\delta_1^{\ell,i} \overset{\varphi_{b,\infty}}{\longrightarrow} \delta_2^{\ell,i} \overset{\varphi_{b,\infty}}{\longrightarrow} \dots \overset{\varphi_{b,\infty}}{\longrightarrow} \delta_{k(\ell,i)}^{\ell,i} \overset{\varphi_{b,\infty}}{\longrightarrow} \delta_1^{\ell,i}.$$

As in the case of pure braids (see also Sect. 4.1 and also Sect. 1.8) we associate to b a nodal surface Y and an isotopy class $\mathfrak{m}_{b,\odot}$ of self-homeomorphisms of Y, using a continuous surjection $w : \mathbb{P}^1 \setminus E_n' \to Y$ whose preimage of each node is a curve $C \in \mathcal{C}$.

Recall that the set of nodes of Y is denoted by $\mathcal{N}$. The surjection w maps $\mathbb{P}^1 \setminus (E_n' \cup \bigcup_{C \in \mathcal{C}} C)$ homeomorphically onto $Y \setminus \mathcal{N}$. Conjugate $\varphi_{b,\infty} \mid \mathbb{P}^1 \setminus (E_n' \cup \bigcup_{C \in \mathcal{C}} C)$ by the inverse of $w \mid \mathbb{P}^1 \setminus (E_n' \cup \bigcup_{C \in \mathcal{C}} C)$ and denote the conjugate by $\varphi_{b,\odot}$. Its isotopy class of self-homeomorphisms of $Y \setminus \mathcal{N}$ is denoted by $\mathfrak{m}_{b,\odot}$. The conjugacy class $\widehat{\mathfrak{m}_{b,\odot}}$ is called the nodal conjugacy class associated to $\mathfrak{m}_{b,\infty}$ and the system of admissible curves $\mathcal{C}$.

Notice that the mapping w extends across the punctures to a continuous surjection from $\mathbb{P}^1$ onto the compact nodal surface Y^c obtained by filling the punctures of Y. The extended mapping is also denoted by w. The elements of the class $\mathfrak{m}_{b,\odot}$ extend across the punctures of the nodal surface Y to self-homeomorphisms of the compact nodal surface Y^c. By an abuse of notation we denote the class of extended mappings also by $\mathfrak{m}_{b,\odot}$, and denote the respective conjugacy class also by $\widehat{\mathfrak{m}_{b,\odot}}$.

Let $Y_j^{\ell,i}$ be the component $w(S_j^{\ell,i} \setminus E_n')$ of $Y \setminus \mathcal{N}$. The mapping $\varphi_{b,\odot}$ permutes the components $Y_j^{\ell,i}$ of $Y \setminus \mathcal{N}$ along cycles corresponding to the cycles of the $S_j^{\ell,i} \setminus E'^{\ell,i}_j$. We denote the cycles of components $Y_j^{\ell,i}$ by $\mathrm{cyc}_{b,\odot}^{\ell,i}$.

The conjugacy classes $\widehat{\mathfrak{m}_{b,\odot}^{\ell,i}}$ of the restrictions $\mathfrak{m}_{b,\odot}^{\ell,i}$ of $\mathfrak{m}_{b,\odot}$ to the cycles $\mathrm{cyc}_{b,\odot}^{\ell,i}$, are called the irreducible nodal components of the class $\widehat{\mathfrak{m}_{b,\infty}}$. The irreducible nodal components determine the class $\widehat{\mathfrak{m}_{b,\infty}}$ up to products of powers of some Dehn twists. This can be seen in the same way as in the case of pure braids. (See later the Remark 5.2.)

Instead of the cycles of components $Y_j^{\ell,i}$, $j = 1, \ldots,$ we may, equivalently, consider the respective cycles of compact Riemann surfaces $(Y_j^{\ell,i})^c$, $j = 1, \ldots,$ and extend the homeomorphisms between the $Y_j^{\ell,i}$, $j = 1, \ldots,$ to homeomorphisms between the compact surfaces. The obtained conjugacy classes will also be denoted by $\widehat{\mathfrak{m}_{b,\odot}^{\ell,i}}$.

If the length of the cycle is bigger than one, the class $\widehat{\mathfrak{m}_{b,\odot}^{\ell,i}}$ is the conjugacy class of a mapping class on a not connected Riemann surface. Lemma 5.1 below states that each irreducible nodal component $\widehat{\mathfrak{m}_{b,\odot}^{\ell,i}}$ is determined by a conjugacy class of mapping classes of self-homeomorphisms of a connected Riemann surface. For instance, for the length $k(\ell, i)$ of the cycle $\mathrm{cyc}_{b,\odot}^{\ell,i}$, we take the $k(\ell, i)$-th power of $\varphi_{b,\odot}$, and restrict it to a component of the cycle, say to $Y_1^{\ell,i}$. The lemma states that the conjugacy class $\widehat{\mathfrak{m}_{b^{k(\ell,i)},\odot} \mid Y_1^{\ell,i}}$ of the restricted mapping determines the irreducible nodal component $\widehat{\mathfrak{m}_{b,\odot}^{\ell,i}}$. Notice that the mappings $\varphi_{b,\odot}^{k(\ell,i)} \mid Y_j^{\ell,i}$, $j = 1, \ldots, k(\ell, i)$, are conjugate, hence, the class $\widehat{\mathfrak{m}_{b^{k(\ell,i)},\odot} \mid Y_1^{\ell,i}}$ is defined by the cycle

$\mathrm{cyc}_{b,\odot}^{\ell,i}$ and is independent on the choice of the set in the cycle to which the mapping $\varphi_{b,\odot}^{k(\ell,i)}$ is restricted.

The key ingredient of the proof of the following Lemma 5.1 is a simple Lemma on Conjugation, which was used by Bers [11]. For convenience of the reader we include the Lemma on Conjugation together with a proof in the Appendix A.2.

Lemma 5.1 *Suppose a self-homeomorphism* $\psi_\odot$ *of* $Y_1^{\ell,i}$ *represents the class* $\widehat{\mathfrak{m}_{b^{k(\ell,i)},\odot} \mid Y_1^{\ell,i}}$ *for the* $k(\ell,i)$*-th power of the braid* b. *Then any self-homeomorphism* $\varphi_\odot$ *of the union* $Y_1^{\ell,i} \cup \ldots \cup Y_{k(\ell,i)}^{\ell,i}$, *that moves the sets* $Y_j^{\ell,i}$ *along the cycle*

$$Y_1^{\ell,i} \xrightarrow{\varphi_\odot} Y_2^{\ell,i} \xrightarrow{\varphi_\odot} \ldots \xrightarrow{\varphi_\odot} Y_{k(\ell,i)}^{\ell,i} \xrightarrow{\varphi_\odot} Y_1^{\ell,i}, \tag{5.3}$$

and satisfies the condition

$$\varphi_\odot^{k(\ell,i)} \mid Y_1^{\ell,i} = \psi_\odot, \tag{5.4}$$

represents $\widehat{\mathfrak{m}_{b,\odot} \mid \mathrm{cyc}_\odot^{\ell,i}}$. *A homeomorphisms* $\varphi_\odot$ *with these properties always exists.*

Vice versa, for any self-homeomorphism $\varphi_\odot$ *of* $Y_1^{\ell,i} \cup \ldots \cup Y_{k(\ell,i)}^{\ell,i}$ *that represents* $\widehat{\mathfrak{m}_{b,\odot} \mid \mathrm{cyc}_\odot^{\ell,i}}$ *the restriction* $\varphi_\odot^{k(\ell,i)} \mid Y_1^{\ell,i}$ *represents* $\widehat{\mathfrak{m}_{b^{k(\ell,i)},\odot} \mid Y_1^{\ell,i}}$.

Proof It is clear that for each $\varphi_\odot \in \widehat{\mathfrak{m}_{b,\odot} \mid \mathrm{cyc}_\odot^{\ell,i}}$ the inclusion $\varphi_\odot^{k(\ell,i)} \mid Y_1^{\ell,i} \in \widehat{\mathfrak{m}_{b^{k(\ell,i)},\odot} \mid Y_1^{\ell,i}}$ holds.

Prove the first part of the lemma. Let $\psi_\odot$ in $\widehat{\mathfrak{m}_{b^{k(\ell,i)},\odot} \mid Y_1^{\ell,i}}$. Define a self-homeomorphisms $\varphi_\odot \in \widehat{\mathfrak{m}_{b,\odot} \mid \mathrm{cyc}_\odot^{\ell,i}}$ as follows. For each $j = 1, \ldots, k(\ell,i) - 1$, we take any homeomorphism $\varphi_{\odot,j}$ from $Y_j^{\ell,i}$ onto $Y_{j+1}^{\ell,i}$. For $j = k(\ell,i)$ we take the homeomorphism $\varphi_{\odot,k(\ell,i)}$ from $Y_{k(\ell,i)}^{\ell,i}$ onto $Y_1^{\ell,i}$ for which

$$\varphi_{\odot,k(\ell,i)} \circ \ldots \circ \varphi_{\odot,1} = \psi_\odot \quad \text{on} \quad Y_1^{\ell,i}. \tag{5.5}$$

Consider the self-homeomorphism $\varphi_\odot$ of $\bigcup_{j=1}^{k(\ell,i)} Y_j^{\ell,i}$ which equals $\varphi_{\odot,j}$ on $Y_j^{\ell,i}$. Then

$$\varphi_\odot^{k(\ell,i)} \mid Y_1^{\ell,i} = \psi_\odot .$$

We have to prove that any mapping $\varphi_\odot$ obtained in this way is in the class $\widehat{\mathfrak{m}_{b,\odot} \mid \mathrm{cyc}_\odot^{\ell,i}}$. Let w be as before the continuous surjection from $\mathbb{P}^1 \setminus E_n'$ onto Y that restricts to a homeomorphism from $\mathbb{P}^1 \setminus (E_n' \cup \bigcup_{C \in \mathcal{C}} C)$ onto $Y \setminus \mathcal{N}$. Let $w^{\ell,i}$ be the restriction of w to $\bigcup_{j=1}^{k(\ell,i)} S_j^{\ell,i} \setminus E_n'$. The mapping $w^{\ell,i}$ takes $S_j^{\ell,i} \setminus E_n'$ homeomorphically onto $Y_j^{\ell,i}$ for each j. Conjugate $\varphi_{b,\infty} \mid \bigcup_{j=1}^{k(\ell,i)} S_j^{\ell,i} \setminus E_n'$ by the inverse of $w^{\ell,i}$ and denote the obtained self-homeomorphism of $\bigcup_{j=1}^{k(\ell,i)} Y_j^{\ell,i}$ by $\tilde{\varphi}_\odot$. Then $\tilde{\varphi}_\odot$ represents the class $\widehat{\mathfrak{m}_{b,\odot} \mid \mathrm{cyc}_\odot^{\ell,i}}$. The mappings $\varphi_\odot$ and $\tilde{\varphi}_\odot$ move the $Y_j^{\ell,i}$ along the same cycle, and both, $\tilde{\varphi}_\odot^{k(\ell,i)} \mid Y_1^{\ell,i}$ and $\varphi_\odot^{k(\ell,i)} \mid Y_1^{\ell,i}$, represent the same conjugacy class $\widehat{\mathfrak{m}_{b^{k(\ell,i)},\odot} \mid Y_1^{\ell,i}}$. Since $\tilde{\varphi}_\odot$ represents $\widehat{\mathfrak{m}_{b,\odot} \mid \mathrm{cyc}_\odot^{\ell,i}}$, by the Lemma on Conjugation (see Appendix A.2) also the mapping $\varphi_\odot$ represents this class. The lemma is proved. □

Remark 5.1 *The restriction $\varphi_{b,\infty}^{k(\ell,i)} \mid S_1^{\ell,i} \setminus E_n'$, equivalently, the mapping $\varphi_{b,\odot}^{k(\ell,i)} \mid Y_1^{\ell,i}$, is irreducible.*

Suppose the contrary. Then there exists an $\mathfrak{M}(\mathbb{P}^1 \setminus E_n')$-isotopy that joins $\varphi_{b,\infty}^{k(\ell,i)}$ with a self-homeomorphism ψ of $\mathbb{P}^1$ and changes the values of $\varphi_{b,\infty}^{k(\ell,i)}$ only on $S_1^{\ell,i}$, such that ψ has the following property. It fixes setwise an admissible system $\gamma_1 \cup \ldots \cup \gamma_k$ of simple closed curves in $S_1^{\ell,i} \setminus E_n'$. Then the mapping $\varphi_{b,\infty}'$, that equals $\varphi_{b,\infty}$ on $\mathbb{P}^1 \setminus S_1^{\ell,i}$ (in particular, on $S_2^{\ell,i} \cup \ldots S_{k(\ell,i)}^{\ell,i}$) and equals $(\varphi_{b,\infty})^{-k(\ell,i)+1} \circ \psi$ on $S_1^{\ell,i}$, fixes the set of loops

$$\gamma_1 \cup \ldots \cup \gamma_k \bigcup \varphi_{b,\infty}'(\gamma_1 \cup \ldots \cup \gamma_k) \bigcup \cdots \bigcup (\varphi_{b,\infty}')^{k(\ell,i)-1}(\gamma_1 \cup \ldots \cup \gamma_k)$$

setwise and is $\mathfrak{M}(\mathbb{P}^1 \setminus E_n')$-isotopic to $\varphi_{b,\infty}$ by the Lemma on conjugation (see the Appendix A.2). This contradicts the fact that the system of curves $\mathcal{C}$ was maximal.

5.2 Irreducible Braid Components. The General Case

We continue to consider non-pure reducible braids b with base point $E_n \subset \mathbb{D}$ together with the associated homeomorphism $\varphi_{b,\infty} \in \mathrm{Hom}^+(\mathbb{P}^1; \mathbb{P}^1 \setminus \mathbb{D}, E_n)$, that fixes setwise an admissible system of curves $\bigcup_{C \in \mathcal{C}} C$ contained in $\mathbb{D}$ and is completely reduced by this system.

Notice that a cycle of discs $\delta_j^{\ell,i}$ of length $k(\ell,i)$ corresponding to a cycle of components of $\mathbb{P}^1 \setminus \underset{C \in \mathbb{P}^1}{C}$ may not contain a cycle of distinguished points of this length. However, the following lemma holds.

Lemma 5.2 *Let $\ell > 1$. Take any φ_b-cycle $\mathrm{cyc}^{\ell,i}$ of sets $S_j^{\ell,i}$, $j = 1, \dots k(\ell, i)$ of length $k(\ell, i)$, and the associated cycles of discs $\delta_j^{\ell,i}$ and of curves $\partial_{\mathfrak{E}} S_j^{\ell,i}$. Then there is an $\mathfrak{M}(\mathbb{P}^1; \mathbb{P}^1 \setminus \mathbb{D}, E_n)$-isotopy of $\varphi_{b,\infty}$, that changes $\varphi_{b,\infty}$ only in a small neighbourhood of $\partial_{\mathfrak{E}} S_1^{\ell,i}$, and provides a homeomorphism $\varphi'_{b,\infty}$ which has a cycle of perhaps non-distinguished points $z_j^{\ell,i} \in S_j^{\ell,i} \subset \delta_j^{\ell,i}$ of length $k(\ell, i)$, $\varphi'_{b,\infty}(z_j^{\ell,i}) = z_{j+1}^{\ell,i}$, $j = 1, \dots, k(\ell, i) - 1$, $\varphi'_{b,\infty}(z_{k(\ell,i)}^{\ell,i}) = z_1^{\ell,i}$. Moreover, the new homeomorphism $\varphi'_{b,\infty}$ has the following property.*

$$\text{For } 1 \le j \le k(\ell, i),\ \varphi'^{k(\ell,i)}_{b,\infty} \text{ fixes a neighbourhood of } \partial_{\mathfrak{E}}\, S_j^{\ell,i} \cup \{z_j^{\ell,i}\} \text{ pointwise.} \tag{5.6}$$

We will apply Lemma 5.2 to all cycles of discs $\delta_j^{\ell,i}$. Recall that the collection of the $\delta_j^{\ell,i}$, $i = 1, \dots, k_\ell$, $j = 1, \dots k(\ell, i)$, is the set of all holes of $Q_{\ell-1}$. Hence, there is a one-to-one correspondence between the set of holes of $Q_{\ell-1}$ and the points $z_j^{\ell,i}$ contained in the holes.

Proof of Lemma 5.2 With the requirement that $\varphi'_{b,\infty} = \varphi_{b,\infty}$ in a neighbourhood of $\partial_{\mathfrak{E}}\, S_j^{\ell,i}$, $j = 2, \dots, k(\ell, i)$, the equation

$$\varphi'_{b,\infty} = \varphi_{b,\infty}^{1-k(\ell,i)} \tag{5.7}$$

in a neighbourhood of $\partial_{\mathfrak{E}}\, S_1^{\ell,i}$ implies $\varphi'^{k(\ell,i)}_{b,\infty} = \mathrm{Id}$ in a neighbourhood of $\partial_{\mathfrak{E}}\, S_1^{\ell,i}$. Then this equation also holds in a neighbourhood of $\partial_{\mathfrak{E}}\, S_j^{\ell,i}$, $j = 2, \dots, k(\ell, i)$, since the restriction of $\varphi'^{k(\ell,i)}_{b,\infty}$ to a neighbourhood of $\partial_{\mathfrak{E}}\, S_j^{\ell,i}$ equals $\varphi_{b,\infty}^{j-2} \circ \varphi'_{b,\infty} \circ \varphi_{b,\infty}^{k(\ell,i)-j+1}$. It is clear that there is an $\mathfrak{M}(\mathbb{P}^1; \mathbb{P}^1 \setminus \mathbb{D}, E_n)$-isotopy of $\varphi_{b,\infty}$ that changes $\varphi_{b,\infty}$ only in a small neighbourhood of $\partial_{\mathfrak{E}} S_1^{\ell,i}$ and provides a homeomorphism $\varphi'_{b,\infty}$ that satisfies (5.7) in a neighbourhood of $\partial_{\mathfrak{E}}\, S_1^{\ell,i}$.

It remains to choose a point $z_1^{\ell,i} \in \delta_1^{\ell,i}$ that is contained in this neighbourhood. The homeomorphism $\varphi'_{b,\infty}$ maps $z_1^{\ell,i} \in \delta_1^{\ell,i}$ along the required cycle and the $k(\ell, i)$-th iterate of $\varphi'_{b,\infty}$ fixes a neighbourhood of $z_1^{\ell,i}$. The lemma is proved. □

In the following the homeomorphism $\varphi_{b,\infty}$ will satisfy all conditions that are obtained for the homeomorphism $\varphi'_{b,\infty}$ of Lemma 5.2.

Let $\varphi_t \in \mathrm{Hom}^+(\mathbb{P}^1; \mathbb{P}^1 \setminus \mathbb{D})$, $t \in [0, 1]$, be a continuous family of self-homeomorphisms of $\mathbb{C}$ such that $\varphi_0 = \mathrm{id}$ and $\varphi_1 = \varphi_{b,\infty}$. The geometric braid $\{(t, \varphi_t(E_n)),\ t \in [0, 1]\}$ is contained in the cylinder $[0, 1] \times \mathbb{D}$ and represents b. Put $g(t) = \varphi_t(E_n)$, $t \in [0, 1]$, $g : [0, 1] \to C_n(\mathbb{C})\diagup \mathcal{S}_n$.

The Outermost Braid The tubular braid associated to the outermost component $S^{1,1}$ can be defined similarly as in Chap. 4. Consider the set of all holes of the outermost component. We give each hole the label of the unique $S_j^{2,i}$ whose exterior boundary is the boundary of the hole, and denote the respective hole by $\delta_j^{2,i}$. Put $E^{1,1} = E \cap S^{1,1}$.

Associate the tubular geometric braid

$$\left\{\left(t, \varphi_t(E^{1,1} \cup \bigcup_{i=1}^{k_2} \bigcup_{j=1}^{k(2,i)} \overline{\delta_j^{2,i}})\right), \quad t \in [0, 1]\right\} \tag{5.8}$$

to b and $S^{1,1}$. The set correctly defines a tubular braid, since φ_0 is the identity and $\varphi_1 = \varphi_{b,\infty}$ maps the set $E^{1,1} \cup \bigcup_{i=1}^{k_2} \bigcup_{j=1}^{k(2,i)} \overline{\delta_j^{2,i}}$ onto itself.

We define a geometric fat braid that is a deformation retract of the tubular braid. For each cycle $\text{cyc}_0^{2,i}$ of sets $S_j^{2,i}$ of length $k(2,i)$ we take the cycle of points $z_j^{2,i} \in S_j^{2,i}$ chosen in Lemma 5.2 (see also relation (5.6)) as initial points of the fat strands. In other words, let $\boldsymbol{E}^{1,1}$ be the collection of all points $z_j^{2,i} \in S_j^{2,i} \subset \delta_j^{2,i}$, $i = 1, \dots, k_2$, $j = 1, \dots, k(2,i)$, assigned to the set of holes of $S^{1,1}$ in Lemma 5.2,

$$\boldsymbol{E}^{1,1} = \bigcup_{i=1}^{k_2} \{z_1^{2,i}, \dots, z_{k(2,i)}^{2,i}\}. \tag{5.9}$$

Associate to $S^{1,1}$ the geometric fat braid

$$\left\{\left(t, \varphi_t(E^{1,1} \cup \boldsymbol{E}^{1,1})\right), \quad t \in [0, 1]\right\}. \tag{5.10}$$

Let $\boldsymbol{B}(1,1)$ be the isotopy class of the geometric fat braid (5.10), and let $\widehat{\boldsymbol{B}(1,1)}$ be its conjugacy class. We call $\widehat{\boldsymbol{B}(1,1)}$ the outermost irreducible braid component of the braid b.

See Fig. 5.1, where $E^{1,1} = \emptyset$. In this figure the connected components of $\mathbb{C} \setminus \bigcup_{C \in \mathcal{C}} C$ of generation 2 are $S_1^{2,1}, S_2^{2,1}, S_3^{2,1}$, which are moved along a 3-cycle by $\varphi_{b,\infty}$. The set $\boldsymbol{E}^{1,1}$ equals $\boldsymbol{E}^{1,1} = \{z_1^{2,1}, z_2^{2,1}, z_3^{2,1}\}$.

Denote by $n(1,1)$ the number of points in $E^{1,1} \cup \boldsymbol{E}^{1,1}$. We write the mapping defining the geometric fat braid (5.10) as a fat map $\boldsymbol{g}^{1,1}$,

$$\boldsymbol{g}^{1,1} : [0, 1] \to C_{n(1,1)}(\mathbb{C})/\mathcal{S}_{n(1,1)}, \ \boldsymbol{g}^{1,1}(t) = \varphi_t(E^{1,1} \cup \boldsymbol{E}^{1,1}). \tag{5.11}$$

The following equality holds

Lemma 5.3

$$\widehat{\mathfrak{m}_{\boldsymbol{B}(1,1),\infty}} = \widehat{\mathfrak{m}_{b,\odot}^{1,1}}. \tag{5.12}$$

Proof of Lemma 5.3 The situation differs from that of Lemma 4.1 by the fact that b is not necessarily a pure braid and, hence, the interior boundary components of $S^{1,1}$ may be permuted by the mapping $\varphi_{b,\infty}$. Again, the mapping $\varphi_{b,\infty} \mid \widehat{S^{1,1} \setminus E'^{1,1}}$ represents the class $\mathfrak{m}^{1,1}_{b,\odot} \in \widehat{\mathfrak{M}}(Y^{1,1})$. The mapping $\varphi_{b,\infty} \mid \mathbb{P}^1 \setminus (\boldsymbol{E}^{1,1} \cup E'^{1,1})$ represents the class $\widehat{\mathfrak{m}_{B(1,1),\infty}} \in \widehat{\mathfrak{M}}(\mathbb{P}^1 \setminus (\boldsymbol{E}^{1,1} \cup E'^{1,1}))$. (Recall that $E'^{1,1} = \boldsymbol{E}^{1,1} \cup \{\infty\}$.)

Similarly as in the proof of Lemma 4.1 we take for each first labeled set $\delta^{2,i}_1$ of a cycle whose exterior boundary is an interior boundary component of $S^{1,1}$ an open annulus $A^{2,i}_1$ that is contained in $S^{1,1} \setminus E'^{1,1}$ and shares a boundary component with the boundary of $\delta^{1,1}_1$. Moreover we require that $A^{2,i}_1$ is contained in the neighbourhood of $\partial\delta^{2,i}_1$ which is fixed by the iterate $\varphi^{k(2,i)}_{b,\infty}$. We put $A^{2,i}_j = \varphi^{j-1}_{b,\infty}(A^{2,i}_1)$, $j = 2, \dots, k(2,1)$. The annuli are taken small enough so that all obtained annuli for $i = 1, \dots, k_2$, $j = 1, \dots, k(2,i)$ are pairwise disjoint.

There is a homeomorphism $\tilde{w}^{1,1}$ from $S^{1,1} \setminus E'^{1,1}$ onto $\mathbb{P}^1 \setminus (E'^{1,1} \cup \boldsymbol{E}^{1,1})$ with the following properties. $\tilde{w}^{1,1}$ is equal to the identical injection on the set $S^{1,1} \setminus (E'^{1,1} \cup \bigcup A^{2,i}_j)$. Moreover, for each j the mapping $\tilde{w}^{1,1}$ takes each annulus $A^{2,i}_j$ onto the punctured disc $\delta'^{2,i}_j \setminus \{z^{2,i}_j\}$, where $\delta'^{2,i}_j \overset{def}{=} A^{2,i}_j \cup \overline{\delta^{2,i}_j}$. Conjugate $\varphi_{b,\infty} \mid S^{1,1} \setminus E'^{1,1}$ with the inverse of $\tilde{w}^{1,1}$. The conjugate $\tilde{\varphi}^{1,1}_{b,\infty}$ is related to $\varphi^{1,1}_{b,\infty} \overset{def}{=} \varphi_{b,\infty} \mid \mathbb{P}^1 \setminus (\boldsymbol{E}^{1,1} \cup E'^{1,1})$ by isotopy and conjugation. Indeed, the two mappings, $\varphi_{b,\infty} \mid S^{1,1} \setminus E'^{1,1}$ and $\tilde{\varphi}^{1,1}_{b,\infty}$, differ only on the punctured discs $\delta'^{2,i}_j \setminus z^{2,i}_j$ around the points $z^{2,i}_j$ of $\boldsymbol{E}^{1,1}$ and map the punctured discs along the same cycles. They are equal on the boundary of each $\delta'^{2,i}_j$. Moreover, for each i the $k(2,i)$-th iterate of both mappings is the identity on each $\partial\delta'^{2,i}_j$, $j = 1, \dots, k(i,j)$, since by (5.6) $\varphi^{k(2,i)}_{b,\infty}$ is the identity on each $A^{2,i}_1$. Apply for each i the Lemma on conjugation (see Appendix A.2) to the restriction of the two mappings $\varphi^{1,1}_{b,\infty}$ and $\tilde{\varphi}^{1,1}_{b,\infty}$ to $\bigcup_j^{k(2,i)} \delta'^{2,i}_j$, and take into account the following fact. If a self-homeomorphism of a closed disc punctured at a point fixes the boundary circle pointwise, then it is isotopic to the identity through self-homeomorphisms that fix the boundary circle pointwise. We proved that $\varphi^{1,1}_{b,\infty}$ and $\tilde{\varphi}^{1,1}_{b,\infty}$ are related by isotopy and conjugation. □

The Irreducible Braid Components $\widehat{\boldsymbol{B}(\ell,i)}$ Associated to the $\varphi_{b,\infty}$-Cycles $\mathrm{cyc}^{\ell,i}$, $\ell > 1$ Recall that, for instance, the ordinary part of the geometric fat braid representing $\boldsymbol{B}(1,1)$ was found by forgetting all strands of the original braid with initial points not in $\boldsymbol{E}^{1,1} = E_n \cap S^{1,1}$. It was important that the set $\boldsymbol{E}^{1,1}$ is invariant under $\varphi_{b,\infty}$.

In order to apply this recipe for getting the irreducible braid component $\widehat{\boldsymbol{B}(\ell,i)}$ for $\ell > 1$ we wish to start with a $\varphi_{b,\infty}$-invariant set of points. The intersection of E_n with the union of the sets $S^{\ell,i}_j$ of the cycle $\mathrm{cyc}^{\ell,i}$ has this property. The problem is that the geometric braid obtained by forgetting all strands with initial point not contained in this set carries also information about the tubular braid $\left(t, \varphi_t(\bigcup_j \delta_j^{\ell,i})\right)$, which is part of the information carried by the geometric fat braids of generation less than ℓ. However, the power $\varphi_{b,\infty}^{k(\ell,i)}$ leaves each $S_j^{\ell,i}$, $j = 1, \ldots, k(\ell, i)$, invariant and, for instance, the geometric braid related to $\varphi_{b,\infty}^{k(\ell,i)}$ by forgetting the strands, whose base points are not contained in $S_1^{\ell,i} \cap E_n'$, does not contain information about the geometric fat braids of generation less than ℓ. We will therefore relate the irreducible braid component $\widehat{B(\ell, i)}$ to the power $b^{k(\ell,i)}$ of the braid b similarly as we related the irreducible nodal component to the power $b^{k(\ell,i)}$. Recall that, the irreducible nodal component $\widehat{\mathfrak{m}_{b,\odot}^{\ell,i}}$ can be identified with the class $\widehat{\mathfrak{m}_{b^{k(\ell,i)},\odot} \mid Y_1^{\ell,i}}$ that is related to the power $b^{k(\ell,i)}$ of the braid b and a set $Y_1^{\ell,i}$ of the cycle $\text{cyc}^{\ell,i}$.

In detail, we extend the family φ_t, $t \in [0, 1]$, to a family defined for all $t \in \mathbb{R}$ by putting $\varphi_t \overset{def}{=} \varphi_{t-k} \circ (\varphi_{b,\infty})^k$, $t \in [k, k+1]$. Since $\varphi_0 = \text{Id}$ and $\varphi_1 = \varphi_{b,\infty}$, the family is correctly defined. The family φ_t, $t \in [0, k(\ell, i)]$, is a parameterizing isotopy for $b^{k(\ell,i)}$ (normalized on the interval $[0, k(\ell, i)]$ rather than on the interval $[0, 1]$). In other words, the geometric braid

$$\left\{\left(t, \varphi_t(E_n)\right), \quad t \in [0, k(\ell, i)]\right\}, \tag{5.13}$$

represents $b^{k(\ell,i)}$. Indeed, $\varphi_0 = \text{Id}$, and $\varphi_{k(\ell,i)} = \varphi_{b,\infty}^{k(\ell,i)}$.

Define a tubular braid in $[0, k(\ell, i)] \times \mathbb{C}$ as follows. Let as before $S_j^{\ell,i}$, $j = 1, \ldots, k(\ell, i)$, be the sets of the cycle $\text{cyc}_0^{\ell,i}$. Put $E_1^{\ell,i} \overset{def}{=} E \cap S_1^{\ell,i}$. Further, consider all interior boundary components of $S_1^{\ell,i}$ and the closed topological discs of generation $\ell+1$ in $\mathbb{C}$ that are bounded by them (in other words, consider all "interior holes" of $S_1^{\ell,i}$). They are of the form $\overline{\delta_{j'}^{\ell+1,i'}}$, where $\delta_{j'}^{\ell+1,i'}$ is the bounded disc in $\mathbb{C}$ that is bounded by the exterior boundary component of $S_{j'}^{\ell+1,i'}$. Let $H(S_1^{\ell,j})$ be the set of "interior" holes of $S_1^{\ell,j}$. Consider the tubular braid

$$\left\{\left(t, \varphi_t\Big(E_1^{\ell,i} \cup \bigcup_{H(S_1^{\ell,j})} \overline{\delta_{j'}^{\ell+1,i'}}\Big)\right), \ t \in [0, k(\ell, i)]\right\} \tag{5.14}$$

in the tube $[0, k(\ell, i)] \times \mathbb{C}$.

For each "interior hole" $\overline{\delta_{j'}^{\ell+1,i'}} \in H(S_1^{\ell,j})$ of $S_1^{\ell,j}$ we consider the point $z_{j'}^{\ell+1,i'} \in S_{j'}^{\ell+1,i'} \subset \overline{\delta_{j'}^{\ell+1,i'}}$ associated to $\overline{\delta_{j'}^{\ell+1,i'}}$ by Lemma 5.2. Let $\boldsymbol{E}_1^{\ell,i}$ be the set of all such points $z_{j'}^{\ell+1,i'}$. The geometric fat braid which is a deformation retract of (5.14) is defined as

$$\left\{\left(t, \varphi_t(E_1^{\ell,i} \cup \boldsymbol{E}_1^{\ell,i})\right), \ t \in [0, k(\ell, i)]\right\} \tag{5.15}$$

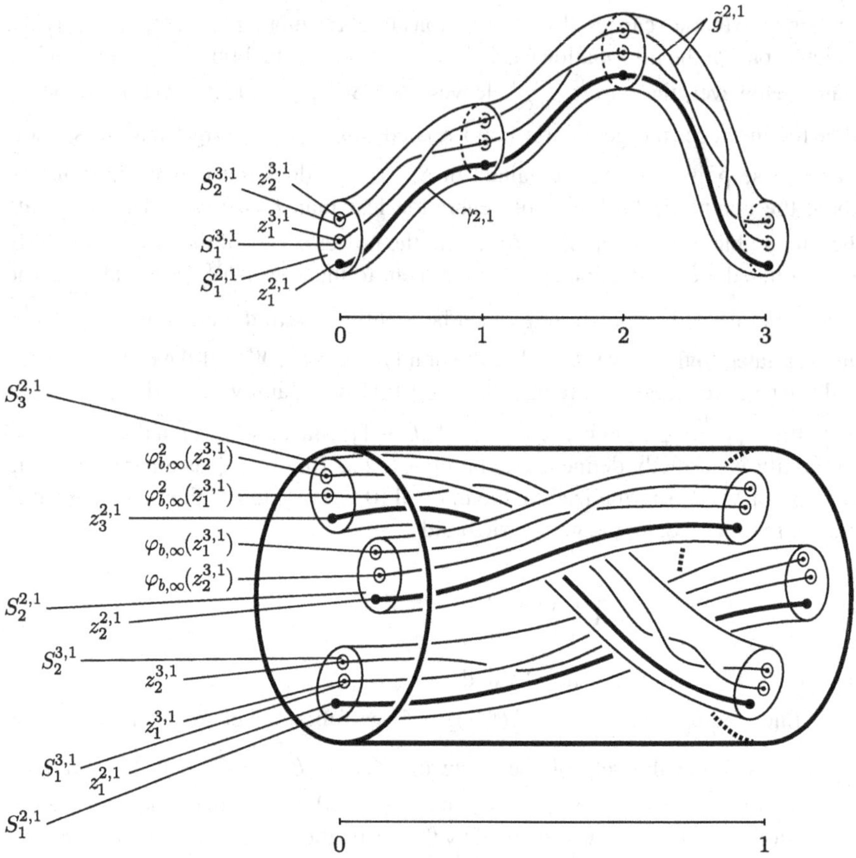

Fig. 5.1 The irreducible braid component $\widehat{\boldsymbol{B}(2,1)}$ of a non-pure braid b

(See Fig. 5.1 where $\boldsymbol{E}_1^{2,1} = \{z_1^{3,1}, z_2^{3,2}\}$, $E_1^{2,1} = \emptyset$.)

Denote by $n(\ell, i)$ the number of points in $E_1^{\ell,i} \cup \boldsymbol{E}_1^{\ell,i}$. We write the fat mapping defining the geometric fat braid (5.15) as

$$\widetilde{\boldsymbol{g}^{\ell,i}}^{k(\ell,i)} : [0, k(\ell,i)] \to C_{n(\ell,i)}(\mathbb{C})\diagup\mathcal{S}_{n(\ell,i)},$$

$$\widetilde{\boldsymbol{g}^{\ell,i}}^{k(\ell,i)}(t) = \varphi_t(E_1^{\ell,i} \cup \boldsymbol{E}_1^{\ell,i}). \tag{5.16}$$

Reparameterize the geometric fat braid $\widetilde{\boldsymbol{g}^{\ell,i}}^{k(\ell,i)}$ to obtain a mapping defined on the unit interval, in other words, put

$$\boldsymbol{g}^{\ell,i}(t) = \widetilde{\boldsymbol{g}^{\ell,i}}^{k(\ell,i)}(k(\ell,i)\,t), \quad \boldsymbol{g}^{\ell,i} : [0,1] \to C_{n(\ell,i)}(\mathbb{C})\diagup\mathcal{S}_{n(\ell,i)}. \tag{5.17}$$

The isotopy class of the geometric fat braid $\boldsymbol{g}^{\ell,i}$ is denoted by $\boldsymbol{B}(\ell, i)$. Its conjugacy class $\widehat{\boldsymbol{B}(\ell, i)}$ is the irreducible braid component of the braid b corresponding to the cycle $\mathrm{cyc}^{\ell,\mathrm{i}}$. The conjugacy class $\widehat{\boldsymbol{B}(\ell, i)}$ can be identified with the free isotopy class of $\boldsymbol{g}^{\ell,i}$. It does not depend on the choice of the set $S_1^{\ell,i}$ of the cycle $\mathrm{cyc}^{\ell,\mathrm{i}}$.

The equality

$$\widehat{\mathfrak{m}_{b^{k(\ell,i)},\odot} \mid Y_1^{\ell,i}} = \widehat{\mathfrak{m}_{B(\ell,i),\infty}} \tag{5.18}$$

follows by applying the arguments proving Eq. (5.12) to the mapping $\varphi_{b,\infty}^{k(\ell,i)} \mid (S_1^{\ell,i} \setminus E_1^{\ell,i})$ representing $\widehat{\mathfrak{m}_{b^{k(\ell,i)},\odot} \mid Y_1^{\ell,i}}$ and the mapping $\varphi_{b,\infty}^{k(\ell,i)} \mid \mathbb{P}^1 \setminus (E'^{\ell,i}_1 \cup \boldsymbol{E}_1^{\ell,i})$ representing $\widehat{\mathfrak{m}_{B(\ell,i),\infty}}$.

5.3 The Building Block of the Recovery. The General Case

We want to describe a procedure that recovers the conjugacy class of a non-pure braid given an admissible system of curves that completely reduces a homeomorphism that represents the mapping class of the braid and knowing the irreducible braid components related to this system. In the case of non-pure braids it is more transparent to work with the associated closed braids rather than with the braids themselves. We obtain a closed geometric fat braid from a geometric fat braid

$$\Big\{(t, \boldsymbol{g}(t)),\ t \in [0, 1]\Big\} \tag{5.19}$$

in the cylinder $[0, 1] \times \mathbb{C}$ by replacing the interval $[0, 1]$ by the quotient $[0, 1]\big/0 \sim 1$ which is diffeomorphic to the unit circle $\partial\mathbb{D}$. In other words, the closed geometric fat braid is obtained from the geometric braid by gluing together the top and the bottom fiber of the cylinder $[0, 1] \times \mathbb{C}$ using the identity mapping of $\mathbb{C}$. We put $\boldsymbol{G}(e^{2\pi i t}) = \boldsymbol{g}(t)$, $t \in [0, 1]$, which is well defined since $\boldsymbol{g}(0) = \boldsymbol{g}(1)$. The associated closed geometric fat braid, denoted by $\boldsymbol{G}$, is defined as the subset of the solid torus $\partial\mathbb{D} \times \mathbb{C}$

$$\Big\{\big(e^{2\pi i t},\ \boldsymbol{g}(t)\big),\ t \in [0, 1]\Big\} = \Big\{\big(e^{2\pi i t},\ \boldsymbol{G}(e^{2\pi i t})\big),\ e^{2\pi i t} \in \partial\mathbb{D}\Big\}. \tag{5.20}$$

The connected components of the closed geometric fat braid (5.20) are simple closed curves in the solid torus $\partial\mathbb{D} \times \mathbb{C}$. For each such loop Γ there is a natural number k such that the loop intersects each fiber $\{e^{2\pi i t}\} \times \mathbb{C}$ along k points. We call k the covering multiplicity of Γ.

Consider the k-fold covering $\widetilde{\partial\mathbb{D}}^k \times \mathbb{C} \ni (\zeta, z) \xrightarrow{\tilde{p}^k} (\zeta^k, z) \in \partial\mathbb{D} \times \mathbb{C}$ of $\partial\mathbb{D} \times \mathbb{C}$. Notice that $\widetilde{\partial\mathbb{D}}^k \times \mathbb{C}$ is isomorphic to $\partial\mathbb{D} \times \mathbb{C}$. We may identify $\partial\mathbb{D} \times \mathbb{C}$ with

$\left\{(e^{2\pi it}, z),\ t \in [0,1], z \in \mathbb{C}\right\}$, and $\widetilde{\partial\mathbb{D}}^k \times \mathbb{C}$ with $\left\{(e^{2\pi i\frac{t}{k}}, z),\ t \in [0,k], z \in \mathbb{C}\right\}$. Then $\tilde{p}^k((e^{2\pi i\frac{t}{k}}, z)) = (e^{2\pi it}, z)$ for $(e^{2\pi i\frac{t}{k}}, z) \in \widetilde{\partial\mathbb{D}}^k \times \mathbb{C}$.

A loop of covering multiplicity k can be written as

$$\varGamma = \left\{\left(e^{2\pi it},\ \gamma(t)\right),\ t \in [0,k]\right\} \tag{5.21}$$

for a function γ with $\gamma(0) = \gamma(k)$. If $\varGamma$ has covering multiplicity k, the lift $\widetilde{\varGamma}^k$ of $\varGamma$ to the k-fold covering,

$$\widetilde{\varGamma}^k = \left\{\left(e^{2\pi i\frac{t}{k}},\ \gamma(t)\right),\ t \in [0,k], z \in \mathbb{C}\right\} \tag{5.22}$$

has covering multiplicity 1 as subset of $\widetilde{\partial\mathbb{D}}^k \times \mathbb{C}$.

A solid torus T in $\partial\mathbb{D} \times \mathbb{C}$ that intersects each fiber $\{e^{2\pi it}\} \times \mathbb{C}$ along a set of k disjoint closed discs is called a solid torus of covering multiplicity k. A solid torus T of covering multiplicity k can be written as $\bigcup_{t\in[0,k]} \{e^{2\pi it}\} \times U_t$, where for each t the sets $U_t, \ldots, U_{t+k-1}$ are disjoint topological discs in $\mathbb{C}$. Take the lift $\widetilde{T}^k = \bigcup_{t\in[0,k]} \{e^{2\pi i\frac{t}{k}}\} \times U_t$ under $\tilde{p}^k$ of the solid torus T of covering multiplicity k. The torus $\widetilde{T}^k$ has covering multiplicity 1 as subset of the k-fold covering space $\widetilde{\partial\mathbb{D}}^k \times \mathbb{C} \cong \partial\mathbb{D} \times \mathbb{C}$.

The following procedure is the basis for the recovery of conjugacy classes of not necessarily pure reducible braids. Let $\boldsymbol{G} : \partial\mathbb{D} \to C_n(\mathbb{C})\diagup\mathcal{S}_n$ be a closed geometric fat braid in $\partial\mathbb{D}\times\mathbb{C}$, written as $\left\{\left(e^{2\pi it},\ \boldsymbol{g}(t)\right),\ t \in [0,1]\right\}$ with $\boldsymbol{G}(e^{2\pi it}) = \boldsymbol{g}(t),\ t \in [0,1]$ (see Eq. (5.20)). Let $\boldsymbol{\varGamma}$ be a fat loop of covering multiplicity k which is a part of the closed geometric fat braid (5.20),

$$\boldsymbol{\varGamma} = \left\{\left(e^{2\pi it},\ \boldsymbol{\gamma}(t)\right),\ t \in [0,k]\right\} \tag{5.23}$$

for a fat function $\boldsymbol{\gamma}$ with $\boldsymbol{\gamma}(0) = \boldsymbol{\gamma}(k)$. The lift $\widetilde{\boldsymbol{\varGamma}}^k$ can be written as

$$\widetilde{\boldsymbol{\varGamma}}^k = \left\{(e^{2\pi i\frac{t}{k}}, \boldsymbol{\gamma}(t)) : t \in [0,k]\right\}. \tag{5.24}$$

$\widetilde{\boldsymbol{\varGamma}}^k$ is the graph of a function depending on $e^{2\pi i\frac{t}{k}} \in \widetilde{\partial\mathbb{D}}^k$.

Take a solid torus T in $\partial\mathbb{D} \times \mathbb{C}$ of covering multiplicity k, that surrounds the fat loop $\boldsymbol{\varGamma}$ of covering multiplicity k (more precisely, $\boldsymbol{\varGamma}$ is contained in T and is a deformation retract of T), such that T does not intersect any other loop of the closed geometric fat braid $\boldsymbol{G}$. Replacing $\boldsymbol{\varGamma}$ by T we obtain a closed geometric tubular fat braid. (A closed geometric tubular fat braid is defined in a similar way as a geometric tubular fat braid.)

Let $\widetilde{F}^k$ be a closed geometric fat braid in $\partial\widetilde{\mathbb{D}}^k \times \mathbb{C}$ that is contained in the lift $\widetilde{T}^k$. We write

$$\widetilde{F}^k = \left\{(e^{2\pi i \frac{t}{k}}, \widetilde{f}^k(t)) : t \in [0, k]\right\} \tag{5.25}$$

with $\widetilde{f}^k(t) \overset{def}{=} \widetilde{F}^k(e^{2\pi i t})$, $t \in [0, k]$, $\widetilde{f}^k : [0, k] \to C_n(\mathbb{C})\diagup\mathcal{S}_n$ for some natural number n. We define the following operation that replaces the fat loop $\boldsymbol{\Gamma}$ of the closed fat braid $\boldsymbol{G}$ by "a copy of the closed fat braid $\boldsymbol{F} = \tilde{p}^k(\widetilde{F}^k) = \left\{(e^{2\pi i t}, \widetilde{f}^k(t)) : t \in [0, k]\right\}$ inserted into T":

$$\boldsymbol{G} \sqcup_{\boldsymbol{\Gamma}} \widetilde{\boldsymbol{F}}^k \overset{def}{=} \left\{\left(e^{2\pi i t}, \boldsymbol{g}(t) \sqcup_{\boldsymbol{\gamma}(t)} \widetilde{\boldsymbol{f}}^k(t)\right), t \in [0, k]\right\}. \tag{5.26}$$

The operation $\boldsymbol{g}(t) \sqcup_{\boldsymbol{\gamma}(t)} \tilde{\boldsymbol{f}}^k(t) \overset{def}{=} \left(\boldsymbol{g}(t) \setminus \{\boldsymbol{\gamma}(t)\}\right) \cup \tilde{\boldsymbol{f}}^k(t)$ for every $t \in [0, k]$ is defined as in Chap. 4 Eq. (4.11), namely, for each t we replace the point $\boldsymbol{\gamma}(t)$ of the unordered tuple of points $\boldsymbol{g}(t)$ by the unordered tuple of points $\widetilde{\boldsymbol{f}}^k(t)$.

Since the closed geometric fat braid defined by $\widetilde{\boldsymbol{F}}^k$ is contained in $\widetilde{T}^k$, the closed geometric fat braid $\left\{\left(e^{2\pi i t}, \widetilde{\boldsymbol{f}}^k(t)\right), t \in [0, k]\right\}$ is contained in T. Hence, (5.26) defines a closed geometric fat braid.

We need the following analog of Lemma 4.3.

Lemma 5.4 *Let $\boldsymbol{G}^0$ and $\boldsymbol{G}^1$ be (not necessarily pure) closed geometric fat braids in $\partial\mathbb{D} \times \mathbb{C}$, such that $\boldsymbol{G}^0$ is free isotopic to $\boldsymbol{G}^1$. Let $\boldsymbol{\Gamma}_0$ be a fat loop of $\boldsymbol{G}^0$ of covering multiplicity k, and let $\boldsymbol{\Gamma}_1$ be the fat loop of $\boldsymbol{G}^1$ of covering multiplicity k that corresponds to $\boldsymbol{\Gamma}_0$ under the isotopy joining $\boldsymbol{G}^0$ and $\boldsymbol{G}^1$.*

For $j = 0, 1$ we let $T(\boldsymbol{\Gamma}_j)$ be a tube around $\boldsymbol{\Gamma}_j$ which does not meet the other loops of $\boldsymbol{G}^j$ and has $\boldsymbol{\Gamma}_j$ as a deformation retract. Suppose for $j = 0, 1$ there are closed geometric fat braids $\widetilde{(\boldsymbol{F}^j)}^k$ in $\partial\widetilde{\mathbb{D}}^k \times \mathbb{C}$ that are contained in the lift $\widetilde{T(\boldsymbol{\Gamma}_j)}^k$ of $T(\boldsymbol{\Gamma}_j)$.

If in addition $\widetilde{(\boldsymbol{F}^0)}^k$ and $\widetilde{(\boldsymbol{F}^1)}^k$ are free isotopic in $\partial\widetilde{\mathbb{D}}^k \times \mathbb{C}$, then $\boldsymbol{G}^0 \sqcup_{\boldsymbol{\Gamma}_0} \widetilde{(\boldsymbol{F}^0)}^k$ and $\boldsymbol{G}^1 \sqcup_{\boldsymbol{\Gamma}_1} \widetilde{(\boldsymbol{F}^1)}^k$ are free isotopic closed geometric fat braids.

Proof Let $\boldsymbol{G}^s$, $s \in [0, 1]$, be an isotopy of closed geometric fat braids joining $\boldsymbol{G}^1$ with $\boldsymbol{G}^2$, and let $\boldsymbol{\Gamma}_s$ be the family of loops in $\boldsymbol{G}^s$ that is obtained by this isotopy and joins $\boldsymbol{\Gamma}_0$ with $\boldsymbol{\Gamma}_1$. Choose ε so small that for each $s \in [0, 1]$ the ε-neighbourhood in $\partial\mathbb{D} \times \mathbb{C}$ of the closed geometric fat braid $\left\{\left(e^{2\pi i t}, \boldsymbol{G}^s(e^{2\pi i t})\right), t \in [0, 1]\right\}$ is a closed geometric tubular braid.

For $j = 0, 1$ the closed geometric fat braid $\widetilde{(\boldsymbol{F}^j)}^k = \left\{\left(e^{2\pi i \frac{t}{k}}, \widetilde{(\boldsymbol{f}^j)}^k(t)\right), t \in [0, k], \right\}$ in $\partial\widetilde{\mathbb{D}}^k \times \mathbb{C}$ is contained in $\widetilde{T(\boldsymbol{\Gamma}_j)}^k$. By the proof of Lemma 4.2 for $j = 0, 1$ there exists an isotopy of closed geometric fat braids $\widetilde{(\boldsymbol{F}^j_s)}^k$, $s \in [0, 1]$, such that for $j = 0, 1$, each $\widetilde{(F_s^j)}^k$ is contained in $\widetilde{T(\Gamma_j)}^k$, and for each j the family joins the closed geometric fat braid $\widetilde{(F^j)}^k$ with a closed geometric fat braid $\widetilde{(F_1^j)}^k$ contained in the ε-neighbourhood $T_\varepsilon(\widetilde{\Gamma}_j^k)$ of the lift $\widetilde{\Gamma}_j^k$ of Γ_j to the k-fold covering of $\partial\mathbb{D} \times \mathbb{C}$. Write $\widetilde{\Gamma}_j^k$ as in formula (5.24).

Since $G^j \sqcup_{\Gamma_j} \widetilde{(F_s^j)}^k$, $s \in [0, 1]$, is a free isotopy of closed geometric fat braids for $j = 0, 1$, we may assume from the beginning, that $\widetilde{F^j}^k$ is contained in the ε-tube $T_\varepsilon(\widetilde{\Gamma}_j^k)$ around $\widetilde{\Gamma}_j^k$. Then $\widetilde{F^j}^k$, $j = 0, 1$, has the form $\widetilde{\Gamma}_j^k \boxplus \varepsilon \widetilde{\mathring{F}^j}^k \overset{def}{=} \{(e^{2\pi i \frac{t}{k}}, \gamma_j(t) \boxplus \varepsilon \widetilde{\mathring{f}_j}^k(t)),\ t \in [0, k]\}$ for a closed geometric fat braid $\widetilde{\mathring{F}_j}^k$, that is defined by $\{(e^{2\pi i \frac{t}{k}}, \widetilde{\mathring{f}_j}^k(t)),\ t \in [0, k]\}$ and contained in $\widetilde{\partial\mathbb{D}}^k \times \mathbb{D}$. Recall that for each positive number ε the set $\varepsilon \widetilde{\mathring{f}_j}^k(t)$ is obtained from $\widetilde{\mathring{f}_j}^k(t)$ (considered as subset of $\mathbb{C}$) by multiplying each point of the latter set by ε, and for each $t \in [0, k]$ the set $\gamma(t) \boxplus \varepsilon \widetilde{\mathring{f}_j}^k(t)$ is obtained by adding $\gamma(t)$ to each point of the set $\varepsilon \widetilde{\mathring{f}_j}^k(t)$. Since the closed geometric fat braids $\widetilde{\Gamma}_0^k \boxplus \varepsilon \widetilde{\mathring{F}^0}^k$ and $\widetilde{\Gamma}_1^k \boxplus \varepsilon \widetilde{\mathring{F}^1}^k$ are free isotopic, also the closed geometric fat braids $\widetilde{\mathring{F}_0}^k$ and $\widetilde{\mathring{F}_1}^k$ are free isotopic. Let $\widetilde{\mathring{F}^s}^k$, $s \in [0, 1]$, be an isotopy of closed geometric fat braids contained in $\widetilde{\partial\mathbb{D}}^k \times \mathbb{D}$ and joining $\widetilde{\mathring{F}^0}^k$ and $\widetilde{\mathring{F}^1}^k$.

By the choice of ε the family $G^s \sqcup_{\Gamma_s} \widetilde{\Gamma}_s^k \boxplus \varepsilon \widetilde{\mathring{F}^s}^k$ is a free isotopy of closed geometric fat braids joining $G^0 \sqcup_{\Gamma_0} \widetilde{\Gamma}_0^k \boxplus \varepsilon \widetilde{\mathring{F}^0}^k$ and $G^1 \sqcup_{\Gamma_1} \widetilde{\Gamma}_1^k \boxplus \varepsilon \widetilde{\mathring{F}^1}^k$. We assumed that the latter two closed geometric fat braids are equal to $G^0 \sqcup_{\Gamma_0} \widetilde{(F^0)}^k$ and $G^1 \sqcup_{\Gamma_1} \widetilde{(F^1)}^k$, respectively. The Lemma is proved. □

Notice that for any closed geometric fat braid given by a mapping F and any diffeomorphism $\sigma : \partial\mathbb{D} \to \partial\mathbb{D}$ the two closed fat braids

$$\left\{(e^{2\pi it}, F(e^{2\pi it})),\ t \in [0, 1]\right\} \text{ and } \left\{(e^{2\pi it}, F(e^{2\pi i\sigma(t)})),\ t \in [0, 1]\right\}$$

are free isotopic.

5.4 Recovery of Conjugacy Classes of Braids from the Irreducible Braid Components. The General Case

Suppose that we know the irreducible braid components $\widehat{B(\ell, i)}$ of the reducible non-pure braid b with respect to the admissible system of curves $\mathcal{C}$ that completely reduces a homeomorphism that represents the class $\widehat{\mathfrak{m}_{b,\infty}}$ of the braid b. Represent each conjugacy class of fat braids $\widehat{\boldsymbol{B}(\ell,i)}$ by an arbitrary closed geometric fat braid $\widetilde{\boldsymbol{F}^{\ell,i}}^{k(\ell,i)}$ in the bounded solid torus $\widetilde{\partial\mathbb{D}}^{k(\ell,i)} \times \mathbb{D}$ which is the $k(\ell,i)$-fold covering space of $\partial\mathbb{D} \times \mathbb{D}$. Here $k(\ell,i)$ is the length of the cycle of $S_j^{\ell,i}$ to which $\widehat{\boldsymbol{B}(\ell,i)}$ is associated. We write $\widetilde{\boldsymbol{F}^{\ell,i}}^{k(\ell,i)}$ as $\left\{\left(e^{2\pi i \frac{t}{k(\ell,i)}}, \widetilde{\boldsymbol{f}^{\ell,i}}^{k(\ell,i)}(t)\right), \, t \in [0, k(\ell,i)]\right\}$ for a continuous mapping $\widetilde{\boldsymbol{f}^{\ell,i}}^{k(\ell,i)}$ from $[0, k(\ell,i)]$ into the symmetrized configuration space of suitable dimension. By induction on the generation ℓ we will (after multiplication by a small positive number and applying the operation $\boxplus$) "insert the closed geometric fat braids $\widetilde{\boldsymbol{F}^{\ell,i}}^{k(\ell,i)}$ into tubes around respective fat loops of the closed geometric fat braid obtained by induction at step $\ell - 1$". To show that indeed the geometric fat braids, that are obtained by the inductive construction with any choice of the representatives of $\widehat{\boldsymbol{B}(\ell,i)}$, represent the class $\hat{b}$, we will compare at each step the inductively constructed closed geometric fat braids with a closed geometric fat braid related to (5.20). We will use Lemma 5.4.

Start with the closed geometric fat braid $\boldsymbol{F}^{1,1} = \widetilde{\boldsymbol{F}^{1,1}}^{1}$ written as $\left\{\left(e^{2\pi i t}, \widetilde{\boldsymbol{f}^{1,1}}^{1}(t)\right), \, t \in [0,1]\right\}$. Let $\varepsilon_2 > 0$ be so small that the ε_2-neighbourhood in $\partial\mathbb{D} \times \mathbb{C}$ of $\boldsymbol{F}^{1,1}$ is the union of disjoint solid tori that retract to the loops of the closed geometric fat braid. Let $\boldsymbol{F}_{2,i}$, $i = 1, \ldots, k_2$, be the fat loops of $\boldsymbol{F}^{1,1}$. The covering multiplicity of $\boldsymbol{F}_{2,i}$ is equal to $k(2,i)$.

For each $i = 1, \ldots, k_2$, we will "insert a suitable closed geometric braid representing $\widehat{\boldsymbol{B}(2,i)}$ into the ε_2-neighbourhood of the loop $\boldsymbol{F}_{2,i}$". For $i = 1$ this means that we consider the closed geometric fat braid given by

$$\widetilde{\boldsymbol{F}^{1,1}}^{1} \sqcup_{\boldsymbol{F}_{2,1}} (\boldsymbol{F}_{2,1} \boxplus \varepsilon_2 \widetilde{\boldsymbol{F}^{2,1}}^{k(2,1)}). \tag{5.27}$$

More detailed, equality (5.27) means the following. Write $\boldsymbol{F}_{2,1}$ as

$$\left\{(e^{2\pi i t}, \boldsymbol{f}_{2,1}(t)), t \in [0, k(2,1)]\right\}.$$

Then $\boldsymbol{F}_{2,1} \boxplus \varepsilon_2 \widetilde{\boldsymbol{F}^{2,1}}^{k(2,1)}$ is the closed geometric fat braid defined by

$$\left\{\left(e^{2\pi i t}, \boldsymbol{f}_{2,1}(t) \boxplus \varepsilon_2 \widetilde{\boldsymbol{f}^{2,1}}^{k(2,1)}(t)\right), \, t \in [0, k(2,1)]\right\}.$$

The closed geometric fat braid (5.27) is defined by

$$\left\{\left(e^{2\pi i t}, \widetilde{\boldsymbol{f}^{1,1}}^{1}(t) \sqcup_{\boldsymbol{f}_{2,1}(t)} \left(\boldsymbol{f}_{2,1}(t) \boxplus \varepsilon_2 \widetilde{\boldsymbol{f}^{2,1}}^{k(2,1)}(t)\right)\right), \, t \in [0, k(2,1)]\right\}. \tag{5.28}$$

We define successively the closed geometic braids

$$\widetilde{\boldsymbol{F}^{1,1}}^{1} \sqcup_{\boldsymbol{F}_{2,1}} (\boldsymbol{F}_{2,1} \boxplus \varepsilon_2 \widetilde{\boldsymbol{F}^{2,1}}^{k(2,1)}) \sqcup \ldots \sqcup_{\boldsymbol{F}_{2,i}} (\boldsymbol{F}_{2,i} \boxplus \varepsilon_2 \widetilde{\boldsymbol{F}^{2,i}}^{k(2,i)}) \tag{5.29}$$

by "inserting the closed geometric fat braid $\boldsymbol{F}_{2,i+1} \boxplus \varepsilon_2 \boldsymbol{F}^{2,i+1}$ representing $\widehat{\boldsymbol{B}(2, i+1)}$ into the ε_2-neighbourhood of the fat loop $\boldsymbol{F}_{2,i+1}$ of the closed geometric fat braid (5.29)" to obtain the closed geometric fat braid (5.29) with i replaced $i+1$.

On the other hand we recall that $\widehat{\boldsymbol{B}(1, 1)}$ is the free isotopy class of the geometric fat braid

$$\Big\{\big(t, \varphi_t(E^{1,1} \cup \boldsymbol{E}^{1,1})\big), \quad t \in [0, 1]\Big\}$$

(see (5.10)). Here $E^{1,1} = E \cap S^{1,1}$, and $\boldsymbol{E}^{1,1}$ is the collection of all points $z_j^{2,i} \in S_j^{2,i} \subset \delta_j^{2,i}$, $i = 1, \ldots, k_2$, $j = 1, \ldots, k(2, i)$, assigned to the set of holes of $S^{1,1}$ in Lemma 5.2, $\boldsymbol{E}^{1,1} = \bigcup_{i=1}^{k_2} \{z_1^{2,i}, \ldots, z_{k(2,i)}^{2,i}\}$. With $n(1, 1)$ being the number of points in $E^{1,1} \cup \boldsymbol{E}^{1,1}$, we write the mapping defining the geometric fat braid (5.10) as a fat map $\boldsymbol{g}^{1,1}$,

$$\boldsymbol{g}^{1,1} : [0, 1] \to C_{n(1,1)}(\mathbb{C})/\mathcal{S}_{n(1,1)},$$

(see (5.11)). With $\widetilde{\boldsymbol{g}^{1,1}}^1 = \boldsymbol{g}^{1,1}$ we consider the closed geometric fat braid $\widetilde{\boldsymbol{G}^{1,1}}^1$ in $\partial\widetilde{\mathbb{D}}^1 \times \mathbb{C}$, defined by

$$\Big\{\Big(e^{2\pi i t}, \widetilde{\boldsymbol{g}^{1,1}}^1(t)\Big), \, t \in [0, 1]\Big\}.$$

The connected components $\boldsymbol{\Gamma}_{2,i}$ of the closed fat braid $\widetilde{\boldsymbol{G}^{1,1}}^1$ that are fat loops correspond to the cycles $\mathrm{cyc}_0^{2,i}$, $\boldsymbol{\Gamma}_{2,i} \overset{def}{=} \Big\{\big(e^{2\pi i t}, \varphi_t(\boldsymbol{E}_1^{2,i})\big), \, t \in [0, k(2, i)]\Big\} \subset \partial\mathbb{D} \times \mathbb{C}$. We also write this fat loop as $\boldsymbol{\Gamma}_{2,i} = \Big\{\big(e^{2\pi i t}, \boldsymbol{\gamma}_{2,i}(t)\big), \, t \in [0, k(2, i)]\Big\}$.

Since by the definition of the extension of φ_t to $\mathbb{R}$ the equalities $\varphi_{k(\ell,i)}(z_1^{2,1}) = \varphi_{b,\infty}^{k(\ell,i)}(z_1^{2,1}) = z_1^{2,1}$, and $\varphi_j(z_1^{2,1}) = \varphi_{b,\infty}^{j}(z_1^{2,1}) = z_{j+1}^{2,1}$ hold for $j = 1, \ldots, k(\ell, i) - 1$, the fat loop $\boldsymbol{\Gamma}_{2,i}$ intersects the fiber $\{1\} \times \mathbb{C}$ at the points $(1, z_1^{2,i}), \ldots, (1, z_{k(\ell,i)}^{2,i})$. It intersects each fiber $\{e^{2\pi i t}\} \times \mathbb{C}$ along $k(2, i)$ points. For each i the fat loop $\boldsymbol{\Gamma}_{2,i}$ is contained in the tube

$$\left\{\Big(e^{2\pi i t}, \varphi_t(E^{1,1} \cup \bigcup_{j=1}^{k(2,i)} \overline{\delta_j^{2,i}})\Big), t \in [0, 1]\right\}. \tag{5.30}$$

of the closed geometric tubular braid

$$\left\{\Big(e^{2\pi i t}, \varphi_t(E^{1,1} \cup \bigcup_{i=1}^{k_2} \bigcup_{j=1}^{k(2,i)} \overline{\delta_j^{2,i}})\Big), t \in [0, 1]\right\}.$$

Further, we recall that for each cycle $\text{cyc}^{2,i}$ of generation 2 the irreducible braid component $\widehat{\boldsymbol{B}(2,i)}$ associated to this cycle is, after reparametrizing, equal to the free isotopy class of the geometric braid, that is defined by

$$\left\{\left(t, \varphi_t(E_1^{2,i} \cup \boldsymbol{E}_1^{2,i})\right),\ \ t \in [0, k(2,i)]\right\}$$

(see (5.15)) and is denoted by

$$\widetilde{g^{2,i}}^{k(2,i)} : [0, k(2,i)] \to C_{n(2,i)}(\mathbb{C}) \diagup \mathcal{S}_{n(2,i)}$$

$$\widetilde{g^{2,i}}^{k(2,i)}(t) = \varphi_t(E_1^{2,i} \cup \boldsymbol{E}_1^{2,i})$$

(see (5.16)). Here $E_1^{2,i} = E \cap S_1^{2,i}$, and the set $\boldsymbol{E}_1^{2,i}$ contains for each interior hole of $S_1^{2,i}$ the point associated to it by Lemma 5.2. We consider for each $(2,i)$ the closed geometric fat braid $\widetilde{\boldsymbol{G}^{2,i}}^{k(2,i)}$ in $\widetilde{\partial\mathbb{D}}^{k(2,i)} \times \mathbb{C}$, that is defined as

$$\left\{\left(e^{2\pi i \frac{t}{k(2,i)}}, \widetilde{g^{2,i}}^{k(2,i)}(t)\right),\ t \in [0, k(2,i)]\right\} \tag{5.31}$$

where $\widetilde{g^{2,i}}^{k(2,i)}$ is defined by (5.15) and (5.16) with $\ell = 2$. The closed geometric fat braid $\widetilde{\boldsymbol{G}^{2,i}}^{k(2,i)}$ is contained in the tube (5.30).

The closed geometric fat braids $\widetilde{\boldsymbol{G}^{1,1}}^{1}$ and $\widetilde{\boldsymbol{F}^{1,1}}^{1}$ are free isotopic, and the fat loops $\boldsymbol{\Gamma}_{2,1}$ and $\boldsymbol{F}_{2,1}$ correspond to each other under the isotopy. By the choice of ε_2 the closed geometric fat braids $\widetilde{\boldsymbol{G}^{2,1}}^{k(2,1)}$ and $\boldsymbol{F}_{2,1} \boxplus \varepsilon_2 \widetilde{\boldsymbol{F}^{2,1}}^{k(2,1)}$, respectively, are contained in suitable tubes around $\boldsymbol{\Gamma}_{2,1}$ and $\boldsymbol{F}_{2,1}$, respectively. By Lemma 5.4 the closed geometric fat braid (5.27) is free isotopic to the closed geometric fat braid $\widetilde{\boldsymbol{G}^{1,1}}^{1} \sqcup_{\boldsymbol{\Gamma}_{2,1}} \widetilde{\boldsymbol{G}^{2,1}}^{k(2,1)}$. By induction the closed geometric fat braid

$$\widetilde{\boldsymbol{G}^{1,1}}^{1} \sqcup_{\boldsymbol{\Gamma}_{2,1}} \widetilde{\boldsymbol{G}^{2,1}}^{k(2,1)} \sqcup \ldots \sqcup_{\boldsymbol{\Gamma}_{2,k_2}} \widetilde{\boldsymbol{G}^{2,k_2}}^{k(2,k_2)} \tag{5.32}$$

is free isotopic to (5.29) with $i = k_2$.

We claim that the obtained closed geometric fat braid

$$\widetilde{\boldsymbol{G}^{1,1}}^{1} \sqcup_{\boldsymbol{\Gamma}_{2,1}} \widetilde{\boldsymbol{G}^{2,1}}^{k(2,1)} \sqcup \ldots \sqcup_{\boldsymbol{\Gamma}_{2,k_2}} \widetilde{\boldsymbol{G}^{2,k_2}}^{k(2,k_2)} \tag{5.33}$$

is associated to the domain $Q_2 = S^{1,1} \cup \bigcup_{i=1}^{k_2} \bigcup_{j=1}^{k(2,i)} \left(S_j^{2,i} \cup C_j^{2,i}\right)$ in the sense that the set of its ordinary strands intersects the fiber $\{1\} \times \mathbb{C}$ of $\partial\mathbb{D} \times \mathbb{C}$ along the set $\{1\} \times (E \cap Q_2)$, and the set of intersection points of the fat strands with $\{1\} \times \mathbb{C}$ contains exactly one point in $\{1\} \times \delta_j^{3,i}$ for each hole $\delta_j^{3,i}$ of Q_2, and contains no other point.

Indeed, the closed geometric fat braid $\widetilde{\boldsymbol{G}^{1,1}}^{1} \sqcup_{\boldsymbol{\Gamma}_{2,1}} \widetilde{\boldsymbol{G}^{2,1}}^{k(2,1)}$ can be written as

$$\left\{\left(e^{2\pi it}, \widetilde{\boldsymbol{g}^{1,1}}^{1}(t) \sqcup_{\boldsymbol{\gamma}_{2,1}(t)} \widetilde{\boldsymbol{g}^{2,1}}^{k(2,1)}(t)\right),\ t \in [0, k(2,1)]\right\}, \tag{5.34}$$

where $\widetilde{\boldsymbol{g}^{2,1}}^{k(2,1)}$ is defined by (5.15), or equivalently by (5.16), with $\ell = 2, i = 1$. Since the closed geometric fat braid $\widetilde{\boldsymbol{G}^{1,1}}^{1}$ can be written by (5.11), the closed geometric fat braid (5.34) is equal to

$$\left\{\left(e^{2\pi it}, \varphi_t(E^{1,1} \cup \boldsymbol{E}^{1,1}) \sqcup_{\varphi_t(z_1^{2,1})} \varphi_t(E_1^{2,1} \cup \boldsymbol{E}_1^{2,1})\right),\ t \in [0, k(2,1)]\right\}. \tag{5.35}$$

The following equalities hold:
$\varphi_{t+j} = \varphi_t \circ \varphi_{b,\infty}^j$ for $t \in [0,1]$,
$\varphi_{b,\infty}^j(\boldsymbol{E}^{1,1}) = \boldsymbol{E}^{1,1}$,
$\varphi_{b,\infty}^j(z_1^{2,1}) = z_{j+1}^{2,1}$ for $j = 1, \dots, k(2,1) - 1$, and $\varphi_{b,\infty}^{k(2,1)}(z_1^{2,1}) = z_1^{2,1}$,
$\varphi_{b,\infty}^j(E_1^{2,1} \cup \boldsymbol{E}_1^{2,1}) = E_{j+1}^{2,1} \cup \boldsymbol{E}_{j+1}^{2,1}$ for $j = 1, \dots, k(2,1) - 1$, and
$\varphi_{b,\infty}^{k(2,1)}(E_1^{2,1} \cup \boldsymbol{E}_1^{2,1}) = E_1^{2,1} \cup \boldsymbol{E}_1^{2,1}$.
Hence, the closed geometric fat braid (5.35) is equal to

$$\left\{\left(e^{2\pi it}, \varphi_t(E^{1,1} \cup (\boldsymbol{E}^{1,1} \setminus \bigcup_{j=1}^{k(2,1)} \{z_j^{2,1}\}) \cup \bigcup_{j=1}^{k(2,1)} (E_j^{2,1} \cup \boldsymbol{E}_j^{2,1}))\right)\ t \in [0,1]\right\}. \tag{5.36}$$

Notice that $\boldsymbol{E}^{1,1} = \bigcup_{i=1}^{k_2} \bigcup_{j=1}^{k(2,i)} \{z_j^{2,i}\}$. Induction over the closed geometric fat braids

$$\widetilde{\boldsymbol{G}^{1,1}}^{1} \sqcup_{\boldsymbol{\Gamma}_{2,1}} \widetilde{\boldsymbol{G}^{2,1}}^{k(2,1)} \sqcup \dots \sqcup_{\boldsymbol{\Gamma}_{2,i}} \widetilde{\boldsymbol{G}^{2,i}}^{k(2,i)} \tag{5.37}$$

for $i = 1, \dots, k_2$, shows that the closed geometric fat braid (5.33) equals

$$\left\{\left(e^{2\pi it}, \varphi_t\left(E^{1,1} \cup \bigcup_{i=1}^{k_2} \bigcup_{j=1}^{k(2,i)} (E_j^{2,i} \cup \boldsymbol{E}_j^{2,i})\right)\right),\ t \in [0,1]\right\}.$$

Hence, the closed geometric fat braid (5.33) is associated to the domain Q_2.

The interior boundary of the domain Q_N is empty. Induction over the number of the generation $\ell = 1, \ldots, N$ shows that the closed geometric fat braid

$$\widetilde{G^{1,1}}^{1} \sqcup_{\boldsymbol{\Gamma}_{2,1}} \widetilde{G^{2,1}}^{k(2,1)} \sqcup \ldots \sqcup_{\boldsymbol{\Gamma}_{2,k_2}} \widetilde{G^{2,k_2}}^{k(2,k_2)} \sqcup$$
$$\ldots \sqcup_{\boldsymbol{\Gamma}_{N,1}} \widetilde{G^{N,1}}^{k(N,1)} \sqcup \ldots \sqcup_{\boldsymbol{\Gamma}_{N,k_N}} \widetilde{G^{N,k_N}}^{k(N,k_N)} \tag{5.38}$$

is an ordinary closed geometric braid which is equal to the closed geometric braid G given by $\left\{\left(e^{2\pi i t},\, \varphi(t)\right),\ t \in [0,1]\right\}$.

By induction over the number of the generation $\ell = 1, \ldots, N$, we find small enough numbers $\varepsilon_{\ell+1}$ and "insert the representatives $\boldsymbol{F}_{\ell+1,i'} \boxplus \varepsilon_{\ell+1} \widetilde{\boldsymbol{F}^{\ell+1,i'}}^{\ell+1,i'}$ of $\widehat{\boldsymbol{B}(\ell+1,i')}$ into the ε_ℓ-neighbourhoods of the fat loops $\boldsymbol{F}_{\ell+1,i'}$ that correspond to the cycles of holes of Q_ℓ". For $\ell = N$ we arrive at the closed geometric fat braid

$$\widetilde{\boldsymbol{F}^{1,1}}^{1} \sqcup_{\boldsymbol{F}_{2,1}} (\boldsymbol{F}_{2,1} \boxplus \varepsilon_2 \widetilde{\boldsymbol{F}^{2,1}}^{k(2,1)}) \sqcup \ldots \sqcup_{\boldsymbol{F}_{2,k_2}} (\boldsymbol{F}_{2,k_2} \boxplus \varepsilon_2 \widetilde{\boldsymbol{F}^{2,k_2}}^{k(2,k_2)})$$
$$\sqcup \ldots \sqcup_{\boldsymbol{F}_{N,1}} (\boldsymbol{F}_{N,1} \boxplus \varepsilon_N \widetilde{\boldsymbol{F}^{N,1}}^{k(N,1)}) \sqcup \ldots \sqcup_{\boldsymbol{F}_{N,k_N}} (\boldsymbol{F}_{N,k_N} \boxplus \varepsilon_N \widetilde{\boldsymbol{F}^{N,1}}^{k(N,k_N)}) . \tag{5.39}$$

An inductive application of Lemma 5.4 shows, that the closed geometric fat braid (5.39) is an ordinary closed braid and is free isotopic to the closed braid (5.38) which is equal to G. We provided a procedure which gives for each choice of closed geometric fat braids, that represent the irreducible braid components $\widehat{\boldsymbol{B}(\ell,i)}$, the closed braid (5.39) that represents $\hat{b}$. We recovered $\hat{b}$ knowing the irreducible braid components.

5.5 Proof of the Main Theorem for Reducible Braids. The General Case

The following lemma describes the entropy of the nodal conjugacy class $\widehat{\mathfrak{m}_{b,\odot}}$, that is associated to $\mathfrak{m}_{b,\infty}$ and an admissible system of curves $\mathcal{C} \subset \mathbb{D}$ that completely reduces $\mathfrak{m}_{b,\infty}$, in terms of the entropy of irreducible nodal components.

Recall that we identify mapping classes on (possibly not connected) punctured Riemann surfaces (or on punctured nodal surfaces) with mapping classes on (possibly not connected) closed Riemann surfaces (or on closed nodal surfaces) with distinguished points. If for notational convenience we write $h(\mathfrak{m})$ for a mapping class $\mathfrak{m}$ on a punctured surface, we mean the entropy of the respective mapping class on the closed surface with distinguished points. Similarly, we call a self-homeomorphism of a punctured Riemann surface non-periodic absolutely extremal if its extension to the closed Riemann surface with the respective set of distinguished points is so.

Lemma 5.5

$$h\left(\widehat{\mathfrak{m}_{b,\odot}}\right) = \max_{\mathrm{cyc}_\odot^{\ell,i}} h\left(\widehat{\mathfrak{m}_{b,\odot} \mid \mathrm{cyc}_\odot^{\ell,i}}\right) = \max_{\mathrm{cyc}_\odot^{\ell,i}} \frac{1}{k(\ell,i)} h\left(\widehat{\mathfrak{m}_{b^{k(\ell,i)},\odot} \mid Y_1^{\ell,i}}\right). \tag{5.40}$$

Proof The first equality is an easy consequence of Theorem 4 of [1].

To obtain the second equality it is enough to prove that the equation

$$h\left(\widehat{\mathfrak{m}_{b,\odot} \mid \mathrm{cyc}_\odot^{\ell,i}}\right) = \frac{1}{k(\ell,i)} h\left(\widehat{\mathfrak{m}_{b^{k(\ell,i)},\odot} \mid Y_1^{\ell,i}}\right) \tag{5.41}$$

holds for each nodal cycle $\mathrm{cyc}_\odot^{\ell,i}$. For each mapping $\varphi_\odot$ representing the class $\widehat{\mathfrak{m}_{b,\odot} \mid \mathrm{cyc}_\odot^{\ell,i}}$ the equality $h(\varphi_\odot^{k(\ell,i)}) = k(\ell,i)h(\varphi_\odot)$ holds. Since $\varphi_\odot^{k(\ell,i)}$ fixes each $Y_j^{\ell,i}$ and for each j the mapping

$$\varphi_\odot^{k(\ell,i)} \mid Y_j^{\ell,i} = (\varphi_\odot^{j-1} \mid Y_1^{\ell,i}) \circ (\varphi_\odot^{k(\ell,i)} \mid Y_1^{\ell,i}) \circ (\varphi_\odot^{j-1} \mid Y_1^{\ell,i})^{-1}$$

is a conjugate of $\varphi_\odot^{k(\ell,i)} \mid Y_1^{\ell,i}$, the equality

$$h(\varphi_\odot^{k(\ell,i)} \mid Y_1^{\ell,i}) = k(\ell,i)h(\varphi_\odot) \tag{5.42}$$

holds. The entropy $h\left(\widehat{\mathfrak{m}_{b,\odot} \mid \mathrm{cyc}_\odot^{\ell,i}}\right)$ is equal to the infimum

$$\begin{aligned} h\left(\widehat{\mathfrak{m}_{b,\odot} \mid \mathrm{cyc}_\odot^{\ell,i}}\right) &= \inf\left\{h(\varphi_\odot) : \varphi_\odot \in \widehat{\mathfrak{m}_{b,\odot} \mid \mathrm{cyc}_\odot^{\ell,i}}\right\} \\ &= \inf\left\{\frac{1}{k(\ell,i)} \{h\left(\varphi_\odot^{k(\ell,i)} \mid Y_1^{\ell,i}\right) : \varphi_\odot \in \widehat{\mathfrak{m}_{b,\odot} \mid \mathrm{cyc}_\odot^{\ell,i}}\right\}. \end{aligned} \tag{5.43}$$

On the other hand,

$$h(\widehat{\mathfrak{m}_{b^{k(\ell,i)},\odot} \mid Y_1^{\ell,i}}) = \inf\left\{h(\psi) : \psi \in \widehat{\mathfrak{m}_{b^{k(\ell,i)},\odot} \mid Y_1^{\ell,i}}\right\}. \tag{5.44}$$

By Lemma 5.1 the inclusion $\psi \in \widehat{\mathfrak{m}_{b^{k(\ell,i)},\odot} \mid Y_1^{\ell,i}}$ holds if and only if $\psi = \varphi_\odot^{k(\ell,i)} \mid Y_1^{\ell,i}$ for a mapping $\varphi_\odot \in \widehat{\mathfrak{m}_{b,\odot} \mid \mathrm{cyc}_\odot^{\ell,i}}$. The lemma is proved. □

By Lemma 5.5 for the proof of the following theorem it is enough to relate the entropy of a mapping class on a surface to the entropy of its nodal mapping class.

Theorem 5.1 *For an arbitrary braid the following equality holds*

$$h\left(\widehat{\mathfrak{m}_b}\right) = h(\widehat{\mathfrak{m}_{b,\odot}}) = \max_{\mathrm{cyc}_{\odot}^{\ell,i}} \left(h\left(\overline{\mathfrak{m}_{b^{k(\ell,i)},\odot} \mid Y_1^{\ell,i}} \right) \cdot \frac{1}{k(\ell,i)} \right). \tag{5.45}$$

Moreover, there exists a mapping with this entropy that represents $\widehat{\mathfrak{m}_b}$.

We split the proof of Theorem 5.1 into two parts. The first part is the lower bound for the entropy of the braid.

Lemma 5.6

$$h\left(\widehat{\mathfrak{m}_b}\right) \geq \max_{\mathrm{cyc}_{\odot}^{\ell,i}} \left(h\left(\overline{\mathfrak{m}_{b^{k(\ell,i)},\odot} \mid Y_1^{\ell,i}} \right) \cdot \frac{1}{k(\ell,i)} \right).$$

Proof Let N be the smallest natural number for which b^N is a pure braid. Notice that

$$h(\mathfrak{m}_{b^N}) \leq N \cdot h(\mathfrak{m}_b). \tag{5.46}$$

Indeed, for a chosen homeomorphism $\varphi_b \in \mathfrak{m}_b$

$$h(\mathfrak{m}_{b^N}) = \inf\left\{ h(\varphi) : \varphi \text{ is } \mathrm{Hom}^+(\overline{\mathbb{D}}; \partial\, \mathbb{D}, E)\text{-isotopic to } \varphi_b^N \right\}, \tag{5.47}$$

and

$$N \cdot h(\mathfrak{m}_b) \;=\; \inf\left\{ h(\psi^N) = N \cdot h(\psi) : \psi \text{ is } \mathrm{Hom}^+(\overline{\mathbb{D}}; \partial\, \mathbb{D}, E)\text{-isotopic to } \varphi_b \right\}. \tag{5.48}$$

Since the infimum in (5.48) is taken over a smaller class than in (5.47), we obtain (5.46).

Suppose $\mathrm{cyc}_{b,\odot}^{\ell,i} = \left(Y_1^{\ell,i}, \ldots, Y_{k(\ell,i)}^{\ell,i}\right)$ is a nodal cycle on the nodal surface Y such that the mapping class $\overline{\mathfrak{m}_{b^{k(\ell,i)},\odot} \mid Y_1^{\ell,i}}$ is pseudo-Anosov, i.e. there exists a self-homeomorphism of $Y_1^{\ell,i}$ that represents $\mathfrak{m}_{b^{k(\ell,i)},\odot} \mid Y_1^{\ell,i}$ and is conjugate to a non-periodic absolutely extremal mapping $\tilde{\varphi}^{\ell,i}$. Any power of $\tilde{\varphi}^{\ell,i}$ is again pseudo-Anosov. Hence the class $\overline{\mathfrak{m}_{b^N,\odot} \mid Y_1^{\ell,i}}$ is pseudo-Anosov, since N is divisible by $k(\ell,i)$, i.e.

$$N = k(\ell,i) \cdot m(\ell,i) \tag{5.49}$$

for an natural number $m(\ell, i)$ Moreover,

$$h\left(\widehat{\mathfrak{m}_{b^N,\odot} \mid Y_1^{\ell,i}}\right) = h\left((\tilde{\varphi}^{\ell,i})^{m(\ell,i)}\right) = m(\ell,i)\, h(\tilde{\varphi}^{\ell,i}) \tag{5.50}$$

since $(\tilde{\varphi}^{\ell,i})^{m(\ell,i)}$ represents $\mathfrak{m}_{b^N,\odot} \mid Y_1^{\ell,i}$. We obtained

$$h\left(\widehat{\mathfrak{m}_{b^N,\odot} \mid Y_1^{\ell,i}}\right) = m(\ell,i)\, h\left(\widehat{\mathfrak{m}_{b^{k(\ell,i)},\odot} \mid Y_1^{\ell,i}}\right). \tag{5.51}$$

By Proposition 4.2 for the pure braid b^N the inequality $h(\widehat{\mathfrak{m}_{b^N}}) \geq h(\widehat{\mathfrak{m}_{b^N,\odot} \mid Y_1^{\ell,i}})$ holds.

By (5.46) and (5.51) we obtain the inequality

$$N \cdot h(\mathfrak{m}_b) \geq h\left(\widehat{\mathfrak{m}_{b^N}}\right) \geq h\left(\widehat{\mathfrak{m}_{b^N,\odot} \mid Y_1^{\ell,i}}\right) = m(\ell,i)\, h\left(\widehat{\mathfrak{m}_{b^{k(\ell,i)},\odot} \mid Y_1^{\ell,i}}\right) \tag{5.52}$$

for all ℓ and i for which $\mathfrak{m}_{b^{k(\ell,i)},\odot} \mid Y_1^{\ell,i}$ is pseudo-Anosov. Hence, by (5.49)

$$h(\mathfrak{m}_b) \geq \frac{1}{k(\ell,i)} h\left(\widehat{\mathfrak{m}_{b^{k(\ell,i)},\odot} \mid Y_1^{\ell,i}}\right) \tag{5.53}$$

if $\mathfrak{m}_{b^{k(\ell,i)},\odot} \mid Y_1^{\ell,i}$ is pseudo-Anosov.

If $\mathfrak{m}_{b^{k(\ell,i)},\odot} \mid Y_1^{\ell,i}$ is not pseudo-Anosov, it is elliptic, since it is irreducible. In this case the entropy equals zero and inequality (5.53) is trivially satisfied. Lemma 5.6 is proved. □

The remaining inequality that is needed for the proof of Theorem 5.1 is stated in the following lemma.

Lemma 5.7 *The following inequality holds*

$$h\left(\widehat{\mathfrak{m}_b}\right) \leq \max_{(\ell,i)} \left(h\left(\widehat{\mathfrak{m}_{b^{k(\ell,i)},\odot} \mid Y_1^{\ell,i}}\right) \cdot \frac{1}{k(\ell,i)} \right).$$

Proof of Lemma 5.7 The goal is to represent $\mathfrak{m}_b$ by an element $\varphi \in \mathrm{Hom}^+(\mathbb{P}^1; \infty, E_n)$ that maps the curve system $\mathcal{C} \subset \mathbb{D}$ onto itself and for each ℓ and i the restriction $\varphi^{k(\ell,i)}|S_1^{\ell,i}$ has entropy equal to $h(\widehat{\mathfrak{m}_{b^{k(\ell,i)},\odot} \mid Y^{\ell,i}})$, equivalently, the restriction $\varphi|\mathrm{cyc}_0^{\ell,i}$ has entropy $\frac{1}{k(\ell,i)} h(\widehat{\mathfrak{m}_{b^{k(\ell,i)},\odot} \mid Y^{\ell,i}})$. We will use Theorem 2.10.

Suppose $\varphi_{b,\infty} \in \mathrm{Hom}^+(\mathbb{P}^1; \mathbb{P}^1 \setminus \mathbb{D}, E_n)$ represents $\mathfrak{m}_{b,\infty}$ and maps the admissible set of curves $\mathcal{C}$ onto itself. After a $\mathrm{Hom}^+(\mathbb{P}^1; \mathbb{P}^1 \setminus \mathbb{D}, E_n)$-isotopy of $\varphi_{b,\infty}$ that changes the values only in a small neighbourhood of $\mathcal{C}$ (see Lemma 5.2) we may assume that for each $\ell > 1$ and $i = 1, \dots k_\ell$ the iterate $\varphi_{b,\infty}^{k(\ell,i)}$ equals the identity on a small neighbourhood of the exterior boundary $\partial_{\mathfrak{E}} S_1^{\ell,i}$ of the first labeled set of each cycle $\text{cyc}_0^{\ell,i}$. Here $k(\ell, i)$ is the length of the cycle. For uniformity of notation we write $S_1^{1,1}$ for $S^{1,1}$.

1. A Representative $\psi_{\circledcirc}^{\ell,i}$ of the Isotopy Class of $\varphi_{b,\infty}^{k(\ell,i)} \mid S_1^{\ell,i} \setminus E_n'$

By Corollary 1.1 for each (ℓ, i) the class $\widehat{\mathfrak{m}_{b^{k(\ell,i)},\odot}|Y^{\ell,i}}$ can be represented by an absolutely extremal self-homeomorphism $\tilde{\psi}_{\odot}^{\ell,i}$ of a punctured Riemann surface $\tilde{Y}^{\ell,i}$. The mapping $\tilde{\psi}_{\odot}^{\ell,i}$ has the following properties. $\tilde{\psi}_{\odot}^{\ell,i}$ minimizes the entropy in the class $\widehat{\mathfrak{m}_{b^{k(\ell,i)},\odot}|Y^{\ell,i}}$. Moreover, the continuous extension of $\tilde{\psi}_{\odot}^{\ell,i}$ to the closed Riemann surface $(\tilde{Y}^{\ell,i})^c$ (also denoted by $\tilde{\psi}_{\odot}^{\ell,i}$) fixes the node corresponding to the exterior boundary component of $S_1^{\ell,i}$ (in the case of $S_1^{1,1}$ it fixes ∞), and it moves the other nodes in $\mathcal{N} \cap (\tilde{Y}_1^{\ell,i})^c$ of $(\tilde{Y}_1^{\ell,i})^c$ along cycles corresponding to the $\varphi_{b^{k(\ell,i)},\infty}$-cycles of the connected components of the interior boundary $\partial_{\mathfrak{I}} S_1^{\ell,i}$.

If the continuous extension of the homeomorphism $\tilde{\psi}_{\odot}^{\ell,i}$ is pseudo-Anosov, then by Theorem 2.10 there exists a self-homeomorphism $\tilde{\psi}_{\circledcirc}^{\ell,i}$ of $(\tilde{Y}_1^{\ell,i})^c$ that is isotopic to $\tilde{\psi}_{\odot}^{\ell,i}$ by an isotopy that changes the mapping only in a punctured neighbourhood of the nodes, that has the same entropy as $\tilde{\psi}_{\odot}^{\ell,i}$, and has the following properties.

$\tilde{\psi}_{\circledcirc}^{\ell,i}$ fixes pointwise a topological disc around the node corresponding to the exterior boundary of $(\tilde{Y}_1^{\ell,i})^c$ (if $(\ell, i) = (1, 1)$ it fixes $\mathbb{P}^1 \setminus \mathbb{D}$). For each $\tilde{\psi}_{\odot}^{\ell,i}$-cycle of nodes there is a $\tilde{\psi}_{\circledcirc}^{\ell,i}$-cycle of closed topological discs around the nodes of the cycle, and, for the length of the cycle being equal to $k \geq 1$, the iterate $(\tilde{\psi}_{\circledcirc}^{\ell,i})^k$ fixes each disc pointwise.

If the continuous extension of $\tilde{\psi}_{\odot}^{\ell,i}$ is a periodic self-mapping of the Riemann sphere $(\tilde{Y}_1^{\ell,i})^c$ with distinguished points, i.e. $(\tilde{\psi}_{\odot}^{\ell,i})^k = \text{Id}$ for a natural number k, then for each cycle of nodes of length k the extension of $\tilde{\psi}_{\odot}^{\ell,i}$ itself maps a collection of topological discs around the distinguished points along a cycle of length k, and the restriction of the extension of $(\tilde{\psi}_{\odot}^{\ell,i})^k$ to each of the discs is the identity. The mapping $\tilde{\psi}_{\odot}^{\ell,i}$ may fix a node. In this case $\tilde{\psi}_{\odot}^{\ell,i}$ rotates a disc around the node. Lemma 2.14 allows to change $\tilde{\psi}_{\odot}^{\ell,i}$ without changing it outside the disc and without increasing the entropy, so that the new mapping is the identity on a smaller disc around the node.

Let $\mathring{Y}_1^{\ell,i}$ denote the complement in $\tilde{Y}_1^{\ell,i}$ of the collection of all respective open discs. Notice that $\mathring{Y}_1^{\ell,i}$ is a bordered surface with punctures. There is a homeomorphism $w_1^{\ell,i} : \mathring{Y}_1^{\ell,i} \to \overline{S_1^{\ell,i}} \setminus E_n'$, for which the conjugate $\psi_{\circledcirc}^{\ell,i} \overset{def}{=} (w_1^{\ell,i}) \circ (\tilde{\psi}_{\odot}^{\ell,i} \mid \mathring{Y}_1^{\ell,i}) \circ (w_1^{\ell,i})^{-1}$ is a self-homeomorphism of $\overline{S_1^{\ell,i}} \setminus E'^{\ell,i}_1$, whose restriction to $S_1^{\ell,i} \setminus E'^{\ell,i}_1$ is isotopic to $\varphi_{b,\infty}^{k(\ell,i)}|S_1^{\ell,i} \setminus E'^{\ell,i}_1$. Moreover, $\psi_{\circledcirc}^{\ell,i} \mid S_1^{\ell,i} \setminus E'^{\ell,i}_1$

also minimizes the entropy among mappings representing $\widehat{\mathfrak{m}_{b^{k(\ell,i)},\odot} \mid Y_1^{\ell,i}}$. In other words,

$$h(\psi_{\circledcirc}^{\ell,i}) = h(\widehat{\mathfrak{m}_{b^{k(\ell,i)},\odot} \mid Y_1^{\ell,i}}) . \tag{5.54}$$

For all (ℓ, i) we found a homeomorphism $\psi_{\circledcirc}^{\ell,i}$ that represents $\widehat{\mathfrak{m}_{b^{k(\ell,i)},\odot} \mid Y_1^{\ell,i}}$, satisfies (5.54), and is equal to the identity at all points of $\overline{S_1^{\ell,i}}$ that are close to the exterior boundary $\partial_{\mathbb{C}} S_1^{\ell,i}$ (on $\mathbb{P}^1 \setminus \mathbb{D}$ in case of $S_1^{1,1}$). Moreover, it moves the interior boundary components along $\psi_{\circledcirc}^{\ell,i}$-cycles. If the length of a $\psi_{\circledcirc}^{\ell,i}$-cycle equals k then $(\psi_{\circledcirc}^{\ell,i})^k$ is the identity near each boundary component of the cycle.

2. The Homeomorphism $\psi_{\circledcirc}^{1,1}$ and an Isotopy of $\varphi_{b,\infty}$
Recall that for $l = 1, \ldots, N$ we denoted by $\overline{Q_l}$ the following closed subset of $\mathbb{P}^1$

$$\overline{Q_l} = \bigcup_{l'=1,\ldots,l} \; \bigcup_{i=1,\ldots,k_{l'}} \; \bigcup_{j=1,\ldots,k(l',i)} \overline{S_j^{l',i}} .$$

$\overline{Q_l}$ is the closure of a domain Q_l in $\mathbb{P}^1$. Notice that $Q_1 = S_1^{1,1}$.

Let first $l = 1$, i.e. $Q_1 = S_1^{1,1}$. Both mappings, $\varphi_{b,\infty}|S_1^{1,1}$ and $\psi_{\circledcirc}^{1,1}|S_1^{1,1}$ represent $\widehat{\mathfrak{m}_{b,\odot} \mid Y_1^{1,1}}$. Both mappings are equal to the identity on $\mathbb{P}^1 \setminus \mathbb{D}$. Take any $\psi_{\circledcirc}^{1,1}$-cycle of interior boundary components of $S_1^{1,1}$. Let k be the length of the cycle. Then the mapping $(\psi_{\circledcirc}^{1,1})^k$ equals the identity on each boundary component of the cycle (and on a small neighbourhood of it in $\overline{S_1^{1,1}}$). Moreover, $h(\psi_{\circledcirc}^{1,1}) = h(\widehat{\mathfrak{m}_{b,\odot} \mid Y_1^{1,1}})$.

Consider a $\mathrm{Hom}^+(\mathbb{P}^1; (\mathbb{P}^1 \setminus \mathbb{D}), E_n)$-isotopy of $\varphi_{b,\infty}$, which changes the mapping only in a small annulus around each interior boundary component of $S_1^{1,1}$, so that the new self-homeomorphism $\varphi_{b,\infty,1}$ of $\mathbb{P}^1$ with distinguished points E_n' coincides with $\psi_{\circledcirc}^{1,1}$ on $\mathbb{P}^1 \setminus \mathbb{D}$ and on all boundary components of $S_1^{1,1}$. Then the mapping $\psi_{\circledcirc}^{1,1} \circ \varphi_{b,\infty,1}^{-1}$ is the identity on $\mathbb{P}^1 \setminus \mathbb{D}$ and on each boundary component of $S_1^{1,1}$. Since both mappings, $\varphi_{b,\infty}|S_1^{1,1}$ and $\psi_{\circledcirc}^{1,1}|S_1^{1,1}$, represent $\widehat{\mathfrak{m}_{b,\odot} \mid Y^{1,1}}$, the isotopy classes of their continuous extensions to $\overline{S_1^{1,1}}$ with distinguished points differ by a product of commuting Dehn twists around curves that are homologous to the boundary circles of $S_1^{1,1}$ or to $\partial\mathbb{D}$. Lemma 2.14 allows to change the isotopy class of $\psi_{\circledcirc}^{1,1}$ by any power of a Dehn twist around the circle $\partial\mathbb{D}$ by changing $\psi_{\circledcirc}^{1,1}$ in a small neighbourhood of this circle without increasing entropy. By a further isotopy of $\varphi_{b,\infty,1}$ on $\mathbb{P}^1$ that does not change the mapping outside small neighbourhoods of the interior boundary of $S_1^{1,1}$ we may change the isotopy class of the restriction $\varphi_{b,\infty,1} \mid \overline{S_1^{1,1}}$ by any product of Dehn twists around curves that are homologous to the interior boundary circles of $S_1^{1,1}$. After changing $\psi_{\circledcirc}^{1,1}$ and

$\varphi_{b,\infty,1}$ in the described way and keeping previous notation, we may assume that $\psi_{\circledcirc}^{1,1} \circ (\varphi_{b,\infty,1}|\overline{S_1^{1,1}})^{-1}$ is $\mathrm{Hom}^+(\overline{S_1^{1,1}}; (S_1^{1,1} \cap E_n'^{\,1,1}) \cup (\mathbb{P}^1 \setminus \mathbb{D}) \cup \partial \overline{S_1^{1,1}})$-isotopic to the identity.

3. The Inductive Construction of the Entropy Minimizing Element of $\widehat{\mathfrak{m}_b}$
By induction on l we will find a self-homeomorphism φ_{Q_l} of $\overline{Q_l}$ with distinguished points $E_n' \cap Q_l$ and a self-homeomorphism $\varphi_{b,\infty,l}$ of $\mathbb{P}^1$ for which the following requirements hold.

The entropy of the self-homeomorphism φ_{Q_l} equals

$$h(\varphi_{Q_l}) = \max_{(l',i):l' \le l, i \le k_{l'}} \Big(h(\widehat{\mathfrak{m}_{b^{k(l',i)},\odot} \mid Y_1^{l',i}}) \cdot \frac{1}{k(l',i)}\Big).$$

Moreover, φ_{Q_l} equals the identity on $\mathbb{P}^1 \setminus \mathbb{D}$ and moves the boundary components of Q_l along cycles so that for each φ_{Q_l}-cycle of length k the homeomorphism $\varphi_{Q_l}^k$ is the identity near each component of the boundary belonging to the cycle.

The self-homeomorphism $\varphi_{b,\infty,l}$ of $\mathbb{P}^1$ is $\mathrm{Hom}^+(\mathbb{P}^1; \infty, E_n)$-isotopic to $\varphi_{b,\infty}$ by an isotopy that changes $\varphi_{b,\infty}$ only in a small neighbourhood of the interior boundary of the $Q_{l'}$ with $l' \le l$ and coincides with φ_{Q_l} on ∂Q_l.

Moreover, the self-homeomorphism φ_{Q_l} of $\overline{Q_l}$ is isotopic to $\varphi_{b,\infty,l} \mid \overline{Q_l}$ by an isotopy that does not change the mappings in a small neighbourhood of $\partial Q_l \cup (E_n' \cap Q_l)$ and on $\mathbb{P}^1 \setminus \mathbb{D}$ (in other words, the mapping $\varphi_{Q_l} \circ (\varphi_{b,\infty,l} \mid \overline{Q_l})^{-1}$ is $\mathrm{Hom}\big(\overline{Q_l}; (E_n' \cap Q_l) \cup \partial Q_l \cup (\mathbb{P}^1 \setminus \mathbb{D})\big)$-isotopic to the identity on $\overline{Q_l}$).

We already found φ_{Q_1} and $\varphi_{b,\infty,1}$. Suppose for some $l \le N-1$ we found a self-homeomorphisms φ_{Q_l} with distinguished points $E_n' \cap Q_l$ and a self-homeomorphism $\varphi_{b,\infty,l}$ of $\mathbb{P}^1$ with distinguished points E_n' that satisfy the requirements. We have to find these objects for the number $l+1$.

The boundary components of Q_l are exactly the curves in $\mathcal{C}$ that are of generation $l+1$. Each such curve is the exterior boundary of a connected component of generation $l+1$ of $\mathbb{P}^1 \setminus \mathcal{C}$. Take any $\varphi_{b,\infty,l}$-cycle $\mathrm{cyc}_0^{l+1,i} = (S_1^{l+1,i}, \dots, S_{k(l+1,i)}^{l+1,i})$ of generation $l+1$ of length $k(l+1,i)$, $1 \le i \le k_{l+1}$. We first consider the set $\overline{Q_{l,i}} \stackrel{def}{=} \overline{Q_l} \cup \overline{S_1^{l+1,i}} \cup \dots \cup \overline{S_{k(l+1,i)}^{l+1,i}}$ instead of $\overline{Q_{l+1}}$, and will associate to $\overline{Q_{l,i}}$ a self-homeomorphism $\varphi_{Q_{l,i}}$ of $\overline{Q_{l,i}}$ and a homeomorphism $\varphi_{b,\infty,l,i} \in \mathrm{Hom}^+(\mathbb{P}^1; \mathbb{P}^1 \setminus \mathbb{D}, \mathcal{C} \cup E_n')$.

We put

$$\varphi_{Q_{l,i}} = \begin{cases} \varphi_{Q_l} & \text{on } \overline{Q_l}\,, \\ \varphi_{b,\infty,l} & \text{on } \overline{S_j^{l+1,i}} \text{ with } j = 2, \dots, k(l+1,i)\,, \\ (\varphi_{b,\infty,l})^{-k(l+1,i)+1} \circ \psi_{\circledcirc}^{l+1,i} & \text{on } \overline{S_1^{l+1,i}}\,. \end{cases} \tag{5.55}$$

Since the homeomorphism $\psi_{\circledcirc}^{l+1,i}$ is equal to the identity at all points of $\overline{S_1^{l+1,i}}$ that are close to the exterior boundary $\partial_{\mathfrak{E}} S_1^{l+1,i}$ and also $(\varphi_{b,\infty,l})^{k(l+1,i)}$ is the identity there, the equality $\varphi_{Q_{l,i}} = \varphi_{b,\infty,l}$ holds on $\partial_{\mathfrak{E}} S_1^{l+1,i}$. Hence, by the induction hypothesis $\varphi_{Q_{l,i}} = \varphi_{Q_l}$ on $\partial_{\mathfrak{E}} S_j^{l+1,i}$, $j = 1, \ldots, k(l+1,i)$, and $\varphi_{Q_{l,i}}$ is a well-defined self-homeomorphism of $\overline{Q_{l,i}}$.

We consider a self-homeomorphism $\varphi_{b,\infty,l,i}$ of $\mathbb{P}^1$ that is obtained from $\varphi_{b,\infty,l}$ by a $\mathrm{Hom}^+(\mathbb{P}^1; \infty, E_n)$-isotopy, that changes the mappings only in a small neighbourhood of the interior boundary components of $S_1^{l+1,i}$ and satisfies the condition $\varphi_{b,\infty,l,i} = \varphi_{Q_{l,i}}$ on all interior boundary components of $S_1^{l+1,i}$. Hence, $\varphi_{b,\infty,l,i} = \varphi_{Q_{l,i}}$ on ∂Q_l and on all boundary components of all $\overline{S_j^{l+1,i}}$.

By the induction hypothesis the restrictions of $\varphi_{Q_{l,i}}$ (see (5.55)) and of $\varphi_{b,\infty,l,i}$ to $\overline{Q_l}$ are isotopic by an isotopy that does not change the mappings in a small neighbourhood of $\partial Q_l \cup (E_n' \cap Q_l)$ and on $\mathbb{P}^1 \setminus \mathbb{D}$.

Since the self-homeomorphism $\psi_{\circledcirc}^{l+1,i} \mid S_1^{l+1,i} \setminus E_n'$ is isotopic to $\varphi_{b,\infty,l}^{k(l+1,i)} \mid S_1^{l+1,i} \setminus E_n'$, and the continuous extensions to $\overline{S_1^{l+1,i}} \setminus E_n'$ of the two mappings coincide on all boundary components of $\overline{S_1^{l+1,i}}$, the isotopy classes of their extensions differ by a product of Dehn twists around curves that are homologous to the boundary components of $S_1^{l+1,i}$.

If necessary, we may change $\psi_{\circledcirc}^{l+1,i}$ without increasing entropy by a power of a Dehn twist about a circle in $S_1^{l+1,i}$ that is homologous to the exterior boundary $\partial_{\mathfrak{E}} S_1^{l+1,i}$ of $\overline{S_1^{l+1,i}}$, and build the mapping $\varphi_{Q_{l,i}}$ from the new $\psi_{\circledcirc}^{l+1,i}$ (see Eq. (5.55)), using the previous notation for the changed objects. On the other hand, we may change the $\mathrm{Hom}^+(\mathbb{P}^1; \infty, E_n)$-isotopy joining $\varphi_{b,\infty,l}$ and $\varphi_{b,\infty,l,i}$, so that the isotopy class of the new $\varphi_{b,\infty,l,i}^{k(l+1,i)} \mid \overline{S_1^{l+1,i}} \setminus E_n'$ differs from that of the previous one by an a priori given product of Dehn twists around curves that are homologous to the interior boundary components of $S_1^{l+1,i}$. Put $\overline{\Omega^{l+1,i}} \overset{def}{=} (\overline{S_1^{l+1,i}} \cup \ldots \cup \overline{S_{k(l+1,i)}^{l+1,i}})$. Since for $j = 2, \ldots, n$ the equality $\varphi_{Q_{l,i}} = \varphi_{b,\infty,l,i} = \varphi_{b,\infty,l}$ holds on $\overline{S_j^{l+1,i}} \setminus E_n'$, we can choose the Dehn twists so that the self-homeomorphism $\varphi_{Q_{l,i}} \mid \overline{\Omega^{l+1,i}}$ of $\overline{\Omega^{l+1,i}}$ (see (5.55)) and the mapping $\varphi_{b,\infty,l,i} \mid \overline{\Omega^{l+1,i}}$ are isotopic on $\overline{\Omega^{l+1,i}}$ with set of distinguished points $E_n' \cap \overline{\Omega^{l+1,i}}$ by an isotopy that does not change the mappings on the boundary of the set. Hence, the self-homeomorphism $\varphi_{Q_{l,i}}$ of $\overline{Q_{l,i}}$ and the mapping $\varphi_{b,\infty,l,i} \mid \overline{Q_{l,i}}$ are isotopic by an isotopy that fixes the boundary of $Q_{l,i}$ and the set $E \cap Q_{l,i}$.

For the obtained self-homeomorphism $\varphi_{Q_{l,i}}$ of $\overline{Q_l} \cup \overline{\Omega^{l+1,i}} = (\overline{Q_l} \cup \overline{S_1^{l+1,i}} \cup \ldots \cup \overline{S_{k(l+1,i)}^{l+1,i}})$ the following equality on entropies holds

$$h(\varphi_{Q_{l,i}} \mid \overline{\Omega^{l+1,i}}) = \frac{1}{k(l+1,i)} h(\mathfrak{m}_{b^{k(l+1,i)},\odot} \mid \overline{Y_1^{l+1,i}}). \tag{5.56}$$

Moreover, the self-homeomorphism $\varphi_{b,\infty,l,i}$ of $\mathbb{P}^1$ with set of distinguished points E_n', is isotopic to $\varphi_{b,\infty}$ by an isotopy that changes the mapping only in a small neighbourhood of the curves of the system $\mathcal{C}$ of generation at most l and in a neighbourhood of the interior boundary components of the set $S_1^{l+1,i}$. The homeomorphisms $\varphi_{Q_{l,i}}$ and $\varphi_{b,\infty,l,i} \mid \overline{Q_{l,i}}$ are isotopic by an isotopy that does not change the values on $E_n' \cap Q_{l,i}$, on $\mathbb{P}^1 \setminus \mathbb{C}$, and on the boundary of $Q_{l,i}$.

We add successively to $\overline{Q_{l,i}}$ the sets $\overline{S_{j'}^{l+1,i'}}$, $j' = 1, \ldots, k(l+1, i')$, corresponding to all other $\varphi_{b,\infty,l}$-cycles $\mathrm{cyc}_0^{l+1,i'}$ of components of $\mathbb{P}^1 \setminus \mathbb{C}$ of generation $l+1$. In the same way as $\varphi_{Q_{l,i}}$ was constructed (see (5.55)) we extend φ_{Q_l} to each cycle $\bigcup_{j'=1,\ldots,k(l+1,i')} \overline{S_{j'}^{l+1,i'}}$ and, similarly as before, we make further isotopies of $\varphi_{b,\infty,l}$. We obtain a self-homeomorphism $\varphi_{Q,l+1}$ of $\overline{Q_{l+1}}$ and a self-homeomorphism $\varphi_{b,\infty,l+1}$ of $\mathbb{P}^1$ that is $\mathrm{Hom}^+(\mathbb{P}^1; \infty, E_n')$-isotopic to $\varphi_{b,\infty}$ by an isotopy that changes $\varphi_{b,\infty}$ only in a small neighbourhood of the interior boundary of the $Q_{l'}$ with $l' \le l+1$. Moreover, the self-homeomorphism $\varphi_{Q_{l+1}}$ of $\overline{Q_{l+1}}$ is isotopic to $\varphi_{b,\infty,l+1} \mid \overline{Q_{l+1}}$ by an isotopy that does not change the mapping on $(E_n' \cap Q_{l+1}) \cup \mathbb{P}^1 \setminus \mathbb{D}$ and in a small neighbourhood of ∂Q_{l+1}.

Using the induction hypothesis, and also the equality (5.56) for all cycles $\bigcup_{j'=1,\ldots,k(l+1,i')} \overline{S_{j'}^{l+1,i'}}$, as well as Theorem 4 of [1] we obtain the estimate for the entropy

$$h(\varphi_{Q_{l+1}}) \le \max_{(l',i):l' \le l+1, i \le k_{l'}} \left(\widehat{h(\mathfrak{m}_{b^{k(l',i)},\odot} \mid Y_1^{l',i})} \cdot \frac{1}{k(l',i)} \right). \tag{5.57}$$

We will prove now that $\varphi_{Q_{l+1}}$ moves each boundary component of Q_{l+1} along a $\varphi_{Q_{l+1}}$-cycle of curves so that if the length of the cycle equals k then the homeomorphism $\varphi_{Q_{l+1}}^k$ is the identity near each component of the cycle.

Each boundary component of Q_{l+1} is an interior boundary component of an element $S_j^{l+1,i}$ of some cycle $\mathrm{cyc}_0^{l+1,i}$.

Consider first an interior boundary component $\partial_{j'}^{l+1,i}$ of the first labeled element $S_1^{l+1,i}$ of an arbitrary cycle $\mathrm{cyc}_0^{l+1,i}$ of sets of generation $l+1$. The mapping $\psi_{\circledcirc}^{l+1,i}$ moves the boundary component along a $\psi_{\circledcirc}^{l+1,i}$-cycle of interior boundary components of $S_1^{l+1,i}$. Denote the length of this cycle by k. Since $\psi_{\circledcirc}^{l+1,i} = \varphi_{Q_{l+1}}^{k(l+1,i)}$ on $\overline{S_1^{l+1,i}}$, the $\psi_{\circledcirc}^{l+1,i}$-cycle of length k is part of a $\varphi_{Q_{l+1}}$-cycle of length $k \cdot k(l+1,i)$, consisting of interior boundary components of $S_1^{l+1,i} \cup \ldots \cup S_{k(l+1,i)}^{l+1,i}$. Since by the choice of $\psi_{\circledcirc}^{l+1,i}$ the iterate $(\psi_{\circledcirc}^{l+1,i})^k$ is the identity on $\partial_{j'}^{l+1,i}$, the iterate $\varphi_{Q_{l+1}}^{k(l+1,i)\cdot k}$ is the identity on $\partial_{j'}^{l+1,i}$.

An arbitrary boundary component $\partial_{j''}^{l+1}$ of Q_{l+1} is an interior boundary component of a component $S_{j_0}^{l+1,i}$ of $\mathbb{P}^1 \setminus \mathcal{C}$. Since $\varphi_{Q_{l+1}}$ moves $\overline{S_{j_0}^{l+1,i}}$ along the cycle

$$\overline{S_{j_0}^{l+1,i}} \to \overline{S_{j_0+1}^{l+1,i}} \to \ldots \to \overline{S_{k(l+1,i)}^{l+1,i}} \to \overline{S_1^{l+1,i}} \to \ldots \overline{S_{j_0}^{l+1,i}}$$

of length $k(l+1,i)$, an iterate of $\varphi_{Q_{l+1}}$ takes $\partial_{j''}^{l+1}$ to an interior boundary component of $S_1^{l+1,i}$. Hence, $\varphi_{Q_{l+1}} \mid \partial_{j''}^{l+1}$ is conjugate to the restriction of $\varphi_{Q_{l+1}}$ to an interior boundary component of $S_1^{l+1,i}$. Hence, the iterate $\varphi_{Q_{l+1}}^{k(l+1,i)\cdot k}$ is the identity on $\partial_{j''}^{l+1}$.

For $l \leq N-1$ we found the mappings $\varphi_{Q_{l+1}}$ and $\varphi_{b,\infty,l+1}$ with the required properties, provided we have these mappings for the number l.

If $l = N-1$ with N being equal to the number of the generations, the sets $S_j^{N,i}$ of the next generation have no interior boundaries. The process stops and we obtain a mapping φ_{Q_N} that represents $\mathfrak{m}_b$ and has entropy equal to the right hand side of the equality in the statement of the lemma. The lemma is proved. □

Remark 5.2 *The construction of the proof of Lemma 5.7 allows to recover, up to a product of commuting Dehn twists, the conjugacy class of a mapping class from its irreducible nodal components.*

Indeed, when we only know the irreducible nodal components of the mapping, we make the construction of the proof of Lemma 5.7 except the corrections related to the Dehn twists about simple closed loops that are homologous to the boundary components of the $S_1^{\ell,i}$.

The Proof of Theorem 5.1 is a consequence of Lemmas 5.6 and 5.7. □

The counterpart of Theorem 5.1 for the irreducible braid components is the following theorem.

Theorem 5.2

$$\mathcal{M}(\hat{b}) = \min_{\ell,j} \left(\mathcal{M}\left(\widehat{B(\ell,j)}\right) \cdot k(\ell,j)\right). \tag{5.58}$$

The following lemma is the lower bound for the conformal module of $\hat{b}$.

Lemma 5.8

$$\mathcal{M}(\hat{b}) \geq \min_{\ell,j} \left(\mathcal{M}\left(\widehat{B(\ell,j)}\right) \cdot k(\ell,j)\right). \tag{5.59}$$

Proof of Lemma 5.8 Let

$$M_0 \overset{\text{def}}{=} \min_{\ell,j} \left(\mathcal{M}\left(\widehat{B(\ell,j)}\right) \cdot k(\ell,j)\right) \tag{5.60}$$

be the right hand side of the inequality (5.59) in the lemma. Let r_0 be a positive number such that $r_0 < e^{2\pi M_0}$. Consider the annulus

$$A_0 = \left\{ z \in \mathbb{C} : \frac{1}{\sqrt{r_0}} < |z| < \sqrt{r_0} \right\},$$

whose conformal module equals

$$m(A_0) = \frac{1}{2\pi} \log r_0 < M_0 ,$$

and an annulus $A \supset \overline{A_0}$, whose conformal module is also less than M_0.

Consider for each ℓ and i the conjugacy class $\widehat{\boldsymbol{B}(\ell, i)}$ corresponding to the cycle $\mathrm{cyc}^{\ell,i}$. Recall that $k(\ell, i)$ is the length of the cycle. The conjugacy class is represented by closed geometric fat braids in the product $\widetilde{\partial \mathbb{D}}^{k(\ell,i)} \times \mathbb{C}$ of the $k(\ell, i)$-fold covering of the unit circle with the complex plane, equivalently, by continuous maps from $\widetilde{\partial \mathbb{D}}^{k(\ell,i)}$ to a symmetrized configuration space. Recall, that we may consider the points of a symmetrized configuration space as subsets of the complex plane $\mathbb{C}$.

We will define fat holomorphic maps similarly as we defined fat geometric braids. A holomorphic map $\boldsymbol{F}$ from an annulus A into $C_m(\mathbb{C})\diagup \mathcal{S}_m$ is called a holomorphic fat map if some of the connected components of the relatively closed complex curve $\{(z, \boldsymbol{F}(z)),\ z \in \mathrm{A}\}$ in $\mathrm{A} \times \mathbb{C}$ are declared to be fat.

We will represent now the conjugacy class $\widehat{\boldsymbol{B}(\ell, i)}$ by a holomorphic fat map $\widetilde{\boldsymbol{F}^{\ell,i}}^{k(\ell,1)}$ of the $k(\ell, i)$-fold covering $\tilde{A}^{k(\ell,1)}$ of the annulus A into the respective symmetrized configuration space $C_{n(\ell,i)}(\mathbb{D})\diagup \mathcal{S}_{n(\ell,i)}$. We choose the representatives so that we obtain complex curves contained in $\tilde{A}^{k(\ell,1)} \times \mathbb{D}$, not merely in $\tilde{A}^{k(\ell,1)} \times \mathbb{C}$. Since $\mathcal{M}(\widehat{\boldsymbol{B}(\ell, i)}) \geq \frac{1}{k(\ell,i)} \cdot M_0 > \frac{1}{k(\ell,i)} m(A)$ and $\frac{1}{k(\ell,i)} m(A)$ is the conformal module of the $k(\ell, i)$-fold covering of the annulus A, such a mapping $\widetilde{\boldsymbol{F}^{\ell,i}}^{k(\ell,1)}$ exists.

We will put together these mappings proceeding similarly as in the process of recovery of conjugacy classes of braids from their irreducible braid components. Start with the holomorphic fat map $\widetilde{\boldsymbol{F}^{1,1}}^{1}$ on A. Choose $\varepsilon_2 > 0$ so that

$$\{(z, \widetilde{\boldsymbol{F}^{1,1}}^{1}(z)), z \in \overline{A_0}\} \tag{5.61}$$

is a deformation retract of its ε_2-neighbourhood in $\overline{A_0} \times \mathbb{C}$. Consider all fat connected components of (5.61). They correspond to the cycles $\mathrm{cyc}^{2,i}$ and are defined by the fat mappings $\boldsymbol{F}_{2,i}$. Define the holomorphic fat map

$$\widetilde{\boldsymbol{F}^{1,1}}^{1} \sqcup_{\boldsymbol{F}_{2,1}} (\boldsymbol{F}_{2,1} \boxplus \varepsilon_2 \widetilde{\boldsymbol{F}^{2,1}}^{k(2,1)}) \sqcup \ldots \sqcup_{\boldsymbol{F}_{2,k_2}} (\boldsymbol{F}_{2,k_2} \boxplus \varepsilon_2 \widetilde{\boldsymbol{F}^{2,k_2}}^{(k(2,k_2)}) \tag{5.62}$$

in a neighbourhood of $\overline{A_0}$ in the same way as we defined the maps that determine the respective geometric fat braids. By induction on the number $\ell \leq N-1$ of the generation we find a small positive number $\varepsilon_{\ell+1}$ so that the set defined by

$$\widetilde{\boldsymbol{F}^{1,1}}^{1} \sqcup_{\boldsymbol{F}_{2,1}} (\boldsymbol{F}_{2,1} \boxplus \varepsilon_2 \widetilde{\boldsymbol{F}^{2,1}}^{k(2,1)}) \sqcup \ldots \sqcup_{\boldsymbol{F}_{2,k_2}} (\boldsymbol{F}_{2,k_2} \boxplus \varepsilon_2 \widetilde{\boldsymbol{F}^{2,k_2}}^{(k(2,k_2)}) \\ \sqcup \ldots \sqcup_{\boldsymbol{F}_{\ell,1}} (\boldsymbol{F}_{\ell,1} \boxplus \varepsilon_\ell \widetilde{\boldsymbol{F}^{\ell,1}}^{k(\ell,1)}) \sqcup \ldots \sqcup_{\boldsymbol{F}_{\ell,k_\ell}} (\boldsymbol{F}_{\ell,k_\ell} \boxplus \varepsilon_\ell \widetilde{\boldsymbol{F}^{\ell,k_\ell}}^{k(\ell,k_\ell)}) \tag{5.63}$$

is a deformation retract of its $\varepsilon_{\ell+1}$-neighbourhood in $\overline{A_0} \times \mathbb{C}$. This allows to define the set (5.63) with ℓ replaced by $\ell+1$. By the same arguments as used for the recovery procedure of conjugacy classes of braids we obtain for $\ell = N$ a holomorphic map from A_0 into the symmetrized configuration space of suitable dimension that extends continuously to the closure $\overline{A_0}$ and represents $\hat{b}$. Recall that for an arbitrary number $r_0 < e^{2\pi M_0}$ we have chosen an annulus A_0 of conformal module $\frac{1}{2\pi} \log r_0$ and represented b by a holomorphic mapping of A_0 into symmetrized configuration space. Hence, $\mathcal{M}(\hat{b}) > M_0$, and Lemma 5.8 is proved. □

The proof of the opposite inequality in the case of pure braids was straightforward, since in that case each mapping that represents a conjugacy class $\hat{b}$ provides also a mapping that represents a given irreducible braid component by forgetting suitable strands. This does not work in the case of non-pure braids.

We will first prove the Main Theorem and then prove the remaining inequality in Theorem 5.2. For the upper bound of the conformal module of the braid in the Main Theorem we will use the estimate of the conformal module by the translation length of the modular transformation which is provided by Proposition 3.1 in the general case. Further we will use the relation between the translation length of the modular transformation and the quasiconformal dilatation of the absolutely extremal mapping on the nodal Riemann surface, and its relation to the quasiconformal dilatation of the mappings of cycles of parts of the nodal Riemann surface. This will imply the Main Theorem. The proof of the remaining inequality of Theorem 5.2 will follow from relations obtained in the proof of the Main Theorem.

One of the inequalities of the Main Theorem is stated in the following corollary of Lemma 5.8.

Corollary 5.1

$$\mathcal{M}(\widehat{b}) \geq \frac{\pi}{2} \frac{1}{h(\widehat{\mathfrak{m}_b})} = \frac{\pi}{2} \frac{1}{h(\widehat{b})} .$$

Proof of Lemma 5.7 By the equality (5.18), and the Main Theorem in the irreducible case we obtain for each ℓ and i

$$k(\ell,i)\,\mathcal{M}\left(\widehat{B(\ell,i)}\right) = \frac{\pi}{2}\,\frac{k(\ell,i)}{h\left(\widehat{\mathfrak{m}_{b^{k(\ell,i)},\odot} \mid Y_1^{\ell,i}}\right)}\,. \tag{5.64}$$

Lemma 5.8 and Theorem 5.1 imply the inequality

$$\mathcal{M}(\widehat{b}) \geq \frac{\pi}{2}\,\frac{1}{h(\widehat{\mathfrak{m}}_b)} = \frac{\pi}{2}\,\frac{1}{h(\widehat{b})}$$

that is stated in the corollary. □

To prove the Main Theorem in the general case it remains to prove the opposite inequality that is stated in Lemma 5.9 below.

Lemma 5.9

$$\mathcal{M}(\widehat{b}) \leq \frac{\pi}{2}\,\frac{1}{h(\widehat{b})}\,.$$

Proof of Lemma 5.7 By Proposition 3.1

$$\mathcal{M}(\widehat{b}) \leq \frac{\pi}{2}\,\frac{1}{L(\varphi^*_{b,\infty})}\,.$$

Here again $\varphi_{b,\infty}$ is a self-homeomorphism of $\mathbb{P}^1$ with the following properties. It is equal to the identity outside $\overline{\mathbb{D}}$ and on $\overline{\mathbb{D}}$ it is equal to a homeomorphism φ_b representing the mapping class $\mathfrak{m}_b$. As before $\varphi^*_{b,\infty}$ denotes the modular transformation of $\varphi_{b,\infty}$. We have

$$L(\varphi^*_{b,\infty}) = \frac{1}{2}\log I(\varphi_{b,\infty})$$

(see (1.46) and (1.47)). As in Chap. 1, $I(\varphi_{b,\infty})$ is the infimum of quasiconformal dilations in the class obtained from $\varphi_{b,\infty}$ by isotopy and conjugation.

We may assume that $\varphi_{b,\infty}$ is completely reduced by an admissible system of curves $\mathcal{C}$ that are contained in the disc $\mathbb{D}$. Let Y be a nodal surface associated to the system $\mathcal{C}$. Let $\tilde{w} : Y \to \tilde{w}(Y) = \tilde{Y}$ be the conformal structure of part (1) of Theorem 1.13 and let $\tilde{\varphi}$ be the absolutely extremal self-homeomorphism of $\tilde{Y}$ which appears in (1), Theorem 1.13. We have

$$\frac{1}{2}\log K(\tilde{\varphi}) = \frac{1}{2}\log I(\varphi_{b,\infty})\,.$$

Then for the $\tilde{\varphi}$-cycles $\mathrm{cyc}_{\odot}^{\ell,i}$ of $\tilde{Y}$ we have

$$\frac{1}{2}\log K(\tilde{\varphi}) = \max_{\ell,i}\left(\frac{1}{2}\log K(\tilde{\varphi} \mid \mathrm{cyc}_{\odot}^{\ell,i})\right). \tag{5.65}$$

By Lemma 1.8

$$\frac{1}{2}\log K(\tilde{\varphi}\mid \mathrm{cyc}_{\odot}^{\ell,i}) = \frac{1}{2}\frac{1}{k(\ell,i)}\log K(\tilde{\varphi}^{k(\ell,i)}\mid \tilde{Y}_1^{\ell,i}). \tag{5.66}$$

By the Theorem of Fathi-Shub for irreducible self-homeomorphisms of connected Riemann surfaces of first kind (see Theorem 2.1) we obtain

$$\frac{1}{2}\frac{1}{k(\ell,i)}\log K(\tilde{\varphi}^{k(\ell,i)}\mid \tilde{Y}_1^{\ell,i}) = \frac{1}{k(\ell,i)}h(\tilde{\varphi}^{k(\ell,i)}\mid \tilde{Y}_1^{\ell,i}). \tag{5.67}$$

Notice that

$$h(\tilde{\varphi}^{k(\ell,i)}\mid \tilde{Y}_1^{\ell,i}) = h\left(\overline{\mathfrak{m}_{b^{k(\ell,i)},\odot}\mid \tilde{Y}_1^{\ell,i}}\right), \tag{5.68}$$

since $\tilde{\varphi}^{k(\ell,i)}$ is an absolutely extremal element of the class $\mathfrak{m}_{b^{k(\ell,i)},\odot}$ induced by $\varphi_{b,\infty}$ on $\tilde{Y}$. By Theorem 5.1

$$\max_{\ell,i}\frac{1}{k(\ell,i)}h\left(\overline{\mathfrak{m}_{b^{k(\ell,i)},\odot}\mid \tilde{Y}_1^{\ell,i}}\right) = h\left(\widehat{\mathfrak{m}_{b,\odot}}\right) = h(\widehat{\mathfrak{m}_b}) = h(\hat{b}). \tag{5.69}$$

Hence, by (5.65), (5.66), (5.67), (5.68) and (5.69) we have

$$\frac{1}{2}\log I(\varphi_{b,\infty}) = \frac{1}{2}\log K(\tilde{\varphi}) = h\left(\widehat{\mathfrak{m}_{b,\odot}}\right).$$

Hence

$$\mathcal{M}(\hat{b}) \le \frac{\pi}{2}\frac{1}{h(\widehat{\mathfrak{m}_{b,\odot}})} = \frac{\pi}{2}\frac{1}{h(\widehat{\mathfrak{m}_b})} = \frac{\pi}{2}\frac{1}{h(\hat{b})}. \tag{5.70}$$

Lemma 5.9, and, hence, the general case of the Main Theorem is proved. □

The following corollary of Lemma 5.9 states the remaining inequality that is needed for the proof of Theorem 5.2.

Corollary 5.2

$$\mathcal{M}(\widehat{b}) \le \min_{\ell,i}\left(\mathcal{M}\left(\widehat{B(\ell,i)}\right)\cdot k(\ell,i)\right).$$

Proof of Lemma 5.7 The corollary follows from inequality (5.70) and Eqs. (5.69) and (5.64). □

Theorem 5.1 together with the Main Theorem imply the following corollary.

Corollary 5.3 *For each $\hat{b} \in \hat{\mathcal{B}}_n$ $(n \geq 2)$ and each nonzero integer ℓ*

$$\mathcal{M}(\widehat{b^{\ell}}) = \frac{1}{|\ell|}\mathcal{M}(\hat{b})\,. \tag{5.71}$$

Proof of Corollary 5.3 By the Main Theorem Corollary 5.3 is equivalent to the equality

$$h\left(\widehat{\mathfrak{m}_{b^\ell}}\right) = |\ell|\, h\left(\widehat{\mathfrak{m}_b}\right)$$

for each braid and each non-zero integer ℓ. The equality is easy for irreducible braids. Indeed, if the class $\widehat{\mathfrak{m}_b}$ is elliptic, the class $\widehat{\mathfrak{m}_{b^\ell}}$ is so and both sides are zero. If the class $\widehat{\mathfrak{m}_b}$ is pseudo-Anosov and $\tilde{\varphi}$ is a pseudo-Anosov representative then $\tilde{\varphi}^\ell$ is a pseudo-Anosov representative of $\widehat{\mathfrak{m}_{b^\ell}}$. The equality follows from

$$h(\tilde{\varphi}^\ell) = |\ell|\, h(\tilde{\varphi})\,.$$

(See [1], Theorem 2 or Sect. 2.1).

If b is reducible one has to use Theorem 5.2 and equality (5.71) for each irreducible braid component. □

Problem 5.1 *Define irreducible nodal components of any mapping class on a Riemann surface of genus g with m punctures, and prove the analogue of Theorem 5.1.*

Chapter 6
The Conformal Module and Holomorphic Families of Polynomials

In this Chapter we discuss the first result [33] which uses the concept of the conformal module of conjugacy classes of braids. The paper [33] was motivated by the interest of the authors in Hilbert's Thirteen's Problem and treats global reducibility of holomorphic families of polynomials. In Sect. 6.1 we give a historical account, in Sect. 6.2 we give a conceptual alternative proof of a slightly improved version of the result of [33].

6.1 Historical Remarks

The conformal module of braids and first applications of this concept to families of polynomials, depending holomorphically on a parameter, appeared in a line of research that was initiated by the 13th Hilbert problem, posed by Hilbert in his famous talk on the International Congress of Mathematicians 1900. Hilbert asked whether each holomorphic function in three variables can be written as finite composition of continuous functions of two variables. The question was answered affirmatively by Kolmogorov and Arnold. In the Proceedings of the Symposium [17] devoted to "Mathematical developments arising from Hilbert Problems", the problem is commented as follows.

> "Hilbert posed this question especially in connection with the solution of a general algebraic equation of degree 7. It is reasonable to presume that he formulated it in terms of continuous functions partly because he had an interest in nomography and partly because he expected a negative answer. Now that it is settled affirmatively, one can ask an equally fundamental, and perhaps more interesting, question with algebraic functions instead of continuous functions."

B. Jöricke, *Braids, Conformal Module, Entropy, and Gromov's Oka Principle*, Lecture Notes in Mathematics 2360, https://doi.org/10.1007/978-3-031-67288-0_6

This latter question goes back to mathematics of the seventeenth century. Consider an algebraic equation of degree n.

$$z^n + a_{n-1} z^{n-1} + \ldots + a_0 = 0 . \tag{6.1}$$

Put $a = (a_0, \ldots, a_{n-1}) \in \mathbb{C}^n$ and consider

$$P(z, a) = z^n + a_{n-1} z^{n-1} + \ldots + a_0$$

as a monic polynomial of degree n whose coefficients are polynomials in a. P is also called an algebraic function of degree n on $\mathbb{C}^n$. Recall that $\overline{\mathfrak{P}}_n$ is the space of monic polynomials of degree n (maybe, with multiple zeros). The adjective monic refers to the property that the coefficient by the highest order of the variable equals 1. Parameterizing the space $\overline{\mathfrak{P}}_n$ by unordered n-tuples of zeros of polynomials, we get a map which assigns to each $a \in \mathbb{C}^n$ the n-tuple $\{z_1(a), \ldots, z_n(a)\}$ of solutions of the equation $P(z, a) = 0$, i.e. what is called an "n-valued algebraic function".

Tschirnhaus suggested to "substitute" the unknown variable z in (6.1) by an algebraic function of a new variable w. In other words, put

$$w = z^k + b_{k-1} z^{k-1} + \ldots + b_0 , \tag{6.2}$$

and let $b = (b_0, \ldots, b_{k-1}) \in \mathbb{C}^k$. To eliminate z from the two Eqs. (6.1) and (6.2) consider P and Q, $Q(z, w) \overset{\text{def}}{=} z^k + b_{k-1} z^{k-1} + \ldots + b_0 - w$, as polynomials in z with coefficients being polynomials in w. They are relatively prime. The resultant R is a polynomial in w (and in the coefficients a and b). It can be written as

$$R = pP + qQ$$

where p and q are polynomials in z and w (and in the coefficients $a \in \mathbb{C}^n, b \in \mathbb{C}^k$). If z satisfies (6.1) and (6.2) then $R(w) = 0$. Vice versa, if $Q(z, w) = 0$ and $R(z) = 0$, then $P(z) = 0$. This allows to think about solutions of (6.1) as part of zeros of a "composition of algebraic functions" determined by the equations $Q(z, w) = 0$, respectively by $R(w) = 0$. The method consists now in choosing the coefficients b depending on a so that the "algebraic functions" depend on a smaller number of variables. This method brings the general equation of degree 7 to the equation

$$z^7 + a\, z^3 + b\, z^2 + c\, z + 1 = 0 . \tag{6.3}$$

The 7-tuple of solutions of (6.3) is an algebraic function of degree 7 depending on three complex variables. Hilbert was interested in the "complexity" of the general 7-valued algebraic function, in other words, he wanted to know how far one can go in this process to get formulas beyond radicals.

There are two difficulties. First, the "function-composition" may have more zeros than P, and, secondly, the "algebraic functions" obtained by choosing the coefficients b are actually determined by polynomials in one variable with coefficients being rational functions in several complex variables, that may have indeterminacy sets. The composition problem was rigorously formulated for entire algebraic functions (i.e. for polynomials in one variable with coefficients being entire functions of complex variables—the restricted composition problem) and was considered first for the case when the "function-composition" coincides with the original function (the faithful composition problem). Arnold's interest in the topological invariants of the space $\mathfrak{P}_n$ of monic polynomials of degree n without multiple zeros was motivated by finding cohomological obstructions for the restricted faithful composition problem. The restricted composition problem is now answered negatively in a series of papers by several authors. The last step was done by Lin (see [61]). But the original question allowing polynomials in one variable with coefficients being rational functions is widely open. Compare also with the more classical account [60]. See also the papers [82, 83].

On the other hand, $\mathfrak{P}_n$ is a complex manifold, in fact, the complement of the algebraic hypersurface $\{\mathsf{D}_n = 0\}$ in complex Euclidean space $\mathbb{C}^n$. Here D_n denotes the discriminant of polynomials in $\mathfrak{P}_n$. Recall that the function D_n is a polynomial in the coefficients of elements of $\mathfrak{P}_n$. While, in connection with his interest in the Thirteen's Hilbert Problem, Arnold studied the topological invariants of the space $\mathfrak{P}_n$, the conformal invariants of $\mathfrak{P}_n$ (i.e. invariants that are preserved under biholomorphic mappings) are of interest in connection with Gromov's Oka Principle. The collection of conformal modules of conjugacy classes of elements of the fundamental group of the space of monic polynomials of degree n without multiple zeros is a collection of conformal invariants of $\mathfrak{P}_n \cong \mathbb{C}_n \setminus \mathsf{D}_n$. Notice, that for any complex manifold the conformal module of conjugacy classes of its fundamental group can be defined. The collection of conformal modules of all conjugacy classes is a biholomorphic invariant of the manifold. We will focus here on the respective invariants of the space $\mathfrak{P}_n$, in other words, on the conformal module of conjugacy classes of braids.

In whatever sense one understands "algebraic functions", for each algebraic function there is a Zariski open set in $\mathbb{C}^n$ such that the restriction of the algebraic function to this set is a separable algebroid function. The restriction of the algebraic function to each loop in this set defines a conjugacy class of braids. Moreover, this conjugacy class is represented by a holomorphic mapping of an annulus into the space of polynomials—any annulus which is mapped holomorphically into the Zariski open set and represents the homotopy class of the loop may serve. We do not know at the moment whether further progress related to the concept of the conformal module of braids may have some impact on the open problems related to the 13th Hilbert problem or stimulated by it. I am grateful to M. Zaidenberg who asked me this question.

6.2 Families of Polynomials, Solvability and Reducibility

The concept of the conformal module of conjugacy classes of braids appeared (without name) in the paper [33]. The paper was motivated by the interest of the authors in Hilbert's Thirteen's Problem and treats global reducibility of holomorphic families of polynomials.

The following objects related to $\mathfrak{P}_n$ have been considered in this connection. A continuous mapping from a topological space X into the set of monic polynomials of fixed degree $\overline{\mathfrak{P}_n}$ (maybe, with multiple zeros) is a quasipolynomial. It can be written as a function in two variables $x \in X$, $\zeta \in \mathbb{C}$: $f(x,\zeta) = a_0(x) + a_1(x)\zeta + ... + a_{n-1}(x)\zeta^{n-1} + \zeta^n$, for continuous functions a_j, $j = 1, ..., n$, on X. If X is a complex manifold and the mapping is holomorphic it is called an algebroid function. If the image of the map is contained in the space $\mathfrak{P}_n$ of monic polynomials of degree n without multiple zeros, it is called separable. A separable quasipolynomial is called solvable if it can be globally written as a product of quasipolynomials of degree 1. It is called irreducible if it can not be written as product of two quasipolynomials of positive degree, and reducible otherwise. We also call a solvable quasipolynomial a globally solvable family of polynomials, and an irreducible quasipolynomial a globally irreducible family of polynomials. Two separable quasipolynomials are isotopic if there is a continuous family of separable quasipolynomials joining them. An algebroid function on $\mathbb{C}^n$ whose coefficients are polynomials is called an algebraic function. In this terminology the result of the paper [33] concerns reducibility of holomorphic quasipolynomials.

The space $\mathfrak{P}_n$ of monic polynomials of degree n without multiple zeros coincides with the symmetrized configuration space $C_n(\mathbb{C})\diagup\mathcal{S}_n$. The space $\mathfrak{P}_n$ can be parameterized either by the coefficients or by the unordered tuple of zeros of polynomials.

A separable quasipolynomial of degree n on a connected closed Riemann surface or a connected bordered Riemann surface X is a continuous mapping from X into the symmetrized configuration space $C_n(\mathbb{C})\diagup\mathcal{S}_n$. The restriction of the quasipolynomial to a loop in X with a base point x_0 is a geometric braid. Hence, the monodromy of the mapping to $C_n(\mathbb{C})\diagup\mathcal{S}_n$ along each element of the fundamental group $\pi_1(X, q_0)$ of X is a braid. Theorem 1.1 in this situation can be stated as follows (see also [36, 76])

The (free) isotopy classes of separable quasipolynomials of degree n on X are in one-to-one correspondence to conjugacy classes of homomorphisms from $\pi_1(X, q_0)$ into $\mathcal{B}_n$.

We start with a simple Lemma on global solvability of holomorphic families of polynomials.

Lemma 6.1 *Let X be a closed Riemann surface of positive genus with a closed smoothly bounded topological disc removed. Suppose f is an irreducible separable algebroid function of degree 3 on X. Suppose X contains a domain A, one of whose boundary components coincides with the boundary circle of X, such that*

A is conformally equivalent to an annulus of conformal module strictly larger than $\frac{\pi}{2}\,(\log(\frac{3+\sqrt{5}}{2}))^{-1}$. *Then f is solvable over A.*

We postpone the proof of Lemma 6.1, and state some simple lemmas which will be needed in the sequel. Let A be an annulus equipped with an orientation (called positive) of simple dividing closed curves. (A simple closed curve is dividing if its complement is not connected.) Let γ be a positively oriented simple closed dividing curve in A. For a continuous mapping f from an annulus A to the symmetrized configuration space $C_n(\mathbb{C})\diagup\mathcal{S}_n$ the restriction $f \mid \gamma$ represents a conjugacy class of braids which is denoted by $\widehat{b}_{f,A}$. (It does not depend on the choice of the positively oriented simple closed dividing curve.)

Lemma 6.2 *Suppose the separable quasipolynomial f of degree n on an annulus A is irreducible and n is prime. Then the induced conjugacy class of braids $\widehat{b}_{f,A}$ is irreducible.*

Notice that, on the other hand, conjugacy classes of irreducible pure braids define solvable, hence reducible, quasipolynomials on the circle.

Proof Recall that $\tau_n : \mathcal{B}_n \to \mathcal{S}_n$ is the natural projection from the braid group to the symmetric group. τ_n maps conjugacy classes of braids to conjugacy classes of permutations. If f is irreducible the conjugacy class of braids $\widehat{b}_{f,A}$ projects to a conjugacy class $\tau_n(\widehat{b}_{f,A})$ of n-cycles. The lemma is now a consequence of the following known lemma (see e.g. [24]). □

Lemma 6.3 *If n is prime then any braid $b \in \mathcal{B}_n$, for which $\tau_n(b)$ is an n-cycle, is irreducible.*

For convenience of the reader we give the short argument.

Proof of Lemma 6.3 If b was reducible then a homeomorphism φ which represents the mapping class corresponding to b would fix setwise an admissible system of curves $\mathcal{C}$. Let C_1 be one of the curves in $\mathcal{C}$ and let δ_1 be the topological disc contained in $\mathbb{D}$, bounded by C_1. δ_1 contains at least two distinguished points z_1 and z_2. Since φ permutes the distinguished points along an n-cycle there is a power φ^k of φ which maps z_1 to z_2. Since n is prime, φ^k also permutes the distinguished points along an n-cycle. Hence it maps some distinguished point in δ_1 to the complement of δ_1. We obtained that $\varphi^k(\delta_1)$ intersects both, δ_1 and its complement. Hence $\varphi^k(C_1) \neq C_1$ but $\varphi^k(C_1)$ intersects C_1. This contradicts the fact that the system of curves $\mathcal{C}$ was admissible and invariant under φ. □

Recall that $\Delta_3 = \sigma_1\sigma_2\sigma_1 = \sigma_2\sigma_1\sigma_2$ and the group $\langle\Delta^2\rangle$ generated by Δ^2 coincides with the center $\mathcal{Z}_3$ of the braid group $\mathcal{B}_3$. We need the following lemma.

Lemma 6.4 *A mapping class in $\mathfrak{M}(\overline{\mathbb{D}};\,\partial\mathbb{D}, E_3)$ is reducible if and only if the braid corresponding to it is conjugate to $\sigma_1^k\Delta_3^{2\ell}$ for integers k and ℓ.*

A mapping class in $\mathfrak{M}(\mathbb{P}^1;\,\{\infty\}, E_3)$ is reducible if and only if it corresponds to a conjugate of $\sigma_1^k\diagup\mathcal{Z}_3$.

Proof Any admissible system of curves in $\overline{\mathbb{D}}$ with three distinguished points consists of one curve that divides $\overline{\mathbb{D}}$ into two connected components. One of these components contains $\partial\mathbb{D}$ and one of the distinguished points, the other one contains two distinguished points. For a reducible mapping class $\mathfrak{m}$ in $\mathfrak{M}(\overline{\mathbb{D}}; \partial\mathbb{D}, E_3)$ we take an admissible curve γ that reduces $\mathfrak{m}$ and a representing homeomorphism $\varphi \in \mathfrak{m}$ that fixes γ pointwise. Suppose the points ζ_1, ζ_2, and ζ_3 in E_3 are labelled by using a homomorphism from $\pi_1(C_3(\mathbb{C})/\mathcal{S}_3, E_3)$ to $\pi_1(C_3(\mathbb{C})/\mathcal{S}_3, E_3^0)$ (which is defined up to conjugation) so that one of the connected components $\mathcal{C}_1$ of the complement of γ contains ζ_1 and ζ_2, the other connected component $\mathcal{C}_2$ contains ζ_3 and $\partial\mathbb{D}$. Since φ fixes $\partial\mathbb{D}$ and γ pointwise, it also fixes the annulus $\overline{\mathcal{C}}_2$ setwise and fixes the point ζ_3. A self-homeomorphism of an annulus that fixes the boundary pointwise is a twist. After an isotopy of φ on $\mathbb{D}$ that fixes γ setwise and fixes each distinguished point we may assume that φ fixes pointwise a simple arc that joins ζ_3 with γ. Then the restriction $\varphi \mid \overline{\mathcal{C}}_2$ represents a power of a Dehn twist in $\mathbb{D}$ about a curve in $\mathcal{C}_2$ that is homotopic in $\overline{\mathcal{C}}_2 \setminus \{\zeta_3\}$ to $\partial\mathbb{D}$. Such a Dehn twist corresponds to the braid Δ_3^2. The restriction $\varphi \mid \overline{\mathcal{C}}_1$ represents the class in $\mathfrak{M}(\overline{\mathcal{C}}_1; \partial\mathcal{C}_1, \{\zeta_1, \zeta_2\})$, corresponding to σ^k for some integer k. Hence the class represented by φ in $\mathfrak{M}(\overline{\mathbb{D}}; \partial\mathbb{D}, E_3)$ corresponds to a braid that is conjugate to $\sigma_1^k \Delta_3^{2\ell}$ for some integer ℓ. Vice versa, it is clear that the mapping class $\mathfrak{m}_{\sigma_1^{2k}\Delta_3^{2\ell}}$ is reducible.

The statement concerning the reducible elements of $\mathfrak{M}(\mathbb{P}^1; \{\infty\}, E_3)$ is proved in the same way. □

We want to point out that the identity in the braid group $\mathcal{B}_n$ with $n \geq 3$ is reducible, the identity in $\mathcal{B}_2$ is irreducible. Recall that by Example 3.8 in Sect. 3.2 the value $\eta = \frac{\pi}{2}\left(\log \frac{3+\sqrt{5}}{2}\right)^{-1}$ is the largest finite conformal module among irreducible 3-braids, and hence, by the Main Theorem the value $\log \frac{3+\sqrt{5}}{2}$ is the smallest non-vanishing entropy among 3-braids. For the latter fact see also [74].

Lemma 6.5 *Suppose for a conjugacy class of braids $\widehat{b} \in \widehat{\mathcal{B}_3}$ the conformal module $\mathcal{M}(\widehat{b})$ satisfies the inequality $\mathcal{M}(\widehat{b}) > \eta$. Then the following holds.*

(a) $\mathcal{M}(\widehat{b}) = \infty$.
(b) If $\widehat{b}$ is irreducible then $\widehat{b}$ is the conjugacy class of a periodic braid, i.e. either of $(\sigma_1\sigma_2)^\ell$ for an integer ℓ not divisible by 3, or of $(\sigma_1\sigma_2\sigma_1)^\ell$ for an integer ℓ not divisible by 2.
(c) If $\widehat{b}$ is reducible then $\widehat{b}$ is the conjugacy class of $\sigma_1^k\Delta_3^{2\ell}$ for integers k and ℓ. Here $\Delta_3^2 = (\sigma_1\sigma_2\sigma_1)^2 = (\sigma_1\sigma_2)^3$.
(d) If $\widehat{b}$ is the conjugacy class of a commutator, then it is represented by a pure braid b.

Proof Suppose $\widehat{b}$ is irreducible. Since $\frac{\pi}{2}\left(\log \frac{3+\sqrt{5}}{2}\right)^{-1}$ is the largest finite conformal module among irreducible 3-braids the equality $\mathcal{M}(\widehat{b}) = \infty$ holds. (Hence, also $h(\widehat{b}) = 0$.) This proves (a) in the irreducible case.

Let $b \in \mathcal{B}_3$ represent the irreducible class $\widehat{b}$, let $\mathfrak{m}_b \in \mathfrak{M}(\overline{\mathbb{D}}; \partial\mathbb{D}, E_3)$ be the mapping class corresponding to b, and let $\mathcal{H}_\infty(\mathfrak{m}_b) \in \mathfrak{M}(\mathbb{P}^1; \infty, E_3)$ be the corresponding mapping class on $\mathbb{P}^1$ with distinguished points. Since $\mathcal{M}(\widehat{b}) = \infty$, the class $\mathcal{H}_\infty(\mathfrak{m}_b)$ is represented by a periodic mapping of the complex plane $\mathbb{C}$ with three distinguished points (see Proposition 3.1, Theorem 1.10, and Theorem 1.11). By a conjugation we may assume that the periodic mapping fixes zero and hence is a rotation by a root of unity. If zero is not a distinguished point, the three distinguished points are equidistributed on a circle with center zero and the mapping is rotation by a power of $e^{\frac{2\pi i}{3}}$. The mapping representing $\mathcal{H}_\infty(\mathfrak{m}_b)$ corresponds to $\widehat{(\sigma_1\,\sigma_2)^\ell/\langle\Delta_3^2\rangle}$ and the mapping representing $\mathfrak{m}_b$ corresponds to $\widehat{(\sigma_1\,\sigma_2)^\ell}$ for an integer number ℓ.

If zero is a distinguished point, the other two distinguished points are equidistributed on a circle with center zero. The mapping representing $\mathcal{H}_\infty(\mathfrak{m}_b)$is a rotation, precisely, a multiplication by a power of -1, and corresponds to $\widehat{(\sigma_1\,\sigma_2\,\sigma_1)^\ell/\langle\Delta_3^2\rangle}$ for an integer number ℓ. (Note that powers of Δ_3^2 are reducible and the mapping class in $\mathfrak{M}(\mathbb{P}^1; \{\infty\}, E_3)$ corresponding to Δ_3^2 is the identity.) The mapping representing $\mathfrak{m}_b$ corresponds to $\widehat{(\sigma_1\,\sigma_2\,\sigma_1)^\ell}$. We proved (b).

Suppose $\widehat{b}$ is reducible, i.e. the mapping class $\mathfrak{m}_b$ is reducible for $b \in \widehat{b}$. Then by Lemma 6.4 the class $\widehat{b}$ is the conjugacy class of $\sigma_1^k\Delta_3^{2\ell}$ for integers $k \neq 0$ and ℓ. We obtained (c). The braid $b = \sigma_1^k\,\Delta_3^{2\ell}$ has infinite conformal module. This gives (a) in the reducible case.

Suppose $\widehat{b}$ is the conjugacy class of a commutator $b \in \mathcal{B}_3$. Then b has exponent sum zero (more detailed, the sum of exponents of the generators in a representing word equals zero). If $\mathcal{M}(\widehat{b}) = \infty$ then this is possible only if b is conjugate to $\sigma_1^k\,\Delta_3^{2\ell}$ with $k + 6\ell = 0$. Hence, k is even and the braid b is pure. This proves (d). The lemma is proved. □

Proof of Lemma 6.1 Let $\mathfrak{S}_f = \{(z,\zeta) \in X \times \mathbb{C} : f(z,\zeta) = 0\}$ be the zero set of the algebroid function f. Since f is irreducible, this set is connected. We obtain an unramified covering $\mathfrak{S}_f \to X$ of degree 3. Hence, for the Euler characteristics $\chi(X)$ and $\chi(\mathfrak{S}_f)$ the relation $\chi(\mathfrak{S}_f) = 3\chi(X)$ holds. Let $m(X)$ and $m(\mathfrak{S}_f)$ be the number of boundary components of X and $\mathfrak{S}_f$. Then $2 - 2g(\mathfrak{S}_f) - m(\mathfrak{S}_f) = 3(2 - 2g(X) - m(X))$. Hence, since $m(X) = 1$, $m(\mathfrak{S}_f)$ equals either 1 or 3. This means that $f \mid A$ is either irreducible or solvable. If $f \mid A$ is irreducible, it represents an irreducible conjugacy class of braids $\widehat{b}_{f,A}$. By Lemma 6.5 $\widehat{b}_{f,A}$ must be the conjugacy class of a periodic braid which corresponds to a 3-cycle. This is impossible for conjugacy classes of products of braid commutators. □

We will discuss in more detail the result of Gorin, Lin and Zjuzin ([33, 84]) and give an alternative proof of a slight improvement. For simplicity we will restrict ourselves to the case when the degree of the quasipolynomial is a prime number. (The general case has been considered by Lin and later by Zjuzin. An alternative proof of the general case can be given as well.) In [33] Gorin and Lin proved that for each prime number n there exists a number r_n such that any separable algebroid function of degree n on an annulus A of conformal module strictly larger than r_n is reducible provided the index of its discriminant $z \to \mathsf{D}_n(f_z)$, $z \in A$, is divisible by n. Recall that the value at a point $z \in A$ of a separable algebroid function of degree n on A is a polynomial $f_z \in \mathfrak{P}_n$. $\mathsf{D}_n(f_z)$ denotes its discriminant. Recall that D_n is a function on the space of all polynomials $\overline{\mathfrak{P}}_n$ of degree n which vanishes exactly on the set of polynomials with multiple zeros. Explicitly, for a polynomial $\mathsf{p} \in \overline{\mathfrak{P}}_n$, $\mathsf{p}(\zeta) = \prod_{j=1}^{n} (\zeta - \zeta_j)$, its discriminant equals

$$\mathsf{D}_n(\mathsf{p}) = \prod_{i<j} (\zeta_i - \zeta_j)^2 . \tag{6.4}$$

The index of the mapping

$$A \ni z \to \mathsf{D}_n(f_z) \in \mathbb{C} \setminus \{0\}, \quad z \in A,$$

is the degree of the map

$$z \to \frac{\mathsf{D}_n(f_z)}{|\mathsf{D}_n(f_z)|}$$

from $\{|z| = 1\}$ to itself. Zjuzin [84] proved that one can take $r_n = n \cdot \rho_0$ for an absolute constant ρ_0, and Petunin showed that $\rho_0 = 10^7$ works. We give a short conceptual proof of the theorem of these authors with an improved constant.

Theorem 6.1 *Suppose n is a prime number. If $A = \{z \in \mathbb{C} : \frac{1}{\sqrt{r}} < |z| < \sqrt{r}\}$ is an annulus of conformal module $m(A) > \frac{2\pi}{\log 2} n$, and f is a separable algebroid function on A of degree n such that the index of its discriminant is divisible by n, then f is reducible.*

The present proof of the theorem uses the Main Theorem (see also Theorem 1 of [43]) and the following result of Penner [70] on the smallest non-vanishing entropy among irreducible n-braids.

Theorem 6.2 (Penner) *Denote by $h_{\mathsf{g}}^{\mathsf{m}}$ the smallest non-vanishing entropy among irreducible self-homeomorphisms of Riemann surfaces of genus g with m distinguished points ($2\mathsf{g} - 2 + \mathsf{m} > 0$). Then*

$$h_{\mathsf{g}}^{\mathsf{m}} \geq \frac{\log 2}{12\mathsf{g} - 12 + 4\mathsf{m}} .$$

Penner's theorem implies that the smallest non-vanishing entropy among irreducible n-braids, $n \geq 3$, is bounded from below by

$$\frac{\log 2}{4(n+1)-12} = \frac{\log 2}{4n-8} \geq \frac{\log 2}{4} \cdot \frac{1}{n}, \qquad n \geq 3 .$$

By the Main Theorem the largest finite cònformal module among irreducible conjugacy classes of n-braid does not exceed $\frac{\pi}{2} \frac{4}{\log 2} \cdot n$ for $n \geq 3$.

Proof of Theorem 6.1 Suppose f is an irreducible separable algebroid function on the annulus A and

$$m(A) > \frac{\pi}{2} \frac{4}{\log 2} \cdot n = \frac{2\pi}{\log 2} \cdot n .$$

By Lemma 6.2 the conjugacy class $\widehat{b}_{f,A} \in \widehat{\mathcal{B}}_3$ induced by f is irreducible. By the Main Theorem the entropy $h(\widehat{b}_{f,A})$ of the conjugacy class of braids $\widehat{b}_{f,A} \in \widehat{\mathcal{B}}_n$ induced by f is strictly smaller than $\frac{\log 2}{4} \cdot \frac{1}{n}$. By Penner's Theorem $h(\widehat{b}_{f,A}) = 0$, i.e. $\mathcal{M}(\widehat{b}_{f,A}) = \infty$.

This implies that $\widehat{b}_{f,A}$ is the conjugacy class of a periodic braid corresponding to an n-cycle, hence it is the conjugacy class of $(\sigma_1 \sigma_2 \ldots \sigma_n)^k$ for an integer k which is not divisible by n. The isotopy class of the algebroid function $\tilde{f}(z, \zeta) = \zeta^n - z^k$, $z \in A$, induces this conjugacy class of braids. Indeed, for $z = e^{2\pi i t}$, $t \in [0, 1]$, the set of solutions of the equation $\tilde{f}(z, \zeta) = 0$ is

$$E_n(t) \overset{def}{=} \left\{ e^{\frac{2\pi i k}{n} t}, e^{\frac{2\pi i k}{n} + \frac{2\pi i k}{n} t}, \ldots, e^{\frac{2\pi i k (n-1)}{n} + \frac{2\pi i k}{n} t} \right\}, \qquad t \in [0, 1] .$$

The path $t \to E_n(t)$, $t \in [0, 1]$, in $\mathfrak{P}_n$ defines a geometric braid in the conjugacy class of $(\sigma_1 \cdot \sigma_2 \cdot \ldots \cdot \sigma_n)^k$.

Compute the discriminant $\mathsf{D}_n(\tilde{f}_z)$:

$$\begin{aligned} \mathsf{D}_n(\tilde{f}_z) &= \prod_{0 \leq m < \ell < n} \left(e^{\frac{2\pi i k}{n} \cdot m + \frac{2\pi i k}{n} \cdot t} - e^{\frac{2\pi i k}{n} \cdot \ell + \frac{2\pi i k}{n} \cdot t} \right)^2 \\ &= e^{\frac{2\pi i k}{n} \cdot t \cdot n \cdot (n-1)} \cdot \prod_{0 \leq m < \ell < n} \left(e^{\frac{2\pi i k}{n} m} - e^{\frac{2\pi i k}{n} \ell} \right)^2 . \end{aligned} \tag{6.5}$$

Hence

$$\mathsf{D}_n(\tilde{f}_z) = e^{2\pi i k \cdot (n-1) \cdot t} \cdot c_n , \quad z = e^{2\pi i t} , \quad t \in [0, 1] ,$$

where c_n is a non-zero constant depending only on n. (It is equal to the product in the last expression of (6.5).) The index of the discriminant equals $k \cdot (n-1)$ which is not divisible by n. We saw that the condition for the discriminant excludes the only possibility for a separable algebroid function on annuli of the given conformal module to be irreducible. Hence under the conditions of Theorem 6.1 the algebroid function f must be reducible. □

Chapter 7
Gromov's Oka Principle and Conformal Module

We study the negating side of the Gromov-Oka Principle, namely the failure and limited validity of this important principle. The obstructions to the principle are based on the relation between conformal invariants of the source and the target. The conformal invariant used in this chapter is the collection of conformal modules of conjugacy classes of elements of the fundamental group of the manifold.

We call a mapping $f : X \to Y$ from a finite open oriented smooth surface to a complex manifold Y a Gromov-Oka mapping if for each orientation preserving conformal structure on X with only thick ends the mapping is homotopic to a holomorphic mapping.

We will confirm Gromov's prediction, that mappings from annuli into a complex manifold play a special role for understanding the homotopy problem of continuous mappings to holomorphic mappings. More precisely, it will be proved, that there are finitely many annuli in an open oriented smooth surface X of finite positive genus, that may be chosen effectively, such that a continuous mapping to the twice punctured complex plane is a Gromov-Oka mapping, if and only if its restriction to each of the annuli is a Gromov-Oka mapping. This allows a complete description of the Gromov-Oka mappings from finite open oriented smooth surfaces to the twice punctured Riemann surface. Similarly, the Gromov-Oka mappings from a smooth oriented surface of genus one with one hole into $\mathfrak{P}_3$ will be completely described.

7.1 Gromov's Oka Principle and the Conformal Module of Conjugacy Classes of Elements of the Fundamental Group

Gromov [35] formulated his Oka Principle as "an expression of an optimistic expectation with regard to the validity of the h-principle for holomorphic maps in

B. Jöricke, *Braids, Conformal Module, Entropy, and Gromov's Oka Principle*, Lecture Notes in Mathematics 2360, https://doi.org/10.1007/978-3-031-67288-0_7

the situation when the source manifold is Stein". Holomorphic maps $X \to Y$ from a complex manifold X to a complex manifold Y are said to satisfy the h-principle if each continuous map from X to Y is homotopic to a holomorphic map. We call a target manifold Y a Gromov-Oka manifold if the h-principle holds for holomorphic maps from any Stein manifold to Y. Gromov [35] gave a sufficient condition on a complex manifold Y to be a Gromov-Oka manifold.

The question of understanding Gromov-Oka manifolds received a lot of attention. It turned out to be fruitful to strengthen the requirement on the target Y by combining the h-principle for holomorphic maps with a Runge type approximation property. Manifolds Y satisfying the stronger requirement are called Oka manifolds. It has been proved by Forstneric that a manifold is an Oka manifold iff Runge approximation holds for holomorphic mappings from neighbourhoods of compact convex subsets of $\mathbb{C}^n$ into Y (see [28]). For more details, examples of Oka manifolds and an account on modern development of Oka theory based on Oka manifolds see [29]. Notice that among the connected Riemann surfaces exactly the non-hyperbolic ones, i.e. the Riemann sphere $\mathbb{P}^1$, the complex plane $\mathbb{C}$, the punctured complex plane $\mathbb{C}^*$, and the tori are Oka manifolds.

On the other hand, in the same paper [35] Gromov mentioned the case of maps from annuli to the twice punctured complex plane as the simplest interesting example where the h-principle for holomorphic mappings fails. He proposed the conformal invariants of the twice complex plane which are called here the conformal modules of conjugacy classes of elements of the fundamental group and ascribed to them a role in capturing certain "conformal rigidity" of the twice punctured complex plane. He supposed that for a complex manifold Y, which is not a Gromov-Oka manifold, the holomorphic mappings from annuli to Y play a special role for understanding obstructions to his Oka Principle.

Recall the definition of these invariants that were introduced in Chap. 3 for general complex manifolds. Let $\mathcal{X}$ be a topological space, and $\hat{e}$ a conjugacy class of elements of the fundamental group of $\mathcal{X}$. $\hat{e}$ can be interpreted as a free homotopy class of loops in $\mathcal{X}$. A continuous map g from an annulus $A = \{z \in \mathbb{C} : r < |z| < R\}$ into $\mathcal{X}$ is said to represent the conjugacy class $\hat{e}$ if for some (and hence for any) $\rho \in (r, R)$ the map $g : \{|z| = \rho\} \to \mathcal{X}$ represents $\hat{e}$. If the mapping is a homeomorphism we will speak about an annulus contained in $\mathcal{X}$ and representing $\hat{e}$. If $\mathcal{X}$ is a complex manifold and the mapping g is holomorphic, we will speak about a holomorphic annulus representing $\hat{e}$.

Definition 7.1 *For any complex manifold $\mathcal{X}$ and any conjugacy class $\hat{e}$ of elements of the fundamental group of $\mathcal{X}$ the conformal module $\mathcal{M}(\hat{e})$ of $\hat{e}$ is defined as the supremum of the conformal modules of annuli $A = \{z \in \mathbb{C} : r < |z| < R\}$ that admit a holomorphic mapping into $\mathcal{X}$ representing $\hat{e}$.*

Obstructions to Gromov's Oka Principle are based on the relation between conformal invariants of the source and the target. Let X and Y be topological spaces. For an element $e \in \pi_1(X, q_0)$ we let $\hat{e}$ be its free homotopy class. Further, for a continuous mapping $f : X \to Y$ we let $f_* : \pi_1(X, q_0) \to \pi_1(Y, f(q_0)$ be the induced homomorphism and $\widehat{f_*(e)}$ the free homotopy class of $f_*(e)$. The following observation provides an obstruction to Gromov's Oka Principle.

Let $f : X \to Y$ be a holomorphic mapping from a complex manifold X to a complex manifold Y. Then $\mathcal{M}(\widehat{f_*(e)}) \geq \mathcal{M}(\hat{e})$ for each free homotopy class $\hat{e}$ of loops in X. Indeed, if a holomorphic mapping $g : A \to X$ from an annulus A to X represents $\hat{e}$, then the composition $f \circ g : A \to Y$ represents $\widehat{f_*(e)}$. Hence, $\mathcal{M}(\widehat{f_*(e)}) \geq m(A)$. Taking annuli of conformal module close to $\mathcal{M}(\hat{e})$, we obtain the statement.

If we know effective estimates of conformal modules of free homotopy classes of loops in the source and in the target, we get effective information on the obstructions.

Here and in the next chapters we will study the following basic problem.

Problem 7.1 *Study obstructions for Gromov' Oka Principle. In particular, look for effective estimates of the conformal modules of conjugacy classes of elements of the fundamental group of complex manifolds, if possible, in algebraic terms related to the fundamental group.*

If the source space is an annulus A, the conformal module of conjugacy classes of elements of the target Y almost describe, which continuous mappings $A \to X$ are homotopic to holomorphic mappings. Namely, with a generator e of the fundamental group of A with some base point chosen, f is homotopic to a holomorphic mapping, if $m(A) < \mathcal{M}(\widehat{f_*(e)})$ and is not homotopic to a holomorphic mapping if $m(A) > \mathcal{M}(\widehat{f_*(e)})$. The case $m(A) = \mathcal{M}(\widehat{f_*(e)})$ is more subtle. For the case when the target is the twice punctured complex plane effective estimates of the conformal modules of conjugacy classes of elements of the fundamental group will be given in Chap. 9, so that the obstructions to Gromov's Oka Principle can be made effective in this case.

Consider the target $Y = \mathfrak{P}_n$. Recall that $\mathfrak{P}_n \cong C_n(\mathbb{C})\diagup \mathcal{S}_n$ is the space of monic polynomials without multiple zeros, parameterized either by the coefficients or by the unordered tuple of the zeros of the polynomials (see Sect. 1.3 and Chap. 6). Note that $\mathfrak{P}_n = \mathbb{C}^n \setminus \{\mathsf{D}_n = 0\}$ is the complement of an algebraic hypersurface in $\mathbb{C}^n$. Here D_n is the discriminant (see also equality (6.4)). A continuous mapping $f : A \to \mathfrak{P}_n$ represents a conjugacy class $\widehat{b_{f,A}}$ of braids. For $n = 3$ there are conjugacy classes of braids that cannot be represented by a holomorphic mapping from an annulus of conformal module bigger than $\frac{\pi}{2}\left(\log \frac{3+\sqrt{5}}{2}\right)^{-1}$ (see Lemma 6.5). The respective fact with another constant is true for any $n > 3$. Hence, $\mathfrak{P}_n$ is not a Gromov-Oka manifold for $n \geq 3$.

On the other hand, any continuous map from an oriented connected open smooth surface X into $\mathfrak{P}_2$ has the Gromov-Oka property, moreover, any such map is homotopic to a holomorphic map for any conformal structure on X. This statement can be proved in the same way as Statement (2) of Theorem 8.2 below.

By the definition of the conformal module of a conjugacy class of n-braids (see Definition 7.1) a continuous mapping f from an annulus A to the space $\mathfrak{P}_n$ is homotopic to a holomorphic mapping if $m(A) < \mathcal{M}(\widehat{b_{f,A}})$ and is not homotopic to a holomorphic mapping if $m(A) > \mathcal{M}(\widehat{b}_{f,A})$. In other words, the relation of the conformal module $\mathcal{M}(\widehat{b}_{f,A})$ to the conformal invariant $m(A)$ of A "almost" determines whether f is homotopic to a holomorphic mapping. Recall that $\mathcal{M}(\widehat{b}_{f,A})$ is inversely proportional to the entropy of $\widehat{b}_{f,A}$. It is reasonable to interpret the entropy as "complexity": The complexity grows with increasing entropy, while the chance for the mapping to be homotopic to a holomorphic one decreases. For $n = 3$ there are effective upper and lower bounds for the conformal module (equivalently, for the entropy) of conjugacy classes of 3-braids that differ by multiplicative constants. We will address this topic in Chap. 9 following [44, 45].

Notice that each element of the fundamental group $\pi_1(\mathbb{C} \setminus \{-1, 1\}, 0)$ can be interpreted as a pure 3-braid and we can associate to its conjugacy class a mapping class. A multiple of the inverse of the conformal modules of a conjugacy class of elements of the fundamental group $\pi_1(\mathbb{C} \setminus \{-1, 1\}, 0)$ (equivalently a multiple of its extremal length) is equal to the entropy of the associated mapping class.

The problem of understanding obstructions to Gromov's Oka Principle is interesting for other target spaces Y as well. In this respect the following problem occurs.

Problem 7.2 *Study the Gromov-Oka Principle in case the target is the complement of an arbitrary algebraic hypersurface of $\mathbb{C}^n$ or the quotient of the n-dimensional round complex ball by a subgroup of its automorphism group which acts freely and properly discontinuously. Study the conformal module of conjugacy classes of elements of the fundamental group of such spaces.*

Let Y be a complex manifold. The collection of conformal modules of all conjugacy classes of elements of its fundamental group $\pi_1(Y, y_0)$ is a biholomorphic invariant of Y, which is expected to play a special role for understanding the Gromov-Oka Principle. In the case of the space $\mathfrak{P}_n$ this may be opposed to the fact, that in connection with his interest in the Thirteen's Hilbert Problem Arnold [7] considered the topological (cohomological) invariants of $\mathfrak{P}_n$.

The facts known for mappings from annuli to $\mathfrak{P}_n$ motivate the following two problems related to the restricted validity of Gromov's Oka Principle in general.

Problem 7.3 *Fix a connected open (i.e. non-compact) complex manifold X and a complex manifold Y. Obtain information about the set of continuous or smooth mappings $X \to Y$ that are homotopic to holomorphic mappings.*

For a smooth manifold X an orientation preserving homeomorphism $\omega : X \to \omega(X)$ from X to a complex manifold $\omega(X)$ is called a complex structure on X. If X is a smooth surface and $\omega(X)$ is a Riemann surface, we will also call ω a conformal structure on X. A mapping f from X to a complex manifold Y is said to be holomorphic for the complex structure ω if the mapping $f \circ \omega^{-1}$ is holomorphic on $\omega(X)$.

The following statement is known as "soft Oka Principle" (see [30]).

Soft Oka Principle *Any continuous mapping from a Stein manifold X to a complex manifold Y gets holomorphic after changing both, the mapping and the complex structure of X, by a homotopy.* (Forstneric, Slapar)

If X is a finite open Riemann surface and $Y = \mathbb{C} \setminus \{-1, 1\}$, the soft Oka principle is just Runge's approximation theorem on compact subsets of Riemann surfaces. Indeed, take any continuous function $f : X \to \mathbb{C} \setminus \{-1, 1\}$. Let Q be a standard bouquet of circles for X. The continuous function $f \mid Q$ can be approximated uniformly on Q by holomorphic functions on X. See [55] for Mergelyan type approximation (approximation of continuous functions on Q by holomorphic functions in a neighbourhood of Q) and [72] for Runge approximation by meromorphic functions on closed Riemann surfaces (here, approximation of holomorphic functions in a neighbourhood of Q by holomorphic functions on X). We obtain approximating holomorphic functions defined on X which do not necessarily omit -1 and 1. But their restrictions to Q are close to $f \mid Q$, hence their restrictions to Q have their image in $\mathbb{C} \setminus \{-1, 1\}$ and are homotopic to $f \mid Q$ as mappings into $\mathbb{C} \setminus \{-1, 1\}$. Hence an approximating holomorphic function F maps a neighbourhood V of Q into $\mathbb{C} \setminus \{-1, 1\}$ and $F|V$ is homotopic to $f \mid V$ through mappings into $\mathbb{C} \setminus \{-1, 1\}$. We may take V to be smoothly bounded and such that there exists a continuous family of mappings $\varphi_t : X \to X$, $t \in [0, 1]$, that map X homeomorphically onto a domain X_t in X such that φ_0 is the identity and $X_1 = V$. The homeomorphisms $\varphi_t : X \to X_t$ are considered as new complex structures on X. The mapping $F \circ \varphi_1$ is holomorphic on X for the complex structure φ_1.

Definition 7.2 *Let X be an oriented finite open smooth surface and Y a complex manifold. A continuous mapping $f : X \to Y$ is said to be homotopic to a holomorphic mapping for a conformal structure $\omega : X \to \omega(X)$ on X, if $f \circ w^{-1}$ is homotopic to a holomorphic mapping from $\omega(X)$ to Y.*

The "soft Oka Principle" motivates the following problem.

Problem 7.4 *Consider an oriented connected smooth open manifold X, a complex manifold Y, and a continuous mapping $f : X \to Y$. Obtain information on the set of orientation preserving complex structures $\omega : X \to \omega(X)$ on X such that f is homotopic to a holomorphic mapping for ω.*

We restrict ourselves to surfaces X as source manifolds. Recall the terminology introduced in Chap. 1. A surface is called finite if its fundamental group is finitely generated. Each finite open Riemann surface X is conformally equivalent to a domain (denoted again by X) on a closed Riemann surface X^c such that each connected component of the complement $X^c \setminus X$ is either a point or a closed topological disc with smooth boundary [77]. The connected components of the complement will be called holes. A Riemann surface is called of first kind, if it is closed, or it is obtained from a closed Riemann surface by removing finitely many points (called punctures). Otherwise the connected Riemann surface is called of second kind. If all holes of a finite open Riemann surface are closed topological discs, the Riemann surface is said to have only thick ends.

Definition 7.3 *Let X be a connected oriented finite open smooth surface, and let Y be a complex manifold. We will say that a continuous mapping $f : X \to Y$ has the Gromov-Oka property if it is homotopic to a holomorphic mapping for any orientation preserving conformal structure $\omega : X \to \omega(X)$ with only thick ends (i.e. with $\omega(X)$ having only thick ends).*

For short we will call mappings with the Gromov-Oka property Gromov-Oka mappings.

We look first at the Gromov-Oka property for mappings from annuli into the spaces $\mathfrak{P}_n$. Recall that if $n = 2$ then for an annulus A each continuous mapping $f : A \to \mathfrak{P}_2$ has the Gromov-Oka property, moreover, it is homotopic to a holomorphic mapping for any conformal structure on A. For the case $n = 3$ the following lemma holds.

Lemma 7.1 *A continuous mapping f from an annulus $A = \{r_1 < |z| < r_2\}$ to $\mathfrak{P}_3$ is homotopic to a holomorphic mapping for each orientation preserving conformal structure of second kind on A if and only if it has the Gromov-Oka property. This happens if and only if the inequality $\mathcal{M}(\widehat{b}_{f,A}) > \frac{\pi}{2}(\log \frac{3+\sqrt{5}}{2})^{-1}$ holds.*

Recall that $\log \frac{3+\sqrt{5}}{2}$ is the smallest non-vanishing entropy among 3-braids. Recall also that an annulus is of second kind if and only if it has at least one thick end, and a mapping $f : A \to Y$ is a Gromov-Oka mapping if it is homotopic to a holomorphic one for any conformal structure on A with only thick ends.

Recall that $\mathfrak{P}_n$ is the space of monic polynomials of degree n without multiple zeros. It can be identified with $C_n(\mathbb{C}_n \diagup \mathcal{S}_n)$ (the zero sets of such polynomials). A mapping into $\mathfrak{P}_n$ can be considered as a quasipolynomial or as mapping with values in $C_n(\mathbb{C}_n \diagup \mathcal{S}_n)$.

Proof of Lemma 7.1 If f is homotopic to a holomorphic mapping for any orientation preserving conformal structure of second kind on A or for any conformal structure with only thick ends, then $\mathcal{M}(\widehat{b}_{f,A}) = \infty$.

Suppose on the other hand that $\mathcal{M}(\widehat{b}_{f,A}) > \frac{\pi}{2}(\log(\frac{3+\sqrt{5}}{2}))^{-1}$. Then Lemma 6.5 implies that $\mathcal{M}(\widehat{b}_{f,A}) = \infty$ and, therefore, f is homotopic to a holomorphic mapping for any orientation preserving conformal structure on A of finite conformal module. There are two orientation preserving conformal structures of infinite conformal module, one conformally equivalent to the punctured disc $\mathbb{D} \setminus \{0\}$, it is of second kind, and the other conformally equivalent to the punctured complex plane, it is of first kind. By Lemma 6.5 a conjugacy class of braids of infinite conformal module is either the class of a periodic braid, or it is the class of $\sigma_1^k \Delta_3^{2\ell}$. We claim that in the case $\widehat{b}_{f,A}$ is a periodic conjugacy class of braids, there is an integer number k, such that f is isotopic on the punctured plane to the quasipolynomial $f_{1,k}(z,\zeta) = z^k - \zeta^3$, $z \in \mathbb{C} \setminus \{0\}$, $\zeta \in \mathbb{C}$, or to $f_{2,k}(z,\zeta) = \zeta(z^k - \zeta^2)$, $z \in \mathbb{C} \setminus \{0\}$, $\zeta \in \mathbb{C}$, respectively. If $\widehat{b}_{f,A}$ is the conjugacy class of σ_1^k, we take any constant $R > 1$. We claim that the mapping f is isotopic on the punctured disc to $f_{3,k}(z,\zeta) = (z^k - \zeta^2)(\zeta - R)$, $z \in \mathbb{D} \setminus \{0\}$, $\zeta \in \mathbb{C}$ for some integer number k.

Indeed, for $z = \frac{1}{2}e^{2\pi i t}, t \in [0, 1]$, the set of solutions of the equation $f_{1,k}(z, \zeta) = 0$ is

$$E_1^k(t) \overset{def}{=} 2^{-\frac{k}{3}} e^{\frac{2\pi i k}{3}t} \left\{1, e^{\frac{2\pi i k}{3}}, e^{\frac{4\pi i k}{3}}\right\}, \quad t \in [0, 1].$$

The path $t \to E_1^k(t)$, $t \in [0, 1]$, in $\mathfrak{P}_3$ defines a geometric braid in the conjugacy class of $(\sigma_1 \cdot \sigma_2)^k$.

The set of solutions of the equation $f_{2,k}(z, \zeta) = 0$, $z = \frac{1}{2}e^{2\pi i t}, t \in [0, 1]$, is

$$E_2^k(t) \overset{def}{=} 2^{-\frac{k}{2}} e^{\frac{2\pi i k}{2}t} \{-1, 0, 1\}, \quad t \in [0, 1].$$

The respective path represents Δ_3^k.

Finally, the set of solutions of the equation $f_{3,k}(z, \zeta) = 0$, $z = \frac{1}{2}e^{2\pi i t}, t \in [0, 1]$, is

$$E_3^k(t) \overset{def}{=} \left\{2^{-\frac{k}{2}} e^{\frac{2\pi i k}{2}t}, -2^{-\frac{k}{2}} e^{\frac{2\pi i k}{2}t}, R\right\}, \quad t \in [0, 1].$$

The respective path represents σ_1^k.

The quasipolynomial f_3 is not isotopic to a holomorphic one on the punctured plane. This follows from Lemma 7.2 below. □

Lemma 7.2 *Let $\hat{b}$ be a reducible conjugacy class of n-braids of infinite conformal module which is not periodic.*

Then there is no holomorphic mapping from $\mathbb{C}^* = \mathbb{C}\backslash\{0\}$ into $\mathfrak{P}_n$ representing $\hat{b}$.

Proof We have to prove that there is no holomorphic mapping from $\mathbb{C}^*$ into $\mathfrak{P}_n$ which represents $\hat{b}$. Assume the contrary. We may assume that $\hat{b}$ is represented by a pure braid b. Indeed, if there is a holomorphic mapping from $\mathbb{C}^*$ into $\mathfrak{P}_n$ representing $\hat{b}$, then there is also such a holomorphic mapping representing $\widehat{b^k}$ for any $k \in \mathbb{Z}\backslash\{0\}$. The number k can be chosen so that b^k is pure. By our assumption b^k is not a power of Δ_n^2. Since b^k is pure, there is a lift of the mapping that represents $\widehat{b^k}$ to a mapping $g : \mathbb{C}^* \to C_n(\mathbb{C})$. Associate to each point $z = (z_1, z_2, \ldots, z_n) \in C_n(\mathbb{C})$ the complex affine mapping $\mathfrak{a}_z$ on $\mathbb{C}$, $\mathfrak{a}_z(\zeta) = \frac{\zeta - z_1}{z_2 - z_1}$., that maps z_1 to 0 and z_2 to 1. Consider the holomorphic mapping $\mathbb{C}^* \ni \xi \to \mathfrak{a}_{g(\xi)}(g(\xi))$ that maps a point $\xi \in \mathbb{C}^*$ to the point $\left(\mathfrak{a}_{g(\xi)}(g_1(\xi)), \mathfrak{a}_{g(\xi)}(g_2(\xi)), \ldots, \mathfrak{a}_{g(\xi)}(g_n(\xi))\right)$, where $g(\xi) = \left(g_1(\xi), g_2(\xi), \ldots, g_n(\xi)\right) \in C_n(\mathbb{C})$. Then for the first two coordinates of $\mathfrak{a}_{g(\xi)}(g(\xi))$ the equalities $(\mathfrak{a}_{g(\xi)}(g(\xi)))_1 = \mathfrak{a}_{g(\xi)}(g_1(\xi)) \equiv 0$ and $(\mathfrak{a}_{g(\xi)}(g(\xi)))_2 = \mathfrak{a}_{g(\xi)}(g_2(\xi)) \equiv 1$ hold. For $j \geq 3$ the coordinate function $(\mathfrak{a}_{g(\xi)}(g(\xi)))_j$ is a mapping from $\mathbb{C}^*$ to $\mathbb{C} \setminus \{0, 1\}$. Its lift to the universal covering $\mathbb{C} \cong \tilde{\mathbb{C}}^*$ maps the complex plane into $\mathbb{C}_+$. Hence by Liouville's Theorem the mapping $\xi \to \mathfrak{a}_{g(\xi)}(g(\xi))$ is constant on $\mathbb{C}^*$. Therefore, $\widehat{b^k}$ can be represented by a mapping of the form $\xi \to \mathcal{P}_{\mathrm{sym}}((\mathfrak{a}_{g(\xi)})^{-1}(c_1, \ldots, c_n)))$ for constants $c_1, \ldots, c_n$. This means that b^k is a power of Δ_n^2 in contrary to the assumption. □

We obtained the following facts for conjugacy classes $\hat{b}$ of n-braids with infinite conformal module. Each annulus of finite conformal module admits a holomorphic mapping into $\mathfrak{P}_n$, representing such a class $\hat{b}$. The conjugacy class of each periodic braid (irreducible or not) can be represented by a holomorphic mapping on $\mathbb{C}^*$ (the proof given in Lemma 7.1 for $n = 3$ works for arbitrary n). The conjugacy class of non-periodic reducible mappings cannot be represented by a holomorphic mapping on $\mathbb{C}^*$, but in some cases there exists a holomorphic representing mapping from $\mathbb{D}^* = \mathbb{D}\setminus\{0\}$.

We saw that only very few mappings f with infinite $\mathcal{M}(\widehat{b_f})$ are homotopic to a holomorphic mapping on an annulus of first kind (i.e. on the punctured plane). This fact together with Lemma 7.1 motivates the use of conformal structures with only thick ends in the definition of the Gromov-Oka property.

The following question arises, which is part of Problem 7.4.

Problem 7.4a *Given a connected oriented finite open smooth surface X and a complex manifold Y with a set of generators whose conjugacy classes have infinite conformal module, which mappings $f : X \to Y$ have the Gromov-Oka property? In particular, which mappings $f : X \to \mathfrak{P}_n$ have the Gromov-Oka property? Which mappings from X into the n-punctured complex plane have the Gromov-Oka property?*

We will use the notion of reducible elements of the fundamental group of the n-punctured complex plane. The notion is an analog of the notion of reducible braids, or reducible mapping classes, respectively.

Definition 7.4 *Let E' be a finite subset of the Riemann sphere $\mathbb{P}^1$ which consists of $n+1 \geq 3$ points. Let X be a connected finite open Riemann surface with non-trivial fundamental group. A non-contractible continuous map $f : X \to \mathbb{P}^1 \setminus E'$ is called reducible if it is free homotopic (as a mapping to $\mathbb{P}^1 \setminus E'$) to a mapping whose image is contained in $D \setminus E'$ for an open topological disc $D \subset \mathbb{P}^1$ with $E' \setminus D$ containing at least two points of E'. Otherwise the mapping is called irreducible.*

Let $E = \{z_1, z_2, \ldots, z_n\}$ be a subset of the complex plane that contains exactly n points and let $E' = E \cup \{\infty\}$. We assign to each continuous mapping $f : [0, 1] \to \mathbb{C} \setminus E$ with $f(0) = f(1)$ the pure geometric $(n + 1)$-braid $F(t) = \{f(t), z_1, z_2, \ldots, z_n\}$, $t \in [0, 1]$. Approximating we may assume that the mapping f is smooth. Consider a parameterizing isotopy φ_t, $t \in [0, 1]$, for the geometric braid F that is contained in the cylinder $[0, 1] \times R\mathbb{D}$ for a large number R. The equality $\varphi_t(\{f(0), z_1, z_2, \ldots, z_n\}) = \{f(t), z_1, z_2, \ldots, z_n\}$, $t \in [0, 1]$ holds. Let e_f be the element of the fundamental group $\pi_1(\mathbb{C} \setminus E, f(0))$ with base point $f(0)$, that is represented by f. We associate to e_f the braid b_f, that is represented by F, and the mapping class $\mathfrak{m}_{f,\infty} \in \mathfrak{M}(\mathbb{P}^1; E \cup \{f(0), \infty\})$ that is represented by the self-homeomorphism φ_1 of $\mathbb{P}^1$, belonging to the family φ_t . We may assign to the homotopy class e_f of f an entropy, namely the entropy of the mapping class $\mathfrak{m}_f$, which is a measure of the complexity of the mapping f.

The following lemma holds.

Lemma 7.3 *For a subset E of the complex plane that contains exactly $n \geq 2$ points a continuous non-contractible mapping $f : [0, 1] \to \mathbb{C} \setminus E$ is reducible if and only if the associated braid b_f is reducible, equivalently, if the associated mapping class $\mathfrak{m}_{f,\infty}$ is reducible.*

Proof We may assume that the mapping f is smooth and all related objects are smooth. Suppose that the mapping f is reducible. Then there exists a free homotopy f_s, $s \in [0, 1]$, of mappings from $[0, 1]$ into $\mathbb{C} \setminus E$ with $f_s(0) = f_s(1)$ and $f_0 = f$, such that for all $t \in [0, 1]$ the inclusion $f_1(t) \subset D$ holds for a disc D with $(E \cup \{\infty\}) \setminus D$ containing at least two points. The pure geometric braids F_s, $s \in [0, 1]$, with varying base point, that are defined by f_s, are free isotopic, hence, represent the same conjugacy class of braids $\widehat{b_f}$. Let E be the set $\{z_1, \ldots, z_n\}$, where the label is chosen so that $z_1, \ldots, z_k$ are the points of E that are contained in D. Let φ_t, $t \in [0, 1]$ be a parameterizing isotopy for the geometric braid $\{f_1(t), z_1, \ldots, z_k\}$, $\varphi_t(\{f_1(0), z_1, \ldots, z_k\}) = \{f_1(t), z_1, \ldots, z_k\}$, $t \in [0, 1]$. Since $f_1(t) \in D$ for all t, the parameterizing isotopy can be chosen so that all φ_t, $t \in [0, 1]$, are equal to the identity on $\mathbb{C} \setminus D$. In particular φ_1 maps ∂D onto itself. Since $(E \cup \{\infty\}) \setminus D$ contains at least two points and $E \cap D$ contains at least one point (since f_1 is not contractible), the mapping φ_1 of $\mathbb{P}^1$ with set of distinguished points $\{f(0), z_1, \ldots, z_n, \infty\}$ is reducible, and hence the associated mapping class $\mathfrak{m}_{f_1,\infty}$ is reducible. Hence $\widehat{b_f}$, and thus b_f, is reducible.

Vice versa, suppose $\mathfrak{m}_{f,\infty}$ is reducible. Since b_f is a pure braid, there exists a mapping $\varphi \in \mathfrak{m}_{f,\infty}$ that fixes each of the points $f(0), z_1, \ldots, z_n$, and fixes a closed curve C pointwise. Each of the components of $\mathbb{P}^1 \setminus C$ contains at least two points among the $f(0), z_1, \ldots, z_n, \infty$. After applying a Möbius transformation and perhaps relabeling the points, we may assume that $f(0), z_1, \ldots, z_k$ are contained in the bounded connected component D of $\mathbb{C} \setminus C$, and $z_{k+1}, \ldots, z_n$ are contained in $\mathbb{C} \setminus \bar{D}$. Take a continuous family φ_t, $t \in [0, 1]$, of self-homeomorphisms of $\mathbb{P}^1$ that fix C pointwise and join the identity with φ. Then the pure geometric braid $\mathcal{F}(t) \overset{def}{=} \varphi_t\big(\{f(0), z_1, \ldots, z_n\}\big)$, $t \in [0, 1]$, is isotopic (through geometric braids with fixed base point) to $\{f(t), z_1, \ldots, z_n\}$, $t \in [0, 1]$, and the strands of $\mathcal{F}$ with initial points $f(0), z_1, \ldots, z_k$, are contained in the domain D.

We prove now, that the geometric braid $\mathcal{F}(t) = \varphi_t\big(\{f(0), z_1, \ldots, z_n)\}\big)$, $t \in [0, 1]$, is isotopic (through geometric braids with fixed base point) to a geometric braid of the form $G(t) = \{g(t), z_1, \ldots, z_n\}$, $t \in [0, 1]$, with $g(t) \in D$ for $t \in [0, 1]$. We look at the geometric braid $\varphi_t\big(\{z_1, \ldots, z_n\}\big)$, $t \in [0, 1]$, that is obtained from $\mathcal{F}(t)$, $t \in [0, 1]$, by forgetting the strand with initial point $f(0)$. This geometric braid $\varphi_t\big(\{z_1, \ldots, z_n\}\big)$, $t \in [0, 1]$, is isotopic with fixed base point to the constant geometric braid $\{z_1, \ldots z_k, z_{k+1}, \ldots, z_n\}$, $t \in [0, 1]$, by an isotopy that fixes C pointwise.

By Remark 1.1 there exists a smooth family $\psi_{t,s}$ of diffeomorphisms of $\mathbb{P}^1$ that are equal to the identity on $\mathcal{C}$, such that $\psi_{t,s}\big(\{z_1, \ldots, z_n\}\big)$ realizes an isotopy of geometric braids with fixed base point that joins $\varphi_t\big(\{z_1, \ldots, z_n\}\big)$, $t \in [0, 1]$, with the constant geometric braid, and has the following property. For a small positive number ε we have $\psi_{t,s} = \varphi_t$ for $0 \le s \le \varepsilon$, $\psi_{t,1} = \mathrm{Id}$, and $\psi_{0,s} = \mathrm{Id}$. Then the equality $\psi_{t,s}\big((z_1, \ldots, z_n)\big) = (z_1, \ldots, z_n)$ holds for (t, s) in the boundary of the square $[0, 1] \times [0, 1]$ except for $\{s = 1, t \in [0, 1]\}$.

We look now at $\psi_{t,s}\big(\{f(0), z_1, \ldots, z_n)\}\big)$, $t, s \in [0, 1]$. Notice that the mapping $[\varepsilon, 1] \ni s \to \psi_{1,s}(f(0))$ defines a curve with base point $f(0)$ in $D \setminus \{z_1, \ldots, z_k\}$. Let h be an orientation preserving self-homeomorphism of $[0, 1] \times [0, 1]$ that equals the identity on $[0, 1] \times \{0\}$ and on $\{0\} \times [0, 1]$, maps $\{1\} \times [0, 1]$ onto $\{1\} \times [0, \varepsilon]$ by a homeomorphism that fixes the point $(1, 0)$, and takes $[0, 1] \times \{1\}$ homeomorphically onto the remaining part of the boundary. The mapping $\psi_{h(t,s)}\big((f(0), z_1, \ldots, z_k)\big)$ equals $\varphi_t\big((f(0), z_1, \ldots, z_k)\big)$ for $t \in [0, 1]$ and $s = 0$, it equals $(f(0), z_1, \ldots, z_k)$ for $t = 0$ and $t = 1$, and is of the form $t \to (g(t), z_1, \ldots, z_k)$, for $s = 1$. Notice that the mapping g is homotopic to the inverse of the mapping $[\varepsilon, 1] \ni s \to \psi_{1,s}(f(0))$. We found an isotopy (through geometric braids with fixed base point) joining $\mathcal{F}(t) = \varphi_t\big((f(0), z_1, \ldots, z_n)\big)$; $t \in [0, 1]$, with a geometric braid of the form $G(t) = (g(t), z_1, \ldots, z_n)$, $t \in [0, 1]$, with $g(t) \in D$ for $t \in [0, 1]$.

It remains to prove the following statement. If two pure geometric braids $(f(t), z_1, \ldots, z_n)$, $t \in [0, 1]$, and $(g(t), z_1, \ldots, z_n)$, $t \in [0, 1]$, with $z_1, \ldots, z_n$ being distinct complex numbers, are isotopic through geometric braids with fixed base point $(f(0), z_1, \ldots, z_n) = (g(0), z_1, \ldots, z_n)$, then g and f are homotopic mappings (with fixed base point) into $\mathbb{C} \setminus \{z_1, \ldots, z_n\} = \mathbb{P}^1 \setminus \{z_1, \ldots, z_n, \infty\}$. This statement is again a consequence of Remark 1.1. Let $f_{t,s}$ be the isotopy of pure geometric braids with fixed base point (contained in the cylinder $[0, 1] \times R\mathbb{D}$ for a large number R). We let $\tilde{f}_{t,s}$ be the isotopy of geometric braids obtained from $f_{t,s}$ by forgetting the first strand. We take the family $\varphi_{t,s}$, $(t, s) \in [0, 1] \times [0, 1] \times R\overline{\mathbb{D}}$ of Remark 1.1, for which $\varphi_{t,s}\big((z_1, \ldots, z_n)\big)$ is equal to $\tilde{f}_{t,s}$, and $\varphi_{0,s} = \varphi_{t,0} = \varphi_{t,1} = \mathrm{Id}$. The family $\varphi_{t,s}^{-1}(f_{t,s})$ defines an isotopy of the form $(f_s(t), z_1, \ldots, z_n)$ that joins $(f(t), z_1, \ldots, z_n)$ and $(g(t), z_1, \ldots, z_n)$. The family $f_s(t)$ is the required homotopy. The lemma is proved. □

We will mostly consider the case when $E = \{-1, 1, \infty\}$. We will often refer to $\mathbb{P}^1 \setminus \{-1, 1, \infty\}$ as the thrice punctured Riemann sphere or the twice punctured complex plane $\mathbb{C} \setminus \{-1, 1\}$. By Definition 7.4 a continuous mapping from a Riemann surface to the twice punctured complex plane is reducible, iff it is homotopic to a mapping with image in a once punctured disc contained in $\mathbb{P}^1 \setminus E$. (The puncture may be equal to ∞.)

To describe the reducible mappings into $\mathbb{P}^1 \setminus \{-1, 1, \infty\}$, we recall, that we denoted by a_1 (a_2, respectively) the generator of $\pi_1(\mathbb{C} \setminus \{-1, 1\}, 0)$ that is represented by a curve with base point 0 that surrounds -1 (1, respectively) positively. We called a_1 and a_2 the standard generators of $\pi_1(\mathbb{C} \setminus \{-1, 1\}, 0)$. By Definition 7.4 a mapping from a finite open Riemann surface into $\mathbb{P}^1 \setminus \{-1, 1, \infty\}$ is reducible if and only if its monodromies are contained in a subgroup generated by a single element which is a conjugate of a power of an element of the form

$$a_1,\ a_2,\ \text{or}\ (a_1a_2)^{-1}. \tag{7.1}$$

Indeed, the images under the mapping of all loops on the Riemann surface represent elements of such a subgroup of $\mathbb{P}^1 \setminus \{-1, 1, \infty\}$ iff the mapping is homotopic to a mapping with image in a punctured neighbourhood of -1 ($+1$, or ∞, respectively).

Lemma 7.4 *Let X be a connected finite open Riemann surface with only thick ends equipped with base point q_0. For each homomorphism $h : \pi_1(X, q_0) \to \pi_1(\mathbb{C} \setminus \{-1, 1\}, 0)$, whose image is contained in a subgroup of $\pi_1(\mathbb{C} \setminus \{-1, 1\}, 0)$ generated by a power of one of the elements of the form* (7.1)*, there exists a holomorphic mapping $f : X \to \mathbb{C} \setminus \{-1, 1\}$, whose monodromy homomorphism $f_* : \pi_1(X, q_0) \to \pi_1(\mathbb{C} \setminus \{-1, 1\}, 0)$ is conjugate to h.*

Proof of Lemma 7.4 Let X be a connected finite open Riemann surface with only thick ends, and let $h : \pi_1(X, q_0) \to \Gamma$ be a homomorphism into the group Γ generated by an element of $\pi_1(\mathbb{C} \setminus \{-1, 1\}, 0)$ of the form (7.1). The Riemann surface X is conformally equivalent to a domain on a closed Riemann surface X^c such that all connected components of its complement are closed topological discs. Hence, there exists an open Riemann surface $X_1 \subset X^c$ such that X is relatively compact in X_1. Let $\omega : X \to X_1$ be a homeomorphism from X onto X_1 that is homotopic on X to the inclusion $X \hookrightarrow X_1$ for which $\omega(q_0) = q_0$. Identify the fundamental groups $\pi_1(X, q_0)$ and $\pi_1(X_1, q_0)$ by the mapping induced by ω.

By Lemma 1.1 X_1 is diffeomorphic to a standard neighbourhood of a standard bouquet B of circles for X_1. Hence, X_1 can be written as union $X_1 = D \cup \bigcup_j V_j$ of an open disc D and half-open bands V_j attached to D (see Sect. 1.2). Let c_j be the circle of the bouquet that corresponds to the generator e_j of the fundamental group $\pi_1(X_1, q_0)$. We consider an open cover of X_1 as follows. Put $U_0 = D$. Then U_0 is an open topological disc in X_1 that contains the base point $\omega(q_0)$. Cover each V_j by two simply connected open subsets U_{j+} and U_{j-} of X_1 with connected and simply connected intersection, so that the $U_{j\pm}$ are disjoint from the $U_{k\pm}$ for $j \neq k$, and for each $U_{j\pm}$ its intersection with U_0 is connected and simply connected. We may also assume that the following holds. The intersection of three different sets among the U_0, U_{j+} and U_{j-} is empty. The intersections of each c_j with U_{j+} and U_{j-} are connected, and each c_j is disjoint from $U_{k+} \cup U_{k-}$ for $k \neq j$. Label the $U_{j\pm}$ so that walking on $c_j \setminus U_0$ (which is contained in V_j) in the direction of the orientation of c_j we first meet U_{j-}. For each j the set $U_0 \cup U_{j-} \cup U_{j+}$ is an annulus that contains c_j.

Define a Cousin I distribution as follows. For each j we put $\mathsf{F}_{0,j-} = 0$ on $U_0 \cap U_{j-}$, $\mathsf{F}_{j-,j+} = 0$ on $U_{j-} \cap U_{j+}$, and $\mathsf{F}_{j+,0} = k_j$ on $U_{j+} \cap U_0$. Since open Riemann surfaces are Stein manifolds (see e.g. [27]), the respective Cousin Problem has a solution (see [31, 40], or Appendix A.1). This means that there are holomorphic functions F_0 on U_0, F_{j+} on U_{j+} and F_{j-} on U_{j-} such that

$$\begin{aligned} F_0 \ - F_{j-} &= 0 \ \text{ on } U_0 \ \cap U_{j-}, \\ F_{j-} - F_{j+} &= 0 \ \text{ on } U_{j-} \cap U_{j+}, \\ F_{j+} - F_0 \ &= k_j \text{ on } U_{j+} \cap U_0 \ . \end{aligned} \tag{7.2}$$

The function

$$F = \begin{cases} e^{2\pi i F_0} & \text{on } U_0, \\ e^{2\pi i F_{j-}} & \text{on } U_{j-} \\ e^{2\pi i F_{j+}} & \text{on } U_{j+} \end{cases} \tag{7.3}$$

is well-defined and holomorphic in X_1. Let e be the generator of the fundamental group of $\mathbb{C} \setminus \{0\}$ with base point $F(q_0)$. The restriction of F to each c_j represents e^{k_j}.

The restriction of F to $X \Subset X_1$ is bounded. Hence, for a positive number C the mapping $-1 + \frac{1}{C}F$ maps X into $\mathbb{C} \setminus \{-1, 1\}$ so that the monodromy along each e_j equals $a_1^{k_j}$ (after identifying $\pi_1\Big(\mathbb{C}\setminus\{-1,1\}, -1+\frac{1}{C}F(q_0)\Big)$ with $\pi_1(\mathbb{C}\setminus\{-1,1\},0)$ by a suitable isomorphism). The case when the monodromies are powers of other elements of form (7.1) is treated by composing F with a suitable conformal self-mapping of the Riemann sphere that permutes the points -1, 1, and ∞. □

In Sect. 7.2 we will address Problem 7.4a for the case when the target manifold is the twice punctured complex plane, and in Sect. 7.3 for the case when $Y = \mathfrak{P}_3$ and X is a smooth surface of genus one with a hole. In Chap. 11 we address Problem 7.3. For instance, we give an upper bound for the number of homotopy classes of irreducible mappings from a finite open Riemann surface to the twice punctured complex plane that contain a holomorphic mapping.

The estimates can be regarded as a quantitative information concerning the restricted validity of Gromov's Oka Principle for some cases when the target manifold is not a Gromov-Oka manifold.

In Sects. 8.3–8.9 we will consider the Gromov-Oka Principle for fiber bundles over a smooth torus with a hole.

7.2 Description of Gromov-Oka Mappings from Open Riemann Surfaces to $\mathbb{C} \setminus \{-1, 1\}$

Let $\mathcal{X}$ be a topological space and let $\hat{e}$ be a conjugacy class of elements of the fundamental group $\pi_1(\mathcal{X}, x_0)$ of $\mathcal{X}$. Choose a loop γ in $\mathcal{X}$ that represents $\hat{e}$. Consider a complex manifold $\mathcal{Y}$ and a continuous mapping $f : \mathcal{X} \to \mathcal{Y}$. The conjugacy class of elements of the fundamental group of $\mathcal{Y}$ represented by the restriction $f \circ \gamma$ depends only on f and on $\hat{e}$ and is denoted by $\hat{f}_e$. Recall that $\mathcal{M}(\hat{f}_e) = \infty$ iff the restriction of f to an annulus representing $\hat{e}$ has the Gromov-Oka property.

Lemma 7.5 *Let X be a connected oriented smooth finite open surface and let $e \in \pi_1(X, q_0)$ be an element whose conjugacy class can be represented by a simple closed curve in X. Then for any $a > 0$ there exists an orientation preserving conformal structure $\omega : X \to \omega(X)$ on X such that the Riemann surface $\omega(X)$ contains a holomorphic annulus A that represents $\hat{e}$ and has conformal module $m(A) > a$.*

Notice that the conjugacy class of any element of a standard basis of $\pi_1(X, q_0)$ and the commutator of two elements of a standard basis of $\pi_1(X, q_0)$ corresponding to a handle can be represented by a simple closed curve.

Proof of Lemma 7.5 Take any conformal structure $\omega' : X \to \omega'(X)$ on X. Let γ be a smooth simple closed curve that represents the conjugacy class $\hat{e}$ of e. Cut $\omega'(X)$ along $\omega'(\gamma)$. Take a neighbourhood of $\omega'(\gamma)$ that is cut by $\omega'(\gamma)$ into connected components X_+ and X_-, each being conformally equivalent to an annulus. Take an annulus A of conformal module $m(A) > a$, and glue the $X_\pm$ conformally to annuli $A_\pm$ in A that are disjoint and adjacent to different boundary components of A. We obtain a Riemann surface and a homeomorphism ω from X onto this Riemann surface. The mapping $\omega : X \to \omega(X)$ gives the required complex structure. □

We will consider now the target space $Y = \mathbb{C} \setminus \{-1, 1\}$. Let A be an annulus in the complex plane. By Lemmas 1.2 and 7.1 a mapping $f : A \to \mathbb{C} \setminus \{-1, 1\}$ has the Gromov-Oka property if and only if for a generator e of the fundamental group of the annulus the conjugacy class $\widehat{f_*(e)}$ has infinite conformal module. By Lemmas 1.2, 6.5, and 7.1 a conjugacy class of elements of $\pi_1(\mathbb{C} \setminus \{-1, 1\}, 0)$ has infinite conformal module iff it is represented by an integer power of one the elements (7.1). By Lemma 7.3 this happens exactly if the mapping f is reducible.

Notice that a continuous mapping f from an annulus $A = \{r^{-1} < |z| < r\}$ into $\mathbb{C} \setminus \{-1, 1\}$ with the Gromov-Oka property is also homotopic to a holomorphic mapping for any orientation reversing conformal structure with only thick ends on the annulus. Indeed, the mappings $z \to f(z)$ and $z \to f(\frac{1}{\bar{z}})$ are homotopic on A since they coincide on $\{|z| = 1\}$.

The following theorem concerns Problem 7.4a for smooth mappings from oriented finite open smooth surfaces X of positive genus to the twice punctured complex plane and confirms the special role of the conformal module of conjugacy classes of elements of the fundamental group of the target with respect to this problem. It states the existence of finitely many annuli contained in X, such that a smooth orientation preserving mapping $X \to \mathbb{C} \setminus \{-1, 1\}$ has the Gromov-Oka property if and only if its restriction to each of the mentioned annuli has this property.

Theorem 7.1 *Let X be a connected smooth oriented surface of positive genus g with $m \geq 1$ holes. There exists a subset $\mathcal{E}'$ of the fundamental group $\pi_1(X, q_0)$ of X*

consisting of at most $(2g+m-1)^3$ *elements such that the conjugacy class* $\hat{e}$ *of each element of* $\mathcal{E}'$ *can be represented by a simple closed curve and the following holds.*

A continuous mapping $f : X \to \mathbb{C} \setminus \{-1, 1\}$ *has the Gromov-Oka property if and only if for each* $e \in \mathcal{E}'$ *the restriction of the mapping to an annulus in X that represents* $\hat{e}$ *has the Gromov-Oka property.*

The following theorem describes all continuous maps $X \to \mathbb{C} \setminus \{-1, 1\}$ with the Gromov-Oka property. In this theorem we allow the surface X to have any genus.

Theorem 7.2 *A continuous mapping* $f : X \to \mathbb{C} \setminus \{-1, 1\}$ *from a connected smooth oriented surface X of genus* $g \geq 0$ *with* $m \geq 1$ *holes,* $2g + m \geq 3$*, into the twice punctured complex plane has the Gromov-Oka property, if and only if f is either reducible, or X is a smooth oriented two-sphere* S^2 *with at least three holes and the mapping f is homotopic to a mapping that extends to an orientation preserving diffeomorphism* $F : S^2 \to \mathbb{P}^1$ *that maps three points of* S^2 *contained in pairwise different holes to the three points* -1, 1, *and* ∞*, respectively.*

Let $\mathcal{E}$ be a standard system of generators of the fundamental group $\pi_1(X, q_0)$ of the smooth surface X of genus g with m holes with base point $q_0 \in X$. Before proving the Theorems 7.1 and 7.2 we will describe a set $\mathcal{E}'$ of elements of $\pi_1(X, q_0)$ such that the conjugacy class of each element of $\mathcal{E}'$ can be represented by a simple closed curve. Later we will compare the Gromov-Oka property of a mapping $X \to \mathbb{C} \setminus \{-1, 1\}$ with the Gromov-Oka property of its restriction to each annulus in X that represents $\hat{e}$ for an element $e \in \mathcal{E}'$.

Let First $g > 0$. For the j-th handle we choose three elements e_{2j-1}, e_{2j}, and $[e_{2j-1}, e_{2j}]$. The conjugacy class of each of them can be represented by a simple closed curve. For the ℓ-th hole, $\ell = 1, \ldots, m-1$, we take the element $e_{2j+\ell}$. There is a simple closed curve that represents its conjugacy class. The obtained $3g+m-1$ elements of the fundamental group will be contained in $\mathcal{E}'$.

We convert each unordered pair of different elements of $\mathcal{E}$, that is different from a pair $\{e_{2j-1}, e_{2j}\}$ corresponding to a handle, into an ordered pair (e', e''). There is a simple closed curve representing $\widehat{e'\,e''} = \widehat{e''\,e'}$. The products $e'e''$ constitute a set of $\frac{1}{2}(2g+m-1)(2g+m-2) - g$ elements of the fundamental group. They will be contained in $\mathcal{E}'$.

For each pair e_{2j-1}, e_{2j} of elements of $\mathcal{E}$ that corresponds to a handle, and each other element e' among the chosen generators (if there is any) we consider the elements $e_{2j-1}^2\, e_{2j}\, e'$, $e_{2j-1}^3\, e_{2j}\, e'$, $e_{2j-1}\, e_{2j}^2\, e'$, and $e_{2j-1}\, e_{2j}^3\, e'$. For the conjugacy class of each such element there is a simple closed curve in X representing it. The obtained $4g\,(2g+m-3)$ elements of the fundamental group will be contained in $\mathcal{E}'$.

Suppose $m > 2$. Let (e_1, e_2) be the pair of elements of $\mathcal{E}$ that corresponds to the first handle. We convert each unordered pair of different elements of $\mathcal{E}$ corresponding to holes to an ordered pair (e', e''), so that the conjugacy class of the element $e'e_1e'e_2e''$ can be represented by a simple closed curve in X. (See Fig. 7.1.)

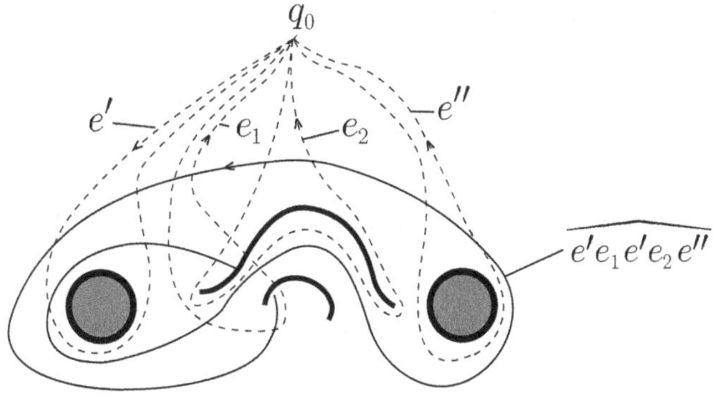

Fig. 7.1 A simple closed curve that represents the free homotopy class of $e'e_1e'e_2e''$

We obtain no more than $\frac{1}{2}(m-1)(m-2)$ elements of $\pi_1(X, q_0)$. They will be contained in $\mathcal{E}'$.

Suppose now $g(X) = 0$. If $m = 1$ the fundamental group is trivial. If $m = 2$ it has one generator. The set $\mathcal{E}'$ will be equal to the set $\mathcal{E}$ consisting of the generator. Let $m \geq 3$. The element $e_m \overset{def}{=} (\prod_{j=1}^{m-1} e_j)^{-1}$ is represented by a simple closed curve that surrounds $\mathcal{C}_m$ positively (i.e. walking along the curve the set $\mathcal{C}_m$ is on the left). All elements $e \in \mathcal{E} \cup \{e_m\}$ will be contained in $\mathcal{E}'$. Further, each unordered pair of different elements of $\mathcal{E} \cup \{e_m\}$ is converted into an ordered pair (e', e'') and the product $e'e''$ will be contained in $\mathcal{E}'$. Finally, we convert each unordered triple of different elements of $\mathcal{E} \cup \{e_m\}$ into an ordered triple (e', e'', e''') so that $\widehat{e'\, e''\, e'''}$ can be represented by a simple closed curve. The product $e'e''e'''$ will be contained in $\mathcal{E}'$.

We described all elements of $\mathcal{E}'$. Recall that the conjugacy class of each of them can be represented by a simple closed curve. We estimate now the total number N of elements of $\mathcal{E}'$.

Let first $g > 0$. With $x = 2g + m - 1 \geq 2$ the number N does not exceed

$$\begin{cases} x + g + \frac{1}{2}x(x-1) - g + 4g(x-2), & 0 < m \leq 2, \\ x + g + \frac{1}{2}x(x-1) - g + 4g(x-2) + \frac{1}{2}(m-1)(m-2), & m > 2. \end{cases}$$

Since $2g \leq x$, for $m \leq 2$ the sum does not exceed $x + x(x-1) + 2x(x-2)$. Hence, $N \leq 3x^2 - 4x < x^3$. For $m > 2$ we use the inequality $m - 1 = x - 2g \leq x - 2$, hence $\big(4g + \frac{1}{2}(m-2)\big)(x-2) \leq 2x(x-2)$. Again $N \leq x^3$.

Let $g = 0$. If $m = 2$ the set $\mathcal{E}'$ contains a single element, and $N = 1 = (m-1)^3 = x^3$. If $x = m - 1 \geq 2$ the number N is not bigger than $x + 1 + \frac{1}{2}(x+1)x + \frac{1}{6}(x+1)x(x-1) = \frac{1}{2}(x+1)(x+2) + \frac{1}{6}x(x^2-1) = \frac{1}{6}x^3 + \frac{1}{2}x^2 + (\frac{3}{2} - \frac{1}{6})x + 1$ and since $x \geq 2$, also in this case $N \leq x^3$. We estimated the number of elements of $\mathcal{E}'$ from above by $x^3 = (2g + m - 1)^3$.

For the proof of Theorem 7.1 we need the following lemma. Its proof will given after the proof of the theorems.

Lemma 7.6 *Let X be a connected oriented smooth open surface of genus g with m holes, $2g + m \geq 0$, and $\mathcal{E}'$ the set of elements of the fundamental group $\pi_1(X, q_0)$ chosen above. Assume the restriction of a continuous mapping $f : X \to \mathbb{C}\setminus\{-1, 1\}$ to each annulus in X representing the conjugacy class $\hat{e}$ of an element $e \in \mathcal{E}'$ has the Gromov-Oka property.*
Then one of the following statements holds.

1. *The image of the monodromy homomorphism $f_* : \pi_1(X, q_0) \to \pi_1(\mathbb{C}\setminus\{-1, 1\}, f(q_0)) \cong \pi_1(\mathbb{C}\setminus\{-1, 1\}, 0)$ is contained in the group generated by a conjugate of a power of one of the elements of the form* (7.1). *In this case the mapping is reducible.*
2. *X is equal to the smooth oriented 2-sphere denoted by S^2 with at least three holes removed and the mapping f is homotopic to a mapping that extends to a diffeomorphism $F : S^2 \to \mathbb{P}^1$ that maps three points of S^2 contained in pairwise different holes to the three points -1, 1, and ∞, respectively.*

If F is orientation preserving, than F is homotopic to a holomorphic map for any orientation preserving conformal structure on X, including structures of first kind. Moreover, the restriction $F \mid X$ is irreducible.

If F is orientation reversing, then for any orientation preserving conformal structure on X the mapping F is homotopic to an antiholomorphic mapping, but F is not a Gromov-Oka mapping.

Proof of Theorem 7.1 We let $\mathcal{E}'$ be the set of elements of $\pi_1(X, q_0)$ chosen above. If a mapping $f : X \to \mathbb{C} \setminus \{-1, 1\}$ has the Gromov-Oka property then by Lemma 7.5 the restriction of f to each annulus in X representing the conjugacy class of an element of $\mathcal{E}'$ has the Gromov-Oka property.

Vice versa, suppose X has positive genus and the restriction of a continuous mapping $f : X \to \mathbb{C} \setminus \{-1, 1\}$ to each annulus in X representing the conjugacy class $\hat{e}$ of an element $e \in \mathcal{E}'$ has the Gromov-Oka property. Then by Lemma 7.6 the monodromy homomorphism f_* is contained in the group generated by a conjugate of a power of an element of the form (7.1), equivalently f is reducible. By Lemma 7.4 each reducible map $X \to \mathbb{C} \setminus \{-1, 1\}$ is homotopic to a holomorphic mapping for each conformal structure on X with only thick ends. Hence, the mapping f has the Gromov-Oka property. □

Proof of Theorem 7.2 As before, $\mathcal{E}'$ is the set of elements of $\pi_1(X, q_0)$ chosen above. Suppose first that $g > 0$. If a continuous mapping $f : X \to \mathbb{C} \setminus \{-1, 1\}$ has the Gromov-Oka property, then by Lemma 7.5 the restriction of f to each annulus represented by a conjugacy class $\hat{e}$ for an element $e \in \mathcal{E}'$ has the Gromov-Oka property. By Lemma 7.6 the monodromy homomorphism f_* is contained in the group generated by a conjugate of a power of an element of the form (7.1) (equivalently f is reducible).

Vice versa, let X be an oriented smooth finite surface of any genus. By Lemma 7.4 each reducible map $f : X \to \mathbb{C} \setminus \{-1, 1\}$ is homotopic to a holomorphic mapping for each conformal structure on X with only thick ends, i.e. reducible maps have the Gromov-Oka property.

Let $g = 0$. Then X is a smooth oriented 2-sphere denoted by S with at least 3 holes removed. If an irreducible continuous mapping $f : X \to \mathbb{C} \setminus \{-1, 1\}$ has the Gromov-Oka property, then by Lemmas 7.5 and 7.6 the mapping f is homotopic to a mapping that extends to a diffeomorphism $F : S \to \mathbb{P}^1$ that maps three points contained in different holes to the points $-1, 1$, and ∞. If F is orientation preserving then by Lemma 7.6 for any conformal structure on S it is homotopic to a conformal mapping from $\mathbb{P}^1$ onto itself that takes an ordered tuple of three distinct points in $\mathbb{P}^1$ to the tuple $(-1, 1, \infty)$.

If the mapping F is not orientation preserving, F is not homotopic to a holomorphic mapping by Lemma 7.6. □

The theorem shows that for oriented finite open surfaces of positive genus only the reducible mappings $X \to \mathbb{C} \setminus \{-1, 1\}$ have the Gromov-Oka property. In case X is the oriented two-sphere with at least three holes there are also irreducible homotopy classes of mappings with the Gromov-Oka property. Each consists of mappings with the following property. For each orientation preserving homeomorphism $\omega : X \to \omega(X)$ onto a Riemann surface $\omega(X)$ (maybe, of first kind) $f \circ \omega^{-1}$ is homotopic to a holomorphic mapping that extends to a conformal mapping $\mathbb{P}^1 \to \mathbb{P}^1$ that maps three points in different holes to $-1, 1$ and ∞, respectively (in some order depending on the class). These homotopy classes are the only irreducible homotopy classes with the Gromov-Oka property, and they are the only homotopy classes of mappings $X \to \mathbb{C} \setminus \{-1, 1\}$ that contain a holomorphic mapping for any conformal structure on X including conformal structures of first kind. Indeed, if the image of a non-trivial monodromy homomorphism f_* is generated by a conjugate of a power of one of the elements of the form (7.1) the mapping f cannot be homotopic to a holomorphic mapping for a conformal structure of first kind (see the proof of Corollary 11.1).

Proof of Lemma 7.6

7.6.1 Identifying the fundamental groups $\pi_1(\mathbb{C} \setminus \{-1, 1\}, 0)$ and $\pi_1(\mathbb{C} \setminus \{-1, 1\}, f(q_0))$ by a fixed isomorphism, we obtain for each $e \in \mathcal{E}'$ the equation $f_*(e) = w_e^{-1}\, b_e\, w_e$ for an element $w_e \in \pi_1(\mathbb{C} \setminus \{-1, 1\}, 0)$ and an element b_e that is a power of an element of the form (7.1). The images $f_*(e_{2j-1})$ and $f_*(e_{2j})$ of two elements corresponding to a handle commute. This is a particular case of the situation treated in the proof of Theorem 7.4 below, but can be seen easily directly. Indeed, the sum of exponents of the terms of a word representing a commutator is equal to zero. Since the commutator $[f_*(e_{2j-1}), f_*(e_{2j})] = f([e_{2j-1}, e_{2j-1}])$ must be a power of a conjugate of an element of the form (7.1) it must be equal to the identity.

Recall that a word in the generators of a free group is called reduced, if neighbouring terms are powers of different generators. We will identify elements of a free group (in particular of $\pi_1(\mathbb{C} \setminus \{-1, 1\}, 0)$) with reduced words in generators of the group. A reduced word is called cyclically reduced, if either the word consists of a single term, or it has at least two terms and the first and the last term of the word are powers of different generators.

7.6.2 Suppose for two generators $e^{(1)}, e^{(2)} \in \mathcal{E} \subset \mathcal{E}'$ we have $f_*(e^{(\ell)}) = w_{e^{(\ell)}}^{-1} b_{e^{(\ell)}} w_{e^{(\ell)}}$ with $b_{e^{(\ell)}} = a_{j_\ell}^{k_\ell}$, $\ell = 1, 2$, where each a_{j_ℓ} is one of the generators a_1, a_2 of $\pi_1(\mathbb{C} \setminus \{-1, 1\}, 0)$. We prove now that there exists an element $w_{e^{(1)}, e^{(2)}}$ of $\pi_1(\mathbb{C} \setminus \{-1, 1\}, 0)$ which conjugates both $f_*(e^{(\ell)})$ to $a_{j_\ell}^{k_\ell}$, $\ell = 1, 2,$. Put $w = w_{e^{(2)}} w_{e^{(1)}}^{-1}$. Since either $e^{(1)} e^{(2)} \in \mathcal{E}'$ or $e^{(2)} e^{(1)} \in \mathcal{E}'$, the monodromy $f_*(e^{(1)} e^{(2)}) = f_*(e^{(1)}) f_*(e^{(2)})$ has infinite conformal module and is conjugate to $a_{j_1}^{k_1} w^{-1} a_{j_2}^{k_2} w$. The element $a_{j_1}^{k_1} w^{-1} a_{j_2}^{k_2} w$ is conjugate to $a_{j_1}^{k_1} w'^{-1} a_{j_2}^{k_2} w'$ for an element $w' = a_{j_2}^{-k_2'} w a_{j_1}^{-k_1'} \in \pi_1(\mathbb{C} \setminus \{-1, 1\}, 0)$ that is either equal to the identity or it can be written as a reduced word that starts with a power of a_{j_1} and ends with a power of a_{j_2}. (In the case $a_{j_1} = a_{j_2}$ we do not exclude so far that w' is equal to a power of the other generator of $\pi_1(\mathbb{C} \setminus \{-1, 1\}, 0)$.) If w' is not the identity, then $a_{j_1}^{k_1} w'^{-1} a_{j_2}^{k_2} w'$ can be written as cyclically reduced word that contains powers of different sign of generators of $\pi_1(\mathbb{C} \setminus \{-1, 1\}, 0)$. Any cyclically reduced word that represents a conjugate of a power of an element of the form (7.1) contains only powers of equal sign of the generators. Hence $w' = \mathrm{Id}$, and $w = a_{j_2}^{k_2'} a_{j_1}^{k_1'}$, equivalently,

$$a_{j_2}^{-k_2'} w_{e^{(2)}} = a_{j_1}^{k_1'} w_{e^{(1)}} . \tag{7.4}$$

Put $w_{e^{(1)}, e^{(2)}} = a_{j_2}^{-k_2'} w_{e^{(2)}} = a_{j_1}^{k_1'} w_{e^{(1)}}$. Then $w_{e^{(1)}, e^{(2)}} f_*(e^{(1)}) (w_{e^{(1)} e^{(2)}})^{-1} = a_{j_1}^{k_1}$ and $w_{e^{(1)}, e^{(2)}} f_*(e^{(2)}) (w_{e^{(1)}, e^{(2)}})^{-1} = a_{j_2}^{k_2}$.

7.6.3 As a corollary we see that $f_*(e^{(1)} e^{(2)})$ is conjugate to $a_{j_1}^{k_1} a_{j_2}^{k_2}$. Hence, either at least one of the monodromies $f_*(e^{(\ell)})$, $\ell = 1, 2$, is the identity, or $j_1 = j_2$, or $k_1 = k_2 = \pm 1$.

Further, if the monodromies $f_*(e^{(\ell)})$, $\ell = 1, \ldots, k$, along a collection of elements of $\mathcal{E}$ are conjugate to powers of a common single generator a_j, there is a single element of $\pi_1(\mathbb{C} \setminus \{-1, 1\}, 0)$ that conjugates the monodromy along each element of this collection to a power of a_j. Indeed, suppose $w_{e^{(\ell)}}$ conjugates the monodromy $f_*(e^{(\ell)})$ to a power of a_j. Then equality (7.4) implies to each pair $e^{(1)}$, $e^{(\ell)}$, $\ell = 2, \ldots, k$, of elements of this collection. By this equation $w_{e^{(\ell)}} = a_j^{k_\ell'} w_{e^{(1)}}$ and, hence $w_{e^{(1)}}^{-1}$ conjugates each $f_*(e^{(\ell)})$ to a power of a_j.

Moreover, if each monodromy of a collection $f_*(e^{(1)}), \ldots, f_*(e^{(k)})$, $e^{(\ell)} \in \mathcal{E}, \ell = 1, \ldots, k$, is conjugate to a power of a common element of form (7.1) and the monodromy $f_*(e^{(k+1)})$, $e^{(k+1)} \in \mathcal{E}$, is conjugate to a different element of form (7.1), then there exists a single element $w \in \pi_1(\mathbb{C} \setminus \{-1, 1\}, 0)$, that conjugates all these monodromies to powers of elements of form (7.1). Indeed, such a w exists for the pair $f_*(e^{(1)})$ and $f_*(e^{(k+1)})$, and by the preceding arguments w also conjugates $f_*(e^{(2)}), \ldots, f_*(e^{(k)})$ to powers of elements of form (7.1).

7.6.4 The same arguments apply if for two elements $e^{(1)}, e^{(2)} \in \mathcal{E}$ the monodromy $f_*(e^{(1)})$ is conjugate to a power of a_j and the monodromy $f_*(e^{(2)})$ is conjugate to a power of $a_1 a_2$. We replace the generators a_1, a_2 of the free group $\pi_1(\mathbb{C} \setminus \{-1, 1\}, 0)$ by the generators $A_1 = a_j$ and $A_2 = (a_1 a_2)^{-1}$. Notice that A_2 can be considered as element of the fundamental group of the thrice punctured plane that is represented by loops surrounding ∞ positively.
The arguments above imply the following.
If the monodromy $f_*(\tilde{e}) = w^{-1} v^k w$ along an element $\tilde{e} \in \mathcal{E}$ is a conjugate to the k-th power of an element v among $a_1, a_2, a_1 a_2$, with $|k| > 1$, then for any $e' \in \mathcal{E}$, $e' \neq \tilde{e}$, the monodromy $f_*(e')$ equals $w^{-1} v^{k'} w$ for an integer k'.

7.6.5 We consider now the case when there is no element in $\mathcal{E}$ the monodromy along which is conjugate to the k-th power of an element of form (7.1) with $|k| > 1$. We claim that if $g > 0$ and the monodromy along an element $e_{2j} \in \mathcal{E}$, $j \leq g$, (i.e. e_{2j} belongs to a pair corresponding to a handle) is not trivial and, hence, is equal to a conjugate $w^{-1} v^{\pm 1} w$ for an element v of form (7.1), then all monodromies are powers of the same element $w^{-1} v w$ or $w^{-1} v^{-1} w$, respectively. Indeed, by our assumption all non-trivial monodromies along elements e of $\mathcal{E}$ are conjugate to $v_e^{\pm 1}$ for an element v_e of the form (7.1). Consider the element e_{2j-1} for which the pair $e_{2j-1}, e_{2j} \in \mathcal{E}$ corresponds to a handle. Conjugating all monodromies by a single element, we suppose for instance that $f_*(e_{2j}) = a_1$. Since the monodromies corresponding to a pair of handles commute (see part 8.6.1 of the proof), the monodromy $f_*(e_{2j-1})$ is a power of a_1, hence, by our assumption $f_*(e_{2j-1})$ is equal to $a_1^{\pm 1}$, or it is the identity. Since for an element $e' \in \mathcal{E}$ different from e_{2j-1}, e_{2j} the products $e_{2j-1}\, e_{2j}^2\, e'$ and $e_{2j-1}\, e_{2j}^3\, e'$ are contained in $\mathcal{E}'$, the monodromies $f_*(e_{2j-1}\, e_{2j}^2\, e')$ and $f_*(e_{2j-1}\, e_{2j}^3\, e')$ must be conjugate to a power of an element of the form (7.1). If $f_*(e_{2j-1})$ equals a_1 or the identity, then the monodromy $f_*(e_{2j-1}\, e_{2j}^2\, e')$ can only be conjugate to a power of an element of the form (7.1), if $f_*(e')$ is a power of a_1. If $f_*(e_{2j-1})$ is equal to a_1^{-1}, then $f_*(e_{2j-1}\, e_{2j}^3\, e')$ can only be conjugate to a power of an element of the form (7.1), if $f_*(e')$ is a power of a_1.

The case when $f_*(e_{2j})$ equals a_1^{-1}, $a_2^{\pm 1}$, or $(a_1 a_2)^{\pm 1}$, or when the role of e_{2j-1} and e_{2j} is interchanged, can be treated similarly. The claim is proved.

7.6.6 Suppose $g(X) > 0$ but all monodromies along elements of $\mathcal{E}$ corresponding to handles are trivial. We claim that still all monodromies are powers of the same conjugate of an element of form (7.1). If the monodromies along all but possibly one element of $\mathcal{E}$ are trivial, there is nothing to prove. Hence, we may assume that $m > 2$. By our assumption all non-trivial monodromies along elements of $\mathcal{E}$ are conjugate to an element of the form (7.1) or are inverse to a conjugate of an element of the form (7.1). Let e_1 and e_2 be the elements of $\mathcal{E}$ corresponding to the first labeled handle. Suppose there is an unordered pair $\{e', e''\}$ of different elements of $\mathcal{E}$ so that the monodromy along each element of the pair is non-trivial. Note that none of the elements equals e_1 or e_2. Convert the unordered pair into an ordered pair (e', e''), so that the conjugacy class $\widehat{e'e_1e'e_2e''}$ can be represented by a simple closed curve. Assume that $f_*(e') = a_1$. If $f_*(e'')$ is not a power of a_1 then $f_*(e'e_1e'e_2e'') = f_*(e')^2 f_*(e'')$ (see Fig. 7.1) cannot be a power of an element of the form (7.1). The remaining cases in which $f_*(e')$ is of the form (7.1) or is inverse to an element of the form (7.1) are treated in the same way. The claim is proved.

7.6.7 Let $g(X) = 0$, i.e. X equals the smooth oriented 2-sphere S^2 with holes. Suppose $f : X \to \mathbb{C} \setminus \{-1, 1\}$ is a continuous mapping whose restriction to each annulus representing an element $e \in \mathcal{E}'$ has the Gromov-Oka property. Assume that the monodromies are not all conjugate to powers of a single common element of the form (7.1). Then by our assumption no monodromy is conjugate to the k-th power of an element of the form (7.1) with $|k| > 1$, and there are at least three holes, and at least two elements $e_{j'}$ and $e_{j''}$ in $\mathcal{E}$ with monodromies $f_*(e_{j'})$ and $f_*(e_{j''})$ being non-trivial powers of absolute value 1 of conjugates of different elements of the form (7.1).

Recall that the element $e_m = (\prod_{j=1}^{m-1} e_j)^{-1} \in \pi_1(X, q_0)$ is represented by a loop with base point q_0 that surrounds the last hole C_m counterclockwise. The product $\prod_{j=1}^{m} f_*(e_j)$ is equal to the identity. The product of two non-trivial powers of different elements of the form (7.1) cannot be equal to the identity. Hence, by part 7.6.3. of the proof there must be an integer number $j''' \in [1, m]$ different from j' and j'', for which $f_*(e_{j'''}) \neq \mathrm{Id}$.

The monodromies along two different elements from $\mathcal{E} \cup \{e_m\}$ cannot be non-trivial powers of the same element of the form (7.1). Otherwise by part 7.6.3. of the proof there would be an ordered couple or an ordered triple of different elements of $\mathcal{E} \cup \{e_m\}$ whose product is in $\mathcal{E}'$ but the monodromy along the product cannot be a conjugate of a power of an element of the form (7.1). Indeed, assume for instance that two monodromies $f_*(e')$ and $f_*(e'')$ are conjugate to powers of a_1 and a third momodromy $f_*(e''')$ is conjugate to a power of a_2. By part 7.6.3. of the proof we may assume that $f_*(e') = a_1^{k'}$, $f_*(e'') = a_1^{k''}$, and $f_*(e''') = a_2^{k'''}$ with $|k'| = |k''| = |k'''| = 1$. If k' and k'' have different sign, then one of the monodromies $f_*(e'e''')$ or $f_*(e''e''')$ cannot be conjugate to a power of an element of form (7.1). If k' and k'' have equal sign, the monodromy along the product of the three elements (e', e'', e''') in any order is not a power of an element of form (7.1). But the product of the three elements (e', e'', e''') for some order is an element of $\mathcal{E}'$. For other possible combinations of powers of elements of form (7.1) the arguments are the same.

7.6.8 Part 7.6.7 of the proof shows that the monodromy along at most three elements of $\mathcal{E} \cup \{e_m\}$ is nontrivial. Hence, there are exactly three elements $e_{j'}$, $e_{j''}$, and $e_{j'''}$ among the e_k, $k = 1, \ldots, m$, with non-trivial monodromy. After conjugating all monodromies by a single element of $\pi_1(\mathbb{C} \setminus \{-1, 1\})$ the monodromies along two of them are equal to either a_1 and a_2, respectively, or to a_1^{-1} and a_2^{-1}, respectively. (The combinations a_1, a_2^{-1} or a_1^{-1}, a_2 are impossible.) Order the three elements by (e', e'', e''') so that the product is in $\mathcal{E}'$. After a cyclic permutation which does not change the conjugacy class of the product, we may assume that the monodromy along e''' is not equal to a power of an a_j. Then the ordered triple of monodromies along (e', e'', e''') is either $(a_1, a_2, (a_1a_2)^{-1})$, or $(a_2, a_1, (a_2a_1)^{-1})$, or $(a_1^{-1}, a_2^{-1}, a_2a_1)$, or $(a_2^{-1}, a_1^{-1}, a_1a_2)$. The elements a_1, a_2, and $(a_1a_2)^{-1}$, respectively, of the fundamental group of the twice punctured complex plane are represented by curves that surround positively the points -1, 1, and ∞, respectively, the elements $a_1^{-1}, a_2^{-1}, a_2a_1$ are represented by curves that surround $-1, 1, \infty$ negatively.

The case $(a_1, a_2, (a_1a_2)^{-1})$ for the monodromies along (e', e'', e''') corresponds to the homotopy class that contains the following mapping. Let $\mathcal{C}', \mathcal{C}'', \mathcal{C}'''$ be the holes of X corresponding to e', e'', e''', respectively. Take points $p' \in \mathcal{C}'$, $p'' \in \mathcal{C}''$, and $p''' \in \mathcal{C}'''$, respectively. Denote by F an orientation preserving diffeomorphism from S^2 onto $\mathbb{P}^1$, that maps p' to -1, p'' to 1, and p''' to ∞. It is straightforward to check that F has the required monodromies along all generators. Hence, f is homotopic to $F \mid X$. The case $(a_2, a_1, (a_2a_1)^{-1})$ for the monodromies along (e', e'', e''') is similar. In these two cases for any orientation preserving conformal structure $\omega : X \to \omega(X)$, including conformal structures of first kind, the mapping $(F|X) \circ \omega^{-1}$ is homotopic to a mapping hat extends to $\mathbb{P}^1 = \omega(X)^c$ as a conformal self-diffeomorphism of $\mathbb{P}^1$ that maps X into $\mathbb{C} \setminus \{-1, 1\}$.

If f has monodromies $(a_1^{-1}, a_2^{-1}, a_2a_1)$ along (e', e'', e'''), then f is homotopic to $F \mid X$ for an orientation reversing diffeomorphism F from S^2 onto $\mathbb{P}^1$, that maps p' to -1, p'' to 1, and p''' to ∞. The case $(a_2^{-1}, a_1^{-1}, a_1a_2)$ is similar. In these cases for any orientation preserving conformal structure $\omega : X \to \omega(X)$, including conformal structures of first kind, the mapping $(F|X) \circ \omega^{-1}$ is homotopic to a mapping hat extends to $\mathbb{P}^1 = \omega(X)^c$ as an anti-conformal self-homeomorphism of $\mathbb{P}^1$ that maps X into $\mathbb{C} \setminus \{-1, 1\}$. But the mapping f does not have the Gromov-Oka property.

The latter fact can be seen as follows. Assume the contrary. Let $\omega_n : X \to \omega_n(X)$ be a sequence of conformal structures of second kind on X such that $X_n \overset{def}{=} \omega_n(X)$ can be identified with an increasing sequence of domains in $\mathbb{C} \setminus \{-1, 1\}$ whose union equals $\mathbb{C} \setminus \{-1, 1\}$. We may choose the ω_n uniformly converging on compact subsets of X. We identify the fundamental groups of X_n and of $\mathbb{C} \setminus \{-1, 1\}$ by the isomorphism induced by inclusion.

Consider the case when the monodromies of f along (e', e'', e''') are equal to $(a_1^{-1}, a_2^{-1}, a_2a_1)$. By our assumption for each n the mapping $f \circ \omega_n^{-1}$ is homotopic to a holomorphic mapping $f_n : X_n \to \mathbb{C} \setminus \{-1, 1\}$ with monodromies a_1^{-1} along e' and a_2^{-1} along e''. By Montel's Theorem there is a subsequence f_{n_k} that converges locally uniformly on $\mathbb{C} \setminus \{-1, 1\}$. The limit function F cannot be a constant (including the values of the constant -1, 1, or ∞), since for curves γ' and γ'' representing e' and e'', respectively, the set $f_{n_j}(\gamma') \cup f_{n_j}(\gamma'')$ separates -1, 1 and ∞. The limit function F is a holomorphic mapping from $\mathbb{C} \setminus \{-1, 1\}$ to itself. Hence it extends to a meromorphic function on $\mathbb{P}^1$, and defines therefore a branched covering of $\mathbb{P}^1$. This is impossible, since the curves γ' and γ'' are mapped to curves that surround -1 and 1, respectively, negatively. □

7.3 Gromov-Oka Mappings from Tori with a Hole to $\mathfrak{P}_3$

A Riemann surface of genus 1 will be called a torus and a Riemann surface of genus 1 with a hole will be called a torus with a hole. A smooth oriented surface of genus 1 will be called a smooth torus.

We consider now a connected smooth oriented surface X of genus one with a hole (i.e. a connected smooth oriented closed surface $X$of genus one with a point or a closed disc removed) and address the question which mappings $X \to \mathfrak{P}_n$ have the Gromov-Oka property. Recall that such a mapping can be considered as a separable quasipolynomial of degree n.

Recall that any continuous map from an oriented connected open smooth surface X into $\mathfrak{P}_2$ has the Gromov-Oka property.

We consider Problem 7.4 and Problem 7.4a for the case when the target manifold equals $\mathfrak{P}_3$.

We will say that a separable quasipolynomial on an open Riemann surface X is isotopic to a holomorphic quasipolynomial, if the corresponding mapping to $\mathfrak{P}_n$ is homotopic to a holomorphic one. We will say that a separable quasipolynomial f on a finite open oriented smooth surface X is holomorphic for the conformal structure $\omega : X \to \omega(X)$ (with $\omega(X)$ being a Riemann surface) if $f \circ \omega^{-1}$ is a holomorphic quasipolynomial on $\omega(X)$.

The soft Oka Principle states that for each separable quasipolynomial f of degree n on a finite open oriented smooth surface X there exists an orientation preserving conformal structure on X for which the quasipolynomial is isotopic to a holomorphic quasipolynomial.

In the following theorem we consider quasipolynomials of degree 3 on a torus X with a hole. The theorem shows that the obstructions for a separable quasipolynomial to be isotopic to a holomorphic quasipolynomial are discrete.

Theorem 7.3 *Let X be a torus with a hole, and let $\mathcal{E} = \{e_1, e_2\}$ be a standard system of generators of the fundamental group of X with base point q_0. Denote by $\mathcal{E}'$ the set $\{e_1, e_2, e_2e_1^{-1}, e_2e_1^{-2}, e_1e_2e_1^{-1}e_2^{-1}\}$.*

Let f be a separable quasipolynomial of degree 3 on X such that for each $e \in \mathcal{E}'$ the quasipolynomial f is isotopic to an algebroid function for an orientation preserving conformal structure w_e on X with a holomorphic annulus representing e of conformal module larger than $\frac{\pi}{2}(\log \frac{3+\sqrt{5}}{2})^{-1}$. Suppose X is a Riemann surface of second kind. Then the quasipolynomial is isotopic to an algebroid function on X.

The following statement is formally slightly stronger. It follows from the proof of Theorem 7.3.

Let X, $\mathcal{E}$ and $\mathcal{E}'$ be as in the theorem. A continuous mapping $f : X \to \mathfrak{P}_3$ has the Gromov-Oka property if and only if for each $e \in \mathcal{E}'$ the restriction of the mapping to an annulus in X that represents $\hat{e}$ has the Gromov-Oka property.

The crucial step for the proof of Theorem 7.3 is Theorem 7.4 below. The rest of this chapter is devoted to the proof of Theorem 7.4. Theorem 7.3 and results related to bundles will be proved in the next chapter. We first provide two lemmas that are needed for the proof of Theorem 7.4.

The proof of the following lemma on permutations can be extracted for instance from [81]. For convenience of the reader we give the short argument.

Lemma 7.7 *Let n be a prime number. Any Abelian subgroup of the symmetric group $\mathcal{S}_n$ which acts transitively on a set consisting of n points is generated by an n-cycle.*

Proof Let $\mathcal{S}_n$ be the group of permutations of elements of the set $\{1, \ldots, n\}$. Suppose the elements $s_j \in \mathcal{S}_n$, $j = 1, \ldots, m$, commute and the subgroup $\langle s_1, \ldots, s_m \rangle$ of $\mathcal{S}_n$ generated by the s_j, $j = 1, \ldots, m$, acts transitively on the set $\{1, \ldots, n\}$. Let $A_{s_1} \subset \{1, \ldots, n\}$ be a minimal s_1-invariant subset. Then $s_1 \mid A_{s_1}$ is a cycle of length $k_1 = |A_{s_1}|$. (The order $|A|$ of a set A is the number of elements of this set.) For any integer ℓ the set $s_2^{\ell}(A_{s_1})$ is minimal s_1-invariant. Hence, two such sets are either disjoint or equal. Take the minimal union $A_{s_1 s_2}$ of sets of the form $s_2^{\ell}(A_1)$ for some integers ℓ, which contains A_{s_1} and is invariant under s_2. Then s_2 moves the $s_2^{\ell}(A_{s_1})$ along a cycle of length k_2 such that $|A_{s_1 s_2}| = k_1 \cdot k_2$. $A_{s_1 s_2}$ is a minimal subset of $\{1, \ldots, n\}$ which is invariant for both, s_1 and s_2. Continue in this way. We obtain a minimal set $A = A_{s_1 \ldots s_m} \subset \{1, \ldots, n\}$ which is invariant under all s_j. By the transitivity condition $A = \{1, \ldots, n\}$. The order $|A|$ equals $k_1 \cdot \ldots \cdot k_m$. Since n is prime, exactly one of the factors, k_{j_0} equals n, the other factors equal 1. Then for some number j_0 the element s_{j_0} is a cycle of length n and, if $j_0 \neq 1$, then $|A_{s_1 \ldots s_{j_0 - 1}}| = 1$. Since each s_j commutes with s_{j_0}, each s_j is a power of s_{j_0}. Indeed, consider a bijection of $\{1, \ldots, n\}$ onto the set of n-th roots of unity so that s_{j_0} corresponds to rotation by the angle $\frac{2\pi}{n}$. In other words, put $\zeta = e^{\frac{2\pi i}{n}}$. The permutation s_{j_0} acts on $\{1, \zeta, \zeta^2, \ldots, \zeta^{n-1}\}$ by multiplication by ζ. Consider an arbitrary s_j. Then for some integer ℓ_j, $s_j(\zeta) = \zeta^{\ell_j}$, i.e. $s_j(\zeta) = \zeta^{\ell_j - 1} \cdot \zeta = (s_{j_0})^{\ell_j - 1}(\zeta)$. Then for any other n-th root of unity ζ^m

$$
\begin{aligned}
s_j(\zeta^m) &= s_j((s_{j_0})^{m-1}(\zeta)) \\
&= (s_{j_0})^{m-1}(s_j(\zeta)) = \zeta^{m-1} \cdot \zeta^{\ell_j} = \zeta^{\ell_j - 1} \cdot \zeta^m = (s_{j_0})^{\ell_j - 1}(\zeta^m).
\end{aligned}
$$

Hence $s_j = (s_{j_0})^{\ell_j - 1}$. □

Lemma 7.8 *Let e_1, e_2 be generators of a free group F_2. Let $\mathcal{E}' \subset F_2$ be the finite subset $\mathcal{E}' = \{e_1, e_2, e_2 e_1^{-1}, e_2 e_1^{-2}\}$ of primitive elements of the group F_2. Put $\mathcal{E}'^{-1} \stackrel{def}{=} \{e^{-1} : e \in \mathcal{E}'\}$. Suppose $\Psi : F_2 \to \mathcal{S}_3$ is a homomorphism from F_2 into the symmetric group $\mathcal{S}_3$ whose image is an Abelian subgroup of $\mathcal{S}_3$ which acts transitively on the set of three elements. Then there are elements $\mathsf{e}_1, \mathsf{e}_2 \in \mathcal{E}' \cup \mathcal{E}'^{-1}$ that generate F_2 such that $\Psi(\mathsf{e}_1)$ is a 3-cycle, $\Psi(\mathsf{e}_2) = \mathrm{id}$ and the commutator $[\mathsf{e}_1, \mathsf{e}_2]$ is conjugate to $[e_1, e_2]$.*

Proof By Lemma 7.7 the image of one of the original generators of F_2 is a 3-cycle. If $\Psi(e_1)$ is a 3-cycle, then $\Psi(e_2 e_1^{-q})$ is the identity for q being either 0 or 1 or 2.

(Recall that $\Psi(e_2)$ is a power of $\Psi(e_1)$ and $\Psi(e_1)^3 = \text{id}$.) Also,

$$[e_1, e_2 e_1^{-q}] = e_1 e_2 e_1^{-q} e_1^{-1} e_1^{q} e_2^{-1} = [e_1, e_2].$$

We may take the pair $\mathsf{e}_1 = e_1, \mathsf{e}_2 = e_2 e_1^{-q}$.

If $\Psi(e_1)$ is the identity then $\Psi(e_2)$ is a 3-cycle. The pair $\mathsf{e}_1 = e_2, \mathsf{e}_2 = e_1^{-1}$ generates F_2 and the commutator $[\mathsf{e}_1, \mathsf{e}_2] = e_2 e_1^{-1} e_2^{-1} e_1$ is conjugate to the commutator $[e_1, e_2]$. □

Let X be a torus with a disc removed. Its fundamental group $\pi_1(X, q_0)$ with base point q_0 is isomorphic to the free group F_2 with two generators.

Theorem 7.4 *Suppose X is a torus with a hole. Let $\mathcal{E} = \{e_1, e_2\}$ be a standard system of generators of the fundamental group of X with base point q_0, and $\mathcal{E}' = \{e_1, e_2, e_2 e_1^{-1}, e_2 e_1^{-2}, e_1 e_2 e_1^{-1} e_2^{-1}\}$. Let f be a separable quasipolynomial of degree 3 on X such that for each $e \in \mathcal{E}'$ the quasipolynomial f is isotopic to an algebroid function for an orientation preserving conformal structure w_e on X with a holomorphic annulus representing $\hat{e}$ of conformal module larger than $\frac{\pi}{2}(\log \frac{3+\sqrt{5}}{2})^{-1}$. Then the isotopy class of f corresponds to the conjugacy class of a homomorphism*

$$\Phi : \pi_1(X, q_0) \to \Gamma \subset \mathcal{B}_3$$

for a subgroup Γ of $\mathcal{B}_3$ which is generated either by $\sigma_1 \sigma_2$, or by $\sigma_1 \sigma_2 \sigma_1$, or by σ_1 and Δ_3^2.

In particular, in all cases the image $\Phi([e_1, e_2])$ of the commutator of the generators of Γ is the identity.

For the proof of the theorem we need two lemmas which we state and prove before proving the theorem.

Lemma 7.9 *Let $\mathsf{b}_1 = (\sigma_1 \sigma_2)^{\pm 1}$ and $\mathsf{b}_2 = w^{-1} \sigma_1^{2k} w$ for an integer k and a braid $w \in \mathcal{B}_3$. If the commutator $[\mathsf{b}_1, \mathsf{b}_2]$ is conjugate to $\sigma_1^{2k^*} \Delta_3^{2\ell^*}$ for integers k^* and ℓ^* then b_2 is the identity and hence also the commutator is the identity.*

Proof of Lemma 7.9 Let $b \in \mathcal{B}_n$ be a pure braid with base point $E_n \in C_n(\mathbb{C})/\mathcal{S}_n$. Fix a point $(x_1, \ldots, x_n) \in C_n(\mathbb{C})$ that projects to E_n. Label each strand of b by the number j of its initial point x_j. Let $\{i, j\}$ be an unordered pair of distinct integer numbers between 1 and n. The linking number $\ell_{\{i,j\}}$ of the i-th and j-th strand is defined as follows. Discard all strands except the i-th and the j-th strand. We obtain a pure braid σ^{2m}. We call the integral number m the linking number of the two strands and denote it by $\ell_{\{i,j\}}$.

Note that the linking numbers $\ell^*_{\{i,j\}}$ of the braid $\sigma_1^{2k^*} \Delta_3^{2\ell^*}$ are equal to $\ell^*_{\{12\}} = k^* + \ell^*, \ell^*_{\{23\}} = \ell^*, \ell^*_{\{13\}} = \ell^*$. Since the braid is conjugate to a commutator, the sum of exponents of generators in a word representing it must be zero. This means that $2k^* + 6\ell^* = 0$. Hence, the (unordered) collection of linking numbers of $\sigma_1^{2k^*} \Delta_3^{2\ell^*}$

is $\{-2\ell^*, \ell^*, \ell^*\}$. Since conjugation only permutes linking numbers between pairs of strands of a pure braid, the unordered collection of linking numbers of pairs of strands of $[\mathrm{b}_1, \mathrm{b}_2]$ equals $\{-2\ell^*, \ell^*, \ell^*\}$ for the integer ℓ^*. For the pure braid σ_1^{2k} the linking number between the first and the second strand equals k, the linking numbers of the remaining pairs of strands equal zero.

Consider a not necessarily pure braid $w \in \mathcal{B}_3$ with base point E_3 and fix a lift (x_1, x_2, x_3) of E_3. Let S_w be the permutation $S_w = \tau_3(w)$. The permutation S_w defines a permutation s_w acting on the set $(1, 2, 3)$, so that $S_w((x_1, x_2, x_3)) = (x_{s_w(1)}, x_{s_w(2)}, x_{s_w(3)})$. Order the linking numbers of pairs of strands of a pure braid $b \in \mathcal{B}_3$ as $(\ell_{\{2,3\}}, \ell_{\{1,3\}}, \ell_{\{1,2\}})$. Notice that the complement of the set $\{2, 3\}$ (of the set $\{1, 3\}$, or $\{1, 2\}$, respectively) in $\{1, 2, 3\}$ is $\{1\}$ ($\{2\}$, or $\{3\}$, respectively). The ordered tuple of linking numbers of pairs of strands of $w^{-1}\, b\, w$ equals

$$S_w((\ell_{\{2,3\}}, \ell_{\{1,3\}}, \ell_{\{1,2\}})) = (\ell_{\{s_w(2)s_w(3)\}}, \ell_{\{s_w(1)s_w(3)\}}, \ell_{\{s_w(1)s_w(2)\}}).$$

Hence, the ordered tuple of linking numbers between pairs of strands of $\mathrm{b}_2 = w^{-1}\,\sigma_1^{2k}\, w$ equals $S_w(0, 0, k)$. Similarly, the ordered tuple of linking numbers between pairs of strands of b_2^{-1} is equal to $S_w(0, 0, -k)$.

Consider the commutator $\mathrm{b}_1\, \mathrm{b}_2\, \mathrm{b}_1^{-1}\, \mathrm{b}_2^{-1}$.

The ordered tuple of linking numbers of pairs of strands of the pure braid $\mathrm{b}_1\, \mathrm{b}_2\, \mathrm{b}_1^{-1}$ equals $(S')^{-1} \circ S_w(0, 0, k)$ for the permutation $S' = \tau_3(\mathrm{b}_1)$ (see Fig. 7.2). Hence the ordered tuple of linking numbers of pairs of strands of the commutator $\mathrm{b}_1\, \mathrm{b}_2\, \mathrm{b}_1^{-1} \circ \mathrm{b}_2^{-1}$ equals $(S')^{-1} \circ S_w(0, 0, k) + S_w(0, 0, -k)$. Since S' is a 3-cycle, the mapping $S'(x_1, x_2, x_3) = (x_{s'(1)}, x_{s'(2)}, x_{s'(3)})$ does not fix (setwise) any pair of points among the x_1, x_2, x_3. Hence the unordered 3-tuple of linking numbers of $\mathrm{b}_1\, \mathrm{b}_2\, \mathrm{b}_1^{-1}\, \mathrm{b}_2^{-1}$ is $\{k, -k, 0\}$. It can coincide with an unordered 3-tuple of the form $\{-2\ell^*, \ell^*, \ell^*\}$ only if $k = \ell^* = 0$. Hence b_2 is the identity and the commutator is the identity. □

Lemma 7.10 *The centralizer of* σ_1^k *in* $\mathcal{B}_3$, $k \neq 0$ *an integral number, equals* $\{\sigma_1^{k'}\, \Delta_3^{2\ell'} : k', \ell' \in \mathbb{Z}\}$.

Proof Consider for each $b \in \mathcal{B}_3$ the modular transformation T_b on the Teichmüller space $\mathcal{T}(0, 4) \cong \mathbb{C}_+$ that is associated to a mapping $\varphi_{b,\infty} \in \mathfrak{m}_{b,\infty}$. Recall that

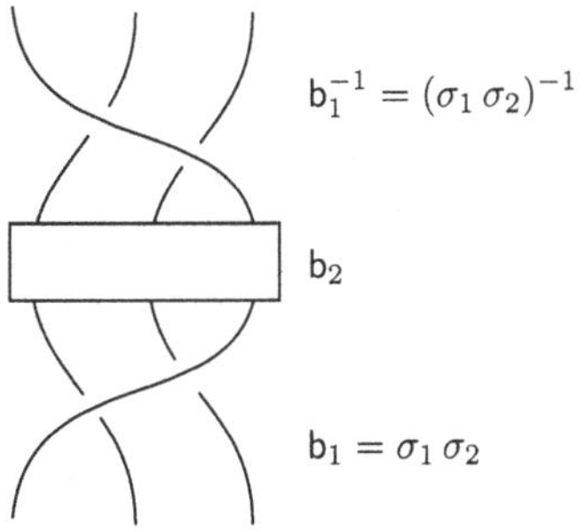

Fig. 7.2 The braid $b_1 b_2 b_1^{-1} = (\sigma_1\,\sigma_2) b_2 (\sigma_1\,\sigma_2)^{-1}$

the mapping $\mathcal{B}_3/\mathcal{S}_3 \ni b/\mathcal{S}_3 \to T_b$ is a bijection that satisfies equality (1.42) (see Sect. 1.7). The group of modular transformations on $\mathbb{C}_+$ is isomorphic to $SL_2(\mathbb{Z})/\pm \text{Id}$. Recall that $T_{\sigma_1}(z) = \frac{z}{1-z}$ corresponds to $\begin{pmatrix} 1 & 0 \\ -1 & 1 \end{pmatrix} / \pm \text{Id}$ (see Sect. 3.2, Example 3.8). Suppose the braid b commutes with σ_1^k for some non-zero integer k and corresponds to $V/\pm \mathrm{I}$ where $\mathrm{V} = \begin{pmatrix} v_{11} & v_{12} \\ v_{21} & v_{22} \end{pmatrix}$. Then

$$\begin{pmatrix} 1 & 0 \\ -k & 1 \end{pmatrix} \begin{pmatrix} v_{11} & v_{12} \\ v_{21} & v_{22} \end{pmatrix} = \begin{pmatrix} v_{11} & v_{12} \\ v_{21} & v_{22} \end{pmatrix} \begin{pmatrix} 1 & 0 \\ -k & 1 \end{pmatrix}$$

i.e.

$$\begin{pmatrix} v_{11} & v_{12} \\ -k\, v_{11} + v_{21} & -k\, v_{12} + v_{22} \end{pmatrix} = \begin{pmatrix} v_{11} - k\, v_{12} & v_{12} \\ v_{21} - k\, v_{22} & v_{22} \end{pmatrix} .$$

Since $k \neq 0$ we have $v_{12} = 0$ and $v_{11} = v_{22}$. Since $\det \mathrm{V} = 1$ we obtain $\mathrm{V} = \pm \begin{pmatrix} 1 & 0 \\ -m & 1 \end{pmatrix}$ for an integer m. Thus $b = \sigma_1^m$ for some integer m. The lemma is proved. □

Proof of Theorem 7.4 Let $\Phi : \pi_1(X, x_0) \to \mathcal{B}_3$ be a homomorphism whose conjugacy class corresponds to the isotopy class of f. Denote by $\Psi = \tau_3 \circ \Phi : \pi_1(X, x_0) \to \mathcal{S}_3$ the related homomorphism into the symmetric group. Since the quasipolynomial F is isotopic to an algebroid function for the conformal structure $\omega_{[e_1,e_2]}$ with an annulus of conformal module larger than $\frac{\pi}{2}(\log \frac{3+\sqrt{5}}{2})^{-1}$ representing the commutator $[e_1, e_2]$, by Lemma 6.5 the conformal module of the commutator $[b_1, b_2]$ equals infinity. Hence, by Lemma 6.5 $\Psi([e_1, e_2]) = \text{id}$.

Consider first the case when the subgroup $\Psi(\pi_1(X, x_0))$ of $\mathcal{S}_3$ acts transitively on the set of three points. Apply Lemma 7.8 to the free group $\pi_1(X, x_0)$ with generators e_1 and e_2 and to the homomorphism Ψ. We obtain new generators $\mathrm{e}_1, \mathrm{e}_2 \in \mathcal{E}' \cup \mathcal{E}'^{-1}$ of $\pi_1(X, x_0)$ (with $\mathcal{E}'$ and $\mathcal{E}'^{-1}$ being the sets of the lemma) such that $\Psi(\mathrm{e}_1)$ is a 3-cycle and $\Psi(\mathrm{e}_2) = \text{id}$. Put $s_1 = \Psi(\mathrm{e}_1)$, $s_2 = \Psi(\mathrm{e}_2)$, $b_1 = \Phi(\mathrm{e}_1)$, $b_2 = \Phi(\mathrm{e}_2)$. Then b_2 and $[b_1, b_2]$ are pure braids (the latter holds since $[\mathrm{e}_1, \mathrm{e}_2]$ is conjugate to $[e_1, e_2]$). Since f is isotopic to an algebroid function for the conformal structure w_{e_1}, Lemma 6.5 implies $\mathcal{M}(b_1) = \infty$. Since $\tau_3(b_1)$ is a 3-cycle, by the same lemma the braid b_1 must be conjugate to an integral power of $\sigma_1 \sigma_2$.

After conjugating Φ we may assume that $b_1 = (\sigma_1 \sigma_2)^{3\ell \pm 1} = (\sigma_1 \sigma_2)^{\pm 1} \cdot \Delta_3^{2\ell}$ for an integer ℓ. By Lemma 6.5 we also have $\mathcal{M}(b_2) = \infty$ for the pure braid b_2. Hence, $b_2 = w^{-1} \sigma_1^{2k} \Delta_3^{2\ell'} w$ for integers k and ℓ' and a conjugating braid $w \in \mathcal{B}_3$. Since $\mathcal{M}([b_1, b_2]) = \infty$ the commutator $[b_1, b_2]$ is conjugate to $\sigma_1^{2k^*} \Delta_3^{2\ell^*}$ for integers k^* and ℓ^*.

Take into account that $\Delta_3^{2\ell}$ commutes with each 3-braid and apply Lemma 7.9 to $\mathsf{b}_1 = (\sigma_1\sigma_2)^{\pm 1}$ and $\mathsf{b}_2 = w^{-1}\sigma_1^{2k^*}w$. This gives the statement of Theorem 7.4 for the case when the subgroup $\Psi(\pi_1(X, x_0))$ acts transitively on the set of three points.

Consider now the case when the subgroup $\Psi(\pi_1(X, x_0))$ of $\mathcal{S}_3$ does not act transitively on the set of three points. Then $\Psi(\pi_1(X, x_0))$ is generated either by a transposition (we may assume the transposition to be (12) by choosing the mapping Φ in its conjugacy class) or is equal to the identity in $\mathcal{S}_3$.

There are generators $\mathsf{e}_1, \mathsf{e}_2 \in \mathcal{E}' \cup \mathcal{E}'^{-1}$ such that $[\mathsf{e}_1, \mathsf{e}_2]$ is conjugate to $[e_1, e_2]$ and $\Psi(\mathsf{e}_2) = \text{id}$. Indeed, this is clear if $\Psi(\pi_1(X, x_0))$ is the identity. Suppose $\Psi(\pi_1(X; x_0))$ is generated by the transposition (12) and neither $\Psi(e_1)$ nor $\Psi(e_2)$ is the identity. Then $\Psi(e_1) = \Psi(e_2) = (12)$. Hence, $\Psi(e_2 e_1^{-1}) = \text{id}$, and for the generators $\mathsf{e}_1 \overset{def}{=} e_1$ and $\mathsf{e}_2 \overset{def}{=} e_2 e_1^{-1}$ the claimed statements hold.

Use the notation $b_1 = \Phi(\mathsf{e}_1)$, $b_2 = \Phi(\mathsf{e}_2)$. By Lemma 6.5 we may assume, that b_1 is conjugate to either $\sigma_1^{k_1}\,\Delta_3^{2\ell_1}$ or to $(\sigma_1\sigma_2\sigma_1)^{k_1} = \Delta_3^{k_1}$ for an odd integer k_1 and an integer ℓ_1, and b_2 is conjugate to $\sigma_1^{2k_2}\,\Delta_3^{2\ell_2}$ for integers k_2 and ℓ_2. Conjugating Φ, we may assume that $b_1 = B_1\,\Delta_3^{2\ell_1}$ with either $B_1 = \sigma_1^{k_1}$ or $B_1 = \sigma_2\,\sigma_1^2 = \sigma_1^{-1}(\sigma_1\sigma_2\sigma_1)\sigma_1$. Respectively, b_2 is conjugate to $B_2\,\Delta_3^{2\ell_2}$ with $B_2 = \sigma_1^{2k_2}$. Since $[b_1, b_2]$ is a pure braid with infinite conformal module, Lemma 6.5 implies that $[b_1, b_2]$ is conjugate to $\sigma_1^{2k^*}\,\Delta_3^{2\ell^*}$. Since the commutator has degree zero the equality $2k^* + 6\ell^* = 0$ holds and the unordered tuple of linking numbers of $[b_1, b_2]$ equals $\{\ell^*, \ell^*, -2\ell^*\}$.

This implies that the commutator $[b_1, b_2]$ and the braid B_2 are equal to the identity. Indeed, suppose $b_2 = w^{-1}\,\sigma_1^{2k_2}\,w\,\Delta_3^{2\ell_1}$ for a braid $w \in \mathcal{B}_3$. The ordered tuple of linking numbers of pairs of strands of $B_2 = w^{-1}\,\sigma_1^{2k_2}\,w$ equals $S_w(0, 0, k_2)$ and the respective ordered triple for B_2^{-1} is $S_w(0, 0, -k_2)$. Here $S_w = \tau_3(w)$. The ordered tuple of linking numbers of pairs of strands of $B_1\,B_2\,B_1^{-1}$ equals $(S')^{-1} \circ S_w(0, 0, k_2)$, where $S' = \tau_3(B_1)$. Since Δ_3^2 is in the center of $\mathcal{B}_3$, the unordered tuple of linking numbers of pairs of strands of $[b_1, b_2]$ is equal to that of $[B_1, B_2]$, i.e. it equals either $\{k_2, -k_2, 0\}$ or $\{0, 0, 0\}$ (in dependence on S_w). Either of these unordered tuples can be equal to $\{\ell^*, \ell^*, -2\ell^*\}$ for some integer ℓ^* only if $\ell^* = 0$. We obtained that $[b_1, b_2] = \text{id}$ also in this case. By Lemma 7.10 the group $\Phi(\pi_1(X, x_0))$ is generated either by (a power of) σ_1 and Δ_3^2, or by (a power of) $\sigma_2\,\sigma_1^2$ and Δ_3^2. The theorem is proved. □

We do not know the answer to the following problem.

Problem 7.5 *Can two braids in $\mathcal{B}_3$ of zero entropy have a non-trivial commutator of zero entropy?*

If the answer is negative, then Theorem 7.4 holds with $\mathcal{E}' = \mathcal{E}$. The proof of Theorem 7.4 gives the following corollary which is weaker than a negative answer to Problem 7.5.

Corollary 7.1 *Let b_1 and b_2 be braids in $\mathcal{B}_3$ with $h(\mathsf{b}_1) = h(\mathsf{b}_2) = h([\mathsf{b}_1, \mathsf{b}_2]) = 0$. If one of the braids is pure or*

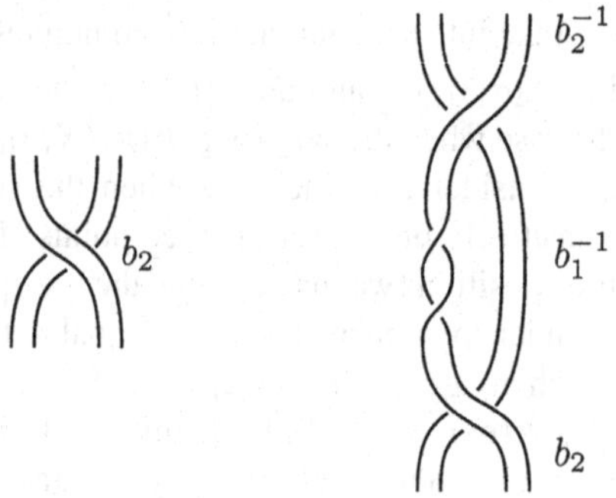

Fig. 7.3 Two 4-braids of vanishing entropy with non-trivial commutator

$$h(b_2\, b_1^{-1}) = h(b_2\, b_1^{-2}) = 0$$

then $[b_1, b_2] = \mathrm{id}$.

On the other hand D.Calegari and A.Walker suggested the following example of two elements of the braid group $\mathcal{B}_3$ with non-trivial commutator of zero entropy.

Example 7.1 *The commutator of the non-commuting braids* $b_1 = (\sigma_2)^{-1}\,\sigma_1$ *and* $b_2 = \sigma_2\,(\sigma_1)^{-1}$ *has entropy zero.*

It is enough to show that the commutator of the two braids equals $[b_1, b_2] = (\sigma_2)^{-6}\,\Delta_3^2$. Since

$$\begin{aligned} b_1^{-1}\,b_2^{-1} &= (\sigma_1)^{-1}\,\sigma_2\,\sigma_1\,(\sigma_2)^{-1} &&= (\sigma_1)^{-1}\,\sigma_2\,\sigma_1\,\sigma_2\,(\sigma_2)^{-2} \\ &= (\sigma_1)^{-1}\,\sigma_1\,\sigma_2\,\sigma_1\,(\sigma_2)^{-2} &&= \sigma_2\,\sigma_1\,(\sigma_2)^{-2}, \\ b_1\,b_2 &= (\sigma_2)^{-1}\,\sigma_1\,\sigma_2\,(\sigma_1)^{-1} &&= (\sigma_2)^{-2}\,\sigma_2\,\sigma_1\,\sigma_2\,(\sigma_1)^{-1} \\ &= (\sigma_2)^{-2}\,\sigma_1\,\sigma_2\,\sigma_1(\sigma_1)^{-1} &&= (\sigma_2)^{-2}\,\sigma_1\,\sigma_2, \end{aligned}$$

we obtain

$$\begin{aligned} [b_1, b_2] = b_1 b_2 b_1^{-1} b_2^{-1} &= (\sigma_2)^{-3}\,\sigma_2\,\sigma_1\,\sigma_2 \cdot \sigma_2\,\sigma_1\,\sigma_2\,(\sigma_2)^{-3} \\ &= (\sigma_2)^{-3}\,\Delta_3^2\,(\sigma_2)^{-3} = (\sigma_2)^{-6}\,\Delta_3^2. \end{aligned}$$

It is clear now that the commutator $[b_1, b_2]$ has entropy zero.

Problem 7.5 has a positive answer for braid groups on more than 3 strands. Consider the braids $b_1 = \sigma_1^{-2}$ and $b_2 = (\sigma_2)^{-1}\,(\sigma_1)^{-1}\,(\sigma_3)^{-1}\,(\sigma_2)^{-1}$ in $\mathcal{B}_4$ of zero entropy. Their commutator equals $[b_1, b_2] = \sigma_1^{-2} \cdot \sigma_3^2 \neq \mathrm{id}$ and is a non-trivial braid of zero entropy (see Fig. 7.3).

Notice the following fact. Let X be a torus with a disc removed. Consider the separable quasipolynomial of degree four on X whose isotopy class corresponds to the conjugacy class of the homomorphism $\Phi : \pi_1(X, x_0) \to \mathcal{B}_4$ with $\Phi(e_1) = b_1$

and $\Phi(e_2) = b_2$ for the just defined 4-braids b_2 and b_2. By a similar argument as in the proof of Theorem 7.3 the quasipolynomial is isotopic to an algebroid function for each conformal structure of second kind on X.

Problem 7.6 *Consider separable quasipolynomials of degree $n \geq 4$ on a torus with a hole. Are the obstructions for them to be isotopic to holomorphic quasipolynomials discrete like in the case $n = 3$?*

Is there an analogue of Corollary 7.1 for braid groups on more than three strands?

Chapter 8
Gromov's Oka Principle for (g, m)-Fiber Bundles

In this chapter we prove theorems related to the failure and the restricted validity of the Gromov-Oka Principle for bundles whose fibers are Riemann surfaces of type $(1, 1)$ (in other words, the fibers are once punctured tori). The problem can be reduced to the respective problem for $(0, 4)$-bundles (and, hence, for mappings into $\mathfrak{P}_3$) by considering double branched coverings.

8.1 (g, m)-Fiber Bundles. Statement of the Problem

We will consider bundles whose fibers are connected closed oriented surfaces of genus $\mathsf{g} \geq 0$ with $\mathsf{m} \geq 0$ distinguished points. Recall that a connected smooth closed oriented surface of genus g with m distinguished points is called a surface of type (g, m). In case the surface is equipped with a complex structure we call it a Riemann surface of type (g, m).

Definition 8.1 (Smooth Oriented (g, m) Fiber Bundles) *Let X be a smooth oriented manifold of dimension k, let $\mathcal{X}$ be a smooth (oriented) manifold of dimension $k + 2$ and $\mathcal{P} : \mathcal{X} \to X$ an orientation preserving smooth proper submersion such that for each point $x \in X$ the fiber $\mathcal{P}^{-1}(x)$ is a smooth closed oriented surface of genus g. Let $\boldsymbol{E}$ be a smooth submanifold of $\mathcal{X}$ that intersects each fiber $\mathcal{P}^{-1}(x)$ along a set E_x of m distinguished points. Then the tuple $\mathfrak{F}_{g,m} = (\mathcal{X}, \mathcal{P}, \boldsymbol{E}, X)$ is called a smooth (oriented) fiber bundle over X with fibers being smooth closed oriented surfaces of genus g with m distinguished points (for short, a smooth oriented (g, m)-bundle).*

If $\mathsf{m} = 0$ the set $\boldsymbol{E}$ is the empty set and we will often denote the bundle by $(\mathcal{X}, \mathcal{P}, X)$. If $\mathsf{m} > 0$ the mapping $x \to E_x$ locally defines m smooth sections of the $(\mathsf{g}, 0)$-bundle $(\mathcal{X}, \mathcal{P}, X)$ that is obtained by forgetting all distinguished points. A section of the bundle $(\mathcal{X}, \mathcal{P}, X)$ is a continuous mapping $s : X \to \mathcal{X}$ with $\mathcal{P} \circ s$

B. Jöricke, *Braids, Conformal Module, Entropy, and Gromov's Oka Principle*, Lecture Notes in Mathematics 2360, https://doi.org/10.1007/978-3-031-67288-0_8

being the identity on X. (g, 0)-bundles will also be called genus g fiber bundles. For $\mathsf{g} = 1$ and $\mathsf{m} = 0$ the bundle is also called an elliptic fiber bundle. We will consider mostly the case when $2\mathsf{g} - 2 + \mathsf{m} > 0$.

Let S be a smooth reference surface of genus g and $E \subset S$ a set of m distinguished points. For an open subset U of X we consider the trivial bundle (also called product bundle) $\big(U \times S, \mathrm{pr}_1, U \times E, U\big)$ with set $\{x\} \times E$ of distinguished points in the fiber $\{x\} \times S$ over x. Here $\mathrm{pr}_1 : U \times S \to U$ is the projection onto the first factor. By Ehresmann's Fibration Theorem each smooth (g, m)-bundle $\mathfrak{F}_{\mathsf{g},\mathsf{m}} = (\mathcal{X}, \mathcal{P}, \boldsymbol{E}, X)$ with set of distinguished points $E_x \overset{def}{=} \boldsymbol{E} \cap \mathcal{P}^{-1}(x)$ in the fiber over x is locally smoothly trivial, i.e. each point in X has a neighbourhood $U \subset X$ such that for a surjective diffeomorphism $\varphi_U : \mathcal{P}^{-1}(U) \to U \times S$ the diagram

$$\begin{array}{ccc} \mathcal{P}^{-1}(U) & \xrightarrow{\ \varphi_U\ } & U \times S \\ {\scriptstyle \mathcal{P}}\big\downarrow & \swarrow{\scriptstyle \mathrm{pr}_1} & \\ U & & \end{array}$$

is commutative and φ_U maps $\boldsymbol{E} \cap \mathcal{P}^{-1}(U)$ onto $U \times E$, equivalently, for all $x \in U$ the diffeomorphism φ_U maps the set of distinguished points $E_x = \boldsymbol{E} \cap \mathcal{P}^{-1}(x)$ in the fiber $\mathcal{P}^{-1}(x)$ to the set of distinguished points $\{x\} \times E$ in the fiber $\{x\} \times S$.

The idea of the proof of Ehresmann's Theorem is the following. Choose smooth coordinates on U by a mapping from a rectangular box in $\mathbb{R}^n$ to U. Consider smooth vector fields v_j on U, which form a basis of the tangent space of U at each point of U. Take smooth vector fields V_j on $\mathcal{P}^{-1}(U)$ that are tangent to $\boldsymbol{E}$ at points of this set and are mapped to v_j by the differential of $\mathcal{P}$. Such vector fields can easily be obtained locally. To obtain the globally defined vector fields V_j on $\mathcal{P}^{-1}(U)$ one uses partitions of unity. The required diffeomorphism φ_U is obtained by composing the flows of the vector fields V_j (in any fixed order).

In this way a trivialization of the bundle can be obtained over any simply connected smooth manifold.

In the case when the base manifold is a Riemann surface, a holomorphic (g,m) fiber bundle over X is defined as follows.

Definition 8.2 *Let X be a Riemann surface, let $\mathcal{X}$ be a complex surface, and $\mathcal{P}$ a holomorphic proper submersion from $\mathcal{X}$ onto X, such that each fiber $\mathcal{P}^{-1}(x)$ is a closed Riemann surface of genus* g*. Suppose* $\boldsymbol{E}$ *is a complex one-dimensional submanifold of $\mathcal{X}$ that intersects each fiber $\mathcal{P}^{-1}(x)$ along a set E_x of* m *distinguished points. Then the tuple $\mathfrak{F}_{\mathsf{g},\mathsf{m}} = (\mathcal{X}, \mathcal{P}, \boldsymbol{E}, X)$ is called a holomorphic* (g, m) *fiber bundle over X.*

Notice that the mapping $x \to E_x$ locally defines m holomorphic sections of the (g, 0)-bundle $(\mathcal{X}, \mathcal{P}, X)$ that is obtained by forgetting all distinguished points.

We will call two smooth oriented (holomorphic, respectively) (g, m) fiber bundles, $\mathfrak{F}^0 = (\mathcal{X}^0, \mathcal{P}^0, \boldsymbol{E}^0, X^0)$ and $\mathfrak{F}^1 = (\mathcal{X}^1, \mathcal{P}^1, \boldsymbol{E}^1, X^1)$, smoothly isomorphic (holomorphically isomorphic, respectively) if there are smooth (holomorphic, respectively) homeomorphisms $\Phi : \mathcal{X}^0 \to \mathcal{X}^1$ and $\phi : X^0 \to X^1$ such that for each $x \in X^0$ the mapping Φ takes the fiber $(\mathcal{P}^0)^{-1}(x)$ onto the fiber $(\mathcal{P}^1)^{-1}(\phi(x))$ and the set of distinguished points in $(\mathcal{P}^0)^{-1}(x)$ to the set of distinguished points in $(\mathcal{P}^1)^{-1}(\phi(x))$. Holomorphic bundles that are holomorphically isomorphic will be considered the same holomorphic bundles.

For a smooth (g, m)-bundle $\mathfrak{F} = (\mathcal{X}, \mathcal{P}, \boldsymbol{E}, X)$ over X and a homeomorphism $\omega : X \to \omega(X)$ onto a surface $\omega(X)$ we denote by $\mathfrak{F}_\omega$ the bundle

$$\mathfrak{F}_\omega = (\mathcal{X}, \omega \circ \mathcal{P}, \boldsymbol{E}, \omega(X)). \tag{8.1}$$

The bundles $\mathfrak{F}$ and $\mathfrak{F}_\omega$ are isomorphic. Indeed, we may consider the homeomorphisms $\omega : X \to \omega(X)$ and $\mathrm{Id} : \mathcal{X} \to \mathcal{X}$.

We are interested in smooth deformations of a smooth bundle to a holomorphic one. Two smooth (oriented) (g, m) fiber bundles over the same oriented smooth base manifold X, $\mathfrak{F}^0 = (\mathcal{X}^0, \mathcal{P}^0, \boldsymbol{E}^0, X)$, and $\mathfrak{F}^1 = (\mathcal{X}^1, \mathcal{P}^1, \boldsymbol{E}^1, X)$, are called (free) isotopic if for an open interval I containing $[0, 1]$ there is a smooth (g, m) fiber bundle $(\mathcal{Y}, \mathcal{P}, \boldsymbol{E}, I \times X)$ over the base $I \times X$ (called an isotopy) with the following property. For each $t \in [0, 1]$ we put $\mathcal{Y}^t = \mathcal{P}^{-1}(\{t\} \times X)$ and $\boldsymbol{E}^t = \boldsymbol{E} \cap \mathcal{P}^{-1}(\{t\} \times X)$. The bundle $\mathfrak{F}^0$ is equal to $\left(\mathcal{Y}^0, \mathcal{P} \mid \mathcal{Y}^0, \boldsymbol{E}^0, \{0\} \times X\right)$, and the bundle $\mathfrak{F}^1$ is equal to $\left(\mathcal{Y}^1, \mathcal{P} \mid \mathcal{Y}^1, \boldsymbol{E}^1, \{1\} \times X\right)$.

Notice that for each $t \in I$ the tuple $\left(\mathcal{Y}^t, \mathcal{P} \mid \mathcal{Y}^t, \boldsymbol{E}^t, \{t\} \times X\right)$ is automatically a smooth (g, m)-fiber bundle.

Lemma 8.1 *Isotopic bundles are isomorphic.*

Proof Let $I \supset [0, 1]$ be an open interval, and let $(\mathcal{Y}, \mathcal{P}, \boldsymbol{E}, I \times X)$ be a smooth (g, m)-bundle over $I \times X$ such that the restrictions to $\{0\} \times X$ and to $\{1\} \times X$ are equal to given smooth (g, m)-bundles over X. Here X is a smooth finite surface. Consider the vector field v on $I \times X$ that equals the unit vector in positive direction of I at each point of $I \times X$, and let V be a smooth vector field on $\mathcal{Y}$ that projects to v under $\mathcal{P}$ and is tangent to $\boldsymbol{E}$ at points of this set. For each $t \in I$ we let φ_t be the time t map of the flow of V, more precisely,

$$\frac{\partial}{\partial t}\varphi_t(y) = V(\varphi_t(y)), \quad \varphi_0(y) = y, \quad t \in I, \ y \in \mathcal{P}^{-1}(\{0\} \times X).$$

Then φ_1 is a diffeomorphism from $\mathcal{P}^{-1}(\{0\} \times X)$ onto $\mathcal{P}^{-1}(\{1\} \times X)$, that maps the fiber over $(0, x)$ onto the fiber over $(1, x)$ and maps distinguished points to distinguished points. □

In Sect. 8.2 we will prove that vice versa, isomorphic bundles are isotopic.

The following problem on isotopies of smooth objects to the respective holomorphic objects concerns another version of the restricted validity of Gromov's Oka Principle.

Problem 8.1 *Let X be a finite open Riemann surface. Can a given smooth (g,m)-fiber bundle over X be smoothly deformed to a holomorphic fiber bundle over X?*

More precisely,, does there exist an isotopy $(\mathcal{Y}, \mathcal{P}, \boldsymbol{E}, X \times I)$ over the base $X \times I$ with $[0, 1] \subset I$, for which the restriction to $\{0\} \times X$ is equal to the given bundle, and the restriction to $\{1\} \times X$ can be equipped with the structure of a holomorphic bundle?

The answer is in general negative. There is a similar notion that describes obstructions for the existence of such isotopies for (g,m)-bundles as for the existence of isotopies of smooth separable quasipolynomials to holomorphic ones. It is called the conformal module of isotopy classes of (g,m)-fiber bundles over the circle, and is defined as follows. Consider a smooth oriented (g, m)-fiber bundle $\mathfrak{F} = \mathfrak{F}_{(g,m)} = (\mathcal{X}, \mathcal{P}, \boldsymbol{E}, \partial\mathbb{D})$ over the circle. Denote by $\widehat{\mathfrak{F}}$ the isotopy class of fiber bundles over $\partial\mathbb{D}$ that contains $\mathfrak{F}$. Let $A_{r,R} = \{z \in \mathbb{C} : r < |z| < R\}$, $r < 1 < R$, be an annulus containing the unit circle. A smooth (g, m)-fiber bundle on $A_{r,R}$ is said to represent $\widehat{\mathfrak{F}}$ if its restriction to the unit circle $\partial\mathbb{D}$ is an element of $\widehat{\mathfrak{F}}$.

Definition 8.3 (The Conformal Module of Isotopy Classes of (g, m)**-Fiber Bundles)** *Let $\widehat{\mathfrak{F}} = \widehat{\mathfrak{F}}_{g,m}$ be the isotopy class of an oriented* (g, m)*-fiber bundle over the circle $\partial\mathbb{D}$. Its conformal module is defined as*

$$\mathcal{M}(\widehat{\mathfrak{F}}) = \sup\{m(A_{r,R}) : \textit{there exists a holomorphic fiber bundle} \\ \textit{on } A_{r,R} \textit{ that represents } \widehat{\mathfrak{F}}\}. \tag{8.2}$$

8.2 The Monodromy of (g, m)-Bundles. Isotopy and Isomorphism of Bundles

Monodromy and Maping Torus Consider a smooth (g, m)-bundle $\mathfrak{F}_{g,m} = (\mathcal{X}, \mathcal{P}, \boldsymbol{E}, \partial\mathbb{D})$ over the unit circle $\partial\mathbb{D}$. Denote as before the set of distinguished points $\boldsymbol{E} \cap \mathcal{P}^{-1}(x)$ in the fiber over x by E_x. Let v be the unit tangent vector field to $\partial\mathbb{D}$. The argument used for the proof of Ehresmann's Theorem provides a smooth vector field V on $\mathcal{X}$ which is tangent to $\boldsymbol{E}$ at points of this set and projects to v, i.e. $(d\,\mathcal{P})(V) = v$. Cover $\partial\mathbb{D}$ by its universal covering $\mathbb{R} \xrightarrow{p} \partial\mathbb{D}$, using the mapping $p(t) = e^{2\pi i t}$, $t \in \mathbb{R}$. Lift the bundle $\mathfrak{F}_{g,m}$ over $\partial\mathbb{D}$ to a bundle $\widetilde{\mathfrak{F}}_{g,m} = (\widetilde{\mathcal{X}}, \widetilde{\mathcal{P}}, \widetilde{\boldsymbol{E}}, \mathbb{R})$ over $\mathbb{R}$.

The fiber $\widetilde{\mathcal{P}}^{-1}(t)$ with set of distinguished points $\tilde{E}_t$ equals $\mathcal{P}^{-1}(e^{2\pi i t})$ with set of distinguished points $E_{e^{2\pi i t}}$, hence for $k \in \mathbb{Z}$ and each $t \in \mathbb{R}$ the sets $\widetilde{\mathcal{P}}^{-1}(t+2\pi k)$ and $\widetilde{\mathcal{P}}^{-1}(t)$, are equal, and $\tilde{E}_{t+2\pi k}$ is equal to $\tilde{E}_t$. Lift the vector field V to a vector field $\widetilde{V}$ on $\widetilde{\mathcal{X}}$ with the following property. The equality $\widetilde{V}(z_1) = \widetilde{V}(z_2)$ holds, if $z_1, z_2 \in \widetilde{\mathcal{X}}$ are mapped to the same point $z \in \mathcal{X}$ under the projection $\widetilde{\mathcal{X}} \xrightarrow{\tilde{p}} \mathcal{X}$.

For $t \in \mathbb{R}$, $\zeta \in \widetilde{\mathcal{X}}$, we let $\widetilde{\varphi}_t(\zeta) \in \widetilde{\mathcal{X}}$ be the solution of the differential equation

$$\frac{\partial}{\partial t}\,\widetilde{\varphi}_t(\zeta) = \widetilde{V}(\widetilde{\varphi}_t(\zeta)),\ \widetilde{\varphi}_0(\zeta) = \zeta. \tag{8.3}$$

Put $S \overset{def}{=} \mathcal{P}^{-1}(1) \cong \widetilde{\mathcal{P}}^{-1}(0)$ and $E \overset{def}{=} \boldsymbol{E} \cap \mathcal{P}^{-1}(1) \subset S$. Let $\zeta \in S$. Then $\widetilde{\varphi}_t(\zeta) \in \widetilde{\mathcal{P}}^{-1}(t)$. The time t map $\widetilde{\varphi}_t(\zeta)$, $\zeta \in S$, of the vector field $\widetilde{V}$ defines a homeomorphism from the fiber $S \cong \widetilde{\mathcal{P}}^{-1}(0)$ onto the fiber $\widetilde{\mathcal{P}}^{-1}(t)$, that maps the set of distinguished points $E \subset S$ to the set of distinguished points $\widetilde{E}_t \subset \widetilde{\mathcal{P}}^{-1}(t)$. In the same way $\widetilde{\varphi}_t$ defines a homeomorphism from the fiber $\widetilde{\mathcal{P}}^{-1}(t_1)$ onto the fiber $\widetilde{\mathcal{P}}^{-1}(t_1+t)$ that maps the distinguished points $\tilde{E}_{t_1}$ to the distinguished points $\tilde{E}_{t_1+t}$. The mappings $\widetilde{\varphi}_t$ form a group:

$$\widetilde{\varphi}_{t_1+t_2}(\zeta) = \widetilde{\varphi}_{t_1}(\widetilde{\varphi}_{t_2}(\zeta)),\ \zeta \in \widetilde{\mathcal{X}},\ t_1, t_2 \in \mathbb{R}. \tag{8.4}$$

Let $\big(\mathbb{R} \times S,\ \mathrm{pr}_1, \mathbb{R} \times E,\ \mathbb{R}\big)$ be the trivial bundle. Here $\mathrm{pr}_1 : \mathbb{R} \times S \to \mathbb{R}$ is the projection onto the first factor. The mapping Φ,

$$\mathbb{R} \times S \ni (t, \zeta) \to \Phi(t, \zeta) \overset{def}{=} \widetilde{\varphi}_t(\zeta) \in \widetilde{\mathcal{X}} \tag{8.5}$$

provides an isomorphism from the trivial bundle $\big(\mathbb{R} \times S,\ \mathrm{pr}_1, \mathbb{R} \times E,\ \mathbb{R}\big)$ to the lifted bundle $\widetilde{\mathfrak{F}}_{\mathrm{g,m}}$. Define by φ_t the projection

$$\tilde{\mathcal{X}} \supset \widetilde{\mathcal{P}}^{-1}(t) \ni \widetilde{\varphi}_t(\zeta) \xrightarrow{\ \tilde{p}\ } \varphi_t(\zeta) = \tilde{p}(\widetilde{\varphi}_t(\zeta)) \in \mathcal{P}^{-1}(e^{2\pi i t}) \in \mathcal{X}\,. \tag{8.6}$$

Since $\widetilde{\mathcal{P}}^{-1}(t) = \mathcal{P}^{-1}(e^{2\pi i t})$ and $\tilde{E}_t = E_{e^{2\pi i t}}$, we obtain a smooth family of homeomorphisms $\varphi_t : \mathcal{P}^{-1}(1) \to \mathcal{P}^{-1}(e^{2\pi i t})$ that map distinguished points to distinguished points. We call the family φ_t, $t \in \mathbb{R}$, a trivializing family of homeomorphisms. Consider the time-1 map

$$\varphi_1 : \mathcal{P}^{-1}(1) \to \mathcal{P}^{-1}(e^{2\pi i}) = \mathcal{P}^{-1}(1) \tag{8.7}$$

which is a self-homeomorphism of the fiber over 1 that maps the set of distinguished points $E \subset S \cong \mathcal{P}^{-1}(1)$ to itself. For $k \in \mathbb{Z}$ we have $\varphi_{t+k} = \varphi_t \circ \varphi_1^k$. For each $n \in \mathbb{Z}$ the projection (8.6) maps the points $\Phi(t, \zeta) = \widetilde{\varphi}_t(\zeta)$ and $\Phi(t+n, \varphi_1^{-n}(\zeta)) = \widetilde{\varphi}_{t+n}(\varphi_1^{-n}(\zeta))$ to the same point $\varphi_t(\zeta) \in \mathcal{P}^{-1}(e^{2\pi i t})$. The group $\mathbb{Z}$ of integer numbers acts on $\Phi(\mathbb{R} \times S) = \widetilde{\mathcal{X}}$ by

$$\Phi(\mathbb{R} \times S) \ni \Phi(t, \zeta) \to \Phi(t+n, \varphi_1^{-n}(\zeta)), \quad n \in \mathbb{Z}. \tag{8.8}$$

The quotient

$$\mathcal{X}_1 \overset{def}{=} \mathbb{R} \times S \big/ \big((t, \zeta) \sim (t+1, \varphi_1^{-1}(\zeta))\big) \tag{8.9}$$

is a smooth manifold that is diffeomorphic to $\mathcal{X}$. Indeed, the mapping $\tilde{p} \circ \Phi : \mathbb{R} \times S \to \mathcal{X}$, $\tilde{p} \circ \Phi(t, \zeta) = \varphi_t(\zeta) \in \mathcal{P}^{-1}(e^{2\pi it})$, satisfies the condition $\tilde{p} \circ \Phi(t+1, \varphi_1^{-1}(\zeta)) = \tilde{p} \circ \Phi(t, \zeta)$. Hence, this mapping descends to a mapping from $\mathcal{X}_1$ to $\mathcal{X}$, that is a local diffeomorphism and is bijective by the construction. Hence, it is a diffeomorphism from $\mathcal{X}_1$ onto $\mathcal{X}$. Put

$$\boldsymbol{E}_1 \overset{def}{=} \mathbb{R} \times E \Big/ \Big((t, \zeta) \sim (t+1, \varphi_1^{-1}(\zeta))\Big), \tag{8.10}$$

and recall that $\tilde{p}(\Phi(t, E)) = \tilde{p}(\widetilde{\varphi}_t(E)) = \boldsymbol{E} \cap \mathcal{P}^{-1}(e^{2\pi it})$, since $\widetilde{\varphi}_t$ maps the distinguished points $E \subset S = \mathcal{P}^{-1}(1)$ to the distinguished points $\boldsymbol{E} \cap \mathcal{P}^{-1}(e^{2\pi it})$ in the fiber $\widetilde{\mathcal{P}}^{-1}(t) = \mathcal{P}^{-1}(e^{2\pi it})$. Define the projection $\mathcal{P}_1 : \mathcal{X}_1 \to \partial\mathbb{D}$, so that $\mathcal{P}_1$ takes the value $e^{2\pi it}$ on the class containing (t, ζ). We obtain a bundle $(\mathcal{X}_1, \mathcal{P}_1, \boldsymbol{E}_1, \partial\mathbb{D})$ that is smoothly isomorphic to $(\mathcal{X}, \mathcal{P}, \boldsymbol{E}, \partial\mathbb{D})$. We will also say for short that $\mathfrak{F}_{\mathrm{g,m}}$ is smoothly isomorphic to the mapping torus

$$([0,1] \times S) \Big/ \Big((0, \zeta) \sim (1, \varphi(\zeta))\Big) \tag{8.11}$$

where $\varphi = \varphi_1^{-1}$ is a smooth orientation preserving self-homeomorphism of the fiber $S = \mathcal{P}^{-1}(1)$ with set of distinguished points $E \subset S$. The mapping φ depends on the trivializing vector field. However, the mapping class of φ is independent on this vector field, it is merely determined by the bundle. We denote it by $\mathfrak{m}_{\mathfrak{F}}$ (with $\mathfrak{F} = \mathfrak{F}_{\mathrm{g,m}}$). The mapping class $\mathfrak{m}_{\mathfrak{F}} \in \mathfrak{M}(S; \emptyset, E)$ is called the monodromy mapping class of the bundle $\mathfrak{F}$ over the circle, or the monodromy, for short.

For diffeomorphic closed surfaces S_1 and S_2 and sets of distinguished points E_1 and E_2 of the same cardinality the groups $\mathfrak{M}(S_1; \emptyset, E_1)$ and $\mathfrak{M}(S_2; \emptyset, E_2)$ are isomorphic. Any diffeomorphism $\varphi : S_1 \to S_2$ with $\varphi(E_1) = E_2$ induces an isomorphism $\mathrm{Is}_\varphi : \mathfrak{M}(S_1; \emptyset, E_1) \to \mathfrak{M}(S_2; \emptyset, E_2)$. Two such isomorphisms $\mathfrak{M}(S_1; \emptyset, E_1) \to \mathfrak{M}(S_2; \emptyset, E_2)$ differ by conjugation with an element of $\mathfrak{M}(S_2; \emptyset, E_2)$. Hence, there is a canonical one-to-one correspondence between the sets of conjugacy classes $\widehat{\mathfrak{M}(S_1; \emptyset, E_1)}$ and $\widehat{\mathfrak{M}(S_2; \emptyset, E_2)}$, and we will identify conjugacy classes according to this correspondence. Similarly, let $h : F_k \to \mathfrak{M}(S_1; \emptyset, E_1)$ be a homomorphism from a free group F_k of rank k. Any diffeomorphism $\varphi : S_1 \to S_2$ with $\varphi(E_1) = E_2$ induces another homomorphism $h_\varphi : F_k \to \mathfrak{M}(S_2; \emptyset, E_2)$. Replacing φ by another diffeomorphism we arrive at a conjugate homomorphism. We obtain a canonical bijection between conjugacy classes of homomorphisms $F_k \to \mathfrak{M}(S_j; \emptyset, E_j)$, $j = 1, 2$.

Problem 8.2 *Prove that for each isotopy class* $\widehat{\mathfrak{F}}_{\mathrm{g,m}}$ *of* (g, m)*-bundles over the circle and the associated conjugacy class* $\widehat{\mathfrak{m}}_{\mathfrak{F}_{\mathrm{g,m}}}$ *of elements of the mapping class of the fiber over the point* 1 *the equality*

$$\mathcal{M}(\widehat{\mathfrak{F}}_{\mathrm{g,m}}) = \frac{\pi}{2} \frac{1}{h(\widehat{\mathfrak{m}}_{\mathfrak{F}_{\mathrm{g,m}}})}$$

holds.

In the sequel we will use the mapping class group $\mathfrak{M}(S; \emptyset, E)$ on a reference (Riemann) surface of type (g, m) and the set of its conjugacy classes $\widehat{\mathfrak{M}(S; \emptyset, E)}$, or the set of conjugacy classes of mappings from a free group to $\mathfrak{M}(S; \emptyset, E)$, respectively. Recall that the group $\mathfrak{M}(S; \emptyset, E)$ is isomorphic to the modular group $\mathrm{Mod}(\mathrm{g}, \mathrm{m})$.

The following theorem holds. (See, e.g. [24] for the case of closed fibers of genus $\mathrm{g} \geq 2$ and paracompact Hausdorff spaces X.)

Theorem 8.1 *Let X be a connected smooth finite open oriented surface with base point q_0. The set of isomorphism classes of smooth oriented* (g, m)*-bundles on X is in one-to-one correspondence to the set of conjugacy classes of homomorphisms from the fundamental group $\pi_1(X, q_0)$ into the modular group* Mod(g, m).

Proof Let first $\mathfrak{F}_0 = (\mathcal{X}_0, \mathcal{P}_0, \boldsymbol{E}_0, \partial\mathbb{D})$ and $\mathfrak{F}_1 = (\mathcal{X}_1, \mathcal{P}_1, \boldsymbol{E}_1, \partial\mathbb{D})$ be isomorphic bundles over the circle $\partial\mathbb{D}$. Suppose the isomorphism is given by a diffeomorphism $\phi : \partial\mathbb{D}\circlearrowleft$ and a diffeomorphism $\Phi : \mathcal{X}_0 \to \mathcal{X}_1$. We may assume that ϕ preserves the base point 1, hence, Φ maps each fiber $\mathcal{P}_0^{-1}(e^{2\pi it})$ of the first bundle onto the fiber $\mathcal{P}_1^{-1}(e^{2\pi i\phi_1(t)})$ of the second bundle for a homeomorphism $\phi_1 : [0, 1]\circlearrowleft$. Let $\varphi_t : S \to \mathcal{P}_0^{-1}(e^{2\pi it})$ be a trivialising family of homeomorphisms for the bundle $\mathfrak{F}_0$. Then $\psi_t \overset{def}{=} \Phi|\mathcal{P}_0^{-1}(e^{2\pi it}) \circ \varphi_t \circ (\Phi|\mathcal{P}_0^{-1}(1))^{-1}$ is a trivialising family of homeomorphisms of the second bundle. We obtain $\psi_1 = \Phi|\mathcal{P}_0^{-1}(1) \circ \varphi_1 \circ (\Phi|\mathcal{P}_0^{-1}(1))^{-1}$. This means that ψ_1 and ϕ_1 represent the same element of $\widehat{\mathfrak{M}(\mathcal{P}_0^{-1}(1))}$. We saw that the monodromies of isomorphic (g, m)-bundles over the circle belong to the same conjugacy class $\widehat{\mathrm{Mod}(\mathrm{g}, \mathrm{m})}$.

Vice versa, suppose the monodromies of two (g, m)-bundles $\mathfrak{F}_0$ and $\mathfrak{F}_1$ over the circle with base point 1 represent the same element of $\widehat{\mathrm{Mod}(\mathrm{g}, \mathrm{m})}$. We will prove that the bundles are isomorphic. The assumption means that the monodromies $\mathfrak{m}_j$ of $\mathfrak{F}_j$, $j = 0, 1$ are represented by mappings $\varphi_j \in \mathfrak{m}_j$, such that $\varphi_1 = \varphi \circ \varphi_0 \circ \varphi^{-1}$ for a homeomorphism $\varphi : \mathcal{P}_0^{-1}(1) \to \mathcal{P}_1^{-1}(1)$ that maps the set of distinguished points of $\mathcal{P}_0^{-1}(1)$ onto the set of distinguished points of $\mathcal{P}_1^{-1}(1)$. The bundle $\mathfrak{F}_0$ is isomorphic to the mapping torus of φ_0. This mapping torus has total space $\mathcal{X}_0 = (\mathbb{R} \times S)\big/\big((t, \zeta) \sim (t + 1, \varphi_0(\zeta))\big)$ and set of distinguished points $E_0 = (\mathbb{R} \times E)\big/\big((t, \zeta) \sim (t + 1, \varphi_0(\zeta))\big)$.

The mapping torus $\mathfrak{F}_0'$ with total space

$$\mathcal{X}_0' = \big(\mathbb{R} \times \varphi(S)\big)\big/\big((t, \zeta) \sim (t + 1, \varphi \circ \varphi_0 \circ \varphi^{-1}(\zeta))\big)$$

and set of distinguished points

$$E_0' = \big(\mathbb{R} \times \varphi(E)\big)\big/\big((t, \zeta) \sim (t + 1, \varphi \circ \varphi_0 \circ \varphi^{-1}(\zeta))\big)$$

is isomorphic to $\mathfrak{F}_0$ and has monodromy represented by $\varphi \circ \varphi_0 \circ \varphi^{-1}$. Indeed, the mapping $\mathbb{R} \times S \ni (t, \zeta) \to (t, \varphi(\zeta)) \in \mathbb{R} \times S$ descends to a diffeomorphism $\mathcal{X}_0 \to \mathcal{X}_0'$ that takes fibers to fibers.

The mapping torus $\mathfrak{F}_0'$ with total space $\mathcal{X}_0'$ is isomorphic to $\mathfrak{F}_1$. Indeed, the monodromy mappings of $\mathfrak{F}_0'$ and $\mathfrak{F}_1$ coincide, and the mapping tori of isotopic mappings are isotopic, and, hence, isomorphic. Therefore, the mapping torus $\mathfrak{F}_0'$ is isomorphic to the mapping torus of $\mathfrak{F}_1$. Hence, $\mathfrak{F}_0$ and $\mathfrak{F}_1$ are isomorphic.

Let now X be a connected smooth oriented open surface of genus g with m holes with base point q_0. Suppose $\mathfrak{F} = (\mathcal{X}, \mathcal{P}, \boldsymbol{E}, X)$ is a smooth (g, m)-bundle over X. Denote the fiber $\mathcal{P}^{-1}(q_0)$ over q_0 by S and the set of distinguished points $\boldsymbol{E} \cap \mathcal{P}^{-1}(q_0)$ by E. We assign to $\mathfrak{F}$ a homomorphism $\pi_1(X, q_0) \to \mathfrak{M}(S, \emptyset, E)$ as follows. Take smooth closed curves in X parameterized by $\gamma_j : [0, 1] \to X$ that represent the elements e_j of a standard system of generators of the fundamental group $\pi_1(X, q_0)$, $j = 1, \ldots, 2g+m-1$. Associate to each generator e_j of $\pi_1(X, q_0)$ the monodromy mapping class of the restricted bundle $\mathfrak{F}|\gamma_j$. This is a mapping class on the fiber $S = \mathcal{P}^{-1}(q_0)$ over q_0 with distinguished points $E = \boldsymbol{E} \cap \mathcal{P}^{-1}(q_0)$, that depends only on the e_j and on the bundle. We obtain a well-defined mapping that associates to each generator of the fundamental group $\pi_1(X, q_0)$ a mapping class. This mapping extends to a homomorphism from the fundamental group $\pi_1(X, q_0)$ to the mapping class group $\mathfrak{M}(S, \emptyset, E)$. In the same way as in the case of bundles over the circle we may prove that the respective homomorphisms corresponding to isomorphic bundles represent the same conjugacy class of homomorphisms from $\pi_1(X, q_0)$ to Mod(g, m).

Vice versa, take a homomorphism h from the fundamental group $\pi_1(X, q_0)$ to the modular group Mod(g, m). A bundle over X whose monodromy homomorphism coincides with h can be obtained as follows.

By Lemma 1.1 the surface X is diffeomorphic to a standard neighbourhood of a standard bouquet B of circles for X. Hence, it can be written as union $X = D \cup \bigcup_j V_j$ of an open disc D and half-open bands V_j attached to D. The disc D contains the base point q_0. The boundary ∂D of D in X is smooth. The set $B \setminus D$ is the union of disjoint closed arcs s_j with endpoints on ∂D. Each circle c_j of the bouquet contains exactly one of the s_j. For each j the set V_j is a neighbourhood in $X \setminus D$ of s_j.

Consider the universal covering $\tilde{X} \xrightarrow{\mathsf{P}} X$. Take a point $\tilde{q}_0 \in \tilde{X}$ with $\mathsf{P}(\tilde{q}_0) = q_0$. For each j we take the lifts $\widetilde{c}_j$ of c_j to $\tilde{X}$ with initial point $\tilde{q}_0$, and the lift $\widetilde{V}_j$ of V_j that intersects $\widetilde{c}_j$. Let $\widetilde{D}_0$ be the lift of D to $\tilde{X}$, that contains $\tilde{q}_0$, and let $\widetilde{D}_j$, $j = 1, \ldots, 2g+m-1$, be the lift of D, that contains the endpoint $\tilde{x}_j \stackrel{def}{=} \widetilde{c}_j(1)$ of $\widetilde{c}_j$. Let $\tilde{U}$ be the domain in $\tilde{X}$ that is the union of $\tilde{D}_j$, $j = 0, 1, \ldots, 2g+m-1$, and $\tilde{V}_j$, $j = 1, \ldots, 2g+m-1$.

Choose for each $j \geq 1$ a mapping φ_j in the mapping class $h(e_j)$, and define a bundle over X by taking the trivial bundle over $\tilde{U}$ and making the following identifications. For each $x' \in D$ and each j we glue the fiber over the point $\tilde{x}_j' \in \widetilde{D}_j$ for which $p(\tilde{x}_j') = x'$ to the fiber over the point $\tilde{x}_0' \in \widetilde{D}_0$ for which $p(\tilde{x}_0') = x'$ using the mapping φ_j. We obtained a bundle with the given monodromy homomorphism.

As in the case of bundles over the circle one can see that each bundle is isomorphic to a bundle that is constructed in the just described way, and bundles constructed in this way with equal conjugacy class of monodromy homomorphisms are isomorphic. We obtained a bijective correspondence of isomorphism classes of smooth (g, m)-bundles over a connected finite open smooth oriented surface X of genus g with $m > 0$ holes and conjugacy classes of homomorphisms from the fundamental group of X to Mod(g, m). □

Isotopy Classes and Isomorphism Classes of Bundles We will consider now the relation between isotopy classes and isomorphism classes of (g, m)-bundles, and their relation to the monodromies of the bundles.

Lemma 8.1 says that isotopic smooth (g, m)-bundles over the circle are smoothly isomorphic. The following lemma states in particular, that, vice versa, isomorphic bundles are isotopic.

Lemma 8.2 *Two smooth* (g, m)*-bundles over a connected smooth finite open oriented surface X are isomorphic if and only of they are isotopic.*

Proof By Lemma 8.1 and Theorem 8.1 it remains to prove that two smooth (g, m)-bundles over X are isotopic if their monodromy homomorphisms represent the same conjugacy class of homomorphisms.

We first consider two smooth (g, m)-bundles $\mathfrak{F}_0$ and $\mathfrak{F}_1$ over the circle with fiber and distinguished points over the base point 1 being equal. Suppose the bundles (g, m)-bundles $\mathfrak{F}_0$ and $\mathfrak{F}_1$ have the same monodromy. We will prove that they are isotopic by an isotopy that fixes the fiber over the base point and the set of distinguished points in this fiber (by a based isotopy, for short). Denote the fiber over 1 by S and the set of distinguished points in S by E. Lift each bundle $\mathfrak{F}_j$, $j = 0, 1$, to a bundle over the vertical line $\{j\} \times \mathbb{R} \subset \mathbb{R}^2$ by the covering $\{j\} \times \mathbb{R} \ni (j, t) \to e^{2\pi i t}$. Let ε be a small positive number. Restrict the lift of the bundle $\mathfrak{F}_j$ to the open interval $\{j\}\times(-\varepsilon, 1+\varepsilon) \supset \{j\}\times[0, 1]$, $\varepsilon > 0$, and extend for each j the restricted bundle to a bundle over the rectangle $(j-\varepsilon, j+\varepsilon)\times(-\varepsilon, 1+\varepsilon)$, so that the restriction of the extended bundle $\mathfrak{F}'_j$ to each horizontal segment is a product bundle.

Consider the product bundle with fiber S and set of distinguished points E over the horizontal line segments $(-\varepsilon, 1 + \varepsilon) \times \{k\}$, $k = 0, 1$. Take an extension of the product bundle over each of the horizontal line segments $(-\varepsilon, 1 + \varepsilon) \times \{k\}$ to a smooth (g, m)-bundle over the open rectangle $(-\varepsilon, 1 + \varepsilon) \times (k - \varepsilon, k + \varepsilon)$ around the segment, so that for $y \in (-\varepsilon, \varepsilon)$ and $x \in (-\varepsilon, 1 + \varepsilon)$ the fibers and the distinguished points over (x, y) and $(x, y + 1)$ are the same, and the bundles over the four rectangles $(-\varepsilon, 1 + \varepsilon) \times (k - \varepsilon, k + \varepsilon)$, $k = 0, 1$ and $(j - \varepsilon, j + \varepsilon) \times (-\varepsilon, 1 + \varepsilon)$, $j = 0, 1$, match to a smooth bundle $\tilde{\mathfrak{F}}_0$ over the neighbourhood $V \overset{def}{=} \Big((-\varepsilon, 1 + \varepsilon) \times (-\varepsilon, 1 + \varepsilon)\Big) \setminus \Big((\varepsilon, 1 - \varepsilon) \times (\varepsilon, 1 - \varepsilon)\Big)$ of the boundary of the square $[0, 1] \times [0, 1]$. Let $(0, 0)$ be the base point of V. The monodromy of the obtained bundle along the generator of the fundamental group $\pi_1(V, 0)$ of V is the identity. Hence, the bundle is smoothly isomorphic to the trivial bundle.

Let φ be a diffeomorphism of the total space of the bundle $\tilde{\mathfrak{F}}_0$ to the total space of the trivial bundle over V, that maps the fiber of the bundle $\tilde{\mathfrak{F}}_0$ over each point in V to the fiber of the trivial bundle over this point, and maps distinguished points to distinguished points. The trivial bundle over V extends to the trivial bundle over the neighbourhood $(-\varepsilon, 1+\varepsilon) \times (-\varepsilon, 1+\varepsilon)$ of the square. Glue the trivial bundle over the square $(\frac{\varepsilon}{2}, 1-\frac{\varepsilon}{2}) \times (\frac{\varepsilon}{2}, 1-\frac{\varepsilon}{2})$, that is relatively compact in the open unit square, to the bundle $\tilde{\mathfrak{F}}_0$ over V along $V \cap \big((\frac{\varepsilon}{2}, 1-\frac{\varepsilon}{2}) \times (\frac{\varepsilon}{2}, 1-\frac{\varepsilon}{2})\big)$ using the mapping φ. We obtain a smooth bundle over the neighbourhood $(-\varepsilon, 1+\varepsilon) \times (-\varepsilon, 1+\varepsilon)$ of the unit square, that coincides with $\tilde{\mathfrak{F}}_0$ over V. For $(x, y) \in (-\varepsilon, 1+\varepsilon) \times (-\varepsilon, \varepsilon)$ we glue the fiber over (x, y) to the fiber over $(x, y+1)$ using the identity mapping. We get a bundle over the product of the interval $(-\varepsilon, 1+\varepsilon)$ with the circle, which provides an isotopy with fixed fiber over the base point and fixed set of distinguished points (a based isotopy for short).

If two bundles over the circle with the same fiber over the point 1 and the same set of distinguished points over the base point have conjugate monodromy, we consider instead of the product bundle on each of the two horizontal line segments $(-\varepsilon, 1+\varepsilon) \times \{k\}$, $k = 0, 1$, the bundle obtained from the product bundle by the mapping Φ, $\Phi(t, \zeta) = \varphi_t(\zeta)$, $t \in (-\varepsilon, 1+\varepsilon)$, $\zeta \in S$, between total spaces, where φ_t is a smooth family of diffeomorphisms with $\varphi_0 = \mathrm{Id}$ and φ_1 representing the conjugating mapping class. Along the same lines as above one can prove that the bundles are free isotopic.

If two bundles over the circle have different fibers S_1 and S_2 over the point 1 with sets of distinguished points E_1 and E_2, respectively, and (after an isomorphism) their monodromy maps are conjugate, then the bundles are free isotopic. Indeed, the condition means that the monodromies of the two bundles differ by "conjugation" by a homeomorphism $\psi : S_1 \to S_2$ that maps E_1 onto E_2. Instead of the product bundle on the horizontal segments $(-\varepsilon, 1+\varepsilon) \times \{k\}$, $k = 0, 1$, we use in this case the bundle obtained from the product bundle by the mapping Φ', $\Phi'(t, \zeta) = \varphi'_t(\zeta)$, $t \in (-\varepsilon, 1+\varepsilon)$, $\zeta \in S$, between total spaces, where $\varphi'_t : S \to S_t$ is a smooth family of diffeomorphisms with $\varphi'_0 = \mathrm{Id}$ and $\varphi'_1 = \psi$.

The proof for bundles over a smooth finite open oriented surface X follows along the same lines using the fact that such a surface X is diffeomorphic to a standard neighbourhood of a standard bouquet of circles.

We obtain a bijective correspondence between smooth isomorphism classes and free isotopy classes of (g, m)-bundles over connected oriented smooth finite open surfaces. □

A smooth (g, m)-bundle $(\mathcal{X}, \mathcal{P}, \boldsymbol{E}, X)$ over a connected smooth finite open oriented surface X admits a smooth section, i.e. a smooth mapping $X \ni x \to s(x) \in \mathcal{X} \setminus \boldsymbol{E}$ for which $\mathcal{P} \circ s(x) = x$ for all $x \in X$. Indeed, let $N(B)$ be a neighbourhood of a standard bouquet of circles for X as in Sect. 1.2, and let D, V_j and s_j be as in Sect. 1.2. There is a smooth section over $D \cup_j \mathsf{s}'_j$, where s'_j are the arcs of ∂D to which the horizontal sides of the V_j are glued. This follows from the fact, that there is a simply connected neighbourhood of this set, and the bundle is smoothly trivial over any simply connected domain. Extend the smooth section to a smooth section over a neighbourhood of $D \cup_j (\mathsf{s}'_j \cup \mathsf{s}_j)$. Since a neighbourhood of each V_j is simply connected and there is a smooth deformation retraction of a neighbourhood of V_j

to a neighbourhood of $\mathsf{s}_j' \cup \mathsf{s}_j$, we obtain a smooth section on the whole $N(B)$. By Lemma 1.1 we obtain a smooth section on the whole surface X.

Reducible Bundles and Irreducible Components Thurston's notion of reducible mapping classes takes over to families of mapping classes on a surface of type (g, m), and therefore to (g, m)-bundles. Namely, an admissible system of curves on a connected oriented closed surface S of genus g with set of m distinguished points E is said to reduce a family of mapping classes $\mathfrak{m}_j \in \mathfrak{M}(S; \emptyset, E)$, $j = 1, \ldots, k$, if it reduces each $\mathfrak{m}_j$. If the family consists of a single mapping class we arrive at the notion of a reduced mapping class (see Sect. 1.8).

Similarly, a (g, m)-bundle over a smooth finite open connected oriented surface X with fiber S over the base point q_0 and set of distinguished points $E \subset S$ is called reducible if there is an admissible system of curves in the fiber over the base point that reduces all monodromy mapping classes simultaneously. Otherwise the bundle is called irreducible.

Consider any isotopy class of reducible (g, m)-bundles with fiber S and set of distinguished points $E \subset S$ over the base point q_0. Take an admissible system $\mathcal{C}$ of simple closed curves on $S \setminus E$ that reduces each monodromy mapping class and is maximal in the sense that there is no strictly larger admissible system with this property. For each monodromy mapping class $\mathfrak{m}_j$ we choose a representing homeomorphism φ_j that fixes $\mathcal{C}$ setwise. Then all homeomorphisms φ_j fix $S \setminus \mathcal{C}$. Decompose $S \setminus \mathcal{C}$ into disjoint open sets S_k, each of which being a minimal union of connected components of $S \setminus \mathcal{C}$ that is invariant under each φ_j. Each S_k is a union of surfaces with holes and possibly with distinguished points and is homeomorphic to a union $\overset{\circ}{S}_k$ of closed surfaces with a set E_k of finitely many punctures and, possibly, a set E_k' of finitely many distinguished points. The restrictions of the φ_j to each S_k (possibly, with distinguished points) are conjugate to self-homeomorphisms of $\overset{\circ}{S}_k$ (possibly with set of distinguished points E_k').

For each k we denote by S_k^c the (possibly non-connected) closed surface which is obtained from $\overset{\circ}{S}_k$ by filling the punctures. The surface S_k^c is equipped with distinguished points $E_k \cup E_k'$. For each k we obtain a conjugacy class of homomorphisms from the fundamental group $\pi_1(X, x_0)$ to the modular group of the possibly non-connected Riemann surface S_k^c with set of distinguished points $E_k \cup E_k'$. By Theorem 8.1 (more precisely, by its analog for possibly non-connected fibers) we obtain for each k an isotopy class of bundles over X with fiber a union of closed surfaces with distinguished points. The obtained isotopy classes of bundles are irreducible and are called irreducible bundle components of the (g, m)-bundle with fiber S. (There may be different decompositions into irreducible bundle components.) The reducible bundle can be recovered from the irreducible bundle components up to commuting Dehn twists in the fiber over the base point. (See the proof of Theorem 8.2 below for the case of $(0, 4)$-bundles.)

Locally Holomorphically Trivial and Isotrivial Bundles Let X be a Riemann surface and $2\mathsf{g} - 2 + \mathsf{m} > 0$. A holomorphic (g, m)-bundle over X is called locally holomorphically trivial if it is locally holomorphically isomorphic to the trivial (g, m)-bundle. All fibers of a locally holomorphically trivial (g, m)-bundle $\mathfrak{F} = (\mathcal{X}, \mathcal{P}, \boldsymbol{E}, X)$ are conformally equivalent to each other.

For a locally trivial holomorphic (g, m)-bundle there exists a finite unramified covering $\hat{\mathsf{P}} : \hat{X} \to X$ and a lift $\hat{\mathfrak{F}} = (\hat{\mathcal{X}}, \hat{\mathcal{P}}, \hat{\boldsymbol{E}}, \hat{X})$ of $\mathfrak{F}$ to $\hat{X}$ such that $\hat{\mathfrak{F}}$ is holomorphically isomorphic to the trivial bundle. This can be seen as follows. All fibers of the bundle $\mathfrak{F}$ are conformally equivalent and for each fiber there are only finitely many different conformal self-homeomorphisms. Consider the lift $\tilde{\mathfrak{F}}$ of the bundle $\mathfrak{F}$ to the universal covering $\mathsf{P} : \tilde{X} \to X$ of X, i.e. $\tilde{\mathfrak{F}} = (\tilde{\mathcal{X}}, \tilde{\mathcal{P}}, \tilde{\boldsymbol{E}}, \tilde{X})$, where the fiber $\tilde{\mathcal{P}}^{-1}(\tilde{x})$ with distinguished points $\tilde{\boldsymbol{E}} \cap \tilde{\mathcal{P}}^{-1}(\tilde{x})$ is conformally equivalent to the fiber $\mathcal{P}^{-1}(x)$ with distinguished points $\boldsymbol{E} \cap \mathcal{P}^{-1}(x)$ with $x = \mathsf{P}(\tilde{x})$. Let $\tilde{P} : \tilde{\mathcal{X}} \to \mathcal{X}$ be the respective fiber preserving projection. The bundle $\tilde{\mathfrak{F}}$ is locally holomorphically trivial. Since $\tilde{X}$ is simply connected, $\tilde{\mathfrak{F}}$ is holomorphically trivial on $\tilde{X}$, i.e., there is a biholomorphic mapping $\Phi : \tilde{\mathcal{X}} \to \tilde{X} \times S$ that maps $\tilde{\mathcal{P}}^{-1}(\tilde{x})$ to $\{\tilde{x}\} \times S$ for each $\tilde{x} \in \tilde{X}$ and maps $\tilde{\boldsymbol{E}}$ to $\tilde{X} \times E$. Here S is the fiber $\tilde{\mathcal{P}}^{-1}(\tilde{q}_0) \cong \mathcal{P}^{-1}(q_0)$ over a chosen point $\tilde{q}_0 \in \tilde{X}$ over the base point $q_0 \in X$ and $E = \tilde{\boldsymbol{E}} \cap \mathcal{P}^{-1}(\tilde{x})$. The mapping Φ^{-1} provides a holomorphic family of conformal mappings $\tilde{\varphi}_{\tilde{x}} : S = \tilde{\mathcal{P}}^{-1}(\tilde{q}_0) \to \tilde{\mathcal{P}}^{-1}(\tilde{x})$, $\tilde{x} \in \tilde{X}$. Each mapping of the family takes the set of distinguished points in the fiber S to the set of distinguished points in the respective fiber of the bundle $\tilde{\mathfrak{F}}$. The total space $\mathcal{X}$ of the bundle $\mathfrak{F}$ is holomorphically equivalent to the quotient of $\tilde{X} \times S$ by the following equivalence relation. Two points $(\tilde{x}_1, \zeta_1)$ and $(\tilde{x}_2, \zeta_2)$ in $\tilde{X} \times S$ are equivalent if $\tilde{P}\Phi^{-1}(\tilde{x}_1, \zeta_1) = \tilde{P}\Phi^{-1}(\tilde{x}_2, \zeta_2)$, i.e. if $\mathsf{P}(\tilde{x}_1) = \mathsf{P}(\tilde{x}_2)$ and $(\tilde{P}\varphi_{\tilde{x}_1})(\zeta_1) = (\tilde{P}\varphi_{\tilde{x}_2})(\zeta_2)$. Put $\varphi_{\tilde{x}} = \tilde{P}\tilde{\varphi}_{\tilde{x}} : S \to \mathcal{P}^{-1}(x)$ with $x = \mathsf{P}(\tilde{x})$. If $\mathsf{P}(x_1) = \mathsf{P}(x_2)$, the mapping $\varphi_{\tilde{x}_2}^{-1}\varphi_{\tilde{x}_1}$ is a holomorphic self-homeomorphism of the fiber S. The set of such self-homeomorphims is finite.

Let $\tilde{x} = \sigma(\tilde{q}_0)$ for a covering transformation σ. The mapping $\varphi_{\sigma(\tilde{q}_0)}$ maps S to $\mathcal{P}^{-1}(q_0) \cong S$. For two covering transformations σ_1 and σ_2 the equality

$$\varphi_{\sigma_1\sigma_2(\tilde{q}_0)} = \varphi_{\sigma_2(\tilde{q}_0)}\varphi_{\sigma_1(\tilde{q}_0)} \tag{8.12}$$

holds.

As before $(\mathrm{Is}^{\tilde{q}_0})^{-1}$ denotes the isomorphism from the fundamental group to the group of covering transformations. Consider the set N of elements $e \in \pi_1(X, q_0)$ for which $\varphi_{(\mathrm{Is}^{\tilde{q}_0})^{-1}(e)(\tilde{q}_0)}$ is the identity. By (8.12) the set N is a normal subgroup of the fundamental group. It is of finite index, since two cosets $e_1 N$ and $e_2 N$ are equal if $\tilde{P}\varphi_{(\mathrm{Is}^{\tilde{q}_0})^{-1}(e_2 e_1^{-1})(\tilde{q}_0)} = \mathrm{Id}$, and there are only finitely many distinct holomorphic self-homeomorphisms of $S \cong \mathcal{P}^{-1}(q_0)$. Hence, $\hat{X} \overset{def}{=} \tilde{X} \diagup (\mathrm{Is}^{\tilde{q}_0})^{-1}(N)$ is a finite unramified covering of X and the lift of the bundle $\mathfrak{F}$ to $\hat{X}$ has the required property.

Vice versa, if for a holomorphic (g, m)-bundle $\mathfrak{F}$ there exists a finite unramified covering $\hat{\mathsf{P}} : \hat{X} \to X$, such that the lift $\hat{\mathfrak{F}} = (\hat{\mathcal{X}}, \hat{\mathcal{P}}, \hat{\boldsymbol{E}}, \hat{X})$ of $\mathfrak{F}$ to $\hat{X}$ is holomorphically isomorphic to the trivial bundle, then $\mathfrak{F}$ is locally holomorphically trivial.

A smooth (holomorphic, respectively) bundle is called isotrivial, if it has a finite covering by the trivial bundle. If all monodromy mapping classes of a smooth bundle are periodic, then the bundle is isotopic (equivalently, smoothly isomorphic) to an isotrivial bundle. This can be seen by the same arguments as above.

8.3 (0, 4)-Bundles Over Genus 1 Surfaces with a Hole

The $(0, n)$-bundles over a manifold X are closely related to the separable quasipolynomials on X. In this section we will state a theorem that concerns $(0, 4)$-bundles and is related to Theorem 7.3. We will prove it together with Theorem 7.3.

Let X be a connected finite open Riemann surface (a connected oriented finite open smooth surface, respectively). By a holomorphic (smooth, respectively) $(0, n)$-bundle with a section over X we mean a holomorphic (smooth, respectively) $(0, n + 1)$-bundle $(\mathcal{X}, \mathcal{P}, \boldsymbol{E}, X)$, such that the complex manifold (smooth manifold, respectively) $\boldsymbol{E} \subset \mathcal{X}$ is the disjoint union of two complex manifolds (smooth manifolds, respectively) $\mathring{\boldsymbol{E}}$ and $\boldsymbol{s}$, where $\mathring{\boldsymbol{E}} \subset \mathcal{X}$ intersects each fiber $\mathcal{P}^{-1}(x)$ along a set $\mathring{E}_x$ of n points, and $\boldsymbol{s} \subset \mathcal{X}$ intersects each fiber $\mathcal{P}^{-1}(x)$ along a single point s_x. In other words, the mapping $x \to s_x$, $x \in X$, is a (holomorphic, smooth, respectively) section of the $(0, n)$-bundle with set of distinguished points $\mathring{E}_x$ in the fiber over x. Two smooth $(0, n)$-bundles with a section are called isomorphic if they are isomorphic as smooth $(0, n + 1)$-bundles and the diffeomorphism between the total spaces takes the section of one bundle to the section of the other bundle. Two smooth $(0, n)$-bundles with a section are called isotopic if they are isotopic as $(0, n+1)$-bundles with an isotopy that joins the sections of the bundles. Two smooth $(0, n)$-bundles with a section are isotopic iff they are isomorphic.

We define a special smooth (holomorphic, respectively) $(0, n + 1)$-bundle over X as a smooth (holomorphic, respectively) bundle over X of the form $(X \times \mathbb{P}^1, \mathrm{pr}_1, \boldsymbol{E}, X)$, where $\mathrm{pr}_1 : X \times \mathbb{P}^1 \to X$ is the projection onto the first factor, and the smooth submanifold $\boldsymbol{E}$ of $X \times \mathbb{P}^1$ is equal to the disjoint union $\mathring{\boldsymbol{E}} \cup \boldsymbol{s}^\infty$, where $\mathring{\boldsymbol{E}}$ is a smooth submanifold of $X \times \mathbb{P}^1$ that intersects each fiber along n finite points and $\boldsymbol{s}^\infty$ is the smooth submanifold of $X \times \mathbb{P}^1$ that intersects each fiber $\{x\} \times \mathbb{P}^1$ along the point $\{x\} \times \{\infty\}$. A special $(0, n + 1)$-bundle is, in particular, a $(0, n)$-bundle with a section.

Each smooth mapping $f : X \to C_n(\mathbb{C})\diagup\mathcal{S}_n$ from X to the symmetrized configuration space defines a smooth special $(0, n + 1)$-bundle over X. The smooth submanifold $\boldsymbol{E}$ of $X \times \mathbb{P}^1$ is equal to the union $\bigcup_{x \in X} \big(x, f(x) \cup \{\infty\}\big)$. Vice versa, for each special $(0, n + 1)$-bundle the mapping $X \ni x \to \mathring{\boldsymbol{E}} \cap (\mathrm{pr}_1)^{-1}(x)$ defines a smooth separable quasi-polynomial f of degree n on X. The special $(0, n + 1)$-bundle is holomorphic if and only if the mapping f is holomorphic.

Choose a base point $x_0 \in X$. Denote by f_* the mapping from $\pi_1(X, x_0)$ to the fundamental group $\pi_1(C_n(\mathbb{C})\diagup\mathcal{S}_n, f(x_0)) \cong \mathcal{B}_n$ induced by f. Two smooth mappings f_1 and f_2 from X to $C_n(\mathbb{C})\diagup\mathcal{S}_n$ define special $(0, n+1)$-bundles that are isotopic through special $(0, n+1)$-bundles if and only if the quasipolynomials f_1 and f_2 are free isotopic. They define bundles that are isomorphic (equivalently, they are isotopic through $(0, n)$-bundles with a section) if and only if for a set of generators e_j of the fundamental group $\pi_1(X, x_0)$, for a braid $w \in \mathcal{B}_n$ and integer numbers k_j the equalities $(f_1)_*(e_j) = w^{-1}(f_2)_*(e_j)w\Delta_n^{2k_j}$ hold. Indeed, the bundles are isomorphic iff their monodromy homomorphisms are conjugate. The monodromy mapping classes of the bundles are elements of $\mathfrak{M}(\mathbb{P}^1; \{\infty\}, f(x_0))$ and the braid group on n strands modulo its center $\mathcal{B}_n/\mathcal{Z}_n$ is isomorphic to the mapping class group $\mathfrak{M}(\mathbb{P}^1; \{\infty\}, f(x_0))$.

Each smooth $(0, n)$-bundle $\mathfrak{F} = (\mathcal{X}, \mathcal{P}, \mathring{E} \cup s, X)$ with a section over an oriented smooth finite open surface X is isomorphic (equivalently, isotopic through $(0, n)$-bundles with a section) to a smooth special $(0, n+1)$-bundle. Indeed, put $\mathring{E} = \mathring{E} \cap \mathcal{P}^{-1}(x_0)$, $s = s \cap \mathcal{P}^{-1}(x_0)$ for the base point $x_0 \in X$. A homeomorphism $\varphi : \mathcal{P}^{-1}(x_0) \to \{x_0\} \times \mathbb{P}^1$ with $\varphi(s) = (x_0, \infty)$ and $\varphi(\mathring{E}) = \{x_0\} \times \mathring{E}'$ for a set $\mathring{E}' \subset C_n(\mathbb{C})/\mathcal{C}_n$ induces an isomorphism Is : $\mathfrak{M}(\mathcal{P}^{-1}(x_0); s, \mathring{E}) \to \mathfrak{M}(\mathbb{P}^1; \{\infty\}, \mathring{E}')$. The bundle $\mathfrak{F}$ corresponds to a conjugacy class of homomorphisms $\pi_1(X, x_0) \to \mathfrak{M}(\mathcal{P}^{-1}(x_0); s, \mathring{E})$. The isomorphism between mapping class groups gives us a conjugacy class of homomorphisms $\pi_1(X, x_0) \to \mathfrak{M}(\mathbb{P}^1; \{\infty\}, \mathring{E}')$. There is a special $(0, n+1)$-bundle that corresponds to the latter conjugacy class of homomorphisms. This bundle is isomorphic to $\mathfrak{F}$.

Lemma 8.3 *Each holomorphic $(0, n)$-bundle $\mathfrak{F} \stackrel{def}{=} (\mathcal{X}, \mathcal{P}, \mathring{E} \cup s, X)$ with a holomorphic section over a connected finite open Riemann surface X is holomorphically isomorphic to a special holomorphic $(0, n+1)$-bundle.*

Proof Choose D, $\tilde{D}_j$, $j = 0, \ldots$, and $\tilde{U}$ as in the proof of Theorem 8.1 in Sect. 8.2 (see also Sect. 1.2). Lift the bundle $\mathfrak{F}$ to a bundle on the universal covering $\tilde{X}$ and restrict the lift to a bundle $\tilde{\mathfrak{F}} = (\tilde{\mathcal{P}}^{-1}(\tilde{U}), \tilde{\mathcal{P}}, \mathring{\tilde{E}} \cup \tilde{s}, \tilde{U})$ on $\tilde{U}$. The lift $\mathring{\tilde{E}}$ of $\mathring{E}$ to $\tilde{U}$ is the union of n connected components $\mathring{\tilde{E}}_{j'}$, $j' = 1, \ldots, n$, each intersecting each fiber along a single point.

Consider the holomorphic $(0, 0)$-bundle $(\tilde{\mathcal{P}}^{-1}(\tilde{U}), \tilde{\mathcal{P}}, \tilde{U})$ that is obtained from $\tilde{\mathfrak{F}}$ by forgetting the distinguished points and the section. All fibers of the $(0, 0)$-bundle $(\tilde{\mathcal{P}}^{-1}(\tilde{U}), \tilde{\mathcal{P}}, \tilde{U})$ are conformally equivalent compact Riemann surfaces. Hence, by a theorem of Fischer and Grauert [26] the bundle is locally holomorphically trivial. This means that for each point $\tilde{x} \in \tilde{U}$ there exists a simply connected open subset $U_{\tilde{x}} \subset \tilde{U}$ containing $\tilde{x}$, and a biholomorphic map $\varphi_{U_{\tilde{x}}} : (\tilde{\mathcal{P}})^{-1}(U_{\tilde{x}}) \to U_{\tilde{x}} \times \mathbb{P}^1$ such that the diagram

$$\begin{array}{ccc} \tilde{\mathcal{P}}^{-1}(U_{\tilde{x}}) & \xrightarrow{\varphi_{U_{\tilde{x}}}} & U_{\tilde{x}} \times \mathbb{P}^1 \\ {\scriptstyle \tilde{\mathcal{P}}} \downarrow & \swarrow {\scriptstyle \mathrm{pr}_1} & \\ U_{\tilde{x}} & & \end{array}$$

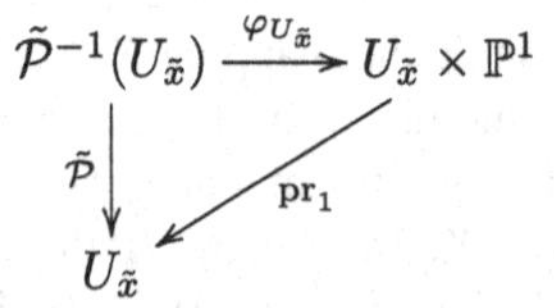

commutes. For each $\tilde{x}' \in U_{\tilde{x}}$ and $j' = 1, \ldots, n$, the mapping $\varphi_{U_{\tilde{x}}}$ takes the point $\mathring{\tilde{E}}_{j'} \cap (\tilde{\mathcal{P}})^{-1}(\tilde{x}')$ to a point denoted by $(\tilde{x}', \zeta_{j'}(\tilde{x}'))$, and takes the point of the section

$s \cap (\tilde{\mathcal{P}})^{-1}(\tilde{x}')$ to a point denoted by $(\tilde{x}', \zeta_{n+1}(\tilde{x}'))$. The $\zeta_{j'}$, $j' = 1, \ldots, n+1$, are holomorphic. Composing $\varphi_{U_{\tilde{x}}}$ with a holomorphic mapping from $U_{\tilde{x}} \times \mathbb{P}^1$ onto itself which preserves each fiber, we may assume that $\varphi_{U_{\tilde{x}}}$ maps no point in $\tilde{\mathcal{P}}^{-1}(\tilde{x}) \cap (\widetilde{\overset{\circ}{\boldsymbol{E}}} \cup \boldsymbol{s})$ to $(\tilde{x}, \infty)$.

Maybe, after shrinking the $U_{\tilde{x}}$, the mapping

$$U_{\tilde{x}} \times \mathbb{P}^1 \ni (\tilde{x}', \zeta) \to w_{\tilde{x}}(\tilde{x}', \zeta) = \left(\tilde{x}', -1 + 2\frac{(\zeta_2(\tilde{x}') - \zeta)(\zeta_{n+1}(\tilde{x}') - \zeta_1(\tilde{x}'))}{(\zeta_2(\tilde{x}') - \zeta_1(\tilde{x}'))(\zeta_{n+1}(\tilde{x}') - \zeta)}\right)$$

provides a biholomorphic map $U_{\tilde{x}} \times \mathbb{P}^1 \circlearrowleft$, that takes for each $\tilde{x}' \in U_{\tilde{x}}$ the point $\zeta_1(\tilde{x}')$ to 1, it maps $\zeta_2(\tilde{x}')$ to -1, and $\zeta_{n+1}(\tilde{x}')$ to ∞.

Denote by $\psi_{\tilde{x}}$ the mapping $w_{\tilde{x}} \circ \varphi_{U_{\tilde{x}}}$ from $\tilde{\mathcal{P}}^{-1}(U_{\tilde{x}})$ onto $U_{\tilde{x}} \times \mathbb{P}^1$. If for two points $\tilde{x}, \tilde{y} \in \tilde{U}$ the intersection $U_{\tilde{x}} \cap U_{\tilde{y}}$ is not empty and connected, then the mappings $\psi_{\tilde{x}}$ and $\psi_{\tilde{y}}$ coincide on $\tilde{\mathcal{P}}_1^{-1}(U_{\tilde{x}} \cap U_{\tilde{y}})$, since in each fiber the ordered triple of points, corresponding to $\widetilde{\overset{\circ}{\boldsymbol{E}}}_1, \widetilde{\overset{\circ}{\boldsymbol{E}}}_2$ and the section, is mapped by each of the mappings to the ordered triple $(-1, 1, \infty)$. We get a holomorphic isomorphism $\tilde{\psi}$ from the total space $\tilde{\mathcal{U}}$ of the bundle $\tilde{\mathfrak{F}}$ onto $\tilde{U} \times \mathbb{P}^1$ that maps the first component of $\widetilde{\overset{\circ}{\boldsymbol{E}}}$ to $\tilde{U} \times \{+1\}$, the second to $\tilde{U} \times \{-1\}$, and the section to $\tilde{U} \times \{\infty\}$. The isomorphism takes the set $\widetilde{\overset{\circ}{\boldsymbol{E}}}$ to a set $\widetilde{\overset{\circ}{\boldsymbol{E}}'} \subset \tilde{U} \times \mathbb{P}^1$, and the section $\tilde{\mathfrak{s}}$ to the set $\tilde{\mathfrak{s}}^\infty = \tilde{U} \times \{\infty\}$. The identity mapping from $\tilde{U}$ onto itself together with the mapping $\tilde{\psi}$ define a holomorphic isomorphism from the bundle $\tilde{\mathfrak{F}}$ onto the bundle $(\tilde{U} \times \mathbb{P}^1, \mathrm{pr}_1, \widetilde{\overset{\circ}{\boldsymbol{E}}'} \cup \tilde{\mathfrak{s}}^\infty, \tilde{U})$.

Recall that the original bundle $\mathfrak{F}$ is obtained from the bundle $\tilde{\mathfrak{F}}$ by gluing for each j the restriction $\tilde{\mathfrak{F}} | \tilde{D}_j$ to the restriction $\tilde{\mathfrak{F}} | \tilde{D}_0$ using a holomorphic isomorphism φ_j from $\tilde{\mathcal{P}}^{-1}(\tilde{D}_0)$ onto $\tilde{\mathcal{P}}^{-1}(\tilde{D}_j)$. A bundle that is holomorphically isomorphic to $\mathfrak{F}$ is obtained as follows. Take the bundle $(\tilde{U} \times \mathbb{P}^1, \mathrm{pr}_1, \widetilde{\overset{\circ}{\boldsymbol{E}}'} \cup \tilde{\mathfrak{s}}^\infty, \tilde{U})$ and glue for each $j = 1, \ldots, 2g + m - 1$, its restrictions to $\tilde{D}_j$ and $\tilde{D}_0$ together using the biholomorphic mapping $\tilde{\psi}_{0,j} \overset{def}{=} \tilde{\psi} \circ \varphi_j \circ (\tilde{\psi})^{-1} \mid \tilde{D}_0 \times \mathbb{P}^1$ between the subsets $\tilde{D}_0 \times \mathbb{P}^1$ and $\tilde{D}_1 \times \mathbb{P}^1$ of the total space $\tilde{U} \times \mathbb{P}^1$ of the bundle $(\tilde{U} \times \mathbb{P}^1, \mathrm{pr}_1, \widetilde{\overset{\circ}{\boldsymbol{E}}'} \cup \tilde{\mathfrak{s}}^\infty, \tilde{U})$.

Since $\tilde{\psi}_{0,j}$ maps the fiber of $\tilde{\psi}(\tilde{\mathfrak{F}})$ over $(\tilde{x}_0, \zeta) \in \tilde{D}_0 \times \mathbb{P}^1$ conformally onto the fiber over $(\tilde{x}_j, \zeta) \in \tilde{D}_j \times \mathbb{P}^1$, the following equality holds

$$\tilde{\psi}_{0,j}(\tilde{x}_0, \zeta) = \left(\tilde{x}_j, \tilde{a}_j(\tilde{x}_0) + \tilde{\alpha}_j(\tilde{x}_0)\zeta\right), \ (\tilde{x}_0, \zeta) \in \tilde{D}_0 \times \mathbb{C}, \tag{8.13}$$

for holomorphic functions $\tilde{a}_j$ and $\tilde{\alpha}_j$ with $\tilde{\alpha}_j$ nowhere vanishing on $\tilde{D}_0$.

We want to find a biholomorphic fiber preserving holomorphic mapping $v : \tilde{U} \times \mathbb{P}^1 \circlearrowleft$ such that for each j the mappings $v \circ \tilde{\psi}_{0,j} \circ v^{-1}$ have the form

$$v \circ \tilde{\psi}_{0,j} \circ v^{-1}(\tilde{x}_0, \zeta) = (\tilde{x}_j, \zeta), \ (\tilde{x}_0, \zeta) \in \tilde{D}_0 \times \mathbb{P}^1, \tag{8.14}$$

where $\tilde{x}_j \in \tilde{D}_j$ is the point for which $\mathsf{P}(\tilde{x}_0) = \mathsf{P}(\tilde{x}_j)$ for the projection $\mathsf{P} : \tilde{X} \to X$. Then the bundle $(v \circ \tilde{\psi})(\tilde{\mathfrak{F}})$ provides a special holomorphic $(0, n+1)$-bundle that is holomorphically isomorphic to $\mathfrak{F}$ by gluing the fibers over $\tilde{x}_j \in \tilde{D}_j$ to the fibers over $\tilde{x}_0 \in \tilde{D}_0$ using the mapping $(\tilde{x}_0, \zeta) \to (\tilde{x}_j, \zeta)$.

For $(\tilde{x}, \zeta) \in \tilde{U} \times \mathbb{C}$ we may write

$$v^{-1}(\tilde{x}, \zeta) = \big(\tilde{x}, a(\tilde{x}) + \alpha(\tilde{x})\zeta\big), \ (\tilde{x}, \zeta) \in \tilde{U} \times \mathbb{C}. \tag{8.15}$$

Condition (8.14) is by (8.13) and (8.15) equivalent to the following requirements on a and α:

$$\alpha(\tilde{x}_j) = \tilde{\alpha}_j(\tilde{x}_0) \cdot \alpha(\tilde{x}_0), \qquad j = 1, \ldots, 2g + m - 1, \tag{8.16}$$

$$a(\tilde{x}_j) = \tilde{a}_j(\tilde{x}_0) + \tilde{\alpha}_j(\tilde{x}_0)\, a(\tilde{x}_0), \quad j = 1, \ldots, 2g + m - 1. \tag{8.17}$$

The Eq. (8.16) leads to a Second Cousin Problem for α_j. If Eq. (8.16) is granted, Eq. (8.17) can be rewritten as

$$\frac{a(\tilde{x}_j)}{\alpha(\tilde{x}_j)} = \frac{a(\tilde{x}_0)}{\alpha(\tilde{x}_0)} + \frac{\tilde{a}_j(\tilde{x}_0)}{\alpha(\tilde{x}_j)}. \tag{8.18}$$

For convenience of the reader we provide the reduction of Eq. (8.16) to the Second Cousin Problem. Let D, V_j and s_j be the same objects as in Sect. 8.2. We consider an open cover of X as follows. Put $U_0 = D$. U_0 is an open topological disc in X that contains the base point x_0. Cover each V_j by two simply connected open sets U_{j+} and U_{j-} with connected and simply connected intersection, so that the $U_{j\pm}$ are disjoint from the $U_{k\pm}$ for $j \neq k$, and for each $U_{j\pm}$ its intersection with U_0 is connected and simply connected. We may also assume that the intersection of at least three of the sets is empty, the intersections of each s_j with U_{j+} and U_{j-} are connected, and each s_j is disjoint from $U_{k+} \cup U_{k-}$ for $k \neq j$. Label the $U_{j\pm}$ so that walking on $\mathsf{s}_j \setminus U_0$ (which is contained in V_j) in the direction of the orientation of s_j we first meet U_{j-} (see Fig. 8.1).

We consider the following subsets of $\tilde{X}$. Let $(U_0 \cup_{j-} U_{j-})^{\sim}$ be a lift of $U_0 \cup_{j-} U_{j-}$ to $\tilde{X}$. Let $(\tilde{x}_0)_0$ be the point in the lifted set that projects to x_0. For each $j \geq 1$ we consider a point $q_j \in U_{j-} \cap U_{j+}$ and the point $\tilde{q}_j$ in $(U_0 \bigcup \cup_{j-} U_{j-})^{\sim}$ that projects to q_j. For each j we denote by $(U_{j+} \cup U_0)^{\sim}$ the lift of $U_{j+} \cup U_0$ to $\tilde{X}$ that contains $\tilde{q}_j$. Let $\widetilde{U_0^{j+}}$ be the subset of $(U_{j+} \cup U_0)^{\sim}$ that projects to U_0.

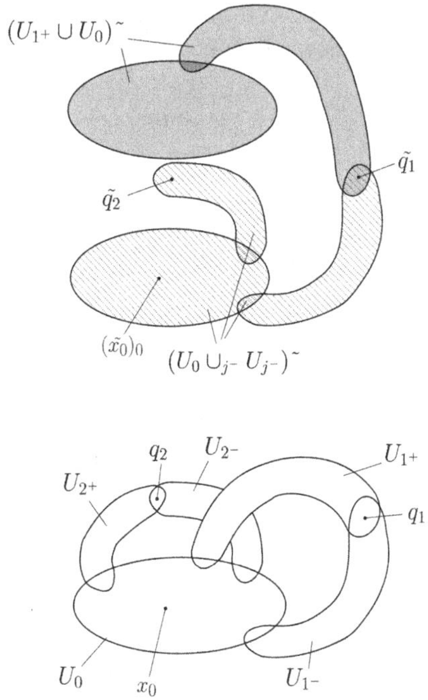

Fig. 8.1 The open cover of X and its lift to $\tilde{X}$

The sets U_0, $U_{j\pm}$ cover X. The intersection of three distinct sets of the cover is empty. On the intersection of pairs of covering sets we define a non-vanishing holomorphic function as follows. Put

$$\begin{aligned}
\alpha_{0,j^-} &= (\alpha_{j^-,0})^{-1} = 1 && \text{on } U_0 \cap U_{j^-} \quad j = 1, \ldots, 2g+m-1\,, \\
\alpha_{j^-,j^+} &= (\alpha_{j^+,j^-})^{-1} = 1 && \text{on } U_{j^-} \cap U_{j^+} \quad j = 1, \ldots, 2g+m-1\,, \\
\alpha_{j^+,0} &= (\alpha_{0,j^+})^{-1} = \tilde{\alpha}_{j,0} && \text{on } U_{j^+} \cap U_0 \quad j = 1, \ldots, 2g+m-1\,,
\end{aligned} \tag{8.19}$$

where for $U_{j^+} \cap U_0$, the function $\tilde{\alpha}_{j,0}$ is defined as $\tilde{\alpha}_{j,0}(x) = \tilde{\alpha}_j(\tilde{x}_0)$ for the lift $\tilde{x}_0$ of the point $x \in U_0$ to $\widetilde{U_0^{j^+}}$.

Since there are no triple intersections, the Eq. (8.19) define a Cousin II distribution. A second Cousin distribution on a complex manifold defines a holomorphic line bundle. The Second Cousin Problem has a solution if and only if the line bundle is holomorphically trivial. Since on an open Riemann surface each holomorphic line bundle is holomorphically trivial the Cousin Problem has a solution (see e.g. [27]).

The solution of the Cousin Problem for the Cousin distribution (8.19) provides non-vanishing holomorphic functions α_0 on U_0 and $\alpha_{j\pm}$ on $U_{j\pm}$ such that

$$\begin{aligned} \alpha_0 &= \alpha_{j-} && \text{on } U_0 \cap U_{j-}, \\ \alpha_{j-} &= \alpha_{j+} && \text{on } U_{j-} \cap U_{j+}, \\ \alpha_{j+}\alpha_0^{-1} &= \tilde{\alpha}_{j,0} && \text{on } U_{j+} \cap U_0 \;. \end{aligned} \tag{8.20}$$

Put

$$\alpha(\tilde{x}) = \begin{cases} \alpha_0(x) & \tilde{x} \in \tilde{U}_0 \\ \alpha_{j-}(x) & \tilde{x} \in \tilde{U}_{j-} \\ \alpha_{j+}(x) & \tilde{x} \in \tilde{U}_{j+} \\ \tilde{\alpha}_{j,0}(x) \cdot \alpha_0(x) & \tilde{x} \in \tilde{U}_0^{j+} \;. \end{cases} \tag{8.21}$$

Here $x = \mathsf{P}(\tilde{x})$ for the projection $\mathsf{P} : \tilde{X} \to X$. The function α is a well-defined holomorphic function on $\tilde{U}$ that satisfies (8.16).

The reduction of (8.18) to a First Cousin Problem for the $\frac{a(\tilde{x}_j)}{\alpha(\tilde{x}_j)}$, $j = 0, \dots 2g + m - 1$, is similar. The First Cousin Problem is solvable since open Riemann surfaces are Stein manifolds (see e.g. [27] and [40], and also Appendix A.1). □

8.4 (0, 4)-Bundles Over Tori with a Hole. Proof of Theorem 7.4

We will consider now deformations of a smooth (0, 3)-bundle with a section over a connected finite open Riemann surface X to a holomorphic bundle. Statement (1) of Theorem 8.2 below is the analog of Theorem 7.3 for (0, 3)-bundles with a section.

Definition 8.4 *We will say that a smooth* (g, m)*-bundle* $\mathfrak{F}$ *over a connected smooth finite open oriented surface* X *is isotopic to a holomorphic bundle for the conformal structure* $\omega : X \to \omega(X)$*, if the pushed forward bundle* $\mathfrak{F}_\omega$ *(see Eq.* (8.1)*) is isotopic to a holomorphic bundle on* $\omega(X)$*. If the bundle is isotopic to a holomorphic bundle for each conformal structure with only thick ends on* X*, then* $\mathfrak{F}$ *is said to have the Gromov-Oka property.*

A $(0, n)$*-bundle on* X *with a section, in particular, a special* $(0, n+1)$*-bundle on* X*, is said to have the Gromov-Oka property, if and only if for each conformal structure on* X *with only thick ends the bundle is isotopic through* $(0, n)$*-bundles with a section to a holomorphic bundle.*

In the following theorem we start with a smooth special (0, 4)-bundle, since each smooth (0, 3)-bundle with a section is isotopic to a special (0, 4)-bundle. Recall that each connected finite open Riemann surface $\mathcal{X}$ is conformally equivalent to a domain in a connected closed Riemann surface $\mathcal{X}^c$. Notice that a special $(0,4)$-bundle is reducible if and only if all monodromies are powers of a single conjugate of $\sigma_1/\mathcal{Z}_3$.

Theorem 8.2

(1) Let X be a connected smooth oriented surface of genus one with a hole with base point x_0, and with a chosen set $\mathcal{E} = \{e_1, e_2\}$ of generators of $\pi_1(X, x_0)$. Define the set $\mathcal{E}_0 = \{e_1, e_2, e_1e_2^{-1}, e_1e_2^{-2}, e_1e_2e_1^{-1}e_2^{-1}\}$ as in Theorem 7.3. Consider a smooth special $(0,4)$-bundle $\mathfrak{F}$ over X. Suppose for each $e \in \mathcal{E}_0$ the restriction of the bundle $\mathfrak{F}$ to an annulus in X representing $\hat{e}$ has the Gromov-Oka property. Then the bundle over X has the Gromov-Oka property.

(2) If a bundle $\mathfrak{F}$ as in (1) is irreducible, then it is isotopic to an isotrivial bundle, and, hence, for any conformal structure ω on X the bundle $\mathfrak{F}_\omega$ is isotopic to a bundle that extends to a holomorphic bundle on $\omega(X)^c$. In particular, the bundle is isotopic to a holomorphic bundle for any conformal structure on X (including conformal structures of first kind).

(3) Let $\mathfrak{F}$ be any smooth reducible special $(0,4)$-bundle over X (without any further requirement). Then each irreducible bundle component of $\mathfrak{F}$ is isotopic to an isotrivial bundle. There exists a Dehn twist in the fiber over the base point such that the bundle $\mathfrak{F}$ can be recovered from the irreducible bundle components up to composing each monodromy with a power of this Dehn twist. The bundle $\mathfrak{F}$ itself is isotopic to a holomorphic bundle for each conformal structure of second kind on X.

(4) Any reducible holomorphic $(0,3)$-bundle with a holomorphic section over a punctured Riemann surface is holomorphically trivial.

Recall that a bundle is called isotrivial if its lift to a finite covering of X is the trivial bundle. We will prove now Theorems 7.3 and 8.2 using Theorem 7.4.

Proof of Theorem 8.2 The special $(0,4)$-bundle $\mathfrak{F}$ defines a smooth quasipolynomial f on X. Under the conditions of Statement (1) for each $e \in \mathcal{E}_0$ the restriction of the bundle to an annulus $A_{\hat{e}}$ in X representing $\hat{e}$ has the Gromov-Oka property. By Lemma 8.3 we may assume that for a conformal structure $\omega(A_{\hat{e}})$ of conformal module bigger than $\frac{\pi}{2}\log(\frac{3+\sqrt{5}}{2})^{-1}$ on $A_{\hat{e}}$ the bundle $(\mathfrak{F} \mid A_{\hat{e}})_\omega$ is isotopic (through $(0,3)$-bundles with a section on $\omega(A_{\hat{e}})$) to a special holomorphic $(0,4)$-bundle $\mathfrak{F}_{\hat{e}}$ (not merely to a holomorphic $(0,3)$-bundle with a section). The bundle $\mathfrak{F}_{\hat{e}}$ defines a holomorphic quasipolynomial $f_{\hat{e}}$ on $\omega(A_{\hat{e}})$. We may assume that $\omega(A_{\hat{e}})$ has the form $\{z \in \mathbb{C} : \frac{1}{r} < |z| < r\}$. The monodromies of the isomorphic bundles $(\mathfrak{F} \mid A_{\hat{e}})_\omega$ and $\mathfrak{F}_{\hat{e}}$ differ by conjugation (after identification of the mapping class groups on the fiber over the base point by an isomorphism). Hence, there exists an integer number k_e, such that the quasipolynomial $f \circ \omega^{-1} \mid \omega(A_{\hat{e}})$ corresponding to $(\mathfrak{F} \mid A_{\hat{e}})_\omega$ is (free) isotopic to the holomorphic quasipolynomial $z \to e^{2\pi k_e z} f_{\hat{e}}(z)$, $z \in \omega(A_{\hat{e}})$. This holds for each $e \in \mathcal{E}_0$. We proved that the quasipolynomial f satisfies the conditions of Theorem 7.4.

The monodromy mapping classes of the quasipolynomial f are elements of the braid group $\mathcal{B}_3$, the monodromy mapping classes of the bundle $\mathfrak{F}$ can be identified with elements of $\mathcal{B}_3/\mathcal{Z}_3$. In terms of the (0, 4)-bundle $\mathfrak{F}$ the Theorem 7.4 implies the following. The isotopy class of the special (0, 4)-bundle $\mathfrak{F}$ on X corresponds to the conjugacy class of a homomorphism $\boldsymbol{\Phi} : \pi_1(X, x_0) \to \boldsymbol{\Gamma} \subset \mathcal{B}_3/\mathcal{Z}_3$, where $\boldsymbol{\Gamma}$ is generated either by $\sigma_1 \sigma_2/\mathcal{Z}_3$, or by $\Delta_3/\mathcal{Z}_3$, or by $\sigma_1/\mathcal{Z}_3$.

The group $\boldsymbol{\Gamma}$ is Abelian. Let ω be any conformal structure on X (maybe, of first kind). The fundamental group $\pi_1(\omega(X)^c, \omega(x_0))$ of the closed torus $\omega(X)^c$ (that contains a conformal copy of $\omega(X)$) is the Abelianization of $\pi_1(\omega(X), \omega(x_0)) \cong \pi_1(X, x_0)$. Hence $\boldsymbol{\Phi}$ defines a homomorphism from $\pi_1(\omega(X)^c, \omega(x_0))$ to $\boldsymbol{\Gamma}$ which we also denote by $\boldsymbol{\Phi}$.

Consider first the case when the generator of $\boldsymbol{\Gamma}$ is either $\sigma_1\sigma_2/\mathcal{Z}_3$, or $\Delta_3/\mathcal{Z}_3$. By Lemma 6.4 this is exactly the case when the bundle $\mathfrak{F}$ is irreducible. The generator of $\boldsymbol{\Gamma}$ corresponds to an element of the mapping class group $\mathfrak{M}(\mathbb{P}^1; \{\infty\}, \mathring{E}_3)$, that is represented by a periodic mapping $\zeta \to \theta \cdot \zeta$, $\zeta \in \mathbb{P}^1$ (see [11, 53]). In the first case $\theta = e^{\frac{2\pi i}{3}}$ and $\mathring{E}_3$ consists of 3 equidistributed points on a circle with center zero, in the second case $\theta = e^{\frac{2\pi i}{2}}$ and $\mathring{E}_3$ consists of the origin and two equidistributed points on a circle with center zero.

Consider the homomorphism from $\boldsymbol{\Gamma}$ to the group of complex linear transformations of $\mathbb{P}^1$, which assigns to the generator $\boldsymbol{b} \in \boldsymbol{\Gamma}$ the complex linear transformation $\zeta \to \theta \cdot \zeta$, $\zeta \in \mathbb{P}^1$. Composing $\boldsymbol{\Phi}$ with this homomorphism we obtain a homomorphism Ξ from $\pi_1(\omega(X)^c, \omega(x_0))$ to the group of complex linear transformations of $\mathbb{P}^1$.

Represent the closed torus $\omega(X)^c$ as quotient $\omega(X)^c = \mathbb{C}/\Lambda$ for a lattice $\Lambda = \{\lambda_1 k_1 + \lambda_2 k_2 : k_1, k_2 \in \mathbb{Z}\}$ with complex numbers $\lambda_1, \lambda_2 \in \mathbb{C}$ that are linearly independent over $\mathbb{R}$. Denote by p_1 the covering map $p_1 : \mathbb{C} \to \mathbb{C}/\Lambda$. Each element $\lambda \in \Lambda$ defines a covering transformation $z \to z + \lambda$, $z \in \mathbb{C}$. We identify Λ with the group of covering transformations. Recall that the fundamental group $\pi_1(\omega(X)^c, \omega(x_0))$ is isomorphic to the group of covering transformations of $\omega(X)^c \cong \mathbb{C}/\Lambda$.

The group Λ acts on $\mathbb{C} \times \mathbb{P}^1$ as follows. Associate to $\lambda \in \Lambda$ the mapping

$$(z, \zeta) \xrightarrow{\lambda} (z + \lambda, \Xi(\lambda)(\zeta)), \quad (z, \zeta) \in \mathbb{C} \times \mathbb{P}^1. \tag{8.22}$$

The thus defined mapping λ takes the set $\mathbb{C} \times \mathring{E}_3$ to itself and fixes the set $\mathbb{C} \times \{\infty\}$. The action is free and properly discontinuous. Hence the quotient $(\mathbb{C} \times \mathbb{P}^1)/\Lambda$ is a complex manifold. For each element of the quotient its projection to $\mathbb{C}/\Lambda$ is well-defined. Indeed, the following diagram commutes:

$$\begin{array}{ccc} (z, \zeta) & \sim & (\lambda + z, \Xi(\lambda)(\zeta)) \\ \downarrow & & \downarrow \\ z & \sim & \lambda + z \end{array}$$

The equivalence relation in the upper line gives the quotient $(\mathbb{C} \times \mathbb{P}^1)\diagup\Lambda$, the relation in the lower line gives $\mathbb{C}\diagup\Lambda$.

Denote by $\mathcal{P}$ the projection $\mathcal{P} : (\mathbb{C}\times\mathbb{P}^1)\diagup\Lambda \to \mathbb{C}\diagup\Lambda$. We obtain a holomorphic (0, 3)-bundle

$$\mathfrak{F}_\omega \overset{def}{=} \Big((\mathbb{C} \times \mathbb{P}^1)\diagup\Lambda,\ \mathcal{P},\ \big((\mathbb{C} \times \mathring{E}_3)\diagup\Lambda\big) \cup \big((\mathbb{C} \times \{\infty\})\diagup\Lambda\big),\ \mathbb{C}\diagup\Lambda\Big)$$

with a holomorphic section over the closed torus $\omega(X)^c$. (See also (8.1) for notation.) By construction the restriction of this bundle to a representative of each generator e_j of $\pi_1(\omega(X)^c, x_0)$ gives the mapping torus corresponding to $\Xi(e_j)$. Hence, the monodromy mapping classes of the bundle are equal to $\boldsymbol{\Phi}(e_j)$. We proved that in the irreducible case of Theorem 8.2 for any conformal structure ω on X the bundle $\mathfrak{F}_\omega$ is smoothly isomorphic (equivalently, isotopic) to a holomorphic bundle that extends to a holomorphic bundle $(\mathfrak{F}_\omega)^c$ on the closed torus $\omega(X)^c$.

Since the monodromy mapping classes of the bundle $\mathfrak{F}_\omega$ are periodic, the lift of the bundle $(\mathfrak{F}_\omega)^c$ to a finite covering of X is isotopic to an isotrivial bundle. Statement (2) is proved and also statement (1) in the irreducible case.

To prove Statement (3), we suppose $\mathfrak{F}$ is an arbitrary reducible smooth special (0, 4)-bundle. Let $\mathring{E}_3$ be the set of finite distinguished points in the fiber over the base point. There is an admissible curve $\gamma \subset \mathbb{P}^1 \setminus (\mathring{E}_3 \cup \{\infty\})$ that reduces each monodromy mapping class of the bundle. Hence, by (the proof of) Lemma 6.4 each monodromy mapping class of the bundle is a non-trivial power of a single conjugate of the mapping class $\mathfrak{m}(\sigma_1\diagup\mathcal{Z}_3)$ in $\mathfrak{M}(\mathbb{P}^1; \infty, \mathring{E}_3)$ that corresponds to $\sigma_1\diagup\mathcal{Z}_3$. We are in the case when $\boldsymbol{\Gamma}$ is generated by $\sigma_1\diagup\mathcal{Z}_3$. Label the points of $\mathring{E}_3$ as in the proof of Lemma 6.4 so that the simple closed curve γ separates $\{\zeta_1, \zeta_2\}$ from $\{\zeta_3, \infty\}$. Let C_1 be the connected component of the complement of γ that contains $\{\zeta_1, \zeta_2\}$, and let C_2 be the connected component of the complement of γ that contains $\{\zeta_3, \infty\}$.

There are two irreducible components of the mapping class $\mathfrak{m}(\sigma_1\diagup\mathcal{Z}_3)$. The irreducible component that corresponds to C_1 is described as follows. The connected component C_1 is homeomorphic to $\mathbb{P}^1 \setminus \{\infty\}$ with set of distinguished points $\{\zeta_1, \zeta_2\}$. Each self-homeomorphism of $\mathbb{P}^1 \setminus \{\infty\}$ with these two distinguished points extends to a self-homeomorphism of $\mathbb{P}^1$ that maps the set $\{\zeta_1, \zeta_2, \infty\}$ of three distinguished points to itself. The irreducible component of the mapping class $\mathfrak{m}(\sigma_1\diagup\mathcal{Z}_3)$ corresponding to C_1 is an element of the mapping class group $\mathfrak{M}(\mathbb{P}^1; \{\infty\}, \{\zeta_1, \zeta_2\})$. This mapping class group is generated by the element $\boldsymbol{\sigma} = \sigma\diagup\mathcal{Z}_2$ which corresponds to the braid σ in the braid group $\mathcal{B}_2$ on two strands. The square of $\boldsymbol{\sigma}$ is the identity. We put $\boldsymbol{\Gamma}_1 = \mathfrak{M}(\mathbb{P}^1; \{\infty\}, \{\zeta_1, \zeta_2\})$.

To describe the irreducible component corresponding to C_2 of the mapping class $\mathfrak{m}(\sigma_1\diagup\mathcal{Z}_3)$, we notice that C_2 is homeomorphic to $\mathbb{P}^1 \setminus \{\zeta_4\}$ with distinguished points $\{\zeta_3, \infty\}$ for a point $\zeta_4 \in \mathbb{P}^1$, $\zeta_4 \neq \zeta_3, \infty$. Each self-homeomorphism of $\mathcal{P}^1 \setminus \{\zeta_4\}$ with distinguished points $\{\zeta_3, \infty\}$ extends to a self-homeomorphism of $\mathbb{P}^1$ with set of distinguished points $\{\zeta_3, \zeta_4, \infty\}$. The respective irreducible component of $\mathfrak{m}(\sigma_1\diagup\mathcal{Z}_3)$ is the identity in the group $\mathfrak{M}(\mathbb{P}^1; \{\zeta_4, \infty\}, z_3)$.

All monodromy mapping classes of a bundle that is isomorphic to $\mathfrak{F}$ are contained in $\boldsymbol{\Gamma}$, hence are powers of $\mathfrak{m}(\sigma_1/\mathcal{Z}_3)$. This implies that there are two irreducible bundle components, corresponding to C_1 and to C_2, respectively. They are described as follows.

The irreducible bundle component of the bundle $\mathfrak{F}$, corresponding to C_2, is associated to the trivial homomorphism. Hence, this irreducible bundle component is smoothly isomorphic to the trivial (0, 3)-bundle over X.

The irreducible bundle component of $\mathfrak{F}$ related to C_1 corresponds to the conjugacy class of the homomorphism $\boldsymbol{\Phi}_1 : \pi_1(X, x_0) \to \boldsymbol{\Gamma}_1$ induced by the homomorphism $\boldsymbol{\Phi}$. The same proof as in the case, when $\boldsymbol{\Gamma}$ is generated either by $\sigma_1\sigma_2/\mathcal{Z}_3$, or by $\Delta_3/\mathcal{Z}_3$, shows that for any conformal structure ω on X there is a holomorphic (0, 2)-bundle $(\mathfrak{F}_1')^\omega$ with a holomorphic section on the closed torus $\omega(X)^c$ whose isomorphism class corresponds to the conjugacy class of the homomorphism $\pi_1(\omega(X)^c, \omega(x_0)) \to \boldsymbol{\Gamma}_1$ that is induced by $\boldsymbol{\Phi}_1$. The restriction of $(\mathfrak{F}_1')^\omega$ to a punctured torus $\omega(X)'$, $\omega(X) \Subset \omega(X)'$, is isomorphic to a special holomorphic (0, 3)-bundle $(\mathfrak{F}_1)^\omega$ with set of distinguished points $\{\zeta_1(x), \zeta_2(x), \infty\}$ in the fiber $\{\omega(x)\} \times \mathbb{P}^1$ over $\omega(x)$. For $\omega(x) \in \omega(X) \Subset \omega(X)'$ the points $\zeta_1(x), \zeta_2(x)$ are contained in a large closed disc in $\mathbb{C}$ centered at the origin.

Take a point $\zeta_3 \in \mathbb{C}$ outside this closed disc. Consider the special holomorphic (0, 4)-bundle $(\mathfrak{F})^\omega \overset{def}{=} \left(X \times \mathbb{P}^1, \omega \circ \mathrm{pr}_1, \mathring{\boldsymbol{E}}_3 \cup \boldsymbol{s}^\infty, X\right)$, where the set of finite distinguished points $\mathring{\boldsymbol{E}}_3 \cap (\{\omega(x)\} \times \mathbb{P}^1)$ in the fiber over $\omega(x)$ equals $\{\omega(x)\} \times \{\zeta_1(x), \zeta_2(x), \zeta_3\}$ and $\boldsymbol{s}^\infty \cap (\{\omega(x)\} \times \mathbb{P}^1) = (\omega(x), \infty)$. The monodromy mapping class of the bundle $(\mathfrak{F})^\omega$ along e_j differs from the monodromy mapping class $\mathfrak{F}_\omega$, which is the push-forward to $\omega(X)$ of the bundle $\mathfrak{F}$, by the k_j-th power of a Dehn twist about the curve γ. Consider again the punctured torus $\omega(X)' \supset \omega(X)$, and assume that $\omega(X)' = (\mathbb{C} \setminus \Lambda)/\Lambda$ with $\Lambda = \{\lambda_1 n_1 + \lambda_2 n_2,\ n_1, n_2 \in \mathbb{Z}\}$, and $\omega(x_0) = \frac{\lambda_1+\lambda_2}{2}/\Lambda$, and that the segments $\frac{\lambda_1+\lambda_2}{2} + [0, \lambda_1]$ and $\frac{\lambda_1+\lambda_2}{2} + [0, \lambda_2]$ are lifts to $\mathbb{C} \setminus \Lambda$ of curves representing a pair of generators of $\pi_1(\omega(X)', \omega(x_0))$.

For all integers ℓ_1 and ℓ_2 there is a holomorphic function F on $\mathbb{C}\setminus\Lambda$ such that

$$F(y + \lambda_1) = \ell_1 \quad \text{and} \quad F(y + \lambda_2) = \ell_2,\ y = \omega(x) \in \mathbb{C} \setminus \Lambda. \tag{8.23}$$

This follows from the proof of Lemma 7.4. A quick way to see this for $\ell_2 = 0$ and any $\ell_1 \in \mathbb{Z}$ is the following. For each $y \in \mathbb{C}$ there are uniquely determined real numbers $t_1(y)$ and $t_2(y)$ depending smoothly on y such that $y = \lambda_1 t_1(y) + \lambda_2 t_2(y)$. Consider a function χ_0 on the real axis which vanishes near zero such that $\chi_0(t+1) = \chi_0(t) + \ell_1, t \in \mathbb{R}$. Consider the function $\chi(y) \overset{def}{=} \chi_0(t_1(y)),\ y \in \mathbb{C}$. The 1-form $\overline{\partial}\chi$ descends to a smooth 1-form δ on the torus $\mathbb{C}/\Lambda$. Let $\mathring{g}$ be a solution of the equation $\overline{\partial}\,\mathring{g} = \delta$ on the punctured torus $\mathbb{C}\setminus\Lambda/\Lambda$. Let g be the lift of $\mathring{g}$ to $\mathbb{C}\setminus\Lambda$. Put $F = \chi - g$.

Let F be the holomorphic function on $\mathbb{C}\setminus\Lambda$ that satisfies Eq. (8.23) for the integers $\ell_1 = -k_1$ and $\ell_2 = -k_2$. Then the monodromy mapping classes along the e_j of the special holomorphic (0, 4)-bundle $(\mathfrak{F}')^\omega$ over $\omega(X)$ with set of distinguished points $\{\zeta_1(x), \zeta_2(x), \zeta_3\, e^{2\pi i F(\omega(x))}, \infty\}$ in each fiber $\{\omega(x)\} \times \mathbb{P}^1$,

$x \in X$, are the same as the monodromy mapping classes of the bundle $\mathfrak{F}_\omega$. Statement (3) and, hence, statement (1) in the reducible case are proved.

It remains to prove Statement (4), namely the fact that any reducible holomorphic $(0, 3)$-bundle $\mathfrak{F}$ with a holomorphic section over a *punctured* Riemann surface X is holomorphically isomorphic to the trivial bundle. The respective statement is known in more general situations, but the proof of the present statement is a simple reduction to Picard's Theorem and we include it.

By Lemma 8.3 we may assume that the bundle is equal to a special $(0, 4)$-bundle $\mathfrak{F} = (X \times \mathbb{P}^1, \mathrm{pr}_1, \mathring{\boldsymbol{E}} \cup s^\infty, X)$. There is a finite unramified covering $\hat{X}$ of the punctured Riemann surface X so that for the lift $\hat{\mathfrak{F}} = \big(\hat{X} \times \mathbb{P}^1, \mathrm{pr}_1, \widehat{\mathring{\boldsymbol{E}}} \cup \hat{s}^\infty, \hat{X}\big)$ of the bundle $\mathfrak{F}$ to $\hat{X}$ the set $\widehat{\mathring{\boldsymbol{E}}} \cup \hat{s}^\infty$ is the union of four disjoint complex curves each intersecting each fiber along a single point. After a holomorphic isomorphism (see the proof of Lemma 8.3) we may assume that the connected components of $\widehat{\mathring{\boldsymbol{E}}}$ are $\hat{X} \times \{-1\}$, $\hat{X} \times \{1\}$, and a component that intersects each fiber $\hat{x} \times \mathbb{P}^1$ along a point $(\hat{x}, \hat{g}(\hat{x}))$ for a holomorphic function $\hat{g}$ on $\hat{X}$ that omits the values $-1,\ 1$, and ∞.

Since the bundle $\mathfrak{F}$ is reducible, the bundle $\hat{\mathfrak{F}}$ is reducible. Hence, $\hat{g} : \hat{X} \to \mathbb{C} \setminus \{-1, 1\}$ is free homotopic to a mapping whose image is contained in a punctured disc around one of the points $-1,\ 1$, or ∞. We may assume, after applying to each fiber the same conformal self-mapping of $\mathbb{P}^1$, that this point is -1.

By Picard's Theorem the mapping $\hat{g}$ extends to a meromorphic mapping $\hat{g}^c$ from the closure $\hat{X}^c$ of $\hat{X}$ to $\mathbb{P}^1$. Then the meromorphic extension $\hat{g}^c$ omits the value 1. Indeed, if $\hat{g}^c$ was equal to 1 at some puncture of $\hat{X}$, then $\hat{g}$ would map a simple closed curve on $\hat{X}$ that surrounds the puncture positively to a loop in $\mathbb{C} \setminus \{-1, 1\}$ that surrounds 1 with positive winding number. This contradicts the fact that $\hat{g}$ is homotopic to a mapping into a disc punctured at -1 and contained in $\mathbb{C} \setminus \{-1, 1\}$. Hence, $\hat{g}^c$ is a meromorphic function on a compact Riemann surface that omits a value, and, hence $\hat{g}$ is constant.

This means that the bundle $\hat{\mathfrak{F}}$ over $\hat{X}$ is holomorphically isomorphic to the trivial bundle. Hence, the monodromy mappings of the original bundle $\mathfrak{F}$ are periodic. By Lemma 6.4 this is possible for a reducible bundle only if the monodromies are equal to the identity. Repeat the same reasoning as above for the bundle $\mathfrak{F}$ instead of $\hat{\mathfrak{F}}$, we see that the bundle $\mathfrak{F}$ is holomorphically trivial. Theorem 8.2 is proved. □

Proof of Theorem 7.3 Consider the special $(0, 4)$-bundle $\mathfrak{F}$ that is associated to the quasipolynomial. By the conditions of Theorem 7.3 the bundle $\mathfrak{F}$ satisfies the assumptions of Statement (1) of Theorem 8.2. Theorem 8.2 implies, that for any a priori given conformal structure ω of second kind on X the isotopy class (with respect to isotopies through $(0, 3)$-bundles with a section) of the bundle $\mathfrak{F}_\omega$ on $\omega(X)$ contains a holomorphic $(0, 3)$-bundle with a section. Hence, by Lemma 8.3, the isotopy class contains a special holomorphic $(0, 4)$-bundle $\mathfrak{F}'_\omega$ on $\omega(X)$. The monodromies of the two bundles $\mathfrak{F}'_\omega$ and $\mathfrak{F}_\omega$ differ by conjugation by a self-homeomorphism of $\mathbb{P}^1$ that fixes the set of distinguished points setwise and fixes $\{\infty\}$. There is an isotopy through special $(0, 4)$-bundles that joins the bundle $\mathfrak{F}'_\omega$ with a special $(0, 4)$-bundle $\mathfrak{F}^\omega$ whose monodromies coincide with those of $\mathfrak{F}_\omega$.

The special holomorphic (0, 4)-bundle $\mathfrak{F}^{\omega}$ determines a holomorphic quasipolynomial f^{ω} for this conformal structure by assigning to each $x \in \omega(X)$ the triple of finite distinguished points in the fiber $\{x\} \times \mathbb{P}^1$ and associating to it the monic polynomial with this set of zeros. The monodromies of the quasipolynomial f^{ω} differ from the monodromies of the push forward f_{ω} of the original quasipolynomial by powers of Δ_3^2. Using the function F that satisfies (8.23) for suitable integers ℓ_1 and ℓ_2 we obtain a holomorphic quasipolynomial $f^{\omega}(x)e^{2\pi i F(x)}$, $x \in \omega(X)$, that is isotopic to f_{ω}. Theorem 7.3 is proved. □

8.5 Smooth Elliptic Fiber Bundles and Differentiable Families of Complex Manifolds

Recall that Kodaira (see [54]) defines a complex analytic family (also called holomorphic family) $\mathcal{M}_t$, $t \in \mathcal{B}$, of compact complex manifolds over the base $\mathcal{B}$ as a triple $(\mathcal{M}, \mathcal{P}, \mathcal{B})$, where $\mathcal{M}$ and $\mathcal{B}$ are complex manifolds, and $\mathcal{P}$ is a proper holomorphic submersion with $\mathcal{P}^{-1}(t) = \mathcal{M}_t$. Thus, each holomorphic genus g fiber bundle over a Riemann surface is a complex analytic family of compact Riemann surfaces of genus g.

We will consider holomorphic (g, m)-bundles as complex analytic families of Riemann surfaces of genus g equipped with m distinguished points, for short, as complex analytic (or holomorphic) families of Riemann surfaces of type (g, m), or holomorphic (g, m)-families.

Kodaira (see [54]) defines a differentiable family of compact complex manifolds as a smooth fiber bundle with compact fibers and the following additional structure. The fibers are equipped with complex structures that depend smoothly on the parameter and induce on each fiber the smooth fiber structure. The formal definition of a differentiable family of Riemann surfaces of type (g, m) is the following.

Definition 8.5 *A differentiable family of Riemann surfaces $\mathcal{M}_t$, $t \in \mathcal{B}$, of type* (g, m) *is a tuple $(\mathcal{M}, \mathcal{P}, \mathbf{E}, \mathcal{B})$, where $\mathcal{M}$ and $\mathcal{B}$ are oriented C^{∞} manifolds, and $\mathcal{P}$ is a smooth proper orientation preserving submersion with the following property. For each $t \in \mathcal{B}$ the fiber $\mathcal{P}^{-1}(t)$, denoted by $\mathcal{M}_t$, is equipped with the structure of a compact Riemann surface of genus* g, *such that the complex structure induces the differentiable structure defined on $\mathcal{P}^{-1}(t)$ as a submanifold of $\mathcal{M}$. $\mathbf{E} \subset \mathcal{M}$ is a smooth submanifold of $\mathcal{M}$ that intersects each fiber $\mathcal{M}_t$ along a set E_t of* m *distinguished points. Moreover, the complex structures of the $\mathcal{M}_t$ depend smoothly on the parameter t, more precisely, there is an open locally finite cover U_j, $j = 1, 2, \ldots,$ of $\mathcal{M}$ and complex valued C^{∞} functions $(z_1^j, \ldots, z_n^j)$ on each U_j (with n the complex dimension of $\mathcal{M}_t$) such that for each j and t these functions define holomorphic coordinates on $U_j \cap \mathcal{M}_t$.*

We will also call such families smooth families of Riemann surfaces of type (g, m), or smooth (g, m)-families.

Notice that each smooth special $(0, n+1)$-bundle over a finite oriented smooth surface carries automatically the structure of a smooth family of Riemann surfaces of type $(0, n+1)$.

Two smooth families of Riemann surfaces of type (g, m) will be called isomorphic (as families of Riemann surfaces) if they are isomorphic as smooth bundles (smoothly isomorphic, for short) and there is a bundle isomorphism that is holomorphic on each fiber.

We will show in this section that each smooth elliptic fiber bundle with a section is smoothly isomorphic (equivalently, isotopic) to a bundle that carries the structure of a differentiable family of Riemann surfaces of type $(1, 1)$, also called a differentiable family of closed Riemann surfaces of genus 1 with a smooth section.

Recall that each compact Riemann surface of genus 1 is conformally equivalent to the quotient $\mathbb{C}/\Lambda$ for a lattice $\Lambda = a\mathbb{Z} + b\mathbb{Z}$ where a and b are two complex numbers that are independent over the real numbers. A torus given in the form $\mathbb{C}/\Lambda$ will be called here a canonical torus. The torus $\mathbb{C}/(\mathbb{Z}+i\mathbb{Z})$ will be called standard.

A family of lattices $\Omega \ni x \to \Lambda(x)$ on a smooth manifold Ω is called smooth if for each $x_0 \in X$ there is a neighbourhood $U(x_0) \subset \Omega$ of x_0 such that the lattices $\Lambda(x)$ can be written as

$$\Lambda(x) = a(x)\mathbb{Z} + b(x)\mathbb{Z}. \tag{8.24}$$

for smooth functions a and b on $U(x_0)$. If Ω is a complex manifold, the family is called holomorphic if it can be locally represented by (8.24) with holomorphic functions a and b.

If on $U(x_0)$ we have $\widetilde{\Lambda}(x) = \Lambda(x)$, with $\widetilde{\Lambda}(x) = \widetilde{a}(x)\mathbb{Z} + \widetilde{b}(x)\mathbb{Z}$ for other smooth functions $\widetilde{a}$ and $\widetilde{b}$, then there is a matrix $A = \begin{pmatrix} \alpha & \beta \\ \gamma & \delta \end{pmatrix} \in \mathrm{SL}_2(\mathbb{Z})$ not depending on x, such that

$$\widetilde{a}(x) = \alpha\, a(x) + \gamma\, b(x), \quad \widetilde{b}(x) = \beta\, a(x) + \delta\, b(x). \tag{8.25}$$

Indeed, $\Lambda(x)$ and $\widetilde{\Lambda}(x)$ are equal lattices and $(a(x), b(x))$ and $(\widetilde{a}(x), \widetilde{b}(x))$ are pairs of generators of the lattice. Put $B(x) = \begin{pmatrix} \mathrm{Re}\, a(x) & \mathrm{Re}\, b(x) \\ \mathrm{Im}\, a(x) & \mathrm{Im} b(x) \end{pmatrix}$ and $\widetilde{B}(x) = \begin{pmatrix} \mathrm{Re}\, \widetilde{a}(x) & \mathrm{Re}\, \widetilde{b}(x) \\ \mathrm{Im}\, \widetilde{a}(x) & \mathrm{Im}\, \widetilde{b}(x) \end{pmatrix}$. Then the real linear self-map of $\mathbb{C}$, defined by $B(x)$ (by $\widetilde{B}(x)$, respectively) takes 1 to $a(x)$ and i to $b(x)$ (1 to $\widetilde{a}(x)$ and i to $\widetilde{b}(x)$, respectively). Hence, both maps take the standard lattice $\mathbb{Z} + i\mathbb{Z}$ onto $\Lambda(x) = \widetilde{\Lambda}(x)$. Then $B^{-1}(x) \circ \widetilde{B}(x)$ maps the standard lattice onto itself, hence, it is in $\mathrm{SL}_2(\mathbb{Z})$. Since $B(x)$ and $\widetilde{B}(x)$ depend continuously on x, the matrices $B^{-1}(x) \circ \widetilde{B}(x)$ depend continuously on x and, hence, are equal to a matrix $A = \begin{pmatrix} \alpha & \beta \\ \gamma & \delta \end{pmatrix} \in \mathrm{SL}_2(\mathbb{Z})$ that does not depend on x. Equation (8.25) is satisfied for the entries of this matrix.

Two smooth families of lattices $\Lambda_0(x)$ and $\Lambda_1(x)$ depending on a parameter x in a smooth manifold Ω are called isotopic, if for an interval $I \supset [0, 1]$ there is a smooth family of lattices $\Lambda(x,t)$, $(x,t) \in \Omega \times I$, such that $\Lambda(x,0) = \Lambda_0(x)$ and $\Lambda(x,1) = \Lambda_1(x)$ for all $x \in \Omega$.

Let X be a finite open Riemann surface and $\Lambda(x)$, $x \in X$, a smooth family of lattices on X. Take an open subset U of X on which there are smooth function $a(x)$ and $b(x)$ such that $\Lambda(x) = a(x)\mathbb{Z}+b(x)\mathbb{Z}$, $x \in U$. Then the family $\Lambda = \Lambda(x)$, $x \in U$, defines a free and properly discontinuous group action (see e.g. [58], III, 3K)

$$U \times \mathbb{C} \ni (x,\zeta) \to \big(x, \zeta + a(x)n + b(x)m\big),\ n,m \in \mathbb{Z}, \tag{8.26}$$

on $U \times \mathbb{C}$. The quotient of $U \times \mathbb{C}$ by this action depends only on Λ, not on the choice of the generators $a(x)$ and $b(x)$ of the lattice. Denote by $(U \times \mathbb{C})/\Lambda$ the quotient of $U \times \mathbb{C}$ by this action (8.26). Let

$$\mathcal{P}_{U,\Lambda} : (U \times \mathbb{C})/\Lambda \to U \tag{8.27}$$

be the mapping whose value at the equivalence class of (x,ζ) equals x. Consider in each fiber $\mathcal{P}_\Lambda^{-1}(x)$ the distinguished point $s_{\Lambda,x} = (x,0)/\Lambda(x)$. The mapping $x \to s_{\Lambda,x}$, $x \in U$, is smooth, hence defines a smooth section. Put $s_{\Lambda,U} \overset{def}{=} \bigcup_{x\in U}\{s_{\Lambda,x}\}$. Then the tuple $\mathfrak{F}_{\Lambda,U} \overset{def}{=} \big((U \times \mathbb{C})/\Lambda, \mathcal{P}_{U,\Lambda}, s_{\Lambda,U}, U\big)$ defines a smooth (1, 1)-bundle over U. Moreover, $\mathfrak{F}_{\Lambda,U}$ is equipped with the structure of a differentiable family of Riemann surfaces of type (1, 1). If the family Λ of lattices on X is holomorphic, then $\mathfrak{F}_{\Lambda,U}$ is a holomorphic (1, 1)-bundle, equivalently, a holomorphic family of Riemann surfaces of type (1, 1). The bundle $\mathfrak{F}_{\Lambda,U}$ depends only on $\Lambda \mid U$, not on the choice of the generators $a(x)$ and $b(x)$ of the lattice $\Lambda(x)$, $x \in U$.

Take an open cover of X by sets U_j on which there are smooth functions $a_j(x)$ and $b_j(x)$ such that $\Lambda(x) = \Lambda_j(x) = a_j(x)\mathbb{Z} + b_j(x)\mathbb{Z}$, $x \in U_j$. Consider the set $\big\{(x,\zeta)/\Lambda(x),\ x \in X,\ \zeta \in \mathbb{C}\big\}$. The family $(U_j \times \mathbb{C})/\Lambda_j$ provides a system of smooth coodinates on this set that are holomorphic in ζ for fixed value of x. We denote the set $\{(x,\zeta)/\Lambda(x),\ x \in X,\ \zeta \in \mathbb{C}\}$ with the thus obtained structure by $\mathcal{X}_\Lambda = (X \times \mathbb{C})/\Lambda$. Define a mapping $\mathcal{P}_\Lambda$ by the equalities $\mathcal{P}_\Lambda = \mathcal{P}_{U_j,\Lambda}$ on U_j, and define s_Λ by $s_\Lambda \cap (\mathcal{P}_\Lambda^{-1}(U_j)) = s_{\Lambda,U_j}$ for each j. We obtain a smooth family of Riemann surfaces of type (1, 1) on X, which we denote by $\mathfrak{F}_{\Lambda,X} = (\mathcal{X}_\Lambda, \mathcal{P}_\Lambda, s_\Lambda, X)$.

If the family Λ of lattices on X is holomorphic, then $\mathfrak{F}_{\Lambda,X}$ is a holomorphic (1, 1)-bundle, equivalently, a holomorphic family of Riemann surfaces of type (1, 1).

The following lemma holds.

Lemma 8.4 *Each smooth* $(1, 1)$*-bundle* $\mathfrak{F}$ *over a smooth finite open oriented surface* X *is smoothly isomorphic (equivalently, isotopic) to a bundle of the form*

$$\mathfrak{F}_{\Lambda,X} = (\mathcal{X}_\Lambda, \mathcal{P}_\Lambda, s_\Lambda, X). \tag{8.28}$$

Recall that the bundle (8.28) carries the structure of a smooth family of Riemann surfaces of type $(1, 1)$.

Proof We need to find a smooth bundle of the form (8.28) whose monodromy homomorphism is conjugate to that of the original bundle $\mathfrak{F}$. This can be done as in Sect. 8.2. We represent X as the union of an open disc D containing the base point of X, and a collection of ℓ attached bands V_j that are relatively closed in X and correspond to the generators of the fundamental group of X. Let as in Sect. 8.2 the set $\tilde{U}$ be a simply connected domain on the universal covering of X such that each point of D is covered $\ell + 1$ times and each other point of X is covered once. Label the preimages of D under the projection $\tilde{U} \to X$ by $\tilde{D}_j$, $j = 0, 1, \ldots \ell$, so that the preimage $\tilde{V}_j$ of the band V_j, $j = 1, \ldots, \ell$, is attached to $\tilde{D}_0$ and $\tilde{D}_j$. Denote by $\mathfrak{m}_j$ the monodromy of $\mathfrak{F}$ along the generator e_j of the fundamental group of X that corresponds to the j-th band V_j.

Map the fiber of $\mathfrak{F}$ over the base point diffeomorphically onto the standard torus $\mathbb{C}\diagup(\mathbb{Z} + i\mathbb{Z})$ so that the distinguished point is mapped to $0\diagup(\mathbb{Z} + i\mathbb{Z})$. By an isomorphism between mapping class groups we identify each monodromy mapping class $\mathfrak{m}_j$, $j = 1, \ldots, \ell$, of $\mathfrak{F}$ with a mapping class on the standard torus $\mathbb{C}\diagup(\mathbb{Z}+i\mathbb{Z})$ with distinguished point $0\diagup(\mathbb{Z}+i\mathbb{Z})$. We denote the new mapping class by the same letter $\mathfrak{m}_j$. Represent the new class $\mathfrak{m}_j$ by a mapping $\varphi_j^{-1} : \mathbb{C}\diagup(\mathbb{Z}+i\mathbb{Z})\circlearrowleft$, such that φ_j lifts to a real linear self-mapping $\widetilde{\varphi}_j$ of $\mathbb{C}$ that maps the lattice $\mathbb{Z} + i\mathbb{Z}$ onto itself. In other words, $\widetilde{\varphi}_j$ corresponds to a 2×2 matrix A_j with integer entries and determinant 1, $\widetilde{\varphi}_j(x + iy) = A_j \begin{pmatrix} x \\ y \end{pmatrix}$, $A_j \in \mathrm{SL}_2(\mathbb{Z})$.

Consider on each set $\widetilde{\Omega}_j \overset{def}{=} \tilde{D}_0 \cup \tilde{V}_j \cup \tilde{D}_j$ (which is a simply connected domain) a smooth family of real linear self-maps $(\widetilde{\varphi}_j)_z$, $z \in \widetilde{\Omega}_j$, of the complex plane $\mathbb{C}$ such that

$$(\widetilde{\varphi}_j)_z = \mathrm{Id} \text{ for } z \in \tilde{D}_0 \text{ and } (\widetilde{\varphi}_j)_z = \widetilde{\varphi}_j \text{ for } z \in \tilde{D}_j\,, \tag{8.29}$$

and put

$$\Lambda_j(z) = (\widetilde{\varphi}_j)_z(\mathbb{Z} + i\mathbb{Z}),\; z \in \widetilde{\Omega}_j\,. \tag{8.30}$$

For points $\tilde{z}_j \in \tilde{D}_j$, $j = 0, \ldots \ell$, that project to the same point in D, the lattices $\Lambda_j(\tilde{z}_j)$ coincide. Hence, there exists a well defined smooth family of lattices $\Lambda(x)$, $x \in X$, that lifts to the family of lattices (8.30) on $\widetilde{\Omega}_j$. We obtain a bundle $\mathfrak{F}_{\Lambda,X} = \big((X \times \mathbb{C})\diagup\Lambda, \mathcal{P}_\Lambda, s_\Lambda, X\big)$ over X, which is isomorphic (as a smooth bundle) to the original bundle over X, since the monodromy homomorphisms of the two bundles are conjugate to each other. The lemma is proved. □

By Lemma 8.4 the Problem 8.1 can be reformulated as follows.

Problem 8.1′ *Let X be a finite open Riemann surface. Is a given smooth family of Riemann surfaces of type* $(1, 1)$ *on X isotopic to a complex analytic family of Riemann surfaces of type* $(1, 1)$*?*

8.6 Complex Analytic Families of Canonical Tori

Let now $\mathfrak{F} = (\mathcal{X}, \mathcal{P}, X)$ be a *holomorphic* elliptic fiber bundle over a finite open Riemann surface X. The fiber bundle is, in particular, a smooth elliptic fiber bundle. For each disc $\Delta \subset X$ there is a smooth family of diffeomorphisms $\varphi_t : S \to \mathcal{P}^{-1}(t)$, $t \in \Delta$, from the reference Riemann surface $S = \mathbb{C}/(\mathbb{Z} + i\mathbb{Z})$ of genus one onto the fiber over t. Consider the Teichmüller class $[\varphi_t]$, $t \in \Delta$. The following Lemma 8.5 states that these Teichmüller classes depend holomorphically on the parameter.

Lemma 8.5 *Let $\mathfrak{F}$ be a holomorphic elliptic fiber bundle over a Riemann surface X. For each small enough disc $\Delta \subset X$ there is a holomorphic map $z \to \tau(z)$, $z \in \Delta$, into the Teichmüller space $\mathcal{T}(1, 0)$ of the standard torus, such that each fiber $\mathcal{P}^{-1}(z)$ is conformally equivalent to $\mathbb{C}/(\mathbb{Z} + \tau(z)\mathbb{Z})$.*

For convenience of the reader we will provide a proof. The key ingredient is a lemma of Kodaira which we formulate now.

Let $\mathcal{X}$ and X be complex manifolds and let $\mathcal{P} : \mathcal{X} \to X$ be a proper holomorphic submersion such that the fibers $\mathcal{X}_z = \mathcal{P}^{-1}(z)$, $z \in X$, are compact complex manifolds of complex dimension n. For each $z \in X$ we denote by Θ_z the sheaf of germs of holomorphic tangent vector fields of the complex manifold $\mathcal{X}_z$. Denote by $H^0(\mathcal{X}_z, \Theta_z)$ the space of global sections of the sheaf.

Lemma 8.6 (Kodaira, [54], Lemma 4.1, p. 204.) *If the dimension* $d \overset{def}{=} \dim(H^0(\mathcal{X}_z, \Theta_z))$ *is independent of z then for any small enough (topological) ball Δ in X there is for each $z \in \Delta$ a basis* $(v_1(z), \ldots, v_d(z))$ *of* $H^0(\mathcal{X}_z, \Theta_z)$ *such that* $(v_1(z), \ldots, v_d(z))$ *depends holomorphically on $z \in \Delta$.*

For a proof we refer to [54]. Let m be the dimension of the complex manifold X in the statement of Kodaira's Lemma. The condition that the $v_j(z)$, $j = 1, \ldots, d$, depend holomorphically on z, means the following. Let $U_\alpha \subset \mathcal{X}$ be a small open subset of $\mathcal{X}$ on which there are local holomorphic coordinates $(\zeta_1^\alpha, \ldots, \zeta_n^\alpha, z_1, \ldots, z_m)$, where $z = (z_1, \ldots, z_m) \in \Delta$ and for fixed $z = (z_1, \ldots, z_m)$ the $\zeta^\alpha = (\zeta_1^\alpha, \ldots, \zeta_n^\alpha)$ are local holomorphic coordinates on the fiber over z. In these local coordinates the vector field v_j, $j = 1, \ldots, n$, can be written as

$$v_j(z) = \sum_{k=1}^{n} v_{j\,k}^{\alpha}(\zeta^\alpha, z)\frac{\partial}{\partial \zeta_k^\alpha}, \tag{8.31}$$

where the v^{α}_{jk} are holomorphic in (ζ^{α}, z). The condition does not depend on the choice of local coordinates with the described properties.

We prepare now the proof of Lemma 8.5. Kodaira's Lemma applies in the situation of Lemma 8.5. Indeed, in the situation of Lemma 8.5 the base X has complex dimension one and the fibers are compact Riemann surfaces of genus 1. The space of holomorphic sections $H^0(\mathcal{X}_z, \Theta_z)$ of the sheaf of germs of holomorphic tangent vector fields of each fiber $\mathcal{X}_z$ has complex dimension one. Indeed, let $\mathbb{T}$ be a compact Riemann surface of genus 1. The complex structure of $\mathbb{T}$ may be given by a system of local holomorphic coordinates with transition functions $\zeta(z) = z + c$ for complex constants c (so called flat coordinates). A holomorphic vector field (equivalently, a holomorphic section in the holomorphic tangent bundle) on $\mathbb{T}$ assigns to each chart $\mathbb{C} \supset U_j \to V_j \subset \mathbb{T}$ a function $v_j(z_j)$, $z_j \in U_j$, such that if $V_j \cap V_k \neq \emptyset$ then $v_j(z_j(z_k))\frac{dz_k}{dz_j} = v_k(z_k)$. Since the complex structure on $\mathbb{T}$ is given by a system of flat local coordinates, the equality $\frac{dz_k}{dz_j} \equiv 1$ holds for each j, k with $V_j \cap V_k \neq \emptyset$. Hence, the $v_j(z_j)$ define a holomorphic function v on $\mathbb{T}$. Each holomorphic function on a compact complex manifold is constant. Hence, the space of holomorphic tangent vector fields to any compact Riemann surface of genus one has complex dimension 1. In particular, for each $z \in X$ the space $H^0(\mathcal{X}_z, \Theta_z)$ is generated by a single holomorphic tangent vector field $v(z)$ on $\mathcal{X}_z$. By Kodaira's Lemma $v(z)$ may be chosen to depend holomorphically on $z \in \Delta$ for each small enough disc Δ in X. We obtain a holomorphic vector field v on $\mathcal{X}_\Delta \stackrel{def}{=} \mathcal{P}^{-1}(\Delta)$ whose restriction to each fiber $\mathcal{X}_z$ equals $v(z)$.

Consider the (holomorphic) universal covering space $\tilde{\mathcal{X}}_\Delta$ of $\mathcal{X}_\Delta$. Denote by p the covering map, $p : \tilde{\mathcal{X}}_\Delta \to \mathcal{X}_\Delta$. Consider the triple $(\tilde{\mathcal{X}}_\Delta, \mathcal{P} \circ p, \Delta)$.

Lemma 8.7 $(\tilde{\mathcal{X}}_\Delta, \mathcal{P} \circ p, \Delta)$ *is holomorphically isomorphic to the trivial holomorphic fiber bundle over* Δ *with fiber* $\mathbb{C}$.

Proof of Lemma 8.7 For each $z \in \Delta$ the fiber $\mathcal{P}^{-1}(z)$ of the bundle $(\mathcal{X}_\Delta, \mathcal{P}, \Delta)$ is a compact Riemann surface of genus 1. Its preimage under p is the set $(\mathcal{P} \circ p)^{-1}(z)$ which is the fiber of the bundle $(\tilde{\mathcal{X}}_\Delta, \mathcal{P} \circ p, \Delta)$ over z. The set $\mathcal{P}^{-1}(z)$ is a complex one-dimensional submanifold of $\mathcal{X}_\Delta$ which is the zero set of the holomorphic function $\mathcal{P} - z$. The set $(\mathcal{P} \circ p)^{-1}(z)$ is a complex one-dimensional submanifold of $\tilde{\mathcal{X}}_\Delta$ which is the zero set of the holomorphic function $\mathcal{P} \circ p - z$. Hence, the restriction $p \mid (\mathcal{P} \circ p)^{-1}(z) : (\mathcal{P} \circ p)^{-1}(z) \to \mathcal{P}^{-1}(z)$ defines a holomorphic covering. Indeed, since $p : \tilde{\mathcal{X}}_\Delta \to \mathcal{X}_\Delta$ is a covering, each point z' of $\mathcal{P}^{-1}(z)$ has a neighbourhood $U_{z'} \subset \mathcal{X}_\Delta$ such that $p^{-1}(U_{z'})$ is the disjoint union of open sets $\tilde{U}^k_{z'} \subset \tilde{\mathcal{X}}$ and p maps each $\tilde{U}^k_{z'}$ biholomorphically onto $U_{z'}$. Then p maps each $\tilde{U}^k_{z'} \cap (\mathcal{P} \circ p)^{-1}(z)$ conformally onto $U_{z'} \cap \mathcal{P}^{-1}(z)$.

Each fiber $(\mathcal{P} \circ p)^{-1}(z)$ is simply connected, since any loop in the fiber $(\mathcal{P} \circ p)^{-1}(z)$ is contractible in $\tilde{\mathcal{X}}_\Delta$, and the bundle $(\tilde{\mathcal{X}}_\Delta, \mathcal{P} \circ p, \Delta)$ is smoothly isomorphic to the trivial bundle $(\Delta \times (\mathcal{P} \circ p)^{-1}(z), \mathrm{pr}_1, \Delta)$.

Since for each $z \in \Delta$ the covered manifold $\mathcal{P}^{-1}(z)$ is a Riemann surface of genus 1, the covering manifold $(\mathcal{P} \circ p)^{-1}(z)$ is conformally equivalent to the complex plane $\mathbb{C}$. Hence, the triple $(\widetilde{\mathcal{X}}_\Delta, \mathcal{P}\circ p, \Delta)$ is a holomorphic fiber bundle with fiber $\mathbb{C}$.

Take a holomorphic section $s(z)$, $z \in \Delta$, of the mapping $\widetilde{\mathcal{X}}_\Delta \xrightarrow{\mathcal{P}\circ p} \Delta$. It exists after, perhaps, shrinking Δ. Let v be the holomorphic vector field on $\mathcal{X}_\Delta$ from Kodaira's Lemma, and $\widetilde{v}$ its lift to the universal covering $\widetilde{\mathcal{X}}_\Delta$. Define a mapping $\mathcal{G} : \Delta \times \mathbb{C} \to \widetilde{\mathcal{X}}_\Delta$ by

$$\mathcal{G}(z,\zeta) = \gamma_{s(z)}(\zeta), \quad (z,\zeta) \in \Delta \times \mathbb{C}, \tag{8.32}$$

where for each z the mapping $\gamma_{s(z)}$ is the solution of the holomorphic differential equation

$$\gamma'_{s(z)}(\zeta) = \widetilde{v}(\gamma_{s(z)}(\zeta)), \quad \gamma_{s(z)}(0) = s(z) \in (\mathcal{P}\circ p)^{-1}(z), \quad \zeta \in \mathbb{C}. \tag{8.33}$$

Since $\widetilde{v}$ is tangential to the fibers we have the inclusion

$$\gamma_{s(z)}(\mathbb{C}) \subset (\mathcal{P}\circ p)^{-1}(z) \cong \mathbb{C} \quad \text{for each} \quad z \in \Delta. \tag{8.34}$$

For each z the solution exists for all $\zeta \in \mathbb{C}$ since the restriction of $\widetilde{v}$ to the fiber over z is the lift of a vector field on a closed torus. The mapping

$$(z,\zeta) \to \gamma_{s(z)}(\zeta), \quad z \in \Delta, \quad \zeta \in \mathbb{C}, \tag{8.35}$$

is a local holomorphic diffeomorphism. By the Poincaré-Bendixson Theorem for each z it maps $\mathbb{C}$ one-to-one onto an open subset of the fiber $(\mathcal{P}\circ p)^{-1}(z) \cong \mathbb{C}$, hence, it maps $\mathbb{C}$ *onto* $(\mathcal{P}\circ p)^{-1}(z)$. Hence, $\mathcal{G}$ defines a holomorphic isomorphism of the trivial bundle $(\Delta \times \mathbb{C}, \mathrm{pr}_1, \Delta)$ onto the bundle $(\widetilde{\mathcal{X}}_\Delta, \mathcal{P}\circ p, \Delta)$. Lemma 8.7 is proved. □

Proof of Lemma 8.5 Consider the covering transformations of the covering $\widetilde{\mathcal{X}}_\Delta \to \mathcal{X}_\Delta$. In terms of the isomorphic bundle $(\Delta \times \mathbb{C}, \mathrm{pr}_1, \mathbb{C})$ the restrictions of the covering transformations to each fiber $\{z\} \times \mathbb{C}$, are translations, hence, the covering transformations have the form

$$\psi_{n,m}(z,\zeta) = \big(z, \zeta + n\,a(z) + m\,b(z)\big), \quad (z,\zeta) \in \Delta \times \mathbb{C}, \tag{8.36}$$

for integral numbers n and m. Here $a(z)$ and $b(z)$ are complex numbers which are linearly independent over $\mathbb{R}$ and depend on z. Since the covering transformations $\psi_{1,0}$ and $\psi_{0,1}$ are holomorphic, the numbers $a(z)$ and $b(z)$ can be taken to depend holomorphically on $z \in \Delta$. After perhaps interchanging $a(z)$ and $b(z)$ we may assume that $\mathrm{Re}\frac{a(z)}{b(z)} > 0$. Hence, $\tau(z) = \frac{a(z)}{b(z)}$ depends holomorphically on $z \in \Delta$ and can be interpreted as a holomorphic mapping from Δ into the Teichmüller space (which is identified with the upper half-plane $\mathbb{C}_+$). Lemma 8.5 is proved. □

Corollary 8.1 *Let $\mathfrak{F}$ be a holomorphic elliptic fiber bundle over a connected simply connected Riemann surface X, or let $\mathfrak{F}$ be a holomorphic elliptic fiber bundle over a connected finite open Riemann surface X, such that all monodromies are equal to the identity. Then there is a holomorphic map $z \to \tau(z)$, $z \in X$, into the Teichmüller space $\mathcal{T}(1,0)$, such that each fiber $\mathcal{P}^{-1}(z)$ is conformally equivalent to $\mathbb{C}/(\mathbb{Z}+\tau(z)\mathbb{Z})$.*

Proof Let first X be simply connected. Consider a smooth family of diffeomorphisms $\varphi_z : S \to \mathcal{P}^{-1}(z)$, $z \in X$, where $S = \mathbb{C}/(\mathbb{Z}+i\mathbb{Z})$ is the standard torus. Then the Teichmüller classes $\tau(z) \overset{def}{=} [\varphi_z]$, $z \in X$, depend smoothly on z and $\mathcal{P}^{-1}(z)$ is conformally equivalent to $\mathbb{C}/(\mathbb{Z}+[\varphi_z]\mathbb{Z})$. By Lemma 8.5 $[\varphi_z]$ depends holomorphically on z. Indeed, on any small enough disc $\Delta \subset X$ there is a holomorphic map τ_Δ to the Teichmüller space such that $\mathcal{P}^{-1}(z)$ is conformally equivalent to $\mathbb{C}/(\mathbb{Z}+\tau_\Delta(z)\mathbb{Z})$. Hence, $[\varphi_z] = \varphi_\Delta^*(\tau_\Delta(z))$, $z \in \Delta$, for a modular transformation φ_Δ^* of the Teichmüller space. The corollary follows from the fact that modular transformations are biholomorphic self-maps of the Teichmüller space.

Let now X be a finite open Riemann surface, and let all monodromies of the bundle $\mathfrak{F}$ be equal to the identity. Consider a lift $\tilde{\mathfrak{F}}$ of the bundle to the universal covering $\tilde{X} \xrightarrow{\mathsf{P}} X$. There exists a holomorphic mapping $\tilde{\tau} : \tilde{X} \to \mathcal{T}(1,0)$ such that for each $\tilde{x} \in \tilde{X}$ the value $\tilde{\tau}(\tilde{x})$ represents the conformal class of the fiber of $\tilde{\mathfrak{F}}$ over $\tilde{x}$ (which is equal to the conformal class of the fiber of $\mathfrak{F}$ over $\mathsf{P}(\tilde{x})$). Let e be an element of the fundamental group of X with base point x_0. We let φ_e be a self-homeomorphism of the fiber of $\mathfrak{F}$ over x_0, that represents the monodromy mapping class of the bundle along e. Denote by φ_e^* the modular transformation on $\mathcal{T}(1,0)$ corresponding to φ_e. Then $\tilde{\tau}(e(\tilde{x})) = \varphi_e^*(\tilde{\tau}(\tilde{x}))$ for each e. Since all monodromies of the bundle $\mathfrak{F}$ are trivial, each φ_e^* is equal to the identity, and therefore $\tilde{\tau}$ descends to a well-defined mapping τ on X with the required property. □

The following proposition holds.

Proposition 8.1 *Each holomorphic $(1,1)$-bundle $(\mathcal{X}, \mathcal{P}, s, X)$ over a finite open Riemann surface X is holomorphically isomorphic to a holomorphic bundle of the form*

$$(\mathcal{X}_\Lambda, \mathcal{P}_\Lambda, s_\Lambda, X). \tag{8.37}$$

for a holomorphic family of lattices $\Lambda = \{\Lambda(x)\}_{x \in X}$.

Here as before, $\mathcal{X}_\Lambda = (X \times \mathbb{C})/\Lambda$, and $\mathcal{P}_\Lambda$ assigns to each class $(x,\zeta)/\Lambda$ in the quotient the point $x \in X$. The value at the point $x \in X$ of the holomorphic section s_Λ of the bundle (8.37) is the class in the quotient that contains $(x, 0)$.

Proof For each $x \in X$ we let Δ_x be a topological disc, $x \in \Delta_x \subset X$, for which Kodaira's Lemma holds. As in Lemma 8.7 we let $p_{\Delta_x} : \widetilde{\mathcal{X}_{\Delta_x}} \to \mathcal{X}_{\Delta_x}$ be the universal covering of $\mathcal{X}_{\Delta_x} = \mathcal{P}^{-1}(\Delta_x)$. Let $\tilde{s}_{\Delta_x}$ be a lift of $s_{\Delta_x} \overset{def}{=} s_\Lambda \cap \mathcal{X}_{\Delta_x}$

to $\widetilde{\mathcal{X}}_{\Delta_x}$. By Lemma 8.7 there exists (after perhaps shrinking Δ_x) a holomorphic bundle isomorphism

$$\left(\widetilde{\mathcal{X}}_{\Delta_x}, (\mathcal{P} \circ p_{\Delta_x})|\widetilde{\mathcal{X}}_{\Delta_x}, \tilde{s}_{\Delta_x}, \Delta_x\right) \to \left(\Delta_x \times \mathbb{C}, \text{pr}_1, \Delta_x \times \{0\}, \Delta_x\right). \tag{8.38}$$

The bundle isomorphism is given by the identity mapping on Δ_x and a holomorphic mapping $\tilde{\Phi}_{\Delta_x}$ from $\widetilde{\mathcal{X}}_{\Delta_x}$ to $\Delta_x \times \mathbb{C}$, which maps the fiber of the first bundle over each point $x' \in \Delta_x$ to the fiber of the second bundle over the same point x' and maps the section of the first bundle to the section of the second bundle.

Cover X by a locally finite set of such discs Δ_j. For each Δ_j we consider the covering transformations of the projection $p_{\Delta_j} : \widetilde{\mathcal{X}}_{\Delta_j} \to \mathcal{X}_{\Delta_j}$. The bundle isomorphism conjugates the group of these covering transformations to a group of fiber preserving transformations of $\Delta_j \times \mathbb{C}$ with free and properly discontinuous action. The conjugated group acts on each fiber $\{x\} \times \mathbb{C}$ as a lattice $\Lambda_j(x)$, and the lattices depend holomorphically on $x \in \Delta_j$. Taking the quotient, we obtain for each j a holomorphic bundle isomorphism

$$\left(\mathcal{X}_{\Delta_j}, \mathcal{P}_{\Delta_j}, s_{\Delta_j}, \Delta_j\right) \to \left((\Delta_j \times \mathbb{C})\diagup\Lambda_j, \text{pr}_1, (\Delta_j \times \{0\})\diagup\Lambda_j, \Delta_j\right). \tag{8.39}$$

The bundle isomorphism (8.39) is given by the identity mapping from Δ_j to itself and by a biholomorphic mapping

$$\Phi_j : \mathcal{X}_{\Delta_j} \to (\Delta_j \times \mathbb{C})\diagup\Lambda_j \tag{8.40}$$

that maps the fiber of the first bundle over any point $x \in \Delta_j$ to the fiber of the second bundle over the same point x and maps the section of the first bundle to the section of the second bundle.

Take j and k so that $\Delta_j \cap \Delta_k \neq \emptyset$. Then $\Phi_j \circ (\Phi_k)^{-1}$ is a fiber preserving biholomorphic mapping from $((\Delta_j \cap \Delta_k) \times \mathbb{C})\diagup\Lambda_k$ onto $((\Delta_j \cap \Delta_k) \times \mathbb{C})\diagup\Lambda_j$. Moreover, $\Phi_j \circ (\Phi_k)^{-1}$ maps $((\Delta_j \cap \Delta_k) \times \{0\})\diagup\Lambda_k$ onto $((\Delta_j \cap \Delta_k) \times \{0\})\diagup\Lambda_j$. The mapping $\Phi_j \circ (\Phi_k)^{-1}$ lifts to a fiber preserving biholomorphic mapping $A_{j,k}$ of $(\Delta_j \cap \Delta_k) \times \mathbb{C}$ onto itself, that takes $(\Delta_j \cap \Delta_k) \times \{0\}$ onto itself. Then

$$A_{j,k}(x, \zeta) = (x, \alpha_{j,k}(x) \cdot \zeta), \; (x, \zeta) \in (\Delta_j \cap \Delta_k) \times \mathbb{C} \tag{8.41}$$

for a nowhere vanishing holomorphic function $\alpha_{j,k}$ on $(\Delta_j \cap \Delta_k)$. Moreover, the mapping $A_{j,k}$ takes each $\{x\} \times \Lambda_k(x) \subset \{x\} \times \mathbb{C}$ onto $\{x\} \times \Lambda_j(x) \subset \{x\} \times \mathbb{C}$.

The $\alpha_{j,k}$ form a Cousin II cocycle on X (see [40] and Appendix A.1). Since X is a finite open Riemann surface, the Cousin II problem is solvable (see [27]), i.e. there exist holomorphic nowhere vanishing functions α_j on Δ_j, such that on $(\Delta_j \cap \Delta_k)$ the equality

$$\alpha_{j,k} = \alpha_j \cdot \alpha_k^{-1}$$

holds. Then for each j, k with $\Delta_j \cap \Delta_k \neq \emptyset$ the equality $(\alpha_j(x))^{-1}\Lambda_j(x) = (\alpha_k(x))^{-1}\Lambda_k(x)$, $x \in \Delta_j \cap \Delta_k$ holds. Hence, we obtain a well-defined holomorphic family of lattices Λ on X, $\Lambda(x) = (\alpha_j(x))^{-1}\Lambda_j(x)$, $x \in \Delta_j$. For this family of lattices we consider the bundle $\big((X \times \mathbb{C})/\Lambda, \mathcal{P}_\Lambda, \mathbf{s}_\Lambda, X\big)$ of the form (8.37).

Let $c_j^{-1} : (\Delta_j \times \mathbb{C})/\Lambda_j \to (\Delta_j \times \mathbb{C})/\Lambda$ be the biholomorphic mapping, that lifts to the biholomorphic self-mapping $(x, \zeta) \to (x, (\alpha_j(x))^{-1}\zeta)$ of $\Delta_j \times \mathbb{C}$, which maps each $\{x\} \times \Lambda_j(x)$ onto $\{x\} \times (\alpha_j(x))^{-1}\Lambda_j(x) = \{x\} \times \Lambda(x)$.

Put $\Phi'_j = c_j^{-1}\Phi_j$ on $\mathcal{X}_{\Delta_j}$. Then on each non-empty intersection $(\Delta_j \cap \Delta_k)$ the equality $\Phi'_j(\Phi'_k)^{-1} = c_j^{-1}\Phi_j(c_k^{-1}\Phi_k)^{-1} = c_j^{-1}(\Phi_j(\Phi_k)^{-1})c_k$ holds. The latter mapping lifts to

$$(x, \zeta) \to (x, \alpha_j(x)^{-1}\alpha_{j,k}(x)\alpha_k(x)\zeta) = (x, \zeta), \quad (x, \zeta) \in (\Delta_j \cap \Delta_k) \times \mathbb{C}.$$

Hence, we obtain $\Phi'_j = \Phi'_k$ on $\mathcal{X}_{\Delta_j} \cap \mathcal{X}_{\Delta_k}$. The mapping $\Phi' : \mathcal{X} \to (X \times \mathbb{C})/\Lambda$, for which $\Phi'(x) = \Phi'_j(x)$ for $x \in \mathcal{X}_{\Delta_j}$, is well defined and determines a holomorphic isomorphism from the original bundle to a bundle of the form (8.37). Proposition 8.1 is proved. □

Similar arguments as used in the proof of Proposition 8.1 give the following statement.

Proposition 8.1′ *Each holomorphic elliptic fiber bundle over a finite open Riemann surface X is holomorphically isomorphic to a holomorphic bundle that admits a holomorphic section.*

Proof Cover X by small discs Δ_j on which a holomorphic section $\mathbf{s}_{\Delta_j}$ of the original bundle can be chosen. For each j we consider the holomorphic bundle isomorphism (8.39), and obtain a holomorphic isomorphism Φ_j of the total space of the bundle on the left onto the total space of the bundle on the right of (8.39), that maps fibers of the first bundle to fibers of the second one and the chosen section $\mathbf{s}_{\Delta_j}$ of the first bundle to the section $(\Delta_j \times \{0\})/\Lambda_j$ of the second one.

The isomorphism $\Phi_j : \mathcal{X}_{\Delta_j} \to (\Delta_j \times \mathbb{C})/\Lambda_j$ lifts to a holomorphic isomorphism $\tilde{\Phi}_j : \tilde{\mathcal{X}}_{\Delta_j} \to \Delta_j \times \mathbb{C}$, that maps a lift $\tilde{\mathbf{s}}_{\Delta_j}$ of $\mathbf{s}_{\Delta_j}$ to $\Delta_j \times \{0\}$. For $\Delta_j \cap \Delta_k \neq \emptyset$ the mapping $\Phi_{j,k} = \Phi_j \circ (\Phi_k)^{-1}$ on $\big((\Delta_j \cap \Delta_k) \times \{\mathbb{C}\}\big)/\Lambda_k$ lifts to a fiber preserving biholomorphic mapping $\tilde{\Phi}_{j,k} = \tilde{\Phi}_j \circ \tilde{\Phi}_k^{-1}$ of $(\Delta_j \cap \Delta_k) \times \mathbb{C}$ onto itself. In contrast to the situation in Proposition 8.1 we have no control about the image of $\big((\Delta_j \cap \Delta_k) \times \{0\}\big)/\Lambda_k$ under the mapping $\Phi_{j,k} = \Phi_j \circ (\Phi_k)^{-1}$. Instead of Eq. (8.41) we obtain the equation

$$\tilde{\Phi}_{j,k}(x, \zeta) = (x, a_{j,k}(x) + \alpha_{j,k}(x) \cdot \zeta), \ (x, \zeta) \in (\Delta_j \cap \Delta_k) \times \mathbb{C} \tag{8.42}$$

for a holomorphic function $a_{j,k}$ and a nowhere vanishing holomorphic function $\alpha_{j,k}$ on $\Delta_j \cap \Delta_k$. Since $\tilde{\Phi}_{j,k}\, \tilde{\Phi}_{k,i}\, \tilde{\Phi}_{i,j} = \mathrm{Id}$ on $\Delta_j \cap \Delta_k \cap \Delta_i$, we obtain

$$\begin{aligned} &\Big(a_{j,k}(x) + \alpha_{j,k}(x)a_{k,i}(x) + \alpha_{j,k}(x)\alpha_{k,i}(x)a_{i,j}(x)\Big) \\ +&\Big(\alpha_{j,k}(x)\alpha_{k,i}(x)\alpha_{i,j}(x)\Big) \cdot \zeta = \zeta, \quad x \in \Delta_j \cap \Delta_k \cap \Delta_i,\ \zeta \in \mathbb{C}. \end{aligned} \tag{8.43}$$

Hence, $\alpha_{j,k}(x)\alpha_{k,i}(x)\alpha_{i,j}(x) \equiv 1$, in other words, the $\alpha_{j,k}$ form a Cousin II cocycle. Since X is an open Riemann surface the Cousin Problem has a solution, i.e. there exist nowhere vanishing holomorphic functions α_j on Δ_j such that for $k \neq j$ the equality $\alpha_{j,k} = \frac{\alpha_j}{\alpha_k}$ holds on $(\Delta_j \cap \Delta_k)$. Then by Eq. (8.43)

$$\frac{a_{j,k}}{\alpha_j} + \frac{a_{k,i}}{\alpha_k} + \frac{a_{i,j}}{\alpha_i} = 0\,.$$

In other words, the $\frac{a_{j,k}}{\alpha_j}$ form a Cousin I cocycle. Since X is a Stein manifold the Cousin Problem is solvable, i.e. there exist holomorphic functions a_j on Δ_j such that

$$\frac{a_{j,k}}{\alpha_j} = \frac{a_j}{\alpha_j} - \frac{a_k}{\alpha_k} \tag{8.44}$$

on $(\Delta_j \cap \Delta_k)$. Consider for all j the mapping $(x, \zeta) \to A_j(x, \zeta) = (x, a_j(x) + \alpha_j(x)\zeta)$. The inverse mapping has the form $A_j^{-1}(x, \zeta) = \big(x, -\frac{a_j(x)}{\alpha_j(x)} + \frac{1}{\alpha_j(x)}\zeta\big)$. Then by Eq. (8.44) we obtain

$$\begin{aligned} (A_j\, A_k^{-1})(x, \zeta) &= \Big(x, a_j(x) + \alpha_j(x)\big(-\frac{a_k(x)}{\alpha_k(x)} + \frac{\zeta}{\alpha_k(x)}\big)\Big) \\ &= \Big(x, a_{j,k}(x) + \alpha_{j,k}(x)\zeta\Big) = \tilde{\Phi}_{j,k}(x, \zeta)\,. \end{aligned} \tag{8.45}$$

Consider the biholomorphic mapping $\tilde{\Phi}_j^* : \tilde{\mathcal{X}}_{\Delta_j} \to \Delta_j \times \mathbb{C}$ which is the composition $\tilde{\Phi}_j^* = A_j^{-1} \circ \tilde{\Phi}_j$ of the mapping $A_j^{-1} : \Delta_j \times \mathbb{C} \circlearrowleft$ with $\tilde{\Phi}_j$. By Eq. (8.45) we obtain

$$\tilde{\Phi}_j^*(\tilde{\Phi}_k^*)^{-1}(x, \zeta) = A_j^{-1}\tilde{\Phi}_j\tilde{\Phi}_k^{-1}A_k(x, \zeta) = A_j^{-1}\tilde{\Phi}_{j,k}(A_k A_j^{-1})A_j(x, \zeta) = (x, \zeta)\,.$$

We found biholomorphic mappings $\tilde{\Phi}_j^* : \tilde{\mathcal{X}}_{\Delta_j} \to \Delta_j \times \mathbb{C}$, such that on $\Delta_j \cap \Delta_k$ the mappings $\tilde{\Phi}_j^*$ and $\tilde{\Phi}_k^*$ coincide.

Each $\mathcal{X}_{\Delta_j}$ is the quotient of $\tilde{\mathcal{X}}_{\Delta_j}$ by the group of covering transformations of the covering $p_{\Delta_j} : \tilde{\mathcal{X}}_{\Delta_j} \to \mathcal{X}_{\Delta_j}$. Conjugating for each j the group of covering transformations by the mapping $\tilde{\Phi}_j^*$, we obtain a group acting on $\Delta_j \times \mathbb{C}$ which can be identified with a holomorphic family of lattices Λ_j on Δ_j. The mapping

$\tilde{\Phi}_j^*$ descends to a biholomorphic mapping $\Phi_j^* : \mathcal{X}_{\Delta_j} \to (\Delta_j \times \mathbb{C})/\Lambda_j$. Since on $\Delta_j \cap \Delta_k$ the mappings $\tilde{\Phi}_j^*$ and $\tilde{\Phi}_k^*$ coincide, the lattices Λ_j and Λ_k coincide on $\Delta_j \cap \Delta_k$, and $\Phi_j^* = \Phi_k^*$ on $\Delta_j \cap \Delta_k$. For the lattice Λ on X that equals Λ_j on Δ_j we obtain a bundle $((X \times \mathbb{C})/\Lambda, \mathcal{P}_\Lambda, s_\Lambda, X)$ of the form (8.37), and the mappings Φ_j define a holomorphic bundle isomorphism from the original bundle to $((X \times \mathbb{C})/\Lambda, \mathcal{P}_\Lambda, X)$. The proposition is proved. □

8.7 Special (0, 4)-Bundles and Double Branched Coverings

In Sect. 8.9 we will study the Gromov-Oka Principle for elliptic fiber bundles, or, in other words, for families of Riemann surfaces of type $(1, 1)$ (see Problems 8.1 and Problem 8.1′). This will be done by a reduction to the Gromov-Oka Principle in the case of special $(0, 4)$-bundles, which we will prepare now. We will represent tori with a distinguished point as double branched coverings over $\mathbb{P}^1$ with a set of four branch points. This is done as follows.

Take any element $\mathring{E} = \{z_1, z_2, z_3\} \in C_3(\mathbb{C})/\mathcal{S}_3$, and put $E \overset{def}{=} \mathring{E} \cup \{\infty\}$. The set

$$\mathring{Y}_E \overset{def}{=} \Big\{(z, w) \in \mathbb{C}^2 : w^2 = 4\,(z - z_1)(z - z_2)(z - z_3)\Big\} \tag{8.46}$$

is a one-dimensional complex submanifold of $\mathbb{C}^2$. Each point of $\mathring{Y}_E$ has a neighbourhood in $\mathring{Y}_E$ on which one of the functions, z or w, defines local holomorphic coordinates. The mapping

$$\mathring{Y}_E \ni (z, w) \to z \in \mathbb{C} \tag{8.47}$$

is a branched holomorphic covering of $\mathbb{C}$ with branch locus $\mathring{E}$.

Consider the 1-point compactification Y_E of $\mathring{Y}_E$. The complex structure on it is obtained as follows. Let $\mathring{Y}_E^r$ be the subset of $\mathring{Y}_E$ where $|z| > r$ for a large positive number r. On this set $|\frac{z}{w}|$ is small. Put $(\tilde{z}, \tilde{w}) = (\frac{1}{z}, \frac{z}{w})$ on a small neighbourhood in $\mathbb{C}^2$ of $\mathring{Y}_E^r$. In these coordinates the equation for $\mathring{Y}_E^r$ becomes

$$\tilde{w}^2 = \tilde{z}\frac{1}{4\prod_{j=1}^3 (1 - z_j\tilde{z})}, \tag{8.48}$$

and $\mathring{Y}_E^r$ can be identified with the subset

$$\left\{(\tilde{z}, \tilde{w}) \in \mathbb{C}^2,\ 0 < |\tilde{z}| < \frac{1}{r} : \tilde{w}^2 = \tilde{z}\frac{1}{4\prod_{j=1}^3 (1 - z_j\tilde{z})}\right\} \tag{8.49}$$

of $\mathbb{C}^2$. Adding the point $(\tilde{z}, \tilde{w}) = (0, 0) \in \mathbb{C}^2$ to $\mathring{Y}_E^r$ we obtain a complex manifold Y_E^r. The two manifolds Y_E^r and $\mathring{Y}_E$ form an open cover of the desired compact complex manifold Y_E. Denote the point $(0, 0)$ in coordinates $(\tilde{z}, \tilde{w})$ on Y_E^r by s^∞. Each point of Y_E has a neighbourhood where one of the functions z, w, $\tilde{z}$, or $\tilde{w}$ defines local holomorphic coordinates. The holomorphic projection $Y_E \to \mathbb{P}^1$ is correctly defined by $(z, w) \to z$, $(z, w) \in \mathring{Y}_E$, and $(\tilde{z}, \tilde{w}) \to \tilde{z}$, $(\tilde{z}, \tilde{w}) \in Y_E^r$. We obtain a double branched covering $Y_E \to \mathbb{P}^1$ over $\mathbb{P}^1$ with branch locus equal to $E \overset{def}{=} \mathring{E} \cup \{\infty\}$. The manifold Y_E is a closed Riemann surface of genus 1. We will consider it as a closed Riemann surface of genus 1 with distinguished point being the preimage s^∞ of ∞ under the branched covering. The set $\mathring{E}$ (considered as subset of $\mathbb{C}$) will be called the finite branch locus of the covering $Y_E \to \mathbb{P}^1$.

Let Y be a closed Riemann surface of genus 1 with distinguished point s and let Y_1 be equal to $\mathbb{P}^1$ with set of distinguished points $\mathring{E} \cup \{\infty\}$ for $\mathring{E} \subset C_3(\mathbb{C})/\mathcal{S}_3$. Suppose $\mathsf{Pr} : Y \to Y_1$ is a double branched covering with branch locus $\mathring{E} \cup \infty$ and $\mathsf{Pr}(s) = \infty$. A mapping class $\mathfrak{m} \in \mathfrak{M}(Y; s, \emptyset)$ is called a lift of a mapping class $\mathfrak{m}_1 \in \mathfrak{M}(Y_1; \infty, \mathring{E})$ if there are representing homeomorphisms $\varphi \in \mathfrak{m}$ and $\varphi_1 \in \mathfrak{m}_1$, such that φ lifts φ_1, i.e. $\varphi_1(\mathsf{Pr}(\zeta)) = \mathsf{Pr}(\varphi(\zeta))$, $\zeta \in Y$.

We define double branched coverings of smooth families of Riemann surfaces with distinguished points as follows.

Definition 8.6 *Let X be an oriented smooth surface (a Riemann surface, respectively). Suppose $\mathfrak{F} = \big(\mathcal{X}, \mathcal{P}, \mathbf{s}, X\big)$ is a smooth (complex analytic, respectively) family of Riemann surfaces of type $(1, 1)$ over X. Let $\mathbf{E} \subset X \times \mathbb{P}^1$ be a smooth (complex, respectively) submanifold of $X \times \mathbb{P}^1$, that intersects each fiber $\{x\} \times \mathbb{P}^1$ along a set of distinguished points $E_x = \{x\} \times (\mathring{E}_x \cup \{\infty\})$ with $\mathring{E}_x \subset C_3(\mathbb{C})/\mathcal{S}_3$.*

The family $\mathfrak{F}$ is called a double branched covering of the special smooth (holomorphic, respectively) $(0, 4)$-bundle $\big(X \times \mathbb{P}^1, \mathrm{pr}_1, \mathbf{E}, X\big)$ if there exists a smooth (holomorphic, respectively) mapping $\mathsf{Pr} : \mathcal{X} \to X \times \mathbb{P}^1$ that maps each fiber $\mathcal{P}^{-1}(x)$ of the $(1, 1)$-family $\mathfrak{F}$ onto the fiber $\{x\} \times \mathbb{P}^1$ of the $(0, 4)$-bundle over the same point x, such that the restriction $\mathsf{Pr} : \mathcal{P}^{-1}(x) \to \{x\} \times \mathbb{P}^1$ is a holomorphic double branched covering with branch locus being the set $\{x\} \times (\mathring{E}_x \cup \{\infty\})$ of distinguished points in the fiber $\{x\} \times \mathbb{P}^1$, and Pr maps the distinguished point $s_x = \mathbf{s} \cap \mathcal{P}^{-1}(x)$ in the fiber $\mathcal{P}^{-1}(x)$ over x to the point $\{x\} \times \{\infty\}$ in the fiber $\{x\} \times \mathbb{P}^1$ of the special $(0, 4)$-bundle.

We will also write $(X \times \mathbb{P}^1, \mathrm{pr}_1, \mathbf{E}, X) = \mathsf{Pr}((\mathcal{X}, \mathcal{P}, \mathbf{s}, X))$, and call the family $(\mathcal{X}, \mathcal{P}, \mathbf{s}, X)$ a lift of $(X \times \mathbb{P}^1, \mathrm{pr}_1, \mathbf{E}, X)$.

Lemma 8.8 *Each special holomorphic $(0, 4)$-bundle $\big(X \times \mathbb{P}^1, \mathrm{pr}_1, \mathbf{E}, X\big)$ over a Riemann surface X admits a double branched covering by a complex analytic family of Riemann surfaces of type $(1, 1)$. Each special smooth $(0, 4)$-bundle over an oriented differentiable manifold X has a double branched covering by a differentiable family of Riemann surfaces of type $(1, 1)$.*

Notice that the statement for the smooth case is true also for families of the mentioned complex manifolds over products $X \times I$ where X is an oriented surface and I is an interval, in other words, it is true for isotopies of families of such complex manifolds over Riemann surfaces and for isotopies of families of such complex manifolds over orientable smooth surfaces.

Proof We prove the statement for the holomorphic case. In the smooth case the dependence on the variable x is only smooth, otherwise the smooth case is treated similarly as the holomorphic case. Assume the sets $\mathring{E}_x$, $x \in X$, (with $\boldsymbol{E} \cap (\{x\} \times \mathbb{P}^1) = \{x\} \times (\mathring{E}_x \cup \{\infty\})$) are uniformly bounded in $C_3(\mathbb{C})/\mathcal{S}_3$. Consider the set

$$\mathring{\mathcal{Y}}_{\boldsymbol{E}} = \left\{ (x, z, w) \in X \times \mathbb{C}^2 : w^2 = 4 \prod_{z_j \in \mathring{E}_x} (z - z_j) \right\}, \tag{8.50}$$

equipped with the structure of an embedded complex hypersurface in $X \times \mathbb{C}^2$. In a neighbourhood of a point (x_0, z_0, w_0) on $\mathring{\mathcal{Y}}_{\boldsymbol{E}}$ with $z_0 \notin \mathring{E}_{x_0}$ the pair (x, w) defines holomorphic coordinates. If $z_0 \in \mathring{E}_{x_0}$ the pair (x, z) defines holomorphic coordinates in a neighbourhood of (x_0, z_0, w_0) on $\mathring{\mathcal{Y}}_{\boldsymbol{E}}$. The projection $\mathcal{P} : \mathring{\mathcal{Y}}_{\boldsymbol{E}} \to X$, $\mathcal{P}(x, z, w) = x$ is holomorphic.

For some large positive number r we may define the set

$$\mathcal{Y}^r_{\boldsymbol{E}} \stackrel{def}{=} \left\{ (x, \tilde{z}, \tilde{w}) \in X \times \mathbb{C}^2,\ |\tilde{z}| < \frac{1}{r} : \tilde{w}^2 = \tilde{z} \frac{1}{4 \prod_{z_j \in \mathring{E}_x} (1 - z_j \tilde{z})} \right\}. \tag{8.51}$$

It is a complex hypersurface of complex dimension two of $X \times \mathbb{C}^2$. Each of its points has a neighbourhood on which either $(x, \tilde{z})$ or $(x, \tilde{w})$ defines holomorphic coordinates. The mapping $(x, \tilde{z}, \tilde{w}) \to x$ is holomorphic. The part $\mathring{\mathcal{Y}}^r_{\boldsymbol{E}} \stackrel{def}{=} \big\{(x, \tilde{z}, \tilde{w}) \in \mathcal{Y}^r_{\boldsymbol{E}} : \tilde{z} \neq 0\big\}$ of $\mathcal{Y}^r_{\boldsymbol{E}}$ can be identified with the subset $\big\{(x, z, w) \in \mathring{\mathcal{Y}}_{\boldsymbol{E}} : |z| > r\big\}$ of $\mathring{\mathcal{Y}}_{\boldsymbol{E}}$ using the transition functions $(x, z, w) \to (x, \tilde{z}, \tilde{w}) = (x, \frac{1}{z}, \frac{z}{w})$. The sets $\mathring{\mathcal{Y}}_{\boldsymbol{E}}$ and $\mathcal{Y}^r_{\boldsymbol{E}}$ form an open cover of a complex manifold denoted by $\mathcal{Y}_{\boldsymbol{E}}$ equipped with a proper holomorphic submersion $\mathcal{P} : \mathcal{Y}_{\boldsymbol{E}} \to X$, such that $\mathcal{P}^{-1}(x)$ is a torus for each $x \in X$. We proved that $(\mathcal{Y}_{\boldsymbol{E}}, \mathcal{P}, X)$ is a holomorphic elliptic fiber bundle. Let $\boldsymbol{s}^\infty$ be the submanifold of $\mathcal{Y}_{\boldsymbol{E}}$ that intersects each fiber $\mathcal{P}^{-1}(x)$ along the distinguished point $s^\infty_x \in \mathcal{Y}^r_{\boldsymbol{E}}$ that is written in coordinates on $\mathcal{Y}^r_{\boldsymbol{E}}$ as $(x, 0, 0)$.

Let $\mathsf{Pr} : \mathcal{Y}_{\boldsymbol{E}} \to X \times \mathbb{P}^1$ be the map that assigns to (x, z, w) the point (x, z) (and to $(x, \tilde{z}, \tilde{w})$ the point $(x, \tilde{z})$). By the construction it is clear that the obtained $(1, 1)$-bundle $(\mathcal{Y}_{\boldsymbol{E}}, \mathcal{P}, \boldsymbol{s}^\infty, X)$ is a double branched covering of the $(0, 4)$-bundle $\big(X \times \mathbb{P}^1, \mathrm{pr}_1, \boldsymbol{E}, X\big)$ whose set of finite distinguished points in the fiber over x is equal to the finite branch locus $\{x\} \times \mathring{E}_x$. The double branched covering map $\mathsf{Pr} : \mathcal{Y}_{\boldsymbol{E}} \to X \times \mathbb{P}^1$ maps the distinguished point $s^\infty_x \in \mathcal{P}^{-1}(x)$ to the point $\{x\} \times \infty$.

In the general case (i.e. without the assumption that the sets $\mathring{E}_x$, $x \in X$, are uniformly bounded) the statement is proved by considering an exhaustion of X by relatively compact open sets. □

Let $\big(X \times \mathbb{P}_1, \mathrm{pr}_1, \boldsymbol{E}_j, X\big)$, $j = 0, 1$, be two special holomorphic (smooth, respectively) $(0,4)$-bundles that are isotopic through smooth special $(0,4)$-bundles. Then the families $(\mathcal{Y}_{\boldsymbol{E}_j}, \mathcal{P}_j, \boldsymbol{s}_j^\infty, X)$, $j = 0, 1$, of Riemann surfaces of type $(1,1)$ are isotopic. Indeed, let I be an open interval containing $[0,1]$. The isotopy of the special $(0,4)$-bundles is given by a bundle

$$\big((I \times X) \times \mathbb{P}^1, \mathrm{pr}_1, \boldsymbol{E}, I \times X\big)$$

over $I \times X$. Here $\boldsymbol{E}$ is a smooth submanifold of $(I \times X) \times \mathbb{P}^1$ such that for each $(t,x) \in I \times X$ the intersection of the set $\boldsymbol{E}$ with the fiber over (t,x) equals $\{(t,x)\} \times E(t,x)$ for subsets $E(t,x)$ that are the union of the point ∞ with an element of $C_3(\mathbb{C} \diagup \mathcal{S}_3)$. Moreover, $\boldsymbol{E} \cap ((\{j\} \times X) \times \mathbb{P}^1) = \boldsymbol{E}_j$, $j = 0, 1$. The smooth special $(0,4)$-bundle $\big((I \times X) \times \mathbb{P}^1, \mathrm{pr}_1, \boldsymbol{E}, I \times X\big)$ has a double branched covering by a smooth family of Riemann surfaces of type $(1,1)$ which defines the required isotopy for the families $(\mathcal{Y}_{\boldsymbol{E}_j}, \mathcal{P}_j, \boldsymbol{s}_j^\infty, X)$, $j = 0, 1$, of Riemann surfaces of type $(1,1)$.

Notice that for a non-contractible oriented finite open smooth surface X and a smooth special $(0,4)$-bundle $\big(X \times \mathbb{P}^1, \mathrm{pr}_1, \boldsymbol{E}, X\big)$ over X the obtained double branched covering $(\mathcal{Y}_{\boldsymbol{E}}, \mathcal{P}, \boldsymbol{s}^\infty, X)$ is not the only double branched covering of the $(0,4)$-bundle (see Sect. 8.8). We will call the bundle $(\mathcal{Y}_{\boldsymbol{E}}, \mathcal{P}, \boldsymbol{s}^\infty, X)$ the canonical double branched covering of the special $(0,4)$-bundle $\big(X \times \mathbb{P}^1, \mathrm{pr}_1, \boldsymbol{E}, X\big)$.

8.8 Lattices and Double Branched Coverings

The Weierstraß $\wp$-Function Consider the quotient $\mathbb{C} \diagup \Lambda$ of the complex plane by a lattice Λ. We want to associate to the quotient a double branched covering over the Riemann sphere with covering space being conformally equivalent to $\mathbb{C} \diagup \Lambda$. The standard tool for this purpose is the Weierstraß $\wp$-function $\wp_\Lambda$. Put $\Lambda \overset{def}{=} a\mathbb{Z} + b\mathbb{Z}$, where a and b are real linearly independent complex numbers. The Weierstraß $\wp$-function,

$$\wp_\Lambda(\zeta) = \frac{1}{\zeta^2} + \sum_{\substack{(n,m)\in\mathbb{Z}^2 \\ (n,m)\neq(0,0)}} \left(\frac{1}{(\zeta - an - bm)^2} - \frac{1}{(an+bm)^2} \right), \quad \zeta \in \mathbb{C} \setminus \Lambda, \tag{8.52}$$

is meromorphic on $\mathbb{C}$, has poles of second order at points of Λ and is holomorphic on $\mathbb{C} \setminus \Lambda$. It has periods a and b and principal part $\zeta \to \frac{1}{\zeta^2}$ at 0. The summation is over all non-zero elements of the lattice. Hence, the function depends only on the lattice, not on the choice of the generators a and b of the lattice.

The Weierstraß $\wp$-function satisfies the following differential equation

$$(\wp_\Lambda')^2(\zeta) = 4(\wp_\Lambda)^3(\zeta) - g_2(\Lambda)(\wp_\Lambda)(\zeta) - g_3(\Lambda) \tag{8.53}$$

for complex numbers $g_2(\Lambda)$ and $g_3(\Lambda)$ that depend only on the lattice. This can be proved using the Laurent series expansion near zero of $\wp_\Lambda$ and $\wp'_\Lambda$ and taking a linear combination that has no pole and is therefore constant. For details see [5]. The numbers $g_2(\Lambda)$ and $g_3(\Lambda)$ are called the elliptic invariants of Λ.

Denote the zeros of the equation $4t^3 - g_2(\Lambda)t - g_3(\Lambda) = 0$ by $e_1(\Lambda), e_2(\Lambda), e_3(\Lambda)$. It follows by symmetry considerations that the zeros of this equation are the values of $\wp_\Lambda$ at the half-periods $\frac{a}{2}, \frac{b}{2}, \frac{a+b}{2}$. Indeed, since the doubly periodic function $\wp'_\Lambda$ is odd and has no pole at its half-periods, it vanishes at its half-periods. Since $\wp'_\Lambda$ is of order 3, these are all its zeros in a period-parallelogram. An argument using that the order of $\wp_\Lambda$ is 2, shows that all roots of $\wp_\Lambda$ are distinct. The differential equation for $\wp_\Lambda$ becomes

$$(\wp'_\Lambda)^2(\zeta) = 4(\wp_\Lambda(\zeta) - e_1(\Lambda))(\wp_\Lambda(\zeta) - e_2(\Lambda))(\wp_\Lambda(\zeta) - e_3(\Lambda)), \tag{8.54}$$

where

$$e_1(\Lambda) = \wp_\Lambda\left(\frac{a}{2}\right), \quad e_2(\Lambda) = \wp_\Lambda\left(\frac{b}{2}\right), \quad e_3(\Lambda) = \wp_\Lambda\left(\frac{a+b}{2}\right) \tag{8.55}$$

are the values of $\wp_\Lambda$ at points which are contained in the lattice $\frac{a}{2}\,\mathbb{Z} + \frac{b}{2}\,\mathbb{Z}$ but are not contained in the lattice $a\mathbb{Z} + b\,\mathbb{Z}$.

The half-periods can be found from the elliptic invariants. The integral

$$\int_y^\infty (4t^3 - g_2 t - g_3)^{-\frac{1}{2}} dt \tag{8.56}$$

with constants $g_2 = g_2(\Lambda)$ and $g_3 = g_3(\Lambda)$ defines a holomorphic function in y on any simply connected domain in the complex plane which does not contain a zero of the function $t \to 4t^3 - g_2 t - g_3$. The derivative in z of the function defined by the integral (8.56) with $y = \wp_\Lambda(z)$ equals

$$\frac{\wp'_\Lambda(z)}{\left(4\wp_\Lambda(z)^3 - g_2(\Lambda)\wp_\Lambda(z) - g_3(\Lambda)\right)^{\frac{1}{2}}}, \tag{8.57}$$

which equals either 1 or -1 on each suitable domain for z. This shows that (after multiplying by ± 1) the integral (8.56) provides an inverse of $\wp_\Lambda$. Knowing e_1, e_2, and e_3 we obtain the half-periods.

Put for instance $e_1 = 1$, $e_2 = -1$ and $e_3 = 0$. Then $g_2 = 4$, $g_3 = 0$. The integral $\int_y^\infty (4t^3 - 4t)^{-\frac{1}{2}} dt$ defines a holomorphic function on $\mathbb{C} \setminus \left(i[0,\infty) \cup [-1,1]\right)$. Its continuous extension ω_1 to 1 is positive, and its continuous extension ω_2 to -1 equals $i\omega_1$. Hence, the Weierstrass $\wp$-function corresponding to the periods $2\omega_1$ and $2\omega_2 = i2\omega_1$ takes the values $1, -1$ and 0 at the points $\omega_1, \omega_2, \omega_1 + \omega_2$, respectively.

The lattice $\Lambda_\tau \overset{def}{=} \mathbb{Z}+\tau\,\mathbb{Z}$ often plays a special role. If the pair 1 and τ generates the lattice, then also the pair 1 and $-\tau$ generates it. Hence, we may assume that $\mathrm{Im}\tau > 0$. We will write $\wp_\tau \overset{def}{=} \wp_{\Lambda_\tau}$,

$$\wp_\tau(\zeta) = \frac{1}{\zeta^2} + \sum_{\substack{(n,m)\in\mathbb{Z}^2\\(n,m)\neq(0,0)}} \left(\frac{1}{(\zeta-n-m\,\tau)^2} - \frac{1}{(n+\tau\,m)^2}\right), \quad \zeta\in\mathbb{C}\setminus\Lambda_\tau\,. \tag{8.58}$$

The function $\wp_\tau$ satisfies the differential equation

$$(\wp_\tau')^2(\zeta) = 4(\wp_\tau(\zeta)-e_1(\tau))(\wp_\tau(\zeta)-e_2(\tau))(\wp_\tau(\zeta)-e_3(\tau))\,, \tag{8.59}$$

where

$$e_1(\tau) = \wp_\tau\left(\frac{1}{2}\right), \quad e_2(\tau) = \wp_\tau\left(\frac{\tau}{2}\right), \quad e_3(\tau) = \wp_\tau\left(\frac{1+\tau}{2}\right) \tag{8.60}$$

are the values of $\wp_\tau$ at points which are contained in the lattice $\frac{1}{2}\,\mathbb{Z}+\frac{\tau}{2}\,\mathbb{Z}$ but are not contained in the lattice $\mathbb{Z}+\tau\,\mathbb{Z}$.

An arbitrary lattice Λ can be written as $\Lambda = \alpha(\mathbb{Z}+\tau\mathbb{Z})$. By (8.52) the equality $\wp_\Lambda(\zeta) = \alpha^{-2}\,\wp_\tau\left(\frac{\zeta}{\alpha}\right)$ holds, hence

$$\{e_1(\Lambda), e_2(\Lambda), e_3(\Lambda)\} = \{\alpha^{-2}e_1(\tau), \alpha^{-2}e_2(\tau), \alpha^{-2}e_3(\tau)\}\,. \tag{8.61}$$

The Double Branched Covering Defined by the Weierstraß $\wp$-Function Let Λ be an arbitrary lattice. Put $z(\zeta) = \wp_\Lambda(\zeta) = \alpha^{-2}\,\wp_\tau\left(\frac{\zeta}{\alpha}\right)$ and $w(\zeta) = \wp_\Lambda'(\zeta) = \alpha^{-3}\,\wp_\tau'\left(\frac{\zeta}{\alpha}\right)$. The mapping

$$\mathbb{C}\backslash\Lambda \ni \zeta \to \big(z(\zeta), w(\zeta)\big) = \big(\wp_\Lambda(\zeta), \wp_\Lambda'(\zeta)\big) \in \mathbb{C}^2 \tag{8.62}$$

descends to a conformal mapping ω_Λ from the punctured torus $\mathbb{C}\backslash\Lambda\diagup\Lambda$ onto the complex hypersurface $\mathring{Y}(\Lambda)$ of $\mathbb{C}^2$,

$$\omega_\Lambda : (\mathbb{C}\setminus\Lambda)\diagup\Lambda \to \mathring{Y}(\Lambda) \overset{def}{=} \left\{(z,w)\in\mathbb{C}^2 : w^2 = 4\prod_{j=1}^{3}(z-e_j(\Lambda))\right\}. \tag{8.63}$$

To see this we notice first that the mapping ω_Λ is one-to-one. Indeed, the mapping $\wp_\Lambda$ descends to a holomorphic mapping $\mathsf{p}_\Lambda : (\mathbb{C}\setminus\Lambda)\diagup\Lambda \to \mathbb{C}$, such that

$$\wp_\Lambda = \mathsf{p}_\Lambda \circ p\,. \tag{8.64}$$

Here p denotes the projection $p : \mathbb{C} \setminus \Lambda \to (\mathbb{C} \setminus \Lambda)/\Lambda$. The mapping p_Λ is 2 to 1. Further, $\wp_\Lambda(\zeta) = \wp_\Lambda(-\zeta)$. Hence, the preimage under $\wp_\Lambda$ of each point in $\mathbb{C} \setminus \{0\}$ equals $(\{\zeta\} + \Lambda) \cup (\{-\zeta\} + \Lambda)$ for some $\zeta \in \mathbb{C}$. Since, $\wp'_\Lambda(\zeta) = -\wp'_\Lambda(-\zeta)$, the mapping ω_Λ is one-to-one.

Moreover, the mapping (8.62) is locally conformal. Indeed, if $\wp_\Lambda(\zeta_0) \neq e_j(\Lambda)$, $j = 1, 2, 3$, then $\wp'_\Lambda(\zeta_0) \neq 0$ and the mapping $\zeta \to \wp_\Lambda(\zeta)$ is conformal in a neighbourhood of ζ_0. Suppose $\wp_\Lambda(\zeta_0) = e_j(\Lambda)$. The differential Eq. (8.54) implies that $\wp'_\Lambda \wp''_\Lambda = 2\wp'_\Lambda \sum_{\ell=1}^{3} \prod_{k\neq\ell}(\wp_\Lambda - e_k(\Lambda))$. Hence, $\wp''_\Lambda(\zeta_0) = 2\prod_{k\neq j}(e_j(\Lambda) - e_k(\Lambda)) \neq 0$. Hence, in this case the mapping $\zeta \to \wp'_\Lambda(\zeta)$ is conformal in a neighbourhood of ζ_0.

The set $\mathring{Y}(\Lambda)$ (see (8.63)) is the covering space of the double branched covering

$$\mathring{Y}(\Lambda) \ni (z, w) \to z \in \mathbb{C} \tag{8.65}$$

of $\mathbb{C}$ with branch locus

$$\mathring{BL}(\Lambda) \overset{def}{=} \{e_1(\Lambda), e_2(\Lambda), e_3(\Lambda)\}. \tag{8.66}$$

With $BL(\Lambda) \overset{def}{=} \mathring{BL}(\Lambda) \cup \{\infty\}$ the equality $\mathring{Y}(\Lambda) = \mathring{Y}_{BL(\Lambda)}$ holds.

For a family of lattices $\Lambda(z)$ depending holomorphically on a complex parameter z the sets $\mathring{BL}(\Lambda(z))$ (see (8.66)) depend holomorphically on z. Indeed, the $e_j(\Lambda)$ are the values of $\wp_\Lambda$ at the half-periods of the lattice. The half-periods of the lattice depend holomorphically on z, and also the Weierstraß function $\wp_\Lambda$ depends holomorphically on z (see Eq. (8.52)).

If the family of lattices is merely smooth, then the set of finite branch points in the fiber over x depends smoothly on the point $x \in X$.

The one-point compactification $Y_{BL(\Lambda)}$ of $\mathring{Y}_{BL(\Lambda)}$ (see Sect. 8.7) is the covering manifold of a double branched covering of the Riemann sphere $\mathbb{P}^1$ with branch locus $BL = \mathring{BL}(\Lambda) \cup \{\infty\}$ and finite branch locus $\mathring{BL}(\Lambda)$. We associate to the Riemann surface $\mathbb{C}/\Lambda$ the conformally equivalent Riemann surface $Y_{BL(\Lambda)}$ which we also denote by $Y(\Lambda)$. Notice that ω_Λ extends to a conformal mapping between the closed Riemann surfaces $\mathbb{C}/\Lambda$ and $Y(\Lambda)$.

Lifts of Mappings to the Double Branched Covering and the Involution For each lattice Λ the mapping $\mathbb{C} \ni \zeta \to -\zeta$ takes Λ onto itself. This mapping descends to an involution of $\mathbb{C}/\Lambda$, i.e. to a self-homeomorphism ι_Λ of $\mathbb{C}/\Lambda$ such that $\iota_\Lambda^2 = \mathrm{id}$.

Formula (8.52) implies the following equality

$$\left(\wp_\Lambda(-\zeta), \wp'_\Lambda(-\zeta)\right) = \left(\wp_\Lambda(\zeta), -\wp'_\Lambda(\zeta)\right). \tag{8.67}$$

Conjugate the restriction $\iota_\Lambda \mid (\mathbb{C} \setminus \Lambda)/\Lambda$ by the conformal mapping ω_Λ^{-1}. We obtain a self-homeomorphism of $\mathring{Y}(\Lambda)$ which we denote by ι. The mapping ι satisfies the equality

$$\iota(z, w) = (z, -w). \tag{8.68}$$

The involution ι extends to an involution of $Y(\Lambda)$, denoted also by ι. By (8.68) the involution ι fixes the projection to $\mathbb{P}^1$ of each point of the double branched covering space $Y(\Lambda)$, and interchanges the sheets over each point. Hence, it fixes each of the three finite branch points, it also fixes ∞, and it does not fix any other point.

Let $\widetilde{\varphi}$ be any real linear self-homeomorphism of $\mathbb{C}$ that maps Λ onto itself, and let φ be the induced mapping on $\mathbb{C}/\Lambda$. Then φ commutes with ι_Λ. Indeed, $\widetilde{\varphi}(-\zeta) = -\widetilde{\varphi}(\zeta)$, $\zeta \in \mathbb{C}$.

Each mapping class $\mathfrak{m}$ in $\mathfrak{M}(\mathbb{C}/\Lambda;\ 0/\Lambda, \emptyset)$ can be represented by a self-homeomorphism of $\mathbb{C}/\Lambda$ which commutes with ι_Λ. Indeed, $\mathfrak{M}(\mathbb{C}/\Lambda;\ 0/\Lambda, \emptyset)$ contains a mapping φ that lifts to a real linear self-map $\widetilde{\varphi}$ of $\mathbb{C}$ which maps Λ onto itself. This mapping φ commutes with ι_Λ.

As a corollary, each mapping class $\mathfrak{m}$ in $\mathfrak{M}(Y(\Lambda);\ s^\infty, \emptyset) \cong \mathfrak{M}(\mathbb{C}/\Lambda;\ 0/\Lambda, \emptyset)$ is a lift of a mapping class $\mathfrak{m}_1 \in \mathfrak{M}(\mathbb{P}^1; \{\infty\}, \mathring{E})$ for the set $\mathring{E} \overset{def}{=} \mathring{B}L(\Lambda) \subset C_3(\mathbb{C})/\mathcal{S}_3$. This can be seen as follows. If ψ' represents a mapping class $\mathfrak{m} \in \mathfrak{M}(Y(\Lambda);\ s^\infty, \emptyset)$, then $\omega_\Lambda^{-1} \circ \psi' \circ \omega_\Lambda$ represents a mapping class $\mathfrak{m}'$ in $\mathfrak{M}(\mathbb{C}/\Lambda;\ 0/\Lambda, \emptyset)$. Let φ be the homemorphism in this class that commutes with ι_Λ. Then $\omega_\Lambda \circ \varphi \circ \omega_\Lambda^{-1} \in \mathfrak{m}'$ commutes with ι.

Suppose ψ represents a mapping class $\mathfrak{m} \in \mathfrak{M}(Y(\Lambda);\ s^\infty, \emptyset)$ and commutes with ι. Denote by $\mathring{\psi}$ the restriction of ψ to $\mathring{Y}$. In coordinates (z, w) on $\mathring{Y}(\Lambda)$ we write

$$\mathring{\psi}(z, w) = (\mathring{\psi}_1(z, w), \mathring{\psi}_2(z, w)). \tag{8.69}$$

Since $\mathring{\psi}$ commutes with ι we obtain by (8.68)

$$(\mathring{\psi}_1(z, -w), \mathring{\psi}_2(z, -w)) = \mathring{\psi} \circ \iota(z, w) = \iota \circ \mathring{\psi}(z, w) = (\mathring{\psi}_1(z, w), -\mathring{\psi}_2(z, w)). \tag{8.70}$$

Hence $\mathring{\psi}_1(z, w) = \mathring{\psi}_1(z, -w)$. Since each point in $(z, w) \in \mathring{Y}(\Lambda)$ is determined by z and the sign of w, this means that $\mathring{\psi}_1(z, w)$ depends only on the coordinate $z \in \mathbb{C}$, not on the sheet (determined by w). Further, if ι fixes (z, w), then $w = 0$, and by (8.70) for $w = 0$, $\mathring{\psi}_2(z, w) = \mathring{\psi}_2(z, -w) = -\mathring{\psi}_2(z, w)$. In other words, if ι fixes (z, w), then it also fixes $\mathring{\psi}(z, w)$. Hence, $\mathring{\psi}$ maps the set of finite branch points (the preimage of $\mathring{B}L(\Lambda)$ under the branched covering map) onto itself, and its extension ψ maps the preimage s^∞ of ∞ to itself. We saw, that ψ induces a self-homeomorphism $\mathsf{Pr}(\psi)$ of $\mathbb{P}^1$ in the class $\mathfrak{M}(\mathbb{P}^1; \{\infty\}, \mathring{B}L(\Lambda))$. For this class $\mathsf{Pr}(\psi)|\mathbb{C} = \mathsf{Pr}(\mathring{\psi})$, $\mathsf{Pr}(\mathring{\psi})(z) = \mathring{\psi}_1(z, w)$, $(z, w) \in \mathbb{C}^2$. We call $\mathsf{Pr}(\psi)$ the projection of ψ. The mapping class $\mathfrak{m}$ is a lift of the mapping class $\mathfrak{m}_1$ of ψ_1.

Notice that the provided arguments imply also the following fact. For two mapping classes $\mathfrak{m}, \mathfrak{m}' \in \mathfrak{M}(Y(\Lambda); s^\infty, \emptyset)$ the equality $\mathsf{Pr}(\mathfrak{m}\,\mathfrak{m}') = \mathsf{Pr}(\mathfrak{m})\mathsf{Pr}(\mathfrak{m}')$ holds. Indeed, take representing maps ψ^1 and ψ^2, that commute with the involution ι. Then $\mathring{\psi}_1^1$ and $\mathring{\psi}_1^2$ depend only on the first variable. Since

$$\mathring{\psi}^1 \circ \mathring{\psi}^2(z,w) = \left(\mathring{\psi}_1^1((\mathring{\psi}_1^2(z,w), \mathring{\psi}_2^2(z,w)), \mathring{\psi}_2^1(\mathring{\psi}_1^2(z,w), \mathring{\psi}_2^2(z,w))\right),$$

the first component of $\psi^1 \circ \psi^2$ depends only on the first variable z. The equality $\mathsf{Pr}(\psi^1 \circ \psi^2) = \mathsf{Pr}(\psi^1) \circ \mathsf{Pr}(\psi^2)$ follows.

Let again $\mathring{E}$ be an element of $C_3(\mathbb{C})\diagup \mathcal{S}_3$ and $E = \mathring{E} \cup \{\infty\}$. Any mapping class $\mathfrak{m}_1 \in \mathfrak{M}(\mathbb{P}^1; \{\infty\}, \mathring{E})$ has exactly two lifts $\mathfrak{m}_\pm \in \mathfrak{M}(Y_E; s^\infty, \emptyset)$. Here Y_E is the double branched covering of $\mathbb{P}^1$ with branch locus E, and s^∞ is the branch point over ∞. Indeed, for any representative ψ_1 of $\mathfrak{m}_1$ there are exactly two self-homeomorphisms of the double branched covering that lift ψ_1. They are obtained as follows. Cut $\mathbb{P}^1$ along two disjoint simple curves γ_1 and γ_2, that join disjoint pairs of points of E. ψ_1 maps the pair of curves γ_1 and γ_2 to another pair of curves γ_1' and γ_2' joining (maybe different) disjoint pairs of points of E. The double branched covering over $\mathbb{P}^1$ is obtained in two different ways, either gluing two sheets of $\mathbb{P}^1 \setminus (\gamma_1 \cup \gamma_2)$ crosswise together, or gluing two sheets of $\mathbb{P}^1 \setminus (\gamma_1' \cup \gamma_2')$ crosswise together. ψ_1 maps $\mathbb{P}^1 \setminus (\gamma_1 \cup \gamma_2)$ homeomorphically onto $\mathbb{P}^1 \setminus (\gamma_1' \cup \gamma_2')$. Consider the mapping that takes the first sheet of $\mathbb{P}^1 \setminus (\gamma_1 \cup \gamma_2)$ onto the first sheet of $\mathbb{P}^1 \setminus (\gamma_1' \cup \gamma_2')$ and the second sheet of $\mathbb{P}^1 \setminus (\gamma_1 \cup \gamma_2)$ onto the second sheet of $\mathbb{P}^1 \setminus (\gamma_1' \cup \gamma_2')$ and lifts $\psi_1 \mid \mathbb{P}^1 \setminus (\gamma_1 \cup \gamma_2)$. This mapping extends to a self-homeomorphism ψ of the double branched covering that lifts ψ_1. There is exactly one more self-homeomorphism of the double branched covering that lifts ψ_1. This self-homeomorphism maps the first sheet of $\mathbb{P}^1 \setminus (\gamma_1 \cup \gamma_2)$ onto the second sheet of $\mathbb{P}^1 \setminus (\gamma_1' \cup \gamma_2')$ and the second sheet of $\mathbb{P}^1 \setminus (\gamma_1 \cup \gamma_2)$ onto the first sheet of $\mathbb{P}^1 \setminus (\gamma_1' \cup \gamma_2')$. In other words, there are two lifts of ψ_1 to the double branched covering of $\mathbb{P}^1$ with branch locus E, and they differ by involution.

We proved the following lemma.

Lemma 8.9 *Each mapping class* $\mathfrak{m} \in \mathfrak{M}(\mathbb{C}\diagup\Lambda;\ 0\diagup\Lambda, \emptyset)$ *is the lift of a mapping class* $\mathfrak{m}_1 = \mathsf{Pr}(\mathfrak{m}) \in \mathfrak{M}(\mathbb{P}^1; \{\infty\}, \mathring{BL}(\Lambda))$. *Vice versa, each* $\mathfrak{m}_1 \in \mathfrak{M}(\mathbb{P}^1; \{\infty\}, \mathring{BL}(\Lambda))$ *has two lifts* $\mathfrak{m}_\pm \in \mathfrak{M}(\mathbb{C}\diagup\Lambda;\ 0\diagup\Lambda, \emptyset)$. *The lifts* $\mathfrak{m}_\pm$ *differ by involution. For two mapping classes* $\mathfrak{m}, \mathfrak{m}' \in \mathfrak{M}(\mathbb{C}\diagup\Lambda;\ 0\diagup\Lambda, \emptyset)$ *the equality* $\mathsf{Pr}(\mathfrak{m}\mathfrak{m}') = \mathsf{Pr}(\mathfrak{m})\mathsf{Pr}(\mathfrak{m}')$ *holds.*

Lifts of Special $(0,4)$-Bundles to Elliptic Fiber Bundles with a Section The following proposition relates Theorem 8.2 to the respective Theorem 8.3 for elliptic bundles that will be formulated below.

Proposition 8.2 *Let X be a connected Riemann surface (connected oriented smooth surface, respectively) of genus g with $m \geq 1$ holes with base point x_0 and curves denoted by γ_j that represent a standard system of generators $e_j \in \pi_1(X, x_0)$.*

(1) Each complex analytic (differentiable, respectively) family of Riemann surfaces of type $(1, 1)$ *over* X *is holomorphically (smoothly, respectively) isomorphic to the canonical double branched covering* $\mathfrak{F}$ *of a special holomorphic (smooth, respectively)* $(0, 4)$*-bundle* $\mathfrak{F}_1$ *over* X*. The monodromy of the bundle* $\mathfrak{F}$ *along each* γ_j *is a lift of the respective monodromy of the bundle* $\mathfrak{F}_1$*.*

(2) Vice versa, for each special holomorphic (special smooth, respectively) $(0, 4)$*-bundle over* X *and each collection* $\mathfrak{m}^j$ *of lifts of the* $2g + m - 1$ *monodromy mapping classes* $\mathfrak{m}_1^j$ *of the bundle along the* γ_j *there exists a double branched covering by a complex analytic (differentiable, respectively) family of Riemann surfaces of type* $(1, 1)$ *with collection of monodromy mapping classes equal to the* $\mathfrak{m}^j$*. For each given holomorphic (smooth, respectively) special* $(0, 4)$*-bundle over* X *there are up to holomorphic (smooth, respectively) isomorphisms exactly* 2^{2g+m-1} *holomorphic (smooth, respectively) families of Riemann surfaces of type* $(1, 1)$ *that lift the* $(0, 4)$*-bundle.*

(3) A lift of a special $(0, 4)$*-bundle is reducible if and only if the special* $(0, 4)$*-bundle is reducible.*

Proof We start with the proof of the **first statement** of the proposition. Consider a complex analytic (smooth, respectively) family of Riemann surfaces of type $(1, 1)$ over X. By Lemma 8.4 and Proposition 8.1 we may assume that the family has the form $(\mathcal{X}_\Lambda, \mathcal{P}_\Lambda, s_\Lambda, X)$ for a holomorphic (smooth, respectively) family of lattices $\Lambda(x)$, $x \in X$.

We consider the complex (smooth, respectively) submanifold $\mathring{\boldsymbol{B}}\boldsymbol{L}(\Lambda)$ of $X \times \mathbb{C}$ that intersects each fiber $\{x\} \times \mathbb{C}$ along the set $\{x\} \times \mathring{B}L(\Lambda(x))$ (see Eqs. (8.55) and (8.66)) and the complex (smooth, respectively) submanifold $\boldsymbol{BL}(\Lambda)$ of $X \times \mathbb{P}^1$ that intersects each fiber $\{x\} \times \mathbb{P}^1$ along the set $\{x\} \times BL(\Lambda(x))$ with $BL(\Lambda(x)) = \mathring{B}L(\Lambda(x)) \cup \{\infty\}$.

We prove first that the holomorphic (smooth, respectively) family

$$(\mathcal{X}_\Lambda, \mathcal{P}_\Lambda, s_\Lambda, X)$$

is holomorphically isomorphic (isomorphic as a smooth family of Riemann surfaces, respectively) to the canonical double branched covering

$$(\mathcal{Y}_{\boldsymbol{BL}(\Lambda)}, \mathcal{P}, s^\infty, X)$$

(see Sect. 8.7) of the special $(0, 4)$-bundle $(X \times \mathbb{P}^1, \mathrm{pr}_1, \boldsymbol{BL}(\Lambda), X)$. To see this we recall that for each $x \in X$ the mapping $\omega_{\Lambda_x} : (\mathbb{C} \setminus \Lambda_x)/\Lambda_x \to \mathring{Y}_{BL(\Lambda_x)}$ is a surjective conformal mapping that depends holomorphically (smoothly, respectively) on x (see equalities (8.62) and (8.63)). Hence, the mapping defines a holomorphic (smooth, respectively) homeomorphism from $\mathcal{X}_\Lambda \setminus s_\Lambda$ onto $\mathring{\mathcal{Y}}_{\boldsymbol{BL}(\Lambda)}$, that maps the fiber over x of the first bundle conformally onto the fiber over x of the second bundle.

The set $\mathcal{Y}_{\boldsymbol{BL}(\Lambda)}$ is obtained as in Sect. 8.7 by "adding a point to each fiber". As in Sect. 8.7 the mapping extends to a holomorphic (smooth, respectively) homeomorphism $\mathcal{X}_\Lambda \to \mathcal{Y}_{\boldsymbol{BL}(\Lambda)}$, that maps the fiber over x conformally onto the fiber over

x. The extension maps s_Λ to s^∞. We proved that the bundles $(\mathcal{X}_\Lambda, \mathcal{P}_\Lambda, s_\Lambda, X)$ and $(\mathcal{Y}_{\boldsymbol{BL}(\Lambda)}, \mathcal{P}, s^\infty, X)$ are holomorphically isomorphic (isomorphic as smooth families of Riemann surfaces, respectively). Hence, each holomorphic (smooth, respectively) $(1, 1)$-family over a finite open Riemann surface is holomorphically isomorphic (smoothly isomorphic), and in particular, isotopic, to the canonical double branched covering of a holomorphic (smooth, respectively) $(0, 4)$-bundle.

We identify the mapping class groups in the fiber over the base point x_0 of the bundles $(\mathcal{X}_\Lambda, \mathcal{P}_\Lambda, s_\Lambda, X)$ and $(\mathcal{Y}_{\boldsymbol{BL}(\Lambda)}, \mathcal{P}, s^\infty, X)$ using the extension of the isomorphism ω_{Λ_0} to the closed fiber $\mathbb{C}/\Lambda_0$. Having in mind this identification, we prove now that the monodromy mapping class along each γ_j of the $(1, 1)$-bundle $(\mathcal{X}_\Lambda, \mathcal{P}_\Lambda, s_\Lambda, X)$ is a lift of the monodromy mapping class along γ_j of the special $(0, 4)$-bundle $(X \times \mathbb{P}^1, \mathrm{pr}_1, \boldsymbol{BL}(\Lambda), X)$. Parameterise γ_j by the unit interval $[0, 1]$. Write $\Lambda_t \overset{def}{=} \Lambda(\gamma_j(t)) = a(t)\mathbb{Z} + b(t)\mathbb{Z}$, $t \in [0, 1]$, for smooth functions a and b on $[0, 1]$. For $t \in [0, 1]$ we denote by $\widetilde{\varphi}^t$ the real linear self-homeomorphism of $\mathbb{C}$ that maps $a(0)$ to $a(t)$ and $b(0)$ to $b(t)$. Let φ^t be the homeomorphism from the fiber over $\gamma_j(0)$ onto the fiber over $\gamma_j(t)$ of the $(1, 1)$-bundle that lifts to $\widetilde{\varphi}^t$. Then $\varphi^0 = \mathrm{Id}$ and $(\varphi^1)^{-1}$ represents the monodromy mapping class of the $(1, 1)$-bundle $(\mathcal{X}_\Lambda, \mathcal{P}_\Lambda, s_\Lambda, X)$ along γ_j. Let $\mathring{\psi}^t : \mathring{Y}(\Lambda_0) \to \mathring{Y}(\Lambda_t)$ be obtained from the commutative diagram

$$\begin{array}{ccc}
(\mathbb{C} \setminus \Lambda_0)/\Lambda_0 & \xrightarrow{\ \mathring{\varphi}^t\ } & (\mathbb{C} \setminus \Lambda_t)/\Lambda_t \\
\Big\downarrow{\scriptstyle \omega_{\Lambda_0}} & & \Big\downarrow{\scriptstyle \omega_{\Lambda_t}} \\
\mathring{Y}(\Lambda_0) & \xrightarrow{\ \mathring{\psi}^t\ } & \mathring{Y}(\Lambda_t)
\end{array}$$

where $\mathring{\varphi}^t = \varphi^t | (\mathbb{C} \setminus \Lambda_0)/\Lambda_0$.

We use the restriction to $\mathring{Y}(\Lambda_0)$ of the coordinates (z, w) on $\mathbb{C}^2$. Let ι_0 be the involution on $\mathring{Y}(\Lambda_0)$ and ι_t the involution on $\mathring{Y}(\Lambda_t)$. Choose respective coordinates on $\mathring{Y}(\Lambda_t)$ and write $\mathring{\psi}^t(z, w) = (\mathring{\psi}^t_1(z, w), \mathring{\psi}^t_2(z, w))$, $(z, w) \in \mathring{Y}(\Lambda_0)$. The equality $\widetilde{\varphi}^t(-\zeta) = -\widetilde{\varphi}^t(\zeta)$, $\zeta \in \mathbb{C}$, can be written as $\mathring{\varphi}^t \iota_{\Lambda_0} = \iota_{\Lambda_t} \mathring{\varphi}^t$. From the diagram we obtain with $\iota_t = \omega_{\Lambda_t} \iota_{\Lambda_t} (\omega_{\Lambda_t})^{-1}$, $\iota_0 = \omega_{\Lambda_0} \iota_{\Lambda_0} (\omega_{\Lambda_0})^{-1}$

$$\mathring{\psi}^t \iota_0 = \iota_t \mathring{\psi}^t . \tag{8.71}$$

In coordinates (z, w) this means $(\mathring{\psi}^t_1(z, -w), \mathring{\psi}^t_2(z, -w)) = (\mathring{\psi}^t_1(z, w), -\mathring{\psi}^t_2(z, w))$. Hence, $\mathring{\psi}^t_1(z, w) = \mathring{\psi}^t_1(z, -w)$, so that $\mathring{\psi}^t_1$ depends only on the coordinate $z \in \mathbb{C}$. Moreover, if ι_0 fixes (z, w) (equivalently, if $z \in \mathring{BL}(\Lambda_0)$), then $w = 0$, hence $\mathring{\psi}^t_2(z, w) = \mathring{\psi}^t_2(z, -w) = -\mathring{\psi}^t_2(z, w)$, i.e. ι_t fixes $\mathring{\psi}^t(z, w)$ (equivalently, $\mathring{\psi}^t_1(z, w) \in \mathring{BL}(\Lambda_t)$). We saw that the self-homeomorphism $\mathring{\psi}^t_1$ of $\mathbb{C}$ maps the set $\mathring{BL}(\Lambda_0)$ onto the set $\mathring{BL}(\Lambda_t)$. Each $\mathring{\psi}^t_1$ extends to a self-homeomorphism ψ^t_1 of $\mathbb{P}^1$ that maps the set of distinguished points $BL(\Lambda_0) \overset{def}{=} \mathring{BL}(\Lambda_0) \cup \{\infty\}$ onto the set of distinguished points $BL(\Lambda_t) \overset{def}{=} \mathring{BL}(\Lambda_t) \cup \{\infty\}$. Hence, ψ^1_1 represents the monodromy mapping class along γ_j of the special $(0,4)$-bundle with set of finite distinguished points $\mathring{BL}(\Lambda_t)$ in the fiber over $\gamma_j(t)$. The mapping $\mathring{\psi}^1 = (\mathring{\psi}_1^1, \mathring{\psi}_2^1)$ is equal to $\omega_{\Lambda_0} \circ \mathring{\varphi}^1 \circ \omega_{\Lambda_0}^{-1}$. Identifying the mapping class groups of $\mathbb{C}/\Lambda_0$ with the mapping class group of $Y_{BL(\Lambda_0)}$ by the isomorphism induced by ω_{Λ_0}, we may identify the monodromy of the bundle $(\mathcal{X}_\Lambda, \mathcal{P}_\Lambda, s_\Lambda, X)$ along γ_j with the mapping class of the extension ψ^1 of $\mathring{\psi}^1 = (\mathring{\psi}_1^1, \mathring{\psi}_2^1)$. The mapping ψ_1^1 is the projection of ψ^1 under the double branched covering from $\mathbb{C}/\Lambda_0$ with distinguished point $0/\Lambda_0$ onto $\mathbb{P}^1$ with distinguished points $BL(\Lambda_0) \cup \{\infty\}$. We proved that the monodromy mapping class of the bundle $(\mathcal{X}_\Lambda, \mathcal{P}_\Lambda, s_\Lambda, X)$ along each γ_j is a lift of the respective monodromy mapping class of the bundle $(X \times \mathbb{P}^1, \mathrm{pr}_1, \boldsymbol{BL}(\Lambda), X)$. The first statement is proved.

We prove now the **second statement**. Consider the special holomorphic, or smooth, respectively, $(0,4)$-bundle $\mathfrak{F}_1 \overset{def}{=} \big(X \times \mathbb{P}^1, \mathrm{pr}_1, \boldsymbol{E}, X\big)$. The complex or smooth, respectively, submanifold $\boldsymbol{E}$ of $X \times \mathbb{P}^1$ intersects each fiber $\{x\} \times \mathbb{P}^1$ along the set $\{x\} \times E_x$, with $E_x = (\mathring{E}_x \cup \{\infty\})$, where $\mathring{E}_x \subset C_3(\mathbb{C})/\mathcal{S}_3$. In Sect. 8.7 we obtained the canonical lift $(\mathcal{Y}_{\boldsymbol{E}}, \mathcal{P}, \boldsymbol{s}^\infty, X)$. Denote the monodromy mapping class along γ_j of the canonical lift by $\mathfrak{m}_+^j$. Recall that each mapping class $\mathfrak{m}_1^j$ has two lifts and they differ by involution. Let $\mathfrak{m}_-^j$ be the mapping class whose representatives differ from those of $\mathfrak{m}_+^j$ by composition with the involution ι of the fiber over the base point.

Take any subset J of $\{1, 2, \ldots, 2g + m - 1\}$. The lift of the $(0,4)$-bundle whose monodromy mapping class along γ_j equals $\mathfrak{m}_+^j$ if $j \notin J$, and $\mathfrak{m}_-^j$ if $j \in J$, is obtained as follows. Let $\tilde{U}$ be the subset of the universal covering $\tilde{X} \xrightarrow{\mathsf{P}} X$ that was used in Sect. 8.2 in the proof of Theorem 8.1, and let D, and $\tilde{D}_j$, $j = 0, \ldots,$ be as in Sect. 8.2. Consider the lift of the bundle $(\mathcal{Y}_{\boldsymbol{E}}, \mathcal{P}, \boldsymbol{s}^\infty, X)$ to $\tilde{X}$ and restrict the lifted bundle to $\tilde{U}$. Denote the obtained bundle on $\tilde{U}$ by $\tilde{\mathfrak{F}}$. For each point $x \in D$ we let $\tilde{x}_j \in \tilde{D}_j$, $j = 0, \ldots,$ be the points with $\mathsf{P}(\tilde{x}_j) = x$. We identify the fiber of $\tilde{\mathfrak{F}}$ over $\tilde{x}_j$, $j = 0, \ldots,$ with the fiber Y_x, $x = \mathsf{P}(\tilde{x}_j)$, of the canonical double branched covering of the $(0,4)$-bundle. $\mathring{Y}_x$ is obtained from Y_x by removing the point of $\boldsymbol{s}^\infty$ from Y_x.

Glue for each $x \in D$ and each $j = 1, \ldots,$ the fiber of $\tilde{\mathfrak{F}}$ over $\tilde{x}_j$ to the fiber of $\tilde{\mathfrak{F}}$ over $\tilde{x}_0$ using the identity if $j \notin J$ and the mapping ι if $j \in J$. More detailed, the gluing mapping of the punctured fibers in the case $j \in J$ equals

$$(\tilde{x}_j, z, w) \to (\tilde{x}_0, z, -w), \quad x \in D, \ (z, w) \in \mathring{Y}_x \,. \tag{8.72}$$

Since the gluing mappings are holomorphic (smooth, respectively), we obtain a complex analytic (smooth, respectively) family over X of double branched coverings of $\mathbb{C}$. Extend the family to a complex analytic (smooth, respectively) family of double branched coverings of $\mathbb{P}^1$ over X. This can be done as in Sect. 8.7 by considering an exhausting sequence of relatively compact open subsets X_n of X and constructing the required extension for the restriction of the bundle to each X_n. We obtained for each collection of lifts of the $\mathfrak{m}_1^j$ a complex analytic (smooth, respectively) family of Riemann surfaces of type $(1, 1)$ over X with given fiber over the base point and with monodromy mapping classes equal to this collection. It is easy to see that two holomorphic (smooth, respectively) $(1, 1)$-bundles that lift a given $(0, 4)$-bundle over a finite open Riemann surface (smooth oriented surface, respectively) are holomorphically isomorphic (smoothly isomorphic, respectively), if and only if their monodromies coincide. Hence, for each given holomorphic (smooth, respectively) special $(0, 4)$-bundle over X there are up to holomorphic (smooth, respectively) isomorphisms exactly 2^{2g+m-1} holomorphic (smooth, respectively) families of Riemann surfaces of type $(1, 1)$ that lift the $(0, 4)$-bundle. The second statement is proved.

It remains to prove the **third statement**. Let

$$\mathsf{Pr} : (\mathcal{X}, \mathcal{P}, \boldsymbol{s}, X) \to (X \times \mathbb{P}^1, \mathrm{pr}_1, \boldsymbol{E}, X)$$

be a double branched covering of a $(0, 4)$-bundle by a $(1, 1)$-family of Riemann surfaces. If the special $(0, 4)$-bundle is reducible, then there is a simple closed curve γ that divides the fiber $\mathbb{P}^1$ over the base point x_0 into two connected components, each of which contains two distinguished points, such that the following holds. The curve γ is mapped by each monodromy mapping class of the $(0, 4)$-bundle to a curve that is free homotopic to γ. The preimage of the closed curve γ under the covering map consists of two simple closed curves $\tilde{\gamma}_1$ and $\tilde{\gamma}_2$, that are homotopic to each other in $\mathcal{P}^{-1}(x) \setminus \boldsymbol{s}$, and each of the two curves cuts the torus into an annulus. Each monodromy mapping class of the $(1, 1)$-bundle, being a lift of the respective monodromy mapping class of the $(0, 4)$-bundle, takes $\tilde{\gamma}_1$ to a curve that is free homotopic to $\tilde{\gamma}_1$ (and is also free homotopic to $\tilde{\gamma}_2$). Hence, the $(1, 1)$-bundle is also reducible.

Suppose now that a double branched covering $\mathfrak{F} = (\mathcal{X}, \mathcal{P}, \boldsymbol{s}, X)$ of a special $(0, 4)$-bundle $\mathsf{Pr}(\mathfrak{F}) = (X \times \mathbb{P}^1, \mathrm{pr}_1, \boldsymbol{E}, X)$ is reducible and prove that the $(0, 4)$-bundle is reducible. Let $x_0 \in X$ be the base point of X, and let the fiber of the bundle $\mathfrak{F}$ over x_0 be $\mathbb{C}/\Lambda$ with distinguished point $0/\Lambda$. Any admissible system of curves on the torus $\mathbb{C}/\Lambda$ with a distinguished point contains exactly one curve.

Let γ_0 be a simple closed curve in $(\mathbb{C} \setminus \Lambda)/\Lambda$ that reduces all monodromy mapping classes of the bundle $\mathfrak{F}$. The curve γ_0 is free isotopic in the closed torus $\mathbb{C}/\Lambda$ to a curve γ with base point $0/\Lambda$ that lifts under the covering map $\mathsf{P} : \mathbb{C} \to \mathbb{C}/\Lambda$ to the straight line segment that joins 0 with another lattice point λ in Λ. Since γ_0 is free isotopic in $\mathbb{C}/\Lambda$ to a simple closed curve on the torus, λ is a primitive element of the lattice. Indeed, with $\Lambda = \{na + mb : n, m \in \mathbb{Z}\}$ for real linearly independent complex numbers a and b we get $\lambda = n(\lambda)a + m(\lambda)b$, where $n(\lambda)$ and $m(\lambda)$ are relatively prime integer numbers. Then there is another element $\lambda' = n(\lambda')a + m(\lambda')b \in \Lambda$ such that

$$\begin{vmatrix} n(\lambda) & n(\lambda') \\ m(\lambda) & m(\lambda') \end{vmatrix} = 1 .$$

The two elements λ and λ' generate Λ. Multiplying Λ by a non-zero complex number and changing perhaps λ' to $-\lambda'$, we may assume that $\lambda = 1$ and $\lambda' = \tau$ with $\mathrm{Im}\tau > 0$. After similar changes of all fibers we obtain an isomorphic bundle for which the fiber over x_0 equals $\mathbb{C}/\Lambda_\tau$ with distinguished point $0/\Lambda_\tau$. After a free isotopy in $(\mathbb{C} \setminus \Lambda_\tau)/\Lambda_\tau$ we may assume that the reducing curve γ_0 lifts under the covering map $\mathsf{P} : \mathbb{C} \to \mathbb{C}/\Lambda_\tau$ to the segment $\frac{1+\tau}{2} + [0, 1] \subset \mathbb{C}$. Denote by γ the curve in the closed torus $\mathbb{C}/\Lambda_\tau$ that lifts to $[0, 1]$, and by γ' the curve on the closed torus that lifts to $[0, \tau]$. γ and γ' represent a pair of generators of the fundamental group of the closed torus with base point $0/\Lambda_\tau$.

The complement $(\mathbb{C}/\Lambda_\tau) \setminus \gamma_0$ of γ_0 in the fiber over x_0 is a topological annulus with distinguished point $0/\Lambda_\tau$. Since γ_0 reduces all monodromy mapping classes of the (1, 1)-family $\mathfrak{F}$, each monodromy mapping class has a representative that fixes γ_0 pointwise. These representatives map the topological annulus $(\mathbb{C}/\Lambda_\tau) \setminus \gamma_0$ homeomorphically onto itself, fixing the distinguished point and fixing the boundary pointwise, or fixing the boundary pointwise after an involution. Hence, each monodromy is a power of a Dehn twist about γ_0, maybe, composed with an involution.

The Dehn twist on $\mathbb{C}/\Lambda_\tau$ about γ_0 can be represented by a self-homeomorphism ψ_τ of $\mathbb{C}/\Lambda_\tau$ which lifts under P to the real linear self-map $\widetilde{\psi}_\tau$ of $\mathbb{C}$ which maps 1 to 1 and τ to $1 + \tau$. This can be seen by looking at the action of $\widetilde{\psi}_\tau$ on the lifts of the curves γ and γ'. $\widetilde{\psi}_\tau$ takes $\widetilde{\gamma} = [0, 1]$ to itself, and $\widetilde{\gamma}' = [0, \tau]$ to a curve that is isotopic to $[0, \tau + 1]$ through simple closed arcs in $\mathbb{C}$ with fixed endpoints, the interiors of the arcs avoiding Λ_τ. Note that the real 2×2 matrix corresponding to $\widetilde{\psi}_\tau$ is $\begin{pmatrix} 1 & (\mathrm{Im}\,\tau)^{-1} \\ 0 & 1 \end{pmatrix}$.

We consider the double branched covering $\mathsf{p}_{\Lambda_\tau} : (\mathbb{C}\setminus\Lambda_\tau)/\Lambda_\tau \to \mathbb{C}$ (see (8.64)) with branch locus $\mathring{BL}(\Lambda_\tau) = \{e_1(\tau), e_2(\tau), e_3(\tau)\}$ determined by the Weierstraß $\wp$-function, $\wp_{\Lambda_\tau} = \mathsf{p}_{\Lambda_\tau} \circ p$ (with p being the projection $p : \mathbb{C}\setminus\Lambda_\tau \to (\mathbb{C}\setminus\Lambda_\tau)/\Lambda_\tau$). Consider the continuous extension $\mathsf{p}^c_{\Lambda_\tau} : \mathbb{C}/\Lambda_\tau \to \mathbb{P}^1$ of $\mathsf{p}_{\Lambda_\tau}$. The Weierstrass $\wp$-function extends to a mapping, also denoted by $\wp_{\Lambda_\tau}$, $\wp_{\Lambda_\tau} : \mathbb{C} \to \mathbb{P}^1$, that takes 0 to ∞, $\frac{1}{2}$ to e_1, $\frac{\tau}{2}$ to e_2, and $\frac{1+\tau}{2}$ to e_3. The line segments $[0, \frac{1}{2}]$, $[0, \frac{\tau}{2}]$, $[\frac{1}{2}, \frac{1+\tau}{2}]$, and $[\frac{\tau}{2}, \frac{1+\tau}{2}]$ are mapped under $\wp_{\Lambda_\tau}$ to simple curves γ_{∞,e_1}, γ_{∞,e_2}, γ_{e_1,e_3}, and γ_{e_2,e_3} in $\mathbb{P}^1$, each of which joins the first-mentioned point with the last-mentioned point. (See Fig. 8.2.) The union of the four curves (each with suitable orientation) is a closed curve in $\mathbb{P}^1$ (the union of the real axis with the point ∞ on Fig. 8.2) that divides $\mathbb{P}^1$ into two connected components C_1 and C_2. The Weierstrass $\wp$-function $\wp_{\Lambda_\tau}$ maps the open parallelogram R in the complex plane with vertices 0, 1, $\frac{\tau}{2}$, $1+\frac{\tau}{2}$ conformally onto its image. Indeed, the "double" of this parallelogram, i.e. the parallelogram with vertices 0, 1, τ, $\tau+1$, is the interior of a fundamental polygon for the covering $\mathbb{C} \to \mathbb{C}/\Lambda_\tau$ and the equality $\wp_{\Lambda_\tau}(\zeta) = \wp_{\Lambda_\tau}(-\zeta+n+m\tau)$, $n, m \in \mathbb{Z}$ holds. By the last equality we get in particular

$$\wp_{\Lambda_\tau}(\frac{\tau+1}{2} + t) = \wp_{\Lambda_\tau}(\frac{\tau+1}{2} - t) \ , t \in [0, \frac{1}{2}]. \tag{8.73}$$

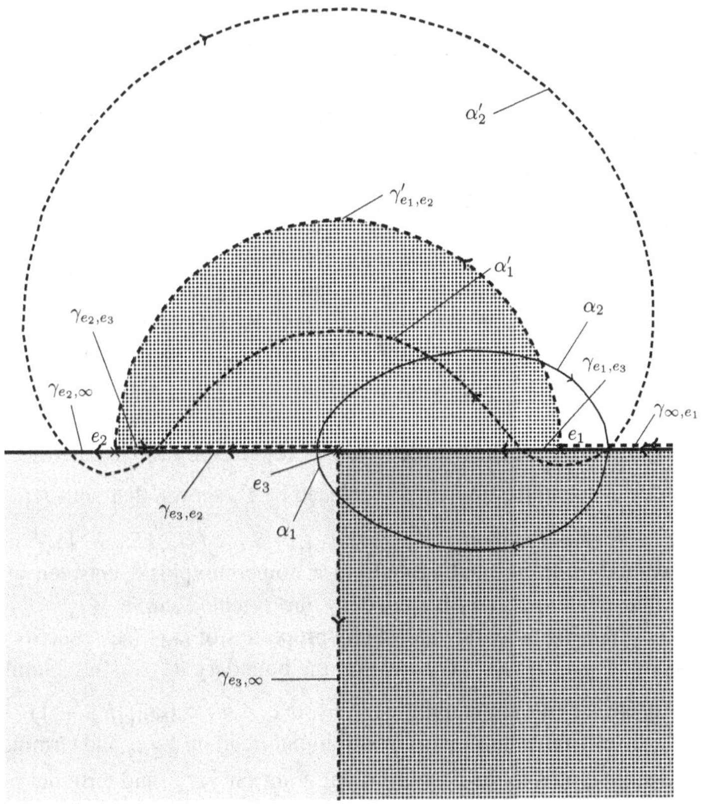

Fig. 8.2 The semi-conjugation of a Dehn twist to a half-twist. In the figure we consider the torus $\mathbb{C}/\Lambda$ for $\Lambda = 2\alpha(\mathbb{Z} + i\mathbb{Z})$ with a real number α such that $e_1 = \wp_\Lambda(\alpha) = 1$, $e_2 = \wp_\Lambda(i\alpha) = -1$, $e_3 = \wp_\Lambda((1+i)\alpha) = 0$

After perhaps relabeling C_1 and C_2, we may assume that $\wp_{\Lambda_\tau}$ maps the open parallelogram R_- with vertices $0, \frac{1}{2}, \frac{\tau}{2}, \frac{1+\tau}{2}$, (the "left half" of R) conformally onto C_1 (the lower half-plane (shadowed) in the case of Fig. 8.2), and then it maps the parallelogram R_+ with vertices $\frac{1}{2}$, 1, $\frac{1+\tau}{2}$, $1+\frac{\tau}{2}$, (the "right half" of R) conformally onto C_2 (the upper half-plane in the case of Fig. 8.2).

The mapping $\tilde{\psi}_\tau$ fixes the segment $[0, \frac{1}{2}]$, and translates the segment $[\frac{\tau}{2}, \frac{1+\tau}{2}]$ by $\frac{1}{2}$. It maps R_- to the parallelogram $\tilde{\psi}_\tau(R_-) \subset R$ whose horizontal sides are $[0, \frac{1}{2}]$ and $[\frac{1+\tau}{2}, \frac{2+\tau}{2}]$.

The Weierstraß $\wp$-function $\wp_{\Lambda_\tau}$ maps $[\frac{1}{2} + \frac{\tau}{2}, 1 + \frac{\tau}{2}]$ to γ_{e_3,e_2} which equals γ_{e_2,e_3} with inverted orientation (see (8.73)), and takes the segment $[0, \frac{1}{2}]$ to a curve γ_{∞,e_1} that joins ∞ with e_1. It maps the segment $[0, \frac{1+\tau}{2}]$ to a curve γ_{∞,e_3} which is contained in C_1 except of its endpoints and joins ∞ and e_3. The segment $[\frac{1}{2}, 1 + \frac{\tau}{2}]$ is mapped to a curve γ'_{e_1,e_2} with initial point e_1 and terminal point e_2, which is contained in C_2 except its endpoints, so that the union of the curves γ_{∞,e_1}, γ'_{e_1,e_2}, γ_{e_2,e_3}, and $\gamma_{e_3,\infty}$ is the oriented boundary of a domain C'_1, i.e. the domain C'_1 is "on the left" when walking along the oriented curve. (In Fig. 8.2 C'_1 is the dotted domain.) Here $\gamma_{e_3,\infty}$ equals γ_{∞,e_3} with inverted orientation. Hence, $\wp_{\Lambda_\tau}$ maps $\tilde{\psi}_\tau(R_-)$ onto C'_1.

The mapping $\wp_{\Lambda_\tau} \circ \tilde{\psi}_\tau$ takes R_- to the domain C'_1. The mapping $\varphi_{\tau,1} \overset{def}{=} \wp_{\Lambda_\tau} \circ \tilde{\psi}_\tau \circ (\wp_{\Lambda_\tau}|R_-)^{-1}$ takes C_1 onto C'_1. It extends continuously to a homeomorphism between closures. The extension fixes γ_{∞,e_1} pointwise, it takes γ_{e_1,e_3} to γ'_{e_1,e_2} fixing e_1, it takes γ_{e_3,e_2} to γ_{e_2,e_3} mapping e_2 to e_3 and e_3 to e_2, and it takes $\gamma_{e_2,\infty}$ to $\gamma_{e_3,\infty}$ fixing ∞. A simple arc α_1 in the closure $\overline{C_1}$ (i.e. an arc without self-intersection) with initial point on γ_{∞,e_1} and terminal point on γ_{e_3,e_2} is mapped by the extension of $\varphi_{\tau,1}$ to a simple arc α'_1 in $\overline{C'_1}$ with initial point on γ_{∞,e_1} and terminal point on γ_{e_2,e_3}.

A similar argument for the parallelogram R replaced by the parallelogram with vertices $\pm\frac{1}{2}$, $\frac{\tau}{2} \pm \frac{1}{2}$ and the domain C_1 replaced by C_2 shows, that with $R_+ - 1 \overset{def}{=} \{z - 1 : z \in R_+\}$ the mapping $\varphi_{\tau,2} \overset{def}{=} \wp_{\Lambda_\tau} \circ \tilde{\psi}_\tau \circ (\wp_{\Lambda_\tau}|R_+ - 1)^{-1}$ takes C_2 homeomorphically onto C'_2 and extends to a homeomorphism between closures. Here C'_2 is the domain that is bounded by the oriented curve $\gamma'_{e_2,e_1} \cup \gamma_{e_1,\infty} \cup \gamma_{\infty,e_3} \cup \gamma_{e_3,e_2}$. Moreover, by the periodicity properties of $\wp_{\Lambda_\tau}$ (see equality (8.73)) the extensions of $\varphi_{\tau,1}$ and $\varphi_{\tau,2}$ coincide on the boundary $\partial C_1 = \partial C_2$. Similarly as before the extension of the mapping $\varphi_{\tau,2} \overset{def}{=} \wp_{\Lambda_\tau} \circ \tilde{\psi}_\tau \circ (\wp_{\Lambda_\tau}|R_+ - 1)^{-1}$ to the closure of C_2 takes simple arcs α_2 in $\overline{C_2}$ with initial point on γ_{e_3,e_2} and terminal point on γ_{∞,e_1} to simple arcs α'_2 in $\overline{C'_2}$ with initial point on γ_{e_2,e_3} and terminal point on γ_{∞,e_1}.

It follows that the mapping φ_τ that is equal to $\varphi_{\tau,j}$ on C_j extends to a self-homeomorphism of $\mathbb{P}^1$, denoted again by φ_τ, whose mapping class in $\mathbb{P}^1$ with distinguished points e_1, e_2, e_3, ∞, is conjugate to $\mathfrak{m}_{\sigma_1}$. Indeed, φ_τ takes a simple closed curve $\alpha_1 \cup \alpha_2$ with α_1 and α_2 as before to the simple closed curve $\alpha'_1 \cup \alpha'_2$ and takes a simple closed curve that surrounds γ_{∞,e_1} and no other point in the branch locus to a simple closed curve that surrounds γ_{∞,e_1} and no other point in the branch locus. The mapping class, that is represented by a half-twist around the interval $[e_2, e_3]$ (see Sect. 1.5), is determined by the following properties. It takes the homotopy class of $\alpha_1 \cup \alpha_2$ to the homotopy class of $\alpha'_1 \cup \alpha'_2$, and it takes the homotopy class of a simple closed curve on $\mathbb{P}^1$, that surrounds γ_{∞,e_1} and no other point in the branch locus, to a curve that is isotopic to this curve on $\mathbb{C} \setminus \{e_1, e_2, e_3\}$.

Since $\tilde{\psi}_\tau|R_- = (p|R)^{-1} \circ \psi_\tau \circ p|R_-$ and $\wp_{\Lambda_\tau} \circ \tilde{\psi}_\tau|R_- = \varphi_{\tau,1} \circ \wp_{\Lambda_\tau}|R_-$ we obtain $\wp_{\Lambda_\tau} \circ (p|R)^{-1} \circ \psi_\tau = \varphi_{\tau,1} \circ \wp_{\Lambda_\tau} \circ (p|R_-)^{-1}$, hence, since $\wp_{\Lambda_\tau} = \mathsf{p}_{\Lambda_\tau} \circ p$ (see (8.64)), the equality $\mathsf{p}_{\Lambda_\tau} \circ \psi_\tau = \varphi_{\tau,1} \circ \mathsf{p}_{\Lambda_\tau} = \varphi_\tau \circ \mathsf{p}_{\Lambda_\tau}$ holds on $p(R_-)$. By the same reason the equality $\mathsf{p}_{\Lambda_\tau} \circ \psi_\tau = \varphi_{\tau,2} \circ \mathsf{p}_{\Lambda_\tau} = \varphi_\tau \circ \mathsf{p}_{\Lambda_\tau}$ holds on $p(R_+)$. Since $p(R_- \cup R_+)$ is dense in $\mathbb{C}/\Lambda_\tau$, the equality $\mathsf{p}_{\Lambda_\tau} \circ \psi_\tau = \varphi_\tau \circ \mathsf{p}_{\Lambda_\tau}$ holds on $\mathbb{C}/\Lambda_\tau$. We proved that the projection of ψ_τ (see Sect. 8.7) is in the homotopy class of a conjugate of $\mathfrak{m}_{\sigma_1}$, which is a reducible mapping class. We proved that the (0, 4)-bundle is reducible. Proposition 8.2 is proved. □

Let $\widehat{\mathfrak{F}}$ be an isomorphism class of smooth $(1,1)$-bundles over the circle. By the analog of Proposition 8.2 for bundles over the circle each smooth $(1,1)$-bundle $\mathfrak{F}$ over $\partial\mathbb{D}$ is isomorphic to the canonical double branched covering of a smooth special $(0,4)$-bundle $\mathfrak{F}_1 = \mathsf{Pr}(\mathfrak{F})$ over $\partial\mathbb{D}$ (the projection of $\mathfrak{F}$). Moreover, by Lemma 8.9 the projections of the monodromy homomorphisms of isomorphic bundles over $\partial\mathbb{D}$ are conjugate, hence the projections of isomorphic bundles are isomorphic. Vice versa, isomorphic special $(0,4)$-bundles over $\partial\mathbb{D}$ lift to isomorphic $(1,1)$-bundles. An isomorphism class of special $(0,4)$-bundles over $\partial\mathbb{D}$ corresponds to a conjugacy class of 3-braids modulo center $\widehat{b/\mathcal{Z}_3}$. By Corollary 2.2 for any braid b that is associated to a bundle representing the class $\mathsf{Pr}(\widehat{\mathfrak{F}})$ the equality $\mathcal{M}(\hat{b}) = \mathcal{M}(\widehat{b\Delta_3^{2k}}) = \mathcal{M}(\widehat{\mathfrak{m}_{b,\infty}}) = \mathcal{M}(\widehat{\mathsf{Pr}(\mathfrak{F})}) = \mathcal{M}(\mathsf{Pr}(\widehat{\mathfrak{F}}))$ holds. Proposition 8.2 implies the following corollary.

Corollary 8.2

$$\mathcal{M}(\widehat{\mathfrak{F}}) = \mathcal{M}(\hat{b})\,.$$

Proof $\mathcal{M}(\widehat{\mathfrak{F}})$ is the supremum of the conformal modules of annuli on which there exists a holomorphic $(1,1)$-bundle that represents $\widehat{\mathfrak{F}}$. $\mathcal{M}(\mathsf{Pr}(\widehat{\mathfrak{F}}))$ is the supremum of the conformal modules of annuli on which there exists a special holomorphic $(0,4)$-bundle that represents $\mathsf{Pr}(\widehat{\mathfrak{F}})$. Statement (1) of Proposition 8.2 implies the inequality $\mathcal{M}(\widehat{\mathfrak{F}}) \le \mathcal{M}(\mathsf{Pr}(\widehat{\mathfrak{F}}))$, Statement (2) implies the opposite inequality. The corollary is proved. □

8.9 The Gromov-Oka Principle for (1, 1)-Bundles over Tori with a Hole

In the following Theorem 8.3 we consider isotopies of smooth $(1,1)$-families over tori with a hole to complex analytic $(1,1)$-families.

Theorem 8.3

(1) Let X be a smooth surface of genus one with a hole with base point x_0, and with a chosen set $\mathcal{E} = \{e_1, e_2\}$ of generators of $\pi_1(X, x_0)$. Define the set $\mathcal{E}_0 = \{e_1, e_2, e_1 e_2^{-1}, e_1 e_2^{-2}, e_1 e_2 e_1^{-1} e_2^{-1}\}$ as in Theorem 7.3. Consider a smooth $(1,1)$-bundle $\mathfrak{F}$ on X. Suppose for each $e \in \mathcal{E}_0$ the restriction of the bundle $\mathfrak{F}$ to an annulus representing $\hat{e}$ has the Gromov-Oka property. Then the bundle $\mathfrak{F}$ has the Gromov-Oka property on X.

(2) If a bundle $\mathfrak{F}$ as in (1) is irreducible, then it is isotopic to an isotrivial bundle, and hence, for each conformal structure ω on X the bundle $\mathfrak{F}_\omega$ is isotopic to a bundle that extends to a holomorphic $(1,1)$-bundle on the closed torus $\omega(X)^c$. In particular, $\mathfrak{F}$ is isotopic to a holomorphic bundle for any conformal structure ω on X (maybe, of first kind).

(3) Any smooth reducible bundle $\mathfrak{F}$ on X has a single irreducible bundle component. This irreducible bundle component is isotopic to an isotrivial bundle. There is a Dehn twist in the fiber over the base point such that the $(1, 1)$*-bundle $\mathfrak{F}$ can be recovered from the irreducible bundle component up to composing each monodromy by a power of the Dehn twist. Any smooth reducible bundle on X is isotopic to a holomorphic* $(1, 1)$*-bundle for each conformal structure of second kind on X.*

(4) A reducible holomorphic $(1, 1)$*-bundle over a punctured Riemann surface is locally holomorphically trivial.*

Proof By Proposition 8.2, Statement (1), we may assume that $\mathfrak{F}$ is the canonical double branched covering of a smooth special $(0, 4)$-bundle, denoted by $\mathsf{Pr}(\mathfrak{F})$. Suppose that for each $e \in \mathcal{E}_0$ the restriction of the smooth $(1, 1)$-bundle $\mathfrak{F}$ to an annulus $A_{\hat{e}}$ representing $\hat{e}$ has the Gromov-Oka property. Then for each $e \in \mathcal{E}_0$ there is a conformal structure $\omega_e : A_{\hat{e}} \to \omega_e(A_{\hat{e}})$ such that $\omega_e(A_{\hat{e}})$ has conformal module bigger than $\frac{\pi}{2}\log(\frac{3+\sqrt{5}}{2})^{-1}$ and the pushed forward bundle $(\mathfrak{F}|A_{\hat{e}})_{\omega_e}$ is isotopic to a holomorphic bundle $\mathfrak{F}_e$ on $\omega_e(A_{\hat{e}})$. By Proposition 8.1 we may assume that $\mathfrak{F}_e$ is of the form (8.37), and by Proposition 8.2 we may assume that $\mathfrak{F}_e$ is the canonical double branched covering of a holomorphic special $(0, 4)$-bundle on $\omega_e(A_{\hat{e}})$, denoted by $\mathsf{Pr}(\mathfrak{F}_e)$. Since the monodromy homomorphisms of $\mathfrak{F}|A_{\hat{e}}$ and $\mathfrak{F}_e$ are conjugate, the monodromy homomorphisms of $\mathsf{Pr}(\mathfrak{F}|A_{\hat{e}})$ and $\mathsf{Pr}(\mathfrak{F}_e)$ are also conjugate. Hence, $\mathsf{Pr}(\mathfrak{F}|A_{\hat{e}})$ and $\mathsf{Pr}(\mathfrak{F}_e)$ are isotopic. We saw that for each $e \in \mathcal{E}$ the restriction of the special $(0, 4)$-bundle $\mathsf{Pr}(\mathfrak{F})$ to an annulus representing $\hat{e}$ is isotopic to a holomorphic bundle for a conformal structure on the annulus of conformal module bigger than $\frac{\pi}{2}\log(\frac{3+\sqrt{5}}{2})^{-1}$. By Theorem 8.2 the special $(0, 4)$-bundle $\mathsf{Pr}(\mathfrak{F})$ is isotopic to a holomorphic special $(0, 4)$-bundle for any conformal structure of second kind on X. Hence, the canonical double branched covering of $\mathsf{Pr}(\mathfrak{F})$ is isotopic to a holomorphic $(1, 1)$-bundle for each conformal structure of second kind on X. Therefore, $\mathfrak{F}$ is isotopic to a holomorphic $(1, 1)$-bundle for each conformal structure of second kind on X. Statement (1) is proved.

Suppose a bundle $\mathfrak{F}$ as in (1) is irreducible. Then by Proposition 8.2 the special $(0, 4)$-bundle $\mathsf{P}(\mathfrak{F})$ is also irreducible. Moreover, by Theorem 8.2, Statement (1), the $(0, 4)$-bundle $\mathsf{P}(\mathfrak{F})$ is isotopic to an isotrivial bundle, i.e. its lift to a finite covering $\hat{X}$ is isotopic to the trivial bundle, and therefore, for each conformal structure ω on X (maybe, of first kind) the $(0, 4)$-bundle $\mathsf{P}(\mathfrak{F})_\omega$ is isotopic to a bundle that extends holomorphically to the closed torus $\omega(X)^c$. Then also the canonical double branched covering of the $(0, 4)$-bundle is isotopic to an isotrivial bundle. Hence, the bundle $\mathfrak{F}$ is isotopic to an isotrivial bundle and extends to a holomorphic bundle on the closed torus. In particular, for each conformal structure (maybe, of first kind) on X the bundle $\mathfrak{F}_\omega$ is isotopic to a holomorphic bundle on $\omega(X)$. This gives statement (2).

Suppose now $\mathfrak{F}$ is any smooth reducible bundle on X. Then by Proposition 8.2 the bundle $\mathsf{Pr}(\mathfrak{F})$ is reducible. By Theorem 8.2 the special $(0, 4)$-bundle $\mathsf{Pr}(\mathfrak{F})$ is isotopic to a holomorphic special $(0, 4)$-bundle for any conformal structure of second kind on X. Hence, for any conformal structure of second kind on X the bundle $\mathfrak{F}$ is isotopic to a holomorphic (1, 1)-bundle which is a double branched covering.

Each admissible set of curves on a punctured torus consists of exactly one curve. Hence, there is an admissible closed curve γ in the punctured fiber $\mathcal{P}^{-1}(x_0) \setminus s^\infty$ over the base point x_0, that is mapped by the monodromy mapping class along each curve γ_j representing an element $e_j \in \mathcal{E}$ to a curve that is isotopic to γ in the punctured fiber over x_0. Choose for each monodromy mapping class a representative that maps the curve γ onto itself.

The complement of γ on the punctured torus $\mathcal{P}^{-1}(x_0) \setminus s^\infty$ is connected and homeomorphic to the thrice punctured Riemann sphere. This implies first that there is a single irreducible component of $\mathfrak{F}$. Further, the restrictions of suitable representatives of the monodromy mapping classes to the complement of γ on the punctured torus are conjugate to self-homeomorphisms of the thrice punctured Riemann sphere. The conjugated homeomorphisms extend to self-homeomorphisms of the Riemann sphere with three distinguished points. The self-homeomorphisms may interchange the two points that come from different edges of γ, but each of them fixes the third distinguished point. Except the identity there is only one isotopy class of such mappings and its square is the identity (in other words, the mapping class is the class of an involution). Hence, the monodromy mapping classes of the irreducible bundle component are powers of a single periodic mapping class. Therefore the irreducible bundle component is isotopic to an isotrivial bundle.

The monodromy mapping classes of the original bundle $\mathfrak{F}$ are powers of Dehn twists about γ, maybe, composed with an involution, and up to powers of Dehn twists about γ in the fiber $\mathcal{P}^{-1}(x_0)$ the bundle can be recovered from the irreducible bundle component. Statement (3) is proved.

We prove now statement (4). Consider a reducible holomorphic (1,1)-bundle $\mathfrak{F}$ over a punctured Riemann surface of genus 1. By Proposition 8.2 it is holomorphically isomorphic to a double branched covering of a holomorphic special (0, 4)-bundle $\mathsf{Pr}(\mathfrak{F})$ over a punctured Riemann surface. By Proposition 8.2 the special holomorphic(0, 4)-bundle $\mathsf{Pr}(\mathfrak{F})$ is also reducible. By Theorem 8.2 the bundle $\mathsf{Pr}(\mathfrak{F})$ is holomorphically trivial. Then the double branched covering $\mathfrak{F}$ of $\mathsf{Pr}(\mathfrak{F})$ is locally holomorphically trivial. Theorem 8.3 is proved. □

Chapter 9
Fundamental Groups and Bounds for the Extremal Length

In Chap. 3 we computed several versions of the extremal length of certain elements of the fundamental group of the twice punctured complex plane. It becomes impracticable to compute the versions of extremal length for all elements regardless of their "complexity". In this chapter we give effective upper and lower bounds for the extremal length with totally real horizontal boundary values of any element of the fundamental group of the twice punctured complex plane. The bounds differ by a multiplicative constant not depending on the element. The estimates are provided in terms of a natural syllable decomposition of the reduced word representing the element. Estimates for the extremal length of conjugacy classes of elements of the fundamental group (i.e. of the invariants mentioned by Gromov) are also obtained. The extremal length with totally real horizontal boundary values is capable to give more subtle information regarding Gromov's Oka Principle and limitations of its validity than the respective invariant of conjugacy classes.

In the last section (Sect. 9.7) we give estimates for the extremal length with totally real horizontal boundary values of 3-braids, and estimates of the extremal length of conjugacy classes of 3-braids (and, hence, of the entropy of 3-braids).

9.1 The Fundamental Group and Extremal Length. Two Theorems

It will be convenient here to normalize the twice punctured complex plane as $\mathbb{C}\backslash\{-1, 1\}$. With this normalization the fundamental group $\pi_1 \overset{def}{=} \pi_1(\mathbb{C}\backslash\{-1, 1\}, 0)$ is a free group in two generators a_j, $j = 1, 2$, where a_1 is represented by simple closed curves that surround -1 counterclockwise, and a_2 is represented by simple closed curves that surround 1 counterclockwise. Recall that there is a canonical isomorphism $\pi_1\big(\mathbb{C} \setminus \{-1, 1\}, 0\big) \to \pi_1^{tr} \overset{def}{=} \pi_1\big(\mathbb{C} \setminus \{-1, 1\}, (-1, 1)\big)$.

B. Jöricke, *Braids, Conformal Module, Entropy, and Gromov's Oka Principle*, Lecture Notes in Mathematics 2360, https://doi.org/10.1007/978-3-031-67288-0_9

The elements of the relative fundamental group $\pi_1(\mathbb{C} \setminus \{-1, 1\}, (-1, 1))$ are homotopy classes of arcs in $\mathbb{C} \setminus \{-1, 1\}$, each of whose endpoints is contained in $(-1, 1)$. The isomorphism $\pi_1(\mathbb{C} \setminus \{-1, 1\}, 0) \to \pi_1^{tr}$ assigns to the element $e \in \pi_1(\mathbb{C} \setminus \{-1, 1\}, 0)$, that is represented by a curve γ with base point 0, the element $e_{tr} \in \pi_1^{tr} = \pi_1(\mathbb{C}\setminus\{-1, 1\}, (-1, 1))$, that is represented by γ. The extremal length $\Lambda(e_{tr})$ of the homotopy class e_{tr} in the sense of Definition 3.2 is called the extremal lengths of e with totally real horizontal boundary values and is sometimes denoted by $\Lambda_{tr}(e)$. Recall that $\Lambda(e_{tr})$ is the infimum of the extremal length of all rectangles that admit a holomorphic mapping into $\mathbb{C} \setminus \{-1, 1\}$ whose restriction to each maximal vertical segment represents e_{tr}. The name "totally real" refers to the notion for 3-braids which is related to the present notion. The conformal module $\mathcal{M}(e_{tr}) = (\Lambda(e_{tr}))^{-1}$ is also denoted by $\mathcal{M}_{tr}(e)$.

We will also consider the relative fundamental group $\pi_1^{pb} \overset{def}{=} \pi_1(\mathbb{C}\setminus\{-1, 1\}, i\mathbb{R})$ whose elements are homotopy classes of arcs in $\mathbb{C} \setminus \{-1, 1\}$ with endpoints on the imaginary axis $i\mathbb{R}$. The canonical group isomorphism $\pi_1(\mathbb{C}\setminus\{-1, 1\}, 0) \to \pi_1^{pb} \overset{def}{=} \pi_1(\mathbb{C} \setminus \{-1, 1\}, i\mathbb{R})$ assigns to each element e of $\pi_1(\mathbb{C} \setminus \{-1, 1\}, 0)$ represented by a curve γ with base point 0 the element e_{pb} of $\pi_1(\mathbb{C} \setminus \{-1, 1\}, i\mathbb{R})$ represented by γ. The extremal length $\Lambda(e_{pb})$ in the sense of Definition 3.2 is called the extremal length with pb boundary values of e and is also denoted by $\Lambda_{pb}(e)$. The conformal module $\mathcal{M}(e_{pb}) = (\Lambda(e_{pb}))^{-1}$ is also denoted by $\mathcal{M}_{pb}(e)$. ("pb" stands for "perpendicular bisector". In fact, the imaginary axis is the perpendicular bisector of the line segment $(-1, 1)$.)

More generally, it will be sometimes convenient to use the fact that $\pi_1(\mathbb{C} \setminus \{-1, 1\}, 0)$ is canonically isomorphic to the relative fundamental group $\pi_1^{+} \overset{def}{=} \pi_1(\mathbb{C}\setminus\{-1, 1\}, i\mathbb{R} \cup (-1, 1))$ whose elements are homotopy classes of arcs in $\mathbb{C} \setminus \{-1, 1\}$ with endpoints in $i\mathbb{R} \cup (-1, 1))$. The group isomorphism assigns to each element e of $\pi_1(\mathbb{C} \setminus \{-1, 1\}, 0)$, represented by a closed curve γ, the element e_+ of $\pi_1(\mathbb{C} \setminus \{-1, 1\}, i\mathbb{R} \cup (-1, 1))$, represented by γ. For each element $e \in \pi_1(\mathbb{C} \setminus \{-1, 1\}, 0)$ all its representatives are contained in both classes, e_{tr} and e_{pb}, associated to e, and all representatives of e_{tr} and e_{pb} are contained in the element e_+ of π_1^+ that is associated to e.

We will identify each element of $\pi_1(\mathbb{C}\setminus\{-1, 1\}, 0)$ with the reduced word in the generators representing it. A word w in the generators is reduced if it is the identity or it has the form $w = w_1^{n_1} \cdot w_2^{n_2} \cdot \ldots,$ where the n_j are non-zero integers and the w_j are alternately equal to either a_1 or a_2. We refer to the $w_j^{n_j}$ as the terms of the word.

Any reduced word w in $\pi_1(\mathbb{C} \setminus \{-1, 1\}, 0)$ can be uniquely decomposed into syllables. They are defined as follows.

Definition 9.1 *The syllables of any reduced word* $w \in \pi_1(\mathbb{C} \setminus \{-1, 1\}, 0)$ *are all its terms* $a_{j_i}^{k_i}$ *with* $|k_i| \geq 2$ *(called syllables of form* (1)*), and all maximal sequences of consecutive terms* $a_{j_i}^{k_i}$ *for which* $|k_i| = 1$ *and all* k_i *have the same sign. A syllable of the latter form is called a syllable of form* (2)*, if it contains more than one term and is called a singleton or a syllable of form* (3) *if it consists of a single term.*

(See also [44, 45]). Define the degree of a syllable deg(syllable) to be the sum of the absolute values of the powers of terms entering the syllable. For example, the syllables of the word $a_2^{-1}\,a_1^2\,a_2^{-3}\,a_1^{-1}\,a_2^{-1}\,a_1^{-1}\,a_2\,a_1^{-1}$ from left to right are the singleton a_2^{-1}, the syllable a_1^2 of form (1) and degree 2, the syllable a_2^{-3} of form (1) and degree 3, the syllable $a_1^{-1}\,a_2^{-1}\,a_1^{-1}$ of form (2) and of degree 3, the singleton a_2 and the singleton a_1^{-1}.

We call words consisting of a single syllable elementary words. Label the syllables of a word from left to right by consecutive integral numbers $j = 1, 2, \dots$. Let d_j be the degree of the j-th syllable $\mathfrak{s}_j$. (We consider each syllable as a reduced word in the elements of the fundamental group.) Put

$$\mathcal{L}(w) \overset{def}{=} \sum_j \log(2d_j + \sqrt{4d_j^2 - 1})\,, \tag{9.1}$$

where the sum runs over the degrees of all syllables of w. If w is the identity we put $\mathcal{L}(w) = 0$. We want to point out that $\mathcal{L}(w^{-1}) = \mathcal{L}(w)$. Notice that for the word consisting of the single syllable $\mathfrak{s}_j$ we have $\mathcal{L}(\mathfrak{s}_j) = \log(2d_j + \sqrt{4d_j^2 - 1})$. Thus, $\mathcal{L}(w) = \sum \mathcal{L}(\mathfrak{s}_j)$ where the sum runs over the syllables of w.

In this chapter we will prove the following theorems. The present proof of the upper bound is shorter and less technical than the proof in [44], but the proof in [44] gives a better constant.

Theorem 9.1 *For any element $w \in \pi_1(\mathbb{C}\setminus\{-1, 1\}, 0)$ the following estimates hold.*

$$\frac{1}{2\pi}\,\mathcal{L}(w) \le \Lambda_{tr}(b) = \frac{1}{\mathcal{M}_{tr}(w)} \le 2^{14} \cdot \mathcal{L}(w), \tag{9.2}$$

except in the following cases: $w = a_1^n$ or $w = a_2^n$ for an integer n. In these cases $\Lambda_{tr}(b) = 0$, i.e $\mathcal{M}_{tr}(b) = \infty$.

Moreover,

$$\frac{1}{2\pi}\,\mathcal{L}(w) \le \Lambda_{pb}(w) = \frac{1}{\mathcal{M}_{pb}(w)} \le 2^{13} \cdot \mathcal{L}(w), \tag{9.3}$$

except in the following case: each term in the reduced word w has the same power, which equals either $+1$ or -1. In these cases $\Lambda_{pb}(w) = 0$, i.e. $\mathcal{M}_{pb}(w) = \infty$.

We will also consider the relative fundamental groups ${}^{pb}\pi_1^{tr}$ and ${}^{tr}\pi_1^{pb}$ with mixed horizontal boundary values. Here the elements of ${}^{pb}\pi_1^{tr}$ are the homotopy classes ${}_{i\mathbb{R}}\mathfrak{h}_{(-1,1)}$ and the elements of ${}^{tr}\pi_1^{pb}$ are the homotopy classes ${}_{(-1,1)}\mathfrak{h}_{i\mathbb{R}}$ in the space $X = \mathbb{C} \setminus \{-1, 1\}$. There is a canonical isomorphism from π_1 onto ${}^{pb}\pi_1^{tr}$ (onto ${}^{tr}\pi_1^{pb}$, respectively) that assigns to an element $e \in \pi_1$ the element ${}_{pb}e_{tr}$ that contains all representatives of e (the element ${}_{tr}e_{pb}$, respectively, that contains all representatives of e).

The statement for mixed boundary values is given in the following Theorem 9.2. Note that for mixed boundary values the estimate of the extremal length holds always while for pb or tr boundary values there are exceptional cases.

Theorem 9.2 *For all $w \in \pi_1$ the following inequalities hold*

$$\frac{1}{2\pi}\,\mathcal{L}(w) \leq \Lambda({}_{tr}w_{pb}) \leq 2^{14} \cdot \mathcal{L}(w), \tag{9.4}$$

$$\frac{1}{2\pi}\,\mathcal{L}(w) \leq \Lambda({}_{pb}w_{tr}) \leq 2^{14} \cdot \mathcal{L}(w). \tag{9.5}$$

Recall that the conjugacy class $\hat{w}$ of an element $w \in \pi_1(\mathbb{C} \setminus \{-1, 1\}, 0)$ can be identified with the free homotopy class of closed curves that contains w.

Definition 9.2 *A word $w \in \pi_1(\mathbb{C} \setminus \{-1, 1\}, 0)$ is called cyclically reduced, if either the word consists of a single term, or it has at least two terms and the first and the last term of the word are powers of different generators.*

Each reduced word which is not the identity is conjugate to a cyclically reduced word.

The following theorem gives upper and lower bounds of the invariant $\mathcal{M}(\hat{w})$ of conjugacy classes of elements of the twice punctured complex plane, that was first mentioned in Gromov's seminal paper [35] in connection with obstructions for the Gromov-Oka principle. Each element of the fundamental group $\pi_1(\mathbb{C} \setminus \{-1, 1\}, 0)$ corresponds to an element of $\mathcal{PB}_3/\mathcal{Z}_3$ and, equivalently, to a mapping class of the twice punctured complex plane. By the Main Theorem the conformal module $\mathcal{M}(\hat{w})$ can be interpreted as a multiple of the inverse of the entropy of the conjugay class $\hat{w}$.

Theorem 9.3 *Let $\hat{w}$ be a conjugacy class of elements of $\pi_1(\mathbb{C} \setminus \{-1, 1\}, 0)$, and let w be a cyclically reduced word representing the conjugacy class $\hat{w}$. Then*

$$\frac{1}{2\pi} \cdot \mathcal{L}(w) \;\leq\; \Lambda(\hat{w}) \;\leq\; 2^{13} \cdot \mathcal{L}(w),$$

with the following exceptions: $\hat{w} = \widehat{a_1^n}$, $\hat{w} = \widehat{a_2^n}$ and $\hat{w} = \widehat{(a_1 a_2)^n}$. In the exceptional cases $\Lambda(\hat{w}) = 0$ and $\mathcal{M}(\hat{w}) = \infty$.

9.2 Coverings of $\mathbb{C} \setminus \{-1, 1\}$ and Slalom Curves

For the proof of the theorems it will be convenient to use instead of the universal covering of $\mathbb{C} \setminus \{-1, 1\}$ two different coverings of $\mathbb{C} \setminus \{-1, 1\}$ by $\mathbb{C} \setminus i\mathbb{Z}$. Though the universal covering is well studied, factorizing the universal covering through two different coverings by $\mathbb{C} \setminus i\mathbb{Z}$ has the advantage to make the contribution of the syllables to the extremal length transparent.

To obtain the first covering $\mathbb{C} \setminus i\mathbb{Z} \to \mathbb{C} \setminus \{-1, 1\}$ we take the universal covering of the twice punctured Riemann sphere $\mathbb{P}^1 \setminus \{-1, 1\}$ (the logarithmic covering) and remove all preimages of ∞ under the covering map. Geometrically the logarithmic covering of $\mathbb{P}^1 \setminus \{-1, 1\}$ can be described as follows. Take copies of $\mathbb{P}^1 \setminus [-1, 1]$ labeled by the set $\mathbb{Z}$ of integer numbers. Attach to each copy two copies of $(-1, 1)$, the $+$-edge (the set of accumulation points contained in $(-1, 1)$ of the upper half-plane) and the $-$-edge (the set of accumulation points contained in $(-1, 1)$ of the lower half-plane). For each $k \in \mathbb{Z}$ we glue the $+$-edge of the k-th copy to the $-$-edge of the $k+1$-st copy (using the identity mapping on $(-1, 1)$ to identify points on different edges). Denote by $U_{\log}$ the set obtained from the described covering by removing all preimages of ∞.

Choose curves $\alpha_j(t), t \in [0, 1], j = 1, 2$, in $\mathbb{C} \setminus \{-1, 1\}$ that represent the generators $a_j, j = 1, 2$, of the fundamental group $\pi_1(\mathbb{C} \setminus \{-1, 1\}, 0\})$ and have the following properties. The initial point and the terminal point of each of the curves are the only points of the curve on the interval $(-1, 1)$ and are also the only points of the curve on the imaginary axis. The following proposition holds.

Proposition 9.1 *The set $U_{\log}$ is conformally equivalent to $\mathbb{C} \setminus i\,\mathbb{Z}$. The mapping $f_1 \circ f_2$, $f_2(z) = \frac{e^{\pi z}-1}{e^{\pi z}+1}$, $z \in \mathbb{C} \setminus i\,\mathbb{Z}$, $f_1(w) = \frac{1}{2}(w + \frac{1}{w})$, $w \in \mathbb{C} \setminus \{-1, 0, 1\}$, is a covering map from $\mathbb{C} \setminus i\,\mathbb{Z}$ to $\mathbb{C} \setminus \{-1, 1\}$. It has period i and takes the punctured strip $\{z \in \mathbb{C} : -\frac{1}{2} < \mathrm{Im} z < \frac{1}{2}\} \setminus \{0\}$ conformally onto $\mathbb{C} \setminus [-1, 1]$. Points of the strip contained in the upper half-plane are mapped to the lower half-plane and points of the strip contained in the lower half-plane are mapped to the upper half-plane. The mapping $f_1 \circ f_2$ takes the set $(-\frac{i}{2}, \frac{i}{2}) \setminus \{0\}$ onto $i\mathbb{R} \setminus \{0\}$, and the line $\mathbb{R} + \frac{i}{2}$ onto $(-1, 1)$.*

For each $k \in \mathbb{Z}$ the lift of α_1 with initial point $\frac{-i}{2} + ik$ is a curve which joins $\frac{-i}{2} + ik$ with $\frac{-i}{2} + i(k + 1)$ and is contained in the closed left half-plane. The only points on the imaginary axis are the endpoints.

The lift of α_2 with initial point $\frac{-i}{2} + ik$ is a curve which joins $\frac{-i}{2} + ik$ with $\frac{-i}{2} + i(k - 1)$ and is contained in the closed right half-plane. The only points on the imaginary axis are the endpoints.

Figure 9.1 shows the curves α_1 and α_2 which represent the generators of the fundamental group $\pi_1(\mathbb{C} \setminus \{-1, 1\}, 0)$ and their lifts under the covering maps f_1 and $f_2 \circ f_1$. For $j = 1, 2$ the curves α'_j and α''_j are the two lifts of α_j under the double covering $f_1 : \mathbb{C} \setminus \{-1, 0, 1\} \to \mathbb{C} \setminus \{-1, 1\}$. The curve $\hat{\alpha}'_1$ is the lift of α'_1 under the mapping f_2 with initial point $\frac{-i}{2}$, the curve $\hat{\alpha}''_1$ is the lift of α''_1 under the mapping f_2 with initial point $\frac{i}{2}$, the curve $\hat{\alpha}'_2$ lifts α'_2 and has initial point $\frac{i}{2}$, and the curve $\hat{\alpha}''_2$ lifts α''_2 and has initial point $-\frac{i}{2}$.

Proof The mapping f_1 is the restriction of the Zhukovsky function to $\mathbb{C} \setminus \{-1, 0, 1\}$. The Zhukovski function defines a double branched covering of the Riemann sphere $\mathbb{P}^1$ with branch locus $\{-1, 1\}$. In particular, the Zhukovsky function provides a conformal mapping from the unit disc $\mathbb{D}$ onto $\mathbb{P}^1 \setminus [-1, 1]$. It maps -1 to -1, 1 to 1

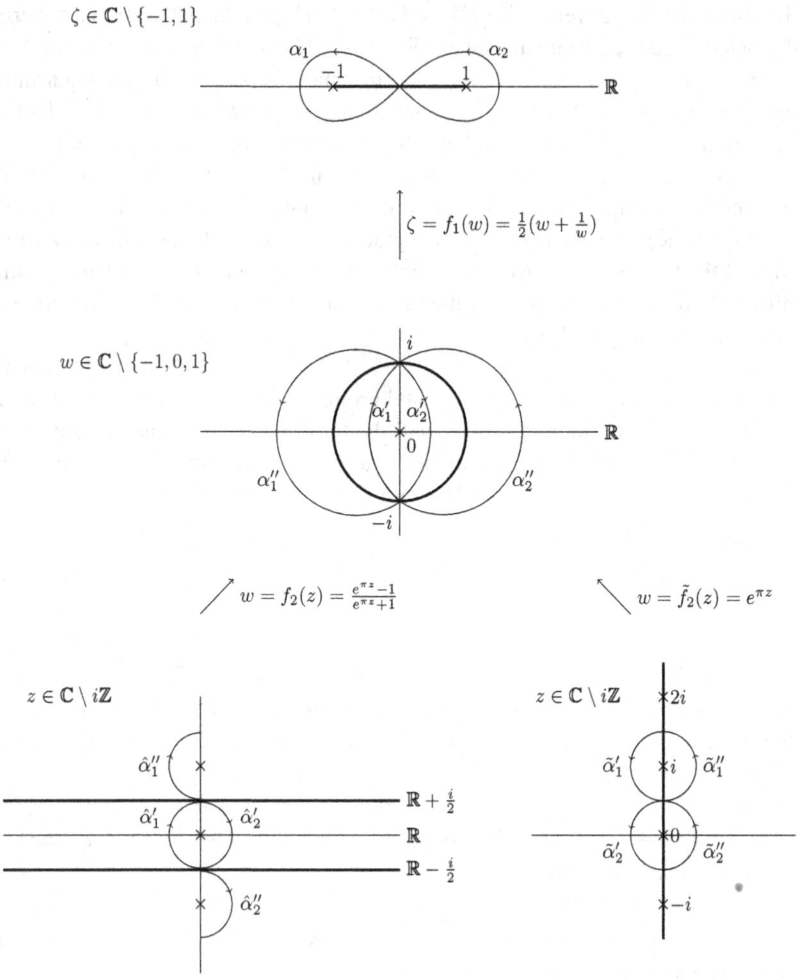

Fig. 9.1 The generators of $\pi_1(\mathbb{C} \setminus \{-1, 1\}, 0)$ and their lifts under two coverings by $\mathbb{C} \setminus i\mathbb{Z}$

and 0 to ∞. The upper half-circle is mapped onto the $-$-edge, the upper half-disc is mapped onto the lower half-plane, the lower half-circle is mapped onto the $+$-edge and the lower half-disc is mapped onto the upper half-plane. Similarly, it provides a conformal mapping of the exterior of the closed unit disc onto $\mathbb{P}^1 \setminus [-1, 1]$ which preserves the upper half-plane and also preserves the lower half-plane.

The mapping f_2 extends through $i\mathbb{Z}$ to an infinite covering of $\mathbb{P}^1 \setminus \{-1, 1\}$ by $\mathbb{C}$. By an abuse of notation we denote this extension also by f_2. (The extension of) f_2 provides a conformal mapping f_2^0 from the strip $\{z \in \mathbb{C} : -\frac{1}{2} < \operatorname{Im} z < \frac{1}{2}\}$ onto the unit disc, which takes the real axis onto the segment $(-1, 1)$, such that $\lim_{x \in \mathbb{R}, x \to -\infty} = -1$, $\lim_{x \in \mathbb{R}, x \to +\infty} = 1$. Further, f_2 maps $\frac{i}{2}$ to i, $-\frac{i}{2}$ to $-i$, and 0 to 0. The line $\{z \in \mathbb{C} : \operatorname{Im} z = \frac{1}{2}\}$ is mapped onto the upper half-circle, the upper half-strip $\{z \in \mathbb{C} : 0 < \operatorname{Im} z < \frac{1}{2}\}$ is mapped onto the upper half-disc, the line $\{z \in \mathbb{C} : \operatorname{Im} z = -\frac{1}{2}\}$ is mapped onto the lower half-circle and the lower half-strip is mapped onto the lower half-disc. It follows that $f_1 \circ f_2$ takes the punctured strip $\{z \in (\mathbb{C} \setminus \{0\}) : -\frac{1}{2} < \operatorname{Im} z < \frac{1}{2}\}$ conformally onto $\mathbb{C} \setminus [-1, 1]$.

The mapping $f = f_1 \circ f_2$ has period i. Indeed, f_2 has period $2i$, $f_2(z+i) = \frac{1}{f_2(z)}$, and $f_1(\frac{1}{w}) = f_1(w)$. The statements concerning the lift of curves under $f_1 \circ f_2$ are now clear.

To see that $U_{\log}$ is conformally equivalent to $\mathbb{C} \setminus i\,\mathbb{Z}$ we identify the set $\mathbb{C} \setminus [-1, 1]$ with the sheet of $U_{\log}$ labeled by 0. The conformal mapping $f \mid \{z \in \mathbb{C} \setminus \{0\} : -\frac{1}{2} < \operatorname{Im} z < \frac{1}{2}\}$ (whose image is $\mathbb{C} \setminus [-1, 1]$) extends by Schwarz's Reflection Principle through the line $\{z \in \mathbb{C} : \operatorname{Im} z = \frac{1}{2}\}$ which is mapped onto the $-$-edge of $\mathbb{C} \setminus [-1, 1]$. The reflected mapping takes $\{z \in \mathbb{C} \setminus \{i\} : \frac{1}{2} < \operatorname{Im} z < \frac{3}{2}\}$ conformally onto $\mathbb{C} \setminus [-1, 1]$. We identify the image of the punctured strip $\{z \in \mathbb{C} \setminus \{i\} : \frac{1}{2} < \operatorname{Im} z < \frac{3}{2}\}$ with the sheet of $U_{\log}$ labeled by -1. Induction on reflection through the lines $\{z \in \mathbb{C} : \operatorname{Im} z = \frac{1}{2} + j\}$, $j \in \mathbb{Z}$, gives the conformal mapping from $C \setminus i\mathbb{Z}$ onto $U_{\log}$.

The remaining statements are easy to see. □

The second covering is given by the mapping $f_1 \circ \tilde{f_2} : \mathbb{C} \setminus i\mathbb{Z} \to \mathbb{C} \setminus \{-1, 1\}$, where f_1 is as before and $\tilde{f_2}$ is the exponential map, $\tilde{f_2}(z) = e^{\pi z}$. Recall that each curve α_j, $j = 1, 2$, has two lifts α'_j and α''_j under f_1. For an illustration of the following proposition see the right part of Fig. 9.1.

Proposition 9.2 *The mapping $\tilde{f_2}$ takes $\mathbb{C} \setminus i\mathbb{Z}$ to $\mathbb{C} \setminus \{-1, 0, 1\}$, and f_1 takes the latter set to $\mathbb{C} \setminus \{-1, 1\}$. The composition $\tilde{f_2} \circ f_1$ is a covering of $\mathbb{C} \setminus \{-1, 1\}$ by $\mathbb{C} \setminus \{-1, 0, 1\}$ and has period $2i$. For each integer k it takes the interval $(ki, (k+1)i)$ to $(-1, 1)$, it maps $\frac{i}{2} + \mathbb{R}$ to $i(0, \infty)$, and maps $-\frac{i}{2} + \mathbb{R}$ to $i(-\infty, 0)$.*

The lifts $\tilde{\alpha}'_j$ and $\tilde{\alpha}''_j$, $j = 1, 2$, of α'_j and α''_j under $\tilde{f_2}$ have the following properties. The lifts $\tilde{\alpha}'_j$, $j = 1, 2$, are contained in the closed left half-plane and are directed downwards (i.e in the direction of decreasing y), the lifts $\tilde{\alpha}''_j$, $j = 1, 2$, are contained in the closed right half-plane and directed upwards. The initial point of $\tilde{\alpha}'_1$ is $i + \frac{1}{2}i$, the initial point of $\tilde{\alpha}''_1$ and $\tilde{\alpha}'_2$ is $\frac{1}{2}i$, the initial point of $\tilde{\alpha}''_2$ is $-\frac{1}{2}i$. All other lifts are obtained by translation by an integer multiple of $2i$.

The straightforward proof is left to the reader.

Consider the curve α_1^n, $n \in \mathbb{Z} \setminus \{0\}$. It runs n times along the curve α_1 if $n > 0$, and $|n|$ times along the curve α_1 with inverted orientation if $n < 0$. It is homotopic in $\mathbb{C} \setminus \{-1, 1\}$ with base point 0 to a curve whose interior (i.e. the complement of its endpoints) is contained in the open left half-plane. We call a representative of a_1^n whose interior is in the open left half-plane a standard representative of a_1^n. In the same way we define standard representatives of a_2^n. For each $k \in \mathbb{Z}$ the curve α_1^n lifts under $f_1 \circ f_2$ to a curve with initial point $\frac{-i}{2} + ik$ and terminal point $\frac{-i}{2} + ik + in$ which is contained in the closed left half-plane and omits the points in $i\mathbb{Z}$. Respectively, α_2^n, $n \in \mathbb{Z} \setminus \{0\}$, lifts under $f_1 \circ f_2$ to a curve with initial point $\frac{\pm i}{2} + ik$ and terminal point $\frac{\pm i}{2} + ik - in$ which is contained in the closed right half-plane and omits the points in $i\mathbb{Z}$. The mentioned lifts are homotopic through curves in $\mathbb{C} \setminus i\mathbb{Z}$ with endpoints on $i\,\mathbb{R} \setminus i\mathbb{Z}$ to curves with interior contained either in the open right half-plane or in the open left half-plane. Standard representatives of a_1^n and a_2^n lift to such curves.

Definition 9.3 *A simple curve in $\mathbb{C} \setminus i\mathbb{Z}$ with endpoints on different connected components of $i\mathbb{R}\setminus i\mathbb{Z}$ is called an elementary slalom curve if its interior is contained in one of the open half-planes $\{z \in \mathbb{C} : Re\, z > 0\}$ or $\{z \in \mathbb{C} : Re\, z < 0\}$.*

A curve in $\mathbb{C}\setminus i\mathbb{Z}$ is called an elementary half slalom curve if one of the endpoints is contained in a horizontal line $\{z \in \mathbb{C} : Im\, z = k+\frac{1}{2}\}$ for an integer k and the union of the curve with its (suitably oriented) reflection in the line $\{z \in \mathbb{C} : Im\, z = k + \frac{1}{2}\}$ is an elementary slalom curve.

A slalom curve in $\mathbb{C} \setminus i\mathbb{Z}$ is a curve which can be divided into a finite number of elementary slalom curves so that consecutive elementary slalom curves are contained in different half-planes.

A curve which is homotopic to a slalom curve (elementary slalom curve, respectively) in $\mathbb{C}\setminus i\mathbb{Z}$ through curves with endpoints in $i\mathbb{R}\setminus i\mathbb{Z}$ is called a homotopy slalom curve (elementary homotopy slalom curve, respectively).

A curve which is homotopic to an elementary half-slalom curve in $\mathbb{C}\setminus i\mathbb{Z}$ through curves with one endpoint in $i\mathbb{R} \setminus i\mathbb{Z}$ and the other endpoint on the line $\{z \in \mathbb{C} : Im\, z = k + \frac{1}{2}\}$ for an integer k is called an elementary homotopy half-slalom curve.

We call an elementary slalom curve non-trivial if its endpoints are contained in intervals $(ik, i(k+1))$ and $(i\ell, i(\ell+1))$ with $|k - \ell| \geq 2$. Note that the union of an elementary half-slalom curve with its reflection in the horizontal line that contains one endpoint is always a non-trivial elementary slalom curve (i.e. an elementary half-slalom curve is "half" of a non-trivial elementary slalom curve). We saw that the lifts under $f_1 \circ f_2$ of representatives of terms ${}_{pb}(a_j^n)_{pb} \in \pi_1^{pb}$ with $|n| \geq 2$ are non-trivial elementary homotopy slalom curves.

The lifts of representatives of elements of π_1 under $f_1 \circ \tilde{f}_2$ look different. The representatives of $(a_j^n)_{tr}$, $|n| \geq 1$, lift to curves which make $|n|$ half-turns around a point in $i\mathbb{Z}$ (positive half-turns if $n > 0$, and negative half-turns if $n < 0$). Each such representative lifts under $f_1 \circ \tilde{f}_2$ to the composition of $|n|$ trivial elementary homotopy slalom curves.

Take any syllable of from (2), i.e. any maximal sequence of at least two consecutive terms of the word which enter with equal power being either 1 or -1. Recall that d denotes the sum of the absolute values of the powers of the terms of the syllable. There is a representing curve that makes d half-turns around the interval $[-1, 1]$ (positive half-turns, if the exponents of terms in the syllable are 1, and negative half-turns otherwise). The lift under $f_1 \circ \tilde{f}_2$ of this representative is a non-trivial elementary slalom curve.

We will call homotopy classes of homotopy slalom curves for short slalom classes.

Fig. 9.2 Lifts of a curve in $\mathbb{C} \setminus \{-1, 1\}$ under two different coverings by $\mathbb{C} \setminus i\mathbb{Z}$

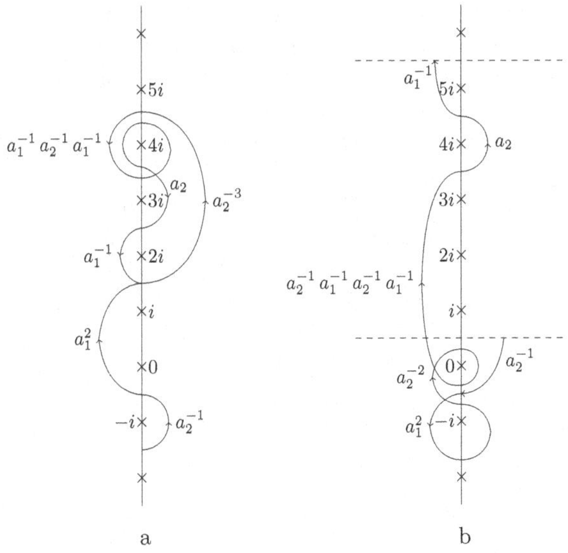

Figure 9.2 shows two slalom curves. The curve in Fig. 9.2a is a lift under $f_1 \circ f_2$ of a curve in $\mathbb{C} \setminus \{-1, 1\}$ with initial point and terminal point equal to $0 \in i\mathbb{R}$ representing the word $a_2^{-1}\, a_1^2\, a_2^{-3}\, a_1^{-1}\, a_2^{-1}\, a_1^{-1}\, a_2\, a_1^{-1}$ in the relative fundamental group $\pi_1(\mathbb{C} \setminus \{-1, 1\}, i\mathbb{R})$. The curve in Fig. 9.2b is a lift under $f_1 \circ \tilde{f}_2$ of a curve with initial and terminal point in $i\mathbb{R} \setminus \{0\}$ representing the same word. For each elementary slalom curve or half-slalom curve in Fig. 9.2 we indicate the element of π_1 which lifts under the considered mapping to the respective elementary slalom class or half-slalom class.

Non-trivial elementary slalom classes and elementary half-slalom classes have positive extremal length (in the sense of Definition 3.2) which can be effectively estimated from above and from below. In this sense homotopy classes of curves in $\mathbb{C} \setminus \{-1, 1\}$ whose lifts under $f_1 \circ f_2$ or $f_1 \circ \tilde{f}_2$ are non-trivial elementary slalom classes or elementary half-slalom classes serve as building blocks. We obtained the following fact. For an element of a relative fundamental group of the twice punctured complex plane the pieces representing syllables of form (1) with *pb* boundary values lift to non-trivial elementary slalom classes under $f_1 \circ f_2$, while the pieces representing syllables of form (2) with *tr* boundary values lift to non-trivial elementary slalom classes under $f_1 \circ \tilde{f}_2$. For a word in π_1 containing a syllable that is a singleton we may select a piece of a representing curve that lifts under $f_1 \circ f_2$ to a non-trivial elementary homotopy half-slalom curve. These facts will be used to obtain a lower bound for the extremal length of elements of the relative fundamental groups. Recall that the extremal length of a homotopy class of curves is equal to the extremal length of the class of their lifts under a holomorphic covering.

The method to obtain the upper bound is roughly to patch together in a quasiconformal way the holomorphic mappings of rectangles representing syllables and to perturb the obtained quasiconformal mapping to a holomorphic mapping.

We conclude this section with relating the two explicitly given coverings of $\mathbb{C} \setminus \{-1, 1\}$ by $\mathbb{C} \setminus i\mathbb{Z}$ to the universal covering of the twice punctured plane. Denote by $\mathsf{P} : U \to \mathbb{C} \setminus i\mathbb{Z}$ the universal covering of $\mathbb{C} \setminus i\mathbb{Z}$. Geometrically the set U is obtained in the following way. Consider the left half-plane $\mathbb{C}_\ell$ and call it the Riemann surface of generation 0. The first step is the following. For each integer k we take a copy of the right half-plane $\mathbb{C}_r$ and glue it to $\mathbb{C}_\ell$ along the interval $(ki, (k+1)i)$ (using the identity mapping for gluing). We obtain a Riemann surface with a natural projection to $\mathbb{C} \setminus i\mathbb{Z}$ called the Riemann surface of first generation. At the second step we consider the Riemann surface of first generation and proceed similarly by gluing the left half-plane along each copy of intervals $(ki, (k+1)i)$ which is an end of the Riemann surface of first generation. By induction we obtain the universal covering of $\mathbb{C} \setminus i\mathbb{Z}$.

Denote by $\widetilde{\mathbb{C}}_\ell$ the lift under P of the left half-plane to the first sheet of U over $\mathbb{C}_\ell$. Let $\mathbb{C}_\ell^{Cl}$ be the closure of $\mathbb{C}_\ell$ in $\mathbb{C} \setminus i\mathbb{Z}$, let $\widetilde{\mathbb{C}}_\ell^{Cl}$ be the closure of $\widetilde{\mathbb{C}}_\ell$ in the universal covering of $\mathbb{C} \setminus i\mathbb{Z}$ and let $(ki, (k+1)i)^\sim \subset \widetilde{\mathbb{C}}_\ell^{Cl}$ be the lift of the intervals $(ki, (k+1)i)$.

For each k we denote by $\mathfrak{D}_k^\ell$ the half-disc $\{z \in \mathbb{C}_\ell : |z - i(k + \frac{1}{2})| < \frac{1}{2}\}$ with diameter $(ik, i(k+1))$ which is contained in the left half-plane and by ρ_k the respective open half-circle $\{z \in \mathbb{C}_\ell : |z - i(k + \frac{1}{2})| = \frac{1}{2}\}$. We call the ρ_k half-circles of generation 0.

Lemma 9.1 *There is a conformal mapping $\varphi : U \to \mathbb{C}_\ell$ that takes $\widetilde{\mathbb{C}}_\ell$ onto $\mathbb{C}_\ell \setminus \bigcup_{k=-\infty}^{\infty} \overline{\mathfrak{D}_k^\ell}$ so that for each k the set $(ki, (k+1)i)^\sim$ is mapped onto ρ_k under the continuous extension of φ.*

Proof Consider the half-strip $\mathfrak{H}_0 \stackrel{def}{=} \{z \in \mathbb{C}_\ell : 0 < \operatorname{Im} z < 1\}$. By a theorem of Caratheodory ([32], Chapter II.3, Theorem 4 and Theorem 4′, and also [62] , Theorem 2.24 and Theorem 2.25) each conformal mapping that takes the lift $\widetilde{\mathfrak{H}}_0$ of $\mathfrak{H}_0$ onto the set $\mathfrak{H}_0 \setminus \overline{\mathfrak{D}_0^\ell}$, extends continuously to a homeomorphism between closures. Let φ_0 be the conformal mapping from $\widetilde{\mathfrak{H}}_0$ onto the set $\mathfrak{H}_0 \setminus \overline{\mathfrak{D}_0^\ell}$ whose extension to the boundary takes the point $\tilde{0}$ over 0 to 0, the point $\tilde{1}$ to 1 and the point $\tilde{\infty}$ (considered as prime end of $\widetilde{\mathfrak{H}}_0$) to ∞. The extension of the conformal mapping to the boundary takes the lift $\{z \in \mathbb{C}_\ell : \operatorname{Im} z = 1\}^\sim$ of $\{z \in \mathbb{C}_\ell : \operatorname{Im} z = 1\}$ onto $\{z \in \mathbb{C}_\ell : \operatorname{Im} z = 1\}$, it takes the lift $\{z \in \mathbb{C}_\ell : \operatorname{Im} z = 0\}^\sim$ of $\{z \in \mathbb{C}_\ell : \operatorname{Im} z = 0\}$ onto $\{z \in \mathbb{C}_\ell : \operatorname{Im} z = 0\}$, and maps the lift $(0, i)^\sim$ of $(0, i)$ onto ρ_0.

By induction we extend φ_0 by Schwarz's Reflection Principle across the half-lines $\{z \in \mathbb{C}_\ell : \operatorname{Im} z = k\}^\sim$ which are the lifts of the respective half-lines in $\mathbb{C}_\ell$. We obtain a conformal mapping of $\widetilde{\mathbb{C}}_\ell$ onto $\mathbb{C}_\ell \setminus \bigcup_{k=-\infty}^{\infty} \overline{\mathfrak{D}_k^\ell}$, denoted again by φ_0, whose extension to $\widetilde{\mathbb{C}}_\ell^{Cl}$ takes for each integer number k the segment $(ik, i(k+1))^\sim$ onto the half-circle ρ_k of generation 0. (See Fig. 9.3.)

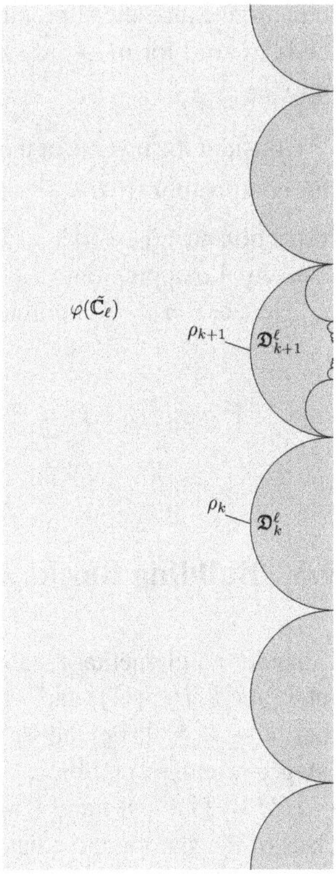

Fig. 9.3 A conformal image of the Riemann surface of generation 2

Schwarz's Reflection Principle across each segment $(ik, i(k+1))^{\sim}$provides an extension of the conformal mapping φ_0 to the Riemann surface of first generation. Note that for each k_0 the image of the half-circles ρ_k, $k \neq k_0$, under reflection of $\mathbb{C}_\ell \setminus \bigcup_{k=-\infty}^{\infty} \overline{\mathfrak{D}^\ell_k}$ across ρ_{k_0} are half-circles with diameter on the imaginary axis. We call them half-circles of the first generation. We obtained a conformal mapping of the Riemann surface of first generation to the unbounded connected component of the left half-plane with the half-circles of first generation removed.

Apply the reflection principle by induction to the Riemann surface of generation n and all copies of intervals $(ik, i(k+1))$ in its "boundary". We obtain the Riemann surface of generation $n+1$, half-circles of generation $n+1$ and a conformal mapping of this Riemann surface onto the unbounded connected component of the left half-plane with the half-circles of generation $n+1$ removed.

The supremum of the diameters of the half-circles of generation n does not exceed 2^{-n}. Indeed, the half-circles of generation n are obtained as follows. Take a half-circle ρ of generation $n-1$ and reflect all other half-circles of generation $n-1$ across it. Each half-circle of generation $n-1$ different from ρ has diameter on one side of the diameter of ρ. Hence, the image of each half-circle of generation $n-1$ different from ρ under reflection across ρ is contained in a quarter disc, hence has diameter not exceeding half of the diameter of ρ. The statement is obtained by induction. Since the supremum of the diameters of the half-circles of generation n tends to zero for $n \to \infty$, we obtain in the limit a conformal mapping φ from U onto $\mathbb{C}_\ell$. □

Consider the inverse of the mapping $\varphi_0 : \widetilde{\mathbb{C}}_\ell \to \mathbb{C}_\ell \setminus \bigcup_{k=-\infty}^{\infty} \overline{\mathfrak{D}_k^\ell}$. By construction the composition $\mathsf{P} \circ \varphi_0^{-1}$ of P with this inverse has the following property: the restriction to $(\mathfrak{H}_0 + ik) \setminus (\mathfrak{D}_0 + ik)$ of $\mathsf{P} \circ \varphi_0^{-1} : \mathbb{C}_\ell \setminus \bigcup_{k=-\infty}^{\infty} \overline{\mathfrak{D}_k^\ell} \to \mathbb{C}_\ell$ is a conformal mapping onto $\mathfrak{H}_0 + ik$ that takes ik to ik, $i(k+1)$ to $i(k+1)$, and ∞ to ∞. Hence, $\mathsf{P} \circ \varphi_0^{-1}$ commutes with translation by i, i.e.

$$\mathsf{P} \circ \varphi_0^{-1}(z+i) = \mathsf{P} \circ \varphi_0^{-1}(z) + i \quad \text{for } z \in \mathbb{C}_\ell \setminus \bigcup_{k=-\infty}^{\infty} \overline{\mathfrak{D}_k^\ell}. \tag{9.6}$$

9.3 Building Blocks and Their Extremal Length

Consider an elementary slalom curve in the left half-plane with initial point in the interval $(ik, i(k+1))$ and with terminal point in the interval $(ij, i(j+1))$. Suppose that $|k-j| \ge 2$, i.e. the slalom curve is non-trivial. After a translation by a (half-integer or integer) multiple of i we obtain a curve that has endpoints in the intervals $(-i(M+1), -iM)$ and $(iM, i(M+1))$ with $M = \frac{|k-j|-1}{2} \ge \frac{1}{2}$. The homotopy class of the slalom curve has the same extremal length as the translated class.

Let M be any positive number. Denote a curve with interior in the open left half-plane, with initial point in $(-i(M+1), -iM)$ and terminal point in $(iM, i(M+1))$, by $\gamma_{\ell,M}$, and the curve with inverted orientation by $\gamma_{\ell,M}^{-1}$. Notice that with the afore mentioned value $M = \frac{|k-j|-1}{2}$ the curve $\gamma_{\ell,M}$ is not an elementary slalom curve for even $|k-j|$, but the curve $\gamma_{\ell,M} + iM$ is always an elementary slalom curve. It will be convenient to work with the normalized curve $\gamma_{\ell,M}$ for computing the extremal length.

For any integer or half-integer number $M \ge \frac{1}{2}$ we assign to the curve $\gamma_{\ell,M}$ the class $\gamma_{\ell,M}^*$ of curves which are homotopic to $\gamma_{\ell,M}$ through curves in $\mathbb{C} \setminus i(\mathbb{Z} - M)$ with initial point in $(-i(M+1), -iM)$ and terminal point in $(iM, i(M+1))$. The class assigned to $\gamma_{\ell,M}^{-1}$ in this way is denoted by $(\gamma_{\ell,M}^*)^{-1}$. Respectively, let $\gamma_{r,M}$ be a curve with interior contained in the right half-plane $\mathbb{C}_r$ with initial point in the interval $(-i(M+1), -iM)$ and terminal point in $(iM, i(M+1)$. Consider the class $\gamma_{r,M}^*$ of curves which are homotopic to $\gamma_{r,M}$ through curves in $\mathbb{C} \setminus i(\mathbb{Z} - M)$ with initial point in $(-i(M+1), -iM)$ and terminal point in $(iM, i(M+1))$.

We will call a homotopy class of elementary slalom curves in $\mathbb{C} \setminus i\mathbb{Z}$ an elementary slalom class with integer or half-integer parameter M, if after a translation by an imaginary integer or half-integer it becomes equal to $(\gamma_{\ell,M}^*)^{\pm 1}$ or $(\gamma_{r,M}^*)^{\pm 1}$. A homotopy class of elementary half-slalom curves in $\mathbb{C} \setminus i\mathbb{Z}$ with endpoints in a real line $\frac{i}{2} + ik + \mathbb{R}$ $(k \in \mathbb{Z})$ and a component of $i\mathbb{R} \setminus i\mathbb{Z}$ is called an elementary half-slalom class with parameter M, if for a representative γ the union $\bigcup_{\gamma \in \Gamma}(\gamma \cup \gamma_{\mathrm{refl}})$ represents an elementary slalom class with parameter M. Here γ_{refl} is the reflection of γ through the real line $\frac{i}{2} + ik + \mathbb{R}$. Notice that for a given M the extremal lengths of all four classes $(\gamma^*_{\ell,M})^{\pm 1}$ or $(\gamma^*_{r,M})^{\pm 1}$ are equal. The same remark concerns the respective half-slalom classes.

Lemma 9.2 *The extremal length of an elementary slalom class γ^* with parameter M equals*

$$\Lambda(\gamma^*) = \frac{4}{\pi}\log(\sqrt{M} + \sqrt{M+1}). \tag{9.7}$$

The extremal length of an elementary half-slalom class $\tilde{\gamma}^*$ with parameter M equals

$$\Lambda(\tilde{\gamma}^*) = \frac{2}{\pi}\log(\sqrt{M} + \sqrt{M+1}). \tag{9.8}$$

Proof It is enough to prove that $\Lambda(\gamma^*_{\ell,M}) = \frac{4}{\pi}\log(\sqrt{M} + \sqrt{M+1})$. Let $\varphi : U \to \mathbb{C}_\ell$ be the mapping considered in the previous section. For a rectangle R we take a holomorphic mapping $f : R \to \mathbb{C} \setminus i\mathbb{Z}$, that represents $\gamma^*_{\ell,M} + iM$. Consider its lift $\tilde{f} : R \to U$ to a mapping to the universal covering U of $\mathbb{C} \setminus i\mathbb{Z}$. Define the mapping $G(\zeta) = -M - i\varphi(\zeta)$, $\zeta \in U$, from U onto the upper half-plane. The composition $G \circ \tilde{f}$ maps R into $\mathbb{C}_+$, and its extension to the closure of R takes the horizontal sides to the circles $\{|z \pm (M + \frac{1}{2})| = \frac{1}{2}\}$. Lemmas 3.2 and 3.4 imply the inequality $\lambda(R) \geq \frac{4}{\pi}\log(\sqrt{M} + \sqrt{M+1})$. (Compare also with the proof of the statement in Example 3.4 of Sect. 3.2.)

Take the conformal mapping f_0 of a rectangle R_0 with $\lambda(R_0) = \frac{4}{\pi}\log(\sqrt{M} + \sqrt{M+1})$ onto $\mathbb{C}_+ \setminus \{|z \pm (M + \frac{1}{2})| \leq \frac{1}{2}\}$ that maps the horizontal sides to the two boundary circles $\{|z \pm (M + \frac{1}{2})| = \frac{1}{2}\}$. Precompose this mapping with the inverse of G, and project the obtained mapping to $\mathbb{C} \setminus i\mathbb{Z}$. We obtain a mapping that represents $\gamma^*_{\ell,M} + iM$. Hence, $\Lambda(\gamma^*_{\ell,M}) = \Lambda(\gamma^*_{\ell,M} + iM) = \frac{4}{\pi}\log(\sqrt{M} + \sqrt{M+1})$.

The statement for the half-slalom class $\tilde{\gamma}^*$ follows in the same way. □

Lemma 9.3 *The extremal length of an element of each of the relative fundamental groups π_1^{pb}, π_1^{tr}, ${}^{tr}\pi_1^{pb}$, and ${}^{pb}\pi_1^{tr}$ is realized on a locally conformal mapping of a rectangle representing the element. The extremal mapping extends locally conformally across the open horizontal sides of the rectangle.*

The extremal length of a conjugacy class of elements of the fundamental group of the twice punctured plane is realized on a locally conformal mapping of an annulus into the twice punctured plane.

Proof of Lemma 9.3 For this proof it is more convenient to consider the fundamental group of the complex plane punctured at 0 and 1 rather than at -1 and 1, and to consider the upper half plane $\mathbb{C}_+$ as universal covering of $\mathbb{C} \setminus \{0, 1\}$.

Each holomorphic mapping of a rectangle into $\mathbb{C} \setminus \{0, 1\}$ that represents an element of one of the relative fundamental groups lifts to the universal covering. The lift takes the open horizontal sides of the rectangle to certain half-circles with diameter on the real axis (maybe, after a conformal self-map of the half-plane) and represents the class of curves that are contained in the half-plane and join the two half-circles. As in the proof of Lemma 3.4 the extremal length is realized on a conformal mapping of a rectangle onto the half-plane with two deleted half-discs such that the horizontal sides are mapped onto the half-circles. Composing with the covering map we obtain a locally conformal mapping that extends locally conformally across the open horizontal sides of the rectangle.

We will now prove the statement concerning conjugacy classes of elements of the fundamental group of $\mathbb{C} \setminus \{0, 1\}$ with base point.

Recall that each element of the fundamental group corresponds to a covering transformation. Each covering transformation is a holomorphic self-homeomorphism of the universal covering $\mathbb{C}_+$ that extends to a Möbius transformation $T(z) = \frac{az+b}{cz+d}$ of the Riemann sphere with integer coefficients a, b, c, d, such that $ad - bc = 1$. Moreover, T is either parabolic (i.e. T has one fixed point and is conjugate to the mapping $z \to z + b'$ for a constant b'), or T is hyperbolic (i.e., T has two fixed points and is conjugate to $z \to \kappa z$ for a positive real number κ), or T is elliptic (i.e., T has two complex fixed points symmetric with respect to the imaginary axis and is conjugate to $z \to e^{i\theta} z$ for a real number θ). See [58], Chapter II, 9D and 9E.

Let $\hat{a}$ be a conjugacy class of elements of the fundamental group of $\mathbb{C} \setminus \{0, 1\}$ and let a be an element of the fundamental group that represents $\hat{a}$. Denote by T_a the covering transformation corresponding to a, and by $\langle T_a \rangle$ the subgroup of the group of covering transformations generated by T_a. Then the quotient $\mathbb{C}_+ \diagup \langle T_a \rangle$ is an annulus. It has extremal length 0 if T_a is parabolic or elliptic and has positive extremal length if T_a is hyperbolic. If $f : A \to \mathbb{C} \setminus \{0, 1\}$ is a holomorphic mapping of an annulus A to $\mathbb{C} \setminus \{0, 1\}$ that represents $\hat{a}$, then f lifts to a holomorphic map of A into $\mathbb{C}_+ \diagup \langle T_a \rangle$. The lift represents the class of a generator of the fundamental group of $\mathbb{C}_+ \diagup \langle T_a \rangle$. The corollary follows from Lemma 3.3. □

The following proposition considers elements of the fundamental group π_1, that are represented by elementary words (i.e. by words consisting of a single syllable), and gives upper and lower bounds for the extremal length of the corresponding classes of curves in the relative fundamental groups with pb, tr or mixed boundary values.

Proposition 9.3 *The following statements hold for elements of the relative fundamental groups that are represented by elementary words.*

1. *Each syllable $\mathfrak{s}$ of form* (1) *and degree $d \geq 2$ with pb horizontal boundary values lifts under $f_1 \circ f_2$ to an elementary slalom class with parameter $M = \frac{d-1}{2}$. The following equality for the extremal length holds*

$$\Lambda({}_{pb}(\mathfrak{s})_{pb}) = \frac{2}{\pi} \log(d + \sqrt{d^2 - 1})\,.$$

2. *Each syllable* $\mathfrak{s}$ *of form* (2) *and degree* $d \geq 2$ *with* tr *horizontal boundary values lifts under* $f_1 \circ \tilde{f_2}$ *to an elementary slalom class with parameter* $M = \frac{d-1}{2}$. *The following equality for the extremal length holds*

$$\Lambda({}_{tr}(\mathfrak{s})_{tr}) = \frac{2}{\pi} \log(d + \sqrt{d^2 - 1}) .$$

3. *Any syllable* $\mathfrak{s}$ *of degree* $d \geq 1$ *with mixed boundary values lifts under* $f_1 \circ f_2$ *to an elementary half-slalom class with parameter* $M = d - \frac{1}{2}$. *For the extremal length the equalities*

$$\Lambda({}_{pb}(\mathfrak{s})_{tr}) = \frac{1}{\pi} \log(2d + \sqrt{4d^2 - 1}) ,$$

$$\Lambda({}_{tr}(\mathfrak{s})_{pb}) = \frac{1}{\pi} \log(2d + \sqrt{4d^2 - 1})$$

hold.

Proof of Statement 1 Assume first that $n = d > 0$. For any integer number k the class ${}_{pb}(a_1^n)_{pb}$ lifts under $f_1 \circ f_2$ to an elementary homotopy slalom class with initial point in $(i(k-1), ik)$ and terminal point in $i(k-1+d), i(k+d))$. There is a representative in the closed left half-plane. The slalom class has parameter $M = \frac{d-1}{2}$. A lift of the class ${}_{pb}(a_1^{-d})_{pb}$ is obtained by inverting the orientation. For $n = d$ the class ${}_{pb}(a_2^n)_{pb}$ lifts under $f_1 \circ f_2$ to an elementary homotopy slalom class, represented by a curve in the closed right half-plane with initial point in $\big(i(k-1), ik\big)$ and terminal point in $\big(i(k-1-d), i(k-d)\big)$. The class has parameter $M = \frac{d-1}{2}$. A lift of the class ${}_{pb}(a_2^{-d})_{pb}$ is obtained by inverting the orientation. See Fig. 9.4a for an elementary slalom curve with $d = 3$ that is the lift under $f_1 \circ f_2$ of a representative of ${}_{pb}(a_2^{-3})_{pb}$ with initial point not equal to 0.

The estimate for the extremal length follows from formula (9.7), since

$$\frac{2}{\pi} \log(\sqrt{M} + \sqrt{M+1})^2 = \frac{2}{\pi} \log(M + M + 1 + 2\sqrt{M(M+1)})$$
$$= \frac{2}{\pi} \log(d + 2\sqrt{\frac{d-1}{2} \cdot \frac{d+1}{2}}) = \frac{2}{\pi} \log(d + \sqrt{d^2 - 1}) . \tag{9.9}$$

Proof of Statement 3 for Syllables of the Form a_j^n, $|n| \geq 1$, ***with Mixed Boundary Values***

Let $n = d > 0$. For each $k \in \mathbb{Z}$ the class ${}_{pb}(a_1^n)_{tr} \in {}^{pb}\pi_1^{tr}$ lifts under the covering $f_1 \circ f_2$ to a class of elementary homotopy half-slalom curves with initial point in $(ik, i(k+1))$ and terminal point in $i(k + d + \frac{1}{2}) + \mathbb{R}$. The half-slalom class has parameter $M = d - \frac{1}{2}$, and can be represented by a curve contained in the closed left half-plane. In the remaining cases ($n < 0$, or a_1 replaced by a_2, or both) a lift of the curve can be obtained from the present one by suitable choices of inverting

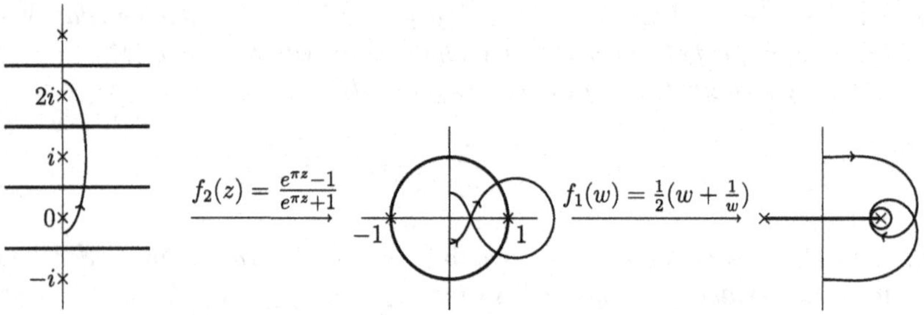

Figure 9.4a

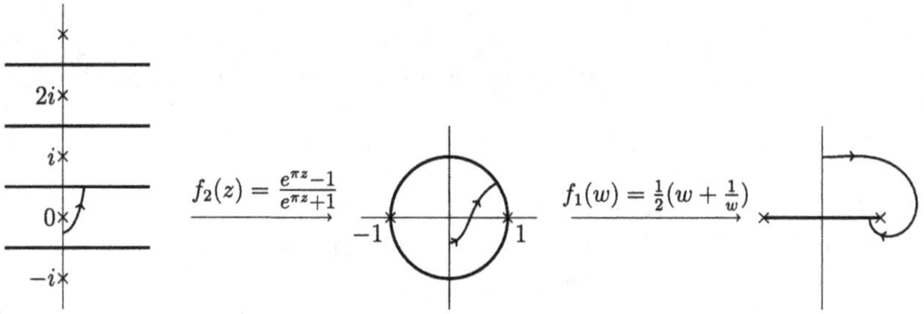

Figure 9.4b

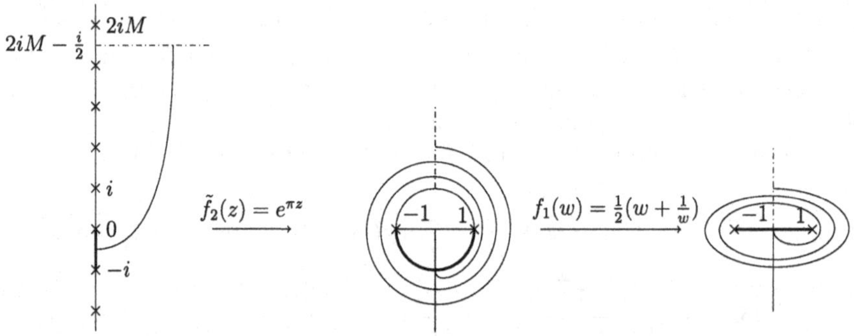

Figure 9.4c

Fig. 9.4 Some elementary slalom curves and their projections under different coverings

orientation or reflection in the imaginary line or both. See Fig. 9.4b for ${}_{pb}(a_2^{-1})_{tr}$ with $d = 1$.

The estimate for the extremal length follows by Lemma 9.2, (9.8).

The case ${}_{tr}(a_j^n)_{pb}$ is treated symmetrically.

Proof of Statement 3 for Syllables of Form (2) with Mixed Boundary Values Recall that for each integer k the mapping $f_1 \circ \tilde{f_2}$ takes $(ik, i(k+1))$ onto $(-1, 1)$, and $\frac{i}{2} + ki + \mathbb{R}$ onto $i(0, \infty)$, if k is even and onto $-i(0, \infty)$, if k is odd. Hence, for each $k \in \mathbb{Z}$ the class ${}_{tr}(\mathfrak{s})_{pb} = {}_{tr}(a_1\, a_2 \ldots)_{pb}$ lifts under $f_1 \circ \tilde{f_2}$ to an elementary homotopy half-slalom class in the closed left half-plane with initial point in $((2k-1)i, 2ki)$ and terminal point in $i(2k - d - \frac{1}{2}) + \mathbb{R}$, and it lifts also under $f_1 \circ \tilde{f_2}$ to an elementary homotopy half-slalom class in the closed right half-plane with initial point in $((2ki, (2k+1)i)$ and terminal point in $i(2k+d+\frac{1}{2})+\mathbb{R}$. Here d is the number of letters in $\mathfrak{s}$. See also Proposition 9.2, as well as Fig. 9.1. The class has parameter $M = d - \frac{1}{2}$.

For each $k \in \mathbb{Z}$ the class ${}_{tr}(\mathfrak{s})_{pb} = {}_{tr}(a_2\, a_1 \ldots)_{pb}$ lifts under $f_1 \circ \tilde{f_2}$ to an elementary homotopy half-slalom class in the closed right half-plane with initial point in $((2k-1)i, 2ki)$ and with terminal point in $i(2k + d - \frac{1}{2}) + \mathbb{R}$, and it also lifts under $f_1 \circ \tilde{f_2}$ to an elementary homotopy half-slalom class in the closed left half-plane with initial point in $(2ki, (2k+1)i)$ and with terminal point in $i(2k-d+\frac{1}{2})+\mathbb{R}$. Again d is the degree of the elementary word. See Fig. 9.4c for ${}_{tr}(a_2\, a_1 a_2 a_1 a_2)_{pb}$ with number of half-turns equal to $d = 5$.

If all generators enter with power -1 the orientation is reversed. In all cases we obtain elementary half-slalom curves with parameter $M = d - \frac{1}{2}$. The estimate of the extremal length follows from (9.8).

Proof of Statement 2 (Syllables of Form (2) with tr Boundary Values) The proof of Statement 2 is related to the proof of Statement 1 in the same way as the proof of Statement 3 for syllables of form (2) is related to the proof of Statement 3 for syllables of form (1). We leave it to the reader.

The proposition is proved. □

9.4 The Extremal Length of Words in π_1. The Lower Bound

Take an element of a relative fundamental group of the twice punctured complex plane. We will break representing curves into elementary pieces. The pieces will be chosen so that we have a good lower bound of the extremal length of the homotopy class of each piece. Ahlfors' Theorem B and Corollary 3.1 will give a lower bound for the extremal length of the element of the relative fundamental group by the sum of the extremal lengths of the classes of the elementary pieces.

We will use the following terminology. Let R be a rectangle. By a curvilinear subrectangle R' of R we mean a simply connected domain in R whose boundary looks as follows. It consist of two vertical segments, one in each vertical side of R, and either two disjoint simple arcs with interior contained in R and endpoints on opposite open vertical sides of R, or one such arc and a horizontal side of the rectangle R. The rectangle R itself may also be considered as a curvilinear subrectangle of R. The vertical segments in the boundary of the curvilinear rectangle

R' are considered as its vertical curvilinear sides, the remaining curvilinear sides are considered as the horizontal curvilinear sides of R'. Recall that each curvilinear rectangle admits a conformal mapping onto a true rectangle whose continuous extension to the boundary takes curvilinear horizontal sides to horizontal sides and curvilinear vertical sides to vertical sides.

We will use the symbol # for the boundary values if we are free to choose either pb or tr boundary values. Let v_1 and v_2 be words in π_1. For the word ${}_\#(v)_\# = {}_\#(v_1\, v_2)_\# \in {}^\#\pi_1{}^\#$ we say that there is a sign change of exponents for the pair (v_1, v_2) if the sign of the exponent of the last term of v_1 is different from the sign of the exponent of the first term of v_2.

The following lemma is a key part for the proof of the lower bound of the extremal length of slalom classes, respectively, of elements of the relative fundamental group of $\mathbb{C} \setminus \{-1, 1\}$.

Lemma 9.4 *Consider a reduced word $w \in \pi_1$. Take a locally conformal mapping g from a neighbourhood of the closure $\bar{R}$ of a rectangle R into the twice punctured plane $\mathbb{C} \setminus \{-1, 1\}$, such that the restriction $g|R$ represents the word w with pb, tr or mixed boundary values.*

The following statements hold.

1. *If w has at least two terms, then R can be divided into a collection of pairwise disjoint curvilinear subrectangles R_j of R which are in bijection to the terms of w in such a way that the restriction of g to R_j represents the j-th term with pb boundary values if the term is not at the (right or left) end of the word and, possibly, with mixed boundary values if the term is at the end of the word.*
2. *If w has at least two syllables and among them there is at least one syllable of form (1) then R contains a collection of mutually disjoint curvilinear subrectangles R'_j of R, $j \in J$, which are in bijective correspondence to the collection of syllables $\mathfrak{s}_j$ of w that are not of the form (1) such that either the restriction of g to R'_j represents the respective syllable $\mathfrak{s}_j$ with mixed boundary values, or the restriction represents an elementary word of form (2) with one more letter than $\mathfrak{s}_j$ and mixed boundary values.*
3. *If the word has at least two syllables and all syllables $\mathfrak{s}_j$, $j = 1, \ldots, N$, of the word w are of form (2) or (3) then there is a division of R into curvilinear subrectangles R'_j of R, $j = 1, \ldots, N$, such that for $j = 1, \ldots, N-1$, the restriction $g \mid R'_j$ represents the j-th syllable with mixed boundary values. In the same way there is a decomposition of R into curvilinear subrectangles R''_j of R, $j = 1, \ldots, N$, such that for $j = 2, \ldots, N$, the restriction $g \mid R''_j$ represents the j-th syllable with mixed boundary values.*

For the proof of Statements 2 and 3 of Lemma 9.4 and we need Statements 2 and 3 of the following lemma.

Lemma 9.5

1. *Let $\hat{w}$ be a conjugacy class of elements of π_1, $\hat{w} \neq \widehat{a_j^k}$ for $j = 1$ or $j = 2$ and integers k. Take a cyclically reduced word w that represents the conjugacy class*

$\hat{w}$, and a mapping $g : A \to \mathbb{C} \setminus \{-1, 1\}$ of an annulus A to $\mathbb{C} \setminus \{-1, 1\}$, that extends to a locally conformal mapping on a neighbourhood of the closure $\bar{A}$ and represents $\hat{w}$. Then there exists a smooth arc $L^0 \subset \bar{A}$ with endpoints on different boundary components of A, such that $g \mid A \setminus L^0$ represents ${}_{pb}w_{pb}$.

2. *Suppose $g : R \to \mathbb{C} \setminus \{-1, 1\}$ extends to a locally conformal mapping in a neighbourhood of the closure $\bar{R}$ of the rectangle R and represents ${}_\#(a_{j_1}^{n_1} a_{j_2}^{n_2})_\#$, where a_{j_k} are different standard generators of π_1, and the integer numbers n_1 and n_2 have different sign. Then there exists a smooth arc $L^0 \subset \bar{R}$ with endpoints on different open vertical sides of R, such that $R \setminus L^0 = R_1 \cup R_2$ for two disjoint open curvilinear rectangles, the restriction $g \mid R_1$ represents ${}_\#(a_{j_1}^{n_1})_{tr}$, and the restriction $g \mid R_2$ represents ${}_{tr}(a_{j_2}^{n_2})_\#$.*
3. *Suppose $g : R \to \mathbb{C} \setminus \{-1, 1\}$ extends to a locally conformal mapping in a neighbourhood of the closure $\bar{R}$ of the rectangle R and represents ${}_\#(a_j^n)_\#$ for a standard generator a_j and an integer number n with $|n| > 1$. Let n_1 and n_2 be non-zero integer numbers of the same sign, such that $n = n_1 + n_2$. Then there is a smooth arc $L^0 \subset \bar{R}$ with endpoints on different open vertical sides of R, such that $R \setminus L^0 = R_1 \cup R_2$ for two disjoint open curvilinear rectangles, $g \mid R_1$ represents ${}_\#(a_j^{n_1})_{tr}$ and $g \mid R_2$ represents ${}_{tr}(a_j^{n_2})_\#$.*

We prove Lemma 9.5 after Lemma 9.4 is proved.

Proof of Lemma 9.4 We start with the **proof of Statement 1.** Notice first that for the reduced word $w = w_1^{n_1} w_2^{n_2} \dots w_k^{n_k} \in \pi_1$ the corresponding element ${}_\#(w)_\#$ can be represented by a smooth curve β' that intersects $\mathbb{R}$ and $i\mathbb{R}$ transversally and lifts under $f_1 \circ f_2$ to a slalom curve (not merely to a homotopy slalom curve) in the case of pb horizontal boundary values, or to the union of a slalom curve with some elementary half-slalom curve(s) if some horizontal boundary values of w are tr. The intersection points of β' with the imaginary axis divide β' into pieces β'_j that represent $w_j^{n_j}$, $j = 1, \dots k$. The β'_j represent elements ${}_\#(w_1^{n_1})_{pb}$ if $j = 1$, ${}_{pb}(w_j^{n_j})_{pb}$ for $j = 2, \dots, k-1$, and ${}_{pb}(w_k^{n_k})_\#$ if $j = k$.

Consider the mapping g of the lemma. By our condition 0 is a regular value of both, $\operatorname{Re} g(z)$ and $\operatorname{Im} g(z)$ in a neighbourhood of the closed rectangle $\bar{R}$. Normalize the rectangle so that we have $R = \{z \in \mathbb{C} : x \in (0, 1),\ y \in (0, \mathrm{a})\}$. Consider the mentioned curve β'.

Take a homotopy that joins the mentioned curve β' with the restriction β^0 of g to the left vertical side of R, i.e. we consider a smooth mapping h from $\overline{R'} \overset{def}{=} \{z \in \mathbb{C} : x \in [-1, 0],\ y \in [0, \mathrm{a}]\}$ to $\mathbb{C} \setminus \{-1, 1\}$, whose restriction to the left side of R' equals β' and whose restriction to the right side of R' equals β_0. Since β' and β^0 are smooth curves that intersect $(-1, 1) \cup i\mathbb{R}$ transversely we may assume that zero is a regular value of both, $\operatorname{Re} h$ and $\operatorname{Im} h$. Indeed, there is a neighbourhood U in $\overline{R'}$ of the two closed vertical sides of the rectangle such that 0 is a regular value of both, $\operatorname{Re} h$ and $\operatorname{Im} h$, on U, and we may assume, by choosing the homotopy and extending it to a neighbourhood of $\overline{R'}$ suitably, that in a neighbourhood U' in $\overline{R'}$ of the closed horizontal sides this is so. Let Ψ be a non-negative smooth function in a neighbourhood of the closed rectangle $\overline{R'}$ which equals 1 outside $U \cup U'$ and equals zero near the boundary of the rectangle. If ε is a sufficiently small complex value such that $\mathrm{Re}\,\varepsilon$ is a regular value of $\mathrm{Re}\,h$ on the closed rectangle, and $\mathrm{Im}\,\varepsilon$ is a regular value of $\mathrm{Im}\,h$ on the closed rectangle, then $h - \varepsilon\Psi$ is another smooth homotopy joining β^0 with β^1, and zero is a regular value of both, the real part and the imaginary part of this mapping on the closed rectangle. We may choose the homotopy h so that we obtain a smooth mapping $\tilde{g}$ from a neighbourhood of the closed rectangle $\overline{R'} \cup \bar{R}$ to $\mathbb{C} \setminus \{-1, 1\}$, which equals g on $\bar{R}$ and h on $\overline{R'}$ such that 0 is a regular value of both, $\mathrm{Re}\,\tilde{g}$ and $\mathrm{Im}\,\tilde{g}$. We defined $\tilde{R}$ as the interior of $\overline{R'} \cup \bar{R}$.

Each connected component of the level set $\{L_0 \overset{def}{=} z \in \tilde{R} : \mathrm{Re}\,h(z) = 0\}$ on the rectangle $\tilde{R}$ is either an open arc (i.e. it is non-compact), and both its end points are on the boundary of $\tilde{R}$, or it is a circle (i.e. it is compact). Consider the closures of all non-compact connected components of L_0. This is a collection of closed arcs. The endpoints of these arcs that are contained in the left side of the rectangle divide the left side into connected components. The restrictions of h to these components are the β'_j. The closure of each non-compact component of L_0 has at most one endpoint on the open left side of the rectangle. Indeed, assume the contrary. Then the restriction of β' to the interval I on the left side of the rectangle between the two endpoints of the arc is equal to the product of at least one or more successive curves ${}_{pb}(\beta'_j)_{pb}$ and, hence is not homotopic to a constant curve through curves with pb boundary values. On the other hand, the existence of an arc in the level set $\mathrm{Re}\,h = 0$ joining the two endpoints of I would provide a relative homotopy in $\mathbb{C} \setminus \{-1, 1\}$ (with endpoints in the imaginary axis) joining the restriction of h to the interval I with a constant curve contained in the imaginary axis, which is a contradiction.

The same reasoning shows that the closure of each non-compact component of L_0 with one endpoint on the open left side of the rectangle cannot have its other endpoint on a closed horizontal side of R. Indeed, otherwise the restriction of h to the interval on the vertical side between the endpoint of the arc and the vertex of the rectangle belonging to the closed horizontal side would be homotopic to a constant through curves with endpoints in $(-1, 1) \cup i\mathbb{R}$, which is impossible.

We may ignore the non-compact components of L_0 whose closures have no endpoint on the open left side of R, and will also ignore the circles contained in L_0. We consider the connected components of L_0 whose closures have one endpoint on the open left side of R, and the other endpoint on the open right side. These arcs divide the rectangle into curvilinear rectangles which are in bijective correspondence to the intervals of division on the left side. We call them dividing arcs. Take the curvilinear rectangle whose left side corresponds to β'_j. The restriction of h to this curvilinear rectangle provides a homotopy with boundary values in the imaginary axes (in $(-1, 1) \cup i\mathbb{R}$, respectively), that joins β'_j to the restriction β^1_j of h to the right side of the curvilinear rectangle $\tilde{R}$. Since the division of the rectangle $\tilde{R}$ into curvilinear rectangles induces a division of the right side of the rectangle into intervals, the curve β^1, which is the restriction of g to the right side of R, is the composition of the curves β^1_j. Denote by E the subset of the right side of R that consists of the endpoints of the mentioned intervals.

Consider all connected components of the level set $\{z \in R : \operatorname{Re} g = 0\}$ in the original rectangle R that have an endpoint on the right side of R contained in E. Each such component is a part of a dividing arc for $\tilde{R}$. Hence, each such component has its other endpoint on the left side of the original rectangle R. We call these components the dividing arcs for R. The dividing arcs for R provide a division of the rectangle R into curvilinear rectangles R_j. The right side of each R_j is a connected component I_j of the complement of E in the right side of R and the restriction of g to I_j equals β_j^1. Since β_j^1 represents the term $w_j^{n_j}$ with the required horizontal boundary conditions, we obtained the required collection of curvilinear rectangles R_j. Statement 1 is proved.

Proof of Statement 2 Choose a syllable $\mathfrak{s}_k$ of form (1). Suppose there are syllables on the left of $\mathfrak{s}_k$ which are not of form (1). Consider them in the order from left to right.

If the left boundary values of w are pb then we consider the most left syllable $\mathfrak{s}_{j_1}$ with $j_1 < k$ which is not of form (1). For the curvilinear rectangle R_{j_1} associated to $\mathfrak{s}_{j_1}$ the restriction $g \mid R_{j_1}$ has pb left boundary values.

Suppose the next syllable $\mathfrak{s}_{j_1+1}$ to the right of $\mathfrak{s}_{j_1}$ is not of form (1). Then $j_1 + 1 < k$ and there is a sign change of exponents for the pair $(\mathfrak{s}_{j_1}, \mathfrak{s}_{j_1+1})$ of consecutive syllables. For the curvilinear rectangles R_j of Statement 1 the restriction of g to $R_{j_1,j_1+1} \overset{def}{=} \operatorname{Int}(\overline{R}_{j_1} \cup \overline{R}_{j_1+1})$ represents ${}_{pb}(\mathfrak{s}_{j_1}\mathfrak{s}_{j_1+1})_{pb}$. (Recall that Int$X$ denotes the interior of a subset X of a topological space.) By Lemma 9.5, Statement 2 the curvilinear rectangle R_{j_1,j_1+1} can be divided into two curvilinear rectangles such that the restriction of g to them represents the syllables ${}_{pb}(\mathfrak{s}_{j_1})_{tr}$ and ${}_{tr}(\mathfrak{s}_{j_1+1})_{pb}$. We obtained the required representation of the two syllables $\mathfrak{s}_{j_1}$ and $\mathfrak{s}_{j_1+1}$ which are both not of form (1).

Suppose the next syllable $\mathfrak{s}_{j_1+1}$ to the right of $\mathfrak{s}_{j_1}$ is of form (1), i.e it equals $w_{j_1+1}^n$ for an integer n, $|n| \geq 2$, with w_{j_1+1} being a standard generator of π_1. We include here the case when $j_1 + 1 = k$. If there is a sign change of exponents for the pair $(\mathfrak{s}_{j_1}, w_{j_1+1}^n)$ the preceding argument applies. Suppose there is no sign change. By Lemma 9.5 Statement 3 the curvilinear rectangle R_{j_1,j_1+1} can be divided into two curvilinear rectangles such that the restrictions of g to them represent the syllables ${}_{pb}(\mathfrak{s}_{j_1} w_{j_1+1}^{\operatorname{sgn}(n)})_{tr}$ and ${}_{tr}(w_{j_1+1}^{n-\operatorname{sgn}(n)})_{pb}$ with mixed boundary values. Since there is no sign change, the word $s_{j_1} w_{j_1+1}^{\operatorname{sgn}(n)}$ is of form (2).

If the left boundary values of w are tr then $g \mid R_1$ represents $\mathfrak{s}_1$ with mixed boundary values. Consider the most left syllable $\mathfrak{s}_{j_1}$ with $1 < j_1 < k$ which is not of form (1) (if there is any) and proceed as in the previous case.

The inductive procedure. We obtained the following facts. If there is a syllable $\mathfrak{s}_{j_1}$ not of form (1) on the left of $\mathfrak{s}_k$ then Statement 2 holds for all syllables not of form (1) with label $j \leq j_1 + 1$. The disjoint curvilinear rectangles contained in R used for representing these syllables (or syllables with one letter more) are contained in the union of the closure of the rectangles R_j, $1 \leq j \leq j_1 + 1$, of Statement 1 of the lemma. Notice that the restriction $g \mid \operatorname{Int}(\bigcup_{j=1}^{j_1+1} \overline{R_j})$ has pb right boundary values.

We proceed by induction as follows. Suppose for some $l < k$ we achieved the following. We found disjoint curvilinear rectangles contained in $\bigcup_{j \le l} \overline{R}_j$ which are in bijective correspondence to all syllables $\mathfrak{s}_j$ with $j \le l$ that are not of form (1) such that the restrictions of g to these rectangles represent the syllable, or a syllable with one more letter, with mixed boundary values. Moreover, the restriction $g \mid \mathrm{Int}(\bigcup_{j=1}^{l} \overline{R_j})$ has pb right boundary values.

Consider the first syllable $\mathfrak{s}_{j_2}$ with $j_2 < k$ on the right of $\mathfrak{s}_l$ with $\mathfrak{s}_{j_2}$ not of form (1) (if there is any). If $j_2 + 1 < k$ we proceed with $\mathfrak{s}_{j_2}$ in the same way as it was done for $\mathfrak{s}_{j_1}$ in the case $j_1 + 1 < k$ and continue the process. If there is no such syllable we stop the process.

Make the same procedure from right to left until each syllable not of form (1) on the right of $\mathfrak{s}_k$ is represented in the desired way. The curvilinear rectangles obtained by the construction which starts from the left do not intersect the curvilinear rectangles obtained by the construction which starts from the right because $\mathfrak{s}_k = w_k^{n_k}$ with $|n_k| \ge 2$. Statement 2 is proved.

Proof of Statement 3 Under the conditions of Statement 3 there is a sign change of exponents for any pair of consecutive syllables.

If the left boundary values of w are pb we may consider curvilinear rectangles $R_{2j-1,2j}$, $j = 1, 2, \ldots$, such that the restriction of g to $R_{2j-1,2j}$, $j = 1, 2, \ldots$, represents ${}_{pb}(\mathfrak{s}_{2j-1}\, \mathfrak{s}_{2j})_{pb}$. By Lemma 9.5 Statement 2 there exists a smooth arc that divides $R_{2j-1,2j}$ into two curvilinear rectangles such that the restriction of g to the first curvilinear rectangle R'_{2j-1} represents the syllable $\mathfrak{s}_{2j-1}$ with right tr boundary values and the restriction of g to the second curvilinear rectangle R'_{2j} represents the syllable with left tr boundary values. In this way each syllable except, maybe, the last one is represented with mixed boundary values by restricting g to a member of the collection of the obtained pairwise disjoint curvilinear rectangles.

If the left boundary values of w are tr then $g \mid R_1$ represents $\mathfrak{s}_1$ with mixed boundary values and we consider instead the rectangles $R_{2j,2j+1}$, $j = 1, \ldots$. We obtained the collection R'_j for both cases of the left boundary values.

Repeating the procedure from right to left gives the rectangles R''_j. This finishes the proof of Statement 3. □

Proof of Lemma 9.5, Statement 1 We may assume that the annulus in the Statement 1 equals $A = \{z \in \mathbb{C} : 2 < |z| < r\}$ for a positive number $r > 2$. Take a smooth closed curve $\beta' : \partial\mathbb{D} \to \mathbb{C} \setminus \{-1, 1\}$ that represents $\hat{w}$, intersects $i\mathbb{R}$ transversally, and has minimal possible number of intersection points with the imaginary axis. Then the intersection points of β' with the imaginary axis divide the curve β' into curves ${}_{pb}(\beta'_l)_{pb}$ whose interiors do not intersect $i\mathbb{R}$. No ${}_{pb}(\beta'_l)_{pb}$ is homotopic with endpoints in $i\mathbb{R}$ to a constant curve, since otherwise the curve β would be homotopic to a curve with smaller number of intersection points with $i\mathbb{R}$. Then the ${}_{pb}(\beta'_l)_{pb}$ represent words of the form ${}_{pb}(a_{j_l})^k_{pb}$, and neighbouring ${}_{pb}(\beta'_l)_{pb}$ on the closed curve β' are powers of different generators $a^k_{j_l}$ among the a_1 and a_2. Since the word w is cyclically reduced, we may suppose (after perhaps relabeling the β'_l) that $w = \beta'_1\, \beta'_2 \ldots \beta'_N$.

The proof follows now along the same lines as the proof of Statement 1 of Lemma 9.4. We consider a smooth homotopy that joins the curves β' and $g \mid \{|z| = 2\}$, more precisely, we consider a smooth mapping $h : \{1 \le |z| \le r\} \to \mathbb{C} \setminus \{-1, 1\}$ for which zero is a regular value of the real part, whose restriction to $\{|z| = 1\}$ coincides with β', and whose restriction to $\bar{A}$ equals g. Similarly as in the proof of Lemma 9.4 the division of β' into ${}_{pb}(\beta'_l)_{pb}$ induces a division of the annulus A into curvilinear rectangles R'_l by smooth arcs in A, each with endpoints on different boundary circles of A and contained in the preimage $g^{-1}(i\mathbb{R})$. The restriction $g \mid R'_l$ represents ${}_{pb}(\beta'_l)_{pb}$. The arc that is the common part of the boundaries of R'_1 and R'_N satisfies the requirement of Statement 1 of the lemma.

Proof of Statement 2. Since there is a sign change of exponents for the two terms of the word, we may represent ${}_{\#}(a_{j_1}^{n_1} a_{j_2}^{n_2})_{\#}$ by a smooth curve that intersects the real axis transversally and equals the product ${}_{\#}(\beta'_1)_{tr}\, {}_{tr}(\beta'_2)_{tr} \cdots {}_{tr}(\beta'_{n_1+n_2})_{\#}$ of curves, where each of the first j_1 curves represents $a_{j_1}^{\text{sgn}(n_1)}$ and the remaining curves represent $a_{j_2}^{\text{sgn}(n_2)}$, and the interior of each curve does not intersect $\mathbb{R}$. The proof follows now along the lines of proof of Statement 1 of Lemma 9.4.

Proof of Statement 3 We choose a representative β' of ${}_{\#}(a_j^n)_{\#}$ that has minimal number of intersection points with the real axis. The curve β' can be written as ${}_{\#}(\beta'_1)_{tr} \cdots {}_{tr}(\beta'_{n-1})_{\#}$ for curves β'_j that represent ${}_{\#}(a_j)_{\#}$ whose interior does not intersect $(-1, 1)$. The proof now follows along the same lines of arguments as in the proof of Statement 1 of Lemma 9.4.
The lemma is proved. □

We will now give the proof of the lower bound in Theorem 9.1 and in Theorem 9.2. The plan is the following. We consider a locally conformal mapping in a neighbourhood of a closed rectangle R, that represents an element $w \in \pi_1^{pb}$. The rectangle will be covered by the closures of curvilinear subrectangles for which the following holds. Each point of R is contained in at most two subrectangles. The restriction of the mapping to each subrectangle R_j represents a syllable $\mathfrak{s}_j$ for which the horizontal boundary values are chosen so that Proposition 9.3 provides a suitable lower bound of the extremal length of the subrectangle in terms of $\mathcal{L}(\mathfrak{s}_j)$.

Figure 9.5 illustrates a mapping g from a rectangle R to $\mathbb{C} \setminus i\mathbb{Z}$ which represents a lift under $f_1 \circ f_2$ of an element $w \in \pi_1$ with pb boundary values. The rectangle R is covered by relatively closed curvilinear rectangles. For each curvilinear rectangle we indicate the word with suitable horizontal boundary conditions (the "building block") whose lift under $f_1 \circ f_2$ is represented by the restriction of g to the curvilinear rectangle. Notice that for some choices of the boundary values the curvilinear rectangles may intersect, but each point of R is contained in no more than two relatively closed curvilinear rectangles.

The right part of the figure shows the image $g(R)$ of R in $\mathbb{C} \setminus i\mathbb{Z}$ and a slalom curve that lifts a representative of w. For the elementary pieces of the slalom curve we indicate the element of π_1 a lift of which the piece represents.

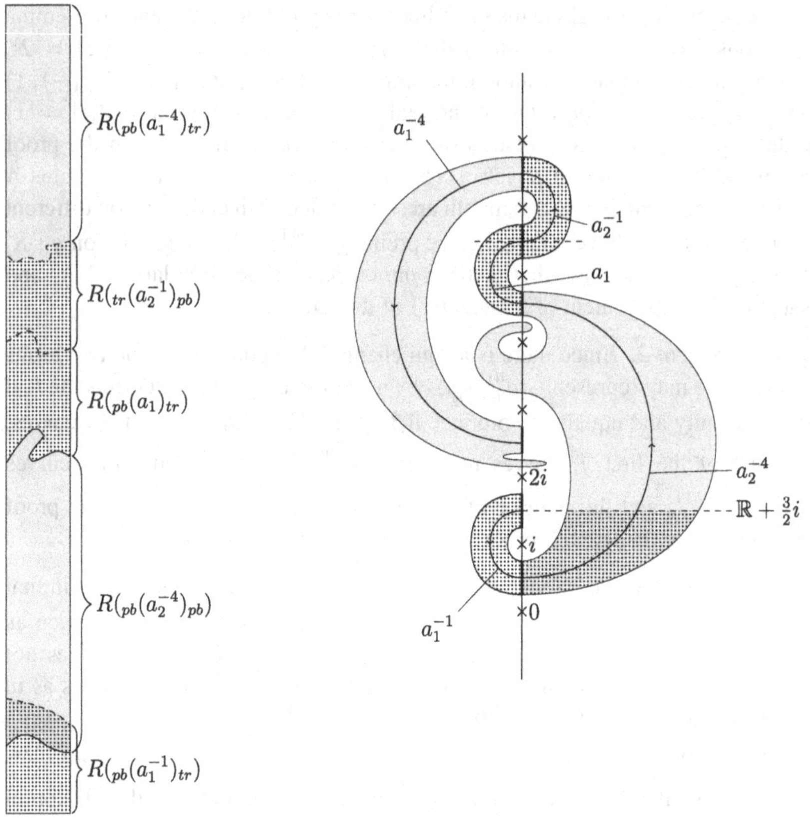

Fig. 9.5 A representative of a lift to $\mathbb{C} \setminus i\mathbb{Z}$ of an element of π_1 and building blocks

Proof of the Lower Bound in Theorem 9.1 and in Theorem 9.2 By Lemma 9.3 the extremal length of an element of a relative fundamental group of $\mathbb{C} \setminus \{-1, 1\}$ is attained on a locally conformal mapping g from an open rectangle R with sides parallel to the axes into $\mathbb{C} \setminus \{-1, 1\}$, which represents the word w in the relative fundamental group of the twice punctured complex plane with tr, pb, or mixed horizontal boundary values. Since the continuous extension of g maps each horizontal side to the real axis or to the imaginary axis, we may extend g holomorphically through the horizontal sides. Shrinking the rectangle slightly in the horizontal direction (and hence, increasing the extremal length slightly), we may assume, that the mapping g is locally conformal in a neighbourhood of the closed rectangle $\bar{R}$.

We assume first that the word has at least two syllables, and the word has at least one syllable of form (1). Let I be the set of the natural numbers j for which the j-th term of w is a syllable of form (1) and let R_j, $j \in I$, be the curvilinear rectangles of Statement (1) of Lemma 9.4.

By Corollary 3.1

$$\sum_{j \in I} \lambda(R_j) \le \lambda(R).$$

The rectangle R_j admits the holomorphic mapping $g \mid R_j$ to $\mathbb{C} \setminus \{-1, 1\}$ which represents the syllable corresponding to R_j with *pb* boundary values if the syllable is not at the end of the word, and, possibly, with mixed boundary values if the syllable is at the end of the word. Let d_j be the degree of the j-th syllable. By Proposition 9.3 the extremal length of R_j is not smaller than $\frac{1}{\pi}\log(2d_j + \sqrt{4d_j^2 - 1})$ if $g \mid R_j$ has mixed boundary values and $d_j \ge 1$, and it is not smaller than

$$\frac{2}{\pi}\log(d_j + \sqrt{d_j^2 - 1}) \ge \frac{1}{\pi}\log(2d_j + \sqrt{4d_j^2 - 1}) \tag{9.10}$$

if $g \mid R_j$ has *pb* boundary values and $d_j \ge 2$.

We obtain

$$\frac{1}{2}\lambda(R) \ge \frac{1}{2}\sum_{j \in I} \lambda(R_j) \ge \frac{1}{2}\sum_{j \in I} \frac{1}{\pi}\log(2d_j + \sqrt{4d_j^2 - 1}). \tag{9.11}$$

Suppose as before that w contains at least one syllable of form (1). Denote by J the set of all natural numbers j for which $\mathfrak{s}_j$ is not of form (1). For $j \in J$ we denote by R'_j the curvilinear rectangle of Statement 2 of Lemma 9.4 corresponding to $\mathfrak{s}_j$ (or to an elementary word of form (2) with one more letter than $\mathfrak{s}_j$).

Since $g \mid R'_j$ has mixed boundary values, Proposition 9.3 implies that the extremal length of R'_j is not smaller than $\frac{1}{\pi}\log(2d_j + \sqrt{4d_j^2 - 1})$. (If R'_j represents an elementary word of form (2) with one letter more than $\mathfrak{s}_j$ then the lower bound is even bigger, namely, d_j may be replaced by $d_j + 1$.) As before $d_j \ge 1$ is the degree of the syllable $\mathfrak{s}_j$. Hence,

$$\frac{1}{2}\lambda(R) \ge \frac{1}{2}\sum_{j \in J} \lambda(R'_j) \ge \sum_{j \in J} \frac{1}{2\pi}\log(2d_j + \sqrt{4d_j^2 - 1}). \tag{9.12}$$

For the case when w contains syllables of form (1) and syllables not of form (1) we add the two inequalities (9.11) and (9.12). We obtain

$$\begin{aligned}\lambda(R) &\ge \sum_{j \in J} \frac{1}{2\pi}\log(2d_j + \sqrt{4d_j^2 - 1}) + \sum_{j \in I} \frac{1}{2\pi}\log(2d_j + \sqrt{4d_j^2 - 1}) \\ &= \sum_{\mathfrak{s}_j} \frac{1}{2\pi}\log(2d_j + \sqrt{4d_j^2 - 1}).\end{aligned} \tag{9.13}$$

The last sum is extended over all syllables $\mathfrak{s}_j$ of the word w, and d_j is the degree of $\mathfrak{s}_j$.

Suppose that the word does not contain syllables of form (1) and the syllables are labeled from left to right by $j = 1, \ldots, N$. Let R_j' be the rectangles from Statement 3 of Lemma 9.4. Since $g \mid R_j'$ represents $\mathfrak{s}_j$ with mixed boundary values we obtain

$$\lambda(R) \geq \sum_{j=1}^{N-1} \lambda(R_j') \geq \sum_{j=1}^{N-1} \frac{1}{\pi} \log(2d_j + \sqrt{4d_j^2 - 1})\,. \tag{9.14}$$

On the other hand, with the curvilinear rectangles R_j'' from Statement 3 of Lemma 9.4 we obtain the inequality

$$\lambda(R) \geq \sum_{j=2}^{N} \lambda(R_j'') \geq \sum_{j=2}^{N} \frac{1}{\pi} \log(2d_j + \sqrt{4d_j^2 - 1})\,. \tag{9.15}$$

It follows that for any word $w \in \pi_1$ with at least two syllables for the respective family Γ the inequality

$$\lambda(R) \geq \sum_{\mathfrak{s}_j} \frac{1}{2\pi} \log(2d_j + \sqrt{4d_j^2 - 1}) \tag{9.16}$$

holds.

If the word consists of a single syllable, Proposition 9.3 implies (9.16) in the case of mixed boundary values as well as in the non-exceptional cases with both boundary values being tr or pb.

Consider the exceptional cases. If $w = a_1^n$ with $n > 0$ the mapping $\zeta \to -1 + e^{\zeta}$, $\zeta \in R$, with $R = \{\xi + i\eta,\ \xi \in (-\infty, 0),\ \eta \in (0, 2\pi n)\}$, represents w_{tr}. Hence, $\Lambda(w_{tr}) = 0$.

If $w = a_1 a_2 \ldots$ and has degree $d \geq 2$ then the mapping $\zeta \to e^{\zeta}$, $\zeta \in R$, with $R = \{\xi + i\eta,\ \xi \in (1, \infty),\ \eta \in (\frac{\pi}{2}, \frac{\pi}{2} + \pi d)\}$, represents w_{pb}. Hence, $\Lambda(w_{pb}) = 0$.

The other exceptional cases are similar. The lower bound in Theorem 9.1 and in Theorem 9.2 is proved. □

It will be useful to have in mind the following corollary.

Corollary 9.1 *Let $\hat{w}$ be a conjugacy class of elements of π_1 that is not among the exceptional cases of Theorem 9.3.*
Then for any cyclically reduced representative $w \in \pi_1$ of $\hat{w}$ the inequality

$$\Lambda(w_{pb}) \leq \Lambda(\hat{w}) \tag{9.17}$$

holds. The inequality may be strict.

Further, let w be any representative of $\hat{w}$, that is obtained by a cyclic permutation of a cyclically reduced representative of $\hat{w}$ and has one of the following properties. Either the first and the last term of w are powers of the same sign of the same standard generator of π, or the first and the last term of w are powers of different sign of different standard generators of π_1. Then

$$\Lambda(w_{tr}) \leq \Lambda(\hat{w}) . \tag{9.18}$$

The inequality may be strict.

Proof Let $\hat{w} \neq \widehat{a_j^k}$ and $\hat{g}$ a locally conformal mapping from an annulus A into the twice punctured complex plane $\mathbb{C} \setminus \{-1, 1\}$ that represents $\hat{w}$. We may assume that $\hat{g}$ is locally conformal in a neighbourhood of $\bar{A}$ and $\lambda(A) < \Lambda(\hat{w}) - \varepsilon$ for an a priori given small positive number ε . By Statement 1 of Lemma 9.5 there is a smooth arc $L_0 \subset \bar{A}$ with endpoints on different boundary circles of A, such that $\hat{g} \mid A \setminus L^0$ represents the cyclically reduced word w with pb boundary values. Hence, by Corollary 3.1 the inequality $\Lambda(w_{pb}) \leq \lambda(A \setminus L^0) \leq \lambda(A) \leq \Lambda(\hat{w}) - \varepsilon$ holds. This proves inequality (9.17).

Inequality (9.18) is proved in the same way using Corollary 3.1 and the analog of Statement 2 or Statement 3 of Lemma 9.5 for annuli instead of rectangles. Examples 3.4 and 3.5 of Section 4.2 show that the inequality may be strict. □

Proof of the Lower Bound of $\Lambda(\hat{w})$ in Theorem 9.3 The lower bound of $\Lambda(\hat{w})$ in Theorem 9.3 for the non-exceptional case follows immediately from Lemma 9.5, Statement 1, Theorem 9.1, and Corollary 3.1.

Similarly as in the proof of the lower bound in Theorem 9.1 we see that in the exceptional cases the extremal length of $\hat{b}$ equals zero. □

9.5 The Upper Bound of the Extremal Length for Words Whose Syllables Have Degree 2

We first obtain the upper bound of the extremal length of reduced words of the form

$$w = a_1^{\pm 2}\, a_2^{\pm 2} \, \dots \tag{9.19}$$

with at least two syllables and pb, tr, or mixed horizontal boundary values, because the proof in this case avoids technical details and gives a better estimate than in the general case. We will use this estimate in Chap. 11. The proof in the general situation is given in Sect. 9.6.

We first represent the lift under $f_1 \circ f_2$ of the elementary word ${}_{pb}(a_2^{-2})_{pb}$ by a holomorphic mapping $g_{a_2^{-2}}$ of a rectangle $R_{a_2^{-2}}$ into $\mathbb{C} \setminus i\mathbb{Z}$.

Consider the rectangle $R_{a_2^{-2}} = \mathrm{R}$ in the plane with vertices $\pm\frac{1}{5} \pm i\frac{\pi}{2}$ and the mapping

$$g_{a_2^{-2}}(\zeta) = \frac{i}{2} + e^{\zeta},\ \zeta \in \mathrm{R}. \tag{9.20}$$

The mapping $g_{a_2^{-2}}$ takes the rectangle R conformally onto a half-annulus in the right half-plane. It maps the point $-\frac{\pi i}{2}$ to $-\frac{i}{2}$, and maps $\frac{\pi i}{2}$ to $\frac{3i}{2}$. The upper side of the rectangle is mapped to the interval $\frac{i}{2} + i(e^{-\frac{1}{5}}, e^{\frac{1}{5}})$. This interval is contained in $(i, 2i)$, and has distance to the endpoints of this interval equal to $\min\{e^{-\frac{1}{5}} - \frac{1}{2}, \frac{3}{2} - e^{\frac{1}{5}}\} \geq 0.25$. The lower side is mapped to the interval $\frac{i}{2} - i(e^{-\frac{1}{5}}, e^{\frac{1}{5}})$, that is contained in $(-i, 0)$ and has distance from the endpoints of $(-i, 0)$ at least equal to 0.25. Hence, the mapping represents a lift of ${}_{pb}(a_2^{-2})_{pb}$ under $f_1 \circ f_2$. Its image has distance at least $\frac{1}{4}$ to $i\mathbb{Z}$. Moreover, $g'_{a_2^{-2}}(-\frac{\pi i}{2}) = -i$, $g'_{a_2^{-2}}(\frac{\pi i}{2}) = i$, $\mathrm{vsl}(\mathrm{R}) = \pi$, $\mathrm{hsl}(\mathrm{R}) = \frac{2}{5}$. The mapping $g_{a_2^{-2}}$ extends holomorphically across the horizontal sides of R. The extension is denoted again by $g_{a_2^{-2}}$.

A lift of the elementary word ${}_{pb}(a_2^{2})_{pb}$ under $f_1 \circ f_2$ is represented by the mapping $g_{a_2^{2}}(\zeta) \overset{def}{=} g_{a_2^{-2}}(-\zeta)$, $\zeta \in \mathrm{R}$, whose derivatives at $\pm\frac{\pi i}{2}$ coincide with those of $g_{a_2^{-2}}$ at these points, and lifts of the words ${}_{pb}(a_1^{\pm 2})_{pb}$ under $f_1 \circ f_2$ are represented by the mappings $g_{a_1^{\pm 2}}(\zeta) \overset{def}{=} -g_{a_2^{2}}(\pm\zeta)$, $\zeta \in \mathrm{R}$, whose derivatives at $\pm\frac{\pi i}{2}$ are $\mp i$.

A lift under $f_1 \circ f_2$ of the syllable ${}_{tr}(a_2^{-2})_{pb}$ with mixed horizontal boundary values can be represented by the mapping $g_{{}_{tr}(a_2^{-2})_{pb}}(\zeta) = -\frac{i}{2} + 2e^{\frac{\zeta}{2}}$ on the rectangle $R_{{}_{tr}(a_2^{-2})_{pb}} = \mathrm{R}_{\mathrm{mix}} \overset{def}{=} \{\zeta \in \mathbb{C} : \mathrm{Im}\zeta \in (0, \pi), |\mathrm{Re}\zeta| < \frac{1}{5}\}$. Indeed, $g_{{}_{tr}(a_2^{-2})_{pb}}$ maps the rectangle $R_{{}_{tr}(a_2^{-2})_{pb}}$ into the right half-plane, 0 is mapped into $-\frac{i}{2} + \mathbb{R}$, and πi is mapped to $\frac{3i}{2}$. Moreover, the lower side of the rectangle is mapped to a relatively compact subset of $-\frac{i}{2} + \mathbb{R}$, and the upper side is mapped to the interval $-\frac{i}{2} + 2i(e^{-\frac{1}{10}}, e^{\frac{1}{10}})$. The interval is contained in $(i, 2i)$ and has distance to the endpoints of $(i, 2i)$ equal to $\min\{-\frac{3}{2} + 2e^{-\frac{1}{10}}, \frac{5}{2} - 2e^{\frac{1}{10}}\} \geq 0.25$.

For the derivative of the mapping the equality $g'_{{}_{tr}(a_2^{-2})_{pb}}(i\pi) = i$ holds. Moreover, $\mathrm{vsl}(R_{{}_{tr}(a_2^{-2})_{pb}}) = \pi$ and $\mathrm{hsl}(R_{{}_{tr}(a_2^{-2})_{pb}}) = \frac{2}{5}$.

Lifts under $f_1 \circ f_2$ of all syllables of the form (9.19) and degree 2 with mixed horizontal boundary values can be represented in a similar way by holomorphic mappings on $\mathrm{R}_{\mathrm{mix}}$.

There exists a universal constant $C' > 1$ such the following holds.

For each $j = 1, 2$, each choice of the sign, and each integer number k the compositions $f_1 \circ f_2 \circ (g_{a_j^{\pm 2}} + ik) \mid \mathsf{R}$, $f_1 \circ f_2 \circ ({}_{tr}(g_{a_j^{\pm 2}})_{pb} + ik) \mid \mathsf{R}_{\text{mix}}$, and $f_1 \circ f_2 \circ ({}_{pb}(g_{a_j^{\pm 2}})_{tr} + ik) \mid \mathsf{R}_{\text{mix}}$ have its image in the domain $\{z \in \mathbb{C} : |z| < C', |z \pm 1| > \frac{1}{C'}\}$. Indeed, $g_{a_j^{\pm 2}}(\mathsf{R})$, ${}_{pb}(g_{a_j^{\pm 2}})_{tr}(\mathsf{R}_{\text{mix}})$ and ${}_{tr}(g_{a_j^{\pm 2}})_{pb}(\mathsf{R}_{\text{mix}})$ have distance at least $\frac{1}{4}$ from $\mathbb{C} \setminus i\mathbb{Z}$, hence the statement is true for $k = 0$. For arbitrary integer numbers k it follows from the periodicity of $f_1 \circ f_2$.

Consider the word ${}_{pb}w_{pb} =_{pb} (a_1^{\pm 2} a_2^{\pm 2} \ldots)_{pb}$ with $N \geq 2$ syllables. Denote the j-th syllable by $\mathfrak{s}_j$. Represent the lift under $f_1 \circ f_2$ of each ${}_{pb}(\mathfrak{s}_j)_{pb}$ by the mapping $g_{\mathfrak{s}_j} : \mathsf{R} \to \mathbb{C} \setminus i\mathbb{Z}$ described above.

Put $R_j \overset{def}{=} \mathsf{R} + \pi i c_j$ for integer numbers c_j, so that for the midpoints ξ_j^- and ξ_j^+, respectively, of the lower and upper side of R_j the equalities $\xi_j^+ = \xi_{j+1}^-$, $j = 1, \ldots N - 1$, hold, where N is the number of syllables of the word w. Take $g_j(\zeta) \overset{def}{=} g_{\mathfrak{s}_j}(\zeta - \pi i c_j) + \pi i b_j$, $\zeta \in \overline{R_j}$, for integer numbers b_j such that $g_j(\xi_j^+) = g_{j+1}(\xi_{j+1}^-)$.

The rectangle $R_w \overset{def}{=} \text{Int}(\cup \overline{R_j})$ has vertical side length $\text{vsl}(R_w) = \pi N$ and horizontal side length $\frac{2}{5}$. The mappings $g_j : R_j \to \mathbb{C} \setminus i\mathbb{Z}$ represent ${}_{pb}(\mathfrak{s}_j)_{pb}$ and extend by the Reflection Principle holomorphically across the horizontal sides to the rectangle of thrice the vertical side length, with the same horizontal side length and the same center as R_j. The extended mappings are denoted again by g_j. The distance of the images of the extended mappings from $i\mathbb{Z}$ is the same as that of the original mappings. We will perform quasiconformal gluing of the g_j using the fact that $g_j(\xi_j^+) = g_{j+1}(\xi_{j+1}^-)$ and $g_j'(\xi_j^+) = g_{j+1}'(\xi_{j+1}^-)$. The latter statement follows from the fact that one of two consecutive syllables is a power of a_1 and the other is a power of a_2.

Consider the C^1-function χ_0 on the interval $[0, 1]$, $\chi_0(t) = 6 \int_0^t \tau(1-\tau)d\tau$. Then $\chi_0(0) = 0$, $\chi_0(1) = 1$, and $0 \leq \chi_0'(t) = 6t(1-t) \leq \frac{3}{2}$. Put $\chi(t) = \chi_0(\frac{5}{2}t)$. Define a function g_w on R_w as follows. Each point ξ in R_w, for which $|\text{Im}\,(\xi - \xi_{j'}^+)| > \frac{1}{5}$ for all $j' < N$, belongs to a single rectangle $\overline{R_j}$ (depending on ξ) and we put $g_w(\xi) \overset{def}{=} g_j(\xi)$ for such a point.

Fix a number $j < N$ and consider the set $Q_j \overset{def}{=} \{\xi \in R_w : |\text{Im}\,(\xi - \xi_j^+)| \leq \frac{1}{5}\}$. Put $\chi_j(\xi) = \chi(\text{Im}\,(\xi - \xi_j^+) + \frac{1}{5})$ for $\xi \in Q_j$. Let $g_w \overset{def}{=} (1 - \chi_j)g_j + \chi_j g_{j+1}$ on Q_j. For $\{\xi \in Q_j : \text{Im}\,\xi = \xi_j^+ - \frac{1}{5}\}$ the equalities $\chi_j(\xi) = \chi_0(0) = 0$ and $\chi_j'(\xi) = \chi_0'(0) = 0$ hold. Hence, the function g_w is C^1 smooth near such points ξ. Further, for $\{\xi \in Q_j : \text{Im}\,\xi = \xi_j^+ + \frac{1}{5}\}$ the equalities $\chi_j(\xi) = \chi_0(1) = 1$ and $\chi_j'(\xi) = \chi_0'(1) = 0$ hold, hence, the function g_w is C^1 smooth near such ξ.

Since $g_{j+1}-g_j$ and $g'_{j+1}-g'_j$ vanish at $\xi_j^+=\xi_{j+1}^-$, and the absolute value of the second derivative of g_j and of g_{j+1} on Q_j does not exceed $e^{\frac{1}{5}} < 1.222$, the estimate

$$|(g_{j+1}-g_j)(\xi)| \le 2e^{\frac{1}{5}}\frac{|\xi-\xi_j^+|^2}{2} \le e^{\frac{1}{5}}\frac{2}{25} < \frac{1}{10} \tag{9.21}$$

holds on Q_j.

Make the same definition for all but the last number j. We obtain a smooth mapping g_w from the rectangle $\overline{R_w}$ to $\mathbb{C}\setminus i\mathbb{Z}$ which represents a lift of ${}_{pb}w_{pb}$ under $f_1\circ f_2$. Since the distance of the image of each mapping g_j from $i\mathbb{Z}$ is not smaller than $\frac{1}{4}$, by inequality (9.21) the distance of the image $g_w(Q_j)$ from $i\mathbb{Z}$ is not smaller than 0.15. It follows that the image $f_1\circ f_2\circ g_w(\overline{R_w})$ is contained in the closure of a domain $\{z\in\mathbb{C}: |z|<C, |z\pm 1|>\frac{1}{C}\}$ with another universal constant C.

Lemma 9.6 *The mapping g_w is a quasiconformal mapping from R_w onto its image. The Beltramy differential μ_{g_w} of g_w has absolute value $|\mu_{g_w}|<\frac{2}{5}$.*

Proof Put $\xi=u+iv$. If $\xi\in R_w$, $|\mathrm{Im}\,(\xi-\xi_j^+)|\le\frac{1}{5}$ for some j, $1\le j<N$, then the Beltrami differential at ξ equals

$$\mu_{g_w}(\zeta)=\frac{\frac{\partial}{\partial\bar\zeta}g_w(\zeta)}{\frac{\partial}{\partial\zeta}g_w(\zeta)}=\frac{\frac{i}{2}\Big(\frac{\partial}{\partial v}\chi_j\cdot(g_{j+1}-g_j)\Big)(\zeta)}{\Big(\frac{-i}{2}\frac{\partial}{\partial v}\chi_j\cdot(g_{j+1}-g_j)+(1-\chi_j)\cdot g_j'+\chi_j\cdot g'_{j+1}\Big)(\zeta)}. \tag{9.22}$$

On the rest of the rectangle R_w the function is analytic. Notice that

$$|\frac{\partial}{\partial v}\chi_j|\le\frac{3}{2}\cdot\frac{5}{2}$$

on Q_j. By inequality (9.21) the numerator in the right hand side of (9.22) has absolute value smaller than $e^{\frac{1}{5}}\cdot\frac{3}{20}<0.1833$.

Since $\max|g_j''|\le e^{\frac{1}{5}}$ on Q_j, the inequality

$$\max\{|g_j'-g_j'(\xi_j^+)|, |g'_{j+1}-g_j'(\xi_j^+)|\}<e^{\frac{1}{5}}\cdot\frac{\sqrt{2}}{5}$$

holds on the $\frac{\sqrt{2}}{5}$-neighbourhood of ξ_j^+. Since $g_j'(\xi_j^+)=g'_{j+1}(\xi_{j+1}^-)$ has absolute value 1, the denominator of the right hand side of (9.21) is not smaller than $1-e^{\frac{1}{5}}\cdot\frac{3}{20}-e^{\frac{1}{5}}\cdot\frac{\sqrt{2}}{5}>1-0.53=0.47$.

We obtain

$$k_w = \sup_{R_w} |\mu_{g_w}(\zeta)| < \frac{0.1833}{0.47} < 0.4 .$$

The quasiconformal dilatation $K_w = \frac{1+k_w}{1-k_w}$ does not exceed $\frac{7}{3}$. □

Let ω_w be the normalized solution of the Beltrami equation

$$\frac{\partial}{\partial \bar{z}}\omega_w = \tilde{\mu}_{g_w}\frac{\partial}{\partial z}\omega_w$$

on the complex plane. Here $\tilde{\mu}_{g_w}$ equals μ_{g_w} on $\overline{R_w}$ and equals 0 outside $\overline{R_w}$. ω_w is a Hölder continuous self-homeomorphism of the complex plane. The mapping $g_w \circ {\omega_w}^{-1}$ is holomorphic on $\omega_w(R_w)$ (see [4], Chapter I C). The image $\omega_w(R_w)$ can be considered as a curvilinear rectangle. The curvilinear sides are the images of the sides of R_w. By [4] (chapter I, Theorem 3) the extremal length of $\omega_w(R_w)$ does not exceed $K_w \cdot \lambda(R_w)$. In other words, there is a conformal mapping ψ_w of a true rectangle $\mathcal{R}_w$ of extremal length not exceeding $K_w \cdot \lambda(R_w)$ onto $\omega_w(R_w)$, which takes the sides of $\mathcal{R}_w$ to the respective curvilinear sides of $\omega_w(R_w)$. The mapping $g_w \circ {\omega_w}^{-1} \circ \psi_w : \mathcal{R}_w \to \mathbb{C} \setminus i\mathbb{Z}$ is a holomorphic mapping from the rectangle $\mathcal{R}_w$ of extremal length not exceeding $K_w \cdot \lambda(R_w)$ to $\mathbb{C} \setminus i\mathbb{Z}$ that represents a lift of ${}_{pb}(w)_{pb}$. Notice that the image of $f_1 \circ f_2 \circ g_w \circ {\omega_w}^{-1} \circ \psi_w$ is contained in $\{z \in \mathbb{C} : |z| < C, |z \pm 1| > \frac{1}{C}\}$.

We obtained the inequality

$$\lambda(\mathcal{R}_w) \le K_w \frac{\pi N}{\frac{2}{5}} \le \frac{7}{3} \cdot \frac{5}{2}\pi N = \frac{35}{6}\pi N$$

Since $\mathcal{L}(w) = N \cdot \log(4 + \sqrt{15})$ and $\frac{35\pi}{6\log(4+\sqrt{15})} < 9$ we obtain

$$\Lambda_{pb}(w) < 9\mathcal{L}(w) .$$

The case of words of the form (9.19) with *tr* or mixed boundary values is treated in the same way. For instance, the first syllable of the word may have totally real left boundary values. But it has always *pb* right boundary values. We described a holomorphic mapping from a rectangle of vertical side length π and horizontal side length $\frac{2}{5}$, that represents the lift of such a syllable under $f_1 \circ f_2$. This representing mapping has the same properties near the upper side of the rectangle as the chosen representing mapping of the lift of the syllable with *pb* boundary values. Therefore the gluing procedure is the same as in the case of *pb* boundary values. The argument concerning the last syllable is the same with left and right boundary values interchanged. We proved the following

Proposition 9.4 *If $w \in \pi_1$ is a word with $N \geq 2$ syllables, all of the form $a_1^{\pm 2}$ or $a_2^{\pm 2}$, then*

$$\Lambda_{pb}(w) \leq \frac{35}{6}\pi N < 9\mathcal{L}(w).$$

The same estimate holds for the extremal length with tr and with mixed boundary values.

The following proposition is obtained in the same way. As in the proof of Proposition 9.4 we represent lifts under $f_1 \circ f_2$ of the terms of a cyclically reduced word w, that represents a class $\hat{w}$, by holomorphic mappings with pb horizontal boundary values from rectangles R_j to $\mathbb{C} \setminus \mathbb{Z}$. We glue the rectangles together so that we obtain an annulus, and perform quasiconformal glueing of the representing holomorphic mappings. We compose the solution of a Beltrami equation on an annulus with a conformal mapping from a round annulus to a curvilinear annulus to obtain a holomorphic mapping on a round annulus that represents $\hat{w}$.

Proposition 9.5 *Let w be a cyclically reduced word with $N \geq 2$ syllables, all of the form $a_1^{\pm 2}$ or $a_2^{\pm 2}$ (in particular, w has an even number of syllables). Then the upper bound of the extremal length of the free homotopy class $\hat{w}$ is given by the following inequality*

$$\Lambda(\hat{w}) \leq \frac{35}{6}\pi N < 9\mathcal{L}(\hat{w}).$$

The proofs of the propositions gave the following slightly more comprehensive statement that will be used later.

Remark 9.1 *Let R be a rectangle in the complex plane with horizontal side length b and vertical side length a. Then for any word w of the form* (9.19) *with N terms and $6N\pi < \frac{\mathsf{a}}{\mathsf{b}}$ there exists a holomorphic map $g_w : R \to \mathbb{C} \setminus \{-1, 1\}$ that represents w with pb horizontal boundary values and has its image in $\{z \in \mathbb{C} : |z| < C, |z \pm 1| > \frac{1}{C}\}$ for a universal constant $C > 1$. Moreover, g_w takes the value zero at the midpoints of the horizontal sides of the rectangle.*

Further, for positive numbers α and δ we consider the annulus $A^{\alpha,\delta} \overset{def}{=} \{z \in \mathbb{C} : |\mathrm{Re} z| < \frac{\delta}{2}\}/(z \sim z + i\alpha)$. If $\lambda(A^{\alpha,\delta}) = \frac{\alpha}{\delta} > 6N\pi$ for an even number N, then for each word w of the form (9.19) *with N terms there exists a holomorphic mapping $\mathfrak{g}_w : (A^{\alpha,\delta}, 0/(z \sim z + i\alpha)) \to (\mathbb{C} \setminus \{-1, 1\}, 0)$ from $A^{\alpha,\delta}$ with base point $0/(z \sim z + i\alpha)$ to $\mathbb{C} \setminus \{-1, 1\}$ with base point 0, that represents $\hat{w}$. Moreover, the image of $\mathfrak{g}_w$ is contained in the domain $\{z \in \mathbb{C} : |z| < C, |z \pm 1| > \frac{1}{C}\}$ for a constant $C > 1$.*

Indeed, for instance to obtain the second statement we consider the holomorphic mapping $g_w \circ {\omega_w}^{-1} \circ \psi_w$ that represents a lift of $\hat{w}$ and was constructed in the proofs of Propositions 9.4 and 9.5. The mapping is defined on an annulus of the form $A^{\alpha,\delta}$ with $\lambda(A^{\alpha,\delta}) = \frac{\alpha}{\delta} > 6N\pi$. The image $f_1 \circ f_2 \circ g_w \circ \omega_w{}^{-1} \circ \psi_w(A^{\alpha,\delta})$ under the composition is contained in $\{z \in \mathbb{C} : |z| < C, |z \pm 1| > \frac{1}{C}\}$. Perhaps after precomposing with a rotation of the annulus $A^{\alpha,\delta}$ it maps $0 \diagup (z \sim z + i\alpha)$ to 0.

9.6 The Extremal Length of Arbitrary Words in π_1. The Upper Bound

We take an arbitrary word w with at least two syllables and write ${}_{\#}w_{\#}$ as product of its syllables ${}_{\#}w_{\#} ={}_{\#} (\mathfrak{s}_1)_{pb} \cdots {}_{pb} (\mathfrak{s}_N)_{\#}$. For each elementary word ${}_{pb}w_{pb}$ with pb horizontal boundary conditions (or for the respective word with other horizontal boundary conditions) we will represent its lift under $f_1 \circ f_2$ by a holomorphic mapping to $\mathbb{C} \setminus i\mathbb{Z}$ from a rectangle of extremal length not exceeding a constant times $\mathcal{L}(w)$. The properties of the representing mappings will allow quasiconformal gluing. The building block is the following lemma.

Lemma 9.7 *For each elementary word ${}_{pb}w_{pb} \in \pi_1^{pb}$ there exists a rectangle R_w with horizontal side length $\mathrm{hsl}(R_w) \geq \frac{1}{16\sqrt{2}}$, and vertical side length $\mathrm{vsl}(R_w) \leq \pi\mathcal{L}(w)$, and a holomorphic mapping $g_w : R_w \to \mathbb{C} \setminus i\mathbb{Z}$ with the following properties.*

There are two points ξ_w^+ and ξ_w^- in the boundary of R_w with $\mathrm{Re}\xi_w^+ = \mathrm{Re}\xi_w^-$, such that the restriction of g_w to the (vertical) line segment that joins ξ_w^- with ξ_w^+ represents a lift of ${}_{pb}w_{pb}$ under $f_1 \circ f_2$. The values of g_w at the points $\xi_w^{\pm}$ are imaginary half-integers that are not imaginary integers. The equalities for the derivatives at these points are $g_w'(\xi_w^-) = i$ if the first letter of the word is $a_1^{\pm 1}$ and $-i$ if the first letter is $a_2^{\pm 1}$, and $g_w'(\xi_w^+) = -i$ if the last letter of the word is $a_1^{\pm 1}$ and i if the last letter is $a_2^{\pm 1}$. Further, the mapping g_w extends holomorphically to the discs $|\zeta - \xi_w^{\pm}| < \frac{1}{16}$ of radius $\frac{1}{16}$ around $\xi_w^{\pm}$ and the inequality $|g_w''(\zeta)| \leq 2^8$ holds for the extended function if $|\zeta - \xi_w^{\pm}| < \frac{1}{16}$.

For each elementary word ${}_{tr}w_{pb}$ or ${}_{pb}w_{tr}$ of the form (1) or (3) with mixed horizontal boundary values there is a lift under $f_1 \circ f_2$ that can be represented by a holomorphic mapping from a rectangle with horizontal side length $\mathrm{hsl}(R_w) \geq \frac{1}{16\sqrt{2}}$, and vertical side length $\mathrm{vsl}(R_w) \leq \pi\mathcal{L}(w)$. Moreover, the horizontal side, that is mapped to $i\mathbb{R}$, contains a point for which the previous statements hold.

Proof of Lemma 9.7

1. Syllables of Form (1) with *pb* Horizontal Boundary Conditions We first consider syllables ${}_{pb}w_{pb}$ of the form ${}_{pb}(a_1^d)_{pb}$, $d \geq 2$. Let $M = \frac{d-1}{2}$. Consider the holomorphic mapping $T^{\mathsf{a},\mathsf{b}}$ from the proof of Lemma 3.4 (see Example 3.4, Sect. 3.2) with $\mathsf{a} = M$ and $\mathsf{b} = M + 1$. Denote the obtained mapping by T_M. Let $t_M = -M + \sqrt{M(M+1)}$ (compare with equality (3.12)). The inverse T_M^{-1} provides a conformal mapping from the half-annulus with center $\frac{1}{2}$ and radii $\frac{1}{2}$ and $\frac{1}{2} - t_M$ to

$\{z \in \mathbb{C} : \mathrm{Im} z > 0, \mathrm{Re} z > 0, |z - \frac{M+1}{2}| > \frac{1}{2}\}$, that takes the half-circle with diameter $(0, 1)$ to the imaginary half-axis and the half-circle with diameter $(t_M, 1 - t_M)$ to the half-circle with diameter $(M, M + 1)$. The mapping takes 0 to 0, 1 to ∞, t_M to M and $1 - t_M$ to $M + 1$. The first two conditions imply that the mapping T_M^{-1} has the form $z \to \frac{a}{c}\frac{z}{z-1}$. The last two conditions imply that $(\frac{a}{c})^2 = M(M + 1)$ and $\frac{a}{c} < 0$. This implies that

$$T_M^{-1}(z) = \sqrt{M(M+1)}\frac{z}{1-z}, \quad \text{and } T_M(z) = \frac{z}{\sqrt{M(M+1)} + z}. \tag{9.23}$$

Further, the entire mapping

$$\omega(z) = \frac{1}{2} + \frac{1}{2}e^{iz} \tag{9.24}$$

takes the rectangle $\mathring{R}_M^+$ with vertices $0, \pi, \pi + i\log\frac{1}{1-2t_M}, i\log\frac{1}{1-2t_M}$ conformally onto the half-annulus with center $\frac{1}{2}$ and radii $\frac{1}{2}$ and $\frac{1}{2} - t_M$. The lower side of the rectangle is mapped onto the larger half-circle, the upper side is mapped onto the smaller half-circle. Let $\mathring{R}_M \supset \mathring{R}_M^+$ be the rectangle that is symmetric with respect to the real axis, has the same horizontal side length $\mathrm{hsl}(\mathring{R}_M) = \mathrm{hsl}(\mathring{R}_M^+) = \pi$, and (see Eqs. (3.13), (9.9) and (9.10)) the double vertical side length

$$\begin{aligned}\mathrm{vsl}(\mathring{R}_M) =& 2\mathrm{vsl}(\mathring{R}_M^+) = 2\log\frac{1}{1-2t_M} = 2\log(\sqrt{M} + \sqrt{M+1})^2\\ =& 2\log(d + \sqrt{d^2-1}) \le 2\log(2d + \sqrt{4d^2-1}).\end{aligned} \tag{9.25}$$

The rectangle $\mathring{R}_M$ has extremal length

$$\lambda(\mathring{R}_M) = \frac{2\log\frac{1}{1-2t_M}}{\pi} \le \frac{2}{\pi}\log(2d + \sqrt{4d^2-1}) = \frac{2}{\pi}\mathcal{L}(w). \tag{9.26}$$

The mapping ω takes $\mathring{R}_M$ onto the half-annulus in $\mathbb{C}_+$ with center $\frac{1}{2}$ and radii of the circles equal to $\frac{1}{2} - t_M$ and $\frac{1}{\frac{1}{2}-t_M}$.

Denote by $\mathfrak{I}$ the mapping, that acts by multiplication with the imaginary unit, $\mathfrak{I}(z) = iz$, $z \in \mathbb{C}$. The mapping $\mathfrak{I} \circ T_M^{-1} \circ \omega = iT_M^{-1} \circ \omega$ takes $\mathring{R}_M$ conformally onto $\mathbb{C}_\ell \setminus (\{|\zeta - i(M + \frac{1}{2})| \le \frac{1}{2}\} \cup \{|\zeta + i(M + \frac{1}{2})| \le \frac{1}{2}\})$.

Recall that $\tilde{\mathbb{C}}_\ell$ is the lift of the left half-plane $\mathbb{C}_\ell$ to the first sheet of the universal covering U of $\mathbb{C} \setminus i\mathbb{Z}$. We consider the conformal mapping φ_0 from $\tilde{\mathbb{C}}_\ell$ onto $\mathbb{C}_\ell \setminus \bigcup_{k=-\infty}^{\infty}\{|z - i(k+\frac{1}{2})| \le \frac{1}{2}\}$, whose continuous extension to the boundary takes each interval $(ki, (k+1)i)^\sim$ to the half-circle $\mathbb{C}_\ell \cap \{|z - i(k + \frac{1}{2})| = \frac{1}{2}\}$ (see Sect. 9.2). Let φ be the extension of φ_0 to a conformal mapping of the universal covering U of $\mathbb{C}\setminus i\mathbb{Z}$ onto $\mathbb{C}_\ell$ (see Lemma 9.1), and let φ^{-1} be its inverse. Denote by P the covering map from the universal covering U of $\mathbb{C} \setminus i\mathbb{Z}$ to the set $\mathbb{C} \setminus i\mathbb{Z}$. The composition $(\mathsf{P}\varphi^{-1})$ of P with the inverse φ^{-1} of φ maps $\mathbb{C}_\ell$ locally conformally to $\mathbb{C} \setminus i\mathbb{Z}$. Moreover, it maps the subset $\mathfrak{H}_0 \setminus \{|z - \frac{i}{2}| < \frac{1}{2}\}$ of the half-strip $\mathfrak{H}_0 \overset{def}{=} \{z \in \mathbb{C}_\ell : 0 < \mathrm{Im} z < 1\}$ conformally onto $\mathfrak{H}_0$. Hence, by reflection in the half-circle $\{z \in \mathbb{C}_\ell : |z - \frac{i}{2}| < \frac{1}{2}\}$ and in the interval $(0, i)$, respectively, it follows that $(\mathsf{P}\varphi^{-1})$ maps $Q \overset{def}{=} \mathfrak{H}_0 \setminus (\{|z - \frac{i}{4}| < \frac{1}{4}\} \cup \{|z - \frac{3i}{4}| < \frac{1}{4}\}$ conformally onto the strip $\mathfrak{H} = \{z \in \mathbb{C} : 0 < \mathrm{Im} z < 1\}$. The mapping $\mathsf{P} \circ \varphi^{-1} \mid Q$ takes $\frac{i-1}{2}$ to $\frac{i}{2}$.

The mapping $(\mathsf{P}\varphi^{-1})_M$, $(\mathsf{P}\varphi^{-1})_M(z) \overset{def}{=} \mathsf{P}\varphi^{-1}(z - iM) + iM$ takes $\mathbb{C}_\ell \setminus \{|z \pm i(M + \frac{1}{2})| < \frac{1}{2}\}$ locally conformally into $\mathbb{C} \setminus (i\mathbb{Z} + iM)$ so that its continuous extension takes the half-circle $\mathbb{C}_\ell \cap \{|z - i(M + \frac{1}{2})| = \frac{1}{2}\}$ to $(iM, i(M + 1))$, and the half-circle $\mathbb{C}_\ell \cap \{|z + i(M + \frac{1}{2})| = \frac{1}{2}\}$ to $(-i(M + 1), -iM)$. The composition

$$\mathcal{G}_M \overset{def}{=} (\mathsf{P} \circ \varphi^{-1})_M \circ \mathfrak{I} \circ T_M^{-1} \circ \omega \tag{9.27}$$

maps $\mathring{R}_M$ locally conformally into $\mathbb{C} \setminus (i\mathbb{Z} + iM)$, and its continuous extension takes the lower side of $\mathring{R}_M$ to $(-(M + 1)i, -Mi)$ and the upper side to the interval $(iM, i(M + 1))$. See Fig. 9.6.

The mapping $\mathcal{G}_M - iM : \mathring{R}_M \to \mathbb{C} \setminus i\mathbb{Z}$ represents a lift under $\mathcal{F} \overset{def}{=} f_1 \circ f_2$ of ${}_{pb}(a_1^d)_{pb}$ (see Sect. 9.2, and also Fig. 9.1). In other words, $\mathcal{F} \circ (\mathcal{G}_M - iM)$ represents ${}_{pb}(a_1^d)_{pb}$. Let $R'_M \supset \mathring{R}_M$ be the rectangle of vertical side length $3 \cdot \mathrm{vsl}(\mathring{R}_M)$ and horizontal side length $\mathrm{hsl}(\mathring{R}_M)$ which is symmetric with respect to the real axis. By the reflection principle $\mathcal{G}_M$ extends holomorphically through each of the horizontal sides of $\mathring{R}$ to a holomorphic mapping from the rectangle R'_M into $\mathbb{C} \setminus i\mathbb{Z}$. The extension is denoted again by $\mathcal{G}_M$. (see Fig. 9.6.)

We put $q_M^{\pm} \overset{def}{=} \pm\frac{2M+1}{2} + \frac{i}{2}$. Notice that $iq_M^{\pm}$ is the midpoint of the half-circle in the left half-plane with diameter $(iM, i(M + 1)$ $((-i(M + 1), -iM)$, respectively). Let $\tilde{\xi}_M^{\pm}$ and $p_M^{\pm}$ be the points for which

$$\tilde{\xi}_M^{\pm} \xrightarrow{\ \omega\ } p_M^{\pm} \xrightarrow{\ \mathfrak{I} \circ T_M^{-1}\ } iq_M^{\pm} .$$

We will consider the inverse of $\mathfrak{I} \circ T_M^{-1} \circ \omega$, $(\mathfrak{I} \circ T_M^{-1} \circ \omega)^{-1} = \omega^{-1} \circ T_M \circ \mathfrak{I}^{-1}$.

By equality (9.23), and since for the branch of the logarithm on $\mathbb{C}_+$ with imaginary part in $(0, \pi)$ the equality $\omega^{-1}(z) = \frac{1}{i}\log(2z - 1)$ holds, we obtain

$$\omega^{-1} \circ T_M(z) = \frac{1}{i}\log(2T_M(z) - 1) = \frac{1}{i}\log\frac{-\sqrt{M(M + 1)} + z}{+\sqrt{M(M + 1)} + z} . \tag{9.28}$$

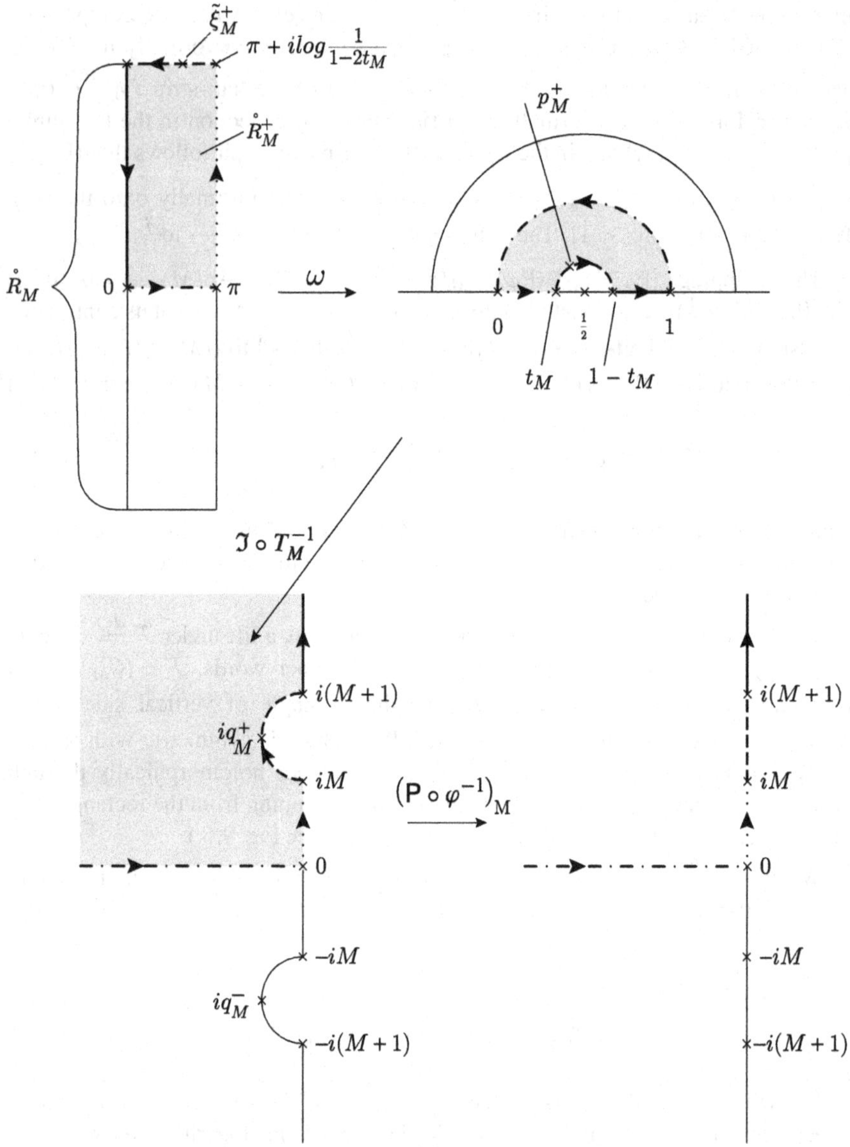

Fig. 9.6 A conformal mapping representing a lift of a syllable of form (1)

For the derivative we get

$$
\begin{aligned}
(\omega^{-1} \circ T_M)'(z) =& \frac{1}{i}\left(\frac{1}{-\sqrt{M(M+1)}+z} - \frac{1}{+\sqrt{M(M+1)}+z}\right) \\
=& i\,\frac{2\sqrt{M(M+1)}}{M(M+1)-z^2} \,.
\end{aligned} \tag{9.29}
$$

Hence,

$$
\begin{aligned}
(T_M^{-1} \circ \omega)'(\tilde{\xi}_M^+) =& \frac{1}{(\omega_M^{-1} \circ T_M)'(M + \frac{1}{2} + \frac{i}{2})} \\
=& \frac{1}{2i} \frac{M(M+1) - (M + \frac{1}{2} + \frac{i}{2})^2}{\sqrt{M(M+1)}} = -\frac{1}{2} \frac{M + \frac{1}{2}}{\sqrt{M(M+1)}} .
\end{aligned}
$$

By Eq. (9.6) the equality $(\mathsf{P} \circ \varphi_0^{-1})'(z + i) = (\mathsf{P} \circ \varphi_0^{-1})'(z)$ holds on $\mathbb{C}_\ell \setminus \bigcup_{k=-\infty}^{\infty} \{|z - i(k + \frac{1}{2})| \le \frac{1}{2}\}$, hence $(\mathsf{P} \circ \varphi^{-1})'(z + i) = (\mathsf{P} \circ \varphi^{-1})'(z)$ holds for $z \in \mathbb{C}_\ell$. This implies that

$$
\begin{aligned}
(\mathcal{G}_M)'(\tilde{\xi}_M^+) =& \big((\mathsf{P} \circ \varphi^{-1})_M \circ (iT_M^{-1}) \circ \omega\big)'(\tilde{\xi}_M^+) = \\
& (\mathsf{P} \circ \varphi^{-1})_M'(iq_M^+) \cdot (iT_M^{-1} \circ \omega)'(\tilde{\xi}_M^+) = \\
& (\mathsf{P} \circ \varphi^{-1})'(\frac{i-1}{2}) \cdot \left(-\frac{i}{2} \frac{M + \frac{1}{2}}{\sqrt{M(M+1)}}\right) .
\end{aligned} \tag{9.30}
$$

We used that $iq_M^+ + iM = -\frac{1}{2} + i(M + \frac{1}{2}) + iM$ differs by an imaginary integer from $-\frac{1}{2} + \frac{1}{2}i$.

The derivative $(\mathsf{P} \circ \varphi^{-1})'(\frac{i-1}{2})$ is a positive real number. Indeed, the composition $\mathsf{P} \circ \varphi^{-1}$ maps $Q = \mathfrak{H}_0 \setminus \big(\{|z - \frac{i}{4}| < \frac{1}{4}\} \cup \{|z - \frac{3i}{4}| < \frac{1}{4}\}\big)$ conformally onto $\mathfrak{H}_0$, and by the symmetry properties of the source and the target it maps the horizontal line $\{\mathrm{Im} z = \frac{1}{2}\}$ to itself, and on this line its real part is increasing in $x = \mathrm{Re} z$. Hence, $\frac{\partial}{\partial z}(\mathsf{P} \circ \varphi^{-1})(\frac{i-1}{2}) = \frac{\partial}{\partial x}(\mathsf{P} \circ \varphi^{-1})(\frac{i-1}{2})$ is a positive real number α_φ that does not depend on M.

We will prove now, that $\alpha_\varphi \le \frac{8}{\pi}$. The set Q contains the disc of radius $\frac{1}{4}$ around the point $\frac{i}{2} - \frac{1}{2}$. The mapping ϱ, $\varrho(z) = \frac{e^{-\pi z}i - 1}{e^{-\pi z}i + 1}$, is a conformal mapping from the strip $\mathfrak{H}$ onto the unit disc. It maps $\frac{i}{2}$ to 0. Hence the composition $\varrho \circ \mathsf{P} \circ \varphi^{-1} : Q \to \mathbb{D}$ maps the disc of radius $\frac{1}{4}$ around the point $\frac{i}{2} - \frac{1}{2}$ conformally onto a subset of the unit disc.

By Cauchy's formula the derivative of the composition $\varrho \circ \mathsf{P} \circ \varphi^{-1}$ at $\frac{i}{2} - \frac{1}{2}$ does not exceed 4. The derivative of ϱ at $\frac{i}{2}$ equals $\varrho'(\frac{i}{2}) = -\frac{\pi}{2}$. Since $(\varrho \circ \mathsf{P} \circ \varphi^{-1})'(\frac{i}{2} - \frac{1}{2}) = \varrho'(\frac{i}{2}) \cdot (\mathsf{P} \circ \varphi^{-1})'(\frac{i}{2} - \frac{1}{2})$, the inequality $|(\mathsf{P} \circ \varphi^{-1})'(\frac{i}{2} - \frac{1}{2})| \le \frac{8}{\pi}$ holds.

We proved that

$$
(\mathcal{G}_M)'(\tilde{\xi}_M^+) = -i \cdot \frac{\alpha_\varphi}{2} \frac{M + \frac{1}{2}}{\sqrt{M(M+1)}} \tag{9.31}
$$

with $0 < \alpha_\varphi \le \frac{8}{\pi}$. By symmetry reasons

$$
(\mathcal{G}_M)'(\tilde{\xi}_M^-) = i \cdot \frac{\alpha_\varphi}{2} \frac{M + \frac{1}{2}}{\sqrt{M(M+1)}} . \tag{9.32}
$$

Since

$$\left(\frac{M+\frac{1}{2}}{\sqrt{M(M+1)}}\right)^2 = 1 + \frac{1}{4M(M+1)} \le \frac{4}{3} \text{ for } M \ge \frac{1}{2},$$

we get

$$|(\mathcal{G}_M)'(\tilde{\xi}_M^{\pm})| \le \frac{8}{\sqrt{3}\pi}. \tag{9.33}$$

We put

$$\begin{aligned} R_M &= |(\mathcal{G}_M)'(\tilde{\xi}_M^{+})| \cdot \mathring{R}_M, \\ \xi_M^{\pm} &= |(\mathcal{G}_M)'(\tilde{\xi}_M^{+})| \cdot \tilde{\xi}_M^{\pm}, \\ g_M(\zeta) &= (\mathcal{G}_M)\left(\frac{\zeta}{|(\mathcal{G}_M)'(\tilde{\xi}_M^{+})|}\right), \ \zeta \in R_M. \end{aligned} \tag{9.34}$$

By Eqs. (9.31) and (9.32) the equality $g'_M(\xi_M^{\pm}) = \mp i$ holds.

We will prove now that g_M extends holomorphically to a disc of radius $\frac{1}{8}$ around $\xi_M^{\pm}$, and for the extended function the inequality $|g''_M(\zeta)| < 2^8$ holds on a disc of radius $\frac{1}{16}$ around $\xi_M^{\pm}$. Let $R_{M,+}$ be the rectangle of twice the vertical side length and the same horizontal side length as R_M, that is symmetric with respect to the upper side of R_M. The mapping g_M takes R_M locally conformally to $\mathbb{C} \setminus (i\mathbb{Z} + iM)$ and its continuous extension to the boundary maps the upper side to the segment $(iM, i(M+1))$. Hence, g_M extends by reflection through the upper side of the rectangle and the segment $(iM, i(M+1))$, respectively, to a locally conformal mapping from $R_{M,+}$ into $\mathbb{C} \setminus (i\mathbb{Z} + iM)$. Denote the extension again by g_M. Put $V = |(\mathcal{G}_M)'(\tilde{\xi}_M^{+})| \cdot (\mathfrak{I} \circ T_M^{-1} \circ \omega)^{-1}(Q)$. Note that $V \subset R_{M,+}$. Since $\mathsf{P} \circ \varphi^{-1} \mid Q$ is a conformal mapping, and $\mathcal{G}_M = \mathsf{P} \circ \varphi^{-1} \circ (\mathfrak{I} \circ T_M^{-1} \circ \omega - iM) + iM$, the restriction $g_M \mid V$ is by (9.34) a conformal mapping onto $\mathfrak{H} + iM$. The image of the mapping $g_M \mid V$ contains the disc of radius $\frac{1}{2}$ around $i(M + \frac{1}{2})$. Let $\mathfrak{f}_M$ be the inverse of $g_M \mid V$.

The restriction of $\mathfrak{f}_M$ to the disc of radius $\frac{1}{2}$ around $i(M + \frac{1}{2})$ is a conformal mapping with image in $R_{M,+}$. The mapping $\zeta \to 2\mathfrak{f}_M(i\frac{M+1}{2} + \frac{\zeta}{2})$ takes the unit disc $\mathbb{D}$ conformally onto its image. It takes 0 to ξ_M^{+}, and has derivative of absolute value equal to 1 at 0. By Koebe's $\frac{1}{4}$ Covering Theorem (see A.2) the image of the unit disc under this mapping contains the disc of radius $\frac{1}{4}$ around ξ_M^{+}. Hence, the image $\mathfrak{f}_M(\{|z - i\frac{M+1}{2}| < \frac{1}{2}\})$ covers the disc of radius $\frac{1}{8}$ around ξ_M^{+}. In particular, the disc of radius $\frac{1}{8}$ around ξ_M^{+} is contained in $R_{M,+}$, hence $\mathrm{hsl}(R_M) \ge \frac{1}{4}$. Moreover, the mapping g_M, being the inverse of $\mathfrak{f}_M$, takes a disc of radius $\frac{1}{8}$ around ξ_M^{+} into a disc of radius $\frac{1}{2}$. By Cauchy's formula the second derivative of g_M on the disc of radius $\frac{1}{16}$ around ξ_M^{+} is estimated by the inequality $|g''_M(\zeta)| \le 2^8$.

The same arguments apply to the point ξ_M^- instead of ξ_M^+ obtained from ξ_M^+ by reflection in the real axis. We obtain

$$|g''_M(\zeta)| \le 2^8 \text{ for } |\zeta - \xi_M^\pm| < \frac{1}{16}. \tag{9.35}$$

For $w = a_1^d$ and $M = \frac{d-1}{2}$ we put $\xi_w^\pm = \xi_M^\pm - \frac{\xi_M^+ + \xi_M^-}{2}$. The rectangle $R_M - \frac{\xi_M^+ + \xi_M^-}{2}$ is symmetric with respect to the imaginary axis and with respect to the real axis.. We define R_w to be the intersection of $R_M - \frac{\xi_M^+ + \xi_M^-}{2}$ with the vertical strip $\{|\text{Re}(z)| < \frac{1}{8}\}$. The points $\xi_w^\pm$ are the midpoints of the horizontal sides of R_w.

Finally we put $g_w(\zeta) = g_M(\zeta + \frac{\xi_M^+ + \xi_M^-}{2}) + iM$. The horizontal boundary values of the mapping g_w are contained in the intervals $(-i, 0)$ and $(2Mi, (2M+1)i) = ((d-1)i, di)$. Hence g_w represents a lift of a_1^d under $f_1 \circ f_2$. By (9.25), (9.33), and (9.34) the vertical side length of R_w is estimated by

$$\text{vsl}(R_w) = \text{vsl}(R_M) \le \frac{16}{\sqrt{3}\pi} \mathcal{L}(w) < \pi \mathcal{L}(w).$$

The lemma is proved for syllables of the form a_1^d with $d \ge 2$.

The statement for syllables of the form a_1^{-d} and $a_2^{\pm d}$ with $d \ge 2$ follows by symmetry. Indeed, the mapping $z \to -z$ takes $R_{a_1^d}$, homeomorphically onto itself, and takes $\mathbb{C} \setminus i\mathbb{Z}$ homeomorphically onto itself. The mapping $z \to -g_{a_1^d}(z)$, $z \in R_{a_1^d}$, represents a lift of a_2^d under $f_1 \circ f_2$. The mapping $z \to g_{a_1^d}(-z)$, $z \in R_{a_1^d}$, represents a lift of a_1^{-d}, and the mapping $z \to -g_{a_1^d}(-z)$, $z \in R_{a_1^d}$ represents a lift of a_2^{-d} under $f_1 \circ f_2$. The lemma is proved for syllables of the form (1) with pb horizontal boundary conditions.

2. Syllables of Form (1) with Mixed Horizontal Boundary Conditions Consider the elementary word ${}_{tr}(a_1^d)_{pb}$ with mixed horizontal boundary values. The other cases are similar. We represent this elementary word as follows. Put $M' = d - \frac{1}{2}$. The mapping $(\mathcal{G}_{M'} - \frac{i}{2}) \mid \mathring{R}_{M'}^+$ represents a lift under $f_1 \circ f_2$ of the word ${}_{tr}(a_1^d)_{pb}$. (Indeed, the restriction to vertical line segments join $-\frac{i}{2} + \mathbb{R}$ with $(i(M' - \frac{1}{2}), i(M' + \frac{1}{2}))$.) The extremal length of the rectangle is

$$\begin{aligned}\lambda(\mathring{R}_{M'}^+) &= \frac{1}{\pi}\log(\sqrt{M'} + \sqrt{M'+1})^2 = \frac{1}{\pi}\log(2M' + 1 + 2\sqrt{M'(M'+1)}) \\ &= \frac{1}{\pi}\log(2d + 2\sqrt{(d - \frac{1}{2})(d + \frac{1}{2})}) = \frac{1}{\pi}\log(2d + \sqrt{4d^2 - 1}) = \frac{1}{\pi}\mathcal{L}(w).\end{aligned}$$

The rest of the proof is the same as above.

3. Singletons We consider syllables of degree $d = 1$. We start with the singleton $w =_{pb} (a_2^{-1})_{pb}$. A lift under $f_1 \circ f_2$ is represented by the mapping $g_0(\zeta) \overset{def}{=} \frac{1}{2}e^{2\zeta}$ from the rectangle $\mathring{R}_0$ with vertices $\pm\frac{1}{3}\log 2 \pm \frac{\pi}{4}i$ into $\mathbb{C} \setminus i\mathbb{Z}$. The points $\xi_0^{\pm} = \pm i\frac{\pi}{4}$ are mapped to $\pm\frac{i}{2}$. The derivative of the mapping at these points equals $(g_0)'(\xi_0^{\pm}) = \pm i$. Moreover, $|g_0''(\zeta)| \le 4$, $\text{hsl}(\mathring{R}_0) = \frac{2}{3}\log 2 > \frac{\sqrt{2}}{16}$, $\mathcal{L}(w) = \log(2d + \sqrt{4d^2 - 1}) = \log(2 + \sqrt{3})$, and $\text{vsl}(\mathring{R}_0) = \frac{\pi}{2} = \frac{\pi}{2\log(2+\sqrt{3})}\mathcal{L}(w) < \frac{\pi}{2}\mathcal{L}(w)$. For $w = a_2^{-1}$ we put $R_w = \mathring{R}_0$, $\xi_w^{\pm} = \xi_0^{\pm}$, and $g_w = g_0$. The points $\xi_w^{\pm}$ are the midpoints of the horizontal sides of R_w. The lemma is proved for ${}_{pb}(a_2^{-1})_{pb}$.

For the syllables a_2 and $a_1^{\pm 1}$ with pb horizontal boundary values the statement is obtained by the symmetry arguments used for syllables of form (1).

Consider a singleton with mixed boundary values. For instance, a lift under $f_1 \circ f_2$ of the word ${}_{tr}(a_2^{-1})_{pb}$ is represented by the mapping $\tilde{g}(\zeta) = -\frac{i}{2} + e^{\zeta}$, $\zeta \in R_0^{\text{mixed}}$, where $R_0^{\text{mixed}} \overset{def}{=} \{\zeta \in \mathbb{C} : \text{Im}\zeta \in (0, \frac{\pi}{2}), |\text{Re}\zeta| < \frac{1}{5}\}$. Indeed, the lower side of the rectangle is mapped to $-\frac{i}{2} + \mathbb{R}$, and the upper side is mapped to $(0, i)$. Further, $g'(\frac{\pi}{2}i) = i$, $|g''| < e^{\frac{1}{5}} < \frac{3}{2}$, $\text{hsl}(R_0^{\text{mixed}}) = \frac{2}{5} > \frac{\sqrt{2}}{16}$, and $\text{vsl}(R_0^{\text{mixed}}) = \frac{\pi}{2}$. The other cases with mixed horizontal boundary values follow by symmetry arguments.

4. Syllables of Form (2) with *pb* Horizontal Boundary Values of Degree At Least 2 We give the proof for the syllables $w =_{pb} (a_2^{-1}a_1^{-1}\dots)_{pb}$. For other syllables of form (2) and degree d the statement follows by symmetry reasons. Put $M = \frac{d+1}{2}$, where $d \ge 2$ is the degree, i.e. the number of letters of the word. The holomorphic function

$$\mathfrak{g}_M(z) = \frac{1}{2}\exp(\pi(z + i(M-1))),\ z \in \mathbb{C}, \tag{9.36}$$

takes the set $\{z \in \mathbb{C} : \text{Re}z < \frac{\log 2}{\pi}, |\text{Im}z| < M\}$ to the punctured unit disc $\{0 < |\zeta| < 1\} \subset \mathbb{C} \setminus i\mathbb{Z}$. The mapping $\mathfrak{g}_M$ takes $\{\text{Im}z = -(M - \frac{1}{2}), \text{Re}z < \frac{\log 2}{\pi}\}$ to the line segment $(-i, 0)$. Further, $\mathfrak{g}_M$ takes $\{\text{Im}z = M - \frac{1}{2}, \text{Re}z < \frac{\log 2}{\pi}\}$ to the line segment $(0, i)$ if $d = 2M - 1$ is odd, and takes it to $(-i, 0)$ if $d = 2M - 1$ is even. The restriction of $\mathfrak{g}_M$ to $(-i(M - \frac{1}{2}), i(M - \frac{1}{2}))$ is the curve $\theta \to \frac{1}{2} \cdot e^{\pi i\theta}, \theta \in (-\frac{1}{2}, d - \frac{1}{2})$. The initial point of this curve is $-\frac{i}{2}$ and the curve makes d half-turns around 0. The curve $\theta \to \frac{1}{2} \cdot e^{\pi i\theta}$, $\theta \in (-\frac{1}{2}, d - \frac{1}{2})$, represents a lift under $f_1 f_2$ of the syllable $a_2^{-1}a_1^{-1}\dots$ of degree d with pb boundary values, and the restriction $\mathfrak{g}_M \mid \{z \in \mathbb{C} : \text{Re}z < \frac{\log 2}{\pi}, |\text{Im}z| < M - \frac{1}{2}\}$ represents a lift of ${}_{pb}(a_2^{-1}a_1^{-1}\dots)_{pb}$ under $f_1 f_2$. The derivative of the mapping $\mathfrak{g}_M$ at $-i(M - \frac{1}{2})$ equals

$$(\mathfrak{g}_M)'(-i(M - \frac{1}{2})) = \pi\mathfrak{g}_M(-i(M - \frac{1}{2})) = -\frac{\pi i}{2}. \tag{9.37}$$

and

$$(\mathfrak{g}_M)'(+i(M-\frac{1}{2})) = \begin{cases} +\frac{\pi i}{2}, & \text{if } d = 2M-1 \text{ is odd}, \\ -\frac{\pi i}{2}, & \text{if } d = 2M-1 \text{ is even}. \end{cases} \tag{9.38}$$

For each natural number k the set $\{z \in \mathbb{C} : \mathrm{Re}z < \frac{\log 2}{\pi}, |\mathrm{Im}z| < M - \frac{1}{2}\}$ contains the rectangle $\mathcal{R}_k \stackrel{def}{=} \{z \in \mathbb{C} : -k < \mathrm{Re}z < \frac{\log 2}{\pi}, |\mathrm{Im}z| < M - \frac{1}{2}\}$ of extremal length $\frac{2M-1}{k+\frac{\log 2}{\pi}}$, which can be made arbitrarily small by choosing large k. The restriction of $\mathfrak{g}_M$ to each of these rectangles represents a lift of $_{pb}(a_2^{-1}a_1^{-1}\ldots)_{pb}$ under $f_1 f_2$. However, we need to represent lifts of syllables of form (2) with parameter M by holomorphic mappings from rectangles with horizontal side length uniformly bounded from above and from below for all M, with vertical side length comparable to $\log M$, and the values of the derivatives of the mappings at the chosen points equal to $\pm i$ where the sign is prescribed by the situation. After normalizing the rectangle and the mapping so that the condition for the derivatives of the mapping is satisfied, the described approach gives rectangles of vertical side length comparable to $d = 2M-1$, which does not give the optimal estimate.

For dealing with this difficulty, we precompose $\mathfrak{g}$ with the mapping used for syllables of form (1). We consider the rectangle $\mathring{R}_M$ and the mapping $\mathfrak{I} \circ T_M^{-1} \circ \omega$ on $\mathring{R}_M$ as in the case of syllables of form (1). Recall that

$$\mathrm{vsl}(\mathring{R}_M) = 2\log(\sqrt{M} + \sqrt{M+1})^2 \le 2\log(2d + \sqrt{4d^2-1}) = 2\mathcal{L}(w).$$

Let $\tilde{R}_M$ be the union of $\mathring{R}_M$ with its open right side and with its reflection through the line $\{\mathrm{Re}z = \pi\}$. The mapping $\mathfrak{I} \circ T_M^{-1} \circ \omega$ extends holomorphically to $\tilde{R}_M$ by Schwarz Reflection Principle. Denote the extended mapping again by $\mathfrak{I} \circ T_M^{-1} \circ \omega$. The extended mapping $\mathfrak{I} \circ T_M^{-1} \circ \omega$ takes $\tilde{R}_M$ conformally onto the complex plane with the discs of radius $\frac{1}{2}$ around $\pm i(M+\frac{1}{2})$ and the rays $(-\infty, -i(M+1)]$ and $[i(M+1), \infty)$ removed. Put

$$\mathring{R}_{M,2} \stackrel{def}{=} (\mathfrak{I} \circ T_M^{-1} \circ \omega)^{-1}\Big(\{z \in \mathbb{C}_\ell : |\mathrm{Im}z| < M - \frac{1}{2}\}\Big),$$
$$R'_{M,2} \stackrel{def}{=} (\mathfrak{I} \circ T_M^{-1} \circ \omega)^{-1}\Big(\{z \in \mathbb{C} : \mathrm{Re}z < \frac{\log 2}{\pi}, |\mathrm{Im}z| < M\}\Big). \tag{9.39}$$

Then $\mathring{R}_{M,2} \subset \mathring{R}_M \subset \tilde{R}_M$, and $\mathfrak{I} \circ T_M^{-1} \circ \omega$ takes $\mathring{R}_{M,2}$ conformally onto $\{z \in \mathbb{C}_\ell : |\mathrm{Im}z| < M - \frac{1}{2}\}$, and takes $R'_{M,2}$ conformally onto $\{z \in \mathbb{C} : \mathrm{Re}z < \frac{\log 2}{\pi}, |\mathrm{Im}z| < M\}$. The composition $\mathcal{G}_M \stackrel{def}{=} \mathfrak{g}_M \circ \mathfrak{I} \circ T_M^{-1} \circ \omega$ takes $R'_{M,2}$ into the punctured disc $\{0 < |z| < 1\}$. We put

$$\tilde{\eta}_M^{\pm} \stackrel{def}{=} (\mathfrak{I} \circ T_M^{-1} \circ \omega)^{-1}(\pm i(M - \frac{1}{2})) = (T_M^{-1} \circ \omega)^{-1}(\pm(M - \frac{1}{2})).$$

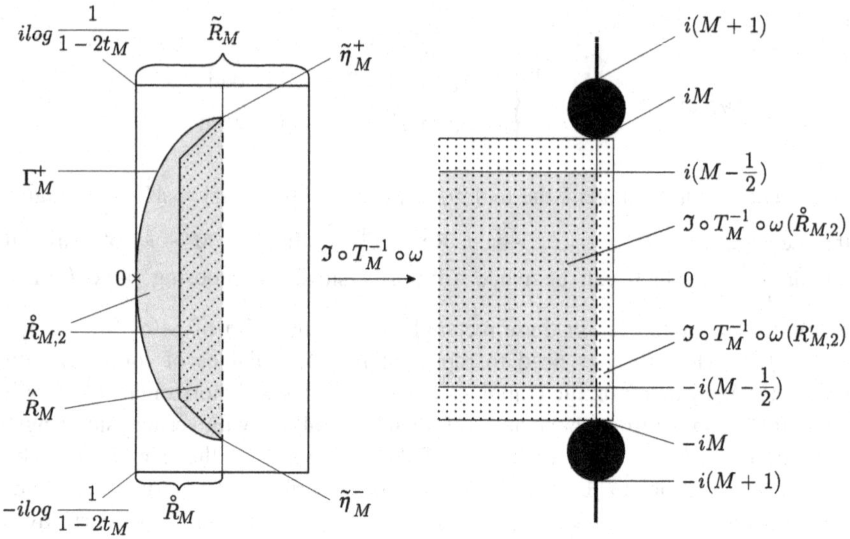

Fig. 9.7 The construction of a rectangle that admits a holomorphic mapping representing a syllable of form (2)

The points $\tilde{\eta}^{\pm}_M$ lie on the boundary of $\mathring{R}_{M,2}$ and on the open right side of $\mathring{R}_M$ (which is contained in $\{\mathrm{Re}z = \pi\} \cap R'_{M,2}$). See Fig. 9.7.

We prove now that the set $\mathring{R}_{M,2}$ contains the set

$$\widehat{R}_M \overset{def}{=} \Big\{z \in \mathbb{C} : -\frac{2}{5} < \mathrm{Re}(z-\pi) < 0,\ \mathrm{Im}(z-\tilde{\eta}^+_M) < \mathrm{Re}(z-\pi),$$
$$\mathrm{Im}(z-\tilde{\eta}^-_M) > -\mathrm{Re}(z-\pi)\Big\}. \tag{9.40}$$

We describe first the curves

$$\Gamma^{\pm}_M \overset{def}{=} (\mathfrak{J}\circ T_M^{-1}\circ\omega)^{-1}\Big(\Big\{z \in \mathbb{C} : -\frac{2}{5} < \mathrm{Re}z < 0,\ \mathrm{Im}z = \pm(M-\frac{1}{2})\}\Big)$$

Since $(\mathfrak{J}\circ T_M^{-1}\circ\omega)^{-1} = \omega^{-1}\circ T_M\circ\mathfrak{J}^{-1}$, we may also write

$$\Gamma^{\pm}_M = \omega^{-1}\circ T_M\Big(\{z \in \mathbb{C} : 0 < \mathrm{Im}z < \frac{2}{5},\ \mathrm{Re}z = \pm(M-\frac{1}{2})\}\Big).$$

To describe the curve $\Gamma^{\pm}_M$, we consider the derivative (see Eq. (9.29))

$$\frac{\partial}{\partial t}(\omega^{-1}\circ T_M)(M-\frac{1}{2}+it) = i(\omega^{-1}\circ T_M)'(M-\frac{1}{2}+it)$$

$$= \frac{2\sqrt{M(M+1)}}{(M-\frac{1}{2}+it)^2 - M(M+1)}, \quad t \geq 0. \tag{9.41}$$

(We used that for a holomorphic function f the equality $\frac{\partial}{z} f = -i\frac{\partial}{y} f$ holds.) The denominator of the last fraction equals $-2M - t^2 + \frac{1}{4} + 2t(M-\frac{1}{2})i$. Hence, (see also Eq. (9.28),) since $M \geq \frac{3}{2}$ we obtain for $t > 0$

$$\mathrm{Re}\frac{\partial}{\partial t}\frac{1}{i}\log(2T_M(M-\frac{1}{2}+it)-1) = \frac{2\sqrt{M(M+1)}(-2M-t^2+\frac{1}{4})}{(-2M-t^2+\frac{1}{4})^2 + (2t(M-\frac{1}{2}))^2} < 0,$$

$$\mathrm{Im}\frac{\partial}{\partial t}\frac{1}{i}\log(2T_M(M-\frac{1}{2}+it)-1) = \frac{-2\sqrt{M(M+1)}2t(M-\frac{1}{2})}{(-2M-t^2+\frac{1}{4})^2 + (2t(M-\frac{1}{2}))^2} < 0. \tag{9.42}$$

We obtain for $t > 0$

$$\frac{\mathrm{Im}\frac{\partial}{\partial t}\frac{1}{i}\log(2T_M(M-\frac{1}{2}+it)-1)}{\mathrm{Re}\frac{\partial}{\partial t}\frac{1}{i}\log(2T_M(M-\frac{1}{2}+it)-1)} = \frac{2t(M-\frac{1}{2})}{2M+t^2-\frac{1}{4}} > 0.$$

For $0 \leq t \leq \frac{1}{2}$

$$\frac{\mathrm{Im}\frac{\partial}{\partial t}\frac{1}{i}\log(2T_M(M-\frac{1}{2}+it)-1)}{\mathrm{Re}\frac{\partial}{\partial t}\frac{1}{i}\log(2T_M(M-\frac{1}{2}+it)-1)} \leq \frac{M-\frac{1}{2}}{2M-\frac{1}{4}} < \frac{1}{2}. \tag{9.43}$$

Further, since $M \geq \frac{3}{2}$, the following estimate holds for $0 \leq t \leq \frac{1}{2}$

$$\begin{aligned}\left|\mathrm{Re}\frac{\partial}{\partial t}\frac{1}{i}\log(2T_M(M-\frac{1}{2}+it)-1)\right| &\geq \frac{2M(2M-\frac{1}{4})}{(2M)^2+(M-\frac{1}{2})^2} \\ &= \frac{4M^2-\frac{M}{2}}{5M^2-M+\frac{1}{4}} > \frac{4}{5}.\end{aligned} \tag{9.44}$$

By inequality (9.44) for $t = \frac{1}{2}$ the point $\frac{1}{i}\log(2T_M(M-\frac{1}{2}+it)-1)$ lies in $\{z \in \mathbb{C} : \mathrm{Re}z < \pi - \frac{2}{5}\}$. Recall that $\frac{1}{i}\log(2T_M(M-\frac{1}{2})-1) = \tilde{\eta}_M^+ \in \{\mathrm{Re}z = \pi\}$. Inequality (9.43) shows that the part of the curve $\Gamma_M^\pm$ that is contained in $\{z \in \mathbb{C} : \pi - \frac{2}{5} < \mathrm{Re}z < \pi\}$, is contained in the set $\{z \in \mathbb{C} : \mathrm{Im}(z - \tilde{\eta}_M^+) > \mathrm{Re}(z-\pi)\}$. A similar fact holds for the curve Γ_M^-. Hence, by the global behaviour of $\Gamma_M^\pm$ (see (9.42)) the set

$$\omega^{-1} \circ T_M\Big(\{z \in \mathbb{C} : \quad 0 < \mathrm{Im}z < \frac{2}{5}, \ |\mathrm{Re}z| < M - \frac{1}{2}\}\Big)$$

$$= (\mathfrak{J} \circ T_M^{-1} \circ \omega)^{-1}\Big(\big\{z \in \mathbb{C} : -\frac{2}{5} < \mathrm{Re} z < 0, \ |\mathrm{Im} z| < M - \frac{1}{2}\big\}\Big)$$

contains the set (9.40).

By Eq. (9.29) we obtain

$$(\omega^{-1} \circ T_M)'(\pm(M - \frac{1}{2})) = i \frac{2\sqrt{M(M+1)}}{-(M-\frac{1}{2})^2 + M(M+1)} = i \frac{\sqrt{M(M+1)}}{M - \frac{1}{8}} \tag{9.45}$$

Since $(T_M^{-1} \circ \omega)'(\tilde{\eta}_M^{\pm}) = \frac{1}{(\omega^{-1} \circ T_M)'(\pm(M-\frac{1}{2}))}$ we obtain

$$(T_M^{-1} \circ \omega)'(\tilde{\eta}_M^{\pm}) = \frac{-i(M - \frac{1}{8})}{\sqrt{M(M+1)}} . \tag{9.46}$$

The derivative of the composition $\mathcal{G}_M^{(2)} \overset{def}{=} \mathfrak{g}_M \circ (iT_M^{-1}) \circ \omega$ at the points $\tilde{\eta}^{\pm}$ equals

$$(\mathcal{G}_M^{(2)})'(\tilde{\eta}_M^{\pm}) = \mathfrak{g}_M'(\pm i(M - \frac{1}{2})) \cdot ((iT_M^{-1}) \circ \omega)'(\tilde{\eta}_M^{\pm}) . \tag{9.47}$$

By Eqs. (9.37), (9.38), and (9.46)

$$\begin{aligned}
(\mathcal{G}_M^{(2)})'(\tilde{\eta}_M^{-}) &= -\frac{\pi}{2}\frac{M - \frac{1}{8}}{\sqrt{M(M+1)}} i , \\
(\mathcal{G}_M^{(2)})'(\tilde{\eta}_M^{+}) &= -\frac{\pi}{2}\frac{M - \frac{1}{8}}{\sqrt{M(M+1)}} i , \ \text{ if } d = 2M - 1 \text{ is even} , \\
(\mathcal{G}_M^{(2)})'(\tilde{\eta}_M^{+}) &= \frac{\pi}{2}\frac{M - \frac{1}{8}}{\sqrt{M(M+1)}} i , \ \text{ if } d \text{ is odd} .
\end{aligned} \tag{9.48}$$

Note that $\frac{M-\frac{1}{8}}{M+1} < \frac{M-\frac{1}{8}}{\sqrt{M(M+1)}} < 1$. Further, the mapping $M \to \frac{M-\frac{1}{8}}{M+1}$ is increasing for M increasing, hence $\frac{M-\frac{1}{8}}{\sqrt{M(M+1)}} > \frac{\frac{3}{2}-\frac{1}{8}}{\frac{5}{2}} > \frac{1}{2}$. It follows that

$$\frac{\pi}{4} < |(\mathcal{G}_M^{(2)})'(\tilde{\eta}_M^{\pm})| < \frac{\pi}{2} . \tag{9.49}$$

Put

$$r_M \overset{def}{=} |(\mathcal{G}_M^{(2)})'(\tilde{\eta}_M^{\pm})| . \tag{9.50}$$

We define

$$
\begin{aligned}
g_{M,2}(\zeta) &=(\mathcal{G}_M^{(2)})(\frac{\zeta}{r_M}), \ \zeta \in r_M R'_{M,2}, \\
\eta_M^{\pm} &= r_M \cdot \tilde{\eta}_M^{\pm}.
\end{aligned}
\tag{9.51}
$$

We will prove now that $r_M R'_{M,2}$ (see (9.39) for the definition) contains a disc of radius $\frac{1}{8}$ around $\eta_M^{\pm}$ and prove the required estimate of the second derive of $g_{M,2}$ in a disc of radius $\frac{1}{16}$ around $\eta_M^{\pm}$. This will be done similarly as for elementary words of form (1). Consider the half-strip $\mathfrak{S}_M^- \stackrel{def}{=} \{z \in \mathbb{C} : \mathrm{Re}z < \frac{\log 2}{\pi}, \ |\mathrm{Im}z + M - \frac{1}{2}| < \frac{1}{2}\} \subset \{z \in \mathbb{C} : \mathrm{Re}z < \frac{\log 2}{\pi}, \ |\mathrm{Im}z| < M\}$. The restriction $\mathfrak{g}_M \mid \mathfrak{S}_M^-$ is a conformal mapping onto its image. The image is the set

$$
\begin{aligned}
\Big\{\frac{1}{2} \cdot \exp\big(\pi(-i(M-\frac{1}{2}) + \zeta) + \pi i(M-1)\big) : \mathrm{Re}\zeta < \frac{\log 2}{\pi}, |\mathrm{Im}\zeta| < \frac{1}{2}\Big\} \\
= \Big\{\frac{1}{2}\exp\big(-\frac{\pi}{2}i + \pi\zeta\big) : \mathrm{Re}\zeta < \frac{\log 2}{\pi}, |\mathrm{Im}\zeta| < \frac{1}{2}\Big\} \\
= \Big\{re^{i\varphi} : 0 < r < 1, -\pi < \varphi < 0\Big\}.
\end{aligned}
$$

This set contains the disc $\{|\zeta + \frac{i}{2}| < \frac{1}{2}\}$. Put $V_M^- = (\mathfrak{g}_M \mid \mathfrak{S}_M^-)^{-1}(\{|\zeta + \frac{i}{2}| < \frac{1}{2}\})$. Let $\tilde{Q}_M^- = (\mathfrak{J} \circ T_M^{-1} \circ \omega)^{-1}(V_M^-)$. We have the following sequence of conformal mappings

$$
\begin{array}{ccccc}
\tilde{\eta}_M^- & \xrightarrow{\mathfrak{J}\circ T_M^{-1}\circ\omega} & -(M-\frac{1}{2})i & \xrightarrow{\mathfrak{g}_M} & -\frac{1}{2}i \\
\Cap & & \Cap & & \Cap \\
\tilde{Q}_M^- & \xrightarrow{\mathfrak{J}\circ T_M^{-1}\circ\omega} & V_M^- & \xrightarrow{\mathfrak{g}_M} & \{|\zeta + \frac{i}{2}| < \frac{1}{2}\}.
\end{array}
\tag{9.52}
$$

The composition $\mathcal{G}_M^{(2)} = \mathfrak{g}_M \circ \mathfrak{J} \circ T_M^{-1} \circ \omega$ maps $\tilde{Q}_M^-$ conformally onto $\{|\zeta + \frac{i}{2}| < \frac{1}{2}\}$. The mapping $g_{M,2}$, $g_{M,2}(\zeta) = \mathcal{G}_M^{(2)}(\frac{\zeta}{r_M})$ takes $Q_M^- \stackrel{def}{=} r_M \tilde{Q}_M^-$ conformally onto $\{|\zeta + \frac{i}{2}| < \frac{1}{2}\}$. Moreover, the derivative of $g_{M,2}$ at η^- has absolute value equal to one (see equality (9.50)). Consider the inverse of the restriction of $g_{M,2}$ to Q_M^-. In the same way as for syllables of form (1) we see that $Q_M^- \subset r_M R'_{M,2}$ contains a disc of radius $\frac{1}{8}$ around η_M^-, and the absolute value of the second derivative of $g_{M,2}$ in a disc of radius $\frac{1}{16}$ around η_M^- does not exceed 2^8. The same arguments apply to η_M^+ instead of η_M^-.

The set $r_M R'_{M,2}$ also contains the set $r_M \widehat{R}_M$, and by (9.40) and inequality (9.49) the set $r_M \widehat{R}_M$ contains the set

$$\{z \in \mathbb{C} : -\frac{2}{5} \cdot \frac{\pi}{2} < \operatorname{Re}(z - \pi r_M) < 0,\ \operatorname{Im} z - \eta_M^+ < \operatorname{Re}(z - \pi r_M),$$
$$\operatorname{Im} z - \eta_M^- > -\operatorname{Re}(z - \pi r_M)\}\,. \tag{9.53}$$

Since $\frac{1}{16} = 0.0625 < 0.628 < \frac{2\pi}{10}$, the union of the discs of radius $\frac{1}{16}$ around $\eta_M^\pm$ with the set (9.53) contains the rectangle of horizontal side length equal to $\frac{1}{16\sqrt{2}}$ whose right vertices are η_M^+ and η_M^-, respectively. For the syllable ${}_{pb}w_{pb} = {}_{pb}(a_2^{-1}a_1^{-1}\ldots)_{pb}$ of degree d and $M = \frac{d+1}{2}$ we let R_w be equal to this rectangle. Further, we put $\xi_w^\pm = \eta_M^\pm = r_M \tilde{\eta}_M^\pm$. The points $\xi_w^\pm$ are the right vertices of the rectangle R_w. Hence, $R_w \subset r_M \mathring{R}_M$. By inequalities (9.25) and (9.49), and by equality (9.50) the inequalities

$$\operatorname{vsl}(R_w) \le r_M \operatorname{vsl}(\mathring{R}_M) \le \frac{\pi}{2} \cdot 2\log(2d + \sqrt{(4d^2 - 1)}) = \pi\,\mathcal{L}(w)\,.$$

hold.

Put $g_w = g_{M,2}$. The conditions of the lemma concerning the first derivatives $g_w'(\xi_w^\pm)$ are satisfied.

Lemma 9.7 is proved for syllables of the form ${}_{pb}w_{pb} = {}_{pb}(a_2^{-1}a_1^{-1}\ldots)_{pb}$ of degree $d \ge 2$. For the other elementary words of form (2) with pb horizontal boundary values and degree $d \ge 2$ the Lemma is obtained by symmetry arguments. The points $\xi_w^\pm$ are either the right vertices or the left vertices of the rectangle R_w.

□

Proof of the Upper Bound in Theorems 9.1, 9.2, and 9.3 Take any word w with pb horizontal boundary values. Write it as product of syllables ${}_{pb}(w_j)_{pb}$ with pb horizontal boundary values. By Lemma 9.7 each ${}_{pb}(w_j)_{pb}$ can be represented by a holomorphic mapping g_{w_j} from the rectangle R_{w_j} into $\mathbb{C} \setminus \{-1, 1\}$. Recall that the vertical side length of R_{w_j} does not exceed $\pi\,\mathcal{L}(w_j)$, the horizontal side length of R_{w_j} is at least $\frac{1}{16\sqrt{2}}$. We shrink the rectangles R_{w_j} in the horizontal direction, so that its horizontal side length is exactly equal to $\frac{1}{16\sqrt{2}}$, and the points $\xi_{w_j}^\pm$ are on the boundary of the shrinked rectangle R_{w_j}. Moreover, they are midpoints of the open horizontal sides of the shrinked rectangles, if they were midpoints of the open horizontal sides of the original rectangle, and they are on the open right (left) side of the shrinked rectangle, if they were on the open right (left) side of the original rectangle.

We may assume that g_{w_j} extends holomorphically to discs of radius $\frac{1}{16}$ around the points $\xi_{w_j}^\pm$.

Take complex numbers a_j, , $j = 1, \ldots, d$, such that for the translated rectangles $R_j \overset{def}{=} R_{w_j} + a_j$ and the translated points $\xi_j^\pm \overset{def}{=} \xi_{w_j}^\pm + a_j$ the equalities $\xi_j^+ = \xi_{j+1}^-$

hold for $j \le N-1$. Choose complex numbers b_j so that for the translated functions $g_j(\zeta) \overset{def}{=} b_j + g_{w_j}(\zeta - a_j)$ the equalities $g_j(\xi_j^+) = g_{j+1}(\xi_{j+1}^-)$, $j = 1, \dots, N-1$, hold.

We will perform quasiconformal gluing of the g_j using the fact that each g_j extends holomorphically to a disc of radius $\frac{1}{16}$ around $\xi_j^{\pm}$. Denote the extended functions again by g_j. The disc of radius $\frac{1}{16}$ around ξ_j^+ contains the square Q_j of side length $\frac{\sqrt{2}}{16}$ with center ξ_j^+. We define a smooth mapping g on

$$\mathfrak{R}'_w \overset{def}{=} \operatorname{Int}\left(\overline{\bigcup_{j=1}^{N} R_j}\right) \cup \bigcup_{j=1}^{N-1} Q_j$$

as follows.

Consider the C^1-function χ_0 on the interval $[0, 1]$, $\chi_0(t) = 6\int_0^t \tau(1-\tau)d\tau$. Then $\chi_0(0) = 0$, $\chi_0(1) = 1$, and $0 \le \chi_0'(t) \le 6t(1-t) \le \frac{3}{2}$. Let $a < 2^{-5}$ be a positive number that will be chosen in a moment.

If $\zeta \in \mathfrak{R}'_w$ and $|\mathrm{Im}\zeta - \xi_j^+| \le \frac{a}{2}$ for some $j = 1, \dots, N-1$, then $\zeta \in Q_j$, and g_j and g_{j+1} are defined and holomorphic in Q_j. In this case we put on Q_j

$$g(\zeta) = (1 - \chi_j)(\mathrm{Im}\zeta) \cdot g_j(\zeta) + \chi_j(\mathrm{Im}\zeta) \cdot g_{j+1}(\zeta)\,.$$

Here $\chi_j(t) = \chi_0(\frac{t - \mathrm{Im}\xi_j^+ + \frac{a}{2}}{a})$. If $|\mathrm{Im}\zeta - \xi_j^{\pm}| > \frac{a}{2}$ for all $j = 1, \dots, N-1$, then ζ is contained in a single rectangle R_j, and we put $g = g_j$. The mapping g is of class C^1. If $|\mathrm{Im}\zeta - \xi_j^+| \le \frac{a}{2}$ for some $j = 1, \dots, N-1$, then for the Beltrami coefficient we obtain with $\zeta = u + iv$

$$\mu_g(\zeta) = \frac{\frac{\partial}{\partial \bar\zeta} g(\zeta)}{\frac{\partial}{\partial \zeta} g(\zeta)} = \frac{\frac{i}{2}(\frac{\partial}{\partial v}\chi_j \cdot (g_{j+1} - g_j))(\zeta)}{(\frac{-i}{2}\frac{\partial}{\partial v}\chi_j \cdot (g_{j+1} - g_j) + (1-\chi_j)\cdot g_j' + \chi_j \cdot g_{j+1}')(\zeta)}.$$

At all other points the Beltrami coefficient equals zero. For ζ in the square of side length a with center ξ_j^+ the inequalities $|g_j'(\zeta) - g_j'(\xi_j^+)| \le 2^8|\zeta - \xi_j^+|$ and $|g_{j+1}'(\zeta) - g_{j+1}'(\xi_j^+)| \le 2^8|\zeta - \xi_j^+|$ hold. Since $g_j(\xi_j^+) = g_{j+1}(\xi_j^+)$, and $g_j'(\xi_j^+) = g_{j+1}'(\xi_j^+)$, we obtain on this square

$$|g_j - g_{j+1}| \le 2 \cdot \frac{1}{2} \cdot 2^8 \cdot \frac{a^2}{2} = 2^7 a^2,$$

and

$$(1-\chi_j)\cdot|g_j' - g_j'(\xi_j^+)| + \chi_j \cdot |g_{j+1}' - (g_{j+1}')(\xi_j^+)| \le 2^7\sqrt{2}a\,.$$

Since $g_j'(\xi_j^+) = g_{j+1}'(\xi_j^+)$ has absolute value equal to 1 and $|\chi_j'| \le \frac{3}{2} \cdot \frac{1}{a}$, we obtain

$$|\mu_g(\xi)| \le \frac{\frac{1}{2} \cdot \frac{3}{2} \cdot 2^7 a}{1 - \frac{1}{2} \cdot \frac{3}{2} \cdot 2^7 a - 2^7 \sqrt{2} a} = \frac{\frac{3}{4} 2^7 a}{1 - (\frac{3}{4} + \sqrt{2}) 2^7 a} .$$

Put $a = \frac{1}{3} 2^{-8}$. Since $\frac{3}{4} + \sqrt{2} < 3$,we obtain the inequality

$$|\mu_g(\zeta)| \le \frac{\frac{3}{4} \cdot \frac{1}{6}}{1 - (\frac{3}{4} + \sqrt{2}) \cdot \frac{1}{6}} < \frac{1}{4} .$$

The quasiconformal dilatation K is less than $\frac{5}{3}$. The set $\mathfrak{R}_w'$ contains a curvilinear rectangle $\mathfrak{R}_w$ of the form $R_{J,\Phi,\mathrm{b}}$ (see Example 3.3, Sect. 3.1) with $|J| \le \pi\, \mathcal{L}(w)$, $\mathrm{b} = a$, and a smooth function Φ. For each positive ε the function Φ can be chosen with $|\Phi'| \le 1 + \varepsilon$. Indeed, replace first each R_j by a rectangle R_j^a contained in R_j of horizontal side length a and vertical side length $\mathrm{vsl}(R_j^a) = \mathrm{Im}(\xi_j^+ - \xi_j^-)$, so that the $\xi_j^{\pm}$ are the midpoints of the horizontal sides of the new rectangle R_j^a if they are midpoints of the horizontal sides of R_j, and they are equal to the endpoints of the right (left) sides of the R_j^a, if they are contained in the open right (left) sides of the R_j. For $j \le N - 1$ we consider the intersection of $\overline{R_j^a} \cup \overline{R_{j+1}^a}$ with the strip $\{|\mathrm{Im}\zeta - \mathrm{Im}\xi_j^+| \le \frac{a}{2}\}$. Replace the intersection by the parallelogram whose horizontal sides are equal to $R_j^a \cap \{\mathrm{Im}\zeta = \mathrm{Im}\xi_j^+ - \frac{a}{2}\}$ and $R_{j+1}^a \cap \{\mathrm{Im}\zeta = \mathrm{Im}\xi_j^+ + \frac{a}{2}\}$, respectively. Doing so for all $j \le N - 1$ we obtain a curvilinear rectangle of the form $R_{J,\Phi_0,\mathrm{b}}$ for a piecewise C^1 function Φ_0 with $|\Phi_0'| \le 1$. For any positive ε we obtain after suitable smoothing a curvilinear rectangle of the required form with a function Φ that satisfies $|\Phi'| \le 1 + \varepsilon$. See Fig. 9.8.

Choose ε and consider the obtained curvilinear rectangle $\mathfrak{R}_w$. The restriction $g \mid \mathfrak{R}_w$ represents ${}_{pb}w_{pb}$. Since $\mathrm{vsl}(R_w) \le \pi\mathcal{L}(w)$, the extremal length $\lambda(\mathfrak{R}_w)$ does not exceed $(1 + (1 + \varepsilon)^2) \cdot \frac{\pi \mathcal{L}(w)}{a}$ (see Lemma 3.1). Using the normalized solution ω of the Beltrami equation on the complex plane with Beltrami coefficient μ_g on $\mathfrak{R}_w$ and 0 else, and the conformal mapping ψ from a true rectangle $\mathcal{R}_w$ onto the curvilinear rectangle $\omega(\mathfrak{R}_w)$, we obtain the holomorphic mapping $g \circ \omega^{-1} \circ \psi : \mathcal{R}_w \to \mathbb{C} \setminus \{-1, 1\}$ representing ${}_{pb}w_{pb}$. The extremal length of $\mathcal{R}_w$ does not exceed

$$\begin{aligned} K \cdot (1 + (1 + \varepsilon)^2) \cdot \frac{\pi\mathcal{L}(w)}{a} &< \frac{5}{3} \cdot (1 + (1 + \varepsilon)^2) \cdot \frac{\pi}{a} \cdot \mathcal{L}(w) \\ =&(1 + (1 + \varepsilon)^2) \cdot 5\pi \cdot 2^8 \cdot \mathcal{L}(w) < 2^{12}(1 + (1 + \varepsilon)^2)\mathcal{L}(w) . \end{aligned}$$

We used the inequality $5\pi < 2^4$. Choosing ε small enough, we get a curvilinear rectangle $\mathcal{R}_w$ with $\Lambda(\mathcal{R}_w) < 2^{13} \cdot \mathcal{L}(w)$, that admits a holomorphic mapping representing w. This proves Theorem 9.1 in case of pb boundary values.

Fig. 9.8 A curvilinear rectangle $\mathfrak{R}_w$ that admits a holomorphic mapping representing ${}_{pb}w_{pb}$

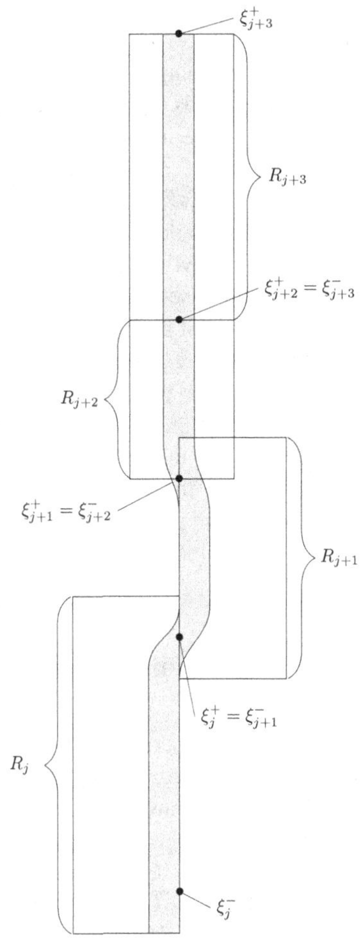

In the case, when w has at least two syllables and totally real right boundary values, or totally real left boundary values, or both boundary values are totally real, the first or last syllable, or both have mixed boundary values. Suppose, the respective syllable $\mathfrak{s}$ is of form (1) or (3). A lift of the syllable $\mathfrak{s}$ with the required mixed boundary values can be represented by a holomorphic mapping from a rectangle with the same estimates for the vertical side length and the horizontal side length as in the case of pb boundary values. Moreover, the value of the derivative of the mapping at the point, in a neighbourhood of which the gluing is performed, is the same as for pb boundary values, and the estimate for the second derivative is the same.

Suppose the respective syllable $\mathfrak{s}$ is of form (2). In this case instead of representing the lift of $\mathfrak{s}$ with mixed boundary values, we represent the lifts under $f_1 \circ f_2$ of a singleton $\mathfrak{s}'$ with mixed boundary values and the lift of a word $\mathfrak{s}''$ with one letter less than $\mathfrak{s}$ with *pb* horizontal boundary values. Let $g_{\mathfrak{s}'} : R_{\mathfrak{s}'} \to \mathbb{C} \setminus i\mathbb{Z}$, and $g_{\mathfrak{s}''} : R_{\mathfrak{s}''} \to \mathbb{C} \setminus i\mathbb{Z}$, be the respective representing mappings. We may choose the mappings so that the derivative at the points in a neighbourhood of which the gluing is performed equals $\pm i$, the estimate for the horizontal side length of each rectangle and the estimate of the second derivative of the mappings is as in the case of *pb* boundary values, and $\text{vsl}(R_{\mathfrak{s}'}) \le \pi\mathcal{L}(\mathfrak{s}') \le \pi\mathcal{L}(\mathfrak{s})$ and $\text{vsl}(R_{\mathfrak{s}''}) \le \pi\mathcal{L}(\mathfrak{s}'') \le \pi\mathcal{L}(\mathfrak{s})$. The quasiconformal gluing and the choice of the curvilinear rectangle $\mathcal{R}$ is done as in the case of *pb* boundary values. The extremal length of the curvilinear rectangle $\mathcal{R}$ differs no more than by a factor 2 from the estimate in the case of *pb* boundary values.

If w consists of a single syllable the upper bound in the case of mixed boundary values (Theorem 9.2) follows directly from Proposition 9.3. The exceptional cases were treated in the proof of the lower bound in Theorem 9.1. Theorems 9.1 and 9.2 are proved.

Theorem 9.3 in the non-exceptional cases is obtained in the same way as Theorem 9.1 in the case of *pb* horizontal boundary values. The only difference is that we have to perform also quasiconformal gluing of the last syllable to the first syllable of the word. This proves Theorem 9.3. □

9.7 The Extremal Length of 3-Braids

In this section we come back to braids. Recall that we may consider n-braids as elements of the fundamental group of $C_n(\mathbb{C})\diagup\mathcal{S}_n$.

In Sect. 3.2 we saw that the totally real subspace $C_n(\mathbb{R})\diagup\mathcal{S}_n$ of the n-dimensional symmetrized configuration space is connected and simply connected, and the fundamental group $\pi_1(C_n(\mathbb{C})\diagup\mathcal{S}_n, E_n)$ of the symmetrized configuration space with base point $E_n \in C_n(\mathbb{R})\diagup\mathcal{S}_n$ is canonically isomorphic to the relative fundamental group $\pi_1(C_n(\mathbb{C})\diagup\mathcal{S}_n, C_n(\mathbb{R})\diagup\mathcal{S}_n)$.

We recall the definition of the extremal length of braids with totally real horizontal boundary values that was given in Sect. 3.2, Definition 3.5. For a braid $b \in \mathcal{B}_n$ the value $\Lambda_{tr}(b) = \Lambda(b_{tr})$ is the infimum of the extremal length of rectangles that admit a holomorphic mapping into $C_n(\mathbb{C})\diagup\mathcal{S}_n$ that represents the image b_{tr} of b under the canonical isomorphism $\pi_1(C_n(\mathbb{C})\diagup\mathcal{S}_n, E_n) \to \pi_1(C_n(\mathbb{C})\diagup\mathcal{S}_n, C_n(\mathbb{R})\diagup\mathcal{S}_n)$.

We will consider now the case $n = 3$. By Lemma 1.2 the group $\mathcal{PB}_3\diagup\langle\Delta_3^2\rangle$ is isomorphic to the fundamental group of $\mathbb{C} \setminus \{-1, 1\}$. Hence for pure braids $b \in \mathcal{PB}$ the estimate of the extremal length $\Lambda(b_{tr})$ with totally real boundary values can be given in terms of the class of b in $\mathcal{PB}_3\diagup\langle\Delta_3^2\rangle$ which can be identified with an element of the fundamental group of the twice punctured complex plane (see Theorem 9.1).

We consider now arbitrary 3-braids (not necessarily pure braids). The following lemma is needed to reduce the case of arbitrary 3-braids to the case of pure 3-braids.

Lemma 9.8 *Any braid $b \in \mathcal{B}_3$ which is not a power of Δ_3 can be written in a unique way in the form*

$$\sigma_j^k \, b_1 \, \Delta_3^\ell \tag{9.54}$$

where $j = 1$ or $j = 2$, $k \neq 0$ is an integer, ℓ is a (not necessarily even) integer, and b_1 is a word in σ_1^2 and σ_2^2 in reduced form. If b_1 is not the identity, then the first term of b_1 is a non-zero even power of σ_2 if $j = 1$, and b_1 is a non-zero even power of σ_1 if $j = 2$.

For an integer $j \neq 0$ we denote by $q(j)$ that even integer neighbour of j, which is closest to zero. In other words, $q(j) = j$ for each even integer $j \neq 0$. For each odd integer j, $q(j) = j - \text{sgn}(j)$, where $\text{sgn}(j)$ for a non-zero integer number j equals 1 if j is positive, and -1 if j is negative. For a braid in form (9.54) we put $\vartheta(b) \overset{def}{=} \sigma_j^{q(k)} \, b_1$.

Theorem 9.4 *Let $b \in \mathcal{B}_3$ be a (not necessarily pure) braid which is not a power of Δ_3, and let w be the word representing the image of $\vartheta(b)$ in $\mathcal{B}_3 / \langle \Delta_3^2 \rangle$. Then*

$$\frac{1}{2\pi} \mathcal{L}(w) \leq \Lambda_{tr}(b) \leq 2^{14} \cdot \mathcal{L}(w) ,$$

except in the case when $b = \sigma_j^k \, \Delta_3^\ell$ where $j = 1$ or $j = 2$, $k \neq 0$ is an integer number, and ℓ is an arbitrary integer. In the exceptional case $\Lambda_{tr}(b) = 0$.

Notice that the periodic 3-braids are among the exceptional cases. We have $\vartheta(\Delta_3^\ell) = \text{id}$, and $\vartheta(\sigma_1\sigma_2\Delta_3^\ell) = \vartheta(\sigma_2^{-1}\Delta_3^{\ell+1}) = id$, and similarly for $(\sigma_1\sigma_2)^{-1}\Delta_3^\ell$ and $(\sigma_2\sigma_1)^{\pm 1}\Delta_3^\ell$.

Proof of Lemma 9.8 Let τ_3 be the natural homomorphism from the braid group $\mathcal{B}_3$ to the symmetric group $\mathcal{S}_3$. When $\tau_3(b)$ is the identity then the braid b is pure and the statement is clear.

Assume $\tau_3(b)$ is not the identity. We consider first the case when $\tau_3(b)$ is a transposition. If $\tau_3(b) = (13)$ then $b\Delta_3^{-1}$ is a pure braid, hence if b is not a power of Δ_3 it can written in the form (9.54) with k even and ℓ odd. If $\tau_3(b) = (12)$ then $\sigma_1^{-1}b$ is a pure braid. Hence, b can be written in the form (9.54) with $j = 1$, k odd and ℓ even. If $\tau_3(b) = (23)$ then $\sigma_2^{-1}b$ is a pure braid. Hence, b can be written in the form (9.54) with $j = 2$, k odd and ℓ even.

It remains to consider the case when $\tau_3(b)$ is a cycle. Suppose, $\tau_3(b) = (123)$. Since $(123)(13) = (12)$ the relation $\tau_3(b)(13)^{-1} = (12)$ holds. Hence, $\sigma_1^{-1}b\Delta_3^{-1}$ is a pure braid and b has the form (9.54) with $j = 1$, and k and ℓ odd. If $\tau_3(b) = (132)$ then, since $(132)(13) = (23)$ the braid $\sigma_2^{-1}b\Delta_3^{-1}$ is a pure braid and b can be written in the form (9.54) with $j = 2$, and k and ℓ odd. □

Proof of the Lower Bound in Theorem 9.4 Since $\Lambda(b_{tr}) = \Lambda((b\Delta_3)_{tr})$ (see Lemma 3.5) we may suppose that $b = \sigma_j^k b_1$ for a pure braid b_1 which is a reduced word in σ_1^2 and σ_2^2. We may suppose that k is an odd integer. The case when k is even follows from Theorem 9.1 and Lemma 1.2. Let $j = 1$. Consider a holomorphic mapping g of a rectangle R to $C_3(\mathbb{C})\diagup\mathcal{S}_3$ that represents the braid and lift it to a mapping $\tilde{g}$ into $C_3(\mathbb{C})$ so that the continuous extension of the mapping to the closed rectangle takes the open lower side to $\{(x_1, x_2, x_3) \in \mathbb{R}^3 : x_2 < x_1 < x_3\}$. Apply the mapping $\mathfrak{C}$. (For notation see also the proof of Lemma 1.2). We obtain a holomorphic mapping $\mathfrak{C}(\tilde{g}) : R \to \mathbb{C}\setminus\{-1, 1\}$, $\mathfrak{C}(\tilde{g})(z) = 2\,\frac{\tilde{g}_2(z)-\tilde{g}_1(z)}{\tilde{g}_3(z)-\tilde{g}_1(z)} - 1$, $z \in R$, whose continuous extension to the closure takes the open lower side of the rectangle to $(-\infty, -1)$.

Consider first the case when $|k| \ge 3$. We represent b_{tr} by a curve γ in $C_3(\mathbb{C})\diagup\mathcal{S}_3$, such that taking the lift $\tilde{\gamma}$ with initial point in $\{(x_1, x_2, x_3) \in \mathbb{R}^3 : x_2 < x_1 < x_3\}$, the curve $\mathfrak{C}(\tilde{\gamma})$ is the union of the following two curves: a half-circle in $\mathbb{C} \setminus \{-1, 1\}$ with endpoints in $\mathbb{R} \setminus \{-1, 1\}$ that is contained in the lower half-plane if $k > 0$, or in the upper half-plane if $k < 0$, and joins a point in $(-\infty, -1)$ with a point in $(-1, 1)$, and a curve representing the element $\mathfrak{C}_*(\vartheta(b)_{tr}) \in \pi_1^{tr}$. It can be proved along the same lines as the proof of Statement 3 of Lemma 9.5, that the rectangle R contains a curvilinear rectangle such that the restriction of the mapping $\mathfrak{C}(\tilde{g})$ to it represents $\mathfrak{C}_*(\vartheta(b)_{tr})$. Therefore $\Lambda(b_{tr}) \ge \Lambda(\vartheta(b)_{tr})$. If $\vartheta(b) \ne \sigma_1^{2k'}$ for an integer k', in other words, if $\vartheta(b)$ is not among the exceptional cases of Theorem 9.1, the lower bound holds.

Suppose $|k| = 1$. If $b \ne \sigma_1^{\pm 1}$ then b_{tr} can be represented by a curve γ for whose lift $\tilde{\gamma}$ with initial point in $\{(x_1, x_2, x_3) \in \mathbb{R}^3 : x_2 < x_1 < x_3\}$ the curve $\mathfrak{C}(\tilde{\gamma})$ is the union of the following two curves: a quarter-circle Γ_0 in $\mathbb{C} \setminus \{-1, 1\}$ in the upper or lower half-plane which joins a point in $(-\infty, -1)$ with a point in $i\mathbb{R}$, and a curve Γ_1 that represents $\mathfrak{C}_*({}_{pb}\vartheta(b)_{tr})$. Following along the lines of proof of Statement 3 of Lemma 9.5, we see that the rectangle R contains a curvilinear rectangle such that the restriction of the mapping $\mathfrak{C}(\tilde{g})$ to it represents $\mathfrak{C}_*({}_{pb}\vartheta(b)_{tr})$. The lower bound for this case follows from Theorem 9.2.

The proof for the case $b = \sigma_2^k b_1$ (with b_1 a reduced word in σ_1^2 and σ_2^2) is similar and is left to the reader. In this case the lift to $C_3(\mathbb{C})$ of a representing mapping for b_{tr} is chosen with initial point in $\{x_1 < x_3 < x_2\}$.

The lower bound of Theorem 9.4 is proved. □

Proof of the Upper Bound of Theorem 9.4 We prove the upper bound for 3-braids that are not pure and not among the exceptional cases. We write again such a braid in the form (9.54), and assume that the braid equals $b = \sigma_1^k b_1$ for an odd natural number k and b_1 a pure braid as in Lemma 9.8 which is not the identity. We want to find a holomorphic mapping g_2 from a rectangle $\mathbb{C} \setminus \{-1, 1\}$, such that the mapping $(-1, g_2, 1)$ from the rectangle into $C_3(\mathbb{C})$ projects under $\mathcal{P}_{\text{sym}}$ to a holomorphic maping that represents b_{tr}. If $k \ge 3$ (or $k \le -3$, respectively) we first represent a lift under $f_1 \circ f_2$ of the homotopy class of a half-circle with center -1 in the lower half-plane (in the upper half-plane, respectively), that joins $(-\infty, -1)$ with $(-1, 1)$, by a suitable holomorphic mapping g from a rectangle to $\mathbb{C} \setminus iZ$. Then we represent a lift of $\mathfrak{C}_*({}_{tr}\vartheta(b)_{tr})$ under $f_1 \circ f_2$ by a suitable holomorphic mapping from a rectangle to $\mathbb{C} \setminus iZ$. Quasiconformal gluing of the two mappings yields a holomorphic mapping from a rectangle to $\mathbb{C}\setminus iZ$. Denote the projection under $f_1 \circ f_2$ of this mapping by g_2. The mapping $(-1, g_2, 1)$ defines a holomorphic mapping of a rectangle into $C_3(\mathbb{C})$ whose projection under $\mathcal{P}_{\text{sym}}$ represents b_{tr}.

If $k = \pm 1$ we represent a lift under $f_1 \circ f_2$ of the homotopy class of a quarter-circle with center -1 that joins $(-\infty, -1)$ with the imaginary axis by a suitable holomorphic mapping from a rectangle to $\mathbb{C}\setminus i\mathbb{Z}$, and represent a lift of $\mathfrak{C}_*({}_{pb}\vartheta(b)_{tr})$ under $f_1 \circ f_2$ by a suitable holomorphic mapping from a rectangle to $\mathbb{C} \setminus iZ$. After quasiconformal gluing of the two mappings we finish the proof of the upper bound as in the previous case. We leave the details to the reader. □

The space $C_3(\mathbb{R}\diagup\mathcal{S}_3)$ has a disadvantage especially when we want to deal with conjugacy classes of 3-braids. Namely, the codimension of $C_3(\mathbb{R}\diagup\mathcal{S}_3)$ in $C_3(\mathbb{C}\diagup\mathcal{S}_3)$ is bigger than one, hence, each curve in generic position avoids this space. It is convenient to consider the set

$$\mathcal{H} \stackrel{def}{=} \{\{z_1, z_2, z_3\} \in C_3(\mathbb{C})\diagup\mathcal{S}_3 : \text{the three points } z_1, z_2, z_3 \\ \text{are contained in a real line in the complex plane}\}. \tag{9.55}$$

The set $\mathcal{H}$ is a smooth real hypersurface of $C_3(\mathbb{C})\diagup\mathcal{S}_3$. Indeed, let $\{z_1^0, z_2^0, z_3^0\}$ be a point of the symmetrized configuration space. Introduce coordinates near this point by lifting a neighbourhood of the point to a connected open set in $C_3(\mathbb{C})$ equipped with coordinates (z_1, z_2, z_3). Since the linear map $M(z) \stackrel{def}{=} \frac{z-z_1}{z_3-z_1}$, $z \in \mathbb{C}$, maps the points z_1 and z_3 to the real axis, the three points z_1, z_2, and z_3 lie on a real line in the complex plane iff the imaginary part of $z_2' \stackrel{def}{=} M(z_2) = \frac{z_2-z_1}{z_3-z_1}$ vanishes. The equation $\text{Im}\frac{z_2-z_1}{z_3-z_1} = 0$ in local coordinates (z_1, z_2, z_3) defines a local piece of a smooth real hypersurface.

The set $\mathcal{H}$ is not simply connected. Nevertheless we may define homotopy classes of curves in $C_3(\mathbb{C})\diagup\mathcal{S}_3$ with endpoints in this set.

For later use we need the following lemmas.

Lemma 9.9 *Let γ be a loop in $\mathcal{H}$ with base point $\{-1, 0, 1\} \in \mathcal{H}$. Then for some $k \in \mathbb{Z}$ the loop γ represents the braid Δ_3^k with this base point.*

Proof Consider the lift $\tilde{\gamma} = (\tilde{\gamma}_1, \tilde{\gamma}_2, \tilde{\gamma}_3) : [0, 1] \to C_3(\mathbb{C})$ of γ with $\tilde{\gamma}(0) = (-1, 0, 1)$. By a homotopy of γ in $\mathcal{H}$ with fixed endpoints we may assume that for the lift $\tilde{\gamma} = (\tilde{\gamma}_1, \tilde{\gamma}_2, \tilde{\gamma}_3)$ for all $t \in [0, 1]$ the equality $\tilde{\gamma}_2(t) = 0$ holds and the segment $[\tilde{\gamma}_1(t), \tilde{\gamma}_3(t)]$ has length 2. Then for each point $\tilde{\gamma}(t)$ there is a unique complex number $e^{i\alpha(t)}$ such that $\alpha(0) = 0$ and $\tilde{\gamma}(t) = e^{i\alpha(t)}\big((-1, 0, 1)\big)$ and the $\alpha(t)$ depend continuously on $t \in [0, 1]$. Since $\tilde{\gamma}(1)$ is equal to $(-1, 0, 1)$ or to $(1, 0, -1)$, the equality $e^{i\alpha(1)} = \pm 1$ holds. The curve $t \to e^{i\alpha(t)}$, $t \in [0, 1]$, is homotopic as a curve in the unit circle with fixed endpoints to the curve $t \to e^{\pi kti}$, $t \in [0, 1]$. Hence, $\tilde{\gamma}$ is homotopic with fixed endpoints to the curve $t \to e^{\pi kti}(-1, 0, 1)$, $t \in [0, 1]$ for an integer number k.

After applying the projection $\mathcal{P}_{\text{sym}}$ we obtain the curve $t \to e^{\pi k t i}\{-1, 0, 1\}$, $t \in [0, 1]$, in $C_3(\mathbb{C})\diagup\mathcal{S}_3$ that is homotopic with fixed base point to γ and represents Δ_3^k. □

Lemma 9.10 *Suppose for a conjugacy class $\hat{b}$ there exists a representing loop that avoids $\mathcal{H}$. Then $\hat{b}$ is the conjugacy class of a periodic braid of the form $(\sigma_1\sigma_2)^k$ for an integer k.*

Proof Let γ be a loop that represents $\hat{b}$ and avoids $\mathcal{H}$. Choose a base point on γ, so that γ represents a braid b. For some natural number k the power b^k of b is a pure braid, and the curve $\gamma^k \overset{def}{=} \underbrace{\gamma \circ \ldots \gamma}_{k}$ represents the pure braid b^k and does not meet $\mathcal{H}$. Then $\mathfrak{C}(\gamma^k) = 1 - 2\frac{\gamma_2^k - \gamma_1^k}{\gamma_3^k - \gamma_1^k}$ is a loop in $\mathbb{C} \setminus \{-1, 1\}$ that avoids $\mathbb{R}$. Here γ_j^k are the coordinate functions of γ^k. This means that $\mathfrak{C}(\gamma^k)$ is contractible, hence γ^k represents $\Delta^{2\ell}$ for an integer ℓ, and hence γ represents a periodic braid.

If a representative $\gamma : [0, 1] \to C_3(\mathbb{C})\diagup\mathcal{S}_3$, $\gamma(0) = \gamma(1)$, of a 3-braid b avoids $\mathcal{H}$, then the associated permutation $\tau_3(b)$ cannot be a transposition. Indeed, assume the contrary. Then there is a lift $\tilde{\gamma}$ of γ to $C_3(\mathbb{C})$, for which $(\tilde{\gamma}_1(1), \tilde{\gamma}_2(1), \tilde{\gamma}_3(1)) = (\tilde{\gamma}_3(0), \tilde{\gamma}_2(0), \tilde{\gamma}_1(0))$. Let L_t be the line in $\mathbb{C}$ that contains $\tilde{\gamma}_1(t)$ and $\tilde{\gamma}_3(t)$, and is oriented so that running along L_t in positive direction we meet first $\tilde{\gamma}_1(t)$ and then $\tilde{\gamma}_3(t)$. The point $\tilde{\gamma}_2(0)$ is not on L_0. Assume without loss of generality, that it is on the left of L_0 with the chosen orientation of L_0. Since for each $t \in [0, 1]$ the three points $\tilde{\gamma}_1(t)$, $\tilde{\gamma}_2(t)$ and $\tilde{\gamma}_3(t)$ in $\mathbb{C}$ are not on a real line, the point $\tilde{\gamma}_2(t)$ is on the left of L_t with the chosen orientation. But the unoriented lines L_0 and L_1 coincide, and their orientation is opposite. This implies $\tilde{\gamma}_2(1) \neq \tilde{\gamma}_2(0)$, which is a contradiction. We proved that the braid b is periodic with period 3 or a power of $\Delta_3^{2\ell}$. □

Lemma 9.11 *Let $A = \{z \in \mathbb{C} : r_1 < |z| < r_2\}$ be an annulus and $g : A \to C_3(\mathbb{C})\diagup\mathcal{S}_3$ a holomorphic mapping that extends holomorphically to a neighbourhood of the closure $\bar{A}$ and represents a conjugacy class $\hat{b}$ of braids. Suppose there exists a smooth arc in a neighbourhood of $\bar{A}$ that intersects A along an arc L_0, with endpoints on different boundary circles, such that $g(L_0) \subset \mathcal{H}$.*

Take any point $q \in L_0$ and a positively oriented dividing curve α in A with initial and terminal points equal to q that does not intersect L_0 at any other point. Let $\mathfrak{A}$ be a complex linear mapping for which $\mathfrak{A}(q) = \{0, x, 1\} \in C_3(\mathbb{R})\diagup\mathcal{S}_3$ with $x \in (0, 1)$. Denote by b_{tr} the element of $\pi_1(C_3(\mathbb{C})\diagup\mathcal{S}_3, C_3(\mathbb{R})\diagup\mathcal{S}_3)$ represented by $\mathfrak{A}(g) \mid \alpha$, and by b the respective braid. Then $b \in \hat{b}$ and $\Lambda(b_{tr}) \leq \lambda(A)$.

If $\hat{b}$ is represented by non-periodic braids and g intersects $\mathcal{H}$ transversally, then an arc L_0 as above exists.

Proof Since the mapping g is free homotopic to the mapping $\mathfrak{A}(g)$ for any complex affine mapping $\mathfrak{A}$, we may assume that for the mapping g itself $g(q) \in C_3(\mathbb{R})\diagup\mathcal{S}_3$, and, moreover, $g(q) = \{0, x, 1\}$. The set $A \setminus L_0$ can be considered as curvilinear rectangle with curvilinear horizontal sides being copies of L_0. Let $\omega(z) : A \setminus L_0 \to R$ be the conformal mapping of the curvilinear rectangle onto a rectangle of the form $R = \{z \in \mathbb{C} : \mathrm{Re} z \in (0, 1),\ \mathrm{Im} z \in (0, \mathsf{a})\}$, that takes the curvilinear side of $A \setminus L_0$, that is attained by moving in $A \setminus L_0$ clockwise towards L_0, to the lower side of R. (Note that the number a is uniquely defined by $A \setminus L_0$.)

Consider the restriction $g \mid A \setminus L_0$ and take the lift $\tilde{g} = (\tilde{g}_1, \tilde{g}_2, \tilde{g}_3) : A \setminus L_0 \to C_3(\mathbb{C})$ of g to a mapping from $A \setminus L_0$ to the configuration space $C_3(\mathbb{C})$, for which the clockwise continous extension to q equals $(0, x, 1)$ with $x \in (0, 1)$. For $z \in A \setminus L_0$ we consider the complex affine mapping

$$\mathfrak{A}_z(\zeta) = \mathfrak{a}(z)\zeta + \mathfrak{b}(z) \stackrel{def}{=} \frac{\zeta - \tilde{g}_1(z)}{\tilde{g}_3(z) - \tilde{g}_1(z)}, \ \zeta \in \mathbb{C}.$$

We define the holomorphic mapping

$$\hat{g}(z) \stackrel{def}{=} \mathfrak{A}_z(\tilde{g}(z)) = (0, \frac{\tilde{g}_2(z) - \tilde{g}_1(z)}{\tilde{g}_3(z) - \tilde{g}_1(z)}, 1), \ z \in A \setminus L_0,$$

and put $\hat{g}_{\mathrm{sym}}(z) \stackrel{def}{=} \mathcal{P}_{\mathrm{sym}}(\hat{g})(z) = \mathfrak{A}_z(g(z)) = \{0, \frac{\tilde{g}_2(z) - \tilde{g}_1(z)}{\tilde{g}_3(z) - \tilde{g}_1(z)}, 1\}, z \in A \setminus L_0$.

The mapping $\hat{g}_{\mathrm{sym}}$ extends continuously to the horizontal sides of $A \setminus L_0$. Since for a triple of complex numbers $\{Z_1, Z_2, Z_2\}$ the ratio $\frac{Z_2 - Z_1}{Z_3 - Z_1}$ is real if and only if the points Z_1, Z_2, and Z_3 lie on a real line and g maps the horizontal sides of $A \setminus L_0$ to $\mathcal{H}$, the mapping $\hat{g}_{\mathrm{sym}}$ takes the horizontal sides of $A \setminus L_0$ into $C_3(\mathbb{R})/\mathcal{S}_3$. Hence, the mapping $\hat{g}_{\mathrm{sym}}$ on $A \setminus L_0$ represents an element $b'_{tr} \in \pi_1(C_3(\mathbb{C})/\mathcal{S}_3, C_3(\mathbb{R})/\mathcal{S}_3)$. We obtain $\Lambda(b'_{tr}) \leq \lambda(A \setminus L_0)$. By Corollary 3.1

$$\Lambda(b'_{tr}) \leq \lambda(A).$$

By construction the mappings $t \to g(\alpha(t))$ and $t \to \hat{g}_{\mathrm{sym}}(\alpha(t))$, $t \in [0, 1]$, take the same value at 0. Moreover, for each $t \in (0, 1)$ the value $\hat{g}_{\mathrm{sym}}(\alpha(t))$ is obtained by applying to the value $g(\alpha(t))$ a complex linear mapping, i.e. $\hat{g}_{\mathrm{sym}}(\alpha(t)) = \mathfrak{a}(\alpha(t)) \cdot g(\alpha(t)) + \mathfrak{b}(\alpha(t))$, $t \in [0, 1]$, for continuous functions $\mathfrak{a}$ and $\mathfrak{b}$ with $\mathfrak{a}$ nowhere vanishing, $\mathfrak{b}(\alpha(0)) = 0$, $\mathfrak{a}(\alpha(0)) = 1$, and $\mathfrak{b}(\alpha(1))$ and $\mathfrak{a}(\alpha(1))$ real valued. The function $\mathfrak{b} \circ \alpha : [0, 1] \to \mathbb{C}$ is homotopic with endpoints in $\mathbb{R}$ to the function that is identically equal to zero. The mapping $\mathfrak{a} \circ \alpha : [0, 1] \to \mathbb{C} \setminus \{0\}$ is homotopic with endpoints in $\mathbb{R}$ to $\frac{\mathfrak{a} \circ \alpha}{|\mathfrak{a} \circ \alpha|}$. Hence, the mappings $\hat{g}_{\mathrm{sym}}(\alpha)$ and $\frac{\mathfrak{a}(\alpha)}{|\mathfrak{a}(\alpha)|} g(\alpha)$ from $[0, 1]$ to $C_3(\mathbb{C})/\mathcal{S}_3$ are homotopic with endpoints in $C_3(\mathbb{R})/\mathcal{S}$.

There is a continous function β on $[0, 1]$ such that $\frac{\mathfrak{a} \circ \alpha(t)}{|\mathfrak{a} \circ \alpha(t)|} = e^{i\beta(t)}$ and $\beta(0) = 0$. Then $\beta(1) = k\pi$ for an integer number k. Put $\mathring{g}(z) = e^{-\frac{\pi k i z}{\mathsf{a}}} \cdot \hat{g}_{\mathrm{sym}}(z)$, $z \in R$. Since the curves $t \to \frac{\mathfrak{a} \circ \alpha(t)}{|\mathfrak{a} \circ \alpha(t)|}$ and $t \to e^{i\beta(t)}$, $t \in [0, 1]$, are homotopic with fixed endpoints, the restrictions of $\mathring{g} \mid \alpha$ and $g \mid \alpha$ are homotopic. Hence, the equality $(b' \Delta_3^{-k})_{tr} = b_{tr}$ holds, and by Lemma 3.5

$$\Lambda(b_{tr}) \leq \lambda(A).$$

To prove the last statement of the lemma we notice that the set $\{z \in A : g(z) \in \mathcal{H}\}$ is a smooth manifold of codimension 1 in a neighbourhood of $\bar{A}$. If it did not contain an arc that joins the two boundary circles there would exist a simple closed dividing curve for the annulus whose image under g avoids $\mathcal{H}$. This would contradict Lemma 9.10. The lemma is proved. □

Chapter 10
Counting Functions

In this chapter we prove that the number of elements of $\mathcal{B}_3/\mathcal{Z}_3$ with positive Λ_{tr} not exceeding a positive number Y grows exponentially. As a corollary we give an alternative proof of the exponential growth of the number of conjugacy classes of elements of $\mathcal{B}_3/\mathcal{Z}_3$ that have positive entropy not exceeding Y. The corollary is a particular case of results of Veech and Eskin-Mirzakhani. Our proof does not use deep techniques from Teichmüller theory.

The first result that states exponential growth of the entropy counting function is due to Veech [80]. More precisely, he considered conjugacy classes of pseudo-Anosov elements of the mapping class group of a closed Riemann surface S, maybe with distinguished points E_n, with hyperbolic universal covering of $S \setminus E_n$, and proved that the number of classes with entropy not exceeding a positive number Y, grows exponentially in Y. Notice that the 3-braids with positive entropy are exactly the 3-braids that correspond to pseudo-Anosov elements of the mapping class group of $\mathbb{P}^1$ with fixed point ∞ and three other distinguished points. The precise asymptotic for the entropy counting function is given by Eskin and Mirzakhani [20] for closed Riemann surfaces of genus at least 2. Both papers, [80] and [20], use deep techniques of Teichmüller theory, in particular, these papers are based on the study of the Teichmüller flow.

We give special attention to the growth estimates for the counting function related to the extremal length Λ_{tr} of braids. The reason is that the extremal length with totally real boudary values is more flexible for applications to Gromov's Oka Principle than the entropy (equivalently the extremal length of conjugacy classes of braids). In particular, estimates of the counting function related to the extremal length Λ_{tr} allow to obtain effective finiteness theorems in the spirit of the Geometric Shafarevich Conjecture for the case when the base manifold is of second kind ([46]). We will address this application of the concept of extremal length in Chap. 11. In this chapter we obtain as a corollary of the results on the counting function related to Λ_{tr} an alternative proof of the exponential growth of the number of conjugacy

B. Jöricke, *Braids, Conformal Module, Entropy, and Gromov's Oka Principle*, Lecture Notes in Mathematics 2360, https://doi.org/10.1007/978-3-031-67288-0_10

classes of elements of $\mathcal{B}_3/\mathcal{Z}_3$ that have positive entropy not exceeding Y. Our proof follows [48] and does not use deep techniques from Teichmüller theory.

Notice that the bounds in the theorems of the present chapter can be improved. For the sake of simplicity of the proof we restrict ourselves to these estimates.

The first theorem of this chapter treats the pure braid group modulo its center, $\mathcal{PB}_3/\mathcal{Z}_3$. Recall that $\mathcal{PB}_3/\mathcal{Z}_3$ is a free group in two generators $\sigma_j^2/\mathcal{Z}_3$, $j = 1, 2$, which is isomorphic to the fundamental group $\pi_1(\mathbb{C} \setminus \{-1, 1\}, 0)$ with standard generators a_j (see Lemma 1.2). The isomorphism takes the generators $\sigma_j^2/\mathcal{Z}_3$ to the generators a_j.

By Lemma 3.5 for each element $\boldsymbol{b} \in \mathcal{PB}_3/\mathcal{Z}_3$ the quantities $\Lambda_{tr}(\boldsymbol{b})$ and $\Lambda(\hat{\boldsymbol{b}})$ are well defined by $\Lambda_{tr}(\boldsymbol{b}) = \Lambda_{tr}(b)$ and $\Lambda(\hat{\boldsymbol{b}}) = \Lambda(\hat{b})$ for any 3-braid b representing the class $\boldsymbol{b}$.

The counting function $N^{\Lambda}_{\mathcal{PB}_3}(Y)$, $Y \in (0, \infty)$, is defined as follows. For each positive parameter Y the value of $N^{\Lambda}_{\mathcal{PB}_3}(Y)$ is equal to the number of elements $\boldsymbol{b} \in \mathcal{PB}_3/\mathcal{Z}_3$ with $0 < \Lambda_{tr}(\boldsymbol{b}) \leq Y$. Note that here we count elements of $\mathcal{PB}_3/\mathcal{Z}_3$ rather than conjugacy classes of such elements. We wish to point out that (with the mentioned choice of the generators) the condition $\Lambda_{tr}(\boldsymbol{b}) > 0$ excludes exactly the classes in $\mathcal{PB}_3/\mathcal{Z}_3$ of the even powers σ_j^{2k} of the standard generators of the braid group $\mathcal{B}_3$ (see Theorem 9.1 in Sect. 9.1). Notice that $N^{\Lambda}_{\mathcal{PB}_3} = N^{\Lambda}_{\mathbb{C}\setminus\{-1,1\}}$ for the counting function $N^{\Lambda}_{\mathbb{C}\setminus\{-1,1\}}(Y)$, that counts the number of elements of the fundamental group $\pi_1(\mathbb{C} \setminus \{-1, 1\}, 0)$ with positive Λ_{tr} smaller than Y.

Theorem 10.1 *For all positive numbers $Y \geq 12\pi$ the inequality*

$$\frac{1}{2}\exp(\frac{\log 2}{6\pi}Y) \leq N^{\Lambda}_{\mathcal{PB}_3}(Y) \leq \frac{1}{2}e^{6\pi Y} \tag{10.1}$$

holds. The upper bound is true for all $Y > 0$.

Consider arbitrary 3-braids. The counting function $N^{\Lambda}_{\mathcal{B}_3}(Y)$, $Y \in (0, \infty)$, is defined as the number of elements $\boldsymbol{b} \in \mathcal{B}_3/\mathcal{Z}_3$ with $0 < \Lambda_{tr}(\boldsymbol{b}) \leq Y$. Note that the condition $\Lambda_{tr}(\boldsymbol{b}) > 0$ excludes exactly the elements $\boldsymbol{b} \in \mathcal{B}_3/\mathcal{Z}_3$ that are represented by $b = \sigma_j^k \Delta_3^\ell$ for $j = 1$ or 2, and $\ell = 0$ or 1. (See Lemma 9.8 and Theorem 9.4.)

Theorem 10.2 *For any positive number $Y \geq 12\pi$ the inequality*

$$\frac{1}{2}\exp(\frac{\log 2}{6\pi}Y) \leq N^{\Lambda}_{\mathcal{B}_3}(Y) \leq 4e^{6\pi Y} \tag{10.2}$$

holds. The upper bound is true for all $Y > 0$.

The entropy counting function $N^{entr}_{\mathcal{B}_n}(Y)$, $Y > 0$, for n-braids is defined as the number of conjugacy classes of pseudo-Anosov elements of $\mathcal{B}_n/\mathcal{Z}_n$ with entropy not exceeding Y. For 3-braids the value $N^{entr}_{\mathcal{B}_3}(Y)$, $Y > 0$, is also equal to the number of conjugacy classes of elements of $\mathcal{B}_3/\mathcal{Z}_3$ with positive entropy not exceeding Y. Indeed, by Lemma 6.4 the reducible elements of $\mathcal{B}_3/\mathcal{Z}_3$ are exactly the conjugates of powers of $\sigma_1/\mathcal{Z}_3$. By the remarks following the proof in Example 3.8 of Sect. 3.2 their entropy equals zero. Since periodic elements of $\mathcal{B}_3/\mathcal{Z}_3$ have zero entropy, the pseudo-Anosov 3-braids are exactly the 3-braids of positive entropy. The following theorem will be obtained as a corollary of the Main Theorem and Theorem 10.2.

Theorem 10.3 *For any number $Y \geq \frac{70}{3}\pi^2$ the estimate*

$$\frac{1}{8}e^{\frac{\log 2}{6\pi^2}Y} \leq N_{\mathcal{B}_3}^{entr}(Y) \leq 4e^{12Y} \tag{10.3}$$

holds. The upper bound holds for all positive Y.

We will prove now Theorem 10.1 using the isomorphism from $\mathcal{PB}_3/\mathcal{Z}_3$ onto $\pi_1(\mathbb{C}\setminus\{-1,1\},0)$ that takes for $j=1,2$ the generator $\sigma_j^2/\mathcal{Z}_3$ to the generator a_j. An element of $\boldsymbol{b} \in \mathcal{PB}_3/\mathcal{Z}_3 \cong \pi_1(\mathbb{C}\setminus\{-1,1\},0)$ will often be identified with the reduced word in the $a_j \cong \sigma_j^2/\mathcal{Z}_3$ that represents $\boldsymbol{b}$.

The proof of Theorem 10.1 uses the syllable decomposition of words $w \in \pi_1(\mathbb{C}\setminus\{-1,1\},0)$ (see Definition 9.1). Recall that for a word $w \in \pi_1(\mathbb{C}\setminus\{-1,1\},0)$ in reduced form $w = a_{j_1}^{k_1} a_{j_2}^{k_2} \ldots$ each term $a_{j_i}^{k_i}$ with $|k_i| \geq 2$ is a syllable of form (1). Further, any maximal sequence of consecutive terms $a_{j_i}^{k_i}$, for which $|k_i| = 1$ and all k_i have the same sign, is a syllable of form (2) (if the number of terms is bigger than 1) or of form (3) (if the number of terms is equal to 1). This gives a uniquely defined decomposition into syllables. Recall that the degree or length of the syllable is the sum of absolute values of the exponents of terms appearing in the syllable. We recall also the convention that the number of syllables of the identity equals zero.

Recall (see (9.1)) that we associated to each word w in the a_j the value $\mathcal{L}(w) \overset{def}{=} \sum_j \log(2d_j + \sqrt{4d_j^2 - 1})$, where the sum runs over the degrees d_j of all syllables of the word. For convenience we estimate the value $\mathcal{L}(w)$ from below by a slightly simpler value. More precisely, for a non-trivial word $w \in \pi_1(\mathbb{C}\setminus\{-1,1\},0) \cong \mathcal{PB}_3/\mathcal{Z}_3$ we put $\mathcal{L}_-(w) \overset{def}{=} \sum \log(3d_k) \leq \mathcal{L}(w) = \sum \log(2d_k + \sqrt{4d_k^2 - 1})$, where each sum runs over the degrees d_k of all syllables of w.

The main ingredient of the proof is the following lemma.

Lemma 10.1 *Let $N_{\mathcal{PB}_3}^{\mathcal{L}_-}$ be the function whose value at any $Y > 0$ is the number of reduced words $w \in \pi_1(\mathbb{C}\setminus\{-1,1\},0)$, $w \neq Id$, for which $\mathcal{L}_-(w) \leq Y$. The following inequality*

$$N_{\mathcal{PB}_3}^{\mathcal{L}_-}(Y) \leq \frac{1}{2}e^{3Y} \tag{10.4}$$

holds.

We need some preparation for the proof of Lemma 10.1. Consider all finite tuples $(d_1, \ldots, d_j)$, where $j \geq 1$ is any natural number (depending on the tuple) and the $d_k \geq 1$ are natural numbers. Before proving Lemma 10.1 we estimate for $Y > 0$ the number $N^*(e^Y)$ of different ordered tuples $(d_1, d_2, \ldots, d_j)$ (with varying j) that may serve as the degrees of the syllables of words (counted from left to right) with $\sum_1^j \log(3d_k) \le Y$. Put $X = e^Y$. Then $N^*(X)$ is the number of distinct tuples with $\prod(3d_k) \le X$.

Fix a natural number j. Denote by $N_j^*(X)$, $X \ge 1$, the number of tuples $(d_1, \dots, d_j)$ for which $\prod_1^j(3d_k) \le X$. The number is not zero if and only if $j \le \frac{\log X}{\log 3}$. For $X \ge 1$ the equality

$$N^*(X) = \sum_{j\in\mathbb{Z}:\, 1\le j\le \frac{\log X}{\log 3}} N_j^*(X) \tag{10.5}$$

holds.

Notice first that $N_1^*(X) = [\frac{X}{3}]$, where $[x]$ denotes the largest integer not exceeding the positive number x. Indeed we are looking for the number of d_1's for which $1 \le d_1 \le \frac{X}{3}$.

The value of N_j^* for $j \ge 2$ is estimated by the following lemma. Notice that the lemma holds also for $j = 1$ if in the inequality (10.6) we define $0! \overset{def}{=} 1$.

Lemma 10.2 *Let $j \ge 2$. Then $N_j^*(X) = 0$ for $X < 3^j$. For $X \ge 3^j$*

$$N_j^*(X) \le \frac{1}{(j-1)!}\frac{1}{3}(\frac{2}{3})^{j-1}X\Big(\log\big(\frac{1}{3}(\frac{2}{3})^{j-1}X\big)\Big)^{j-1}. \tag{10.6}$$

Proof The number $N_2^*(X)$ is the number of tuples (d_1, d_2) for which $3d_1 \cdot 3d_2 \le X$. Since $d_1 d_2 \ge 1$, $N_2^*(X) = 0$ for $X < 3^2$. If $X \ge 3^2$ the inequality $d_1 \le \frac{X}{9}$ holds, and for given d_1 the number d_2 runs through all natural numbers with $1 \le d_2 \le \frac{X}{9d_1}$. Hence, for $X \ge 3^2$

$$N_2^*(X) \le \sum_{k\in\mathbb{Z}:\, 1\le k\le \frac{X}{3^2}} \frac{X}{3^2 k}. \tag{10.7}$$

Put $a = \frac{X}{3^2}$ and $k' = \frac{k}{a}$. Since for positive numbers k' and α with $k' > \alpha$ the inequality $\frac{1}{k'} \le \frac{1}{\alpha}\int_{k'-\alpha}^{k'} \frac{dx}{x}$ holds, we obtain

$$N_2^*(X) \le \sum_{k'\in\frac{1}{a}\mathbb{Z}:\, \frac{1}{a}\le k'\le 1} \frac{1}{k'} \le 2a \int_{\frac{1}{2a}}^{1} \frac{dx}{x} = 2a\log(2a) = \frac{1}{3}\cdot\frac{2}{3}X\cdot\log(\frac{1}{3}\cdot\frac{2}{3}X). \tag{10.8}$$

For $j \geq 3$ we provide induction using the following fact. Let $0 < x < 1$. Then for any positive integer j the value $(\frac{(-\log x)^j}{x})' = -\frac{(-\log x)^{j-1}}{x^2}(j - \log x)$ is negative. Hence, for $k' \in (0, 1)$ and $0 < \alpha < k'$

$$\frac{1}{k'}(-\log(k'))^j \leq \frac{1}{\alpha}\int_{k'-\alpha}^{k'} \frac{(-\log x)^j}{x} dx \,. \tag{10.9}$$

We saw that the lemma is true for $j = 2$. Suppose it is true for j. Prove that then it holds for $j+1$. The number $N^*_{j+1}(X)$ is the number of tuples $(d_1, \ldots, d_{j+1})$ for which $3d_1 \cdot \ldots \cdot 3d_{j+1} \leq X$. Hence, $N^*_{j+1}(X) = 0$ if $X < 3^{j+1}$. If $X \geq 3^{j+1}$, then $d_1 \leq \frac{X}{3^{j+1}}$ and for given d_1 the tuples $(d_2, \ldots, d_{j+1})$ run through all tuples with $3d_2 \cdot \ldots \cdot 3d_{j+1} \leq \frac{X}{3d_1}$. Hence, for $X \geq 3^{j+1}$

$$\begin{aligned} N^*_{j+1}(X) &= \sum_{k \in \mathbb{Z}:\, 1 \leq k \leq \frac{X}{3^{j+1}}} N^*_j\left(\frac{X}{3k}\right) \\ &\leq \sum_{k \in \mathbb{Z}:\, 1 \leq k \leq \frac{X}{3^{j+1}}} \frac{1}{(j-1)!}\frac{1}{3}\left(\frac{2}{3}\right)^{j-1}\frac{X}{3k}\left(\log\left(\frac{1}{3}\left(\frac{2}{3}\right)^{j-1}\frac{X}{3k}\right)\right)^{j-1}. \end{aligned} \tag{10.10}$$

Put $a = \frac{1}{9}(\frac{2}{3})^{j-1}X$ and $k' = \frac{k}{a}$. Then

$$\begin{aligned} N^*_{j+1}(X) \leq& \frac{1}{(j-1)!} \sum_{k' \in \frac{1}{a}\mathbb{Z}:\, \frac{1}{a} \leq k' \leq \frac{1}{2^{j-1}}} \frac{1}{k'}\left(\log\left(\frac{1}{k'}\right)\right)^{j-1} \\ \leq& \frac{1}{(j-1)!} 2a \int_{\frac{1}{2a}}^{\frac{1}{2^{j-1}}} \frac{1}{x}(-\log x)^{j-1} dx \\ \leq& \frac{1}{(j-1)!} 2a \int_{\frac{1}{2a}}^{1} \frac{1}{x}(-\log x)^{j-1} dx \\ =& (-1)^{j-1}\frac{1}{(j-1)!} 2a \frac{1}{j}(\log x)^j \,\Big|_{\frac{1}{2a}}^{1} \,. \end{aligned} \tag{10.11}$$

We obtain

$$N^*_{j+1}(X) \leq \frac{1}{j!}2a\big(\log(2a)\big)^j = \frac{1}{j!}\frac{1}{3}\left(\frac{2}{3}\right)^j X\left(\log\left(\frac{1}{3}\left(\frac{2}{3}\right)^j X\right)\right)^j. \tag{10.12}$$

Lemma 10.2 is proved. □

Lemma 10.2 implies the following upper bound for N^*.

Lemma 10.3 *For $X < 3$ the function $N^*(X)$ vanishes. Moreover, for any positive number X*

$$N^*(X) \le \left(\frac{X}{3}\right)^{\frac{5}{3}}. \tag{10.13}$$

Proof of Lemma 10.3 Since all $N_j^*(X)$ vanish for $X < 3$ the value $N^*(X)$ vanishes for such X. For $X \ge 3$ the equality (10.5) and Lemma 10.2 imply

$$\begin{aligned} N^*(X) &\le \sum_{j\in\mathbb{Z}:\, 1\le j\le \frac{\log X}{\log 3}} \frac{1}{(j-1)!}\frac{1}{3}\left(\frac{2}{3}\right)^{j-1}X\Big(\log\big(\frac{1}{3}(\frac{2}{3})^{j-1}X\big)\Big)^{j-1} \\ &\le \frac{1}{3}X \sum_{j\in\mathbb{Z}:\, 1\le j\le \frac{\log X}{\log 3}} \frac{1}{(j-1)!}\left(\frac{2}{3}\log(\frac{1}{3}X)\right)^{j-1} \\ &\le \frac{1}{3}X\exp\big(\frac{2}{3}\log(\frac{1}{3}X)\big) = \left(\frac{X}{3}\right)^{\frac{5}{3}}. \end{aligned} \tag{10.14}$$

Lemma 10.3 is proved. □

Proof of Lemma 10.1 We assume that $[\frac{Y}{\log 3}] \ge 1$. Otherwise $N^*(e^Y)$ vanishes, and therefore $N_{\mathcal{PB}_3}^{\mathcal{L}_-}(Y) = 0$, and the inequality is satisfied. We will use the notation $N_j^{\mathcal{L}_-}(Y)$, $j \ge 1$, for the number of different reduced words w in $\pi_1(\mathbb{C}\setminus\{-1,1\},0)$ that consist of j syllables and satisfy the inequality $\mathcal{L}_-(w) = \sum_{k=1}^{j}\log(3d_k) \le Y$. Then $N_{\mathcal{PB}_3}^{\mathcal{L}_-}(Y) = \sum_{j=1}^{j_0} N_j^{\mathcal{L}_-}(Y)$ with $j_0 \overset{def}{=} [\frac{Y}{\log 3}]$. We will estimate $N_j^{\mathcal{L}_-}(Y)$ by $N_j^*(X)$ with $X = e^Y$. Recall that $N^*(e^Y)$ is the number of different tuples $(d_1,\dots,d_j)$ with $d_k \ge 1$ for which $\prod_{k=1}^{j}(3d_k) \le e^Y$.

For this purpose we take a tuple $(d_1,\dots,d_j)$ and estimate the number of different reduced words with tuple of lengths of syllables (from left to right) equal to $(d_1,\dots,d_j)$. The first syllable is completely determined by its first letter (which may be $a_1^{\pm 1}$ or $a_2^{\pm 1}$), its length, and the fact which of the following two options hold: either it is of form (1), or of form (2) or (3). Hence, there are at most 8 different choices for the first syllable if we require the syllable to have exactly degree d_1. For all other syllables the first letter of the syllable cannot be $a_i^{\pm 1}$ if the last letter in the preceding syllable is $a_i^{\pm 1}$ for the same a_i. Hence, for all but the first syllable there are at most 4 choices given the degree of the syllable and the preceding syllable.

We showed that for all $j = 1,\dots,j_0 = [\frac{Y}{\log 3}]$, and each tuple $(d_1,\dots,d_j)$, there are at most $2\cdot 4^j$ different reduced words with tuple of lengths of syllables equal to $(d_1,\dots,d_j)$. Hence, for $Y \ge \log 3$ the number $N_{\mathcal{PB}_3}^{\mathcal{L}_-}(Y)$ of reduced words $w \in \pi_1(\mathbb{C}\setminus\{-1,1\},0)$, $w \ne \mathrm{Id}$, with $\prod_1^{j_0}(3d_k) \le \exp(Y)$ equals

$$N_{\mathcal{PB}_3}^{\mathcal{L}_-}(Y) = \sum_{j=1}^{j_0} N_j^{\mathcal{L}_-}(Y) \le \sum_{j=1}^{j_0} 2\cdot 4^{j_0} N_j^*(e^Y) = 2\cdot 4^{j_0}\cdot N^*(e^Y). \tag{10.15}$$

Using the inequality $j_0 \le \frac{Y}{\log 3}$ and Lemma 10.3 with $X = e^Y$ we obtain the requested estimate by the value

$$2 \cdot 3^{-\frac{5}{3}} \exp((\frac{\log 4}{\log 3} + \frac{5}{3})Y) < \frac{1}{2} \exp(3Y). \tag{10.16}$$

We used the inequalities $(\frac{\log 4}{\log 3} + (\frac{5}{3})) < 2.93$ and $4 \cdot 3^{-\frac{5}{3}} < 0.65 < 1$. Lemma 10.1 is proved. □

Proof of Theorem 10.1 By Theorem 9.1 the inequality $\frac{1}{2\pi}\mathcal{L}_-(w) \le \frac{1}{2\pi}\mathcal{L}(w) \le \Lambda_{tr}(w)$ holds for all reduced words w representing elements in $\mathcal{PB}_3/\mathcal{Z}_3 \cong \pi_1(\mathbb{C} \setminus \{-1, 1\}, 0)$ that are not equal to a power of a_1 or of a_2 or to the identity (equivalently, for which $\Lambda_{tr}(w) > 0$). This inequality implies the inclusion $\{w : 0 < \Lambda_{tr}(w) \le Y\} \subset \{w \ne \mathrm{Id} : \mathcal{L}_-(w) \le 2\pi Y\}$. We obtain the inequality $N^{\Lambda}_{\mathcal{PB}_3}(Y) \le N^{\mathcal{L}_-}_{\mathcal{PB}_3}(2\pi Y)$ and by Lemma 10.1 the right hand side of this inequality does not exceed $\frac{1}{2}e^{6\pi Y}$. This gives the upper bound.

The lower bound is obtained as follows. Consider all reduced words in $\pi_1(\mathbb{C} \setminus \{-1, 1\}, 0)$ of the form

$$a_1^{2k_1} a_2^{2k_2} \ldots \tag{10.17}$$

where each k_i is equal to 1 or -1. If j is the number of syllables (i.e the number of the $a_i^{k_i}$) of a word w of the form (10.17), then by Proposition 9.4 the inequality $\Lambda_{tr}(w) \le \frac{35}{6}\pi j$ holds. Consider the words of the mentioned form for which $j = j_0 \stackrel{def}{=} [\frac{Y}{6\pi}]$. Since j_0 must be at least equal to 2 we get the condition $Y \ge 12\pi$. For the chosen j_0 the extremal length of the considered words does not exceed Y. The number of different words of such kind equals $2^{j_0} = 2^{[\frac{Y}{6\pi}]} \ge \exp(\log 2 \cdot (\frac{Y}{6\pi} - 1)) = \frac{1}{2}\exp(\frac{\log 2}{6\pi} Y)$. Theorem 10.1 is proved. □

Consider now arbitrary elements of the braid group modulo its center $\mathcal{B}_3/\mathcal{Z}_3$ and their extremal length with totally real horizontal boundary values.

Proof of Theorem 10.2 We will use the fact that by Lemma 9.8 any $b \in \mathcal{B}_3$ which is not a power of Δ_3 can be uniquely written in the form $\sigma_j^k\, b_1\, \Delta_3^\ell$, where k and ℓ are integers, b_1 is a word in σ_1^2 and σ_2^2 in reduced form whose first term is a non-zero power of $\sigma_{j'} \ne \sigma_j$ if $b_1 \ne \mathrm{id}$. For a braid in this form we defined $\vartheta(b) = \sigma_j^{q(k)}\, b_1$ where $q(j)$ is that even integer neighbour of j (including perhaps j itself), which is closest to zero. For each element $\boldsymbol{b}$ of $\mathcal{B}_3/\mathcal{Z}_3$ the value $\vartheta(\boldsymbol{b})$ is well defined by putting it equal to $\vartheta(b)$ for any $b \in \mathcal{B}_3$ that represents $\boldsymbol{b}$.

Take any element $\boldsymbol{b}$ of $\mathcal{PB}_3/\mathcal{Z}_3$. Choose its unique representative that can be written as a reduced word w in σ_1^2 and σ_2^2. We describe now all elements $\boldsymbol{b}$ of $\mathcal{B}_3/\mathcal{Z}_3$ with $\vartheta(\boldsymbol{b}) = w$. If $w \ne \mathrm{Id}$ these are the elements represented by the following braids. If the first term of w is σ_j^{2k} with $k \ne 0$, then the possibilities are $b = w\Delta_3^\ell$ with $\ell = 0$ or 1, $b = \sigma_j^{\mathrm{sgn}k} w\Delta_3^\ell$ with $\ell = 0$ or 1, or $b = \sigma_{j'}^{\pm 1} w\Delta_3^\ell$ with $\ell = 0$ or 1 and $\sigma_{j'} \neq \sigma_j$. Hence, for $w \neq \mathrm{Id}$ there are 8 possible choices of elements $\boldsymbol{b} \in \mathcal{B}_3/\mathcal{Z}_3$ with $\vartheta(\boldsymbol{b}) = w$. By Theorem 9.4 the set of $\boldsymbol{b} \in \mathcal{B}_3/\mathcal{Z}_3$ with $0 < \Lambda_{tr}(\boldsymbol{b}) \leq Y$ is contained in the set of $\boldsymbol{b} \in \mathcal{B}_3/\mathcal{Z}_3$ with $\vartheta(\boldsymbol{b}) = w \neq \mathrm{Id}$, $\mathcal{L}_-(w) \leq 2\pi Y$. We obtain

$$N_{\mathcal{B}_3}^{\Lambda}(Y) \leq 8 N_{\mathcal{PB}_3}^{\mathcal{L}_-}(2\pi Y). \tag{10.18}$$

By Lemma 10.1 we obtain $N_{\mathcal{B}_3}^{\Lambda}(Y) \leq 4e^{6\pi Y}$.

Since each pure 3-braid is also an element of the braid group $\mathcal{B}_3$ the lower bound of Theorem 10.1 provides also a lower bound for Theorem 10.2. Theorem 10.2 is proved. □

Proof of the Upper Bound of Theorem 10.3 For each conjugacy class $\hat{\boldsymbol{b}}$ of elements of $\mathcal{B}_3/\mathcal{Z}_3$ with $h(\hat{\boldsymbol{b}}) > 0$ and each positive number ε we will find a braid b for which $\widehat{b/\mathcal{Z}_3} = \hat{\boldsymbol{b}}$ and

$$\Lambda_{tr}(b) \leq \frac{2}{\pi} h(\hat{\boldsymbol{b}}) + \varepsilon\,. \tag{10.19}$$

For this purpose we represent a conjugacy class $\hat{b}$ that represents $\hat{\boldsymbol{b}}$ by a holomorphic map $g : A \to C_3(\mathbb{C})/\mathcal{S}_3$ from an annulus A of extremal length

$$\lambda(A) < \frac{2}{\pi} h(\hat{\boldsymbol{b}}) + \varepsilon \tag{10.20}$$

to the symmetrized configuration space. By the Holomorphic Transversality Theorem [51] we may assume, after shrinking A (keeping inequality (10.20)) and approximating g, that g is holomorphic in a neighbourhood A' of the closure $\bar{A}$ of A and is transversal to the smooth real hypersurface $\mathcal{H}$ (see Eq. (9.55)). Since $h(\hat{\boldsymbol{b}}) > 0$, the representing braids of $\hat{\boldsymbol{b}}$ are non-periodic and the conditions of Lemma 9.11 are satisfied. Hence, there exists a braid b for which $b/\mathcal{Z}_3$ represents $\hat{\boldsymbol{b}}$ such that

$$\Lambda(b_{tr}) \leq \lambda(A)\,.$$

By inequality (10.20) we obtain $\Lambda_{tr}(b) \leq \frac{2}{\pi} h(\hat{\boldsymbol{b}}) + \varepsilon$. Using also Lemma 3.5, we see that the number of conjugacy classes $\hat{\boldsymbol{b}}$ of $\mathcal{B}_3/\mathcal{Z}_3$ of positive entropy not exceeding Y does not exceed the number of elements $\boldsymbol{b} \in \mathcal{B}_3/\mathcal{Z}_3$ with $\Lambda_{tr}(\boldsymbol{b}) < \frac{2}{\pi} Y + \varepsilon$. In other words, $N_{\mathcal{B}_3}^{entr}(Y) \leq N_{\mathcal{B}_3}^{\Lambda}(\frac{2}{\pi} Y + \varepsilon)$. Since for ε we may take any a priory given positive number, Theorem 10.2 implies

$$N_{\mathcal{B}_3}^{entr}(Y) \leq N_{\mathcal{B}_3}^{\Lambda}(\frac{2}{\pi} Y) \leq 4e^{12Y}\,. \tag{10.21}$$

We obtained the upper bound. □

For obtaining the lower bound we need the following preparations.

Lemma 10.4 *Suppose b_1 and b_2 are elements of the free group $\mathbb{F}_n$ in n generators, that are both represented by cyclically reduced words. Then b_1 and b_2 are conjugate in $\mathbb{F}_n$ if and only if the word representing b_2 is obtained from the word representing b_1 by a cyclic permutation of terms.*

Proof It is enough to prove the following statement. If under the conditions of the lemma $b_2 = w^{-1} b_1 w$ for an element $w \in \mathbb{F}_n$, $w \neq \mathrm{Id}$, then $b_2 = w'^{-1} b_1' w'$ where b_1' is represented by a cyclically reduced word that is obtained from the reduced word representing b_1 by a cyclic permutation of terms, and w' is represented by a reduced word which has less terms than the word representing w.

This statement is proved as follows. Write $w = w_1 w'$ where $w_1 \in \mathbb{F}_n$ is represented by the first term of the word representing w, and $b_1 = a' B_1' a''$ where a' and a'' are represented by the first and the last term, respectively, of the word that represents b_1. Then $b_2 = w'^{-1} w_1^{-1} a' B_1' a'' w_1 w'$. If both relations $w_1^{-1} a' \neq \mathrm{Id}$ and $a'' w_1 \neq \mathrm{Id}$ were true, then the first and the last term of the reduced word representing b_2 would be a power of the same generator of $\mathbb{F}_n$, which contradicts the fact that b_2 can be represented by a cyclically reduced word. If either $w_1^{-1} a' = \mathrm{Id}$ or $a'' w_1 = \mathrm{Id}$, then the reduced words representing $b_1' \overset{def}{=} w_1^{-1} a' B_1' a'' w_1$ and b_1 are cyclic permutations of each other. Hence, the statement is true. □

Lemma 10.5 *The following equalities hold.*

$$\begin{aligned}
\Delta_3 \sigma_1 &= \sigma_2 \Delta_3 ,\\
\Delta_3 \sigma_2 &= \sigma_1 \Delta_3 ,\\
\sigma_1^{-1} (\sigma_2^{-4} \Delta_3^4) \sigma_1 &= \sigma_2^2 \sigma_1^2 \sigma_2^2 \sigma_1^2 ,\\
\sigma_2^{-1} (\sigma_1^{-4} \Delta_3^4) \sigma_2 &= \sigma_1^2 \sigma_2^2 \sigma_1^2 \sigma_2^2 .
\end{aligned} \tag{10.22}$$

Proof The third equality is obtained as follows

$$\begin{aligned}
&\sigma_1^{-1} (\sigma_2^{-4} \Delta_3^4) \sigma_1 = \sigma_1^{-1} \sigma_2^{-1} \Delta_3^2 \sigma_2^{-2} \Delta_3^2 \sigma_2^{-1} \sigma_1\\
=&(\sigma_1^{-1} \sigma_2^{-1})(\sigma_2 \sigma_1 \sigma_2)(\sigma_2 \sigma_1 \sigma_2) \sigma_2^{-1} \sigma_2^{-1} (\sigma_2 \sigma_1 \sigma_2)(\sigma_2 \sigma_1 \sigma_2) \sigma_2^{-1} \sigma_1\\
=&\sigma_2^2 \sigma_1^2 \sigma_2^2 \sigma_1^2 .
\end{aligned} \tag{10.23}$$

The fourth equality is obtained by conjugating the third equality by Δ_3. □

Consider two elements $\boldsymbol{b}_1$ and $\boldsymbol{b}_2$ of the pure braid group modulo center $\mathcal{PB}_3 / \mathcal{Z}_3$. In particular, $\boldsymbol{b}_1$ and $\boldsymbol{b}_2$ are elements of the full braid group modulo center $\mathcal{B}_3 / \mathcal{Z}_3$. We write a_1 and a_2 for the generators of $\mathcal{PB}_3 / \mathcal{Z}_3$.

Lemma 10.6 *Suppose both elements $\boldsymbol{b}_1$ and $\boldsymbol{b}_2$ of $\mathcal{PB}_3/\mathcal{Z}_3$ are of the form*

$$a_1^{\pm 2}a_2^{\pm 2}\cdots a_1^{\pm 2}a_2^{\pm 2} \quad or \quad a_2^{\pm 2}a_1^{\pm 2}\cdots a_2^{\pm 2}a_1^{\pm 2}, \tag{10.24}$$

with a positive number of terms, and the reduced word representing $\boldsymbol{b}_1$ has at least four terms. Then $\boldsymbol{b}_2$ cannot be conjugated to $\boldsymbol{b}_1$ by an element $\boldsymbol{\beta}$ of $\mathcal{B}_3/\mathcal{Z}_3$ that can be represented by a braid $\beta = \sigma_j\beta_1\Delta_3^\ell$ with some j and ℓ and β_1 being a word in σ_1^2 and σ_2^2.

(See Lemma 9.8 for writing a 3-braid in a special form.)

Proof of Lemma 10.6 Indeed, suppose the contrary,

$$\boldsymbol{b}_1 = \boldsymbol{\beta}^{-1}\boldsymbol{b}_2\boldsymbol{\beta} \tag{10.25}$$

with $\boldsymbol{\beta}$ represented by $\beta = \sigma_j\beta_1\Delta_3^\ell$, where $\ell = 0, 1$, and $\beta_1 \in \mathcal{PB}_3$ is a word in σ_1^2 and σ_2^2. For $j = 1, 2$, we let b_j be the representative of $\boldsymbol{b}_j$ which can be written as reduced word in σ_1^2 and σ_2^2. (In other words, for the representing braids b_j the first and third strand have linking number zero.) By Eq. (10.22) there is an integer number n such that the braid $b_2' \overset{def}{=} \sigma_j^{-1}b_2\sigma_j\Delta_3^{2n}$ is a product of a positive even number of factors which either have alternately the form $\sigma_1^{\pm 4}$ and $(\sigma_2^2\sigma_1^2\sigma_2^2\sigma_1^2)^{\pm 1}$, or they have alternately the form $\sigma_2^{\pm 4}$ and $(\sigma_1^2\sigma_2^2\sigma_1^2\sigma_2^2)^{\pm 1}$. The braid $\beta_1\Delta_3^\ell b_1\Delta_3^{-\ell}\beta_1^{-1}$ can also be written as reduced word in σ_1^2 and σ_2^2. Hence, by Eq. (10.25) the two braids b_2' and $\beta_1\Delta_3^\ell b_1\Delta_3^{-\ell}\beta_1^{-1}$ must be equal.

The element $b_2'/\mathcal{Z}_3$ is the product of at least two factors equal to $a_1^{\pm 2}$ and $(a_1a_2a_1a_2)^{\pm 1}$ alternately, or it is the product of at least two factors equal to $a_2^{\pm 2}$ and $(a_2a_1a_2a_1)^{\pm 1}$ alternately. Hence, the reduced word in a_1 and a_2 representing b_2' contains at least 4 terms, and each sequence of four consecutive terms of this word contains at least two terms that appear with power $+1$ or -1.

Indeed, put $A = a_1a_2a_1a_2$. Replace in the product $A^{\ell_1}a_1^{\pm 2}A^{\ell_2}$ with $\ell_1 = 0, \pm 1$, $\ell_2 = 0, \pm 1$, each factor A^{ℓ_1} by the reduced word in a_1, a_2, representing it. We obtain a (possibly not reduced) word in a_1, a_2.

If $\ell_1 = 0$ or $\ell_1 = 1$, and $\ell_2 = 0$ or -1, the word is already reduced.

If $\ell_1 = -1$, and $\ell_2 = 0$ or -1, the reduced word representing the product, consists of the first three letters of the word representing $A^{-1} = A^{\ell_1}$, a non-trivial power of a_1 and all letters of the word representing A^{ℓ_2}.

If $\ell_1 = 0$ or $\ell_1 = 1$, and $\ell_2 = 1$, the reduced word representing the product, consists of all letters of the word representing A^{ℓ_1}, a non-trivial power of a_1 and the last three letters of the word representing $A = A^{\ell_2}$.

Finally , if $\ell_1 = -1$ and $\ell_2 = 1$ then the reduced word representing the product, consists of the first three letters of the word representing $A^{-1} = A^{\ell_1}$, a non-trivial power of a_1 and the last three letters of the word representing $A = A^{\ell_2}$.

Hence, the reduced word representing $b_2'/\mathcal{Z}_3$ contains for each factor A^{ℓ_j} in the whole product at least the two middle letters of the word representing A^{ℓ_j}. The case, when $b_2'/\mathcal{Z}_3$ is the alternating product of $a_2a_1a_2a_1$ and a_2^2 is obtained by replacing the role of the generators a_1 and a_2.

Since the element $(\Delta_3^\ell b_1 \Delta_3^{-\ell})/\mathcal{Z}_3$ can be represented by a word of the form (10.24) with at least four terms, we arrived at the following conjugation problem in the free group $\mathbb{F}_2$. We have an element $B_1 \in \mathbb{F}_2$ of the from (10.24) with at least 4 terms, and an element $B_2 \in \mathbb{F}_2$ written as reduced word with at least 4 terms, such that each sequence of four consecutive terms of B_2 contains at least two terms that appear with power $+1$ or -1. We have to prove that there is no element of $\mathbb{F}_2$ that conjugates B_1 to B_2.

To prove this statement we assume the contrary. Suppose the reduced word $a_{j_1}^{k_1} \dots a_{j_\ell}^{k_\ell}$ representing B_2 is not cyclically reduced. Identify each term $a_{j_{\ell'}}^{k_{\ell'}}$ of the word with the element of $\mathbb{F}_2$ represented by it. The conjugate $(a_{j_1}^{k_1})^{-1} B_2 a_{j_1}^{k_1}$ can be represented by a reduced word. If this reduced word is not cyclically reduced, then the consecutive sequence of all its terms is also a consecutive sequence of terms of the reduced word representing B_2. Continue by induction in this way with the $a_{j_{\ell'}}^{k_{\ell'}}$ until we arrive at an element $B_2' \in \mathbb{F}_2$ that can be represented by a cyclically reduced word. At each inductive step the consecutive sequence of all terms of the new reduced but not cyclically reduced word is a consecutive sequence of terms of the previously obtained word and, hence, also a consecutive sequence of terms of B_2. Since B_2 is not the identity, the reduced word representing B_2' is not the identity and is cyclically reduced. The sequence of all its consecutive terms except the last one is also a sequence of consecutive terms of the reduced word representing B_2.

Since by our assumption B_2' is conjugate to B_1 by an element of $\mathbb{F}_2$, and both, B_2' and B_1, are represented by cyclically reduced words, the representing words have the same number of terms (see Lemma 10.4). By the assumption for B_1 the number of terms is at least 4. Hence, the words B_2' and B_2 have at least 3 consecutive terms in common. Therefore B_2' contains a term that appears with power ± 1 which contradicts Lemma 10.4. The lemma is proved. □

Proof of the Lower Bound of Theorem 10.3 Consider the elements $\boldsymbol{b} \in \mathcal{PB}_3/\mathcal{Z}_3 \subset \mathcal{B}_3/\mathcal{Z}_3$ which can be represented by words of the form

$$w = a_1^{\pm 2} a_2^{\pm 2} \cdots a_1^{\pm 2} a_2^{\pm 2} \tag{10.26}$$

in the generators $a_1 \cong \sigma_1^2/\mathcal{Z}_3$ and $a_2 \cong \sigma_1^2/\mathcal{Z}_3$ of $\mathcal{PB}_3/\mathcal{Z}_3$ with at least 4 terms. We denote the number of syllables (in this case the number of the terms $a_i^{k_i}$) of the word (10.26) by $2j$. Since each word of the form (10.26) is cyclically reduced, the Main Theorem and Proposition 9.5 imply that the entropy $h(\hat{\boldsymbol{b}})$ of the conjugacy class of the element $\boldsymbol{b} \in \mathcal{PB}_3/\mathcal{Z}_3$ represented by w satisfies the inequality

$$h(\hat{\boldsymbol{b}}) \leq \frac{\pi}{2} \cdot \frac{35}{6} \cdot 2\pi j \tag{10.27}$$

Take $j = j_0' \overset{def}{=} [\frac{Y}{\frac{35}{6}\pi^2}]$. For this choice of j_0' the inequality $h(\hat{\boldsymbol{b}}) \le Y$ holds. Since we required that the number of syllables $2j_0'$ is at least 4, we get the condition $Y \ge \frac{70}{3}\pi^2$. The number of different words of such kind equals $2^{2j_0'}$.

We prove now that the number of different conjugacy classes of $\mathcal{B}_3/\mathcal{Z}_3$ that can be represented by elements in $\mathcal{PB}_3$ corresponding to words (10.26) with $2j_0'$ syllables is not smaller than $\frac{2^{2j_0'}}{2j_0'}$. It is enough to prove the following claim. For each element $\boldsymbol{b} \in \mathcal{PB}_3/\mathcal{Z}_3$ of form (10.26) the number of elements of $\mathcal{PB}_3/\mathcal{Z}_3$ of form (10.26) that are conjugate to $\boldsymbol{b}$ by an element of $\mathcal{B}_3/\mathcal{Z}_3$ does not exceed $2j_0'$.

Suppose two elements $\boldsymbol{b}_1$ and $\boldsymbol{b}_2$ of $\mathcal{PB}_3/\mathcal{Z}_3 \subset \mathcal{B}_3/\mathcal{Z}_3$ are represented by a word of form (10.26) and belong to the same conjugacy class, i.e. $\boldsymbol{b}_2 = \boldsymbol{\beta}\boldsymbol{b}_1\boldsymbol{\beta}^{-1}$ for an element $\boldsymbol{\beta} \in \mathcal{B}_3/\mathcal{Z}_3$. Then by the Lemmas 9.8 and 10.6 for the element $\boldsymbol{\beta}$ the equality $\boldsymbol{\beta} = \boldsymbol{\beta}_1\boldsymbol{\Delta}_3^{\ell}$ holds with $\boldsymbol{\beta}_1$ being a word in $\sigma_1^2/\mathcal{Z}_3$ and $\sigma_2^2/\mathcal{Z}_3$, $\boldsymbol{\Delta}_3 = \Delta_3/\mathcal{Z}_3$ and $\ell = 0, 1$.

Put $\boldsymbol{b}_1' = \boldsymbol{\Delta}_3{}^{\ell}\boldsymbol{b}_1\boldsymbol{\Delta}_3{}^{-\ell}$. Then $\boldsymbol{b}_1'$ is represented by a word of form (10.24). The element $\boldsymbol{b}_2$ is conjugate by an element of $\mathcal{PB}_3/\mathcal{Z}_3$ either to $\boldsymbol{b}_1$ or to $\boldsymbol{b}_1'$. By Lemma 10.4 the reduced word representing the element $\boldsymbol{b}_2$ is obtained from the reduced word representing either $\boldsymbol{b}_1$ or $\boldsymbol{b}_1'$ by a cyclic permutation of terms. The number of cyclic permutations of $2j_0'$ letters, equals $2j_0'$. Hence the number of different elements of $\mathcal{PB}_3/\mathcal{Z}_3$ that can be represented by words of form (10.26) and are obtained from $\boldsymbol{b}_1$ by conjugation with an element of $\mathcal{B}_3/\mathcal{Z}_3$ does not exceed twice the number of cyclic permutations of $2j_0'$ letters, i.e. it does not exceed $2 \cdot 2j_0'$. (Notice, that e.g. by cyclically permuting the terms of a word with symmetries we may sometimes arrive at the same word.)

We proved that the number of different conjugacy classes of $\mathcal{PB}_3/\mathcal{Z}_3$ represented by words of form (10.26) with $2j_0'$ terms is not smaller than $\frac{2^{2j_0'}}{2 \cdot 2j_0'}$.

We obtain

$$N_{\mathcal{B}_3}^{entr}(Y) \ge \frac{2^{2j_0'-1}}{2 \cdot 2j_0'}. \tag{10.28}$$

Notice that $\frac{2^j}{j} \ge 2$ for natural j. Indeed, the function $x \to \frac{2^x}{x}$ increases for $x \ge 2$ (since $(\frac{2^x}{x})' = -\frac{2^x}{x^2} + \frac{2^x \log 2}{x}$, and $\log 2 > 0.6$) and for $j = 1$ and 2 the expression equals 2). Hence, $\frac{2^{2j}}{2j} \ge 2^j$. Hence, for $Y \ge \frac{70}{3}\pi^2$

$$N_{\mathcal{B}_3}^{entr}(Y) \ge 2^{j_0'-2} \ge \frac{1}{8} 2^{\frac{Y}{\frac{35}{6}\pi^2}} \ge \frac{1}{8} e^{\frac{\log 2}{6\pi^2} Y}. \tag{10.29}$$

The lower bound of the theorem is proved. □

Chapter 11
Riemann Surfaces of Second Kind and Finiteness Theorems

This chapter presents our deepest application of the concept of conformal module and extremal length to Gromov's Oka Principle. We address Problem 7.3 that proposes to obtain information on the set of homotopy classes of continuous mappings from a connected open Riemann surface X to a complex manifold Y. We consider the case when the target Y is the twice punctured complex plane and focus on the irreducible homotopy classes, since the reducible classes are well described and have the Gromov Oka property. The information we are interested in is the number of irreducible homotopy classes of mappings $X \to Y$ that contain a holomorphic mapping. More specifically, for any connected finite open Riemann surface X (maybe, of second kind) we give an effective upper bound for the number of irreducible holomorphic mappings up to homotopy from X to the twice punctured complex plane, and an effective upper bound for the number of irreducible holomorphic torus bundles up to isotopy on such a Riemann surface. The bound depends on a conformal invariant of the Riemann surface that is expressed in terms of the conformal module of a finite number of annuli in X. On the other hand, in the proof we use instead of the conformal modules of conjugacy classes of elements of the fundamental group of the twice punctured complex plane the more powerful extremal length with totally real boundary values of elements of the fundamental group themselves.

The estimates are in some sense asymptotically sharp: If X_σ is the σ-neighbourhood of a skeleton of a connected finite open hyperbolic Riemann surface with a Kähler metric, then the number of irreducible holomorphic mappings up to homotopy from X_σ to the twice punctured complex plane grows exponentially in $\frac{1}{\sigma}$.

These statements are analogs for Riemann surfaces of second kind of the Geometric Shafarevich Conjecture and the Theorem of de Franchis, that state the finiteness of the number of certain holomorphic objects on closed or punctured Riemann surfaces.

B. Jöricke, *Braids, Conformal Module, Entropy, and Gromov's Oka Principle*, Lecture Notes in Mathematics 2360, https://doi.org/10.1007/978-3-031-67288-0_11

11.1 Riemann Surfaces of First and of Second Kind and Finiteness Theorems

It seems that the oldest finiteness theorem for mappings between complex manifolds is the following theorem, which was published by de Franchis [19] in 1913.

Theorem 𝔄 (de Franchis) *For closed connected Riemann surfaces X and Y with Y of genus at least 2 there are at most finitely many non-constant holomorphic mappings from X to Y.*

There is a more comprehensive Theorem in this spirit.

Theorem 𝔅 (de Franchis-Severi) *For a closed connected Riemann surface X there are (up to isomorphism) only finitely many non-constant holomorphic mappings $f : X \to Y$ where Y ranges over all closed Riemann surfaces of genus at least 2.*

A finiteness theorem which became more famous because of its relation to number theory was conjectured by Shafarevich [73].

Theorem ℭ (Geometric Shafarevich Conjecture) *For a given compact or punctured Riemann surface X and given non-negative numbers g and m such that $2\mathsf{g} - 2 + \mathsf{m} > 0$ there are only finitely many locally holomorphically non-trivial holomorphic fiber bundles over X with fiber of type (g, m).*

Theorem ℭ was conjectured by Shafarevich [73] in the case of compact base and fibers of type (g, 0). It was proved by Parshin [69] in the case of compact base and fibers of type (g, 0), $\mathrm{g} \geq 2$, and by Arakelov [6] for punctured Riemann surfaces as base and fibers of type (g, 0). Imayoshi and Shiga [41] gave a proof of the quoted version using Teichmüller theory. See also the subsequent research [18] and [37].

The statement of Theorem ℭ "almost" contains the so called Finiteness Theorem of Sections which is also called the Geometric Mordell Conjecture (see [66]), giving an important conceptual connection between geometry and number theory. For more details we refer to the surveys by C.McMullen [66] and B.Mazur [63]. (See also [21, 22], and [34].)

Theorem 𝔄 is a consequence of Theorem ℭ, and Theorem 𝔄 has analogs for the source X and the target Y being punctured Riemann surfaces. Indeed, we may associate to any holomorphic mapping $f : X \to Y$ of Theorem 𝔄 the bundle over X with fiber over $x \in X$ equal to Y with distinguished point $\{f(x)\}$. Thus, the fibers are of type (g, 1). A holomorphic self-isomorphism of a locally holomorphically non-trivial (g, 1)-bundle may lead to a new holomorphic mapping from X to Y, but there are only finitely many different holomorphic self-isomorphisms.

We will consider here analogs of Theorems 𝔄 and ℭ for the case when the base X is a Riemann surface of second kind. Notice that finite hyperbolic Riemann surfaces of second kind are interesting from the point of view of spectral theory of the Laplace operator with respect to the hyperbolic metric (see also [15]). There are interesting relations to scattering theory and (the Hausdoff dimension of) the limit set of the Fuchsian group defining X.

The Theorems $\mathfrak{A}$ and $\mathfrak{C}$ do not hold literally if the base X is of second kind. If the base is a Riemann surface of second kind the problem to be considered is the finiteness of the number of irreducible isotopy classes (homotopy classes, respectively) containing holomorphic objects. In case the base is a punctured Riemann surface this is equivalent to the finiteness of the number of holomorphic objects. For more detail see Sects. 11.3 and 11.4.

We will present here finiteness theorems with effective estimates for the case when the base is a Riemann surface of second kind. The estimates depend on a conformal invariant of the Riemann surface. We will now prepare the definition of this invariant.

Let X be a connected open Riemann surface of genus $g \geq 0$ with $m + 1$ holes, $m \geq 0$, equipped with a base point q_0. Recall that the fundamental group $\pi_1(X, q_0)$ of X is a free group in $2g+m$ generators. We describe now the conformal invariant of the Riemann surface X that will appear in the mentioned estimate. We fix a standard system of generators $\mathcal{E}$ of $\pi_1(X, q_0)$ that is associated to a standard bouquet of circles for X (see Sect. 1.2). The circles are denoted by α_j, β_j and γ_j (see Fig. 1.1). The generators $\mathcal{E} \subset \pi_1(X, q_0)$ are labeled as follows. The elements $e_{2j-1,0} \in \pi_1(X, q_0)$, $j = 1, \ldots, g$, are represented by the curves α_j, the elements $e_{2j,0} \in \pi_1(X, q_0)$, $j = 1, \ldots, g$, are represented by the curves β_j, and the elements $e_{2g+k,0} \in \pi_1(X, q_0)$, $k = 1, \ldots, m$, of $\pi_1(X, q_0)$ are represented by the curves γ_k.

Let $\tilde{X}$ be the universal covering of X. For each element $e_0 \in \pi_1(X, q_0)$ we consider the subgroup $\langle e_0 \rangle$ of $\pi_1(X, q_0)$ generated by e_0. Fix a point $\tilde{q}_0$ in $\tilde{X}$ that projects to the base point $q_0 \in X$, let $(\mathrm{Is}^{\tilde{q}_0})^{-1}(e_0)$ be the covering transformation corresponding to e_0 (see Sect. 1.2), and $\langle (\mathrm{Is}^{\tilde{q}_0})^{-1}(e_0) \rangle$ the group generated by $(\mathrm{Is}^{\tilde{q}_0})^{-1}(e_0)$.

Definition 11.1 *Denote by $\mathcal{E}_j$, $j = 2, \ldots, 10$, the set of primitive elements of $\pi_1(X, q_0)$ which can be written as product of at most j factors with each factor being either an element of $\mathcal{E}$ or an element of $\mathcal{E}^{-1}$, the set of inverses of elements of $\mathcal{E}$. Define $\lambda_j = \lambda_j(X)$ as the maximum over $e_0 \in \mathcal{E}_j$ of the extremal length of the annulus $\tilde{X} \diagup \langle (\mathrm{Is}^{\tilde{q}_0})^{-1}(e_0) \rangle$.*

The quantity $\lambda_7(X)$ (for mappings to the twice punctured complex plane), or $\lambda_{10}(X)$ (for $(1, 1)$-bundles) is the mentioned conformal invariant.

In the following theorem we put $E = \{-1, 1, \infty\}$. We will often refer to $\mathbb{P}^1 \setminus \{-1, 1, \infty\}$ as the thrice punctured Riemann sphere or the twice punctured complex plane $\mathbb{C} \setminus \{-1, 1\}$. Recall that a continuous mapping from a Riemann surface to the twice punctured complex plane is called reducible, iff it is homotopic to a mapping with image in a once punctured disc contained in $\mathbb{P}^1 \setminus E$. (The puncture may be equal to ∞.) See Definition 7.4. For any connected finite open Riemann surface X with only thick ends and non-trivial fundamental group there are countably many non-homotopic reducible holomorphic mappings from X to the twice punctured complex plane (see Lemma 7.4). On the other hand the following theorem holds.

Theorem 11.1 *For each open connected Riemann surface X of genus $g \geq 0$ with $m + 1 \geq 1$ holes there are up to homotopy at most $3(\frac{3}{2}e^{36\pi\lambda_7(X)})^{2g+m}$ irreducible holomorphic mappings from X into $Y \stackrel{def}{=} \mathbb{P}^1 \setminus \{-1, 1, \infty\}$.*

Notice that the Riemann surface X is allowed to be of second kind. If X is a torus with a hole, $\lambda_7(X)$ may be replaced by $\lambda_3(X)$. If X is a planar domain, $\lambda_7(X)$ may be replaced by $\lambda_4(X)$

Recall that a holomorphic $(1, 0)$-bundle is also called a holomorphic torus bundle. A holomorphic torus bundle equipped with a holomorphic section is also considered as a holomorphic $(1, 1)$-bundle. Moreover, a smooth $(1, 0)$-bundle over a finite connected open smooth surface admits a smooth section. A holomorphic torus bundle over a finite connected open Riemann surface is (smoothly) isotopic to a holomorphic torus bundle that admits a holomorphic section. The following theorem for torus bundles holds.

Theorem 11.2 *Let X be an open connected Riemann surface of genus $g \geq 0$ with $m+1 \geq 1$ holes. Up to isotopy there are no more than $\big(2\cdot 15^6\cdot\exp(36\pi\lambda_{10}(X))\big)^{2g+m}$ irreducible holomorphic $(1, 1)$-bundles over X.*

For the definition of irreducible (g, m)-bundles see Sect. 8.2. Since on each finite open Riemann surface with only thick ends and non-trivial fundamental group there are countably many non-homotopic reducible holomorphic mappings with target being the twice punctured complex plane, there are also countably many non-isotopic reducible holomorphic $(1, 1)$-bundles over each such Riemann surface. This follows from Proposition 8.2.

Notice that Caporaso proved the existence of a uniform bound for the number of locally holomorphically non-trivial holomorphic fiber bundles over closed Riemann surfaces of genus g with $m \geq 0$ punctures with fibers being closed Riemann surfaces of genus g ≥ 2. The bound depends only on the numbers g, g and m. Heier gave effective uniform estimates, but the constants are huge and depend in a complicated way on the parameters.

Theorems 11.1 and 11.2 imply effective estimates for the number of locally holomorphically non-trivial holomorphic $(1, 1)$-bundles over punctured Riemann surfaces, however, the constants depend also on the conformal type of the base. More precisely, the following corollaries hold.

Corollary 11.1 *There are no more than $3(\frac{3}{2}e^{36\pi\lambda_7(X)})^{2g+m}$ non-constant holomorphic mappings from a Riemann surface X of type $(g, m + 1)$ to $\mathbb{P}^1 \setminus \{-1, 1, \infty\}$.*

Corollary 11.2 *There are no more than $\big(2 \cdot 15^6 \cdot \exp(36\pi\lambda_{10}(X))\big)^{2g+m}$ locally holomorphically non-trivial holomorphic $(1, 1)$-bundles over a Riemann surface X of type $(g, m + 1)$.*

The following examples demonstrate the different nature of the problem in the two cases, the case when the base is a punctured Riemann surface, and when it is a Riemann surface of second kind.

Example 11.1 *There are no non-constant holomorphic mappings from a torus with one puncture to the twice punctured complex plane.*

Indeed, by Picard's Theorem each such mapping extends to a meromorphic mapping from the closed torus to the Riemann sphere. This implies that the preimage of the set $\{-1, 1, \infty\}$ under the extended mapping must contain at least three points, which is impossible.

The situation changes if X is a torus with a large enough hole. Let $\alpha \geq 1$ and $\sigma \in (0, 1)$. Consider the torus with a hole $T^{\alpha,\sigma}$ that is obtained from $\mathbb{C}/(\mathbb{Z}+i\alpha\mathbb{Z})$, (with $\alpha \geq 1$ being a real number) by removing a closed geometric rectangle of vertical side length $\alpha - \sigma$ and horizontal side length $1 - \sigma$ (i.e. we remove a closed subset that lifts to such a closed rectangle in $\mathbb{C}$). A fundamental domain for this Riemann surface is "the golden cross on the Swedish flag" turned by $\frac{\pi}{2}$ with width of the laths being σ and length of the laths being 1 and α.

Proposition 11.1 *Up to homotopy there are at most* $7e^{2^5 \cdot 3^2 \pi \frac{2\alpha+1}{\sigma}}$ *irreducible holomorphic mappings from* $T^{\alpha,\sigma}$ *to the twice punctured complex plane.*

On the other hand, there are positive constants c, C, *and* σ_0 *such that for any positive number* $\sigma < \sigma_0$ *and any* $\alpha \geq 1$ *there are at least* $ce^{C\frac{\alpha}{\sigma}}$ *non-homotopic holomorphic mappings from* $T^{\alpha,\sigma}$ *to the twice punctured complex plane.*

Example 11.2 *There are only finitely many holomorphic maps from a thrice punctured Riemann sphere to another thrice punctured Riemann sphere.*

Indeed, after normalizing both, the source and the target space, by a Möbius transformation we may assume that both are equal to $\mathbb{C}\setminus\{-1, 1\}$. Each holomorphic map from $\mathbb{C} \setminus \{-1, 1\}$ to itself extends to a meromorphic map from the Riemann sphere to itself, which maps the set $\{-1, 1, \infty\}$ to itself and maps no other point to this set. By the Riemann-Hurwitz formula the meromorphic map takes each value exactly once. Indeed, suppose it takes each value l times for a natural number l. Then each point in $\{-1, 1, \infty\}$ has ramification index l. Apply the Riemann Hurwitz formula for the branched covering $X = \mathbb{P}^1 \to Y = \mathbb{P}^1$ of multiplicity l

$$\chi(X) = l \cdot \chi(Y) - \sum_{x \in Y} (e_x - 1).$$

Here e_x is the ramification index at the point x. For the Euler characteristic we have $\chi(\mathbb{P}^1) = 2$, and $\sum_{x\in Y}(e_x - 1) \geq \sum_{x=-1,1,\infty}(e_x - 1) = 3(l - 1)$. We obtain $2 \leq 2l - 3(l - 1)$ which is possible only if $l = 1$. We saw that each non-constant holomorphic mapping from $\mathbb{C} \setminus \{-1, 1\}$ to itself extends to a conformal mapping from the Riemann sphere to the Riemann sphere that maps the set $\{-1, 1, \infty\}$ to itself. There are only finitely may such maps, each a Möbius transformation permuting the three points.

For Riemann surfaces of second kind the situation changes, as demonstrated in the following proposition. The proposition does not only concern the case when the Riemann surface equals $\mathbb{P}^1$ with three holes. We consider an open Riemann surface X of genus g with $m \geq 1$ holes.

Proposition 11.2 *Let X be a connected finite open hyperbolic Riemann surface of genus g with $m+1$ holes, that is equipped with a Kähler metric. Suppose S is a standard bouquet of piecewise smooth circles for X with base point q_0. We assume that q_0 is the only non-smooth point of the circles, and all tangent rays to circles in S at q_0 divide a disc in the tangent space into equal sectors. Let S_σ be the σ-neighbourhood of S (in the Kähler metric on X).*

Then there exists a constant $\sigma_0 > 0$, and positive constants C', C'', c', c'', depending only on X, S and the Kähler metric, such that for each positive $\sigma < \sigma_0$ the number $N_{S_\sigma}^{\mathbb{C}\setminus\{-1,1\}}$ of non-homotopic irreducible holomorphic mappings from S_σ to the twice punctured complex plane satisfies the inequalities

$$c' e^{\frac{c''}{\sigma}} \le N_{S_\sigma}^{\mathbb{C}\setminus\{-1,1\}} \le C' e^{\frac{C''}{\sigma}}. \tag{11.1}$$

The following proposition demonstrates the role of the conformal module (extremal length, respectively) with suitable horizontal boundary values for the homotopy problem of individual continuous mappings to holomorphic ones. The proposition concerns mappings from small neighbourhoods of standard bouquets of circles for open Riemann surfaces and uses the estimates of the extremal lengths in terms of the $\mathcal{L}$-invariant of the words representing the monodromies along the standard generators of the fundamental group.

Proposition 11.3 *Let X, q_0, S and S_σ be as in Proposition 11.2. For each $j = 1, \dots 2g+m$ we let γ_j be the circle of the standard bouquet S for X that represents the standard generator e_j of $\pi_1(X, q_0)$. Denote by ℓ_j the length of γ_j in the Kähler metric. There exist positive constants σ_0 and C_1, depending only on X, S, q_0 and the Kähler metric, such that for each positive $\sigma < \sigma_0$ any continuous mapping $f : S_\sigma \to \mathbb{C}\setminus\{-1,1\}$ with $f(q_0) = 0$ is homotopic to a holomorphic mapping if the inequalities*

$$\mathcal{L}(f_*(e_j)) \le C_1 \frac{\ell_j}{\sigma} \tag{11.2}$$

hold for all $j = 1, \dots 2g+m$ for the induced mapping f_* on fundamental groups with base points.

Vive versa, there exists a constant C_2 depending only on X, S, q_0 and the Kähler metric, such that for each holomorphic mapping $f : S_\sigma \to \mathbb{C}\setminus\{-1,1\}$ with $f(q_0) = 0$ and each $j = 1, \dots 2g+m$ the inequality

$$\mathcal{L}(f_*(e_j)) \le C_2 \frac{\ell_j}{\sigma} \tag{11.3}$$

holds for the induced mapping f_* on fundamental groups with base points.

Theorem 11.1 and Propositions 11.1, 11.2 and 11.3 contribute to Problem 7.3 on Gromov's Oka Principle. Theorem 11.1 gives an upper bound for the number of homotopy classes of holomorphic mappings from connected finite open Riemann surfaces to the twice punctured complex plane. Theorem 11.2 gives an upper bound for the number of the isotopy classes of (1, 1)-bundles on connected finite open Riemann surfaces that contain holomorphic bundles. Propositions 11.1 and 11.2 consider families of Riemann surfaces Y_σ, $\sigma \in (0, \sigma_0)$, obtained by continuously changing the conformal structure of a fixed Riemann surface, and determine the growth rate for $\sigma \to 0$ of the number of irreducible holomorphic mappings $X_\sigma \to \mathbb{C} \setminus \{-1, 1\}$ up to homotopy. In Proposition 11.1 the family of Riemann surfaces depends also on a second parameter α, and the growth rate is determined in α and σ. The proof of the lower bound in both propositions uses solutions of a $\bar\partial$-problem. The solution in the case of Proposition 11.1 uses a simple explicit formula. Proposition 11.3 concerns the existence of homotopies of individual continuous mappings to holomorphic ones.

For the proof of the upper bound in the theorems and propositions we have to understand what prevents a continuous map from a finite open Riemann surface X to $\mathbb{C}\setminus\{-1, 1\}$ to be homotopic to a holomorphic map. Proposition 11.5 below says that an irreducible map $X \to \mathbb{C} \setminus \{-1, 1\}$ can only be homotopic to a holomorphic map, if the "complexity" of the monodromies of the map are compatible with conformal invariants of the source manifold.

11.2 Some Further Preliminaries on Mappings, Coverings, and Extremal Length

In this section we will prepare the proofs of the Theorems.

Regular Zero Sets We will call a subset of a smooth manifold $\mathcal{X}$ a simple relatively closed curve if it is a connected component of a regular level set of a smooth real-valued function on $\mathcal{X}$.

Let $\mathcal{X}$ be a connected finite open Riemann surface. Suppose the zero set L of a non-constant smooth real valued function on $\mathcal{X}$ is regular. Each component of L is either a simple closed curve or it can be parameterized by a continuous mapping $\ell : (-\infty, \infty) \to \mathcal{X}$. We call a component of the latter kind a simple relatively closed arc in $\mathcal{X}$.

A relatively closed curve γ in a connected finite open Riemann surface $\mathcal{X}$ is said to be contractible to a hole of $\mathcal{X}$, if the following condition holds. Consider $\mathcal{X}$ as domain $\mathcal{X}^c \setminus \cup \mathcal{C}_j$ on a closed Riemann surface $\mathcal{X}^c$. Here the $\mathcal{C}_j$ are the holes, each is either a closed topological disc with smooth boundary or a point. The condition on γ is the following. For each pair U_1, U_2 of open subsets of $\mathcal{X}^c$, $\cup \mathcal{C}_j \subset U_1 \Subset U_2$, there exists a homotopy of γ that fixes $\gamma \cap U_1$ and moves γ into U_2. Taking for U_2 small enough neighbourhoods of $\cup \mathcal{C}_j$ we see that the homotopy moves γ into an annulus adjacent to one of the holes.

For each relatively compact domain $\mathcal{X}' \Subset \mathcal{X}$ in $\mathcal{X}$ there is a finite cover of $L \cap \overline{\mathcal{X}'}$ by open subsets U_k of $\mathcal{X}$ such that each $L \cap U_k$ is connected. Each set $L \cap U_k$ is contained in a component of L. Hence, only finitely many connected components of

L intersect $\overline{\mathcal{X}'}$. Let L_0 be a connected component of L which is a simple relatively closed arc parameterized by $\ell_0 : \mathbb{R} \to \mathcal{X}$. Since each set $L_0 \cap U_k$ is connected it is the image of an interval under ℓ_0. Take real numbers t_0^- and t_0^+ such that all these intervals are contained in (t_0^-, t_0^+). Then the images $\ell\big((-\infty, t_0^-)\big)$ and $\ell\big((t_0^+, +\infty)\big)$ are contained in $\mathcal{X} \setminus \overline{\mathcal{X}'}$, maybe, in different components. Such parameters t_0^- and t_0^+ can be found for each relatively compact deformation retract $\mathcal{X}'$ of $\mathcal{X}$. Hence for each relatively closed arc $L_0 \subset L$ the set of limit points L_0^+ of $\ell_0(t)$ for $t \to \infty$ is contained in a boundary component of $\mathcal{X}$. Also, the set of limit points L_0^- of $\ell_0(t)$ for $t \to -\infty$ is contained in a boundary component of $\mathcal{X}$. The boundary components may be equal or different.

Moreover, if $\mathcal{X}' \Subset \mathcal{X}$ is a relatively compact domain in $\mathcal{X}$ which is a deformation retract of $\mathcal{X}$, and a connected component L_0 of L does not intersect $\overline{\mathcal{X}'}$ then L_0 is contractible to a hole of $\mathcal{X}$. Indeed, $\mathcal{X} \setminus \overline{\mathcal{X}'}$ is the union of disjoint annuli, each of which is adjacent to a boundary component of $\mathcal{X}$, and the connected set L_0 must be contained in a single annulus.

Further, denote by L' the union of all connected components of L that are simple relatively closed arcs. Consider those components L_j of L' that intersect $\mathcal{X}'$. There are finitely many such L_j. Parameterize each L_j by a mapping $\ell_j : \mathbb{R} \to \mathcal{X}$. For each j we let $[t_j^-, t_j^+]$ be a compact interval for which

$$\ell_j(\mathbb{R} \setminus [t_j^-, t_j^+]) \subset \mathcal{X} \setminus \overline{\mathcal{X}'}. \tag{11.4}$$

Let $\mathcal{X}''$, $\mathcal{X}' \Subset \mathcal{X}'' \Subset \mathcal{X}$, be a domain which is a deformation retract of $\mathcal{X}$ such that $\ell_j([t_j^-, t_j^+]) \subset \mathcal{X}''$ for each j. Then all connected components of $L' \cap \mathcal{X}''$, that do not contain a set $\ell_j([t_j^-, t_j^+])$, are contractible to a hole of $\mathcal{X}''$. Indeed, each such component is contained in the union of annuli $\mathcal{X}'' \setminus \overline{\mathcal{X}'}$.

Subgroups of Covering Transformations Let again q_0 be the base point of a Riemann surface X, and let $\tilde{q}_0$ be the base point in the universal covering $\tilde{X}$ with $\mathsf{P}(\tilde{q}_0) = q_0$ for the covering map $\mathsf{P} : \tilde{X} \to X$. Let N be a subgroup of $\pi_1(X, q_0)$, and $(\mathrm{Is}^{\tilde{q}_0})^{-1}(N)$ the respective subgroup of covering transformations.. Denote by $X(N)$ the quotient $\tilde{X} \diagup (\mathrm{Is}^{\tilde{q}_0})^{-1}(N)$. We obtain a covering $\omega_{\mathrm{Id}}^N : \tilde{X} \to X(N)$ with group of covering transformations isomorphic to N. The fundamental group of $X(N)$ with base point $(q_0)_N \overset{def}{=} \omega_{\mathrm{Id}}^N(\tilde{q}_0)$ can be identified with N.

If N_1 and N_2 are subgroups of $\pi_1(X, q_0)$ and N_1 is a subgroup of N_2 (we write $N_1 \le N_2$), then there is a covering map $\omega_{N_1}^{N_2} : \tilde{X} \diagup (\mathrm{Is}^{\tilde{q}_0})^{-1}(N_1) \to \tilde{X} \diagup (\mathrm{Is}^{\tilde{q}_0})^{-1}(N_2)$, such that $\omega_{N_1}^{N_2} \circ \omega_{\mathrm{Id}}^{N_1} = \omega_{\mathrm{Id}}^{N_2}$. Moreover, the diagram Fig. 11.1 is commutative.

Indeed, take any point $x_1 \in \tilde{X} \diagup (\mathrm{Is}^{\tilde{q}_0})^{-1}(N_1)$ and a preimage $\tilde{x}$ of x_1 under $\omega_{\mathrm{Id}}^{N_1}$. There exists a neighbourhood $V(\tilde{x})$ of $\tilde{x}$ in $\tilde{X}$ such that $V(\tilde{x}) \cap \sigma(V(\tilde{x})) = \emptyset$ for all covering transformations $\sigma \in \mathrm{Deck}(\tilde{X}, X)$. Then for $j = 1, 2$ the mapping $\omega_{\mathrm{Id}}^{N_j, \tilde{x}} \overset{def}{=} \omega_{\mathrm{Id}}^{N_j} \mid V(\tilde{x})$ is a homeomorphism from $V(\tilde{x})$ onto its image denoted by

Fig. 11.1 A commutative diagram related to subgroups of the group of covering transformations

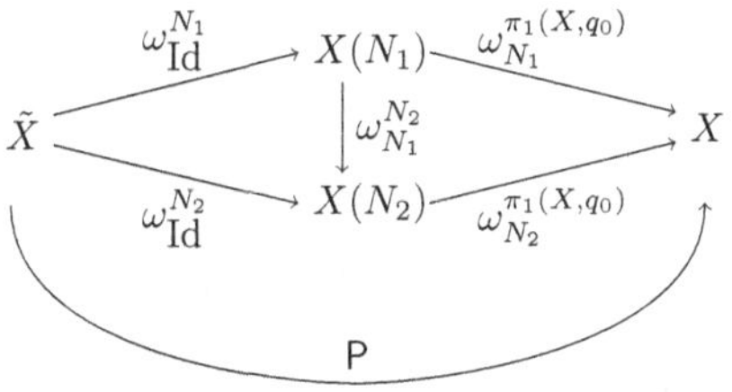

V_j. Put $x_2 = \omega_{\text{Id}}^{N_2,\tilde{x}}(\tilde{x})$. The set $V_j \subset \tilde{X}/(\text{Is}^{\tilde{q}_0})^{-1}(N_j)$ is a neighbourhood of x_j for $j = 1, 2$.

For each preimage $\tilde{x}' \in (\omega_{\text{Id}}^{N_1})^{-1}(x_1)$ there is a covering transformation $\varphi_{\tilde{x},\tilde{x}'}$ in $(\text{Is}^{\tilde{q}_0})^{-1}(N_1)$ which maps a neighbourhood $V(\tilde{x}')$ of $\tilde{x}'$ conformally onto the neighbourhood $V(\tilde{x})$ of $\tilde{x}$ so that on $V(\tilde{x}')$ the equality $\omega_{\text{Id}}^{N_1,\tilde{x}'} = \omega_{\text{Id}}^{N_1,\tilde{x}} \circ \varphi_{\tilde{x},\tilde{x}'}$ holds.

Choose $\tilde{x} \in (\omega_{\text{Id}}^{N_1})^{-1}(x_1)$ and define

$$\omega_{N_1}^{N_2}(y) = \omega_{N_1}^{N_2,\tilde{x}}(y) \overset{def}{=} \omega_{\text{Id}}^{N_2,\tilde{x}}((\omega_{\text{Id}}^{N_1,\tilde{x}})^{-1}(y)) \text{ for each } y \in V_1 . \tag{11.5}$$

We get a correctly defined mapping from V_1 onto V_2. Indeed, since N_1 is a subgroup of N_2, for another point $\tilde{x}' \in (\omega_{\text{Id}}^{N_1})^{-1}(x_1)$ the covering transformation $\varphi_{\tilde{x},\tilde{x}'}$ is contained in $(\text{Is}^{\tilde{q}_0})^{-1}(N_2)$, and we get the equality $\omega_{\text{Id}}^{N_2,\tilde{x}'} = \omega_{\text{Id}}^{N_2,\tilde{x}} \circ \varphi_{\tilde{x},\tilde{x}'}$. Hence, for $y \in V_1(x_1)$

$$\begin{aligned} \omega_{\text{Id}}^{N_2,\tilde{x}'} \circ (\omega_{\text{Id}}^{N_1,\tilde{x}'})^{-1}(y) =& (\omega_{\text{Id}}^{N_2,\tilde{x}} \circ \varphi_{\tilde{x},\tilde{x}'}) \circ (\omega_{\text{Id}}^{N_1,\tilde{x}} \circ \varphi_{\tilde{x},\tilde{x}'})^{-1}(y) \\ =& \omega_{\text{Id}}^{N_2,\tilde{x}} \circ (\omega_{\text{Id}}^{N_1,\tilde{x}})^{-1}(y). \end{aligned} \tag{11.6}$$

Since each mapping $\omega_{\text{Id}}^{N_j,\tilde{x}}$, $j = 1, 2$, is a homeomorphism from $V(\tilde{x})$ onto its image, the mapping $\omega_{N_1}^{N_2}$ is a homeomorphism from $V(x_1)$ onto $V(x_2)$. The same holds for all preimages of $V(x_2)$ under $\omega_{N_1}^{N_2}$. Hence, $\omega_{N_1}^{N_2}$ is a covering map. The commutativity of the part of the diagram that involves the mappings $\omega_{\text{Id}}^{N_1}$, $\omega_{\text{Id}}^{N_2}$, and $\omega_{N_1}^{N_2}$ follows from Eq. (11.5).

The existence of $\omega_{N_1}^{\pi_1(X,q_0)}$ and the equality $\mathsf{P} = \omega_{N_1}^{\pi_1(X,q_0)} \circ \omega_{\text{Id}}^{N_1}$ follows by applying the above arguments with $N_2 = \pi_1(X, q_0)$. The equality $\mathsf{P} = \omega_{N_2}^{\pi_1(X,q_0)} \circ \omega_{\text{Id}}^{N_2}$ follows in the same way. Since

$$\begin{aligned} \mathsf{P} =& \omega_{N_2}^{\pi_1(X,q_0)} \circ \omega_{N_1}^{N_2} \circ \omega_{\text{Id}}^{N_1} \\ =& \omega_{N_1}^{\pi_1(X,q_0)} \circ \omega_{\text{Id}}^{N_1}, \end{aligned}$$

we have

$$\omega_{N_2}^{\pi_1(X,q_0)} \circ \omega_{N_1}^{N_2} = \omega_{N_1}^{\pi_1(X,q_0)}$$

We will also use the notation $\omega^N \stackrel{def}{=} \omega_{\mathrm{Id}}^N$ and $\omega_N \stackrel{def}{=} \omega_N^{\pi_1(X,q_0)}$ for a subgroup N of $\pi_1(X, q_0)$.

Let again $N_1 \leq N_2$ be subgroups of $\pi_1(X, q_0)$. Consider the covering $\omega_{N_1}^{N_2}$: $\tilde{X}/(\mathrm{Is}^{\tilde{q}_0})^{-1}(N_1) \to \tilde{X}/(\mathrm{Is}^{\tilde{q}_0})^{-1}(N_2)$. Let β be a simple relatively closed curve in $\tilde{X}/(\mathrm{Is}^{\tilde{q}_0})^{-1}(N_2)$. Then $(\omega_{N_1}^{N_2})^{-1}(\beta)$ is the union of simple relatively closed curves in $\tilde{X}/(\mathrm{Is}^{\tilde{q}_0})^{-1}(N_1)$ and $\omega_{N_1}^{N_2} : (\omega_{N_1}^{N_2})^{-1}(\beta) \to \beta$ is a covering. Indeed, we cover β by small discs U_k in $\tilde{X}/(\mathrm{Is}^{\tilde{q}_0})^{-1}(N_2)$ such that for each k the restriction of $\omega_{N_1}^{N_2}$ to each connected component of $(\omega_{N_1}^{N_2})^{-1}(U_k)$ is a homeomorphism onto U_k, and U_k intersects β along a connected set. Take any k with $U_k \cap \beta \neq \emptyset$. Consider the preimages $(\omega_{N_1}^{N_2})^{-1}(U_k)$. Restrict $\omega_{N_1}^{N_2}$ to the intersection of any preimage $(\omega_{N_1}^{N_2})^{-1}(U_k)$ with $(\omega_{N_1}^{N_2})^{-1}(\beta)$. We obtain a homeomorphism onto $U_k \cap \beta$. It follows that the map $(\omega_{N_1}^{N_2})$ is a covering from each connected component of $(\omega_{N_1}^{N_2})^{-1}(\beta)$ onto β.

The Extremal Length of Monodromies Let as before X be a connected finite open Riemann surface with base point q_0, and $\tilde{q}_0$ a point in the universal covering $\tilde{X}$ for which $\mathsf{P}(\tilde{q}_0) = q_0$ for the covering map $\mathsf{P} : \tilde{X} \to X$.

Recall that for an arbitrary point $q \in X$ the free homotopy class of an element e of the fundamental group $\pi_1(X, q)$ can be identified with the conjugacy class of elements of $\pi_1(X, q)$ containing e and is denoted by $\widehat{e}$. Notice that for $e_0 \in \pi_1(X, q_0)$ and a curve α in X with initial point q_0 and terminal point q the free homotopy classes of e_0 and of $e = \mathrm{Is}_\alpha(e_0)$ coincide, i.e. $\widehat{e} = \widehat{e_0}$. Consider a simple smooth relatively closed curve L in X. We will say that a free homotopy class of curves $\widehat{e_0}$ intersects L if each representative of $\widehat{e_0}$ intersects L. Choose an orientation of L. The intersection number of $\widehat{e_0}$ with the oriented curve L is the intersection number with L of some (and, hence, of each) smooth loop representing $\widehat{e_0}$ that intersects L transversally. The intersection number of such a loop with the relatively closed curve L is the sum of the intersection numbers over all intersection points. The intersection number at an intersection point equals $+1$ if the orientation determined by the tangent vector to the loop followed by the tangent vector to L is the orientation of X, and equals -1 otherwise.

Let A be an annulus equipped with an orientation (called positive orientation) of simple closed dividing curves in A. Recall that a relatively closed curve γ in a surface X is called dividing, if $X \setminus \gamma$ consists of two connected components. A continuous mapping $\omega : A \to X$ is said to represent a conjugacy class $\widehat{e}$ of elements of the fundamental group $\pi_1(X, q)$ for a point $q \in X$, if the composition $\omega \circ \gamma$ represents $\widehat{e}$ for each positively oriented dividing curve γ in A.

Let A be an annulus with base point p with a chosen positive orientation of simple closed dividing curves in A. Let ω be a continuous mapping from A to a finite

Riemann surface X with base point q such that $\omega(p) = q$. We write $\omega : (A, p) \to (X, q)$. The mapping is said to represent the element e of the fundamental group $\pi_1(X, q)$ if $\omega \circ \gamma$ represents e for some (and hence for each) positively oriented simple closed dividing curve γ in A with base point q.

Let q_0 be the base point of X chosen above and $\tilde{q}_0$ a point in the universal covering $\tilde{X}$ with $\mathsf{P}(\tilde{q}_0) = q_0$. We associate to each element $e_0 \in \pi_1(X, q_0)$ of the free group $\pi_1(X, q_0)$ the annulus $X(\langle e_0 \rangle) = \tilde{X} \diagup (\mathrm{Is}^{\tilde{q}_0})^{-1}(\langle e_0 \rangle)$ with base point $q_{\langle e_0 \rangle} = \omega_{\mathrm{Id}}^{\langle e_0 \rangle}(\tilde{q}_0)$ and the covering map $\omega_{\langle e_0 \rangle} \overset{def}{=} \omega_{\langle e_0 \rangle}^{\pi_1(X,q_0)} : X(\langle e_0 \rangle) \to X$. By the commutative diagram Fig. 11.1 the equality $\omega_{\langle e_0 \rangle}(q_{\langle e_0 \rangle}) = q_0$ holds. We choose the orientation of simple closed dividing curves in $X(\langle e_0 \rangle) = \tilde{X} \diagup (\mathrm{Is}^{\tilde{q}_0})^{-1}(\langle e_0 \rangle)$ so that for a curve $\tilde{\gamma}$ in $\tilde{X}$ with initial point $\tilde{q}_0$ and terminal point $\mathrm{Is}^{\tilde{q}_0}(e_0)$ the curve $\gamma_{\langle e_0 \rangle} \overset{def}{=} \omega^{\langle e_0 \rangle}(\tilde{\gamma})$ is positively oriented. The locally conformal mapping $\omega_{\langle e_0 \rangle}^{\pi_1(X,q_0)} : (X(\langle e_0 \rangle), q_{\langle e_0 \rangle}) \to (X, q_0)$ represents e_0. This follows from the equality $\omega_{\langle e_0 \rangle}(\gamma_{\langle e_0 \rangle}) = \omega_{\langle e_0 \rangle}(\omega^{\langle e_0 \rangle}(\tilde{\gamma})) = \mathsf{P}(\tilde{\gamma}) = \gamma$, since $\mathsf{P}(\tilde{\gamma})$ represents e_0.

Take a curve α in X that joins q_0 and q, and the point $\tilde{q} = \tilde{q}(\alpha) \in \tilde{X}$ such that α and $\tilde{q}$ are compatible, i.e. $\mathrm{Is}^{\tilde{q}} = \mathrm{Is}_\alpha \circ \mathrm{Is}^{\tilde{q}_0}$ (see Eq. (1.2)). Put $e = \mathrm{Is}_\alpha(e_0)$. By Eq. (1.2) $(\mathrm{Is}^{\tilde{q}})^{-1}(e) = (\mathrm{Is}^{\tilde{q}_0})^{-1}(e_0)$, hence, $\tilde{X} \diagup (\mathrm{Is}^{\tilde{q}})^{-1}(\langle e \rangle) = \tilde{X} \diagup (\mathrm{Is}^{\tilde{q}_0})^{-1}(\langle e_0 \rangle) = X(\langle e_0 \rangle)$. The locally conformal mapping $\omega_{\langle e_0 \rangle} : X(\langle e_0 \rangle) \to X$ takes the point $q_{\langle e \rangle} \overset{def}{=} \omega^{\langle e_0 \rangle}(\tilde{q})$ to $q \in X$. Moreover, $\omega_{\langle e_0 \rangle} : (X(\langle e_0 \rangle), q_{\langle e \rangle}) \to (X, q)$ represents $e \in \pi_1(X, q)$. This can be seen by repeating the previous arguments.

Let α be an arbitrary curve in X joining q_0 with q, and $\tilde{q} \in \mathsf{P}^{-1}(q)$ be arbitrary (i.e. α and $\tilde{q}$ are not required to be compatible). Let $e \in \pi_1(X, q)$. Denote the projection $\tilde{X} \to \tilde{X} \diagup (\mathrm{Is}^{\tilde{q}})^{-1}(\langle e \rangle)$ by $\omega^{\langle e \rangle, \tilde{q}}$, and the projection $\tilde{X} \diagup (\mathrm{Is}^{\tilde{q}})^{-1}(\langle e \rangle) \to X$ by $\omega_{\langle e \rangle, \tilde{q}}$. Put $q_{\langle e \rangle, \tilde{q}} = \omega^{\langle e \rangle, \tilde{q}}(\tilde{q})$. For any such choice we choose the orientation of simple closed dividing curves on $\tilde{X} \diagup (\mathrm{Is}^{\tilde{q}})^{-1}(\langle e \rangle)$ so that $\omega^{\langle e \rangle, \tilde{q}}$ maps any curve $\tilde{\gamma}$ in $\tilde{X}$ with initial point $\tilde{q}$ and terminal point $(\mathrm{Is}^{\tilde{q}})^{-1}(\langle e \rangle)(\tilde{q})$ to a positively oriented dividing curve. We will call it the standard orientation of dividing curves in $\tilde{X} \diagup (\mathrm{Is}^{\tilde{q}})^{-1}(\langle e \rangle)$. The mapping $\omega_{\langle e \rangle, \tilde{q}} : \left(\tilde{X} \diagup (\mathrm{Is}^{\tilde{q}})^{-1}(\langle e \rangle), q_{\langle e \rangle, \tilde{q}}\right) \to (X, q)$ represents e.

Since the mapping $\omega_{\langle e_0 \rangle} : (X(\langle e_0 \rangle), (q_0)_{\langle e_0 \rangle}) \to (X, q_0)$ represents e_0, the mapping $\omega_{\langle e_0 \rangle} : X(\langle e_0 \rangle) \to X$ represents the free homotopy class $\widehat{e_0}$. The following simple lemma will be useful.

Lemma 11.1 *The annulus $X(\langle e_0 \rangle)$ has smallest extremal length among annuli which admit a holomorphic mapping to X, that represents the conjugacy class $\widehat{e_0}$.*

In other words, $X(\langle e_0 \rangle)$ is the "thickest" annulus with the property stated in Lemma 11.1.

Proof Take an annulus A with a choice of positive orientation of simple closed dividing curves. Suppose $A \xrightarrow{\omega} X$ is a holomorphic mapping that represents $\widehat{e_0}$. The annulus A is conformally equivalent to a round annulus in the plane, hence, we may assume that A has the form $A = \{z \in \mathbb{C} : r < |z| < R\}$ for $0 \le r < R \le \infty$ and the positive orientation of dividing curves is the counterclockwise one.

Take a positively oriented simple closed dividing curve γ^A in A. Its image $\omega \circ \gamma^A$ under ω represents the class $\widehat{e_0}$. Choose a point q^A in γ^A, and put $q = \omega(q^A)$. Then γ^A represents a generator of $\pi_1(A, q^A)$ and $\gamma = \omega \circ \gamma^A$ represents an element e of $\pi_1(X, q)$ in the conjugacy class $\widehat{e_0}$. Choose a curve α in X with initial point q_0 and terminal point q, and a point $\tilde{q}$ in $\tilde{X}$ so that α and $\tilde{q}$ are compatible, and, hence, for $e = \mathrm{Is}_\alpha(e_0)$ the equality $(\mathrm{Is}^{\tilde{q}_0})^{-1}(e_0) = (\mathrm{Is}^{\tilde{q}})^{-1}(e)$ holds. Let L be the relatively closed arc $\{q^A \cdot r : r \in \mathbb{R}\} \cap A$ in A that contains q^A. After a homotopy of γ^A with fixed base point, we may assume that its base point q^A is the only point of γ^A that is contained in L. The restriction $\omega|(A \setminus L)$ lifts to a mapping $\tilde{\omega} : (A \setminus L) \to \tilde{X}$, that extends continuously to the two strands $L_\pm$ of L. (We assume that L_- is reached by moving clockwise from $A \setminus L$ towards L.) Let $q^A_\pm$ be the copies of q^A on the two strands $L_\pm$. We choose the lift $\tilde{\omega}$ so that $\tilde{\omega}(q^A_-) = \tilde{q}$. Since the mapping $(A, q^A) \to (X, q)$ represents e, we obtain $\tilde{\omega}(q^A_+) = \sigma(\tilde{q})$ for $\sigma = (\mathrm{Is}^{\tilde{q}})^{-1}(e)$. Then for each $z \in L$ the covering transformation σ maps the point $\tilde{z}_- \in \tilde{\omega}(L_-)$ for which $\mathsf{P}(\tilde{z}_-) = z$ to the point $\tilde{z}_+ \in \tilde{\omega}(L_+)$ for which $\mathsf{P}(\tilde{z}_+) = z$. Hence ω lifts to a holomorphic mapping $\iota : A \to X(\langle e_0 \rangle)$. By Lemma 3.3 (see also Lemma 7 of [45]) $\lambda(A) \ge \lambda(X(\langle e_0 \rangle))$. □

For each point $q \in X$ and each element $e \in \pi_1(X, q)$ we denote by $A(\widehat{e})$ the conformal class of the "thickest" annulus that admits a holomorphic mapping into X that represents $\widehat{e}$. We saw that $\lambda(A(\widehat{e_0})) = \lambda(\tilde{X}/(\mathrm{Is}^{\tilde{q}_0})^{-1}(\langle e_0 \rangle))$ for $e_0 \in \pi_1(X, q_0)$. By the same reasoning as before $\lambda(A(\widehat{e})) = \lambda(\tilde{X}/(\mathrm{Is}^{\tilde{q}'})^{-1}(\langle e \rangle))$ for each $\tilde{q}' \in \tilde{X}$ and each element $e \in \pi_1(X, \mathsf{P}(\tilde{q}'))$. Hence, if e_0 and e are free homotopic (equivalently, if they represent to the same conjugacy class,) then $\lambda(\tilde{X}/(\mathrm{Is}^{\tilde{q}_0})^{-1}(\langle e_0 \rangle)) = \lambda(\tilde{X}/(\mathrm{Is}^{\tilde{q}})^{-1}(\langle e \rangle))$ for any $\tilde{q}_0 \in \mathsf{P}^{-1}(q_0) \subset \tilde{X}$ and any $\tilde{q} \in \mathsf{P}^{-1}(q)$. Notice that $A(\widehat{e^{-1}}) = A(\widehat{e})$ for each $e \in \pi_1(X, q)$, $q \in X$.

11.3 Holomorphic Mappings into the Twice Punctured Plane

The following lemma will be crucial for the estimate of the $\mathcal{L}_-$-invariant of the monodromies of holomorphic mappings from a finite open Riemann surface to $\mathbb{C} \setminus \{-1, 1\}$. For a point $q' \in (-1, 1)$ we will identify $\pi_1(\mathbb{C} \setminus \{-1, 1\}, q')$ with $\pi_1(\mathbb{C} \setminus \{-1, 1\}, 0)$ by the canonical isomorphism $\mathrm{Is}_{\alpha_{q'}}$ for the curve $\alpha_{q'}$ that runs along the line segment in $(-1, 1)$ joining 0 with q'.

Lemma 11.2 *Let $f : X \to \mathbb{C} \setminus \{-1, 1\}$ be a non-contractible holomorphic mapping on a connected finite open Riemann surface X, such that 0 is a regular value of $\mathrm{Im} f$. Assume that L_0 is a simple relatively closed curve in X such that $f(L_0) \subset (-1, 1)$. Let $q \in L_0$ and $q' = f(q)$.*

If for an element $e \in \pi_1(X, q)$ the free homotopy class $\widehat{e}$ intersects L_0, then either the reduced word $f_*(e) \in \pi_1(\mathbb{C} \setminus \{-1, 1\}, q')$ is a non-zero power of a standard generator of $\pi_1(\mathbb{C} \setminus \{-1, 1\}, q')$ or the inequality

$$\mathcal{L}_-(f_*(e)) \leq 2\pi\lambda(A(\widehat{e})) \tag{11.7}$$

holds.

Notice that we make a normalization in the statement of the Lemma by requiring that f maps L_0 into the interval $(-1, 1)$, not merely into $\mathbb{R} \setminus \{-1, 1\}$.

Lemma 11.2 will be a consequence of the following lemma.

Lemma 11.3 *Let X, f, L_0, $q \in L_0$ be as in Lemma 11.2, and $e \in \pi_1(X, q)$. Let $\tilde{q}$ be an arbitrary point in $\mathsf{P}^{-1}(q)$. Consider the annulus $A \overset{def}{=} \tilde{X} \diagup (\mathrm{Is}^{\tilde{q}})^{-1}(\langle e \rangle)$ and the holomorphic projection $\omega_A \overset{def}{=} \omega_{\langle e \rangle, \tilde{q}}$. Put $q_A \overset{def}{=} \omega^{\langle e \rangle, \tilde{q}}(\tilde{q})$. The mapping $\omega_A : (A, q_A) \to (X, q)$ represents e.*

Let $L_A \subset A$ be the connected component of $(\omega_A)^{-1}(L_0)$ that contains q_A. If $\widehat{e}$ intersects L_0, then L_A is a relatively closed curve in A that has limit points on both boundary components of A, and the lift $f \circ \omega_A$ is a holomorphic mapping from A to $\mathbb{C} \setminus \{-1, 1\}$ that maps L_A into $(-1, 1)$.

Proof of Lemma 11.3 Let $\gamma : [0, 1] \to X$ be a curve with base point q in X that represents e, and let $\tilde{\gamma}$ be the lift of γ to $\tilde{X}$ with initial point $\tilde{\gamma}(0)$ equal to $\tilde{q}$. Put $\sigma \overset{def}{=} (\mathrm{Is}^{\tilde{q}})^{-1}(e)$. Then the terminal point $\tilde{\gamma}(1)$ equals $\sigma(\tilde{q})$.

All connected components of $\mathsf{P}^{-1}(L_0)$ are relatively closed curves in $\tilde{X} \cong \mathbb{C}_+$ (where $\mathbb{C}_+$ denotes the upper half-plane) with limit points on the boundary of $\tilde{X}$. Indeed, the lift $f \circ \mathsf{P}$ of f to $\tilde{X}$ takes values in $(-1, 1)$ on $\mathsf{P}^{-1}(L_0)$. Hence, $|\exp(\pm i\, f \circ \mathsf{P})| = 1$ on $\mathsf{P}^{-1}(L_0)$. A compact connected component of $\mathsf{P}^{-1}(L_0)$ would bound a relatively compact topological disc in $\tilde{X} = \mathbb{C}_+$, and by the maximum principle $|\exp(\pm i\, f \circ \mathsf{P})| = 1$ on the disc. This would imply that $f \circ \mathsf{P}$ is constant on $\tilde{X}$ in contrary to the assumptions.

Let $\tilde{L}_{\tilde{q}}$ be the connected component of $\mathsf{P}^{-1}(L_0)$ that contains $\tilde{q}$. The point $\sigma(\tilde{q})$ cannot be contained in $\tilde{L}_{\tilde{q}}$. Indeed, assume the contrary. Then the arc $\tilde{\alpha}$ on $\tilde{L}_{\tilde{q}}$ that joins $\tilde{q}$ and $\sigma(\tilde{q})$ is homotopic in $\tilde{X}$ with fixed endpoints to $\tilde{\gamma}$. The projection $\alpha = \mathsf{P}(\tilde{\alpha})$ is contained in L_0 and is homotopic in X with fixed endpoints to γ. Since γ represents e and e is a primitive element of the fundamental group $\pi_1(X, q)$, this is possible only if L_0 is compact and (after orienting it) L_0 represents e. A small translation of α to a side of L_0 gives a curve in X that does not intersect L_0 and represents the free homotopy class $\widehat{e}$ of e. This contradicts the fact that $\widehat{e}$ intersects L_0.

Take any other point $\tilde{q}' \in \tilde{L}_{\tilde{q}}$. If $\sigma(\tilde{q}')$ was in $\tilde{L}_{\tilde{q}}$, then the arc on $\tilde{L}_{\tilde{q}}$ that joins $\tilde{q}$ with $\tilde{q}'$ would project to a loop that represents $\hat{e}$ and is contained in L_0. By the arguments above this is not possible. We proved that the curves $\tilde{L}_{\tilde{q}}$ and $\sigma(\tilde{L}_{\tilde{q}})$ are disjoint.

Each of the two connected components $\tilde{L}_{\tilde{q}}$ and $\sigma(\tilde{L}_{\tilde{q}})$ divides $\tilde{X}$. Let Ω be the domain on $\tilde{X}$ that is bounded by $\tilde{L}_{\tilde{q}}$ and $\sigma(\tilde{L}_{\tilde{q}})$ and parts of the boundary of $\tilde{X}$. After a homotopy of $\tilde{\gamma}$ that fixes the endpoints we may assume that $\tilde{\gamma}((0,1))$ is contained in Ω. Indeed, for each connected component of $\tilde{\gamma}((0,1)) \setminus \Omega$ there is a homotopy with fixed endpoints that moves the connected component to an arc on $\tilde{L}_{\tilde{q}}$ or $\sigma(\tilde{L}_{\tilde{q}})$. A small perturbation yields a curve $\tilde{\gamma}'$ which is homotopic with fixed endpoints to $\tilde{\gamma}$ and has interior contained in Ω. Notice that by the same reasoning as above, $\tilde{\gamma}'((0,1))$ does not meet any $\sigma^k(\tilde{L}_{\tilde{q}})$.

The curve $\omega^{\langle e\rangle,\tilde{q}}(\tilde{\gamma}')$ is a closed curve on A that represents a generator of the fundamental group of A with base point q_A. Moreover, $\omega_A \circ \omega^{\langle e\rangle,\tilde{q}}(\tilde{\gamma}') = \omega_{\langle e\rangle,\tilde{q}} \circ \omega^{\langle e\rangle,\tilde{q}}(\tilde{\gamma}') = \mathsf{P}(\tilde{\gamma}')$ represents e. Hence, the mapping $\omega_A : (A, q_A) \to (X, q)$ represents e.

The curve $\omega^{\langle e\rangle,\tilde{q}}(\tilde{\gamma}')$ intersects $L_{\langle e\rangle} = \omega^{\langle e\rangle,\tilde{q}}(\tilde{L}_{\tilde{q}})$ exactly once. Hence, $L_{\langle e\rangle}$ has limit points on both boundary circles of A for otherwise $L_{\langle e\rangle}$ would intersect one of the components of $A \setminus \omega^{\langle e\rangle,\tilde{q}}(\tilde{\gamma}')$ along a set which is relatively compact in A, and $\tilde{\gamma}'$ would have intersection number zero with $L_{\langle e\rangle}$. It is clear that $f \circ \omega_A(L_A) = f(L_0) \subset (-1,1)$. The lemma is proved. □

Proof of Lemma 11.2 Let $\omega_A : (A, q_A) \to (X, q)$ be the holomorphic mapping from Lemma 11.3 that represents e, and let $L_A \ni q_A$ be the relatively closed curve in A with limit set on both boundary components of A. Consider a positively oriented dividing curve $\gamma_A : [0,1] \to A$ with base point $\gamma(0) = \gamma(1) = q_A$ such that $\gamma_A((0,1)) \subset A \setminus L_A$. The curve $\gamma = \omega_A(\gamma_A)$ represents e. The mapping $f \circ \omega_A$ is holomorphic on A and $f \circ \omega_A(\gamma_A) = f(\gamma)$ represents $f_*(e) \in \pi_1(\mathbb{C} \setminus \{-1,1\}, q')$ with $q' = f \circ \omega_A(q_A) = f(q) \in (-1,1)$. Hence, $f \circ \omega_A(\gamma_A)$ also represents the element $(f_*(e))_{tr} \in \pi_1^{tr}(\mathbb{C} \setminus \{-1,1\})$ in the relative fundamental group $\pi_1(\mathbb{C} \setminus \{-1,1\}, (-1,1)) = \pi_1^{tr}(\mathbb{C} \setminus \{-1,1\})$ corresponding to $f_*(e)$.

We prove now that $\Lambda_{tr}(f_*(e)) \le \lambda(A)$. Let $A_0 \Subset A$ be any relatively compact annulus in A with smooth boundary such that $q_A \in A_0$. If A_0 is sufficiently large, then the connected component L_{A_0} of $L_A \cap A_0$ that contains q_A has endpoints on different boundary components of A_0. The set $A_0 \setminus L_{A_0}$ is a curvilinear rectangle. The open horizontal curvilinear sides are the strands of the cut that are reachable from the curvilinear rectangle moving counterclockwise, or clockwise, respectively. The open vertical curvilinear sides are obtained from the boundary circles of A_0 by removing an endpoint of the arc L_{A_0}. Since $f \circ \omega_A$ maps L_A to $(-1,1)$, the restriction of $f \circ \omega_A$ to $A_0 \setminus L_{A_0}$ represents $(f_*(e))_{tr}$. Hence,

$$\Lambda_{tr}(f_*(e)) \le \lambda(A_0 \setminus L_{A_0}). \tag{11.8}$$

By Corollary 3.1

$$\lambda(A_0 \setminus L_{A_0}) \le \lambda(A_0). \tag{11.9}$$

We obtain the inequality $\Lambda_{tr}(f_*(e)) \le \lambda(A_0)$ for each annulus $A_0 \Subset A$, hence, since A belongs to the class $A(\widehat{e})$ of conformally equivalent annuli,

$$\Lambda_{tr}(f_*(e)) \le \lambda(A(\widehat{e}))\,, \tag{11.10}$$

and the Lemma follows from Theorem 9.1. □

The Monodromies Along Two Generators In the following Lemma we combine the information on the monodromies along two generators of the fundamental group $\pi_1(X, q)$. We allow the situation when the monodromy along one generator or along each of the two generators of the fundamental group of X is a power of a standard generator of $\pi_1(\mathbb{C} \setminus \{-1, 1\}, f(q))$.

Lemma 11.4 *Let $f : X \to \mathbb{C}\setminus\{-1, 1\}$ be a non-contractible holomorphic function on a connected open Riemann surface X such that 0 is a regular value of the imaginary part of f. Suppose f maps a simple relatively closed curve L_0 in X to $(-1, 1)$, and q is a point in L_0. Let $e^{(1)}$ and $e^{(2)}$ be primitive elements of $\pi_1(X, q)$. Suppose that for each $e = e^{(1)}$, $e = e^{(2)}$, and $e = e^{(1)}\,e^{(2)}$, the free homotopy class $\widehat{e}$ intersects L_0. Then either $f_*(e^{(j)})$, $j = 1, 2$, are (trivial or non-trivial) powers of the same standard generator of $\pi_1(\mathbb{C} \setminus \{-1, 1\}, q')$ with $q' = f(q) \in (-1, 1)$, or each of them is the product of at most two elements w_1 and w_2 of $\pi_1(\mathbb{C}\setminus\{-1, 1\}, q')$ with*

$$\mathcal{L}_-(w_j) \le 2\pi\,\lambda_{e^{(1)},e^{(2)}},\ j = 1, 2,$$

where

$$\lambda_{e^{(1)},e^{(2)}} \stackrel{def}{=} \max\{\lambda(A(\widehat{e^{(1)}})),\ \lambda(A(\widehat{e^{(2)}})),\ \lambda(A(\widehat{e^{(1)}\,e^{(2)}}))\}.$$

Hence,

$$\mathcal{L}_-(f_*(e^{(j)})) \le 4\pi\,\lambda_{e^{(1)},e^{(2)}},\ j = 1, 2. \tag{11.11}$$

The $\mathcal{L}_-$ invariant was defined before Lemma 10.1.

Proof If the monodromies $f_*(e^{(1)})$ and $f_*(e^{(2)})$ are not powers of a single standard generator (the identity is considered as zeroth power of a standard generator) we obtain the following. At most two of the elements, $f_*(e^{(1)})$, $f_*(e^{(2)})$, and $f_*(e^{(1)}\,e^{(2)}) = f_*(e^{(1)})\,f_*(e^{(2)})$, are powers of a standard generator, and if two of them are powers of a standard generator, then they are non-zero powers of different standard generators.

If two of them are non-zero powers of different standard generators, then the third has the form $a_\ell^k a_{\ell'}^{k'}$ with a_ℓ and $a_{\ell'}$ being different generators and k and k' being non-zero integers. By Lemma 11.2 the $\mathcal{L}_-$ of the third element does not exceed $2\pi\,\lambda_{e^{(1)},e^{(2)}}$. On the other hand it equals $\log(3|k'|) + \log(3|k|)$. Hence, $\mathcal{L}_-(a_\ell^k) = \log(3|k|) \le 2\pi\,\lambda_{e^{(1)},e^{(2)}}$ and $\mathcal{L}_-(a_{\ell'}^{k'}) = \log(3|k'|) \le 2\pi\,\lambda_{e^{(1)},e^{(2)}}$.

If two of the elements $f_*(e^{(1)})$, $f_*(e^{(2)})$, and $f_*(e^{(1)}\,e^{(2)}) = f_*(e^{(1)})\,f_*(e^{(2)})$, are not powers of a standard generator, then the $\mathcal{L}_-$ of each of the two elements does not exceed $2\pi\,\lambda_{e^{(1)},e^{(2)}}$. Since the $\mathcal{L}_-$ of an element coincides with the $\mathcal{L}_-$ of its inverse, the third element is the product of two elements with $\mathcal{L}_-$ not exceeding $2\pi\,\lambda_{e^{(1)},e^{(2)}}$. Since for $x, x' \geq 2$ the inequality $\log(x+x') \leq \log x + \log x'$ holds, the $\mathcal{L}_-$ of the product does not exceed the sum of the $\mathcal{L}_-$ of the factors. Hence the $\mathcal{L}_-$ of the third element does not exceed $4\pi\,\lambda_{e^{(1)},e^{(2)}}$. Hence, inequality (11.11) holds. □

The following proposition states the existence of suitable connected components of the zero set of the imaginary part of certain analytic functions on tori with a hole and on planar domains. As in Sect. 11.1 for a Riemann surface X we let $\mathcal{E}$ be a standard system of generators of $\pi_1(X, q_0)$ that is associated to a standard bouquet of circles for X. Recall that $\mathcal{E}_j$ is the set of primitive elements of $\pi_1(X, q_0)$ which can be written as product of at most j elements of $\mathcal{E} \cup (\mathcal{E})^{-1}$. Here as before for any subset $\mathcal{E}'$ of $\pi_1(X; q_0)$ we denote by $(\mathcal{E}')^{-1}$ the set of all elements that are inverse to elements in $\mathcal{E}'$.

Proposition 11.4 *Let X be a torus with a hole or a planar domain with base point q_0 and fundamental group $\pi_1(X, q_0)$, and let $\mathcal{E}$ be a standard system of generators of $\pi_1(X, q_0)$ that is associated to a standard bouquet of circles for X. Let $f : X \to \mathbb{C} \setminus \{-1, 1\}$ be a non-contractible holomorphic mapping such that 0 is a regular value of $\operatorname{Im} f$. Then there exist a simple relatively closed curve $L_0 \subset X$ such that $f(L_0) \subset \mathbb{R} \setminus \{-1, 1\}$, and a set $\mathcal{E}_2' \subset \mathcal{E}_2 \subset \pi_1(X, q_0)$ of primitive elements of $\pi_1(X, q_0)$, such that the following holds. Each element $e_{j,0} \in \mathcal{E} \subset \pi_1(X, q_0)$ is the product of at most two elements of $\mathcal{E}_2' \cup (\mathcal{E}_2')^{-1}$. Moreover, for each $e_0 \in \pi_1(X, q_0)$ which is the product of one or two elements from $\mathcal{E}_2'$ the free homotopy class $\widehat{e_0}$ has positive intersection number with L_0 (after suitable orientation of L_0).*

If X is a torus with a hole or X equals $\mathbb{P}^1$ with three holes, we may chose $\mathcal{E}_2'$ consisting of two elements, one of them contained in $\mathcal{E}$, the other is either contained in $\mathcal{E} \cup \mathcal{E}^{-1}$ or is a product of two elements of $\mathcal{E}$.

Notice the following facts. By Theorem 1.1 a mapping $f : X \to \mathbb{C} \setminus \{-1, 1\}$ is contractible if and only if for each $e_0 \in \pi_1(X, q_0)$ the monodromy $f_*(e_0)$ is equal to the identity. The mapping f is reducible if and only if the monodromy mapping $f_* : \pi_1(X, q_0) \to \pi_1(\mathbb{C} \setminus \{-1, 1\}, f(q_0))$ is conjugate to a mapping into a subgroup Γ of $\pi_1(\mathbb{C} \setminus \{-1, 1\}, f(q_0))$ that is generated by a single element that is represented by a curve which separates one of the points $1, -1$ or ∞ from the other points. In other words, Γ is (after identifying fundamental groups with different base point up to conjugacy) generated by a conjugate of one of the elements a_1, a_2 or $a_1 a_2$ of $\pi_1(\mathbb{C} \setminus \{-1, 1\}, 0)$. Recall that for $q' \in (-1, 1)$ the fundamental group $\pi_1(\mathbb{C} \setminus \{-1, 1\}, q')$ is canonical isomorphic to $\pi_1(\mathbb{C} \setminus \{-1, 1\}, 0)$ (by changing the base point along the segment in $(-1, 1)$ that joins q' and 0), and we will identify $\pi_1(\mathbb{C} \setminus \{-1, 1\}, q')$ with $\pi_1(\mathbb{C} \setminus \{-1, 1\}, 0)$.

If f is irreducible, then it is not contractible, and, hence, the preimage $f^{-1}(\mathbb{R})$ is not empty.

Denote by M_1 a Möbius transformation which permutes the points $-1,\ 1,\ \infty$ and maps the interval $(-\infty, -1)$ onto $(-1, 1)$, and let M_2 be a Möbius transformation which permutes the points $-1,\ 1,\ \infty$ and maps the interval $(1, \infty)$ onto $(-1, 1)$. Let $M_0 \stackrel{def}{=} \mathrm{Id}$.

The main step for the proof of Theorem 11.1 is the following Proposition 11.5.

Recall that the invariant $\lambda_j(X)$ was defined in Definition 11.1, Sect. 11.1. Since for $e_0 \in \pi_1(X, q_0)$ the equality $\lambda(\tilde{X}/(\mathrm{Is}^{\tilde{q}_0})^{-1}(\langle e_0 \rangle)) = \lambda(A(\widehat{e_0}))$ holds, $\lambda_j(X)$ is the maximum of $\lambda(A(\widehat{e_0}))$ over $e_0 \in \mathcal{E}_j$.

Proposition 11.5 *Let X be a connected finite open Riemann surface with base point q_0, and let $\mathcal{E}$ be the same system of generators of $\pi_1(X, q_0)$ as in Proposition 11.4. Suppose $f : X \to \mathbb{C} \setminus \{-1, 1\}$ is an irreducible holomorphic mapping, such that 0 is a regular value of $\mathrm{Im} f$. Then for one of the functions $M_l \circ f$, $l = 0, 1, 2$, which we denote by F, there exists a point $q \in X$ (depending on f), such that the point $q' \stackrel{def}{=} F(q)$ is contained in $(-1, 1)$, and there exists a smooth curve α in X joining q_0 with q, such that the following holds. For each element $e_j \in Is_\alpha(\mathcal{E})$ the monodromy $F_*(e_j)$ is the product of at most six elements of $\pi_1(\mathbb{C} \setminus \{-1, 1\}, q')$ (canonically identified with $\pi_1(\mathbb{C} \setminus \{-1, 1\}, 0)$) of $\mathcal{L}_-$ not exceeding $2\pi\lambda_7(X)$ and, hence,*

$$\mathcal{L}_-(F_*(e_j)) \leq 12\pi\lambda_7(X) \ \text{for each}\ j. \tag{11.12}$$

If X is a torus with a hole the proposition holds with $\lambda_7(X)$ replaced by $\lambda_3(X)$. If X is a planar domain the proposition holds with $\lambda_4(X)$ instead of $\lambda_7(X)$.

Notice, that all monodromies of contractible mappings are equal to the identity, hence the inequality (11.12) holds automatically for contractible mappings.

We postpone the proof of the two propositions and prove first the Theorem 11.1.

Proof of Theorem 11.1 Let X be a connected finite open Riemann surface (possibly of second kind) with base point q_0. We need to estimate the number N_X of homotopy classes of irreducible mappings $X \to \mathbb{C} \setminus \{-1, 1\}$ that contain a holomorphic mapping. Take any relatively compact open subset $X^0 \Subset X$ that is a deformation retract of X and contains q_0. We claim that the number N_X does not exceed the number $N^*_{X^0}$ of homotopy classes of irreducible mappings $X^0 \to \mathbb{C} \setminus \{-1, 1\}$ that contain a holomorphic mapping for which 0 is a regular value of the imaginary part.

Notice first that homotopic mappings $X \to \mathbb{C} \setminus \{-1, 1\}$ restrict to homotopic mappings $X^0 \to \mathbb{C}\setminus\{-1, 1\}$. Vice versa, if the restrictions $f_j \mid X^0$ of two mappings $f_j : X \to \mathbb{C} \setminus \{-1, 1\}$, $j = 1, 2$, are homotopic then the f_j are homotopic. Indeed, let $\varphi_t : X \to X$ be a continuous family of mappings $\varphi_t : X \to X$, $t \in [0, 1]$, that map X homeomorphically onto a domain X_t in X such that φ_0 is the identity and $X_1 = X^0$ (see also Lemma 1.1). Then f_j is homotopic to $f_j \circ \varphi_1$, $j = 1, 2$, and $f_1 \circ \varphi_1$ is homotopic to $f_2 \circ \varphi_1$.

For each irreducible holomorphic mapping $f : X \to \mathbb{C} \setminus \{-1, 1\}$ and each small enough complex number ε the holomorphic mapping $(f - i\varepsilon) \mid X^0$ is homotopic to $f \mid X^0$, is irreducible and takes values in $\mathbb{C} \setminus \{-1, 1\}$. Hence, the homotopy class of the restriction to X^0 of any irreducible holomorphic mapping $X \to \mathbb{C} \setminus \{-1, 1\}$ contains an irreducible holomorphic mapping for which zero is a regular value of the imaginary part. This means that the number N_X does not exceed the number $N^*_{X_0}$ of irreducible homotopy classes $X^0 \to \mathbb{C} \setminus \{-1, 1\}$ that contain a holomorphic mapping for which 0 is a regular value of the imaginary part.

We estimate now the number $N^*_{X_0}$. Take any irreducible holomorphic mapping $\mathrm{f} : X^0 \to \mathbb{C} \setminus \{-1, 1\}$ such that 0 is a regular value of the imaginary part $\mathrm{Im\,f}$ and apply Proposition 11.5. We obtain the following objects: a Möbius transformation M_l, that maps one of the components of $\mathbb{R} \setminus \{-1, 1\}$ onto $(-1, 1)$; further, a point $q \in X^0$ and a smooth curve α in X^0 with initial point q_0 and terminal point q, such that for the mapping $F = M_l \circ \mathrm{f}$ the inclusion $q' \overset{def}{=} F(q) \in (-1, 1)$ holds, and for the generators $e_j \overset{def}{=} \mathrm{Is}_\alpha(e_{j,0}),\ e_{j,0} \in \mathcal{E}$, of $\pi_1(X^0, q)$ the inequalities (11.12) hold with X replaced by X^0. Our goal is to find a smooth mapping $\hat{F} : X^0 \to \mathbb{C}\setminus\{-1, 1\}$, that is free homotopic to F, maps q_0 to q', and satisfies $(\hat{F})_*(e_0) = F_*(e)$. Then the inequalities (11.12)) will imply the inequality

$$\mathcal{L}_-((\hat{F})_*(e_0)) \leq 12\pi\lambda_7(X^0) \tag{11.13}$$

for each $e_{j,0} \in \mathcal{E}$.

Write $e = \mathrm{Is}_\alpha(e_0) \in \pi_1(X, q)$ for each $e_0 \in \pi_1(X, q_0)$. Parameterise α by the interval $[0, 1]$. The image of α under the mapping F is the curve $\beta = F \circ \alpha$ in $\mathbb{C} \setminus \{-1, 1\}$ with initial point $F(q_0)$ and terminal point $F(q) = q'$. Then $F_*(e_0) = (\mathrm{Is}_\beta)^{-1}(F_*(e))$. The curve β is parameterized by the interval $[0, 1]$. Choose a free homotopy F_t, $t \in [0, 1]$, of mappings from X^0 to $\mathbb{C} \setminus \{-1, 1\}$, that changes the mappings only in a small neighbourhood of β and joins the mapping $F_0 \overset{def}{=} F$ with a (smooth) mapping F_1 denoted by $\hat{F}$, so that $F_t(q_0) = \beta(t)$, $t \in [0, 1]$. The value $\beta(t)$ moves from the point $\beta(0) = F(q_0)$ to $\beta(1) = q'$ along the curve β. Denote by β_t the curve that runs from $\beta(t)$ to $\beta(1)$ along β. Then $\beta_0 = \beta$ and β_1 is a constant curve. Let γ_0 be a curve that represents e_0. Then $\gamma \overset{def}{=} \mathrm{Is}_\alpha(\gamma_0)$ represents e. The base point of the curve $F_t(\gamma_0)$ equals $F_t(q_0) = \beta(t)$. Hence, we obtain a continuous family of curves $\beta_t^{-1} F_t(\gamma_0)\beta_t$ with base point $\beta(1) = F(q)$. For $t = 1$ the curve is equal to $F_1(\gamma_0) = \hat{F}(\gamma_0)$, for $t = 0$ the curve is equal to $\beta^{-1} F_0(\gamma_0)\beta = F_0(\alpha^{-1}\gamma_0\alpha) = F_0(\mathrm{Is}_\alpha(\gamma_0)) = F_0(\gamma)$. Since the two curves $F_1(\gamma_0)$ and $F_0(\gamma)$ are homotopic and $F_1 = \hat{F}$, $F_0 = F$, we obtain $\hat{F}_*(e_0) = F_*(e)$.

We estimate now the number of free homotopy classes of mappings $X^0 \to \mathbb{C} \setminus \{-1, 1\}$ whose associated conjugacy class of homomorphisms $\pi_1(X^0, q_0) \to \pi_1(\mathbb{C} \setminus \{-1, 1\}, q')$ contains a homomorphism that satisfies the inequalities (11.13). Recall that $\pi_1(\mathbb{C} \setminus \{-1, 1\}, q')$ is canonically isomorphic to $\pi_1(\mathbb{C} \setminus \{-1, 1\}, 0)$ by the isomorphism $\mathrm{Is}_{\beta'}$ where the curve β' runs along the segment that joins q' and 0 and is contained in $(-1, 1)$. By Lemma 10.1 there are at most $\frac{1}{2}e^{36\pi\lambda_7(X^0)} + 1 \leq$

$\frac{3}{2}e^{36\pi\lambda_7(X^0)}$ different reduced words $w \in \pi_1(\mathbb{C}\setminus\{-1,1\}),0)$ (including the identity) with $\mathcal{L}_-(w) \le 12\pi\lambda_7(X^0)$. Hence, there are at most $(\frac{3}{2}e^{36\pi\lambda_7(X^0)})^{2g+m}$ different homomorphisms $h : \pi_1(X, q_0) \to \pi_1(\mathbb{C}\setminus\{-1,1\},0)$ with $\mathcal{L}_-(h(e_0)) \le 12\pi\lambda_7(X^0)$ for each element e_0 of the set of generators $\mathcal{E}$ of $\pi_1(X^0, q_0)$. By Theorem 1.1 there are at most $(\frac{3}{2}e^{36\pi\lambda_7(X^0)})^{2g+m}$ different free homotopy classes of mappings $X^0 \to \mathbb{C}\setminus\{-1,1\}$, whose associated conjugacy class of homomorphisms $\pi_1(X^0, q_0) \to \pi_1(\mathbb{C}\setminus\{-1,1\},0)$ contains a homomorphism that satisfies the inequalities (11.13). It follows that $N^*_{X_0} \le (\frac{3}{2}e^{36\pi\lambda_7(X^0)})^{2g+m}$.

Our arguments show that for each irreducible or contractible holomorphic mapping f on X there are arbitrarily small numbers ε, such that one of the three mappings $M_l(f - i\varepsilon) \mid X^0, l = 0, 1, 2$, belongs to one of at most $(\frac{3}{2}e^{36\pi\lambda_7(X^0)})^{2g+m}$ free homotopy classes of mappings $X^0 \to \mathbb{C}\setminus\{-1,1\}$. Since for small enough numbers ε the mappings $M_l(f) \mid X^0$ and $M_l(f - i\varepsilon) \mid X^0$ are homotopic, the mapping $M_l(f) \mid X^0$ belongs to one of at most $(\frac{3}{2}e^{36\pi\lambda_7(X^0)})^{2g+m}$ free homotopy classes of mappings $X^0 \to \mathbb{C}\setminus\{-1,1\}$. Compose the mappings of each class with each of the three Möbius transformations $M_{l'}^{-1}$. We obtain no more than $3(\frac{3}{2}e^{36\pi\lambda_7(X^0)})^{2g+m}$ free homotopy classes of mappings $X^0 \to \mathbb{C}\setminus\{-1,1\}$, and $f \mid X^0 = M_l^{-1}(M_l(f)) \mid X^0$ belongs to one of them. Identifying $\pi_1(X^0, q_0)$ with $\pi_1(X, q_0)$, the homotopy classes of mappings $X \to \mathbb{C}\setminus\{-1,1\}$ have the same associated conjugacy classes of homomorphism $\pi_1(X, q_0) \to \pi_1(\mathbb{C}\setminus\{-1,1\},0)$ as the homotopy classes of their restrictions. We see that there are no more than $3(\frac{3}{2}e^{36\pi\lambda_7(X^0)})^{2g+m}$ holomorphic mappings $X \to \mathbb{C}\setminus\{-1,1\}$ that are irreducible or contractible. Theorem 11.1 is proved with the upper bound $3(\frac{3}{2}e^{24\pi\lambda_7(X^0)})^{2g+m}$ for an arbitrary relatively compact domain $X^0 \subset X$ that is a deformation retract of X.

It remains to prove that

$$\lambda_7(X) = \inf\{\lambda_7(X^0) : X^0 \Subset X \text{ is a deformation retract of } X\}.$$

We have to prove that for each $e_0 \in \pi_1(X, q_0)$ the quantity $\lambda(\tilde{X}/(\mathrm{Is}^{\tilde{q}_0})^{-1}(\langle e_0\rangle))$ is equal to the infimum of $\lambda(\widetilde{X^0}/(\mathrm{Is}^{\tilde{q}_0})^{-1}(\langle e_0\rangle))$ over all X^0 being open relatively compact subsets of X which are deformation retracts of X. Here $\widetilde{X^0}$ is the universal covering of X^0, and the fundamental groups of X and X^0 are identified. $\widetilde{X^0}$ ($\tilde{X}$, respectively) can be defined as set of homotopy classes of arcs in X^0 (in X, respectively) joining q_0 with a point $q \in X^0$ (in X respectively) equipped with the complex structure induced by the projection to the endpoint of the arcs, and the point $\tilde{q}_0$ corresponds to the class of the constant curve. The isomorphism $(\mathrm{Is}^{\tilde{q}_0})^{-1}$ from $\pi_1(X^0, q_0)$ to the group of covering transformations on $\widetilde{X^0}$ is defined in the same way as it was done for X instead of X^0. These considerations imply that there is a holomorphic mapping from $\tilde{X}^0/(\mathrm{Is}^{\tilde{q}_0})^{-1}(\langle e_0\rangle)$ into $\tilde{X}/(\mathrm{Is}^{\tilde{q}_0})^{-1}(\langle e_0\rangle)$. Hence, the extremal length of the first set is not smaller than the extremal length of the second set, $\lambda(\tilde{X}^0/\mathrm{Is}^{\tilde{q}_0})^{-1}(\langle e_0\rangle) \ge \lambda(\tilde{X}/(\mathrm{Is}^{\tilde{q}_0})^{-1}(\langle e_0\rangle)$.

Vice versa, take any annulus A^0 that is a relatively compact subset of $\widetilde{X}\diagup(\mathrm{Is}^{\tilde{q}_0})^{-1}(\langle e_0\rangle)$ and is a deformation retract of $\widetilde{X}\diagup(\mathrm{Is}^{\tilde{q}_0})^{-1}(\langle e_0\rangle)$. Its projection to X is relatively compact in X, hence, it is contained in a relatively compact open subset X^0 of X that is a deformation retract of X. Hence, A^0 can be considered as subset of $\widetilde{X^0}\diagup(\mathrm{Is}^{\tilde{q}_0})^{-1}(\langle e_0\rangle)$, and, hence, $\lambda(\widetilde{X^0}\diagup(\mathrm{Is}^{\tilde{q}_0})^{-1}(\langle e_0\rangle)) \le \lambda(A^0)$. Since

$$\lambda(\widetilde{X}\diagup(\mathrm{Is}^{\tilde{q}_0})^{-1}(\langle e_0\rangle)) = \inf\{\lambda(A^0) : A^0 \Subset \widetilde{X}\diagup(\mathrm{Is}^{\tilde{q}_0})^{-1}(\langle e_0\rangle)$$
$$\text{is a deformation retract of } \widetilde{X}\diagup(\mathrm{Is}^{\tilde{q}_0})^{-1}(\langle e_0\rangle)\},$$

we obtain $\lambda(\widetilde{X}^0\diagup\mathrm{Is}^{\tilde{q}_0})^{-1}(\langle e_0\rangle) \le \lambda(\widetilde{X}\diagup(\mathrm{Is}^{\tilde{q}_0})^{-1}(\langle e_0\rangle))$. We are done. □

We proved a slightly stronger statement, namely, the number of homotopy classes of mappings $X \to \mathbb{C}\setminus\{-1,1\}$ that contain a contractible holomorphic mapping or an irreducible holomorphic mapping does not exceed $3(\frac{3}{2}e^{36\pi\lambda_7(X)})^{2g+m}$.

Proof of Proposition 11.4 Denote the zero set $\{x \in X : \mathrm{Im} f(x) = 0\}$ by L. Since f is not contractible, $L \neq \emptyset$.

1. A Torus with a Hole Assume first that X is a torus with a hole with base point q_0. For notational convenience we denote by e_0 and e_0' the two elements of the set of generators $\mathcal{E}$ of $\pi_1(X, q_0)$ that is associated to a standard bouquet of circles for X. We claim that there is a connected component L_0 of L such that (after suitable orientation) the intersection number of the free homotopy class of one of the elements of $\mathcal{E}$, say of $\widehat{e_0}$, with L_0 is positive, and the intersection number with one of the classes $\widehat{e_0'}$, or $\widehat{(e_0')^{-1}}$, or $\widehat{e_0\, e_0'}$ with L_0 is positive.

The claim is easy to prove in the case when there is a component L_0 of L which is a simple closed curve that is not contractible and not contractible to the hole of X. Indeed, consider the inclusion of X into a closed torus X^c and the homomorphism on fundamental groups $\pi_1(X, q_0) \to \pi_1(X^c, q_0)$ induced by the inclusion. Denote by e_0^c and ${e_0'}^c$ the images of e_0 and e_0' under this homomorphism. Notice that e_0^c and ${e_0'}^c$ commute. The (image under the inclusion of the) curve L_0 is a simple closed non-contractible curve in X^c. It represents the free homotopy class of an element $(e_0^c)^j({e_0'}^c)^k$ for some integers j and k which are not both equal to zero. Hence, L_0 is not null-homologous in X^c, and by the Poincaré Duality Theorem for one of the generators, say for e_0^c, the representatives of the free homotopy class $\widehat{e_0^c}$ have non-zero intersection number with L_0. After possibly reorienting L_0, we may assume that this intersection number is positive. There is a representative of the class $\widehat{e_0^c}$ which is contained in X, hence, $\widehat{e_0}$ has positive intersection number with L_0.

Suppose all compact connected components of L are contractible or contractible to the hole of X. Consider a relatively compact domain $X'' \Subset X$ in X with smooth boundary which is a deformation retract of X such that for each connected component of L at most one component of its intersection with X'' is not contractible to the hole of X''. (See the paragraph on "Regular zero sets".) There is at least one component of $L \cap X''$ that is not contractible to the hole of X''. Indeed, otherwise the free homotopy class of each element of $\mathcal{E}$ could be represented by a loop avoiding L, and, hence, the monodromy of f along each element of $\mathcal{E}$ would be conjugate to the identity, and, hence, equal to the identity, i.e. contrary to the assumption, $f : X \to \mathbb{C} \setminus \{-1, 1\}$ would be free homotopic to a constant.

Take a component L_0'' of $L \cap X''$ that is not contractible to the hole of X''. There is an arc of $\partial X''$ between the endpoints of L_0'' such that the union $\tilde{L}_0$ of the component L_0'' with this arc is a closed curve in X that is not contractible and not contractible to the hole. Hence, for one of the elements of $\mathcal{E}$, say for e_0, the intersection number of the free homotopy class $\widehat{e_0}$ with the closed curve $\tilde{L}_0$ is positive after orienting the curve $\tilde{L}_0$ suitably. We may take a representative γ of $\widehat{e_0}$ that is contained in X''. Then γ has positive intersection number with L_0''. Denote the connected component of L that contains L_0'' by L_0. All components of $L_0 \cap X''$ different from L_0'' are contractible to the hole of X''. Hence, γ has intersection number zero with each of these components. Hence, γ has positive intersection number with L_0 since $\gamma \subset X''$. We proved that the class $\widehat{e_0}$ has positive intersection number with L_0.

If $\widehat{e_0'}$ also has non-zero intersection number with L_0 we define $e_0'' = (e_0')^{\pm 1}$ so that the intersection number of $\widehat{e_0''}$ with L_0 is positive. If $\widehat{e_0'}$ has zero intersection number with L_0 we put $e_0'' = e_0\, e_0'$. Then again the intersection number of $\widehat{e_0''}$ with L_0 is positive. Also, the intersection number of $\widehat{e_0\, e_0''}$ with L_0 is positive. The set $\mathcal{E}_2' \overset{def}{=} \{e_0, e_0''\}$ satisfies the condition required in the proposition. We obtained Proposition 11.4 for a torus with a hole.

2. A Planar Domain Let X be a planar domain. The domain X is conformally equivalent to a disc with m smoothly bounded holes, equivalently, to the Riemann sphere with $m+1$ smoothly bounded holes, $\mathbb{P}^1 \setminus \bigcup_{j=1}^{m+1} \mathcal{C}_j$, where $\mathcal{C}_{m+1}$ contains the point ∞. As before the base point of X is denoted by q_0, and for each $j = 1, \ldots, m$, the generator $e_{j,0} \in \mathcal{E} \subset \pi_1(X, q_0)$ is represented by a curve that surrounds $\mathcal{C}_j$ once counterclockwise. Since f is not contractible, there must be a connected component of L that has limit points on some $\mathcal{C}_j$ with $j \le m$. Indeed, otherwise the free homotopy class of each generator could be represented by a curve that avoids L. This would imply that all monodromies are equal to the identity. We claim that

there exists a component L_0 of L with limit points on the boundary of two components $\partial\mathcal{C}_{j'}$ and $\partial\mathcal{C}_{j''}$ for some $j', j'' \in \{1, \ldots, m+1\}$ with $j'' \neq j'$.

Indeed, assume the contrary. Then, if a component of L has limit points on a component $\partial\mathcal{C}_j$, $j \le m$, then all its limit points are on $\partial\mathcal{C}_j$. Take a smoothly bounded simply connected domain $\mathcal{C}_j' \Subset X \cup \mathcal{C}_j$ that contains the closed set $\mathcal{C}_j$, so that its boundary $\partial\mathcal{C}_j'$ represents $\widehat{e_{j,0}}$. Then all components L_k' of $L \setminus \mathcal{C}_j'$ with an endpoint on $\partial\mathcal{C}_j'$ have another endpoint on this circle. The two endpoints of each L_k' on $\partial\mathcal{C}_j'$ divide $\partial\mathcal{C}_j'$ into two connected components. The union of each $\overline{L_k'}$ with each of the two components of $\partial\mathcal{C}_j' \setminus \overline{L_k'}$ is a simple closed curve in $\mathbb{C}$, and, hence, by the Jordan Curve Theorem it bounds a relatively compact topological disc in $\mathbb{C}$. One of these discs contains $\mathcal{C}_j'$, the other does not. Assign to each component L_k' of $L \setminus \mathcal{C}_j'$ with both endpoints on $\partial\mathcal{C}_j'$ the closed arc α_k in $\partial\mathcal{C}_j'$ with the same endpoints as L_k', whose union with L_k' bounds a relatively compact topological disc in $\mathbb{C}$ that does not contain C'_j. These discs are partially ordered by inclusion, since the L'_k are pairwise disjoint. Hence, the arcs α_k are partially ordered by inclusion. For an arc α_k which contains no other of the arcs (a minimal arc) the curve $f \circ \alpha_k$ except its endpoints is contained in $\mathbb{C} \setminus \mathbb{R}$. Moreover, the endpoints of $f \circ \alpha_k$ lie on $f(\overline{L'_k})$, which is contained in one connected components of $\mathbb{R} \setminus \{-1, 1\}$, since $\overline{L'_k}$ is connected. Hence, the curve $f \circ \alpha_k$ is homotopic in $\mathbb{C} \setminus \{-1, 1\}$ (with fixed endpoints) to a curve in $\mathbb{R} \setminus \{-1, 1\}$. The function f either maps all points on $\partial C'_j \setminus \alpha_k$ that are close to α_k to the open upper half-plane or maps them all to the open lower half-plane. (Recall, that zero is a regular value of $\operatorname{Im} f$.) Hence, for an open arc $\alpha'_k \subset \partial C'_j$ that contains α_k the curve $f \circ \alpha'_k$ is homotopic in $\mathbb{C} \setminus \{-1, 1\}$ (with fixed endpoints) to a curve in $\mathbb{C} \setminus \mathbb{R}$.

Consider the arcs α_k with the following property. For an open arc α'_k in $\partial C'_j$ which contains the closed arc α_k the mapping $f \circ \alpha'_k$ is homotopic in $\mathbb{C} \setminus \{-1, 1\}$ (with fixed endpoints) to a curve contained in $\mathbb{C} \setminus \mathbb{R}$. Induction on the arcs by inclusion shows that this property is satisfied for all maximal arcs among the α_k and, hence, $f \mid \partial C'_j$ is contractible in $\mathbb{C} \setminus \{-1, 1\}$. Hence, if the claim was not true, then for each hole C_j, $j \le m$, whose boundary contains limit points of a connected component of L, the monodromy along the curve C'_j (with any base point contained in C'_j) that represents $\widehat{e_{j,0}}$ would be trivial. Then all monodromies would be trivial, which contradicts the fact that the mapping is not contractible. The contradiction proves the claim.

With j' and j'' being the numbers of the claim and $j' \le m$ we consider the set $\mathcal{E}'_2 \subset \mathcal{E}_2$ which consists of the following primitive elements: $e_{j',0}$, the element $(e_{j'',0})^{-1}$ provided $j'' \neq m+1$, and $e_{j',0}\, e_{j,0}$ for all $j = 1, \ldots, m$, $j \neq j'$, $j \neq j''$. The free homotopy class of each element of $\mathcal{E}'_2$ has intersection number 1 with L_0 after suitable orientation of the curve L_0. Each product of at most two different elements of $\mathcal{E}'_2$ is a primitive element of $\pi_1(X, q)$ and is contained in $\mathcal{E}_4$. Moreover, the intersection number with L_0 of the free homotopy class of each product of two different elements of $\mathcal{E}'_2$ equals 2. Each element of $\mathcal{E}$ is the product of at most two elements of $\mathcal{E}'_2 \cup (\mathcal{E}'_2)^{-1}$.

The proposition is proved for the case of planar domains X. □

Proof of Proposition 11.5

1. A Torus with a Hole Suppose X is a torus with a hole. Consider the curve L_0 and the set $\mathcal{E}'_2 \subset \pi_1(X, q_0)$ obtained in Proposition 11.4. For one of the functions $M_l \circ f$, denoted by F, the image $F(L_0)$ is contained in $(-1, 1)$. Let e_0, e'_0 be the two elements of $\mathcal{E}$. Move the base point q_0 to a point $q \in L_0$ along a curve α in X, and consider the generators $e = \operatorname{Is}_\alpha(e_0)$ and $e' = \operatorname{Is}_\alpha(e'_0)$ of $\pi_1(X, q)$, and the set $\operatorname{Is}_\alpha(\mathcal{E}'_2) \subset \pi_1(X, q)$. Then e and e' are products of at most two elements of $\operatorname{Is}_\alpha(\mathcal{E}'_2)$. Since the free homotopy class of an element of $\pi_1(X, q_0)$ coincides with the free homotopy class of the element of $\pi_1(X, q)$ obtained by applying Is_α, the free homotopy class of each product of one or two elements of $\operatorname{Is}_\alpha(\mathcal{E}'_2)$ intersects L_0. We may assume as in the proof of Proposition 11.4 that $\operatorname{Is}_\alpha(\mathcal{E}'_2)$ consists of the elements e and e'', where e'' is either equal to $e'^{\pm 1}$, or equals the product of

e and e'. Lemma 11.4 applies to the pair e, e'', the function F, and the curve L_0. Since F is irreducible, the monodromies of F along e and e'' are not powers of a single standard generator of the fundamental group of $\pi_1(\mathbb{C} \setminus \{-1, 1\}, q')$. Hence, the monodromy along each of the e and e'' is the product of at most two elements of $\mathcal{L}_-$ not exceeding $2\pi\lambda_{e,e''}$. Therefore, the monodromy of F along each of the e and e'' has $\mathcal{L}_-$ not exceeding $4\pi\lambda_{e,e''}$. Notice that $\lambda_{e,e''} = \lambda_{e_0 e_0''} \le \lambda_3(X)$, since e_0'' is the product of at most two factors, each an element of $\mathcal{E} \cup \mathcal{E}^{-1}$. Since e' is the product of at most two different elements among the e and e'' and their inverses, we obtain Proposition 11.5 for e and e', in particular $\mathcal{L}_-(F_*(e))$ and $\mathcal{L}_-(F_*(e'))$ do not exceed $8\pi\lambda_3(X)$. Proposition 11.5 is proved for tori with a hole.

2. A Planar Domain Consider the curve L_0 and the set $\mathcal{E}_2'$ of Proposition 11.4. Move the base point q_0 along an arc α to a point $q \in L_0$. Then $f(q) \in \mathbb{R} \setminus \{-1, 1\}$ and for one of the mappings f, $M_1 \circ f$, or $M_2 \circ f$, denoted by F, the inclusion $F(L_0) \subset (-1, 1))$ holds, hence, $q' \overset{def}{=} F(q)$ is contained in $(-1, 1)$. Put $e_j = \text{Is}_\alpha(e_{j,0})$ for each element $e_{j,0} \in \mathcal{E}$. The e_j form the basis $\text{Is}_\alpha(\mathcal{E})$ of $\pi_1(X, q)$. The set $\text{Is}_\alpha(\mathcal{E}_2')$ consists of primitive elements of $\pi_1(X, q)$ such that the free homotopy class of each product of one or two elements of $\text{Is}_\alpha(\mathcal{E}_2')$ intersects L_0. Moreover, each element of $\text{Is}_\alpha(\mathcal{E})$ is the product of one or two elements of $\text{Is}_\alpha(\mathcal{E}_2') \cup (\text{Is}_\alpha(\mathcal{E}_2'))^{-1}$.

By the condition of the proposition not all monodromies $F_*(e)$, $e \in \text{Is}_\alpha(\mathcal{E}_2')$, are (trivial or non-trivial) powers of the same standard generator of $\pi_1(\mathbb{C} \setminus \{-1, 1\}, q')$. Apply Lemma 11.4 to all pairs of elements of $\text{Is}_\alpha(\mathcal{E}_2')$ whose monodromies are not (trivial or non-trivial) powers of the same standard generator of $\pi_1(\mathbb{C} \setminus \{-1, 1\}, q')$. Since the product of at most two different elements of $\text{Is}_\alpha(\mathcal{E}_2')$ is contained in $\text{Is}_\alpha(\mathcal{E}_4)$, Lemma 11.4 shows that the monodromy $F_*(e)$ along each element $e \in \text{Is}_\alpha(\mathcal{E}_2')$ is the product of at most two factors, each with $\mathcal{L}_-$ not exceeding $2\pi\lambda_4(X)$. Since each element of $\text{Is}_\alpha(\mathcal{E})$ is a product of at most two factors in $\mathcal{E}_2' \cup (\mathcal{E}_2')^{-1}$, the monodromy $F_*(e_j)$ along each generator e_j of $\pi_1(X, q)$ is the product of at most 4 factors of $\mathcal{L}_-$ not exceeding $2\pi\lambda_4(X)$, and, hence, each monodromy $F_*(e_j)$ has $\mathcal{L}_-$ not exceeding $8\pi\lambda_4(X)$. Proposition 11.5 is proved for planar domains.

3.1. The General Case. Diagrams of Coverings We will use diagrams of coverings to reduce this case to the case of a torus with a hole or to the case of the Riemann sphere with three holes.

Let as before $\tilde{q}_0$ be the point in $\tilde{X}$ with $\mathsf{P}(\tilde{q}_0) = q_0$ chosen in Sect. 1.2. Let N be a subgroup of the fundamental group $\pi_1(X, q_0)$ and let $\omega^N : \tilde{X} \to \tilde{X}/(\text{Is}^{\tilde{q}_0})^{-1}(N) = X(N)$ be the projection defined in Sect. 1.2. Put $(q_0)_N \overset{def}{=} \omega^N(\tilde{q}_0)$. For an element $e_0 \in N \subset \pi_1(X, q_0)$ we denote by $(e_0)_N$ the element of $\pi_1(X(N), (q_0)_N)$ that is obtained as follows. Take a curve γ in X with base point q_0 that represents $e_0 \in N$. Let $\tilde{\gamma}$ be its lift to $\tilde{X}$ with initial point $\tilde{q}_0$. Then $\gamma_N \overset{def}{=} \omega^N(\tilde{\gamma})$ is a closed curve in $X(N) = \tilde{X}/(\text{Is}^{\tilde{q}_0})^{-1}(N)$ with base point $(q_0)_N$. The element of $\pi_1(X(N), (q_0)_N)$ represented by γ_N is the required element $(e_0)_N$. All curves γ_N' representing $(e_0)_N$ have the form $\omega^N(\tilde{\gamma}')$ for a curve $\tilde{\gamma}'$ in $\tilde{X}$ with initial point $\tilde{q}_0$ and terminal point

$(\mathrm{Is}^{\tilde{q}_0})^{-1}(e_0)(\tilde{q}_0)$. Since $\omega_N \circ \omega^N = \mathsf{P}$, the curve $\omega_N(\gamma'_N) = \mathsf{P}(\tilde{\gamma}') = \gamma'$ represents e_0 for each curve γ'_N in $X(N)$ that represents $(e_0)_N$. We obtain $(\omega_N)_*((e_0)_N) = e_0$. For two subgroups $N_1 \le N_2$ of $\pi_1(X, q_0)$ we obtain $(\omega_{N_1}^{N_2})_*((e_0)_{N_1}) = (e_0)_{N_2}$, $e_0 \in N_1$ (see the commutative diagram Fig. 11.1).

Let $\tilde{q}$ be another base point of $\tilde{X}$ and let $\tilde{\alpha}$ be a curve in $\tilde{X}$ with initial point $\tilde{q}_0$ and terminal point $\tilde{q}$. Let again N be a subgroup of $\pi_1(X, q_0)$. Put $q_N \overset{def}{=} \omega^N(\tilde{q})$. The curve $\alpha_N = \omega^N(\tilde{\alpha})$ in $X(N)$, and the base point $\tilde{q}$ of $\tilde{X}$ are compatible, hence, $X(\mathrm{Is}_{\alpha_N}(N)) \overset{def}{=} \tilde{X} \diagup (\mathrm{Is}^{\tilde{q}})^{-1}(\mathrm{Is}_{\alpha_N}(N)) = \tilde{X} \diagup (\mathrm{Is}^{\tilde{q}_0})^{-1}(N) = X(N)$.

We will use the previous notation $\omega_{N_1}^{N_2}$ also for the projection

$$\tilde{X} \diagup (\mathrm{Is}^{\tilde{q}})^{-1}(\mathrm{Is}_{\alpha_{N_1}}(N_1)) \to \tilde{X} \diagup (\mathrm{Is}^{\tilde{q}})^{-1}(\mathrm{Is}_{\alpha_{N_2}}(N_2)),$$

with $N_1 \le N_2$ being subgroups of $\pi_1(X, q_0)$ (N_1 may be the identity and N_2 may be $\pi_1(X, q_0)$.)

Put $\alpha \overset{def}{=} \mathsf{P}(\tilde{\alpha})$. For an element $e_0 \in \pi_1(X, q_0)$ we put $e \overset{def}{=} \mathrm{Is}_\alpha(e_0) \in \pi_1(X, q)$ and denote by e_N the element of $\pi_1(X(N), q_N)$, that is represented by $\omega^N(\tilde{\gamma})$ for a curve $\tilde{\gamma}$ in $\tilde{X}$ with initial point $\tilde{q}$ and projection $\mathsf{P}(\tilde{\gamma})$ representing e. Again $(\omega_{N_1}^{N_2})_*(e_{N_1}) = e_{N_2}$ for subgroups $N_1 \le N_2$ of $\pi_1(X, q)$ and $e \in N_1$, in particular $(\omega_N)_*(e_N) = e$ for a subgroup N of $\pi_1(X, q)$ and $e \in N$.

3.2. The Estimate for a Chosen Pair of Monodromies Since the mapping $f : X \to \mathbb{C} \setminus \{-1, 1\}$ is irreducible, there exist two elements $e'_0,\ e''_0 \in \mathcal{E} \subset \pi_1(X, q_0)$ such that the monodromies $f_*(e'_0)$ and $f_*(e''_0)$ are not powers of a single conjugate of a power of one of the elements a_1, a_2 or a_1a_2. The fundamental group of the Riemann surface $X(\langle e'_0, e''_0 \rangle)$ is a free group in the two generators $(e'_0)_{\langle e'_0, e''_0 \rangle}$ and $(e''_0)_{\langle e'_0, e''_0 \rangle}$, hence, $X(\langle e'_0, e''_0 \rangle)$ is either a torus with a hole or is equal to $\mathbb{P}^1$ with three holes. Moreover, the system $\mathcal{E}_{\langle e'_0, e''_0 \rangle} = \{(e'_0)_{\langle e'_0, e''_0 \rangle},\ (e''_0)_{\langle e'_0, e''_0 \rangle}\}$ of generators of the fundamental group $\pi_1(X(\langle e'_0, e''_0 \rangle), (q_0)_{\langle e'_0, e''_0 \rangle})$ is associated to a standard bouquet of circles for $X(\langle e'_0, e''_0 \rangle)$. This can be seen as follows. The set of generators $\mathcal{E}$ of $\pi_1(X, q_0)$ is associated to a standard bouquet of circles for X. For each $e_0 \in \mathcal{E}$ we denote the circle of the bouquet that represents e_0 by γ_{e_0}. For each $e_0 \in \mathcal{E}$ we lift the circle γ_{e_0} with base point q_0 to an arc $\widetilde{\gamma_{e_0}}$ in $\tilde{X}$ with initial point $\tilde{q}_0$. Let D be a small disc in X around q_0, and $\widetilde{D_0}$, $\widetilde{D_{e_0}}$, $e_0 \in \mathcal{E}$, be the preimages of D under the projection $\mathsf{P} : \tilde{X} \to X$, that contain $\tilde{q}_0$, or the terminal point of $\widetilde{\gamma_{e_0}}$, respectively. We assume that D is small enough so that the mentioned preimages of D are pairwise disjoint. Put $D_{\langle e'_0, e''_0 \rangle} = \omega^{\langle e'_0, e''_0 \rangle}(\widetilde{D_0})$. For $e_0 \neq e'_0, e''_0$ the image $\omega^{\langle e'_0, e''_0 \rangle}(\widetilde{D_0} \cup \widetilde{\gamma_{e_0}} \cup \widetilde{D_{e_0}})$ is the union of an arc $\omega^{\langle e'_0, e''_0 \rangle}(\widetilde{\gamma_{e_0}})$ in $X(\langle e'_0, e''_0 \rangle)$ with two disjoint discs, each containing an endpoint of the arc and one of them equal to $D_{\langle e'_0, e''_0 \rangle}$. For $e_0 = e'_0, e''_0$ the image $\omega^{\langle e', e'' \rangle}(\widetilde{D_0} \cup \widetilde{\gamma_{e_0}} \cup \widetilde{D_{e_0}})$ is the union of $D_{\langle e'_0, e''_0 \rangle}$ with the loop $(\gamma_{e_0})_{\langle e'_0, e''_0 \rangle} \overset{def}{=} \omega^{\langle e'_0, e''_0 \rangle}(\widetilde{\gamma_{e_0}})$. For $e_0 = e'_0, e''_0$ the loop $(\gamma_{e_0})_{\langle e'_0, e''_0 \rangle}$ in

$X(\langle e_0', e_0''\rangle)$ has base point $(q_0)_{\langle e_0', e_0''\rangle} = \omega^{\langle e_0', e_0''\rangle}(\tilde{q}_0)$ and represents the generator $(e_0)_{\langle e_0', e_0''\rangle}$ of the fundamental group of $\pi_1(X(\langle e_0', e_0''\rangle), (q_0)_{\langle e_0', e_0''\rangle})$.

Since the bouquet of circles $\cup_{e_0 \in \mathcal{E}} \gamma_{e_0}$ is a standard bouquet of circles for X, the union $(\gamma_{e_0'})_{\langle e_0', e_0''\rangle} \cup (\gamma_{e_0''})_{\langle e_0', e_0''\rangle}$ is a standard bouquet of circles in $X(\langle e_0', e_0''\rangle)$. This can be seen by looking at the intersections of the loops with a circle that is contained in $D_{\langle e_0', e_0''\rangle}$ and surrounds $(q_0)_{\langle e_0', e_0''\rangle}$. By the commutative diagram of coverings the intersection behaviour is the same as for the images of these objects under $\omega_{\langle e_0', e_0''\rangle}$. Hence, since $(\gamma_{e_0'})_{\langle e_0', e_0''\rangle}$ and $(\gamma_{e_0''})_{\langle e_0', e_0''\rangle}$ represent the generators $(e_0')_{\langle e_0', e_0''\rangle}$ and $(e_0'')_{\langle e_0', e_0''\rangle}$ of $\mathcal{E}_{\langle e_0', e_0''\rangle}$, the union $(\gamma_{e_0'})_{\langle e_0', e_0''\rangle} \cup (\gamma_{e_0''})_{\langle e_0', e_0''\rangle}$ is a standard bouquet of circles for $X(\langle e_0', e_0''\rangle)$.

The set $X(\langle e_0', e_0''\rangle)$ is either a torus with a hole or is equal to $\mathbb{P}^1$ with three holes. Apply Proposition 11.4 to the Riemann surface $X(\langle e_0', e_0''\rangle)$ with base point $(q_0)_{\langle e_0', e_0''\rangle}$, the holomorphic mapping $f_{\langle e_0', e_0''\rangle} = f \circ \omega_{\langle e_0', e_0''\rangle}$ into $\mathbb{C} \setminus \{-1, 1\}$, and the set of generators $\mathcal{E}_{\langle e_0', e_0''\rangle}$ of the fundamental group $\pi_1(X(\langle e_0', e_0''\rangle), (q_0)_{\langle e_0', e_0''\rangle})$. We obtain a relatively closed curve $L_{\langle e_0', e_0''\rangle}$ on which the function $f_{\langle e_0', e_0''\rangle}$ is real, and a set $(\mathcal{E}_{\langle e_0', e_0''\rangle})_2' = \{(\mathrm{e}_0')_{\langle e_0', e_0''\rangle}, (\mathrm{e}_0'')_{\langle e_0', e_0''\rangle}\}$ which contains one of the elements of $\mathcal{E}_{\langle e_0', e_0''\rangle}$. The second element of $(\mathcal{E}_{\langle e_0', e_0''\rangle})_2'$ is either equal to the second element of $\mathcal{E}_{\langle e_0', e_0''\rangle}$ or to its inverse, or to the product of the two elements (in any order) of $\mathcal{E}_{\langle e_0', e_0''\rangle}$. (We will usually refer to the product $(e_0')_{\langle e_0', e_0''\rangle} (e_0'')_{\langle e_0', e_0''\rangle} = (e_0' \, e_0'')_{\langle e_0', e_0''\rangle}$, but we may change the product $(e_0' \, e_0'')_{\langle e_0', e_0''\rangle}$ to the product $(e_0'' \, e_0')_{\langle e_0', e_0''\rangle}$, without changing the estimate of the $\mathcal{L}_-$ of the monodromies of the elements of $\mathcal{E}_2'$.) The free homotopy classes $\widehat{(\mathrm{e}_0')_{\langle e_0', e_0''\rangle}}$, $\widehat{(\mathrm{e}_0'')_{\langle e_0', e_0''\rangle}}$, and $\widehat{(\mathrm{e}_0')_{\langle e_0', e_0''\rangle} (\mathrm{e}_0'')_{\langle e_0', e_0''\rangle}} = \widehat{(\mathrm{e}_0' \, \mathrm{e}_0'')_{\langle e_0', e_0''\rangle}}$ intersect $L_{\langle e_0', e_0''\rangle}$.

Choose a point $q_{\langle e_0', e_0''\rangle} \in L_{\langle e_0', e_0''\rangle}$ and a point $\tilde{q} \in \tilde{X}$ with $\omega^{\langle e_0', e_0''\rangle}(\tilde{q}) = q_{\langle e_0', e_0''\rangle}$. Let $\tilde{\alpha}$ be a curve in $\tilde{X}$ with initial point $\tilde{q}_0$ and terminal point $\tilde{q}$, and $\alpha_{\langle e_0', e_0''\rangle} = \omega^{\langle e_0', e_0''\rangle}(\tilde{\alpha})$. Put $\mathrm{e}'_{\langle e_0', e_0''\rangle} = \mathrm{Is}_{\alpha_{\langle e_0', e_0''\rangle}}((\mathrm{e}_0')_{\langle e_0', e_0''\rangle})$ and $\mathrm{e}''_{\langle e_0', e_0''\rangle} = \mathrm{Is}_{\alpha_{\langle e_0', e_0''\rangle}}((\mathrm{e}_0'')_{\langle e_0', e_0''\rangle})$. For one out of three Möbius transformations M_l the mapping $F_{\langle e_0', e_0''\rangle} = M_l \circ f_{\langle e_0', e_0''\rangle} = M_l \circ f \circ \omega_{\langle e_0', e_0''\rangle}$ takes $L_{\langle e_0', e_0''\rangle}$ to $(-1, 1)$, and hence $F_{\langle e_0', e_0''\rangle}$ takes a value $q' = F_{\langle e_0', e_0''\rangle}(q_{\langle e_0', e_0''\rangle}) \in (-1, 1)$ at $q_{\langle e_0', e_0''\rangle}$. By Lemma 11.4 each of the $(F_{\langle e_0', e_0''\rangle})_*(\mathrm{e}'_{\langle e_0', e_0''\rangle})$ and $(F_{\langle e_0', e_0''\rangle})_*(\mathrm{e}''_{\langle e_0', e_0''\rangle})$ is the product of at most two elements of $\pi_1(\mathbb{C} \setminus \{-1, 1\}, q')$ of $\mathcal{L}_-$ not exceeding $2\pi \lambda_3(X(\langle e_0', e_0''\rangle))$, hence,

$$\mathcal{L}_-((F_{\langle e_0', e_0''\rangle})_*(\mathrm{e}'_{\langle e_0', e_0''\rangle})) \le 4\pi \lambda_3(X(\langle e_0', e_0''\rangle)),$$

$$\mathcal{L}_-((F_{\langle e_0', e_0''\rangle})_*(\mathrm{e}''_{\langle e_0', e_0''\rangle})) \le 4\pi \lambda_3(X(\langle e_0', e_0''\rangle)) .$$

It follows that each of the $(F_{\langle e_0', e_0'' \rangle})_*(e'_{\langle e_0', e_0'' \rangle})$ and $(F_{\langle e_0', e_0'' \rangle})_*(e''_{\langle e_0', e_0'' \rangle})$ is the product of at most four elements of $\pi_1(\mathbb{C} \setminus \{-1, 1\}, q')$ of $\mathcal{L}_-$ not exceeding $2\pi\lambda_3(X(\langle e_0', e_0'' \rangle))$, hence,

$$\mathcal{L}_-((F_{\langle e_0', e_0'' \rangle})_*(e'_{\langle e_0', e_0'' \rangle})) \leq 8\pi\lambda_3(X(\langle e_0', e_0'' \rangle)),$$

$$\mathcal{L}_-((F_{\langle e_0', e_0'' \rangle})_*(e''_{\langle e_0', e_0'' \rangle})) \leq 8\pi\lambda_3(X(\langle e_0', e_0'' \rangle)).$$

It remains to take into account that for a subgroup N of $\pi_1(X, q_0)$ the equation $(F_N)_*(e_N) = F_*(e)$ holds for each $e \in \mathrm{Is}_\alpha(N)$, and $\lambda_j(X(N)) \leq \lambda_j(X)$ for each natural number j.

3.3. Other Generators. Intersection of Free Homotopy Classes with a Component of the Zero Set Take any element $e \in \mathrm{Is}_\alpha(\mathcal{E})$ that is not in $\langle e', e'' \rangle$. Then the monodromy $F_*(e)$ is either equal to the identity, or one of the pairs $(F_*(e), F_*(\mathrm{e}'))$ or $(F_*(e), F_*(\mathrm{e}''))$ consists of two elements of $\pi_1(\mathbb{C} \setminus \{-1, 1\}, q')$ that are not powers of the same standard generator a_j, $j = 1$ or 2. We assume that the second option holds. Interchanging if necessary e' and e'', we may suppose this option holds for the pair $(F_*(e), F_*(\mathrm{e}'))$. Moreover, changing if necessary e' to its inverse $(\mathrm{e}')^{-1}$, we may assume that e' is either an element of $\mathrm{Is}_\alpha(\mathcal{E})$ or it is a product of two elements of $\mathrm{Is}_\alpha(\mathcal{E})$. The quotient $X(\langle e, \mathrm{e}' \rangle) = \tilde{X} / (\mathrm{Is}^{\tilde{q}})^{-1}(\langle e, \mathrm{e}' \rangle)$ is a Riemann surface whose fundamental group is a free group in two generators. Hence $X(\langle e, \mathrm{e}' \rangle)$ is either a torus with a hole or is equal to $\mathbb{P}^1$ with three holes.

We consider a diagram of coverings as follows. Let first $X(\langle \mathrm{e}' \rangle) = \tilde{X} / (\mathrm{Is}^{\tilde{q}})^{-1}(\langle \mathrm{e}' \rangle)$ be the annulus with base point $q_{\langle \mathrm{e}' \rangle} = \omega^{\langle \mathrm{e}' \rangle}(\tilde{q})$, that admits a mapping $\omega_{\langle \mathrm{e}' \rangle} : X(\langle \mathrm{e}' \rangle) \to X$ that represents e'. By Lemma 11.3 with X replaced by $X(\langle \mathrm{e}', \mathrm{e}'' \rangle)$ the connected component $L_{\langle \mathrm{e}' \rangle}$ of $(\omega^{\langle \mathrm{e}', \mathrm{e}'' \rangle}_{\langle \mathrm{e}' \rangle})^{-1}(L_{\langle \mathrm{e}', \mathrm{e}'' \rangle})$ that contains $q_{\langle \mathrm{e}' \rangle} = \omega^{\langle \mathrm{e}' \rangle}(\tilde{q})$ is a relatively closed curve in $X(\langle \mathrm{e}' \rangle)$ with limit points on both boundary components. The free homotopy class of the generator $\mathrm{e}'_{\langle \mathrm{e}' \rangle}$ of $\pi_1(X(\langle \mathrm{e}' \rangle), q_{\langle \mathrm{e}' \rangle})$ intersects $L_{\langle \mathrm{e}' \rangle}$. The mapping $F_{\langle \mathrm{e}' \rangle} = M_l \circ f \circ \omega_{\langle \mathrm{e}' \rangle}$ maps $L_{\langle \mathrm{e}' \rangle}$ into $(-1, 1)$, and $F_{\langle \mathrm{e}' \rangle}(q_{\langle \mathrm{e}' \rangle}) = F \circ \mathsf{P}(\tilde{q}) = q'$.

Next we consider the quotient $X(\langle \mathrm{e}', e \rangle) = \tilde{X} / (\mathrm{Is}^{\tilde{q}})^{-1}(\langle \mathrm{e}', e \rangle)$ whose fundamental group is again a free group in two generators. The image $L_{\langle \mathrm{e}', e \rangle} \overset{def}{=} \omega^{\langle \mathrm{e}', e \rangle}_{\langle \mathrm{e}' \rangle}(L_{\langle \mathrm{e}' \rangle})$ is a connected component of the preimage of $(-1, 1)$ under $F_{\langle \mathrm{e}', e \rangle} = F \circ \omega_{\langle \mathrm{e}', e \rangle}$. Indeed, $L_{\langle \mathrm{e}', e \rangle}$ is connected as image of a connected set under a continuous mapping, and

$$F_{\langle \mathrm{e}', e \rangle}(L_{\langle \mathrm{e}', e \rangle}) = F_{\langle \mathrm{e}', e \rangle}(\omega^{\langle \mathrm{e}', e \rangle}_{\langle \mathrm{e}' \rangle}(L_{\langle \mathrm{e}' \rangle})) = F \circ \omega_{\langle \mathrm{e}', e \rangle} \circ \omega^{\langle \mathrm{e}', e \rangle}_{\langle \mathrm{e}' \rangle}(L_{\langle \mathrm{e}' \rangle})$$
$$= F_{\langle \mathrm{e}' \rangle}(L_{\langle \mathrm{e}' \rangle}) \subset (-1, 1).$$

Moreover, since the mapping $\omega^{\langle \mathrm{e}', e \rangle}_{\langle \mathrm{e}' \rangle} : X(\langle \mathrm{e}' \rangle) \to X(\langle \mathrm{e}', e \rangle)$ is a covering, its restriction $(\mathrm{Im} F \circ \omega_{\langle \mathrm{e}' \rangle})^{-1}(0) \to (\mathrm{Im} F \circ \omega_{\langle \mathrm{e}', e \rangle})^{-1}(0)$ is a covering. Hence the image under $\omega^{\langle \mathrm{e}',e\rangle}_{\langle \mathrm{e}'\rangle}$ of a connected component of $(\mathrm{Im} F \circ \omega_{\langle \mathrm{e}'\rangle})^{-1}(0)$ is open and closed in $(\mathrm{Im} F \circ \omega_{\langle \mathrm{e}',e\rangle})^{-1}(0)$. Hence, $L_{\langle \mathrm{e}',e\rangle} = \omega^{\langle \mathrm{e}',e\rangle}_{\langle \mathrm{e}'\rangle}(L_{\langle \mathrm{e}'\rangle})$ is a connected component of the preimage of $(-1,1)$ under $F_{\langle \mathrm{e}',e\rangle}$. Put $q_{\langle \mathrm{e}',e\rangle} = \omega^{\langle \mathrm{e}',e\rangle}_{\langle \mathrm{e}'\rangle}(q_{\langle \mathrm{e}'\rangle}) = \omega^{\langle \mathrm{e}',e\rangle}_{\langle \mathrm{e}'\rangle} \circ \omega^{\langle \mathrm{e}'\rangle}(\tilde{q}) = \omega^{\langle \mathrm{e}',e\rangle}(\tilde{q})$. Note that $F_{\langle \mathrm{e}',e\rangle}(q_{\langle \mathrm{e}',e\rangle}) = F \circ \omega_{\langle \mathrm{e}',e\rangle}(q_{\langle \mathrm{e}',e\rangle}) = F(q) = q'$.

We prove now that the free homotopy class $\widehat{\mathrm{e}'_{\langle \mathrm{e}',e\rangle}}$ in $X(\langle \mathrm{e}',e\rangle)$ that is related to e' intersects $L_{\langle \mathrm{e}',e\rangle}$. Consider any loop $\gamma'_{\langle \mathrm{e}',e\rangle}$ in $X(\langle \mathrm{e}',e\rangle)$ with some base point $q'_{\langle \mathrm{e}',e\rangle}$, that represents $\widehat{\mathrm{e}'_{\langle \mathrm{e}',e\rangle}}$. There exists a loop $\gamma'_{\langle \mathrm{e}'\rangle}$ in $X(\langle \mathrm{e}'\rangle)$ which represents $\widehat{\mathrm{e}'_{\langle \mathrm{e}'\rangle}}$ such that $\omega^{\langle \mathrm{e}',e\rangle}_{\langle \mathrm{e}'\rangle}(\gamma'_{\langle \mathrm{e}'\rangle}) = \gamma'_{\langle \mathrm{e}',e\rangle}$. Such a curve $\gamma'_{\langle \mathrm{e}'\rangle}$ can be obtained as follows. There is a loop $\gamma''_{\langle \mathrm{e}',e\rangle}$ in $X(\langle \mathrm{e}',e\rangle)$ with base point $q_{\langle \mathrm{e}',e\rangle}$ that represents $(\mathrm{e}')_{\langle \mathrm{e}',e\rangle}$, and a curve $\alpha'_{\langle \mathrm{e}',e\rangle}$ in $X(\langle \mathrm{e}',e\rangle)$ with initial point $q_{\langle \mathrm{e}',e\rangle}$ and terminal point $q'_{\langle \mathrm{e}',e\rangle}$, such that $\gamma'_{\langle \mathrm{e}',e\rangle}$ is homotopic with fixed endpoints to $(\alpha'_{\langle \mathrm{e}',e\rangle})^{-1}\gamma''_{\langle \mathrm{e}',e\rangle}\alpha'_{\langle \mathrm{e}',e\rangle}$. Consider the lift $\tilde{\gamma}''$ of $\gamma''_{\langle \mathrm{e}',e\rangle}$ to $\tilde{X}$ with initial point $\tilde{q}$, and the lift $\tilde{\alpha}'$ of $\alpha'_{\langle \mathrm{e}',e\rangle}$ with initial point $\tilde{q}$. The terminal point of $\tilde{\gamma}''_{\langle \mathrm{e}',e\rangle}$ equals $\sigma(\tilde{q})$ for the covering transformation $\sigma = (\mathrm{Is}^{\tilde{q}})^{-1}(\mathrm{e}') = (\mathrm{Is}^{\tilde{q}_0})^{-1}(\mathrm{e}'_0)$. (See Eq. (1.2).) The terminal point of the curve $(\tilde{\alpha}')^{-1}\tilde{\gamma}''_{\langle \mathrm{e}',e\rangle}\sigma(\tilde{\alpha}')$ is obtained from its initial point by applying the covering transformation σ. Hence, $\omega^{\langle \mathrm{e}'\rangle}((\tilde{\alpha}')^{-1}\tilde{\gamma}''_{\langle \mathrm{e}',e\rangle}\sigma(\tilde{\alpha}'))$ is a closed curve in $X(\langle \mathrm{e}'\rangle)$ that represents $\widehat{\mathrm{e}'_{\langle \mathrm{e}'\rangle}}$ and projects to $(\alpha'_{\langle \mathrm{e}',e\rangle})^{-1}\gamma''_{\langle \mathrm{e}',e\rangle}\alpha'_{\langle \mathrm{e}',e\rangle}$ under $\omega^{\langle \mathrm{e}',e\rangle}_{\langle \mathrm{e}'\rangle}$. Since $\gamma'_{\langle \mathrm{e}',e\rangle}$ is homotopic to $(\alpha'_{\langle \mathrm{e}',e\rangle})^{-1}\gamma''_{\langle \mathrm{e}',e\rangle}\alpha'_{\langle \mathrm{e}',e\rangle}$ with fixed base point, it also has a lift to $X(\langle \mathrm{e}'\rangle)$ which represents $\widehat{\mathrm{e}'_{\langle \mathrm{e}'\rangle}}$.

Since $\widehat{\mathrm{e}'_{\langle \mathrm{e}'\rangle}}$ intersects $L_{\langle \mathrm{e}'\rangle}$, the loop $\gamma'_{\langle \mathrm{e}'\rangle}$ has an intersection point $p'_{\langle \mathrm{e}'\rangle}$ with $L_{\langle \mathrm{e}'\rangle}$. The point $p'_{\langle \mathrm{e}',e\rangle} = \omega^{\langle \mathrm{e}',e\rangle}_{\langle \mathrm{e}'\rangle}(p'_{\langle \mathrm{e}'\rangle})$ is contained in $\gamma'_{\langle \mathrm{e}',e\rangle}$ and in $L_{\langle \mathrm{e}',e\rangle}$. We proved that the free homotopy class $\widehat{\mathrm{e}'_{\langle \mathrm{e}',e\rangle}}$ in $X(\langle \mathrm{e}',e\rangle)$ intersects $L_{\langle \mathrm{e}',e\rangle}$.

3.4. A System of Generators Associated to a Standard Bouquet of Circles We claim that the system of generators $\mathrm{e}'_{\langle \mathrm{e}',e\rangle}$, $e_{\langle \mathrm{e}',e\rangle}$ of $\pi_1(X(\langle \mathrm{e}',e\rangle), q_{\langle \mathrm{e}',e\rangle})$ is associated to a standard bouquet of circles for $X(\langle \mathrm{e}',e\rangle)$. If $\mathrm{e}' \in \mathcal{E}$ the claim can be obtained as in paragraph 3.2.

Suppose $\mathrm{e}' = e'e''$ for $e', e'' \in \mathcal{E}$. Consider the system $\mathcal{E}'$ of generators of $\pi_1(X,q)$ that is obtained from $\mathcal{E}$ by replacing e' by $e'e''$. If e' and e'' correspond to a handle of X, then $\mathcal{E}'$ is associated to a part of a standard bouquet of circles for X, see Fig. 11.2a for the case when e' is represented by an α-curve and e'' is represented by a β-curve. The situation when e' is represented by a β-curve and e'' is represented by an α-curve is similar. The claim is obtained as in paragraph 3.2. In this case $X(\langle \mathrm{e}',e\rangle)$ is a torus with a hole.

Suppose one of the pairs (e,e') or (e,e'') corresponds to a handle of X. We assume that e corresponds to an α-curve and e' corresponds to a β-curve of a handle of X (see Fig. 11.2b). The remaining cases are treated similarly, maybe, after replacing $e'e''$ by $e''e'$ (see paragraph 3.2). With our assumption $\mathcal{E}'$ is not associated

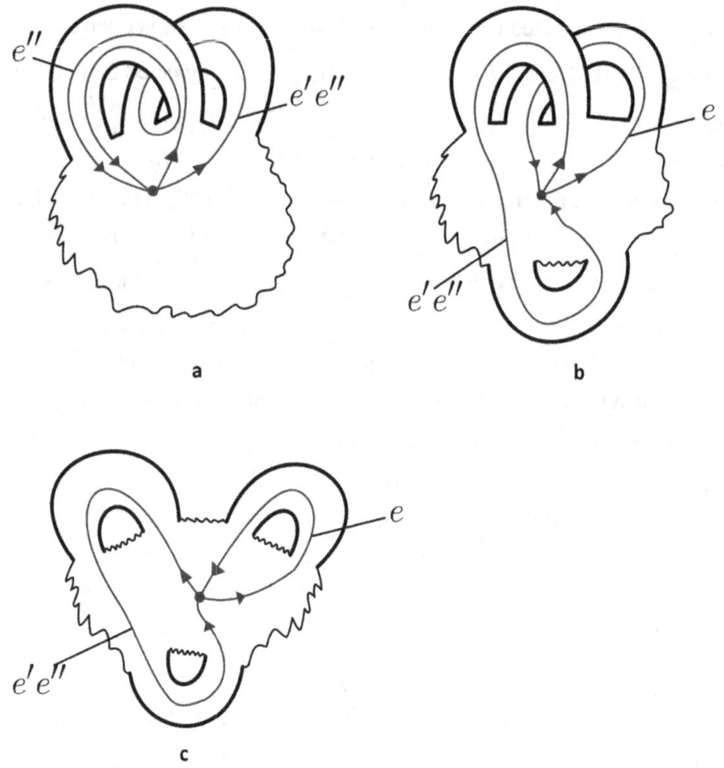

Fig. 11.2 Standard bouquets of circles

to a standard bouquet of circles for X. Nevertheless, the pair $(e_{\langle \mathrm{e}', e\rangle}, \mathrm{e}'_{\langle \mathrm{e}', e\rangle})$ with $\mathrm{e}' = e'e''$ is associated to a standard bouquet of circles for $X(\langle \mathrm{e}', e\rangle)$. This can be seen as before. Consider a (non-standard) bouquet of circles in X corresponding to $\mathcal{E}'$ and take its union with a disc D around q. Lift this set to $\tilde{X}$. We obtain the union of a collection of arcs in $\tilde{X}$ with initial point $\tilde{q}$, with a collection of discs in $\tilde{X}$ around $\tilde{q}$ and around the terminal points of the arcs. Take the union of the arcs and the discs. The image in $X(\langle \mathrm{e}', e\rangle)$ of this union under the projection $\omega^{\langle \mathrm{e}', e\rangle}$ is the union of the two loops $(\gamma_e)_{\langle \mathrm{e}', e\rangle} \cup (\gamma_{\mathrm{e}'})_{\langle \mathrm{e}', e\rangle}$, the disc $D_{\langle \mathrm{e}', e\rangle}$ and a set, that is contractible to $D_{\langle \mathrm{e}', e\rangle}$. Looking at the intersection of the two loops with a small circle contained in $D_{\langle \mathrm{e}', e\rangle}$ and surrounding $q_{\langle \mathrm{e}', e\rangle}$, we see as before that $(\gamma_e)_{\langle \mathrm{e}', e\rangle} \cup (\gamma_{\mathrm{e}'})_{\langle \mathrm{e}', e\rangle}$ is a standard bouquet of circles for $X(\langle \mathrm{e}', e\rangle)$. In this case $X(\langle \mathrm{e}', e\rangle)$ is a torus with a hole.

In the remaining case no pair of generators among e, e', and e'' corresponds to a handle. In this case again $\mathcal{E}'$ does not correspond to a standard bouquet of circles for X. But $\{e_{\langle \mathrm{e}', e\rangle}, (e'e'')_{\langle \mathrm{e}', e\rangle}\}$ (maybe, after changing $e'e''$ to $e''e'$) corresponds to a standard bouquet of circles for $X(\langle \mathrm{e}', e\rangle)$. (See Fig. 11.2c for the case when walking along a small circle around q counterclockwise, we meet the incoming and outgoing rays of representatives of the three elements of $\mathcal{E}$ in the order e, e', e''. If the order is different the situation is similar, maybe, after replacing $e'e''$ by $e''e'$.) In this case $X(\langle \mathrm{e}', e\rangle)$ is a planar domain.

3.5. End of the Proof Consider first the case when $X(\langle \mathrm{e}', e\rangle)$ is a torus with a hole. Since $\widehat{\mathrm{e}'_{\langle \mathrm{e}', e\rangle}}$ intersects $L_{\langle \mathrm{e}', e\rangle}$, we see as in the proof when X itself is a torus with a hole, that the curve $L_{\langle \mathrm{e}', e\rangle}$ cannot be contractible or contractible to the hole, and the intersection number must be different from zero. Then the intersection number with $L_{\langle \mathrm{e}', e\rangle}$ of the free homotopy class of one of the choices $e^{\pm 1}_{\langle \mathrm{e}', e\rangle}$ or $(\mathrm{e}'e)_{\langle \mathrm{e}', e\rangle}$, denoted by $\mathrm{e}'''_{\langle \mathrm{e}', e\rangle}$, is not zero and has the same sign. By Lemma 11.4 each of the $(F_{\langle \mathrm{e}', e\rangle})_*(\mathrm{e}'_{\langle \mathrm{e}', e\rangle})$ and $(F_{\langle \mathrm{e}', e\rangle})_*(\mathrm{e}'''_{\langle \mathrm{e}', e\rangle})$ is the product of at most two elements of $\pi_1(\mathbb{C} \setminus \{-1, 1\}, q')$ with $\mathcal{L}_-$ not exceeding

$$2\pi \lambda_{\mathrm{e}'_{\langle \mathrm{e}', e\rangle}, \mathrm{e}'''_{\langle \mathrm{e}', e\rangle}} \le 2\pi \lambda_5(X), \tag{11.14}$$

since e' is the product of at most two elements of $\mathcal{E} \cup \mathcal{E}^{-1}$ and e''' is the product of at most three elements of $\mathcal{E} \cup \mathcal{E}^{-1}$. The element e is the product of at most two different elements among the e' and e''' or their inverses. Hence, the monodromy $F_*(e) = (F_{\langle \mathrm{e}', e\rangle})_*(e_{\langle \mathrm{e}', e\rangle})$ is the product of at most four elements with $\mathcal{L}_-$ not exceeding (11.14). Hence,

$$F_*(e) \le 8\pi \lambda_5(X)\,. \tag{11.15}$$

Consider now the case when $X(\langle \mathrm{e}', e\rangle)$ equals $\mathbb{P}^1$ with three holes. Since $\mathrm{e}'_{\langle \mathrm{e}', e\rangle}$ and $e_{\langle \mathrm{e}', e\rangle}$ correspond to a standard bouquet of circles for $X(\langle \mathrm{e}', e\rangle)$, the curves representing $\mathrm{e}'_{\langle \mathrm{e}', e\rangle}$ surround counterclockwise one of the holes, denoted by $\mathcal{C}'$, and the curves representing $e_{\langle \mathrm{e}', e\rangle}$ surround counterclockwise another hole, denoted by $\mathcal{C}''$. After applying a Möbius transformation we may assume that the remaining hole, denoted by $\mathcal{C}_\infty$, contains the point ∞. There are several possibilities for the behaviour of the curve $L_{\langle \mathrm{e}', e\rangle}$. Since $\widehat{\mathrm{e}'_{\langle \mathrm{e}', e\rangle}}$ intersects $L_{\langle \mathrm{e}', e\rangle}$, the curve $L_{\langle \mathrm{e}', e\rangle}$ must have limit points on $\mathcal{C}'$. The first possibility is that $L_{\langle \mathrm{e}', e\rangle}$ has limit points on $\partial\mathcal{C}'$ and $\partial\mathcal{C}''$, the second possibility is, $L_{\langle \mathrm{e}', e\rangle}$ has limit points on $\mathcal{C}'$ and $\mathcal{C}_\infty$, the third possibility is, $L_{\langle \mathrm{e}', e\rangle}$ has all limit points on $\mathcal{C}'$, and $\mathcal{C}''$ is contained in the bounded connected component of $\mathbb{C} \setminus (L_{\langle \mathrm{e}', e\rangle} \cup \mathcal{C}')$.

In the first case the free homotopy classes $\widehat{\mathrm{e}'_{\langle \mathrm{e}', e\rangle}}$ and $\widehat{e^{-1}_{\langle \mathrm{e}', e\rangle}}$ have positive intersection number with the suitably oriented curve $L_{\langle \mathrm{e}', e\rangle}$. In the second case the free homotopy classes $\widehat{\mathrm{e}'_{\langle \mathrm{e}', e\rangle}}$ and $\widehat{(\mathrm{e}'e)_{\langle \mathrm{e}', e\rangle}}$ have positive intersection number with the suitably oriented curve $L_{\langle \mathrm{e}', e\rangle}$. In the third case the free homotopy classes of $\mathrm{e}'_{\langle \mathrm{e}', e\rangle}$, $(\mathrm{e}'^2 e)_{\langle \mathrm{e}', e\rangle}$ and of their product intersect $L_{\langle \mathrm{e}', e\rangle}$. The first two cases were treated in paragraph 2 of this section. The statement concerning the third case is proved as follows.

Any curve that is contained in the complement of $\mathcal{C}' \cup L_{\langle \mathrm{e}', e\rangle}$ has either winding number zero around $\mathcal{C}'$ (as a curve in the complex plane $\mathbb{C}$), or its winding number around C' coincides with the winding number around C''. On the other hand the representatives of the free homotopy class of $\mathrm{e}'_{\langle \mathrm{e}',e\rangle}$ have winding number 1 around C' and winding number 0 around C''. The representatives of the free homotopy class of $(\mathrm{e}'^2 e)_{\langle \mathrm{e}',e\rangle}$ have winding number 2 around C', and winding number 1 around C''. Hence, $\mathrm{e}'_{\langle \mathrm{e}',e\rangle}$ and $(\mathrm{e}'^2 e)_{\langle \mathrm{e}',e\rangle}$ intersect $L_{\langle \mathrm{e}',e\rangle}$. By the same argument the free homotopy class of the product of $\mathrm{e}'_{\langle \mathrm{e}',e\rangle}$ and $(\mathrm{e}'^2 e)_{\langle \mathrm{e}',e\rangle}$ intersects $L_{\langle \mathrm{e}',e\rangle}$.

We let $\mathrm{e}'''_{\langle \mathrm{e}',e\rangle}$ be equal to $e^{-1}_{\langle \mathrm{e}',e\rangle}$ in the first case, equal to $(\mathrm{e}'e)_{\langle \mathrm{e}',e\rangle}$ in the second case, and equal to $(\mathrm{e}'^2 e)_{\langle \mathrm{e}',e\rangle}$ in the third case. Then e is the product of at most three elements among the $(\mathrm{e}')^{\pm 1}$ and $(\mathrm{e}''')^{\pm 1}$.

By Lemma 11.4 each of the $(F_{\langle \mathrm{e}',e\rangle})_*(\mathrm{e}'_{\langle \mathrm{e}',e\rangle})$ and $(F_{\langle \mathrm{e}',e\rangle})_*(\mathrm{e}'''_{\langle \mathrm{e}',e\rangle})$ is the product of at most two elements of $\pi_1(\mathbb{C} \setminus \{-1, 1\}, q')$ with $\mathcal{L}_-$ not exceeding

$$2\pi \lambda_{\mathrm{e}'_{\langle \mathrm{e}',e\rangle}, \mathrm{e}'''_{\langle \mathrm{e}',e\rangle}} \le 2\pi \lambda_7(X), \tag{11.16}$$

We used that e' is the product of at most two elements of $\mathcal{E} \cup \mathcal{E}^{-1}$, $e \in \mathcal{E} \cup \mathcal{E}^{-1}$ and e''' is the product of at most five elements of $\mathcal{E} \cup \mathcal{E}^{-1}$. Since e is the product of at most three elements among the $(\mathrm{e}')^{\pm 1}$ and $(\mathrm{e}''')^{\pm 1}$, the monodromy $F_*(e) = (F_{\langle \mathrm{e}',e\rangle})_*(e_{\langle \mathrm{e}',e\rangle})$ is the product of at most six elements with $\mathcal{L}_-$ not exceeding (11.16). Hence,

$$F_*(e) \le 12\pi \lambda_7(X). \tag{11.17}$$

The proposition is proved. □

11.4 (g, m)-Bundles over Riemann Surfaces

Theorem 11.2 is a consequence of the following theorem on (0, 3)-bundles with a section.

Theorem 11.3 *Over a connected Riemann surface of genus g with $m+1$ holes there are up to isotopy no more than $(15 \exp(6\pi \lambda_{10}(X)))^{6(2g+m)}$ irreducible holomorphic $(0, 3)$-bundles with a holomorphic section.*

Theorem 11.1 (with a weaker estimate) is a consequence of Theorem 11.3. Indeed, consider holomorphic (smooth, respectively) bundles whose fiber over each point $x \in X$ equals $\mathbb{P}^1$ with set of distinguished points $\{-1, 1, f(x), \infty\}$ for a function f which depends holomorphically (smoothly, respectively) on the points $x \in X$ and does not take the values -1 and 1. Then we are in the situation of Theorem 11.3. The mapping f is reducible, iff the bundle is reducible (see also Lemma 7.3).

The relation between Theorems 11.2 and 11.3 will be obtained by representing families of (1, 1)-bundles as double branched coverings of special (0, 4)-bundles (see Definition 8.6) and using Proposition 8.2.

Preparation of the Proof of Theorem 11.3 The proof of Theorem 11.3 will go now along the same lines as the proof of Theorem 11.1 with some modifications. Recall that the set

$$\mathcal{H} =\{\{z_1, z_2, z_3\} \in C_3(\mathbb{C})\diagup\mathcal{S}_3 : \text{the three points } z_1, z_2, z_3 \\ \text{are contained in a real line in the complex plane}\}$$

is a smooth real hypersurface of $C_3(\mathbb{C})\diagup\mathcal{S}_3$.

As before, for each complex affine self-mapping M of the complex plane we consider the diagonal action $M\big((z_1, z_2, z_3)\big) = \big(M(z_1), M(z_2), M(z_3)\big)$ on points $(z_1, z_2, z_3) \in C_3(\mathbb{C})$, and the diagonal action $M\big(\{z_1, z_2, z_3\}\big) = \{M(z_1), M(z_2), M(z_3)\}$ on points $\{z_1, z_2, z_3\} \in C_3(\mathbb{C})\diagup\mathcal{S}_3$.

The following two lemmas replace Lemma 11.2 in the case of (0, 3)-bundles with a section.

Lemma 11.5 *Let A be an annulus with an orientation of simple closed dividing curves and $F : A \to C_3(\mathbb{C})\diagup\mathcal{S}_3$ be a holomorphic mapping. Suppose L_A is a simple relatively closed curve in A with limit points on both boundary circles of A, and $F(L_A) \subset \mathcal{H}$. Moreover, we assume that for a point $q_A \in L_A$ the value $F(q_A)$ is in the totally real subspace $C_3(\mathbb{R})\diagup\mathcal{S}_3$. Let $e_A \in \pi_1(A, q_A)$ be the positively oriented generator of the fundamental group of A with base point q_A. If the braid $b \overset{def}{=} F_*(e_A) \in \mathcal{B}_3$ is different from $\sigma_j^k\,\Delta_3^{2\ell'}$ with j equal to 1 or 2, and $k \neq 0$ and ℓ' being integers, then*

$$\mathcal{L}_-(\vartheta(b)) \le 2\pi\lambda(A). \tag{11.18}$$

Notice that the braids $\sigma_j^k\,\Delta_3^\ell$ for odd ℓ are exceptional for Theorem 9.4, but not exceptional for Lemma 11.5. The reason is that the braid in Lemma 11.5 is related to a mapping of an annulus, not merely to a mapping of a rectangle. For $t \in [0, \infty)$ we put $\log_+ t \overset{def}{=} \begin{cases} \log t & t \in [1, \infty) \\ 0 & t \in [0, 1) \end{cases}$.

Lemma 11.6 *If the braid in Lemma 11.5 equals $b = \sigma_j^k\,\sigma_{j'}^{k'}\,\Delta_3^\ell$ with j and j' equal to 1 or to 2, $j' \neq j$, and k and k' being non-zero integers, and ℓ an even integer, then*

$$\log_+(3[\frac{|k|}{2}]) + \log_+(3[\frac{|k'|}{2}]) \le 2\pi\lambda(A). \tag{11.19}$$

As before, for a non-negative number x we denote by $[x]$ the smallest integer not exceeding x.

Proof of Lemma 11.5 By the same argument as in the proof of Lemma 11.2 we may assume that the annulus A has smooth boundary, the mapping F extends continuously to the closure $\overline{A}$, and the curve L_A is a smooth (connected) curve in $\overline{A}$ whose endpoints are on different boundary components of A. By Lemma 9.11 the inequality

$$\Lambda(b_{tr}) \leq \lambda(A) \tag{11.20}$$

holds.

For $b \neq \sigma_j^k \Delta_3^\ell$ with j equal to 1 or 2, and $k \neq 0$ and ℓ being integers, the statement of Lemma 11.5 follows from Theorem 9.4 in the same way as Lemma 11.2 follows from Theorem 9.1. For $b = \sigma_j^k \Delta_3^\ell$ with $k = 0$ the statement is trivial since then $\vartheta(\mathrm{Id}) = \mathrm{Id}$ and $\mathcal{L}_-(\mathrm{Id}) = 0$.

To obtain the statement in the remaining case $b = \sigma_j^k \Delta_3^{2\ell'+1}$ with j equal to 1 or 2, and k and ℓ' being integers, we use Lemma 11.6. Notice that $\sigma_1 \Delta_3 = \Delta_3 \sigma_2$ and $\sigma_2 \Delta_3 = \Delta_3 \sigma_1$. Hence, $b^2 = \sigma_j^k \sigma_{j'}^k \Delta_3^{4\ell'+2}$ with $\sigma_j \neq \sigma_{j'}$. Let $\omega_2 : A^2 \to A$ be the twofold unbranched covering of A by an annulus A^2. The equality $\lambda(A^2) = 2\lambda(A)$ holds. Let q_{A^2} be a point in $\omega_2^{-1}(q_A)$, and let $L_{q_{A^2}}$ be the lift of L_A to A^2 that contains q_{A^2}. Denote by γ_{A^2} the loop $\omega_2^{-1}(\gamma_A)$ with base point q_{A^2}. Then $F \circ \omega_2 \mid \gamma_{A^2}$ represents b^2 and $(b^2)_{tr}$. Lemma 11.6 applied to $\sigma_j^k \sigma_{j'}^k \Delta_3^{4\ell'+2}$ gives the estimate $2\log_+(3[\frac{|k|}{2}]) \leq 2\pi\lambda(A^2) = 4\pi\lambda(A)$. Since $\vartheta(b) = \sigma_j^{2[\frac{|k|}{2}]\mathrm{sgn}(k)}$, the inequality (11.18) follows. The lemma is proved. □

Proof of Lemma 11.6 Lemma 3.5 gives $\Lambda_{tr}(\sigma_j^k \sigma_{j'}^{k'} \Delta_3^\ell) = \Lambda_{tr}(\sigma_j^k \sigma_{j'}^{k'})$. Along the lines of proof of Statement 1 of Lemma 9.4 we see that for $k \cdot k' \neq 0$

$$\Lambda_{tr}(\sigma_j^k \sigma_{j'}^{k'}) \geq \Lambda\big({}_{tr}(\sigma_j^k)_{pb}\big) + \Lambda\big({}_{pb}(\sigma_{j'}^{k'})_{tr}\big).$$

A similar argument as used in the proof of Theorem 9.4 shows that for any positive integer number ℓ and each of the σ_j the inequality $\Lambda\big({}_{tr}(\sigma_j^{\pm\ell})_{pb}\big) \geq \Lambda\big({}_{tr}(\sigma_j^{\pm 2[\frac{\ell}{2}]})_{pb}\big)$ holds. Theorem 9.1 implies $\Lambda\big({}_{tr}(\sigma_j^{\pm 2[\frac{\ell}{2}]})_{pb}\big) \geq \frac{1}{2\pi}\mathcal{L}_-(\sigma_j^{\pm 2[\frac{\ell}{2}]}) = \frac{1}{2\pi}\log_+(3[\frac{\ell}{2}])$. Since by (11.20) the inequality $\Lambda_{tr}(\sigma_j^k \sigma_{j'}^{k'} \Delta_3^\ell) \leq \lambda(A)$ holds, the lemma is proved. □

We want to emphasize that periodic 3-braids are not of the form $\sigma_j^k \Delta_3^{2\ell}$ for non-zero powers σ_j^k of a σ_j and an integer number ℓ, hence, the lemma is true also for periodic braids. For each periodic braid b of the form $\sigma_1\sigma_2 = \sigma_1^{-1}\Delta_3$, $(\sigma_1\sigma_2)^2 = \sigma_1\Delta_3$, $\sigma_2\sigma_1 = \sigma_2^{-1}\Delta_3$, $(\sigma_2\sigma_1)^2 = \sigma_2\Delta_3$, and Δ_3 the $\mathcal{L}_-(\vartheta(b))$ vanishes. However, for instance for the conjugate $\sigma_1^{-2k}\Delta_3\sigma_1^{2k} = \sigma_1^{-2k}\sigma_2^{2k}\Delta_3$ of Δ_3

we have $\mathcal{L}_-(\vartheta(\sigma_1^{-2k}\Delta_3\sigma_1^{2k})) = 2\log(3|k|)$. Another example, for the conjugate $\sigma_2^{-2k}\,\sigma_1\sigma_2\,\sigma_2^{2k}$ of $\sigma_1\sigma_2$ we have

$$\sigma_2^{-2k}\,\sigma_1\sigma_2\,\sigma_2^{2k} = \sigma_2^{-2k-1}\,\Delta_3\,\sigma_2^{2k} = \sigma_2^{-2k-1}\,\sigma_1^{2k}\,\Delta_3\,.$$

and $\mathcal{L}_-(\vartheta(\sigma_2^{-2k}\,\sigma_1\sigma_2\,\sigma_2^{2k}))$ equals $2\log(3|k|)$.

Notice that the lemmas and Theorem 9.4 descend to statements on elements of $\mathcal{B}_3/\mathcal{Z}_3$ rather than on braids. For an element b of the quotient $\mathcal{B}_3/\mathcal{Z}_3$ we put $\vartheta(\mathsf{b}) = \vartheta(b)$ for any representative $b \in \mathcal{B}_3$ of b.

Lemma 11.7 below is an analog of Lemma 11.4. It will follow from Lemma 11.5 in the same way as Lemma 11.4 follows from Lemma 11.2.

Lemma 11.7 *Let X be a connected finite open Riemann surface, and let $F : X \to C_3(\mathbb{C})/\mathcal{S}_3$ be a non-contractible holomorphic map that is transverse to the hypersurface $\mathcal{H}$ in $C_3(\mathbb{C})/\mathcal{S}_3$. Suppose L_0 is a simple relatively closed curve in X such that $F(L_0)$ is contained in $\mathcal{H}$, and for a point $q \in L_0$ the point $F(q)$ is contained in the totally real space $C_3(\mathbb{R})/\mathcal{S}_3$. Let $e^{(1)}$ and $e^{(2)}$ be primitive elements of $\pi_1(X,q)$. Suppose that for $e = e^{(1)}$, $e = e^{(2)}$, and $e = e^{(1)}e^{(2)}$ the free homotopy class $\widehat{e}$ intersects L_0. Then either the two monodromies of F modulo the center $F_*(e^{(j)})/\mathcal{Z}_3$, $j = 1, 2$, are powers of the same element $\sigma_j/\mathcal{Z}_3$ of $\mathcal{B}_3/\mathcal{Z}_3$, or each of them is the product of at most two elements b_1 and b_2 of $\mathcal{B}_3/\mathcal{Z}_3$ with*

$$\mathcal{L}_-(\vartheta(b_j)) \le 2\pi\,\lambda_{e^{(1)},e^{(2)}},\; j = 1, 2, \tag{11.21}$$

where

$$\lambda_{e^{(1)},e^{(2)}} \overset{def}{=} \max\{\lambda(A(\widehat{e^{(1)}})),\; \lambda(A(\widehat{e^{(2)}})),\; \lambda(A(\widehat{e^{(1)}\,e^{(2)}}))\}.$$

Proof Suppose for an element $e \in \pi_1(X,q)$ the free homotopy class $\widehat{e}$ intersects L_0. We claim that all components of $\mathsf{P}^{-1}(L_0)$ have limit points on the boundary of $\tilde{X} \cong \mathbb{C}_+$.

Suppose the claim is not true. Then some connected component $\tilde{L}_0$ of $\mathsf{P}^{-1}(L_0)$ is compact. Lift the holomorphic mapping $F \circ \mathsf{P} : \tilde{X} \cong \mathbb{C}_+ \to C_3(\mathbb{C})/\mathcal{S}_3$ to a holomorphic mapping $\widetilde{F \circ \mathsf{P}} : \tilde{X} \cong \mathbb{C}_+ \to C_3(\mathbb{C})$. For each $z \in \mathbb{C}_+$ we let $\mathfrak{A}_z$ be the complex affine mapping that takes the first coordinate of $\widetilde{F \circ \mathsf{P}(z)}$ to -1, the third to 1, and the second to a point $g(z) \in \mathbb{C} \setminus \{-1, 1\}$. Then $\mathfrak{A}_z(\widetilde{F \circ \mathsf{P}(z)}) = \big(-1, g(z), 1\big)$, $z \in \mathbb{C}_+$. By the conditions of the lemma $g(z)$ is real on $\tilde{L}_0$. Then, since g is holomorphic on $\mathbb{C}_+$, $g \equiv c$ on $\mathbb{C}_+$ for a constant c (see the proof of Lemma 11.3). It follows that $\widetilde{F \circ \mathsf{P}(z)} = \mathfrak{A}_z^{-1}(c)$, $z \in \mathbb{C}_+$, which means that the image of $F \circ \mathsf{P}$ is contained in $\mathcal{H}$. This contradicts the fact that F is not constant and is transverse to $\mathcal{H}$. The claim is proved.

It follows along the lines of the proof of Lemma 11.3 that an analog of Lemma 11.3 holds. More precisely, for any point $\tilde{q}$ in $\mathsf{P}^{-1}(q)$, with $A \overset{def}{=} \tilde{X}/(\mathrm{Is}^{\tilde{q}})^{-1}(\langle e\rangle)$, $\omega_A \overset{def}{=} \omega_{\langle e\rangle,\tilde{q}}$, and $q_A \overset{def}{=} \omega^{\langle e\rangle,\tilde{q}}(\tilde{q})$ the connected component $L_A \subset A$ of $(\omega_A)^{-1}(L_0)$ that contains q_A has limit points on both boundary circles of A.

Put $F_A = F \circ \omega_A$. By the conditions of Lemma 11.7 $F_A(L_A) = F(L_0) \subset \mathcal{H}$ and $F_A(q_A) \in C_3(\mathbb{R})/\mathcal{S}_3$. Let e_A be the generator of $\pi_1(A, q_A)$ for which $\omega_A(e_A) = e$. The mapping $F_A : A \to C_3(\mathbb{C})/\mathcal{S}_3$, the point q_A and the curve L_A satisfy the conditions of Lemma 11.5. Notice that the equality $(F_A)_*(e_A) = F_*(e)$ holds. Hence, if $F_*(e)$ is not equal to $\sigma_j^k \Delta^{2\ell}$, $j = 1$ or 2, $k \neq 0$, $\ell \in \mathbb{Z}$, then inequality (11.18) holds for $F_*(e)$.

Suppose the two monodromies modulo center $F_*(e^{(j)})/\mathcal{Z}_3$, $j = 1, 2$, of Lemma 11.7 are not (trivial or non-trivial) powers of the same element $\sigma_j/\mathcal{Z}_3$ of $\mathcal{B}_3/\mathcal{Z}_3$. Then at most two of the elements, $F_*(e^{(1)})/\mathcal{Z}_3$, $F_*(e^{(2)})/\mathcal{Z}_3$, and $F_*(e^{(1)}e^{(2)})/\mathcal{Z}_3 = F_*(e^{(1)})/\mathcal{Z}_3 \cdot F_*(e^{(2)})/\mathcal{Z}_3$, are powers of an element of the form $\sigma_j/\mathcal{Z}_3$.

If the monodromies modulo center along two elements among $e^{(1)}$, $e^{(2)}$, and $e^{(1)}e^{(2)}$ are not (zero or non-zero) powers of a $\sigma_j/\mathcal{Z}_3$ then by Lemma 11.5 for each of these two monodromies modulo center inequality (11.21) holds, and the third monodromy modulo center is the product of two elements of $\mathcal{B}_3/\mathcal{Z}_3$ for which inequality (11.21) holds. If the monodromies modulo center along two elements among $e^{(1)}$, $e^{(2)}$, and $e^{(1)}e^{(2)}$ have the form $\sigma_j^k/\mathcal{Z}_3$ and $\sigma_{j'}^{k'}/\mathcal{Z}_3$, then the σ_j and the $\sigma_{j'}$ are different and k and k' are non-zero. The third monodromy modulo center has the form $\sigma_j^{\pm k}\sigma_{j'}^{\pm k'}/\mathcal{Z}_3$ (or the order of the two factors interchanged). Lemma 11.6 gives the inequality $\log_+(3[\frac{|k|}{2}]) + \log_+(3[\frac{|k'|}{2}]) \leq 2\pi\lambda_{e^{(1)},e^{(2)}}$. Since $\mathcal{L}_-(\vartheta(\sigma_j^{\pm k})) = \log_+(3[\frac{k}{2}])$ and $\mathcal{L}_-(\vartheta(\sigma_{j'}^{\pm k'})) = \log_+(3[\frac{k'}{2}])$, inequality (11.21) follows for the other two monodromies. The lemma is proved. □

The following lemma is more comprehensive than Lemma 9.10.

Lemma 11.8 *Let X be a connected finite open Riemann surface, and $F : X \to C_3(\mathbb{C})/\mathcal{S}_3$ a smooth mapping. Suppose for a base point q_1 of X each element of $\pi_1(X, q_1)$ can be represented by a curve with base point q_1 whose image under F avoids $\mathcal{H}$. Then all monodromies of F are powers of the same periodic braid of period* 3.

Proof By Lemma 9.10 the monodromy of F along each element of $\pi_1(X, q_1)$ is the power of a periodic braid with period 3. There is a smooth homotopy F_s, $s \in [0, 1]$, of F, such that $F_0 = F$, each F_s is different from F only on a small neighbourhood of q_1, each F_t avoids $\mathcal{H}$ on this neighbourhood of q_1, and $F_1(q_1)$ is the set of vertices of an equilateral triangle with barycenter 0. Since F and F_1 are free homotopic, their monodromy homomorphisms are conjugate, and it is enough to prove the statement of the lemma for F_1.

For notational convenience we will keep the notation F for the new mapping and assume that $F(q_1)$ is the set of vertices of an equilateral triangle with barycenter 0. Let $e \in \pi_1(X, q_1)$ be an arbitrary element. The monodromy $F_*(e)$ along the element e is a power of a periodic braid of period 3. Hence, $\tau_3(F_*(e))$ is a cyclic permutation. Consider the braid b with base point $F(0)$ that corresponds to rotation by the angle $\frac{2\pi}{3}$, i.e. it is represented by the geometric braid $t \to e^{\frac{i2\pi t}{3}} F(0)$, $t \in [0, 1]$, that avoids $\mathcal{H}$. There exists an integer k such that $F_*(e)\, b^k$ is a pure braid that is represented by a mapping that avoids $\mathcal{H}$. Hence, $F_*(e)\,b^k$ represents Δ_3^{2l} for some integer l. We proved that for each $e \in \pi_1(X, q_1)$ the monodromy $F_*(e)$ is represented by rotation of $F(0)$ around the origin by the angle $\frac{2\pi j}{3}$ for some integer j. The Lemma is proved. □

Let as before X be a finite open connected Riemann surface. The following proposition is the main ingredient of the proof of Theorem 11.3. As before $\mathcal{E} \subset \pi_1(X, q_0)$ denotes a standard system of generators of the fundamental group with base point $q_0 \in X$ that is associated to a standard bouquet of circles for X.

Proposition 11.6 *Let $(X \times \mathbb{P}^1, \mathrm{pr}_1, \boldsymbol{E}, X)$ be an irreducible holomorphic special $(0, 4)$-bundle over a finite open Riemann surface X, that is not isotopic to a locally holomorphically trivial bundle. Let $F(x)$, $x \in X$, be the set of finite distinguished points in the fiber over x. Assume that F is transverse to $\mathcal{H}$. Then there exists a complex affine mapping M and a point $q \in X$ such that $M \circ F(q)$ is contained in $C_3(\mathbb{R})\diagup \mathcal{S}_3$, and for an arc α in X with initial point q_0 and terminal point q and each element $e_j \in Is_\alpha(\mathcal{E})$ the monodromy modulo center $(M \circ F)_*(e_j)\diagup \mathcal{Z}_3$ can be written as product of at most 6 elements $\mathrm{b}_{j,k}$, $k = 1, 2, 3, 4, 5, 6$, of $\mathcal{B}_3\diagup \mathcal{Z}_3$ with*

$$\mathcal{L}_-(\vartheta(\mathrm{b}_{j,k})) \le 2\pi\lambda_{10}(X). \tag{11.22}$$

If X is a torus with a hole the monodromy along each e_j is the product of at most 4 elements with $\mathcal{L}_-(\vartheta(\mathrm{b}_{j,k})) \le 2\pi\lambda_3(X)$, and in case of a planar domain the monodromy along each e_j is the product of at most 6 elements with $\mathcal{L}_-(\vartheta(\mathrm{b}_{j,k})) \le 2\pi\lambda_8(X)$.

If X is the sphere with $m = 3$ holes, the monodromy along each e_j is the product of at most 6 elements with $\mathcal{L}_-(\vartheta(\mathrm{b}_{j,k})) \le 2\pi\lambda_5(X)$.

Proof of Proposition 11.6 Since the bundle is not isotopic to a locally holomorphically trivial bundle, it is not possible that all monodromies are powers of the same periodic braid, and by Lemma 11.8 the set

$$L \overset{def}{=} \{z \in X : F(z) \in \mathcal{H}\} \tag{11.23}$$

is not empty.

1. A Torus with a Hole Let X be a torus with a hole and let $\mathcal{E} = \{e_0', e_0''\}$ be a set of generators of $\pi_1(X, q_0)$ that is associated to a standard bouquet of circles for X. There exists a connected component L_0 of L which is not contractible and not contractible to the hole. Indeed, otherwise there would be a base point q_1 and a curve α_{q_1} that joins q_0 with q_1, such that for both elements of $\mathrm{Is}_{\alpha_{q_1}}(\mathcal{E})$ there would be representing loops with base point q_1 which do not meet L, and hence, by Lemma 11.8 the monodromies along both elements would be powers of a single periodic braid of period 3. Hence, as in the proof of Proposition 11.4 there exists a component L_0 of L, which is a simple smooth relatively closed curve in X, such that the free homotopy class of one of the elements of $\mathcal{E}$, say of e_0', has positive intersection number with L_0 after orienting L_0 suitably. Put $\mathrm{e}_0' = e_0'$. Moreover, the intersection number with L_0 is positive for the free homotopy class of one of the elements $e_0''^{\pm 1}$ or $e_0' e_0''$. Denote this element by e_0''. (Since $\widehat{e_0' e_0''} = \widehat{e_0'' e_0'}$ we may also put $\mathrm{e}_0'' = e_0'' e_0'$ if the free homotopy class of $e_0' e_0''$ intersects L_0.) Put $\mathcal{E}_2' = \{\mathrm{e}_0', \mathrm{e}_0''\}$. The free homotopy class of each element of $\mathcal{E}_2'$ and of the product of its two elements intersects L_0. We proved the following claim which we formulate for later use.

Claim 11.1 Let X be a torus with a hole and $\mathcal{E} = \{e_0', e_0''\}$ *a set of standard generators of* $\pi_1(X, q_0)$. *There exists a component* L_0 *of L, which is a simple smooth relatively closed curve in X, and two primitive elements* e_0' *and* e_0'' *in* $\langle \mathcal{E} \rangle$ *such that the free homotopy class of each element of* $\mathcal{E}_2'' \overset{def}{=} \{\mathrm{e}_0', \mathrm{e}_0''\}$ *and of the product of its two elements intersects* L_0. *Moreover, one of the* e_0' *and* e_0'' *is an element of* $\mathcal{E}$, *the other is in* $\mathcal{E} \cup \mathcal{E}^{-1}$ *or is the product of two elements of* $\mathcal{E}$. *Each element of* $\mathcal{E}$ *is the product of at most two elements of* $\mathcal{E}_2' \cup \mathcal{E}_2'^{-1}$.

The proof of the proposition in the case of a torus with a hole is finished as follows. Move the base point q_0 to a point $q \in L_0$ along a curve α, and consider the respective generators $\mathrm{e}' = \mathrm{Is}_\alpha(\mathrm{e}_0')$ and $\mathrm{e}'' = \mathrm{Is}_\alpha(\mathrm{e}_0'')$ of the fundamental group $\pi_1(X, q)$ with base point q. Since $F(L_0) \subset \mathcal{H}$ there is a complex affine mapping M such that $M \circ F(q) \in C_3(\mathbb{R})/\mathcal{S}_3$. Since F is irreducible, the monodromy maps modulo center $(M \circ F)_*(\mathrm{e}')/\mathcal{Z}_3$ and $(M \circ F)_*(\mathrm{e}'')/\mathcal{Z}_3$ are not powers of a single standard generator $\sigma_j/\mathcal{Z}_3$ of $\mathcal{B}_3/\mathcal{Z}_3$ (see Lemma 6.4, or Lemma 7 of [47]). Hence, the second option of Lemma 11.7 occurs. We obtain that each of the $(M \circ F)_*(\mathrm{e}')/\mathcal{Z}_3$ and $(M \circ F)_*(\mathrm{e}'')/\mathcal{Z}_3$ is a product of at most two elements b_j of $\mathcal{B}_3/\mathcal{S}_3$ with $\mathcal{L}_-(\vartheta(\mathrm{b}_j)) \le 2\pi\lambda_3(X)$. Hence, $(M \circ F)_*(e')/\mathcal{Z}_3$ and $(M \circ F)_*(e'')/\mathcal{Z}_3$ are products of at most 4 elements of $\mathcal{B}_3/\mathcal{Z}_3$ with this property. The proposition is proved for tori with a hole.

2. A Planar Domain Let X be a planar domain. Maybe, after applying a Möbius transformation, we represent X as the Riemann sphere with holes $\mathcal{C}_j$, $j = 1, \ldots, m+1$, such that $\mathcal{C}_{m+1}$ contains ∞. Recall, that we have chosen a standard system $\mathcal{E}$ of generators $e_{j,0}$, $j = 1, \ldots, m$, of the fundamental group $\pi_1(X, q_0)$ with base point q_0 that is associated to a standard bouquet of circle for X. Hence, $e_{j,0}$ is represented by a loop with base point q_0 that surrounds $\mathcal{C}_j$ counterclockwise and does not surround any other hole.

We claim that there is a connected component L_0 of L of one of the following kinds. Either L_0 has limit points on the boundary of two different holes (one of them may contain ∞) (first kind), or a component L_0 has limit points on a single hole $\mathcal{C}_j$, $j \le m+1$, and $\mathcal{C}_j \cup L_0$ divides the plane $\mathbb{C}$ into two connected components each of which contains a hole (maybe, only the hole containing ∞) (second kind), or there is a compact component L_0 that divides $\mathbb{C}$ into two connected components each of which contains at least two holes (one of them may contain ∞). Indeed, suppose each non-compact component of L has boundary points on the boundary of a single hole and the union of the component with the hole does not separate the remaining holes of X, and for each compact component of L one of the connected components of its complement in X contains at most one hole. Then there exists a base point q_1, a curve α_{q_1} in X with initial point q_0 and terminal point q_1, and a representative of each element of $\text{Is}_{\alpha_{q_1}}(\mathcal{E}) \subset \pi_1(X, q_1)$ that avoids L. Lemma 11.8 implies that all monodromies modulo center are powers of a single periodic element of $\mathcal{B}_3/\mathcal{Z}_3$ which is a contradiction.

If there is a component L_0 of the first kind we choose the same set of primitive elements $\mathcal{E}_2' \subset \mathcal{E}_2 \subset \pi_1(X, q_0)$ as in the proof of Proposition 11.4 in the planar case. The free homotopy class of each element of $\mathcal{E}_2'$ and of the product of two different elements of $\mathcal{E}_2'$ intersects L_0. Moreover, each element of $\mathcal{E}$ is the product of at most two elements of $\mathcal{E}_2'$.

Suppose there is no component of the first kind but a component L_0 of the second kind. Assume first that all limit points of L_0 are on the boundary of a hole $\mathcal{C}_j$ that does not contain ∞. Put $\mathcal{E}_3' = \{e_{j,0}\} \cup_{1 \le k \le m,\, k \ne j} \{e_{j,0}^2 e_{k,0}\}$. Each element of $\mathcal{E}_3'$ is a primitive element and is the product of at most three generators contained in the set $\mathcal{E}$. Further, each element of $\mathcal{E}$ is the product of at most three elements of $\mathcal{E}_3' \cup \mathcal{E}_3'^{-1}$.

The free homotopy class of each element of $\mathcal{E}_3'$ and of each product of two different elements of $\mathcal{E}_3'$ intersects L_0. Indeed, any curve that is contained in $\mathbb{C} \setminus (\mathcal{C}_j \cup L_0)$ has either winding number zero around $\mathcal{C}_j$ (as a curve in the complex plane $\mathbb{C}$), or its winding number around $\mathcal{C}_j$ coincides with the winding number around each of the holes in the bounded connected component of $\mathbb{C} \setminus (\mathcal{C}_j \cup L_0)$. On the other hand the representatives of the free homotopy class of $e_{j,0}$ have winding number 1 around $\mathcal{C}_j$ and winding number 0 around each other hole that does not contain ∞. The representatives of the free homotopy class of $e_{j,0}^2 e_{k,0}$, $k \le m$, $k \ne j$, have winding number 2 around $\mathcal{C}_j$, winding number 1 around $\mathcal{C}_k$, and winding number zero around each other hole $\mathcal{C}_l$, $l \le m$. The argument for products of two elements of $\mathcal{E}_3'$ is the same.

Assume that the limit points of L_0 are on the boundary of the hole $\mathcal{C}_\infty$ that contains ∞. Let $\mathcal{C}_{j_0}$ and $\mathcal{C}_{k_0}$ be holes that are contained in different components of $\mathbb{C} \setminus (L_0 \cup \mathcal{C}_\infty)$, and let $e_{j_0,0}$ and $e_{k_0,0}$ be the elements of $\mathcal{E}$ whose representatives surround $\mathcal{C}_{j_0}$, and $\mathcal{C}_{k_0}$ respectively. Denote by $\mathcal{E}_3'$ the set that consists of the elements $e_{j_0,0}e_{k_0,0}$, $e_{j_0,0}^2 e_{k_0,0}$, and all elements $e_{j_0,0}e_{k_0,0}\tilde{e}_0$ with $\tilde{e}_0$ running over $\mathcal{E} \setminus \{e_{j_0,0}, e_{k_0,0}\}$. Each element of $\mathcal{E}_3'$ is the product of at most 3 elements of $\mathcal{E}$, and each element of $\mathcal{E}$ is the product of at most 3 elements of $\mathcal{E}_3' \cup (\mathcal{E}_3')^{-1}$.

Each element of $\mathcal{E}_3'$ and each product of at most two different elements of $\mathcal{E}_3'$ intersects L_0. Indeed, if a closed curve is contained in one of the components of $\mathbb{C} \setminus (L_0 \cup \mathcal{C}_\infty)$ then its winding number around each hole contained in the other component is zero. But for all mentioned elements there is a hole in each component of $\mathbb{C} \setminus (L_0 \cup \mathcal{C}_\infty)$ such that the winding number of the free homotopy class of the element around the hole does not vanish.

Notice that in case of $m + 1 = 3$ holes only these two possibilities for the curve L_0 may occur. We proved the following claim which we formulate for later use.

Claim 11.2 Suppose X equals $\mathbb{P}^1$ with $m + 1 = 3$ holes and $\mathcal{E}$ is a standard system of generators of $\pi_1(X, q_0)$. Then there is a set $\mathcal{E}_3' = \{\mathrm{e}_0', \mathrm{e}_0''\} \subset \pi_1(X, q_0)$, such that *one of the elements of $\mathcal{E}_3'$ is the product of at most two elements of $\mathcal{E} \cup \mathcal{E}^{-1}$, and the free homotopy classes of both elements and of their product intersect L_0.*

If in addition L_0 has limit points on a hole C_j with $j \leq m$, then one of the elements of $\mathcal{E}_3'$ can be taken to be an element of $\mathcal{E}$, not merely a product of two such elements.

Moreover, e and e' are products of at most three factors, each an element of $\mathcal{E}_3' \cup \mathcal{E}_3'^{-1}$.

For the general case of planar domains we also need the following considerations. Suppose there are no components of L of the first or the second kind, but there is a connected component L_0 of L of the third kind. Let C_{j_0} be a hole contained in the bounded component of the complement of L_0 in $\mathbb{C}$, and let C_{k_0}, $k_0 \leq m$, be a hole that is contained in the unbounded component of $\mathbb{C} \setminus L_0$. Let $e_{j_0,0}$ and $e_{k_0,0}$ be the elements of $\mathcal{E}$ whose representatives surround C_{j_0}, and C_{k_0} respectively. Consider the set $\mathcal{E}_4'$ consisting of the following elements: $e_{j_0,0}e_{k_0,0}$, $e^2_{j_0,0}e_{k_0,0}$, and $e^2_{j_0,0}e_{k_0,0}\tilde{e}_0$ for each $\tilde{e}_0 \in \mathcal{E}$ different from $e_{j_0,0}$ and $e_{k_0,0}$. Each element of $\mathcal{E}_4'$ is the product of at most 4 elements of $\mathcal{E}$ and each element of $\mathcal{E}$ is the product of at most 3 elements of $\mathcal{E}_4' \cup (\mathcal{E}_4')^{-1}$. The product of two different elements of $\mathcal{E}_4'$ is contained in $\mathcal{E}_8'$.

The free homotopy classes of each element of $\mathcal{E}_4'$ and of each product of two different elements of $\mathcal{E}_4'$ intersect L_0. Indeed, if a loop is contained in the bounded connected component of $\mathbb{C} \setminus L_0$, its winding number around the holes C_j, $j \leq m$, contained in the unbounded component is zero. If a loop is contained in the unbounded connected component of $\mathbb{C} \setminus L_0$, its winding numbers around all holes contained in the bounded connected component are equal. But the winding numbers of $e_{j_0,0}e_{k_0,0}$ and $e^2_{j_0,0}e_{k_0,0}$ around the hole C_{j_0} are positive and the winding numbers around the other holes that are contained in the bounded connected component of $\mathbb{C} \setminus L_0$ vanish, hence the representatives of these two elements cannot be contained in the unbounded component of $\mathbb{C} \setminus L_0$. Since the winding numbers of representatives of these elements around C_{k_0} are positive, the representatives cannot be contained in the bounded component of $\mathbb{C} \setminus L_0$. For representatives of the elements $e^2_{j_0,0}e_{k_0,0}\tilde{e}_0$ the winding numbers around C_{j_0} equal 2, the winding numbers around any other hole in the bounded component of $\mathbb{C} \setminus L_0$ are at most 1, and the winding numbers around C_{k_0} equal 1. Hence, the free homotopy classes of the mentioned elements must intersect both components of $\mathbb{C} \setminus L_0$, hence they intersect L_0.

Representatives of any product of two different elements of $\mathcal{E}_4'$ have winding numbers around C_{j_0} at least 3, the winding numbers around any other hole in the bounded component of $\mathbb{C} \setminus L_0$ are at most 1, and the winding numbers around C_{k_0} equal 2. Hence, the free homotopy classes of these elements intersect L_0. We obtained the following

Claim 11.3 If X is a planar domain and $\mathcal{E}$ is a standard system of generators of $\pi_1(X, q_0)$, then there is a set $\mathcal{E}_4' \subset \pi_1(X, q_0)$, such that each of the elements of $\mathcal{E}_4'$ is the product of at most 4 elements of $\mathcal{E} \cup \mathcal{E}^{-1}$, and the free homotopy classes of the product of at most two elements of $\mathcal{E}_4'$ intersect L_0. Moreover, e and e' are products of at most three factors, each an element of $\mathcal{E}_4' \cup \mathcal{E}_4'^{-1}$.

The proof of the proposition in the planar case is finished as follows. Let α_q be a curve in X with initial point q_0 and terminal point q, and M a complex affine mapping, such that $(M \circ F)(q) \in C_3(\mathbb{R})/\mathcal{S}_3$. Since $M \circ F$ is irreducible, the monodromies modulo center of $M \circ F$ along the elements of $\mathrm{Is}_\alpha(\mathcal{E}_4')$ are not (trivial or non-trivial) powers of a single element $\sigma_j/\mathcal{Z}_3$. Hence, for each element of $\mathrm{Is}_\alpha(\mathcal{E}_4')$ there exists another element of $\mathrm{Is}_\alpha(\mathcal{E}_4')$ so that the second option of Lemma 11.7 holds for this pair of elements of $\mathrm{Is}_\alpha(\mathcal{E}_4')$. Therefore, the monodromy modulo center of $M \circ F$ along each element of $\mathrm{Is}_\alpha(\mathcal{E}_4')$ is the product of at most two elements $\mathrm{b}_j \in \mathcal{B}_3/\mathcal{Z}_3$ of $\mathcal{L}_-$ not exceeding $2\pi\lambda_8(X)$, and the monodromy modulo center of $M \circ F$ along each element $\mathrm{Is}_\alpha(\mathcal{E})$ is the product of at most 6 elements $\mathrm{b}_j \in \mathcal{B}_3/\mathcal{Z}_3$ with $\mathcal{L}_-(\vartheta(\mathrm{b}_j))$ not exceeding $2\pi\lambda_8(X)$. Proposition 11.6 is proved in the planar case.

3. The General Case Since not all monodromies are powers of a single element of $\mathcal{B}_3/\mathcal{Z}_3$ that is either periodic or reducible, there exists a pair of generators e_0', e_0'' in $\mathcal{E}$, such that the monodromies along them are not powers of a single periodic or reducible element. Consider the projection $\omega^{\langle e_0', e_0''\rangle} : \tilde{X} \to X(\langle e_0', e_0''\rangle)$. By the proof for tori with a hole or for $\mathbb{P}^1$ with three holes there exist a relatively closed curve $L_{\langle e_0', e_0''\rangle}$ in $X(\langle e_0', e_0''\rangle)$ and a Möbius transformation M, such that for $F = M \circ f$ the mapping $F_{\langle e_0', e_0''\rangle} = F \circ \omega_{\langle e_0', e_0''\rangle}$ takes $L_{\langle e_0', e_0''\rangle}$ into $\mathcal{H}$, and takes a chosen point $q_{\langle e_0', e_0''\rangle} \in L_{\langle e_0', e_0''\rangle}$ to a point in $C_3(\mathbb{R})/\mathcal{S}_3$.

Choose a point $\tilde{q} \in \tilde{X}$, for which $\omega^{\langle e_0', e_0''\rangle}(\tilde{q}) = q_{\langle e_0', e_0''\rangle}$. Let $\tilde{\alpha}$ be a curve in $\tilde{X}$ with initial point $\tilde{q}_0$ and terminal point $\tilde{q}$. Then $\alpha_{\langle e_0', e_0''\rangle} \stackrel{def}{=} \omega^{\langle e_0', e_0''\rangle}(\tilde{\alpha})$ is a curve in $X(\langle e_0', e_0''\rangle)$ with initial point $(q_0)_{\langle e_0', e_0''\rangle}$ and terminal point $q_{\langle e_0', e_0''\rangle}$, and the curve $\alpha_{\langle e_0', e_0''\rangle}$ in $X(\langle e_0', e_0''\rangle)$ and the point $\tilde{q}$ in the universal covering $\tilde{X}$ of $X(\langle e_0', e_0''\rangle)$ are compatible. Put $\alpha = \omega_{\langle e_0', e_0''\rangle}(\alpha_{\langle e_0', e_0''\rangle})$, and for each $e_0 \in \pi_1(X, q_0)$ we denote as before the element $\mathrm{Is}_\alpha(e_0)$ by e.

Put $\mathcal{E}^* \stackrel{def}{=} \{e_0', e_0''\}$. For an element $e_0 \in \mathcal{E}^*$ we define $(\mathrm{e}_0')_{\langle e_0', e_0''\rangle}$ and $(\mathrm{e}_0'')_{\langle e_0', e_0''\rangle}$ as in the end of paragraph 3.1 of the proof of Proposition 11.5. The Riemann surface $X(\langle e_0', e_0''\rangle)$ is a torus with a hole or $\mathbb{P}^1$ with three holes. This implies the following fact.

There are elements e_0' and e_0'', one of them contained in $\mathcal{E}^*$ or equal to the product of at most two factors among the e_0' and e_0'', the second either contained in $\mathcal{E}^* \cup (\mathcal{E}^*)^{-1}$, or equal to the product of at most three factors among the e_0' and e_0'', such that the free homotopy classes of $(\mathrm{e}_0')_{\langle e_0', e_0''\rangle}$, of $(\mathrm{e}_0'')_{\langle e_0', e_0''\rangle}$, and of their product intersect $L_{\langle e_0', e_0''\rangle}$. Moreover, e_0' and e_0'' are products of at most three factors, each being either $(\mathrm{e}_0')^{\pm 1}$ or $(\mathrm{e}_0'')^{\pm 1}$.

Put $\mathrm{e}'_{\langle e_0', e_0''\rangle} = \mathrm{Is}_{\alpha_{\langle e_0', e_0''\rangle}}((\mathrm{e}_0')_{\langle e_0', e_0''\rangle})$, $\mathrm{e}''_{\langle e_0', e_0''\rangle} = \mathrm{Is}_{\alpha_{\langle e_0', e_0''\rangle}}((\mathrm{e}_0'')_{\langle e_0', e_0''\rangle})$.

Since the monodromies along e' and e'' are not powers of a single periodic or reducible element, by Lemma 11.7 each monodromy $(F_{\langle e_0', e_0''\rangle})_*(\mathrm{e}'_{\langle e_0', e_0''\rangle}) = F_*(\mathrm{e}')$ and $(F_{\langle e_0', e_0''\rangle})_*(\mathrm{e}''_{\langle e_0', e_0''\rangle}) = F_*(\mathrm{e}'')$ is the product of at most two elements $\mathrm{b}_j \in \mathcal{B}_3/\mathcal{Z}_3$ with $\mathcal{L}_-(\vartheta(\mathrm{b}_j)) \le 2\pi\lambda_5(X)$. Since e' and e'' are products of at most three elements among $(\mathrm{e}')^{\pm 1}$ and $(\mathrm{e}'')^{\pm 1}$, each of the monodromies $F_*(\mathfrak{e}')$ and $F_*(\mathfrak{e}'')$ is the product of at most 6 elements $\mathrm{b}_j \in \mathcal{B}_3/\mathcal{Z}_3$ with $\mathcal{L}_-(\vartheta(\mathrm{b}_j)) \le 2\pi\lambda_5(X)$.

Take any element $e_0 \in \mathcal{E} \setminus \{e_0', e_0''\}$. Let $e = \mathrm{Is}_\alpha(e_0)$. Either the pair of monodromies $(F_*(\mathrm{e}'), F_*(e))$ or the pair of monodromies $(F_*(\mathrm{e}''), F_*(e))$ does not consist of two powers of the same element of $\mathcal{B}_3/\mathcal{Z}_3$ that is either periodic or reducible. Suppose this is so for the pair $(F_*(\mathrm{e}'), F_*(e))$.

Let $L_{\langle \mathrm{e}_0' \rangle}$ be the connected component of $(\omega_{\langle \mathrm{e}_0' \rangle}^{\langle \mathrm{e}_0', \mathrm{e}_0'' \rangle})^{-1}(L_{\langle \mathrm{e}_0', \mathrm{e}_0'' \rangle})$ that contains $\omega^{\langle \mathrm{e}_0' \rangle}(\tilde{q})$. By the analog of Lemma 11.3, applied to the holomorphic projection $\tilde{X}/(\mathrm{Is}^{\tilde{q}_0})^{-1}(\langle \mathrm{e}_0' \rangle) \to X(\langle \mathrm{e}_0', \mathrm{e}_0'' \rangle)$, the free homotopy class $\widehat{(\mathrm{e}_0')_{\langle \mathrm{e}_0' \rangle}}$ intersects $L_{\langle \mathrm{e}_0' \rangle}$. (For the definition of $(\mathrm{e}_0')_{\langle \mathrm{e}_0' \rangle}$ see the end of the paragraph 3.1 of the proof of Proposition 11.5.) As in the proof of Proposition 11.5 we consider the Riemann surface $X(\langle e_0, \mathrm{e}_0' \rangle)$ and the curve $L_{\langle e_0, \mathrm{e}_0' \rangle} = \omega_{\langle \mathrm{e}_0' \rangle}^{\langle e_0, \mathrm{e}_0' \rangle}(L_{\langle \mathrm{e}_0' \rangle})$ (see paragraph 3.3. of the proof of Proposition 11.5). As there we see that the free homotopy class $\widehat{(\mathrm{e}_0')_{\langle e_0, \mathrm{e}_0' \rangle}}$ intersects $L_{\langle e_0, \mathrm{e}_0' \rangle}$. The system $\big((e_0)_{\langle e_0, \mathrm{e}_0' \rangle}, (\mathrm{e}_0')_{\langle e_0, \mathrm{e}_0' \rangle}\big)$ is associated to a standard bouquet of circles for $X(\langle e_0, \mathrm{e}_0' \rangle)$ (though the system (e_0, e_0') may not be part of a system of generators that is associated to a standard bouquet of circles for X). This can be seen in the same way as in the proof of Proposition 11.5. We will apply now the arguments, used for $X(\langle e_0', e_0'' \rangle)$ and the generators $(e_0')_{\langle e_0', e_0'' \rangle}, (e_0'')_{\langle e_0', e_0'' \rangle}$ of the fundamental group $\pi_1(X(\langle e_0', e_0'' \rangle), q_{\langle e_0', e_0'' \rangle})$, to $X(\langle e_0, \mathrm{e}_0' \rangle)$ and the generators $(e_0)_{\langle e_0, \mathrm{e}_0' \rangle}, (\mathrm{e}_0')_{\langle e_0, \mathrm{e}_0' \rangle}$ of the fundamental group $\pi_1(X(\langle e_0, \mathrm{e}_0' \rangle), q_{\langle e_0, \mathrm{e}_0' \rangle})$.

In the case when $X(\langle e_0, \mathrm{e}_0' \rangle)$ is a torus with a hole, the intersection number of $\widehat{(\mathrm{e}_0')_{\langle e_0, \mathrm{e}_0' \rangle}}$ with $L_{\langle e_0, \mathrm{e}_0' \rangle}$ is non-zero. Put $\mathfrak{e}_0' = \mathrm{e}_0'$. For one of the choices $e_0^{\pm 1}$, or $\mathrm{e}_0' e_0$, denoted by $\mathfrak{e}_0''$, the free homotopy classes of $(\mathfrak{e}_0')_{\langle e_0, \mathrm{e}_0' \rangle}, (\mathfrak{e}_0'')_{\langle e_0, \mathrm{e}_0' \rangle}$, and of their product intersect $L_{\langle e_0, \mathrm{e}_0' \rangle}$. Since by Claim 11.1 and Claim 11.2 the element e_0' is the product of at most three elements of $\mathcal{E}$, the element $\mathfrak{e}_0''$ is the product of at most $1 + 3 = 4$ elements of $\mathcal{E}$. Moreover, e_0 is the product of at most two factors, each being $(\mathfrak{e}_0')^{\pm 1}$, or $(\mathfrak{e}_0'')^{\pm 1}$. Hence, e_0 is the product of at most 7 elements of $\mathcal{E}$.

In case $X(\langle e_0, \mathrm{e}_0' \rangle)$ is planar, the curve $L_{\langle e_0, \mathrm{e}_0' \rangle}$ must have limit points on the hole that corresponds to the generator $(\mathrm{e}_0')_{\langle e_0, \mathrm{e}_0' \rangle}$ of the fundamental group $\pi_1(X(\langle e_0, \mathrm{e}_0' \rangle), q_{\langle e_0, \mathrm{e}_0' \rangle})$. By Claim 11.2 we find elements $\mathfrak{e}_0'$ and $\mathfrak{e}_0''$ such that $\mathfrak{e}_0' = \mathrm{e}_0'$ and $\mathfrak{e}_0''$ is either equal to e_0^{-1}, or to the product of at most three factors, one being equal to e_0 and the others equal to e_0', and the free homotopy classes of $(\mathfrak{e}_0')_{\langle e_0, \mathrm{e}_0' \rangle}$, $(\mathfrak{e}_0'')_{\langle e_0, \mathrm{e}_0' \rangle}$, and their product intersect $L_{\langle e_0, \mathrm{e}_0' \rangle}$. Moreover, e_0 is the product of at most 3 factors, each being equal to $(\mathfrak{e}_0'')^{\pm 1}$ or $(\mathfrak{e}_0')^{\pm 1}$. Since e_0' is the product of at most three elements of $\mathcal{E}$, e_0'' is the product of at most $1 + 2 \cdot 3 = 7$ elements of $\mathcal{E}$. Hence, the elements $\mathfrak{e}_0'$, $\mathfrak{e}_0''$, and $\mathfrak{e}_0'\mathfrak{e}_0''$ are products of at most 10 elements of $\mathcal{E} \cup \mathcal{E}^{-1}$.

Since for $e = \mathrm{Is}_\alpha(e_0)$, and $\mathrm{e}' = \mathrm{Is}_\alpha(\mathrm{e}_0')$ the monodromies $F_*(e)$ and $F_*(\mathrm{e}_0')$ are not powers of a single periodic or reducible element, for $\mathfrak{e}' = \mathrm{Is}_\alpha(\mathfrak{e}_0')$ and $\mathfrak{e}'' = \mathrm{Is}_\alpha(\mathfrak{e}_0'')$ the monodromies $F_*(\mathfrak{e}')$ and $F_*(\mathfrak{e}'')$ cannot be powers of a single element that is either periodic or reducible. Lemma 11.7 implies, that $F_*(\mathfrak{e}')$ and $F_*(\mathfrak{e}'')$ are products of at most two factors b with $\mathcal{L}_-(\vartheta(\mathsf{b}))$ not exceeding $2\pi\lambda_{10}(X)$. Hence, $F_*(e)$ is the product of at most 6 factors b with $\mathcal{L}_-(\vartheta(\mathsf{b}))$ not exceeding $2\pi\lambda_{10}(X)$. We obtain the statement of Proposition 11.6 in the general case. Proposition 11.6 is proved.

□

Proof of Theorem 11.3 Let X be a connected Riemann surface of genus g with $\mathsf{m}+1 \geq 1$ holes. Since each holomorphic (0, 3)-bundle with a holomorphic section on X is isotopic to a holomorphic special (0, 4)-bundle, we need to estimate the number of isotopy classes of irreducible smooth special (0, 4)-bundles on X, that contain a holomorphic bundle. By Lemma 6.4 the monodromies of an irreducible bundle are not powers of a single element of $\mathcal{B}_3/\mathcal{Z}_3$ which is conjugate to a $\sigma_j/\mathcal{Z}_3$, but they may be powers of a single periodic element of $\mathcal{B}_3/\mathcal{Z}_3$ (equivalently, the isotopy class may contain a locally holomorphically trivial holomorphic bundle).

Consider an irreducible special holomorphic (0, 4)-bundle on X which is not isotopic to a locally holomorphically trivial bundle. Let $F(x)$, $x \in X$, be the set of finite distinguished points in the fiber over x. By the Holomorphic Transversality Theorem [51] the mapping $F : X \to C_3(\mathbb{C})/\mathcal{S}_3$ can be approximated on relatively compact subsets of X by holomorphic mappings that are transverse to $\mathcal{H}$. Similarly as in the proof of Theorem 11.1 we will therefore assume in the following (after slightly shrinking X to a deformation retract of X and approximating F) that F is transverse to $\mathcal{H}$.

By Proposition 11.6 there exists a complex affine mapping M and a point $q \in X$ such that $M \circ F(q)$ is contained in $C_3(\mathbb{R})/\mathcal{S}_3$, and for an arc α in X with initial point q_0 and terminal point q and each element $e_j \in \mathrm{Is}_\alpha(\mathcal{E})$ the monodromy $(M \circ F)_*(e_j)/\mathcal{Z}_3$ of the bundle can be written as product of at most 6 elements $\mathsf{b}_{j,k}$, $k = 1, 2, 3, 4, 5, 6$, of $\mathcal{B}_3/\mathcal{Z}_3$ with

$$\mathcal{L}_-(\vartheta(\mathsf{b}_{j,k})) \leq 2\pi\lambda_{10}(X). \tag{11.24}$$

The mappings F and $M \circ F$ from X into the symmetrized configuration space are free homotopic.

Consider an isotopy class of special (0, 4)-bundles that corresponds to a conjugacy class of homomorphisms $\pi_1(X, q_0) \to \mathcal{B}_3/\mathcal{Z}_3$ whose image is generated by a single periodic element of $\mathcal{B}_3/\mathcal{Z}_3$. Up to conjugacy we may assume that this element is one of the following: $\Delta_3/\mathcal{Z}_3$, $(\sigma_1\sigma_2)/\mathcal{Z}_3$, $(\sigma_1\sigma_2)^{-1}/\mathcal{Z}_3$, or Id. For each of these elements b the equality $\mathcal{L}_-(\vartheta(\mathsf{b})) = 0$ holds. Hence, in this case the isotopy class contains a smooth mapping $\tilde{F}$ such that for each $e_{j,0} \in \mathcal{E}$ the monodromy $(M \circ F)_*(e_{j,0})/\mathcal{Z}_3$ of the bundle can be written as product of at most 6 elements $\mathsf{b}_{j,k}$, $k = 1, 2, 3, 4, 5, 6$, of $\mathcal{B}_3/\mathcal{Z}_3$ satisfying inequality (11.24).

The same argument as in the proof of Theorem 11.1 shows the following fact. Each irreducible free homotopy class of mappings $X \to C_3(\mathbb{C})/\mathcal{S}_3$ that contains a holomorphic mapping contains a smooth mapping $\tilde{F}$ such that for each $e_{j,0} \in \mathcal{E}$

the monodromy $\tilde{F}_*(e_{j,0})/\mathcal{Z}_3$ of the bundle can be written as product of at most 6 elements $\mathsf{b}_{j,k}$, $k = 1, 2, 3, 4, 5, 6$, of $\mathcal{B}_3/\mathcal{Z}_3$ satisfying inequality (11.24).

Using Lemma 10.1 (see also Lemma 1 of [45]) the number of elements $\mathsf{b} \in \mathcal{B}_3/\mathcal{Z}_3$ (including the identity), for which $\mathcal{L}_-(\vartheta(\mathsf{b})) \le 2\pi\lambda_{10}(X)$, is estimated as follows. The element $\mathsf{w} \overset{def}{=} \vartheta(\mathsf{b}) \in \mathcal{PB}_3/\mathcal{Z}_3$ can be considered as a reduced word in the free group generated by $a_1 = \sigma_1^2/\mathcal{Z}_3$ and $a_2 = \sigma_2^2/\mathcal{Z}_3$. By Lemma 10.1 there are no more than $\frac{1}{2}\exp(6\pi\lambda_{10}(X)) + 1 \le \frac{3}{2}\exp(6\pi\lambda_{10}(X))$ reduced words w in a_1 and a_2 (including the identity) satisfying the inequality $\mathcal{L}_-(\mathsf{w}) \le 2\pi\lambda_{10}(X)$.

For a given element $\mathsf{w} \in \mathcal{PB}_3/\mathcal{Z}_3$ (including the identity) we describe now all elements b of $\mathcal{B}_3/\mathcal{Z}_3$ with $\vartheta(\mathsf{b}) = \mathsf{w}$. If $\mathsf{w} \ne \mathrm{Id}$ these are the following elements. If the first term of w equals a_j^k with $k \ne 0$, then the possibilities are $\mathsf{b} = \mathsf{w} \cdot (\Delta_3^\ell/\mathcal{Z}_3)$ with $\ell = 0$ or 1, $\mathsf{b} = (\sigma_j^{\mathrm{sgn}k}/\mathcal{Z}_3)\cdot\mathsf{w}\cdot(\Delta_3^\ell/\mathcal{Z}_3)$ with $\ell = 0$ or 1, or $\mathsf{b} = (\sigma_{j'}^{\pm 1}/\mathcal{Z}_3)\cdot \mathsf{w} \cdot (\Delta_3^\ell/\mathcal{Z}_3)$ with $\ell = 0$ or 1 and $\sigma_{j'} \ne \sigma_j$. Hence, for $\mathsf{w} \ne \mathrm{Id}$ there are 8 possible choices of elements $\mathsf{b} \in \mathcal{B}_3/\mathcal{Z}_3$ with $\vartheta(\mathsf{b}) = \mathsf{w}$.

If $\mathsf{b} = \mathrm{Id}$ then the choices are $\Delta^\ell/\mathcal{Z}_3$ and $(\sigma_j^{\pm 1}\Delta^\ell)/\mathcal{Z}_3$ for $j = 1, 2$, and $\ell = 0$ or $\ell = 1$. These are 10 choices. Hence, there are no more than $15\exp(6\pi\lambda_{10}(X))$ different elements $\mathsf{b} \in \mathcal{B}_3/\mathcal{Z}_3$ with $\mathcal{L}_-(\vartheta(\mathsf{b})) \le 2\pi\lambda_{10}(X)$.

Each monodromy is the product of at most six elements b_j of $\mathcal{B}_3/\mathcal{Z}_3$ with $\mathcal{L}_-(\vartheta(\mathsf{b}_j)) \le 2\pi\lambda_{10}(X)$. Hence, for each monodromy there are no more than $(15\exp(6\pi\lambda_{10}(X)))^6$ possible choices. We proved that there are up to isotopy no more than $(15\exp(6\pi\lambda_{10}(X)))^{6(2g+m)}$ irreducible holomorphic $(0, 3)$-bundles with a holomorphic section over X. Theorem 11.3 is proved. □

Notice that we proved a slightly stronger statement, namely, over a Riemann surface of genus g with $m + 1 \ge 1$ holes there are no more than $(15\exp(6\pi\lambda_{10}(X)))^{6(2g+m)}$ isotopy classes of smooth $(0, 3)$-bundles with a smooth section that contain a holomorphic bundle with a holomorphic section that is either irreducible or isotopic to the trivial bundle.

Proof of Theorem 11.2 Proposition 8.2 and Theorem 11.3 imply Theorem 11.2 as follows. Suppose an isotopy class of smooth $(1, 1)$-bundles over a finite open Riemann surface X contains a holomorphic bundle. By Proposition 8.2 the class contains a holomorphic bundle which is the double branched covering of a holomorphic special $(0, 4)$-bundle. If the $(1, 1)$-bundle is irreducible then also the $(0, 4)$-bundle is irreducible. There are up to isotopy no more than $\big(15(\exp(6\pi\lambda_{10}(X)))\big)^{6(2g+m)}$ holomorphic special $(0, 4)$-bundles over X that are either irreducible or isotopic to the trivial bundle.

Theorem 8.1 and Theorem 11.3 imply, that there are no more than $\big(15\big(\exp(6\pi\lambda_{10}(X))\big)\big)^{6(2g+m)}$ conjugacy classes of monodromy homomorphisms that correspond to a special holomorphic $(0, 4)$-bundle over X that is either irreducible or isotopic to the trivial bundle. Each monodromy homomorphism of the holomorphic double branched covering is a lift of the respective monodromy homomorphism of the holomorphic special $(0, 4)$-bundle. Different lifts of a monodromy mapping class of a special $(0, 4)$-bundle differ by involution, and the fundamental group of X has $2g+m$ generators. Using Theorem 8.1 for (1, 1)-bundles, we see that there are no more than $2^{2g+m}\big(15(\exp(6\pi\lambda_{10}(X)))\big)^{6(2g+m)} = \big(2\cdot 15^6\cdot \exp(36\pi\lambda_{10}(X))\big)^{2g+m}$ isotopy classes of (1, 1)-bundles that contain a holomorphic bundle that is either irreducible or isotopic to the trivial bundle. Theorem 11.2 is proved. □

For convenience of the reader we give the short reduction of the Corollaries 11.1 and 11.2 to Theorems 11.1 and 11.2, respectively. The arguments of the reduction are known in principle, but the case considered here is especially simple.

Proof of Corollary 11.1 We will prove that on a punctured Riemann surface there are no non-constant reducible holomorphic mappings to the twice punctured complex plane and that any non-trivial homotopy class of mappings from a punctured Riemann surface to the twice punctured complex plane contains at most one holomorphic mapping. This implies the corollary.

Recall that a holomorphic mapping f from any punctured Riemann surface X to the twice punctured complex plane extends by Picard's Theorem to a meromorphic function f^c on the closed Riemann surface X^c. Suppose now that X is a punctured Riemann surface and that the mapping $f: X \to \mathbb{C}\setminus\{-1,1\}$ is reducible, i.e. it is homotopic to a mapping into a punctured disc contained in $\mathbb{C}\setminus\{-1,1\}$. Perhaps after composing f with a Möbius transformation we may suppose that this puncture equals -1. Then the meromorphic extension f^c omits the value 1. Indeed, if f^c was equal to 1 at some puncture of X, then f would map the boundary of a sufficiently small disc on X^c that contains the puncture to a loop in $\mathbb{C}\setminus\{-1,1\}$ with non-zero winding number around 1, which contradicts the fact that f is homotopic to a mapping into a disc punctured at -1 and contained in $\mathbb{C}\setminus\{-1,1\}$. Hence, f^c is a meromorphic function on a compact Riemann surface that omits a value, and, hence f is constant. Hence, on a punctured Riemann surface there are no non-constant reducible holomorphic mappings to the twice punctured complex plane.

Suppose f_1 and f_2 are non-constant homotopic holomorphic mappings from the punctured Riemann surface X to $\mathbb{C}\setminus\{-1,1\}$. Then for their meromorphic extensions f_1^c and f_2^c the functions f_1^c-1 and f_2^c-1 have the same divisor on the closed Riemann surface X^c. Indeed, suppose, for instance, that f_1^c-1 has a zero of order $k>0$ at a puncture p. Then for the boundary γ of a small disc in X^c around p the curve $(f_1-1)\circ\gamma$ in $\mathbb{C}\setminus\{-2,0\}$ has index k with respect to the origin. Since f_2-1 is homotopic to f_1-1 as mapping to $\mathbb{C}\setminus\{-2,0\}$, the curve $(f_2-1)\circ\gamma$ is free homotopic to $(f_1-1)\circ\gamma$. Hence, f_2-1 has a zero of order k at p. Applying the same arguments with 0 replaced by ∞, we obtain that f_1^c-1 and f_2^c-1 have the same divisor. Hence, f_1^c-1 and f_2^c-1 differ by a non-zero multiplicative constant. Since the functions are non-constant they must take the value -2. By the same reasoning as above the functions are equal to -2 simultaneously. Hence, the multiplicative constant is equal to 1. We proved that non-constant homotopic holomorphic maps from punctured Riemann surfaces to $\mathbb{C}\setminus\{-1,1\}$ are equal. □

Proof of Corollary 11.2 We need the following fact. For each special (0, 4)-bundle $\mathfrak{F} = (X\times\mathbb{P}^1, \mathrm{pr}_1, E, X)$ there is a finite unramified covering $\hat{\mathsf{P}}: \hat{X}\to X$ of

X, such that $\mathfrak{F}$ lifts to a special $(0, 4)$-bundle $(\hat{X} \times \mathbb{P}^1, \mathrm{pr}_1, \hat{\boldsymbol{E}}, X)$, for which the complex curve $\hat{\boldsymbol{E}}$ is the union of four disjoint complex curves $\hat{\boldsymbol{E}}^k$, $k = 1, 2, 3, 4$, each intersecting each fiber $\{\hat{x}\} \times \mathbb{P}^1$ along a single point $(\hat{x},\ \hat{g}^k(\hat{x}))$. This can be seen as follows. Let q_0 be the base point of X. The monodromy mapping class along each element e of $\pi_1(X, q_0)$ takes the set of distinguished points $\hat{\boldsymbol{E}} \cup (\{q_0\} \times \mathbb{P}^1)$ onto itself, permuting them by a permutation $\sigma(e)$. Consider the set N of elements $e \in \pi_1(X, q_0)$ for which $\sigma(e)$ is the identity. The set N is a normal subgroup of $\pi_1(X, q_0)$. Its index is finite, since two left cosets $e_1 N$ and $e_2 N$ are equal if $\sigma(e_2^{-1} e_1) = \sigma(e_2)^{-1}\sigma(e_1) = \mathrm{Id}$, and there are only finitely many distinct permutations of points of $\hat{\boldsymbol{E}} \cup (\{q_0\} \times \mathbb{P}^1)$. The quotient $\hat{X} \overset{def}{=} \tilde{X}/\mathrm{Is}^{\tilde{q}_0}(N)$ of the universal covering of X by the group of covering transformations corresponding to N and the canonical projection $\hat{X} \to X$ define the required covering.

To prove the corollary, we have to show first, that any reducible holomorphic $(1, 1)$-bundle over a punctured Riemann surface X is locally holomorphically trivial and secondly, that two isotopic (equivalently, smoothly isomorphic) locally holomorphically non-trivial holomorphic $(1, 1)$-bundles over X are holomorphically isomorphic.

The second fact is obtained as follows. Suppose the locally holomorphically non-trivial holomorphic $(1, 1)$-bundles $\mathfrak{F}_j$, $j = 1, 2$, have conjugate monodromy homomorphisms. By Proposition 8.2 each bundle $\mathfrak{F}_j$, $j = 1, 2$, is holomorphically isomorphic to a double branched covering of a special holomorphic $(0, 4)$-bundle $(X \times \mathbb{P}^1, \mathrm{pr}_1, \boldsymbol{E}_j, X) \overset{def}{=} \mathsf{Pr}(\mathfrak{F}_j)$. The bundles $\mathsf{Pr}(\mathfrak{F}_j)$ are isotopic, since they have conjugate monodromy homomorphisms. There is a finite unramified covering $\hat{\mathsf{P}} : \hat{X} \to X$ of X, such that the bundles $\mathsf{Pr}(\mathfrak{F}_j)$ have isotopic lifts $(\hat{X} \times \mathbb{P}^1, \mathrm{pr}_1, \hat{\boldsymbol{E}}_j, X)$ to $\hat{X}$, and for each j the complex curve $\hat{\boldsymbol{E}}_j$ is the union of four disjoint complex curves $\hat{\boldsymbol{E}}_l^k$, $k = 1, 2, 3, 4$, each intersecting each fiber $\{\hat{x}\} \times \mathbb{P}^1$ along a single point $(\hat{x},\ \hat{g}_j^k(\hat{x}))$. We may choose the label of the points so that the isotopy moves the complex curve $\hat{\boldsymbol{E}}_1^k$ to $\hat{\boldsymbol{E}}_2^k$, $k = 1, 2, 3, 4$. The lifted bundles are not isotopic to the trivial bundle. The mappings $\hat{X} \ni \hat{x} \to \hat{g}_j^k(\hat{x}) \in \mathbb{P}^1$ are holomorphic. We may assume that $\hat{g}_j^4(\hat{x}) = \infty$ for each $\hat{x}$. Define for $j = 1, 2$, a holomorphic isomorphism of the bundle $(\hat{X} \times \mathbb{P}^1, \mathrm{pr}_1, \hat{\boldsymbol{E}}_j, X)$ by

$$\{\hat{x}\} \times \mathbb{P}^1 \ni (\hat{x},\ \zeta) \to \left(\hat{x},\ \frac{\hat{g}_j^1(\hat{x}) - \zeta}{\hat{g}_j^1(\hat{x}) - \hat{g}_j^2(\hat{x})}\right).$$

The image $\hat{\boldsymbol{E}}_j'$ of $\hat{\boldsymbol{E}}_j$ under the j-th isomorphism intersects the fiber over each $\hat{x} \in \hat{X}$ along the four points $(\hat{x}, 0)$, $(\hat{x}, 1)$, $(\hat{x}, \infty)$, and $(\hat{x}, \mathring{g}_j(\hat{x}))$ for a holomorphic mapping $\mathring{g}_j : \hat{X} \to \mathbb{P}^1$ that avoids 0, 1 and ∞. The mappings $\mathring{g}_j$, $j = 1, 2$, are homotopic, since the bundles are isotopic. They are not homotopic to a constant mapping since the bundles are not isotopic to the trivial bundle. By the proof of Corollary 11.1 the mappings $\mathring{g}_1$ and $\mathring{g}_2$ coincide. Hence, the bundles $(\hat{X} \times \mathbb{P}^1, \mathrm{pr}_1, \hat{\boldsymbol{E}}_j, X)$ are holomorphically isomorphic to the bundle $\mathfrak{F}' = (\hat{X} \times \mathbb{P}^1, \mathrm{pr}_1, \hat{\boldsymbol{E}}, X)$ with $\hat{\boldsymbol{E}} \cap (\{\hat{x}\} \times \mathbb{P}^1) = \{\hat{x}\} \times (0, 1, \mathring{g}(\hat{x}), \infty)$ for a holomorphic mapping $\mathring{g}(\hat{x}) : \hat{X} \to \mathbb{C} \setminus \{0, 1\}$.

For each $j = 1, 2$, we obtain a bundle that is holomorphically isomorphic to $\mathsf{Pr}(\mathfrak{F}_j)$ by considering the bundle $\mathfrak{F}'$ and gluing for each $x \in X$ the fibers over $\hat{x} \in (\hat{\mathsf{P}})^{-1}(x)$ together using complex affine mappings determined by $\mathsf{Pr}(\mathfrak{F}_j)$. The complex affine mappings depend holomorphically on $\hat{x}$. Each of them takes the unordered triple of finite distinguished points in one fiber to the unordered triple of finite distinguished points in the other fiber over $\hat{x}$. Assign to each finite distinguished point the number k of the curve $\hat{\boldsymbol{E}}^k$ in $\hat{\boldsymbol{E}}$ to which the point belongs. Each of the complex affine mappings is uniquely determined by the permutation of these numbers k, induced by it. Since the bundles $\mathsf{Pr}(\mathfrak{F}_j)$ are smoothly isomorphic, the respective permutations for the two bundles coincide. Hence, the bundles $\mathsf{Pr}(\mathfrak{F}_j)$ are holomorphically isomorphic to a single bundle that is obtained from $\mathfrak{F}'$ by a holomorphic gluing procedure of the fibers over points $\hat{x} \in (\hat{\mathsf{P}})^{-1}(x)$ for $x \in X$.

The bundle $\mathfrak{F}_j$, $j = 1, 2$, is a double branched covering of $\mathsf{Pr}(\mathfrak{F}_j)$ and the $\mathsf{Pr}(\mathfrak{F}_j)$ are holomorphically isomorphic. Moreover, the monodromy homomorphisms of the bundles $\mathfrak{F}_j$ are conjugate. Hence, the bundles $\mathfrak{F}_j$, $j = 1, 2$, are holomorphically isomorphic. The second fact is proved.

The first fact is obtained as follows. After a holomorphic isomorphism we may assume that the reducible holomorphic $(1, 1)$-bundle is a double branched covering of a reducible special $(0, 4)$-bundle $\mathsf{Pr}(\mathfrak{F}) = (X \times \mathbb{P}^1, \mathrm{pr}_1, \mathring{\boldsymbol{E}} \cup s^{\infty}, X)$. After a further isomorphism the bundle $\mathsf{Pr}(\mathfrak{F})$ lifts to a holomorphic bundle $\widehat{\mathsf{Pr}(\mathfrak{F})} = (\hat{X} \times \mathbb{P}^1, \mathrm{pr}_1, \hat{\mathring{\boldsymbol{E}}} \cup \widehat{s^{\infty}}, \hat{X})$, such that $\hat{\mathring{\boldsymbol{E}}}$ intersects each fiber $\{x\} \times \mathbb{P}^1$ along a set of the form $\{\hat{x}\} \times \{0, 1, \mathring{g}(\hat{x})\}$. Since the bundle $\mathfrak{F}$ is reducible, the bundle $\mathsf{Pr}(\mathfrak{F})$ and also $\widehat{\mathsf{Pr}(\mathfrak{F})}$ are reducible. Hence, the mapping $\mathring{g}$ is constant by the proof of Corollary 11.1. Hence, the bundle $\widehat{\mathsf{Pr}(\mathfrak{F})}$ is holomorphically trivial and, hence, $\mathsf{Pr}(\mathfrak{F})$ is locally holomorphically trivial. Since $\mathfrak{F}$ is a double branched covering of $\mathsf{Pr}(\mathfrak{F})$, the bundle $\mathfrak{F}$ is locally holomorphically trivial. The first fact is proved. □

Proof of Proposition 11.1 Denote by S^α a skeleton of $T^{\alpha,\sigma} \subset T^\alpha$ which is the union of two circles each of which lifts under the covering $\mathsf{P} : \mathbb{C} \to T^\alpha$ to a straight line segment which is parallel to an axis in the complex plane. Denote the intersection point of the two circles by q_0. Note that S^α is a standard bouquet of circles for $T^{\alpha,\sigma}$ with base point q_0, and $\mathsf{P}^{-1}(T^{\alpha,\sigma})$ is the $\frac{\sigma}{2}$-neighbourhood of $\mathsf{P}^{-1}(S^\alpha)$. We may assume that $\mathsf{P}^{-1}(q_0)$ is the lattice $\mathbb{Z} + i\alpha\mathbb{Z}$.

Denote by e the generator of $\pi_1(T^{\alpha,\sigma}, q_0)$, that lifts to a vertical line segment and e' the generator of $\pi_1(T^{\alpha,\sigma}, q_0)$, that lifts to a horizontal line segment. Put $\mathcal{E} = \{e, e'\}$. We show first the inequality

$$\lambda_3(T^{\alpha,\sigma}) \leq \frac{4(2\alpha + 1)}{\sigma}. \tag{11.25}$$

For this purpose we take any primitive element e'' of the fundamental group $\pi_1(T^{\alpha,\sigma}, q_0)$ which is the product of at most three factors, each of the factors being an element of $\mathcal{E}$ or the inverse of an element of $\mathcal{E}$. We represent the element e'' by a piecewise C^1 mapping f_1 from an interval $[0, l_1]$ to the skeleton S^α. We may consider f_1 as a piecewise C^1 mapping from the circle $\mathbb{R}/(x \sim x + l_1)$ to the skeleton, and assume that for all points t' of the circle where f_1 is not smooth, $f_1(t') = q_0$. Let $t_0 \in [0, l_1]$ be a point for which $f_1(t_0) \neq q_0$. Let $\tilde{f}_1$ be a piecewise smooth mapping from $[t_0, t_0 + l_1]$ to the universal covering $\mathbb{C}$ of $T^\alpha \subset T^{\alpha,\sigma}$ which projects to f_1. We may take f_1 so that the equality $|\tilde{f}_1'| = 1$ holds for the derivative $\tilde{f}_1'$ of $\tilde{f}_1$. The mapping may be chosen so that $l_1 \leq 2\alpha + 1$. (Recall that $\alpha \geq 1$ and the element e is primitive.)

Take any t' for which f_1 is not smooth. We may assume that f_1 is chosen so that the direction of $\tilde{f}_1'$ changes by the angle $\pm\frac{\pi}{2}$ at each such point. Hence, there exists a neighbourhood $I(t')$ of t' on $(t_0, t_0 + \ell_1)$, such that the restriction $\tilde{f}_1(t')$ covers two sides of a square of side length $\frac{\sigma}{2}$. Denote by $\tilde{q}_0'$ the common vertex $\tilde{f}_1(t')$ of these sides, and by $\tilde{q}_0''$ the vertex of the square that is not a vertex of one of the two sides. Replace the union of the two sides of the square that contain $\tilde{q}_0'$ by a quarter-circle of radius $\frac{\sigma}{2}$ with center at the vertex $\tilde{q}_0''$, and parameterize the latter by $t \to \frac{\sigma}{2}e^{\pm i \frac{2}{\sigma} t}$ so that the absolute value of the derivative equals 1. Notice that the quarter-circle is shorter than the union of the two sides.

Proceed in this way with all such points t'. After a reparameterization we obtain a C^1 mapping $\tilde{f}$ of the interval $[0, l]$ of length l not exceeding $2\alpha + 1$ whose image is contained in the union of $\mathsf{P}^{-1}(S^\alpha)$ with some quarter-circles, such that $|\tilde{f}'| = 1$. The distance of each point of the image of $\tilde{f}$ to the boundary of $\mathsf{P}^{-1}(T^{\alpha,\sigma})$ is not smaller than $\frac{\sigma}{2}$. The mapping $\tilde{f}$ is piecewise of class C^2. The normalization condition $|\tilde{f}'| = 1$ implies $|\tilde{f}''| \leq \frac{2}{\sigma}$.

The projection $f = \mathsf{P} \circ \tilde{f}$ can be considered as a mapping from the circle $\mathbb{R}/(x \sim x + l)$ of length l not exceeding $2\alpha + 1$ to $T^{\alpha,\sigma}$, that represents the free homotopy class $\widehat{e''}$ of the chosen element of the fundamental group.

Consider the mapping $x + iy \to \tilde{F}(x + iy) \stackrel{def}{=} \tilde{f}(x) + i\tilde{f}'(x)y \in \mathbb{C}$, where $x + iy$ runs along the rectangle $R_l = \{x + iy \in \mathbb{C} : x \in [0, l], |y| \leq \frac{\sigma}{4}\}$. The image of this mapping is contained in the closure of $\mathsf{P}^{-1}(T^{\alpha,\sigma})$. Since $2\frac{\partial}{\partial z}\tilde{F}(x + iy) = 2\tilde{f}'(x) + i\tilde{f}''(x)y$ and $2\frac{\partial}{\partial \bar{z}}\tilde{F}(x + iy) = i\tilde{f}''(x)y$, the Beltrami coefficient $\mu_{\tilde{F}}(x + iy) = \frac{\frac{\partial}{\bar{z}}\tilde{F}(x+iy)}{\frac{\partial}{z}\tilde{F}(x+iy)}$ of $\tilde{F}$ satisfies the inequality $|\mu_{\tilde{F}}(x + iy)| \leq \frac{1}{3}$. Hence, for $K = \frac{1+\frac{1}{3}}{1-\frac{1}{3}} = 2$ the mapping $\tilde{F}$ descends to a K-quasiconformal mapping F from the annulus A_l to $T^{\alpha,\sigma}$ of extremal length $\lambda(A_l) = \frac{l}{\frac{\sigma}{2}} \leq 2\frac{(2\alpha+1)}{\sigma}$ that represents the free homotopy class of the element e'' of the fundamental group $\pi_1(T^{\alpha,\sigma}, q_0)$. Realize A_l as an annulus in the complex plane. Let φ be the solution of the Beltrami equation on $\mathbb{C}$ with Beltrami coefficient $\mu_{\tilde{F}}$ on A_l and zero else. Then the mapping $g = F \circ \varphi^{-1}$ is a holomorphic mapping of the annulus $\varphi(A_l)$ of extremal length not exceeding $K\lambda(A_l) \le \frac{4(2\alpha+1)}{\sigma}$ into $T^{\alpha,\sigma}$ that represents the chosen element of the fundamental group $\pi_1(T^{\alpha,\sigma}, q_0)$. Inequality (11.25) is proved.

By Theorem 11.1 for tori with a hole there are up to homotopy no more than $3(\frac{3}{2}e^{36\pi\lambda_3(T^{\alpha,\sigma})})^2 \le \frac{27}{4}e^{2^5\cdot 3^2\pi\frac{2\alpha+1}{\sigma}} < 7e^{2^5\cdot 3^2\pi\frac{2\alpha+1}{\sigma}}$ non-constant irreducible holomorphic mappings from $T^{\alpha,\sigma}$ to the twice punctured complex plane.

We give now the proof of the lower bound. Fix $\alpha \ge 1$ and let $\delta \le \frac{1}{10}$. We consider the annulus $A^{\alpha,5\delta} = \{z \in \mathbb{C} : |\text{Re}z| < \frac{5\delta}{2}\}/(z \sim z + \alpha i)$ of extremal length equal to $\frac{\alpha}{5\delta} \ge 2\alpha$.

For any natural number j we consider all elements of $\pi_1(\mathbb{C} \setminus \{-1, 1\}, 0)$ of the form

$$a_1^{\pm 2}a_2^{\pm 2}\dots a_1^{\pm 2}a_2^{\pm 2} \tag{11.26}$$

containing $2j$ terms, each of the form $a_j^{\pm 2}$. The choice of the sign in the exponent of each term is arbitrary. There are 2^{2j} elements of this kind.

Put $j = [\frac{\alpha}{60\pi\delta}]$, where $[x]$ is the largest integer not exceeding a positive number x. Then $\frac{\alpha}{5\delta} \ge 12j\pi$ and Remark 9.1 implies that for each word w of the form $a_1^{\pm 2}a_2^{\pm 2}\dots a_1^{\pm 2}a_2^{\pm 2}$ with $N = 2j$ terms there exists a holomorphic mapping $\mathfrak{g}_w : \left(A^{\alpha,5\delta}, 0/(z \sim z + i\alpha)\right) \to \left(\mathbb{C} \setminus \{-1, 1\}, 0\right)$ from $A^{\alpha,5\delta}$ with base point $0/(z \sim z + i\alpha)$ to $\mathbb{C} \setminus \{-1, 1\}$ with base point 0, that represents w. Moreover, the image of $\mathfrak{g}_w$ is contained in the domain $\{z \in \mathbb{C} : |z| < C, |z \pm 1| > \frac{1}{C}\}$ for a constant $C > 1$.

The plan is the following. We will embed $A^{\alpha,\delta}$ into $T^{\alpha,\delta}$, restrict $\mathfrak{g}_w$ to a smaller annulus, and extend the pushforward of the restriction to a smooth mapping $\mathring{\mathfrak{g}}_w$ from $T^{\alpha,\delta}$ into $\{z \in \mathbb{C} : |z| < C, |z \pm 1| > \frac{1}{C}\}$ such that (with P being the projection $\mathsf{P} : \mathbb{C} \to T^\alpha$) the monodromy along the circle $\mathsf{P}(\{\text{Re}z = 0\})$ with base point $\mathsf{P}(0)$ is equal to (11.26), and the monodromy along $\mathsf{P}(\{\text{Im}z = 0\}$ with the same base point equals the identity. We will show the existence of a positive number ε depending only on C, such that correcting the smooth mapping $\mathring{\mathfrak{g}}_w$ on $T^{\alpha,\sigma} \stackrel{def}{=} T^{\alpha,\varepsilon\delta}$ by a solution of a $\bar\partial$-problem on $T^{\alpha,\sigma}$ with L^∞-norm smaller than C, we obtain a holomorphic mapping $T^{\alpha,\sigma} \to \mathbb{C} \setminus \{-1, 1\}$ with monodromies w and Id. In this way for $\sigma = \varepsilon\delta$ we will obtain $2^{2j} \ge \frac{1}{4}e^{\frac{2\varepsilon\log 2}{60\pi}\frac{\alpha}{\sigma}}$ non-homotopic holomorphic mappings from $T^{\alpha,\sigma}$ to $\mathbb{C} \setminus \{-1, 1\}$. We will now give the detailed proof following the plan.

Let $\tilde{\mathfrak{g}}_w$ be the lift of $\mathfrak{g}_w$ to a mapping from the strip $\{z \in \mathbb{C} : |\text{Re}z| < \frac{5\delta}{2}\}$ to $\{z \in \mathbb{C} : |z| < C, |z \pm 1| > \frac{1}{C}\}$. On the thinner strip $\{|\text{Re}z| < \frac{3\delta}{2}\}$ the derivative of $\tilde{\mathfrak{g}}_w$ satisfies the inequality $|\tilde{\mathfrak{g}}_w'| \le \frac{C}{\delta}$, since $|\tilde{\mathfrak{g}}_w| \le C$.

Let $F_\alpha = [-\frac{1}{2}, \frac{1}{2}) \times [-\frac{\alpha}{2}, \frac{\alpha}{2}) \subset \mathbb{C}$ be a fundamental domain for the projection $\mathsf{P} : \mathbb{C} \to T^\alpha$. Put $\Delta^{\alpha,\delta} = F_\alpha \cap \mathsf{P}^{-1}(T^{\alpha,\delta})$. Let $\chi_0 : [0, 1] \to \mathbb{R}$ be a non-decreasing function of class C^2 with $\chi_0(0) = 0$, $\chi_0(1) = 1$, $\chi_0'(0) = \chi_0'(1) = 0$

Fig. 11.3 A fundamental domain for a torus with a hole and the poles of the kernel for the $\overline{\partial}$-equation

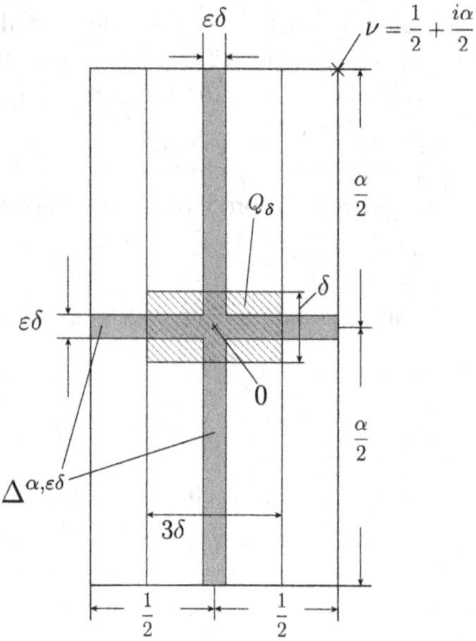

and $|\chi_0'(t)| \leq \frac{3}{2}$ (see also Sect. 9.5). Define $\chi_\delta : [\frac{-3\delta}{2}, \frac{+3\delta}{2}] \to [0, 1]$ by

$$\chi_\delta(t) = \begin{cases} \chi_0(\frac{1}{\delta}t + \frac{3}{2}) & t \in [\frac{-3\delta}{2}, \frac{-\delta}{2}] \\ 1 & t \in [\frac{-\delta}{2}, \frac{+\delta}{2}] \\ \chi_0(-\frac{1}{\delta}t + \frac{3}{2}) & t \in [\frac{\delta}{2}, \frac{3\delta}{2}]. \end{cases} \tag{11.27}$$

Notice that χ_δ is a C^2-function that vanishes at the endpoints of the interval $[\frac{-3\delta}{2}, \frac{+3\delta}{2}]$ together with its first derivative, is non-decreasing on $[\frac{-3\delta}{2}, \frac{-\delta}{2}]$, and non-increasing on $[\frac{\delta}{2}, \frac{3\delta}{2}]$. Put $\mathring{\mathfrak{g}}_w(z) = \chi_\delta(\mathrm{Re} z)\, \tilde{\mathfrak{g}}_w(z) + (1 - \chi_\delta(\mathrm{Re} z))\, \tilde{\mathfrak{g}}_w(0)$ for z in the intersection of $\Delta^{\alpha,\delta}$ with $\{|\mathrm{Re} z| < \frac{3\delta}{2}\}$, and $\mathring{\mathfrak{g}}_w(z) = \tilde{\mathfrak{g}}_w(0)$ for z in the rest of $\Delta^{\alpha,\delta}$.

Put $\varphi_w(z) = \frac{\partial}{\partial \bar{z}} \mathring{\mathfrak{g}}_w(z)$ on $\Delta^{\alpha,\delta}$. Since $\frac{\partial}{\partial \bar{z}} \chi_\delta(\mathrm{Re} z) = 0$ for $|\mathrm{Re} z| < \frac{\delta}{2}$ and for $|\mathrm{Re} z| > \frac{3\delta}{2}$, the function $\varphi_w(z)$ vanishes on $\Delta^{\alpha,\delta} \setminus Q_\delta$ with $Q_\delta \overset{def}{=} ([-\frac{3\delta}{2}, +\frac{3\delta}{2}] \times [-\frac{\delta}{2}, \frac{\delta}{2}])$. On $Q_\delta \cap \Delta^{\alpha,\delta}$ the inequality

$$|\varphi_w(z)| \leq \frac{1}{2} |\chi_\delta'(\mathrm{Re} z)|\, |\tilde{\mathfrak{g}}_w(z) - \tilde{\mathfrak{g}}_w(0)| \leq \frac{3}{4\delta} \cdot \frac{C}{\delta} |z| < \frac{3}{4\delta^2} \cdot C \cdot 2\delta = \frac{3}{2} \frac{C}{\delta} \tag{11.28}$$

holds. Notice that the functions $\mathring{\mathfrak{g}}_w$ and φ_w extend to $\mathsf{P}^{-1}(T^{\alpha,\delta})$ as continuous doubly periodic functions. Hence, we may consider them as functions on $T^{\alpha,\delta}$ (see Fig. 11.3).

We want to find now a small positive number ε that depends only on C, such that for $\sigma \overset{def}{=} \varepsilon\delta \le \frac{\varepsilon}{10}$ there exists a solution f_w of the equation $\frac{\partial}{\partial \bar{z}} f_w(z) = \varphi_w(z)$ on $T^{\alpha,\sigma}$ such that for each z the inequality $|f_w(z)| \le \frac{1}{C}$ holds. Then $\mathfrak{g}_w - f_w$ is a holomorphic mapping from $T^{\alpha,\sigma}$ to $\mathbb{C} \setminus \{-1, 1\}$ whose class has monodromies equal to (11.26), and to the identity, respectively. This provides the $2^{2j} \ge \frac{1}{4} e^{\frac{\varepsilon \log 2}{30\pi} \frac{\alpha}{\sigma}}$ promised different homotopy classes of mappings from $T^{\alpha,\sigma}$ to $\mathbb{C} \setminus \{-1, 1\}$, and, hence proves the lower bound.

To solve the $\bar{\partial}$-problem on $T^{\alpha,\varepsilon\delta} = T^{\alpha,\sigma}$, we consider an explicit kernel function which mimics the Weierstraß $\wp$-function. The author is grateful to Bo Berndtsson who suggested to use this kernel function.

Recall that the Weierstraß $\wp$-function related to the torus T^{α} is the doubly periodic meromorphic function

$$\wp_\alpha(\zeta) = \frac{1}{\zeta^2} + \sum_{(n,m)\in\mathbb{Z}^2\setminus(0,0)} \left(\frac{1}{(\zeta - n - im\alpha)^2} - \frac{1}{(n + im\alpha)^2} \right)$$

on $\mathbb{C}$. It defines a meromorphic function on T^{α} with a double pole at the projection of the origin and no other pole.

Put $\nu = \frac{1}{2} + \frac{\alpha i}{2}$ (See Fig. 11.3). Since for $\zeta \notin (\mathbb{Z} + i\alpha\mathbb{Z}) \cup (\nu + \mathbb{Z} + i\alpha\mathbb{Z})$ the equality

$$\frac{1}{(\zeta - n - im\alpha)} - \frac{1}{(\zeta - n - im\alpha - \nu)} + \frac{\nu}{(\zeta - n - im\alpha)^2} = \frac{-\nu^2}{(\zeta - n - im\alpha)^2(\zeta - n - im\alpha - \nu)}$$

holds, and the series with these terms converges uniformly on compact sets not containing poles, the expression

$$\wp_\alpha^\nu(\zeta) \overset{def}{=} \frac{1}{\zeta} - \frac{1}{\zeta - \nu} + \sum_{(n,m)\in\mathbb{Z}^2\setminus(0,0)} \left(\frac{1}{(\zeta - n - im\alpha)} - \frac{1}{(\zeta - n - im\alpha - \nu)} + \frac{\nu}{(n + im\alpha)^2} \right)$$

defines a doubly periodic meromorphic function on $\mathbb{C}$ with only simple poles. The function descends to a meromorphic function on T^{α} with two simple poles and no other pole.

The support of φ_w is contained in Q_δ. The set Q_δ is contained in the 2δ-disc in $\mathbb{C}$ (in the Euclidean metric) around the origin. If ζ is contained in the 2δ-disc around the origin and $z \in \Delta^{\alpha,\delta}$, then the point $\zeta - z$ is contained in the 2δ-neighbourhood (in $\mathbb{C}$) of $\Delta^{\alpha,\delta}$. By the choice of δ the distance of any such point $\zeta - z$ to any lattice point $n + i\alpha m$ except 0 is larger than $\frac{1}{2} - 2\delta > \frac{1}{4}$. Further, for $z \in \Delta^{\alpha,\delta}$ and ζ in the

2δ-disc around the origin the distance of the point $\zeta - z$ to any point $n + i\alpha m + \nu$ (including the point ν) is not smaller than $\frac{1}{2} - \frac{5\delta}{2} \geq \frac{1}{4}$. Put $Q'_{\varepsilon,\delta} \stackrel{def}{=} Q_\delta \cap \Delta^{\alpha,\varepsilon\delta} = ([-\frac{3\delta}{2}, +\frac{3\delta}{2}] \times [-\frac{\varepsilon\delta}{2}, +\frac{\varepsilon\delta}{2}]) \bigcup ([-\frac{\varepsilon\delta}{2}, +\frac{\varepsilon\delta}{2}] \times [-\frac{\delta}{2}, +\frac{\delta}{2}])$. (See Fig. 11.3). Then the function

$$f_w(z) = -\frac{1}{\pi} \iint_{Q'_{\varepsilon,\delta}} \varphi_w(\zeta) \wp_\alpha^\nu(\zeta - z) dm_2(\zeta) , \tag{11.29}$$

for z in $\Delta^{\alpha,\varepsilon\delta}$ is holomorphic outside $Q'_{\varepsilon,\delta}$ and satisfies the equation $\frac{\partial}{\partial \bar{z}} f_w = \varphi_w$ on $Q'_{\varepsilon,\delta}$. It extends continuously to a doubly periodic function on $\mathsf{P}^{-1}(T^{\alpha,\varepsilon\delta})$ and hence descends to a continuous function on $T^{\alpha,\varepsilon\delta}$. It remains to estimate the supremum norm of the function f_w on $\Delta^{\alpha,\sigma} = \Delta^{\alpha,\varepsilon\delta}$. The following inequality holds for $z \in \Delta^{\alpha,\sigma}$

$$\begin{aligned}
&\left| \iint_{Q'_{\varepsilon,\delta}} \varphi_w(\zeta) \wp_\alpha^\nu(\zeta - z) dm_2(\zeta) \right| = \\
&\left| \frac{1}{\pi} \iint_{Q'_{\varepsilon,\delta}} \varphi_w(\zeta) \Big(\frac{1}{\zeta - z} + (\wp_\alpha^\nu(\zeta - z) - \frac{1}{\zeta - z}) \Big) dm_2(\zeta) \right| \leq \\
&\frac{1}{\pi} \iint_{Q'_{\varepsilon,\delta}} \frac{3C}{2\delta} \Big(|\frac{1}{\zeta - z}| + C' \Big) dm_2(\zeta) .
\end{aligned} \tag{11.30}$$

We used the upper bound (11.28) for φ_w and the fact that for $z \in \Delta^{\alpha,\sigma}$ and ζ in $Q'_{\varepsilon,\delta}$ the expression $|\wp_\alpha^\nu(\zeta - z) - \frac{1}{\zeta - z}|$ is bounded by a universal constant C'. The integral of the second term on the right hand side does not exceed $\frac{3C}{2\delta} \cdot C' \cdot 4\varepsilon\delta^2 = 6CC'\varepsilon\delta$. The integral $\iint_{Q'_{\varepsilon,\delta}} |\frac{1}{\zeta - z}| dm_2(\zeta)$ does not exceed the sum of the two integrals $I_1 = \iint_{(-\frac{3}{2}\delta, \frac{3}{2}\delta) \times (-\frac{1}{2}\varepsilon\delta, \frac{1}{2}\varepsilon\delta)} | \frac{1}{\zeta - z} | \, dm_2(\zeta)$, and $I_2 = \iint_{(-\frac{1}{2}\varepsilon\delta, \frac{1}{2}\varepsilon\delta) \times (-\frac{1}{2}\delta, \frac{1}{2}\delta)} | \frac{1}{\zeta - z} | \, dm_2(\zeta)$. The first integral I_1 is largest when $z = 0$. Hence, it does not exceed

$$\iint_{|\zeta| < (\sqrt{2})^{-1} \varepsilon\delta} | \frac{1}{\zeta} | \, dm_2(\zeta) + 2\varepsilon\delta \int_{\frac{1}{2}\varepsilon\delta}^{\frac{3}{2}\delta} \frac{1}{\eta} \, d\eta \leq \sqrt{2}\pi\varepsilon\delta + 2\varepsilon\delta \log \frac{3}{\varepsilon} . \tag{11.31}$$

The second integral I_2 is smaller. We obtain the estimate

$$\begin{aligned}
|f_w(z)| \leq & \frac{6CC'\varepsilon\delta}{\pi} + \frac{3C}{\pi\delta} (\sqrt{2}\pi\varepsilon\delta + 2\varepsilon\delta \log \frac{3}{\varepsilon}) \\
= & \frac{6CC'\delta}{\pi} \varepsilon + \frac{3C}{\pi} (\sqrt{2}\pi\varepsilon + 2\varepsilon \log \frac{3}{\varepsilon}) .
\end{aligned} \tag{11.32}$$

Recall that we have chosen $\delta \leq \frac{1}{10}$. We may choose $\varepsilon_0 > 0$ depending only on C so that if $\varepsilon < \varepsilon_0$ the supremum norm of f_w is less than $\frac{1}{C}$. The proposition is proved. □

Proof of Proposition 11.2 Let ℓ_0 be the length in the Kähler metric of the longest circle in the bouquet. For each natural number k and each positive $\sigma < \sigma_0$ the value $\lambda_k(S_\sigma)$ satisfies the inequalities

$$C_1' \frac{\ell_0}{\sigma} \leq \lambda_k(S_\sigma) \leq C_1'' \frac{\ell_0}{\sigma} \tag{11.33}$$

for constants C_1' and C_1'' depending on k, X, S and the Kähler metric. This can be seen by the argument used in the proof of Proposition 11.1.

The upper bound in inequalities (11.1) follows from Theorem 11.1.

The proof of the lower bound in (11.1) will follow along the same lines as the proof of Proposition 11.1. It will lead to a $\overline{\partial}$-problem on an open Riemann surface, for which Hörmander's L^2-method can be used. The case of open Riemann surfaces is easier to treat than the general case of pseudo-convex domains. The needed results for Riemann surfaces are explicitly formulated in [68].

To obtain the lower bound we consider the circle of the bouquet S with largest length (in the Kähler metric). After a small deformation of the circle we may assume that it is real analytic outside a small neighbourhood of q. Take a slightly larger neighbourhood of q, so that the part γ_0 of this circle outside this neighbourhood is connected. Consider a conformal mapping ω of a neighbourhood of γ_0 onto a relatively compact domain V in $\mathbb{C}$, such that ω takes γ_0 to an interval on the imaginary axis. We may assume that its length is bigger than $3c\ell_0$ for a positive constant c, and the mapping is normalized so that $\omega(\gamma_0)$ contains the interval $(-\frac{3c\ell_0}{2}, \frac{3c\ell_0}{2})$. Moreover, we may suppose that the orientation of $\omega(\gamma_0)$ coincides with the orientation of the imaginary axis.

For any positive δ we denote by R_δ the rectangle $R_\delta \overset{def}{=} \{x + iy : x \in (-\delta, \delta), y \in (-c\ell_0 - 2\delta, c\ell_0 + 2\delta)\}$ of extremal length $\lambda(R_\delta) = c\frac{\ell_0}{\delta} + 2$ in the complex plane. Let δ_0 be a positive number such that $S_{2\delta_0}$ is relatively compact in X and is a deformation retract of X, and for all positive $\delta \leq 2\delta_0$ the rectangle R_δ is contained in V. We denote by R_δ^X the curvilinear rectangle $\omega^{-1}(R_\delta)$ in X.

Let $\delta \leq \delta_0$. Let $\mathring{R}_\delta \overset{def}{=} \{x + iy : x \in (-\delta, \delta), y \in (-c\ell_0, c\ell_0)\} \subset R_\delta$ be the rectangle in the complex plane with the same center and horizontal side length as R_δ, with vertical side length $2c\ell_0$ and extremal length $\lambda(\mathring{R}_\delta) = \frac{c\ell_0}{\delta}$. Put $j \overset{def}{=} [\frac{c\ell_0}{128\pi}]$. By Remark 9.1 we may represent any word w of the form $a_1^{\pm 2} a_2^{\pm 2} \ldots a_2^{\pm 2}$ in the relative fundamental group $\pi_1(\mathbb{C}\backslash\{-1, 1\}, (-1, 1))$ with $2j$ terms by a holomorphic mapping $g_w : \mathring{R}_\delta \to \{z \in \mathbb{C} : |z| < C, |z \pm 1| > \frac{1}{C}\} \subset \mathbb{C} \setminus \{-1, 1\}$ that vanishes at $\pm ic\ell_0$. The mapping g_w extends by reflection through the horizontal sides of $\mathring{R}_\delta$ to a holomorphic function on R_δ, that we also denote by g_w. Since $|g_w| \leq C$ on $\mathring{R}_\delta$ for the constant C from Remark 9.1, hence, $|g_w| \leq C$ also on R_δ, for any positive $\alpha < 1$ the inequality $|g_w'| \leq \frac{C}{\delta(1-\alpha)}$ holds for the derivative of the mapping g_w on the smaller rectangle $R_{\alpha\delta}$ (defined as R_δ with δ replaced by $\alpha\delta$). This fact implies that $|g_w| \le \frac{\sqrt{2}C\alpha}{1-\alpha}$ on $Q^{\pm}_{\alpha\delta} \stackrel{def}{=} \{x+iy : x \in (-\alpha\delta, \alpha\delta), \pm y \in (c\ell_0, c\ell_0+\alpha\delta)\}$. We took into account that $g_w(\pm ic\ell_0) = 0$. We take α so that $\frac{\sqrt{2}C\alpha}{1-\alpha} < 1 - \frac{1}{C}$.

With the same function χ_0 as in the proof of Proposition 11.1 we define

$$\mathring{\chi}_\delta(t) = \begin{cases} 1 & t \in [-c\ell_0, c\ell_0] \\ \chi_0(\frac{c\ell_0+\alpha\delta-|t|}{\alpha\delta}) & |t| \in (c\ell_0, c\ell_0+\alpha\delta)\,, \\ 0 & t \in \mathbb{R} \setminus [-c\ell_0-\alpha\delta, c\ell_0+\alpha\delta]\,. \end{cases} \tag{11.34}$$

Consider the function $\tilde{g}_w(z) \stackrel{def}{=} g_w(z) \cdot \mathring{\chi}_\delta(\mathrm{Im}(z))$ and the continuous $(0,1)$-form $\varphi_w \stackrel{def}{=} \bar\partial \tilde{g}_w$ on $R_{\alpha\delta}$. The form φ_w vanishes outside $Q^{\pm}_{\alpha\delta}$. Let ε be a small positive number that does not depend on δ and will be chosen later. Consider the measurable $(0,1)$-form $\varphi_{w,\varepsilon}$ on R_δ that equals φ_w on $Q^{\pm}_{\alpha\delta,\varepsilon} \stackrel{def}{=} \{x+iy : x \in (-\alpha\delta\varepsilon, \alpha\delta\varepsilon), \pm y \in (c\ell_0, c\ell_0+\alpha\delta)\}$ and vanishes outside this set. Extend its pullback under the conformal mapping $\omega : R^X_\delta \to R_\delta$ to a measurable $(0,1)$-form on X by putting it equal to zero outside R^X_δ. Denote the obtained form by $\varphi^X_{w,\varepsilon}$.

By Corollary 2.14.2 of [68] there exists a strictly subharmonic exhaustion function ψ on X. The L^2-norm of $\varphi_{w,\varepsilon}$ with respect to the Euclidean metric on the complex plane does not exceed $C_2\sqrt{\varepsilon}$ for an absolute constant C_2. Hence, the weighted L^2-norm on X of $\varphi^X_{w,\varepsilon}$ with respect to the Kähler metric and the weight $e^{-\psi}$ (see Definition 2.6.1 of [68]) does not exceed $C_3C_2\sqrt{\varepsilon}$ for a constant C_3 that depends on ψ and on the Kähler metric on the relatively compact subset $V \supset R^X_\delta$ of X. By Corollary 2.12.6 of [68] there exists a function $f^X_{w,\varepsilon}$ with $\bar\partial f^X_{w,\varepsilon} = \varphi^X_{w,\varepsilon}$ in the weighted L^2-space on X with respect to the Kähler metric and the weight $e^{-\psi}$ (see Definition 2.6.1 of [68]), whose norm in this space does not exceed $C_4C_3C_2\sqrt{\varepsilon}$ for a constant C_4 depending only on X, ψ, and the Kähler metric.

Let $(Q^{\pm}_{\alpha\delta,\varepsilon})^X$ be the preimages of $Q^{\pm}_{\alpha\delta,\varepsilon}$ under ω. The function $f^X_{w,\varepsilon}$ is holomorphic on $X \setminus \big(\overline{(Q^{+}_{\alpha\delta,\varepsilon})^X} \cup \overline{(Q^{-}_{\alpha\delta,\varepsilon})^X}\big)$. For the above chosen constant δ_0 we put $\tilde{Q}^{\pm}_{\delta_0} \stackrel{def}{=} \{x+iy \in R_{\delta_0} : \pm y \in (c\ell_0-\delta_0, c\ell_0+2\delta_0)\}$, and $(\tilde{Q}^{\pm}_{\delta_0})^X = \omega^{-1}(\tilde{Q}^{\pm}_{\delta_0})$. Then $(Q^{\pm}_{\alpha\delta,\varepsilon})^X$ is relatively compact in $(\tilde{Q}^{\pm}_{\delta_0})^X$. On a relatively compact open subset X_0 of X, that contains the closed subset $\overline{S_{2\delta_0}}\setminus((\tilde{Q}^{+}_{\delta_0})^X \cup (\tilde{Q}^{-}_{\delta_0})^X)$ of X, the supremum norm of $|f^X_{w,\varepsilon}|$ is estimated by its weighted L^2-norm: $|f^X_{w,\varepsilon}| < C_5\sqrt{\varepsilon}$ in a neighbourhood of $\overline{S_{2\delta_0}}\setminus((\tilde{Q}^{+}_{\delta_0})^X \cup (\tilde{Q}^{-}_{\delta_0})^X)$ for a constant C_5 that depends on the relatively compact open subset V of X, on the Kähler metric, on ψ and on the constants chosen before (see Theorem 2.6.4 of [68]).

On the other hand the classical Cauchy-Green formula on the complex plane provides a solution $\tilde{f}_{w,\varepsilon}$ of the equation $\bar\partial \tilde{f}_{w,\varepsilon} = \varphi_{w,\varepsilon}$ on the set $\tilde{Q}^{+}_{\delta_0} \cup \tilde{Q}^{-}_{\delta_0}$. The supremum norm of the function $\tilde{f}_{w,\varepsilon}$ is estimated by $C_6\sqrt{\varepsilon}$ for an absolute constant C_6. Let $\tilde{f}^X_{w,\varepsilon}$ be the pullback of $\tilde{f}_{w,\varepsilon}$ to $(\tilde{Q}^{+}_{\delta_0})^X \cup (\tilde{Q}^{-}_{\delta_0})^X$. The function $f^X_{w,\varepsilon} - \tilde{f}^X_w$ is holomorphic on $(\tilde{Q}^{+}_{\delta_0})^X \cup (\tilde{Q}^{-}_{\delta_0})^X$ and satisfies the inequality $|f^X_{w,\varepsilon} - \tilde{f}^X_{w,\varepsilon}| <$

$(C_5+C_6)\sqrt{\varepsilon}$ at all points of the set $(\tilde{Q}^+_{\delta_0})^X \cup (\tilde{Q}^-_{\delta_0})^X$, that are close to its boundary. Hence, the inequality is satisfied on $(\tilde{Q}^+_{\delta_0})^X \cup (\tilde{Q}^-_{\delta_0})^X$. As a consequence,

$$|f^X_{w,\varepsilon}| < (C_5+2C_6)\sqrt{\varepsilon} \text{ on } (\tilde{Q}^+_{\delta_0})^X \cup (\tilde{Q}^-_{\delta_0})^X . \tag{11.35}$$

Choose now ε depending on C_5 and C_6, so that

$$(C_5+2C_6)\sqrt{\varepsilon} < \frac{1}{C} . \tag{11.36}$$

Then

$$|f^X_{w,\varepsilon}| < \frac{1}{C} \text{ on } \overline{S_{2\delta_0}} . \tag{11.37}$$

Put $\sigma_0 = \varepsilon\alpha\delta_0$, and take $\sigma = \varepsilon\alpha\delta \le \sigma_0$. Consider the smooth function $g^X_{w,\sigma}$ on $S_{\alpha\delta\varepsilon} = S_\sigma$ which equals the pullback $\tilde{g}^X_w = \tilde{g}_w \circ \omega$ on $\omega^{-1}(R_{\alpha\delta}) \cap S_\sigma$, and vanishes on the rest of S_σ. The function $g^X_{w,\sigma}$ vanishes on all circles of the bouquet except γ_0, and therefore, the monodromy of its homotopy class along each circle of the bouquet except γ_0 is the identity. The restriction of $g^X_{w,\sigma}$ to $\mathring{R}^X_\delta \cap S_\sigma = \omega^{-1}(\mathring{R}_\delta) \cap S_\sigma$ represents the element $a_1^{\pm 2} a_2^{\pm 2} \dots a_2^{\pm 2} \in \pi_1(\mathbb{C} \setminus \{-1,1\}, (-1,1))$. Moreover, on $\left(R^X_\delta \setminus \mathring{R}^X_\delta\right) \cap S_\sigma$ the inequality $|g^X_{w,\sigma}| < 1 - \frac{1}{C}$ holds. Hence, the monodromy of the homotopy class of $g^X_{w,\sigma}$ along γ_0 equals $a_1^{\pm} a_2^{\pm} \dots a_2^{\pm}$. By inequality (11.37) the restriction $(g^X_{w,\sigma} - f^X_{w,\varepsilon}) \mid S_\sigma$ is a holomorphic mapping into $\mathbb{C} \setminus \{-1,1\}$ and its monodromies along all circles of the bouquet coincide with those of the homotopy class of $g^X_{w,\sigma}$. For each positive $\sigma = \epsilon\alpha\delta < \epsilon\alpha\delta_0 = \sigma_0$ we found no less than $2^{2j} \ge \frac{1}{4} e^{\frac{c\alpha\varepsilon \log 2}{6\pi} \frac{\ell_0}{\sigma}}$ irreducible non-homotopic holomorphic mappings from S_σ to $\mathbb{C} \setminus \{-1,1\}$. The proposition is proved. □

Proof of Proposition 11.3 The first part of the proposition follows along the lines of the proof of the lower bound in Proposition 11.2.

Here is a sketch of the proof of the second part of Proposition 11.3. Let $f : S_\sigma \to \mathbb{C} \setminus \{-1,1\}$ be a holomorphic mapping with $f(q_0) = 0$. For each positive constant $c' < \frac{1}{2}$ there exists a positive constant c depending only on c' and on the Kähler metric, such that f maps the $c\sigma$-neighbourhood $D_{c\sigma}$ of q_0 in S_σ (in the Kähler metric) into the disc of radius c' with center 0 in $\mathbb{C} \setminus \{-1,1\}$ (in the Euclidean metric). Indeed, we take a conformal mapping ω from the unit disc $\mathbb{D}$ in the complex plane onto the σ-neighbourhood D_σ of q_0 with $\omega(0) = q_0$. The holomorphic mapping $f \circ \omega$ takes $\mathbb{D}$ into $\mathbb{C} \setminus \{-1,1\}$ and 0 to 0. It lifts to a holomorphic mapping $\widetilde{f \circ \omega}$ from $\mathbb{D}$ to the right half-plane $\mathbb{C}_r$ that takes 0 to $\frac{1+i}{2}$. Let $\mathfrak{c}$ be a conformal mapping from the right half-plane $\mathbb{C}_r$ onto the unit disc $\mathbb{D}$ that takes $\frac{1+i}{2}$ to 0. The composition $\mathfrak{c} \circ \widetilde{f \circ \omega}$ maps the unit disc to the unit disc and takes 0 to 0. Hence, for the derivative of this mapping the inequality $|(\mathfrak{c} \circ \widetilde{f \circ \omega})'| \le 2$

holds on the disc $\frac{1}{2}\mathbb{D}$ of radius $\frac{1}{2}$ around 0. Using the properties of $\mathfrak{c}$ and of the projection $\mathbb{C}_r \to \mathbb{C} \setminus \{-1, 1\}$ we find the constant c.

We have the isomorphisms of fundamental groups $\pi_1(X, q_0) \cong \pi_1(X, D_{c\sigma})$ and $\pi_1(\mathbb{C}\setminus\{-1, 1\}, 0) \cong \pi_1(\mathbb{C}\setminus\{-1, 1\}, \mathbb{D}_{c'})$. For each j we take a curvilinear rectangle R_j contained in $S_{c\sigma}$ that has horizontal sides in $D_{c\sigma}$ and extremal length not exceeding $C_3\frac{\ell_j}{\sigma}$ for a constant C_3 depending on the Kähler metric, and represents e_j with horizontal boundary values in $D_{c\sigma}$. The restriction $f \mid R_j$ represents $f_*(e_j)$ with horizontal boundary values in $\mathbb{D}_{c'}$.

If c' is small then for the mappings f_1 and f_2 of Proposition 9.1 the preimage $(f_1 \circ f_2)^{-1}(\mathbb{D}_{c'})$ is contained in the union $\bigcup_{k\in\mathbb{Z}} \big(\mathbb{D}_{c''} + ik\big)$ for a positive constant $c'' < \frac{1}{2}$. As a consequence, for each elementary word w (including the exceptional cases) the extremal length $\Lambda({}_{\mathbb{D}_{c'}} w\, {}_{\mathbb{D}_{c'}})$ with boundary values in $\mathbb{D}_{c'}$ is not smaller than const $\mathcal{L}(w)$ for an absolute constant const. Moreover, the extremal lengths with mixed horizontal boundary values $\Lambda({}_{\mathbb{D}_{c'}} w\, {}_{tr})$, $\Lambda({}_{\mathbb{D}_{c'}} w\, {}_{pb})$, $\Lambda({}_{tr} w\, {}_{\mathbb{D}_{c'}})$, $\Lambda({}_{pb} w\, {}_{\mathbb{D}_{c'}})$ are not smaller than const $\mathcal{L}(w)$. These facts can be proved similarly as Proposition 9.3.

The estimates $\Lambda({}_{\mathbb{D}_{c'}}(f_*(e_j))_{\mathbb{D}_{c'}}) \geq C_4\mathcal{L}(f_*(e_j))$, $j = 1, \dots, 2g + m$, for a constant C_4 depending on c'' can now be proved following along the lines of the proof of Theorems 9.1 and 9.2. By the choice of the curvilinear rectangles R_j inequalities (11.3) hold. □

Appendix A

All's well that ends well

A.1 Several Complex Variables

Theorem A.1 ([40], Theorem 5.5.1) *Let Ω be a Stein manifold and Ω_j open subsets of Ω such that $\Omega = \bigcup_{j=1}^{\infty} \Omega_j$. If $g_{j,k}$, $j, k = 1, 2, \ldots$, are holomorphic functions on $\Omega_j \cap \Omega_k$ such that*

$$\begin{aligned} g_{j,k} &= - g_{k,j} \textit{ for all } j, k, \\ g_{j,k} + g_{k,i} + g_{i,j} &= 0 \textit{ on } \Omega_j \cap \Omega_k \cap \Omega_i \textit{ for all } j, k, i, \end{aligned} \tag{A.1}$$

then there are holomorphic functions g_j on Ω_j such that

$$g_{j,k} = g_j - g_k \textit{ for all } j, k. \tag{A.2}$$

A family of holomorphic functions $g_{j,k}$ on $\Omega_j \cap \Omega_k$ that satisfies (A.1) is called a Cousin I distribution. If there are holomorphic functions g_j on Ω_j satisfying (A.2) we say that the first Cousin Problem for the Cousin distribution has a solution.

A family of nowhere vanishing holomorphic functions $g_{j,k}$ on $\Omega_j \cap \Omega_k$ is called a Cousin II distribution if

$$\begin{aligned} g_{j,k} g_{k,j} &= 1 \text{ for all } j, k, \\ g_{j,k} g_{k,i} g_{i,j} &= 1 \text{ on } \Omega_j \cap \Omega_k \cap \Omega_i \text{ for all } j, k, i. \end{aligned} \tag{A.3}$$

If there are nowhere vanishing holomorphic functions g_j on Ω_j satisfying the equation

$$g_{j,k} = g_j g_k^{-1} \text{ for all } j, k, \tag{A.4}$$

we say that the second Cousin Problem for the Cousin II distribution has a solution.

B. Jöricke, *Braids, Conformal Module, Entropy, and Gromov's Oka Principle*, Lecture Notes in Mathematics 2360, https://doi.org/10.1007/978-3-031-67288-0

If the Ω_j cover a Stein manifold and all sets $\Omega_j \cap \Omega_k$ are simply connected, then the respective second Cousin Problem has a solution. Indeed, for each j, k we may consider a branch of the logarithm of $g_{j,k}$ on $\Omega_j \cap \Omega_k$ and solve a Cousin I Problem for these functions.

A.2 A Lemma on Conjugation

Lemma A.1 (Lemma on Conjugation)

1. *Let $Q_1, \dots, Q_k$ be topological spaces, and let φ be a self-homeomorphism of (the formal disjoint union) $Q \overset{def}{=} \bigcup_{1 \le j \le k} Q_j$. Suppose φ permutes the Q_j cyclically, i.e. $\varphi(Q_j) = Q_{j+1}$, $j = 1, \dots, k-1$, $\varphi(Q_k) = Q_1$.*
   Let $\mathring{Q}_1, \dots, \mathring{Q}_k$ be topological spaces, and let $\mathring{\varphi}$ be a self-homeomorphism of (the formal disjoint union) $\mathring{Q} \overset{def}{=} \bigcup_{1 \le j \le k} \mathring{Q}_j$, that permutes the $\mathring{Q}_j$ along the respective cycle, i.e. $\mathring{\varphi}(\mathring{Q}_j) = \mathring{Q}_{j+1}$, $j = 1, \dots, k-1$, $\mathring{\varphi}(\mathring{Q}_k) = \mathring{Q}_1$. Assume that $\varphi^k \mid Q_1$ and $\mathring{\varphi}^k \mid \mathring{Q}_1$ are related by isotopy and conjugation.
   Then φ and $\mathring{\varphi}$ are related by isotopy followed by conjugation, i.e. there exists a homeomorphism σ from $\mathring{Q}$ onto Q, such that each $\sigma_j \overset{def}{=} \sigma | \mathring{Q}_j$ maps $\mathring{Q}_j$ homeomorphically onto Q_j, $j = 0, \dots, k$, and a self-homeomorphism $\tilde{\varphi}$ of Q, which is isotopic to φ and such that $\mathring{\varphi} = \sigma^{-1} \circ \tilde{\varphi} \circ \sigma$.
2. *Suppose in addition, that $\mathring{Q}_j = Q_j$ for each $j = 1, \dots, k$, and there are subsets Q'_j of Q_j, $j = 1, \dots, k$, such that $\varphi = \mathring{\varphi}$ on $Q' \overset{def}{=} \bigcup_{1 \le j \le k} Q'_j$ and $\varphi \mid Q'$ is a self-homeomorphism of Q'. Moreover, assume that $\varphi^k \mid Q_1$ is isotopic to $\mathring{\varphi}^k \mid \mathring{Q}_1$ by an isotopy of the form $\varphi^k \circ \tilde{\chi}_t$, $t \in [0, 1]$, for a* $\mathrm{Hom}^+(Q_1; Q'_1)$*-isotopy $\tilde{\chi}_t$.*
   Then the mapping σ of Statement 1 may be chosen to fix the union $Q' = \bigcup_{1 \le j \le k} Q'_j$ pointwise and the self-homeomorphism $\tilde{\varphi}$ of Statement 1 may be chosen to be isotopic to φ by an isotopy of the form $\varphi \circ \chi_t$, $t \in [0, 1]$, for a $\mathrm{Hom}^+(Q; Q')$*-isotopy χ_t with $\chi_0 =$* Id.

Proof of the Lemma on Conjugation Assume first that in Statement 1

$$\mathring{\varphi}^k \mid \mathring{Q}_1 = \sigma_1^{-1} \circ \varphi^k \circ \sigma_1 \mid \mathring{Q}_1 \tag{A.5}$$

for a homeomorphism σ_1 from $\mathring{Q}_1$ onto Q_1. Put $\varphi_j = \varphi \mid Q_j$ and $\mathring{\varphi}_j = \mathring{\varphi} \mid \mathring{Q}_j$ for $j = 1, \dots, k$. The conditions of the lemma give the following commutative diagram:

$$\begin{array}{ccccccccc}
\mathring{Q}_1 & \xrightarrow{\mathring{\varphi}_1} & \mathring{Q}_2 & \xrightarrow{\mathring{\varphi}_2} & \cdots & \xrightarrow{\mathring{\varphi}_{k-1}} & \mathring{Q}_k & \xrightarrow{\mathring{\varphi}_k} & \mathring{Q}_1 \\
\downarrow{\scriptstyle \sigma_1} & & & & & & & & \downarrow{\scriptstyle \sigma_1} \\
Q_1 & \xrightarrow{\varphi_1} & Q_2 & \xrightarrow{\varphi_2} & \cdots & \xrightarrow{\varphi_{k-1}} & Q_k & \xrightarrow{\varphi_k} & Q_1
\end{array}$$

Let $\sigma_2 : \mathring{Q}_2 \to Q_2$ be the homeomorphism for which the following diagram commutes:

$$
\begin{array}{ccc}
\mathring{Q}_1 & \xrightarrow{\mathring{\varphi}_1} & \mathring{Q}_2 \\
\downarrow{\scriptstyle \sigma_1} & & \downarrow{\scriptstyle \sigma_2} \\
Q_1 & \xrightarrow{\varphi_1} & Q_2
\end{array}
$$

In other words, we put $\sigma_2 = \varphi_1 \circ \sigma_1 \circ \mathring{\varphi}_1^{-1}$. If σ_j is defined for all $j \leq j_0 \leq k-1$, we define $\sigma_{j_0+1} = \varphi_{j_0} \circ \sigma_{j_0} \circ \mathring{\varphi}_{j_0}^{-1}$. Then for all $j \leq k$ the equality

$$
\sigma_j = \left(\prod_{j'=1}^{j-1} \varphi_{j'}\right) \circ \sigma_1 \circ \left(\prod_{j'=1}^{j-1} \mathring{\varphi}_{j'}\right)^{-1} \tag{A.6}
$$

holds. Since $\prod_{j'=1}^{k} \varphi_{j'} = \varphi^k | Q_1$ and $\prod_{j'=1}^{k} \mathring{\varphi}_{j'} = \mathring{\varphi}^k | \mathring{Q}_1$ Eq. (A.5) implies that the following diagram commutes:

$$
\begin{array}{ccccccccc}
\mathring{Q}_1 & \xrightarrow{\mathring{\varphi}_1} & \mathring{Q}_2 & \xrightarrow{\mathring{\varphi}_2} & \cdots & \xrightarrow{\mathring{\varphi}_{k-1}} & \mathring{Q}_k & \xrightarrow{\mathring{\varphi}_k} & \mathring{Q}_1 \\
\downarrow{\scriptstyle \sigma_1} & & \downarrow{\scriptstyle \sigma_2} & & & & \downarrow{\scriptstyle \sigma_k} & & \downarrow{\scriptstyle \sigma_1} \\
Q_1 & \xrightarrow{\varphi_1} & Q_2 & \xrightarrow{\varphi_2} & \cdots & \xrightarrow{\varphi_{k-1}} & Q_k & \xrightarrow{\varphi_k} & Q_1
\end{array}
$$

Let $\sigma : \mathring{Q} \to Q$ be the mapping that equals σ_j on $\mathring{Q}_j$. The commutativity of the diagram means that $\mathring{\varphi} = \sigma^{-1} \circ \varphi \circ \sigma$. Indeed, the equality

$$
\mathring{\varphi} = \sigma^{-1} \circ \varphi \circ \sigma \quad \text{on} \quad \bigcup_{j=1}^{k} Q_j \tag{A.7}
$$

is equivalent to the system of equations

$$
\begin{aligned}
\mathring{\varphi}_1 &= \sigma_2^{-1} \circ \varphi_1 \circ \sigma_1 \\
&\vdots \\
\mathring{\varphi}_{k-1} &= \sigma_k^{-1} \circ \varphi_{k-1} \circ \sigma_{k-1} \\
\mathring{\varphi}_k &= \sigma_1^{-1} \circ \varphi_k \circ \sigma_k\,.
\end{aligned} \tag{A.8}
$$

The Statement 1 of the Lemma A.1 is clear for the case when equality (A.5) holds.

Consider now the general case of Statement 1, i.e. we assume that

$$\mathring{\varphi}^k \mid \mathring{Q}_1 = \sigma_1^{-1} \circ \widetilde{\varphi^k} \circ \sigma_1$$

on $\mathring{Q}_1$ for a homeomorphism σ_1 from $\mathring{Q}_1$ onto Q_1, and a self-homeomorphism $\widetilde{\varphi^k}$ of Q_1 that is isotopic to the self-homeomorphism $\varphi^k \mid Q_1$ of Q_1. To prove the first statement of the lemma, it is enough to find a self-homeomorphism $\tilde{\varphi}$ of Q that is obtained from φ by isotopy and conjugation, for which $\tilde{\varphi}^k = \widetilde{\varphi^k}$ on Q_1. Then by the arguments for the case when equality (A.5) holds $\mathring{\varphi}$ is conjugate to $\tilde{\varphi}$ and the first statement is proved in the general situation.

We put $\tilde{\varphi} = \varphi$ on $\bigcup_{1<j\le k} Q_j$ and $\tilde{\varphi} = (\tilde{\varphi})^{-k+1} \circ \widetilde{\varphi^k} = (\varphi)^{-k+1} \circ \widetilde{\varphi^k}$ on Q_1. Then $(\tilde{\varphi})^k = \widetilde{\varphi^k}$ on Q_1. The isotopy joining φ and $\tilde{\varphi}$ is described as follows. It is equal to $\varphi \circ \chi_t$ on $\bigcup_{1\le j\le k} Q_j$ with χ_t being the identity mapping on $\bigcup_{1<j\le k} Q_j$ for $t \in [0, 1]$, and on Q_1 the χ_t form an isotopy of self-homeomorphisms of Q_1 for which χ_0 is the identity and χ_1 is equal to the mapping $\chi_1 = (\varphi)^{-k} \circ (\widetilde{\varphi^k})$. Then $\varphi \circ \chi_t \mid Q_k$ is equal to φ for $t = 0$ and to $(\varphi)^{-k+1} \circ \widetilde{\varphi^k} = \tilde{\varphi}$ for $t = 1$.

Statement 2 is obtained as follows. We look again for a suitable self-homeomorphism $\tilde{\varphi}$ of Q that is isotopic to φ and satisfies the equality $\tilde{\varphi}^k \mid Q_1 = \mathring{\varphi}^k \mid Q_1$. As in the proof of Statement 1 we put $\tilde{\varphi} = \varphi$ on $\bigcup_{1<j\le k} Q_j$ and $\tilde{\varphi} = (\varphi^{-k+1}) \circ \mathring{\varphi}^k$ on Q_1. Then $\mathring{\varphi}^k \mid Q_1 = \tilde{\varphi}^k \mid Q_1$ and $\mathring{\varphi} = \tilde{\varphi}$ on $\bigcup_{1\le j<k} Q'_j$.

We prove now that φ and $\tilde{\varphi}$ are related by the required isotopy. Put $\chi_t = \mathrm{Id}$ on $\bigcup_{1<j\le k} Q_j$ for all $t \in [0, 1]$, and $\chi_t = \tilde{\chi}_t$ on Q_1. Here $\tilde{\chi}_t$ is the $\mathrm{Hom}^+(Q_1, Q'_1)$-isotopy in the assumption of Statement 2 of the lemma, $\tilde{\chi}_0 = \mathrm{Id}$ and $\tilde{\chi}_1 = \varphi^{-k} \circ \mathring{\varphi}^k$. Then χ_t is a $\mathrm{Hom}^+(Q, Q')$-isotopy. The family $\varphi \circ \tilde{\chi}_t$ joins $\varphi = \varphi \circ \chi_0$ with $\tilde{\varphi} = \varphi \circ \chi_1$. Indeed, this is true on $\bigcup_{1<j\le k} Q_j$, since there $\tilde{\varphi} = \varphi$ and $\chi_t = \mathrm{id}$. On Q_1 $\varphi \circ \chi_0 = \varphi$ and $\varphi \circ \chi_1 = \varphi \circ \varphi^{-k} \circ \mathring{\varphi}^k = \tilde{\varphi}$. Finally $\tilde{\varphi}^k \mid Q_1 = \mathring{\varphi}^k \mid Q_1$.

The homeomorphism σ that conjugates $\tilde{\varphi}$ to $\mathring{\varphi}$ is given by equality (A.6) (see the proof of Statement 1). Since under the present conditions σ_1 is the identity and the $\tilde{\varphi}_j$ and $\mathring{\varphi}_j$ coincide on Q'_j, σ equals the identity on Q'. The lemma is proved. □

A.3 Koebe's Theorem

Theorem A.2 (Koebe's $\frac{1}{4}$-Covering Theorem [32], Theorem 2, II,§4) *Suppose a holomorphic function f on the unit disc $\mathbb{D}$ in the complex plane with $f(0) = 0$ and $f'(0) = 1$ is univalent (in other words, it maps the unit disc conformally onto its image). Then the image $f(\mathbb{D})$ contains the disc of radius $\frac{1}{4}$ with center* 0.

References

1. R.L. Adler, A.G. Konheim, M.H. McAndrew, Topological entropy. Trans. Am. Math. Soc. **114**, 309–319 (1965)
2. L. Ahlfors, On quasiconformal mappings. J. Anal. Math. **3**, 1–58 (1953/1954)
3. L. Ahlfors, Finitely generated Kleinian groups. Am. J. Math. **86**, 413–429 (1964)
4. L. Ahlfors, *Lectures on Quasiconformal Mappings* (Van Nostrand, Princeton, 1966)
5. T. Apostol, *Modular Functions and Dirichlet Series in Number Theory*. Graduate Texts in Mathematics, vol. 41, 2nd edn. (Springer-Verlag, New York, 1990)
6. S.J. Arakelov, Families of algebraic curves with fixed degeneracies (Russian). Izv. Akad. Nauk SSSR Ser. Mat. **35**, 1269–1293 (1971)
7. V. Arnold, On some topological invariants of algebraic functions. Trudy Moskov. Mat. Obsc. **21**, 27–46 (1970). Engl. Transl.: Trans. Moscow Math. Soc. **21**, 30–52 (1970)
8. S. Bell, S. Krantz, Smoothness to the boundary of conformal maps. Rocky Mt. J. Math. **17**(1), 23–40 (1987)
9. L. Bers, *Quasiconformal Mappings and Teichmüller's Theorem.* Analytic Functions (Princeton University Press, Princeton, 1960), pp. 89–119
10. L. Bers, Automorphic forms and general Teichmüller spaces, in *Proceedings of Conference on Complex Analysis, Minneapolis (1964)* (Springer Verlag, 1965), pp. 109–113
11. L. Bers, An extremal problem for quasiconformal mappings and a theorem by thurston. Acta Math. **141**, 73–98 (1978)
12. L. Bers, H.L. Royden, Holomorphic families of injections. Acta Math. **157**(3-4), 259–286 (1986)
13. M. Bestvina, M. Handel, Train tracks for surface homeomorphisms. Topology **34**(1), 109–140 (1995)
14. J. Birman, *Braids, Links and Mapping Class Groups.* Annals of Mathematics Studies, vol. 82 (Princeton University Press, Princeton, 1975)
15. D. Borthwick, *Spectral Theory of Infinite-Area Hyperbolic Surfaces.* Progress in Mathematics, vol. 256 (Birkhäuser, Boston/Basel/Berlin, 2016)
16. G. Bredon, *Topology and Geometry*. Graduate Texts in Mathematics, vol. 139 (Springer-Verlag, New York, 1993)
17. F.E. Browder (ed.), Mathematical developments arising from Hilbert problems, in *Proceedings of Symposia in Pure Mathematics*, vol XXVIII (American Mathematical Society, Providence, 1976)
18. L. Caporaso, On certain uniformity properties of curves over function fields. Compos. Math. **130**(1), 1–19 (2002)

B. Jöricke, *Braids, Conformal Module, Entropy, and Gromov's Oka Principle*, Lecture Notes in Mathematics 2360, https://doi.org/10.1007/978-3-031-67288-0

19. M. de Franchis, Un teorema sulle involuzioni irrationali. Rend. Circ. Mat. Palermo **36**, 368 (1913)
20. A. Eskin, M. Mirzakhani, Counting closed geodesics in moduli space. J. Mod. Dyn. **5**(1), 71–105 (2011)
21. G. Faltings, Endlichkeitssätze für abelsche Varietäten über Zahlkörpern (German) [Finiteness theorems for abelian varieties over number fields]. Invent. Math. **73**(3), 349–366 (1983)
22. G. Faltings, Arakelov's theorem for abelian varieties. Invent. Math. **73**(3), 337–347 (1983)
23. H. Farkas, I. Kra, *Riemann Surfaces*. Graduate Texts in Mathematics, vol. 71 (Springer-Verlag, New York-Berlin, 1980)
24. B. Farb, D. Margalit, *A Primer on Mapping Class Groups*. Princeton Mathematical Series, vol. 49 (Princeton University Press, Princeton, 2012)
25. A. Fathi, F. Laudenbach, V. Poénaru, Travaux de Thurston sur les surfaces. Astérisque, vol. 66–67 (Societe Mathematique de France, Paris, 1979)
26. W. Fischer, H. Grauert, *Lokal-triviale Familien kompakter komplexer Mannigfaltigkeiten* (German). Nachr. Akad. Wiss. G¨ttingen Math.-Phys. Kl. II 89–94 (1965)
27. O. Forster, *Lectures on Riemann surfaces*. Graduate Texts in Mathematics, vol. 81 (Springer-Verlag, New York, 1991). Translated from the 1977 German original by Bruce Gilligan
28. F. Forstneric, Runge approximation on convex sets implies the Oka property. Ann. Math. (2) **163**(2), 689–707 (2006)
29. F. Forstneric, Stein manifolds and holomorphic mappings. *The Homotopy Principle in Complex Analysis*. A Series of Modern Surveys in Mathematics, vol. 56 (Springer, Heidelberg, 2011)
30. F. Forstneric, M. Slapar, Stein structures and holomorphic mappings. Math. Z. 256(3), 615–646 (2007)
31. K. Fritsche, H. Grauert, *From Holomorphic Functions to Complex Manifolds*. Graduate Texts in Mathematics, vol. 213 (Springer, New York/Berlin/Heidelberg, 2002)
32. G.M. Goluzin, *Geometric Theory of Functions of a Complex Variable*. Translations of Mathematical Monographs, vol. 26 (American Mathematical Society, Providence, 1969)
33. E. Gorin, V. Lin, On separable polynomials over commutative Banach algebras. Dokl. Akad. Nauk SSSR **218**(3), 505–508 (1974). Engl. Transl.: Soviet Math. Dokl. **15**(5), 1357–1361 (1974)
34. H. Grauert, Mordells Vermutung über rationale Punkte auf algebraischen Kurven und Funktionenkörper (German). Inst. Hautes Études Sci. Publ. Math. **25**, 131–149 (1965)
35. M. Gromov, Oka's principle for holomorphic sections of elliptic bundles. J. Am. Math. Soc. **2**(4), 851–897 (1989)
36. V.L. Hansen, Polynomial covering spaces and homomorphisms into the braid groups. Pac. J. Math. **81**(2), 399–410 (1979)
37. Ph. Hartman, *Ordinary differential equations*. John Wiley & Sons, Inc., New York-London-Sydney, 1964. xiv+612 pp.
38. G. Heier, Uniformly effective Shafarevich conjecture on families of hyperbolic curves over a curve with prescribed degeneracy locus. J. Math. Pures Appl. **83**(7), 845–867 (2004)
39. Z. He, O. Schramm, Fixed points, Koebe uniformization and circle packings. Ann. Math. **137**(2), 396–406 (1993)
40. L. Hörmander *An Introduction to Complex Analysis in Several Variables*. North-Holland Mathematical Library, vol. 7, 3rd edn. (North-Holland Publishing Co., Amsterdam, 1990)
41. Y. Imayoshi, H. Shiga, A finiteness theorem for holomorphic families of Riemann surfaces. *Holomorphic Functions and Moduli II (Berkeley, CA, 1986)*. Mathematical Sciences Research Institute Publications, vol. 11 (Springer, New York, 1988), pp. 207–219
42. D. Johnson, The structure of the Torelli group I: A finite set of generators of $\mathcal{T}$. Ann. Math. **118**, 423–442 (1983)
43. B. Jöricke, Braids, conformal module and entropy. C.R. Acad. Sci. Paris Ser. I **351**, 289–293 (2013)
44. B. Jöricke, Fundamental groups, slalom curves and extremal length, in *Operator Theory: Advances and Applications*, vol. 261 (Springer, Berlin, 2018), pp. 307–315

45. B. Jöricke, Fundamental groups, 3-braids, and effective estimates of invariants. Math. Z. **294**, 1553–1609 (2020)
46. B. Jöricke, Riemann surfaces of second kind and finiteness theorems. Math. Z. **302**, 73–127 (2022)
47. B. Jöricke, Gromov's Oka principle, fiber bundles and the conformal module. Acta Math. Vietn. **47**, 375–440 (2022)
48. B. Jöricke, Conformal invariants of 3-braids and counting functions. Ann. Fac. Sci. Toulouse **31**(5), 1323–1341 (2022)
49. Š.I. Kaliman, A holomorphic universal covering of the space of polynomials without multiple roots (Russian). Teor. Funkciǐ Funkcional. Anal. i Priložen. **28**(ii), 25–35 (1977)
50. S. Kaliman, The universal holomorphic covers of polynomials without multiple roots. Selecta Math. Soviet **12**(4), 395–405 (1993)
51. S. Kaliman, M. Zaidenberg, A transversality theorem for holomorphic maps and stability of Eisenman-Kobayashi measures. Trans. Am. Mat. Soc. **348**(2), 661–672 (1996)
52. C. Kassel, V.Turaev, *Braid Groups*. Graduate Texts in Mathematics, vol. 247 (Springer, Berlin, 2008)
53. B. von Kerérekjárto, Über die Periodischen Transformationen der Kreisscheibe und der Kugelfläche. Math. Ann. **80**, 3–7 (1919)
54. K. Kodaira, *Complex Manifolds and Deformation of Complex Structures.* Grundlehren der Mathematischen Wissenschaften, vol. 283 (Springer-Verlag, New York, 1986)
55. L. Kodama, Boundary measures of analytic differentials and uniform approximation on a Riemann surface. Pac. J. Math. **15**, 1261–1277 (1965)
56. I. Kra, *Automorphic forms and Kleinian groups.* Mathematics Lecture Note Series (W. A. Benjamin, Inc., Reading, 1972)
57. S. Kravetz, On the geometry of Teichmüller spaces and the structures of their modular groups. Ann. Acad. Sci. Fenn. **278**, 1–35 (1959)
58. J. Lehner, *Discontinuous Groups and Automorpic Functions.* Mathematical Surveys, vol. VIII (American Mathematical Society, Providence, 1964)
59. O. Lehto, *Univalent Functions and Teichmüller Spaces* (Springer, Berlin, 1986)
60. V. Lin, Around the 13th Hilbert problem for algebraic functions, in *Proceedings of the Hirzebruch 65 Conference in Algebraic Geometry (Ramat Gan, 1993), Israel Math Conference Proceedings*, vol. 9 (Bar-Ilan University, Ramat Gan, 1996), pp. 307–327
61. V. Lin, Algebraic functions, configuration spaces, Teichmüller spaces, and new holomorphic combinatorial invariants. Funct. Anal. Appl. **45**(3), 55–78 (2011)
62. A.I. Markushevich, *Theory of Functions of a Complex Variable*, vol. III (Chelsea Publishing Co., New York City, 1977)
63. B. Mazur, Arithmetic on curves. Bull. Am. Math. Soc. (N.S.) **14**(2), 207–259 (1986)
64. J.Milnor, *Topology from the Differentiable Viewpoint.* Princeton Landmarks in Mathematics (Princeton University Press, Princeton, 1997). Based on notes by David W. Weaver. Revised reprint of the 1965 original
65. J. Milnor, A note on curvature and fundamental group. J. Differ. Geom. **2**, 1–7 (1968)
66. C. McMullen, From dynamics on surfaces to rational points on curves. Bull. Am. Math. Soc. (N.S.) **37**(2), 119–140 (2000)
67. S. Nag, *The Complex Analytic Theory of Teichmüller Spaces.* Canadian Mathematical Society Series (John Wiley & Sons, Hoboken, 1988)
68. T. Napier, M. Ramachandran, *An introduction to Riemann surfaces. Cornerstones* (Birkhäuser/Springer, New York, 2011)
69. A.N. Parshin, Algebraic curves over function fields I (Russian). Izv. Akad. Nauk SSSR Ser. Mat. **32**, 1191–1219 (1968)
70. R. Penner, Bounds on least dilatations. Proc. Am. Math. Soc. **113**(2), 443–450 (1991)
71. H. Royden, Automorphisms and isometries of Teichmüller space. *Advances in the Theory of Riemann Surfaces.* Annals of Mathematics Studies, vol. 66 (Princeton University Press, Princeton, 1971), pp. 369–383

72. S. Scheinberg, Uniform approximation by meromorphic functions having prescribed poles. Math. Ann. **243**(1), 83–93 (1979)
73. I. R. Shafarevich, Algebraic number fields (1963) (Russian), in *Proceedings of the International Congress Mathematicians* (Stockholm, 1962) (Inst. Mittag-Leffler, Djursholm, 1962), pp. 163–176
74. W. Song, K. Ko, J. Los, Entropy of Braids. J. Knot Theory Ramifications **11**(4), 647–666 (2002)
75. E.H.Spanier, *Algebraic Topology* (McGraw-Hill Book Company, New York, 1966)
76. N. Steenrod, *The Topology of Fibre Bundles.* Princeton Mathematical Series, vol. 14 (Princeton University Press, Princeton, 1951)
77. E.L. Stout, Bounded holomorphic funtions on finite Riemann surfaces. Trans. Am. Math. Soc. **120**(2), 255-285 (1965)
78. K. Strebel, *Quadratic Differentials* (Springer Verlag, Berlin/Heidelberg/New York/Tokyo, 1984)
79. W. Thurston, On the geometry and dynamics of diffeomorphisms of surfaces. Bull. Am. Math. Soc. (N.S.) **19**(2), 417–431 (1988)
80. W.Veech, *The Teichmüller geodesic flow.* Ann. Math. (2) **124**(3), 441–530 (1986)
81. H. Wielandt, *Finite Permutation Groups.* (Academic Press, New York-London, 1964). Translated from the German by R. Bercov
82. O. Zariski, On the problem of existence of algebraic functions of two variables possessing a given branch curve. Am. J. Math. **52**(2), 305–328 (1929)
83. O. Zariski, On the Poincaré group of rational plane curves. Am. J. Math. **58**(3), 607–619 (1936)
84. J. Zjuzin, Irreducible separable polynomials with holomorphic coefficients on a certain class of complex spaces (Russian). Mat. Sb. (N.S.) 102 **144**(4), 569–591, 632 (1977)

Index

B. Jöricke, *Braids, Conformal Module, Entropy, and Gromov's Oka Principle*, Lecture Notes in Mathematics 2360, https://doi.org/10.1007/978-3-031-67288-0

LECTURE NOTES IN MATHEMATICS Springer

Editors in Chief: J.-M. Morel, B. Teissier;

Editorial Policy

1. Lecture Notes aim to report new developments in all areas of mathematics and their applications – quickly, informally and at a high level. Mathematical texts analysing new developments in modelling and numerical simulation are welcome.

   Manuscripts should be reasonably self-contained and rounded off. Thus they may, and often will, present not only results of the author but also related work by other people. They may be based on specialised lecture courses. Furthermore, the manuscripts should provide sufficient motivation, examples and applications. This clearly distinguishes Lecture Notes from journal articles or technical reports which normally are very concise. Articles intended for a journal but too long to be accepted by most journals, usually do not have this "lecture notes" character. For similar reasons it is unusual for doctoral theses to be accepted for the Lecture Notes series, though habilitation theses may be appropriate.

2. Besides monographs, multi-author manuscripts resulting from SUMMER SCHOOLS or similar INTENSIVE COURSES are welcome, provided their objective was held to present an active mathematical topic to an audience at the beginning or intermediate graduate level (a list of participants should be provided).

   The resulting manuscript should not be just a collection of course notes, but should require advance planning and coordination among the main lecturers. The subject matter should dictate the structure of the book. This structure should be motivated and explained in a scientific introduction, and the notation, references, index and formulation of results should be, if possible, unified by the editors. Each contribution should have an abstract and an introduction referring to the other contributions. In other words, more preparatory work must go into a multi-authored volume than simply assembling a disparate collection of papers, communicated at the event.

3. Manuscripts should be submitted either online at www.editorialmanager.com/lnm to Springer's mathematics editorial in Heidelberg, or electronically to one of the series editors. Authors should be aware that incomplete or insufficiently close-to-final manuscripts almost always result in longer refereeing times and nevertheless unclear referees' recommendations, making further refereeing of a final draft necessary. The strict minimum amount of material that will be considered should include a detailed outline describing the planned contents of each chapter, a bibliography and several sample chapters. Parallel submission of a manuscript to another publisher while under consideration for LNM is not acceptable and can lead to rejection.

4. In general, **monographs** will be sent out to at least 2 external referees for evaluation.

   A final decision to publish can be made only on the basis of the complete manuscript, however a refereeing process leading to a preliminary decision can be based on a pre-final or incomplete manuscript.

   Volume Editors of **multi-author works** are expected to arrange for the refereeing, to the usual scientific standards, of the individual contributions. If the resulting reports can be forwarded to the LNM Editorial Board, this is very helpful. If no reports are forwarded or if other questions remain unclear in respect of homogeneity etc, the series editors may wish to consult external referees for an overall evaluation of the volume.

5. Manuscripts should in general be submitted in English. Final manuscripts should contain at least 100 pages of mathematical text and should always include

   - a table of contents;
   - an informative introduction, with adequate motivation and perhaps some historical remarks: it should be accessible to a reader not intimately familiar with the topic treated;
   - a subject index: as a rule this is genuinely helpful for the reader.
   - For evaluation purposes, manuscripts should be submitted as pdf files.

6. Careful preparation of the manuscripts will help keep production time short besides ensuring satisfactory appearance of the finished book in print and online. After acceptance of the manuscript authors will be asked to prepare the final LaTeX source files (see LaTeX templates online: https://www.springer.com/gb/authors-editors/book-authors-editors/manuscriptpreparation/5636) plus the corresponding pdf- or zipped ps-file. The LaTeX source files are essential for producing the full-text online version of the book, see http://link.springer.com/bookseries/304 for the existing online volumes of LNM). The technical production of a Lecture Notes volume takes approximately 12 weeks. Additional instructions, if necessary, are available on request from lnm@springer.com.

7. Authors receive a total of 30 free copies of their volume and free access to their book on SpringerLink, but no royalties. They are entitled to a discount of 33.3 % on the price of Springer books purchased for their personal use, if ordering directly from Springer.

8. Commitment to publish is made by a *Publishing Agreement*; contributing authors of multiauthor books are requested to sign a *Consent to Publish form*. Springer-Verlag registers the copyright for each volume. Authors are free to reuse material contained in their LNM volumes in later publications: a brief written (or e-mail) request for formal permission is sufficient.

Addresses:

Professor Jean-Michel Morel, CMLA, École Normale Supérieure de Cachan, France
E-mail: moreljeanmichel@gmail.com

Professor Bernard Teissier, Equipe Géométrie et Dynamique,
Institut de Mathématiques de Jussieu – Paris Rive Gauche, Paris, France
E-mail: bernard.teissier@imj-prg.fr

Springer: Ute McCrory, Mathematics, Heidelberg, Germany,
E-mail: lnm@springer.com

Printed in the United States
by Baker & Taylor Publisher Services